ELECTRICITY

Fundamental Concepts and Applications

ELECTRICITY

Fundamental Concepts and Applications

Timothy J. Maloney

NOTICE TO THE READER

Cover photo by Tim Barret
Cover design by Cathy Watterson

Delmar Staff
Administrative Editor: Wendy Welch
Developmental Editor: Mary Ormsbee Clyne
Project Editor: Eleanor Isenhart
Production Coordinator: Helen Yackel
Senior Design Supervisor: Susan Mathews
Art Coordinator: Michael Nelson

For information, address Delmar Publishers Inc.
2 Computer Drive West, Box 15-015
Albany, New York 12212-5015

10 9 8 7 6 5 4 3 2

Printed in the United States of America
Published simultaneously in Canada
by Nelson Canada,
A division of The Thomson Corporation

Library of Congress Cataloging in Publication Data
Maloney, Timothy J.
Electricity: fundamental concepts and applications / Timothy J. Maloney.
p. cm.
Includes index.
ISBN 0-8273-4675-1
1. Electrical engineering. I. Title.
TK146.M357 1992 91-30637
621.3—dc20 CIP

CONTENTS

CHAPTER 5 Measuring Electric Circuits 58

CHAPTER 6 Ohm's Law 76

CHAPTER 7 Power and Energy 92

CHAPTER 8 Safety Devices, Wires, Switches, and Relays 110

CHAPTER 9 Series Circuits 137

PREFACE

A MESSAGE TO STUDENTS

Electronics is an area of human effort that has had a great impact on our lives. For the last two generations, most of the major changes in the way we live can be traced to developments in electronics such as advanced medical testing and surgical procedures, high-fidelity music recording and playback, reliable long-distance telephones, television broadcasting and cable transmission, video recording, higher-efficiency car engines, instant data retrieval on video screens in stores, credit-card processing in restaurants and stores, transfer of funds between banks, exploration of space, microwaved hot meals, computer-aided design and manufacturing, bar-code readers at grocery checkout counters, and home computing and word processing.

Electronics also has profound economic effects. Overall, United States inflation from 1965 to the early 1990s was about 400%. This large inflation rate has been the dominant fact of economic life for your parent's entire adulthood. Were it not for the counter-trend of electronic devices, many of which actually decreased in price, and were it not for the gains in productivity resulting from electronics applications, inflation would have been even higher. It is perhaps not too much to say that the reviving effect of electronics is the reason the American dream is still alive for you.

Some people believe that this period of time is the third revolution in history. Humanity had an agricultural revolution about 10,000 years ago, an industrial revolution in the last century, and is now having an electronic revolution. Naturally, this electronic revolution has created many jobs. Some of the jobs are in research and development; some are in design; some are in manufacturing; some are in sales; some are in installation, maintenance, and repair.

Throughout this book, you will see many photos and descriptions of electrical and electronic applications that affect our lives, as well as many of the employment possibilities. All of these applications, and all of the jobs, have in common the same fundamental concepts. This book is meant to explain the fundamental concepts in a way that is friendly, but not watered-down.

Whenever you set out to become good at something, you must master the fundamentals. This is as true in the study of electricity as it is in sports like baseball or golf. The fundamental electrical ideas are here in this book. You are invited to read carefully, so that you master each one. If any explanation does not come through to you, try to put into words what you do not understand. This effort forces you to really think about the idea, which might by itself clear up things in your mind. Even if this does not make it clear for you, your carefully focused question will make it easier for your instructor to help you.

FEATURES OF THIS TEXTBOOK

Examples

With any new idea, it is always helpful to see examples. Here, there are plenty of realistic example problems that show how the ideas are applied. The best way for you to benefit from the examples is to ask yourself, "How would I answer that question?" Then make an attempt to answer on your own, before looking at the solution. If numerical calculation is required, get out a hand calculator and perform the arithmetic.

If your answer agrees with the book, you can feel satisfied that you are understanding the material. If your answer does not agree, then follow the solution steps carefully to discover where you went astray.

Technical Facts

You will see brief statements or equations highligted in color. These Technical Facts represent ideas of major importance, so they deserve your special attention.

Self-Checks

Self-check questions and problems appear every four to eight pages. Use these Self-Checks to practice what you have learned, and to help uncover any weaknesses in your understanding. The answers to Self-Checks are given at the end of the text.

Safety

You surely know that your physical safety is the first concern of everyone. The Working Safely section after this Preface spells out all the dangers of working in the field of electricity and electronics. Please take that information seriously. It could mean the difference between a long and satisfying career in electricity and electronics, or serious injury.

In addition, specific bits of Safety Advice are placed throughout the book, at locations where they become understandable and necessary.

Troubleshooting

Most people like to be able to apply their electrical knowledge to trouble situations and breakdowns. To help you do this, there are several sections that are devoted specifically to troubleshooting particular types of circuits. Also, there are several Troubleshooting Tips placed throughout the book, at locations where they become understandable and useful.

Boxed Articles

Each chapter contains an article with photographs that describes an interesting application of electrical technology. These articles should give you a feel for some of the exciting and imaginative fields of work that are open to electrical technicians.

Color Inserts

There are two 8-page color sections in the book. The first one, called Applications, contains color photographs that go with the chapter boxed articles. The second one, called Circuits, shows the detailed construction of circuits containing color-coded electrical components. This gives you an opportunity to practice what you are learning by making observations of actual circuit structure and measuring instruments.

Calculator Use

It is expected that you will be doing your calculations with a hand-held calculator. Whenever a new calculator function is needed, it is demonstrated with an exact sequence of keystrokes. In addition, the Appendix, Using a Scientific Handheld Calculator can serve as a quick reference for all the calculator functions used throughout the book.

Chapter Features

As with all books, each chapter introduces a fundamentally new topic. Two things have been done to help you keep track of your progress through each topic:

- *New Terms.* The most important new terms that you will learn are listed at the beginning of the chapter. These are not the only new words that you will see, but they are ones that deserve special attention. When one of these terms is first introduced, it is printed in **bold type**.
- *Objectives.* Each chapter has a list of numbered objectives. These are abilities that you should gain as you work through the chapter. As you are studying each chapter, you will see a symbol showing where each objective is covered. When you have finished that part of the chapter, ask yourself "Can I do what that objective calls for?" If you can answer an honest "yes," then you are on track. Knowing that you are on track gives you some immediate reinforcement, which is good for your confidence. Of course, if you are not on track, it is better to find out right away.

When possible, Self-Check questions and problems are keyed to these chapter objectives. The number of the relevant objective is given in brackets at the end of each question.

Finally, the end of each chapter has these items:

- *A collection of the important equations.* This is a handy reference when you are doing your homework problems.
- *A Summary of the chapter's major ideas.* You can use this summary as another source of feedback about your understanding of the material.
- *Questions and Problems* dealing with the chapter topics. These are keyed to specific chapter objectives, just like Self-Check questions. No doubt your instructor will assign particular questions and problems as homework. You may wish to do additional problems on your own, since the more practice you get at solving problems, the better you will comprehend the ideas.

You should be getting the picture that learning electricity and electronics requires considerable effort on your part. Nevertheless, it can be fun.

A MESSAGE TO INSTRUCTORS

Organization

This text follows a natural topic sequence for students who have no prior knowledge of electricity and electronics. In Chapter 1 atomic structure and the idea of electric charge are introduced first. This leads into the conceptually simplest electrical variable, current.

The conceptually more difficult variable, voltage, is introduced in Chapter 2. It is treated in a separate chapter, to underscore the essential difference between voltage and current. I find myself grinding my teeth at mid-semester whenever one of my students asks about "the voltage going through this resistor."

Resistance is also treated in a separate chapter, Chapter 3. Then we go on to representing complete circuits by schematic diagrams, the subject of Chapter 4. Purposely postponed is any mention of work, energy, or power. These ideas are not necessary for understanding basic circuit action.

All three basic measuring instruments are presented together in Chapter 5. I find it easier to emphasize the connection differences among ammeters, voltmeters, and ohmmeters if they are all right there together, for comparison. However, if you consider it better to explain the use of ammeters at the same time that you cover current, and voltmeters alongside coverage of the voltage idea, and ohmmeters alongside coverage of resistance, the meter material in Chapter 5 can be extracted in thirds.

Scientific notation is introduced in Chapter 6 only when it is really needed for solving Ohm's law with realistic values of resistance and current. We don't want to frighten away those students who are weak in mathematics before we have even given them a chance to develop an interest in our subject.

After dealing with energy and power in Chapter 7, I believe it is appropriate to discuss circuit protection and control. Most beginning students are familiar with fuses and circuit breakers, at least by name. And many of them have a practical curiosity about wire gages, switch configurations and types, and the grounding idea. These topics are covered in Chapter 8, along with the related ideas of ground-fault interruption and relay switching. If you do not have time for formal discussion of this material, it can be skipped. Eventually you will need to cover switch configurations, but that can be done at your convenience.

Unlike most other texts, this book dedicates a separate chapter to each of the three fundamentally different circuit configurations—series (Chapter 9), parallel (Chapter 10), and series-parallel (Chapter 11). Kirchhoff's laws are introduced in the natural way—voltage law with series circuits; current law with parallel circuits.

Capacitance and inductance (Chapters 12–14) are covered in a dc context first, before sine wave ac is ever mentioned. Otherwise, students tend to jumble together time constant analysis with reactance. Chapter 15 deals with series RC and RL time constants, applied to switched dc circuits.

Ac sine waves are presented in Chapter 16, and subsequent ac topics are covered in Chapters 17 through 20. An overview of solid-state electronic devices and circuits is given in Chapter 21.

Level

The math level for the entire text is held at basic algebra with some very limited trigonometry. I have made considerable effort to keep the reading level appropriate for today's students.

To help provide the encouragement that comes from achieving small goals frequently, each chapter objective is called out in the margin when it is encountered.

Frequent reinforcement is also provided by the Self-Checks that appear after every two or three sections. I hope that these features of the text will be useful in encouraging questions from your students and in starting class discussions.

Supplements

A full range of supplement materials, including an Instructor's Guide, Lab Manual, Test Bank, and other items is available by contacting your Delmar sales representative.

ACKNOWLEDGMENTS

I gratefully acknowledge the contributions of those who supplied valuable information for the development of this textbook:

- Leroy Anderson
  Shawnee Mission East High School
- Mark Baumgardner
  Cumberland Valley High School
- George Carr
  Lancaster High School
- David Coeyman
  Brevard Community College
- Pete Curran
  Scott Intermediate High School
- John Davis
  Willowbridge High School
- James T. Deeter
  National Education Center
- Thomas Gibson
  Mohave Community College
- Thomas Harrill
  Woodhaven High School
- Jack Horvatis
  Kenmore East Senior High
- Edward Kaufenberg
  Blackhawk Technical College
- Michael Montgomery
  Central High School
- Dick Shannon
  South Plains College

My sincere thanks go to Wendy Welch and Mary Ormsbee Clyne for their valuable ideas in choosing the substance and developing the form of this book. Thanks also to Eleanor Isenhart, for guiding the production process so smoothly.

Special thanks to my dear wife Pat for her skillful and arduous work in photographing the circuits that appear in the Circuits color section.

Timothy J. Maloney

WORKING SAFELY

Everyone knows that electricity can be dangerous. The most serious threat is electric shock. There are also other dangers that you must beware of when working on electrical and electronic circuits.

ELECTRIC SHOCK

An electric shock is possible whenever a large voltage exists between two points in a circuit. If one part of your body touches one of the points at the same time that another part of your body touches the other circuit point, then the large voltage causes current to flow through your body. If the current's path is through your vital chest organs, then a current as low as 0.01 ampere can be fatal. This is because the current interrupts the nerve signals that maintain the rhythm of your heart and lungs.

There are certain safety practices that you must follow in order to prevent electric shock from occurring. Always assume that the circuit is "live," even though you may have turned the On-Off switch to Off. This is because the switch might malfunction. It might remain electrically closed, or On, even though it has been mechanically moved to Off.

Make sure that your tools and instruments are in good working condition. For instance, if the insulation on a meter probe is worn or cracked, replace it. The same holds for the insulation on tool handles.

Remove metal bracelets, rings, and watches when working on a circuit. Your skin beneath such metal surfaces becomes moist. You are in greater danger when your skin is moist than when it is dry, because moisture permits greater current flow. For the same reason, do not work under wet conditions. This includes working in wet clothing or where there is water on the floor.

Never try to defeat safety features on electrical equipment. For example, the third prong on an electrical plug is designed to guarantee that a circuit's metal enclosure never becomes electrically "hot," even if some internal insulation breaks down. Never cut off this third prong to adapt it to a two-prong receptacle. Instead, the two-prong receptacle should be replaced.

In some situations, you may not be able to keep the On-Off switch in view while you are working. Another person, not knowing that you are working on the circuit, might turn the switch back On. To eliminate this possibility, some switches can be padlocked in the Off position. This padlocking feature is especially useful in high-voltage industrial circuits where the switch can be some distance away from the location where you are working. In those situations, padlock the switch Off, and attach a tag to explain why the switch has been turned Off. Most companies have a firm policy regarding this practice.

A capacitor (studied in Chapter 12) can retain its voltage for a long period of time after the circuit power has been turned off. You must perform a special operation, called discharging, to remove a capacitor's voltage.

A victim of severe electrical shock may be saved by cardiopulmonary resuscitation (CPR). Everyone should learn how to administer CPR.

If a shock victim remains in contact with the live wire, that contact must first be broken. Do not take hold of the victim to begin pulling, or you may be severely shocked yourself. Instead, you must first turn off the power before beginning the rescue. Of course, you must know where the main On-Off disconnect switch is located. In any new work environment, that switch location is the first thing you should find out.

BURNS

Even in low-voltage circuits, many electrical devices operate at a very high temperature. This tends to happen when there is a large value of current in the device. Touching or bumping against such a hot device can cause a serious burn.

Automobile circuits are especially a problem. Although the 12-V battery voltage itself is not dangerous, the current capacity of an auto battery is very great, typically several hundred amperes. If a hand-tool accidentally gets wedged between a hot 12-V terminal and the car's chassis, a huge current passes through the tool, causing it to get very hot. Even more serious is the possibility of wedging a ring or bracelet between a hot 12-V terminal and the car's chassis.

Working on electrical circuits often involves soldering and desoldering electrical connections. Proper soldering techniques are taught in a special project-construction course. But even at this early point in your studies, you should be warned against any technique that causes molten solder to be thrown about. (For example, allowing a springy wire to spring loose from a connection when the solder is melted.) Flying solder is a risk to your eyes. The best practice is to wear eye protection when soldering.

In case of an electrical fire, you should be able to reach the proper fire extinguisher without hesitation. Always check out the locations and types of the fire extinguishers whenever you enter a new work environment. Only a carbon-dioxide (CO_2) or foam-type fire extinguisher can be used on an electrical fire.

OTHER CONCERNS

Take care to protect your eyes whenever you cut wire or strip the insulation from wire. When cutting, point the wire end away from your face, preferably into a wastebasket. When stripping, move the stripping tool away from, not toward, your face.

Some of the chemicals used to make printed circuit boards give off harmful vapors. Always have someone watch you when using these chemicals. Follow the label instructions with regard to proper ventilation.

Pay very close attention to the electrical ratings and restrictions on circuit components. For example, every fuse has a maximum voltage rating (besides its current limitation). You must be careful not to use a 24-V rated fuse, say, in a 120-V circuit. Many other components have maximum voltage ratings as well. Their rated voltage must not be exceeded when they are placed into a circuit.

Some electric devices, including certain kinds of capacitors, have voltage polarity restrictions. They may explode if a reverse-polarity voltage is applied to their leads.

Innovation, invention, and discovery are common in electronics. This color insert will show you some of the newest ways that people are using electronics technology in the real world. As you study this book, you will learn the electrical concepts that make this technology possible.

Most photos shown here are cross-referenced to a chapter in this book. To learn more about these technical applications, turn to the chapter listed and read the Boxed Article.

Water Purification

(See Boxed Article—Water Purification—in Chapter 1.)

The surface of a fishing pond before purification.

The same pond after 48 hours of ion purification. (Those are clouds reflected on the water.) This pond water exceeds the Environmental Protection Agency's standards for drinking water.

Physical Therapy

(See Boxed Article—Physical Therapy—in Chapter 10.)

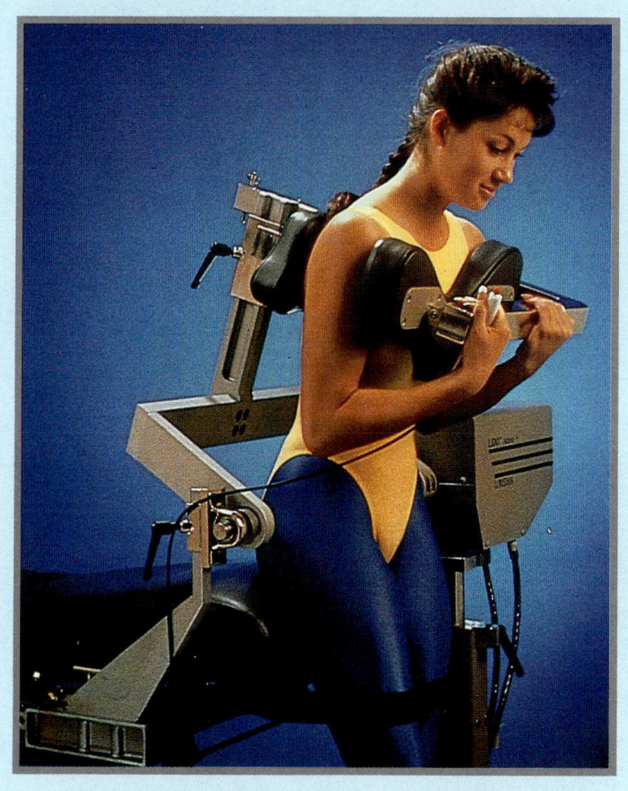

This woman is exercising her hip, back, and shoulder on a physical therapy machine that has electronic sensors to measure her performance.

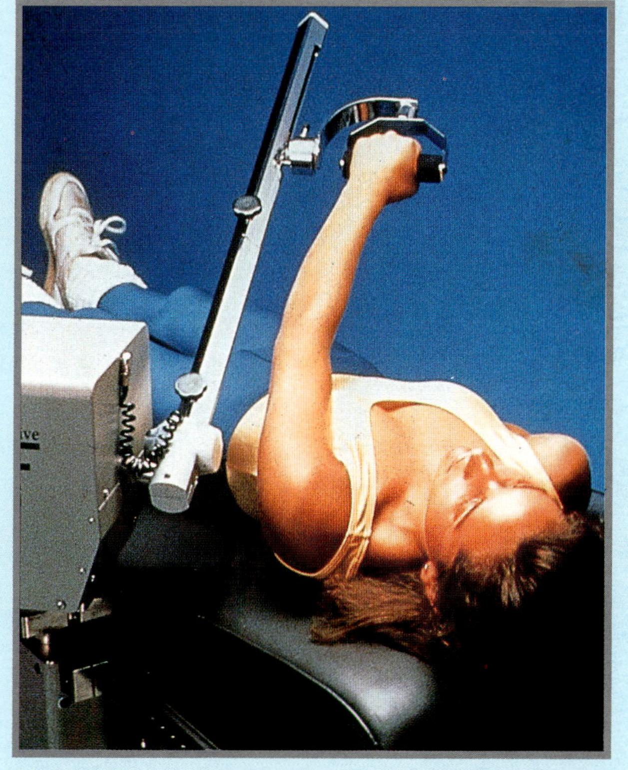

A computer stores the woman's performance data and uses it to tailor the therapy to her needs. Here, she is exercising her wrist and elbow joints.

Courtesy of NASA

Satellite Imaging

(See Boxed Article—Satellite Imaging—in Chapter 2.)

This computer-colored satellite image was created by a LANDSAT satellite and computers on the ground. The satellite carries electronic sensors that measure the radiation information representing the image. The ground computers then process the information to complete the image. The colors that appear in the image are not necessarily the actual colors that appear on earth. However, water is usually colored blue or green by the computer software.

LANDSAT satellite being prepared for rocket launch into space.

Courtesy of NASA

APPLICATIONS

Glass Art (See Boxed Article—Glass Art—in Chapter 3.)

These glass art pieces were created in a furnace that can reach 1800° F. An artist is placing the pieces in an electric oven for slow controlled cooling. The glass must be cooled slowly to maintain the color separation.

Finished pieces.

Medical Thermal Imaging (See Boxed Article—Medical Thermal Imaging—in Chapter 17.)

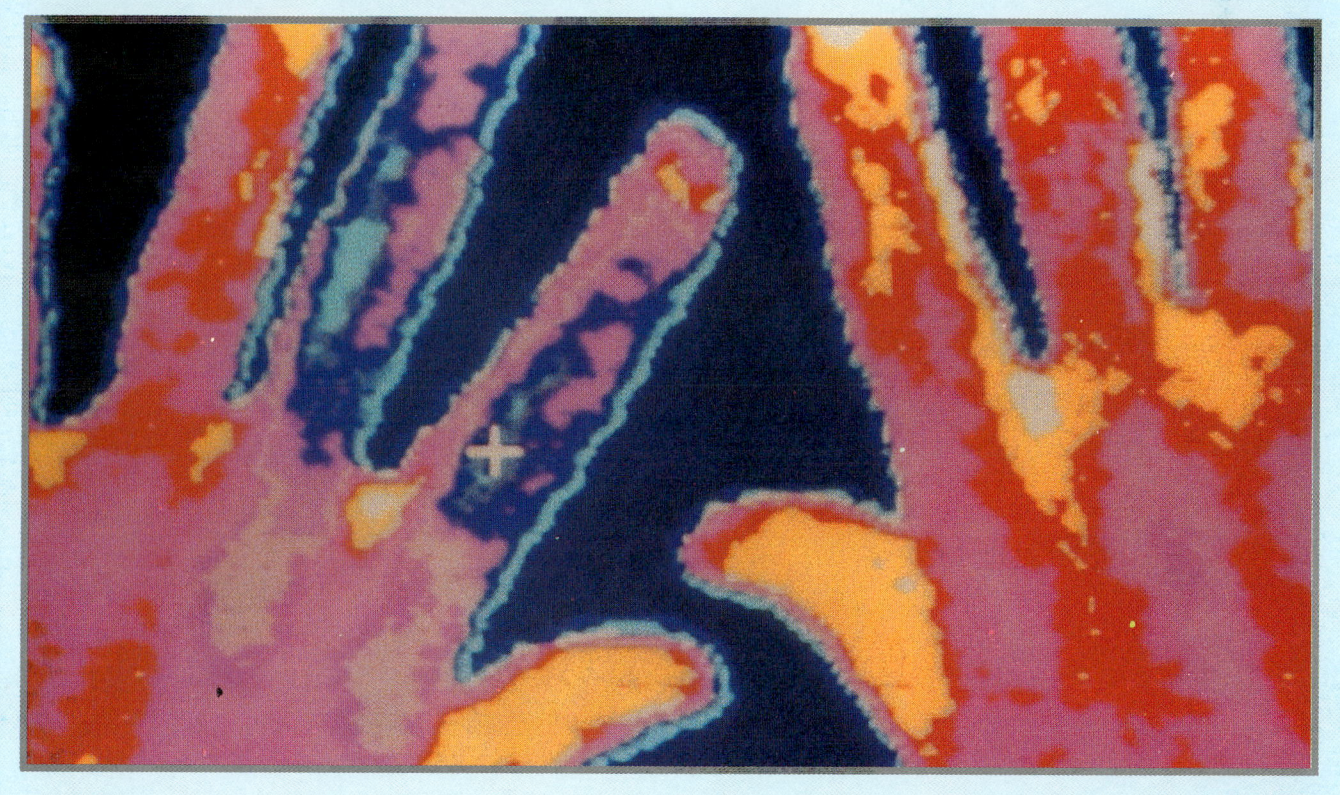

This is a thermal image (thermogram) of two hands. The blue and violet colors in the first two fingers of the left hand indicate subnormal blood circulation.

Courtesy of NASA

APPLICATIONS

Land Survey

(See Boxed Article—Land Survey—in Chapter 6.)

This photo shows a color-enhanced satellite image of about three square miles of land. The red areas are open grazing pasture, light brown are brushlands, green are fields of crops, dark brown are wooded areas, and white are developed areas with buildings, houses and roads.

Weather Forecasting

(See Boxed Article—Weather Forecasting—in Chapter 6.)

A color-enhanced satellite image of a hurricane.

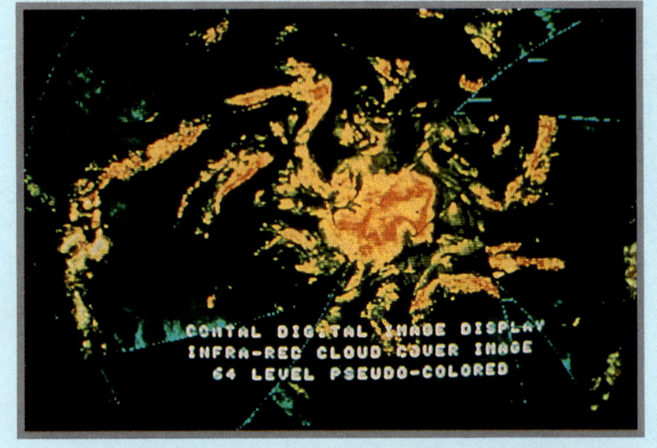

A computer-enhanced infrared photograph of the same hurricane, taken from an airplane. This photograph gives more detailed information regarding the storm's internal temperatures. It enables meteorologists to predict the storm's future development.

Forest Survey

A forest survey image. Green areas are healthy trees. The yellow areas show moderate damage, and orange indicates severe damage.

Courtesy of NASA

APPLICATIONS

Skin Care Testing

(See Boxed Article—Skin Care Testing—in Chapter 12.)

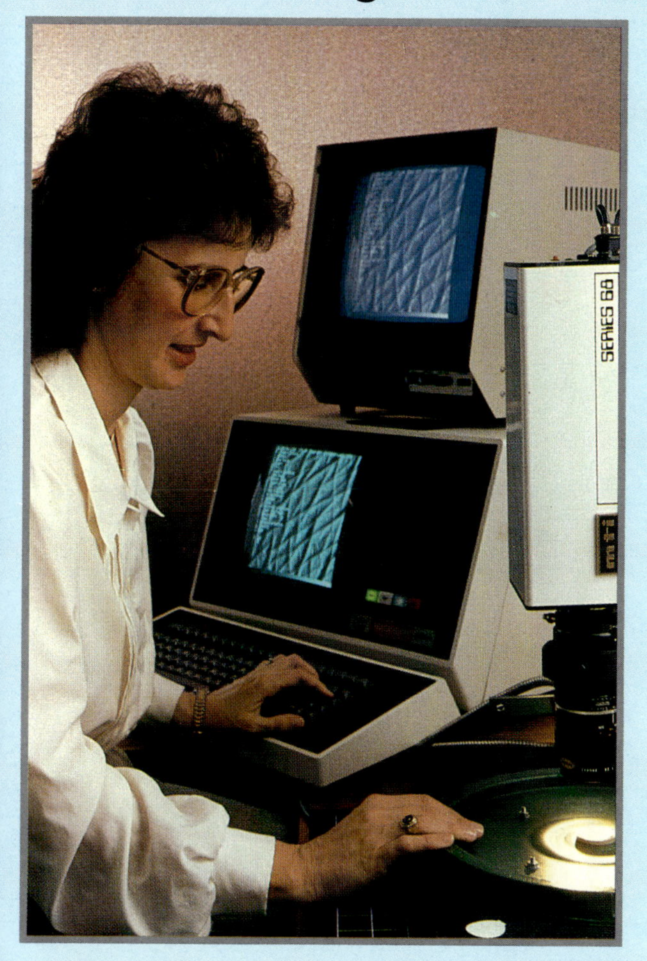

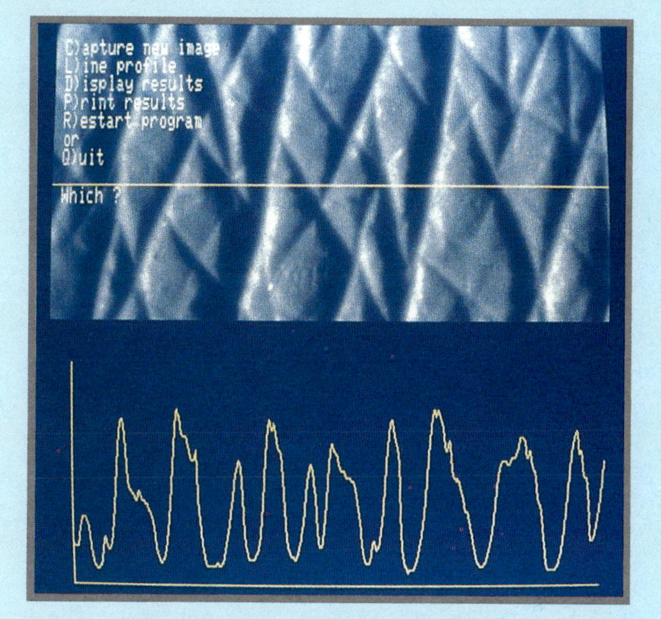

These photos show the procedures used to test the effects of a wrinkle-reducing cream. First, an impression was made of a volunteer's skin. In the photo at left, electronically produced fiber-optic illumination sidelights the skin replica, in order to obtain a magnified video image (above).

A computer-processed color-enhanced view of the skin before (below) and after (right) a treatment program. The skin's wrinkles have clearly been reduced.

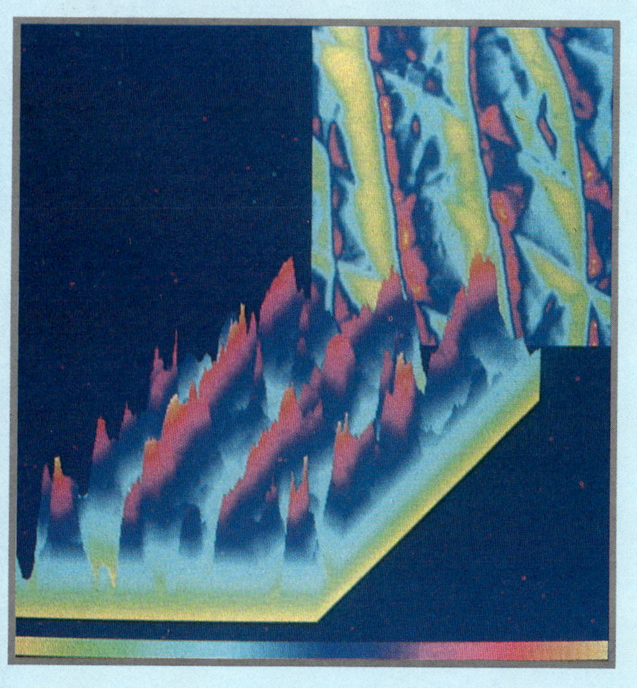

Courtesy of NASA

APPLICATIONS

Wetlands Survey

(See Boxed Article—Wetlands Survey—in Chapter 11.)

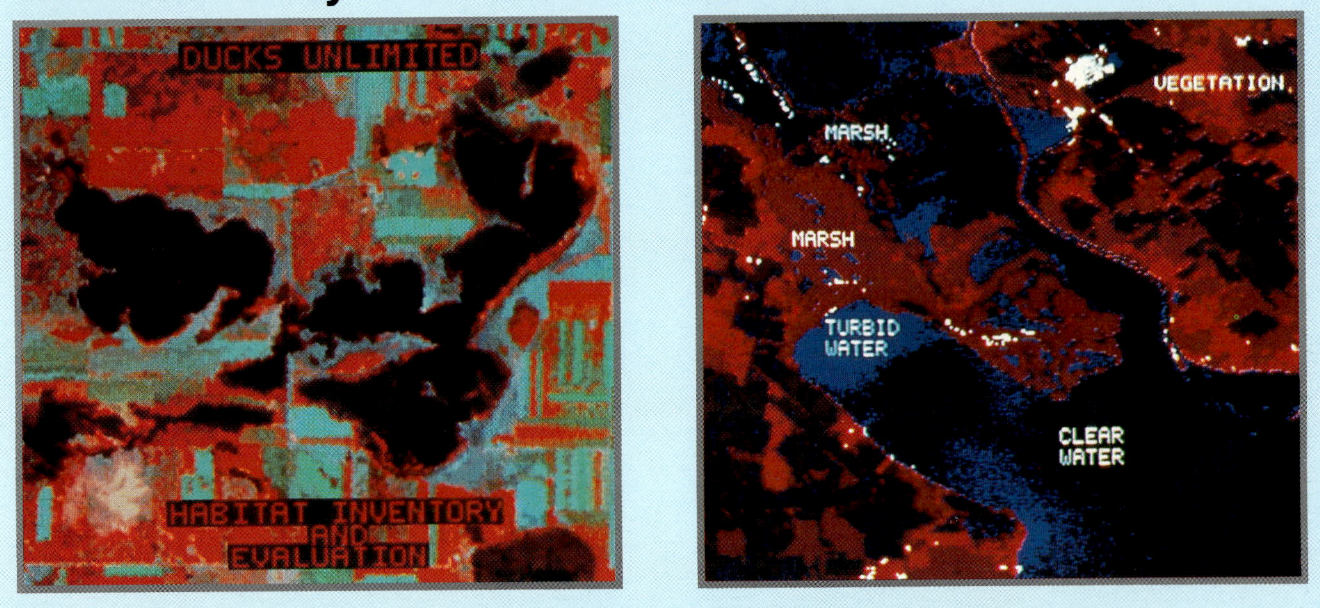

Many state and private environmental organizations monitor the condition of waterfowl habitats throughout North America. The photo images of wetland areas are created electronically from data gathered by the LANDSAT 5 satellite. Such images help researchers study the ability of an area to support waterfowl.

Courtesy of NASA

Film Special Effects (See Boxed Article—Film Special Effects—in Chapter 18.)

Creating the illusion of ghostly powers.

When Patrick Swayze put his arm through a solid door in the movie Ghost, the film's special effects technicians used electronic instrumentation to make it look real. Optical electronic sensors enabled them to make the dupe film's exposure to the background frame perfectly matched with its exposure to the separately shot action frame.

Background.

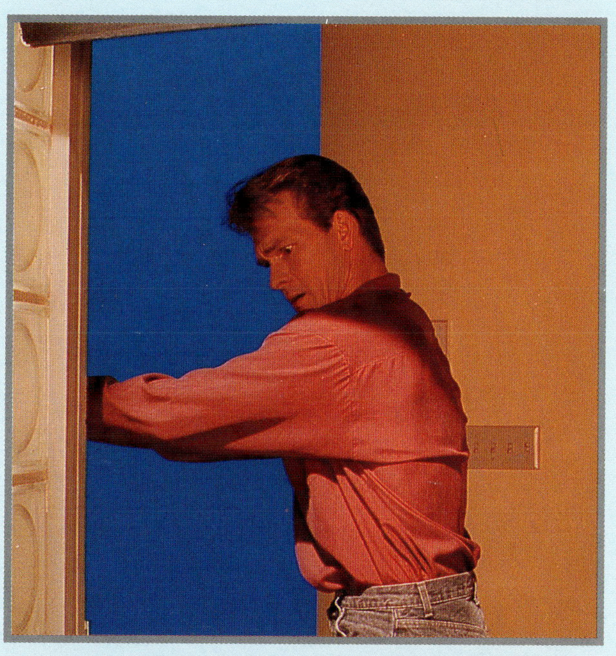

Action photographed against blue screen.

Altered counter-matte.

Final composite frame.

APPLICATIONS

Radar

(See Boxed Article—Radar—in Chapter 21.)

Military radar installations.

Courtesy of Datron Systems, Inc.

CHAPTER 1

CHARGE AND CURRENT

The negative charge stored in a storm cloud must return to the earth. *Courtesy of National Center for Atmospheric Research/National Science Foundation*

OUTLINE

1-1 Structure of Atoms
1-2 Charge
1-3 Concentrating Charge
1-4 Current
1-5 Ampere—The Unit of Current

NEW TERMS TO WATCH FOR

atomic particle
proton
neutron
electron
electric charge
positive charge
negative charge
electrostatic
current
direct current (dc)
alternating current (ac)
ampere
coulomb

In the study of electricity, the idea to start with is electric current. You need to have a good grasp of the meaning of current in order to go forward to other ideas. But current itself is the combination of two other ideas. These ideas are time and electric charge. Everyone understands time, but if this is your first course in electricity, the idea of electric charge is probably new to you.

After studying this chapter, you should be able to:

1. Describe the electrical structure of an atom.
2. Use the rule of charge attraction and repulsion.
3. Explain static electricity and the operation of an electrostatic precipitator.
4. Describe electric current on the atomic level.
5. Specify charge in basic units of coulombs.
6. Calculate current in amperes if charge and elapsed time are known.

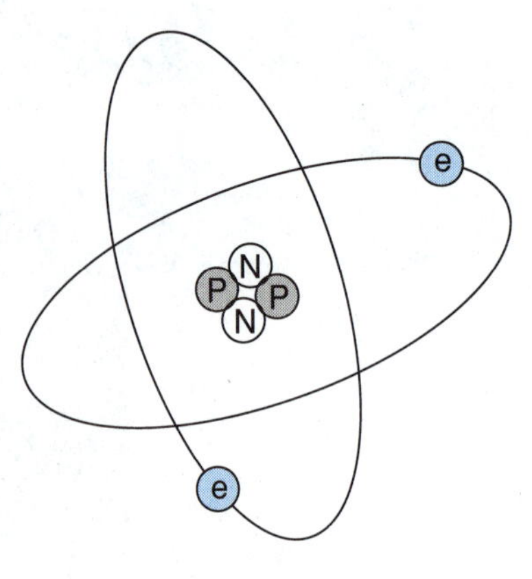

FIGURE 1-1 How we visualize the structure of a helium atom

1-1 STRUCTURE OF ATOMS

All matter is made up of atoms. An atom can be thought of as a collection of three kinds of **atomic particles,** as shown in Figure 1-1.

Two types of particles, **protons** and **neutrons,** crowd together in the center of the atom, called the nucleus. The third type of atomic particle, **electrons,** orbit around the nucleus.

Every intact atom has the same number of electrons as protons. This number is called the atomic number, and it defines the kind of atom that is present. There are 92 naturally occurring kinds of atoms in the universe. The simplest atom, hydrogen, contains one proton and one electron. Helium, shown in Figure 1-1, contains two protons and two electrons. The most complex atom, uranium, contains 92 protons and 92 electrons.

The number of neutrons may or may not be the same as the atomic number. Figure 1-1 shows a helium atom in which it happens to be the same. But oxygen, for example, has an atomic number of 8, with some individual oxygen atoms containing 8 neutrons, some atoms containing 9 neutrons, and some containing 10 neutrons.

1-2 CHARGE

Protons and electrons are important to the study of electricity because they possess a quality called **electric charge.** We say that protons have a **positive charge** (+), and electrons have a **negative charge** (–). Neutrons have no charge. Charge is important because particles that have unlike charges tend to attract each other, and particles that have like charges tend to repel each other. These ideas are shown in Figure 1-2.

An intact atom has zero net charge. This is because the total positive charge in the protons is balanced by an equal amount of negative charge in the electrons. If we focus our attention on *metal* atoms like copper, aluminum, iron, silver, or lithium, the atoms have no effect on each other. The neutral (zero) charge of one atom means that it neither attracts nor repels the electrons in a neighboring atom. For example, if a great number of lithium atoms are packed together in a solid, as shown in Figure 1-3, the atomic particles simply hold their positions.

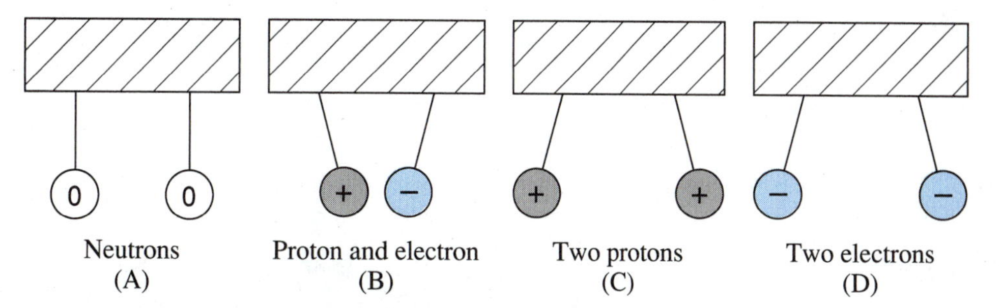

FIGURE 1-2 If we could hang individual subatomic particles on strings, this is how they would react. (A) Neutrons would hang free, since they have no charge. (B) Unlike charges attract. (C) Like charges (both positive) repel. (D) Like charges (both negative) repel.

1-3 CONCENTRATING CHARGE

Figure 1-3 points out that some electrons are physically closer to the nucleus, while others are farther away. The closer electrons are tightly held to the atom by the attraction force of the nucleus. The outer-shell electrons are more loosely held.

For some materials, it is possible to pull loosely held outer-shell electrons away from their atoms. For instance, if a glass object is rubbed with a smooth piece of cloth, outer electrons from the atoms in the glass are pulled loose. These electrons move over to the piece of cloth and crowd into the outer shells of the cloth atoms. Then the two objects are no longer electrically neutral. They are said to be charged, as shown in Figure 1-4. This is the familiar static electricity that everyone has experienced. Of course, we have more effective ways of transferring charge between objects, other than rubbing them together.

3

One important application of static electricity is for removing dust particles from smokestacks. As shown in Figure 1-5(A), a group of positively charged rods are located between negatively charged collecting plates. As the dust particles pass over the rods, they lose some of their excess electrons to the rods. That causes the dust particles themselves to have a net positive charge. The negatively charged plates attract and hold onto the positively charged dust particles, removing them from the air stream. Periodically, the captured dust particles must be cleaned off the plates. This is done by returning the rods and plates to a neutral condition and rinsing the plates. An actual view of an **electrostatic** precipitator is shown in Figure 1-5(B). Electrostatic means using static electricity.

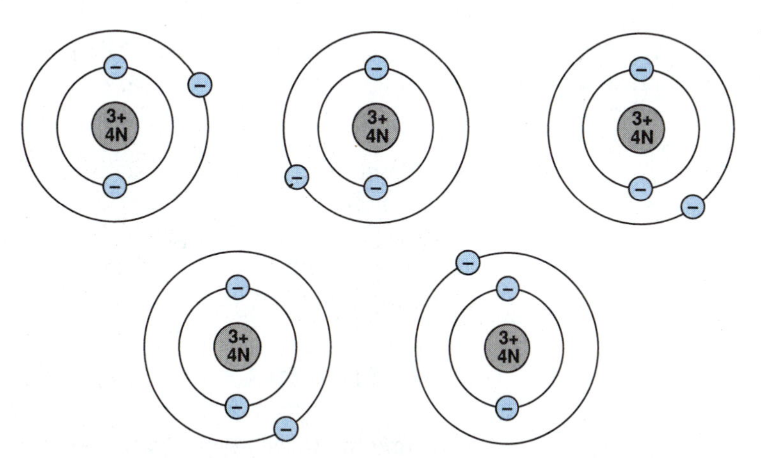

FIGURE 1-3 Lithium has atomic number 3. Two of the electrons orbit closer to the nucleus, in the "inner shell." One of the electrons orbits farther away, in the "outer shell."

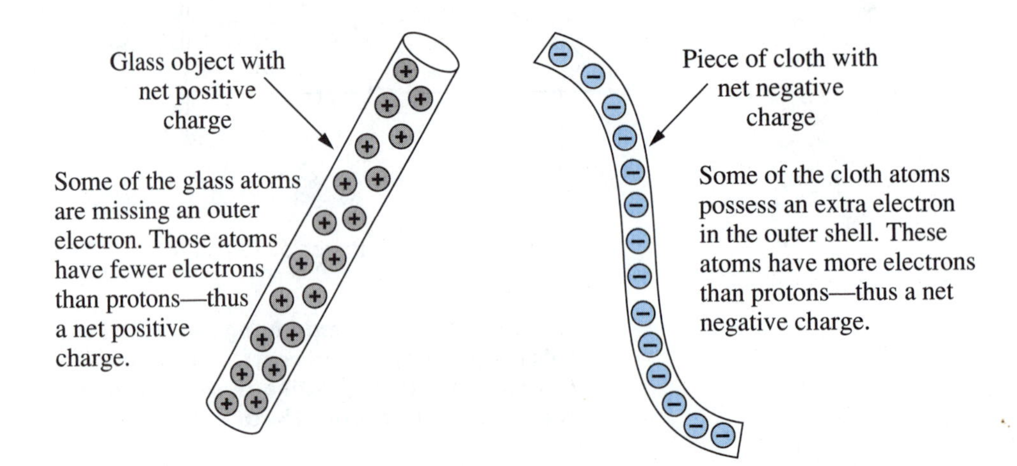

FIGURE 1-4 Static electricity is produced by transferring electrons from one object to another. The object that the electrons are taken from becomes net positive. The object that the electrons are deposited on becomes net negative.

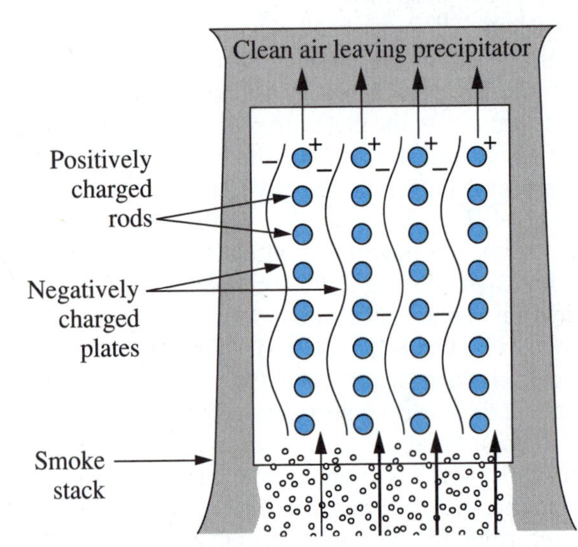

(A)

(B)

FIGURE 1-5 (A) Diagram of an electrostatic precipitator. The positive rods and negative plates are held in a charged state by electrical devices not shown here. (B) An electrostatic precipitator.
Courtesy of American Air Filter

Another application of static electricity is for electroplating. In that process, the solid object to be plated is made positive, while a molten liquid bath of the metal being deposited is held negative. When the positive object is immersed into the negative bath, some of the liquid atoms cling to the positive object and become permanently bound by chemical reaction.

✔ SELF-CHECK FOR SECTIONS 1-1, 1-2, AND 1-3

1. Name the three atomic particles that make up atoms. [1]
2. The nucleus of an atom contains __________ and __________. [1]
3. Which particles are located toward the outside of an atom, orbiting around the nucleus? [1]
4. Describe the electric charge possessed by each atomic particle. [1]

5. Unlike charges _________ each other; like charges _________ each other. [2]
6. Suppose that a large number of electrons are pulled away from the atoms of object A and are placed into the atoms of object B. Object A becomes net _________ charged and object B becomes net _________ charged. [3]
7. If an electron is removed from an atom, does it come from the outer shell or from an inner shell? [1]
8. If a free electron is placed into an atom, does that extra electron go into the outer shell or into an inner shell? [1]
9. In spray-painting a complicated object, the paint droplets can be charged negative by making them flow past negatively charged rods. To make the paint stick, the complicated object must then be charged _________. [2]

1-4 CURRENT

Look again at Figure 1-3, which represents many intact lithium metal atoms packed together as a solid. Let us place that metal solid between two static-charged objects, as shown in Figure 1-6. Under these conditions, outer-shell electrons will tend to move from left to right through the solid block. Here is why.

In those atoms near the left side of the block, electrons are strongly repelled from the negatively charged object, because of its closeness. In those atoms near the right side of the block, electrons are strongly attracted to the positively charged object. In the center of the block, electrons are affected by both charged objects. The inner-shell electrons are tightly held and cannot break away from their parent atoms. But the loosely held outer-shell electrons can break away in reaction to these forces.

When a loosely held electron jumps away, it lands in the outer shell of a neighboring atom to its right. This is illustrated in Figure 1-7. Then that atom has only three protons attempting to hold onto four electrons, so it has an even looser hold on the outer-shell electrons. Therefore, it is even more likely that one of those electrons will jump into the next atom to the right. Figure 1-7 shows this action.

With the huge number of atoms that are present, this process becomes a massive migration of electrons throughout the entire cross section of the solid block. This overall migration of electrons is called electric **current.**

In Figure 1-7, it is true that current cannot continue to flow indefinitely. Eventually the right side of the block will become concentrated with electrons that have no way to escape, since the block is not touching the positive object. Similarly, the left side will have a concentration of excess protons. Therefore , the two sides of the solid block eventually acquire net charge, which is opposite to the

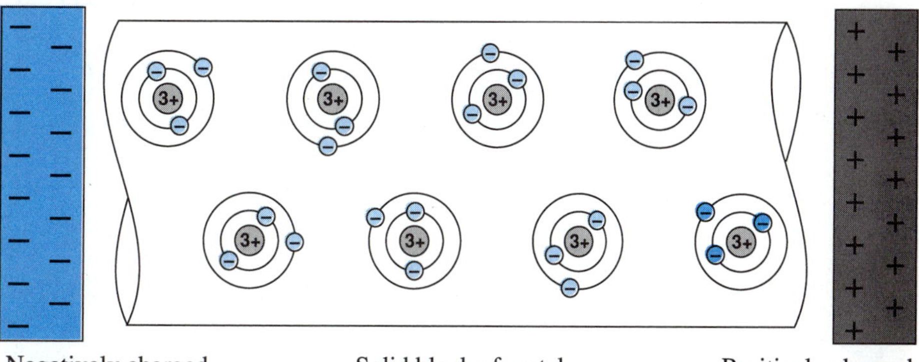

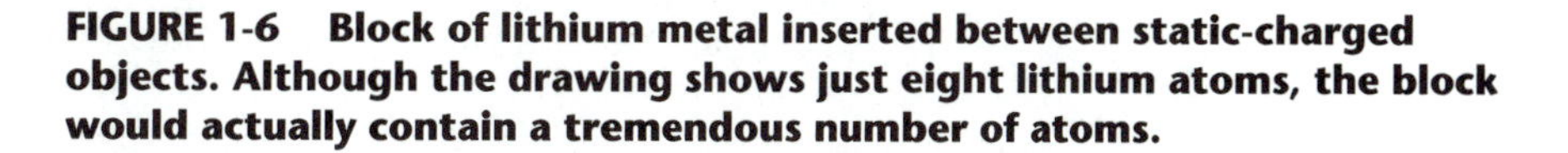

FIGURE 1-6 Block of lithium metal inserted between static-charged objects. Although the drawing shows just eight lithium atoms, the block would actually contain a tremendous number of atoms.

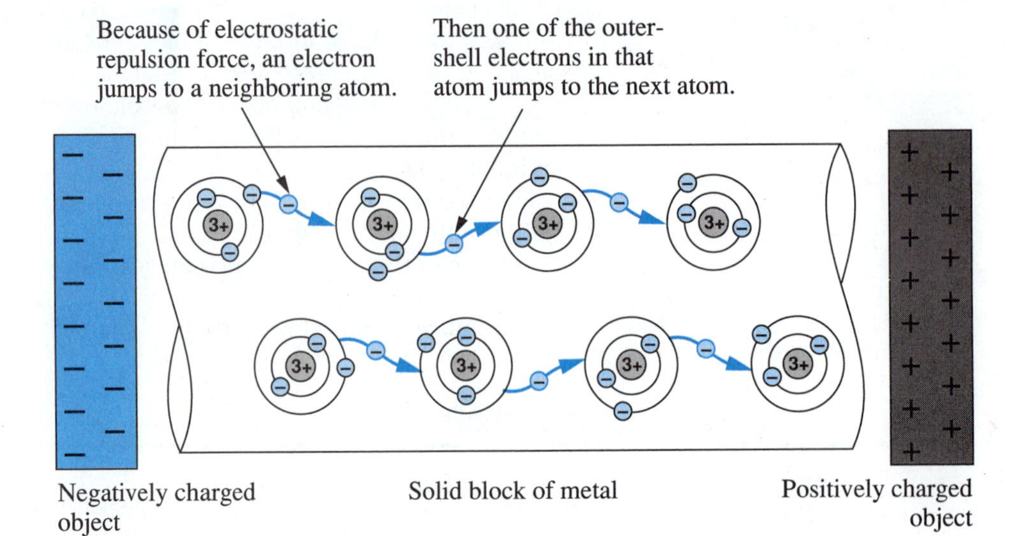

FIGURE 1-7 Migration of loosely held electrons resulting from electrostatic repulsion and attraction

nearby static-charged objects. This cancels their effect and stops the current flow. So Figure 1-7 is not really a practical electric circuit. It is just a useful arrangement for explaining the nature of current. In a practical electric circuit there must be a method for keeping the current flowing indefinitely, or at least for a long time. We will see later how this is accomplished.

In Figures 1-6 and 1-7 we show lithium as the solid material that is carrying electric current. Lithium was chosen for illustration purposes only. With an atomic number of only 3, lithium atoms are simple to draw. Actually, other metals are much more commonly used as current carriers. Copper, atomic number 29, has one loosely held electron in its outermost shell. It is the most commonly used metal for carrying electric current. Aluminum, atomic number 13, has three loosely held electrons. It is also popular for making electric wire. For difficult corrosion applications, silver and gold, both having one loosely held electron, are sometimes used.

Electric current can flow through liquids and gases too. The subatomic flow mechanism is similar to that shown in Figure 1-7. If the direction of electron movement is always the same, as it would be in Figure 1-7, the current is called **direct current**, abbreviated **dc.** This is the kind of current that occurs in a flashlight or an automobile circuit.

Alternating current, abbreviated **ac,** reverses direction periodically. It is the kind of current that occurs in residential circuits. In a residential appliance such

Every manufacturer of instruments or systems needs a staff of skilled service technicians. *Courtesy of Hewlett-Packard Company*

WATER PURIFICATION

The system shown here purifies pool and pond water without the use of chlorine or bromine. Water is pumped through a plastic fitting that holds two electrodes of silver-copper alloy, shown at top. An electronic controller varies the amount of current that is passed through the flowing water between the electrodes. This in turn controls the concentration of silver atoms and copper atoms in the overall body of water. These atoms lose electrons from their outer shells as they enter the water from the metal electrodes. Such altered silver atoms, called silver *ions*, are especially lethal to various bacteria. Copper ions are lethal to algae. The technician shown here is running a comprehensive testing procedure to measure water purity. Page 1 in the Applications color section shows the appearance of a fishing pond before and after purification with the silver-copper ion system.

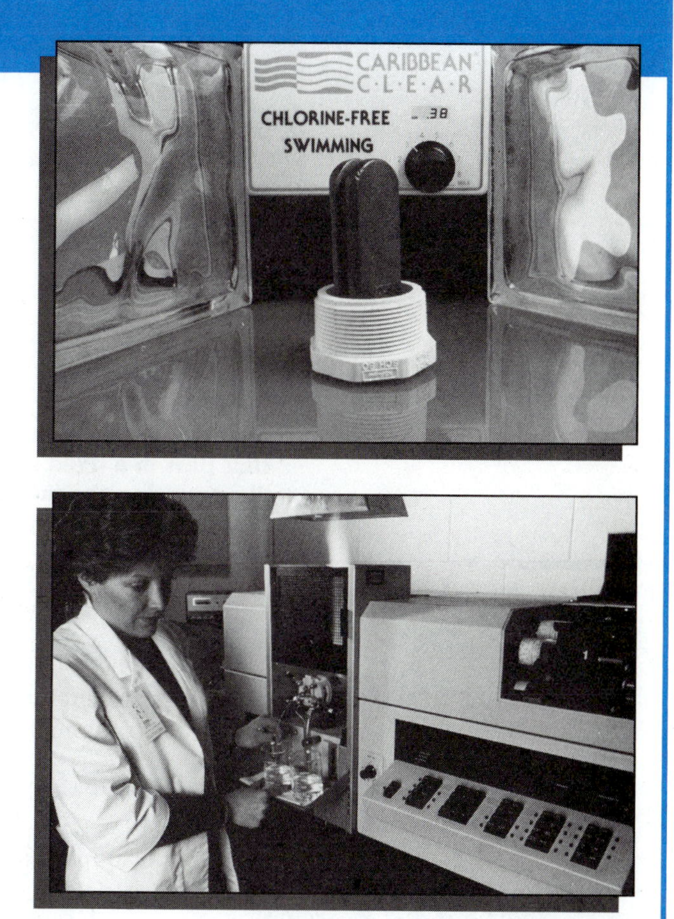

Courtesy of NASA

as a reading lamp, the current reverses and then reverses back again every 1 / 60 second. In a radio circuit, the current may reverse and then reverse back again one hundred million times per second. Because of this rapid reversal in an ac circuit, individual electrons don't actually travel much distance through the material. However, that doesn't matter as far as the proper operation of the circuit is concerned. The only thing that is necessary for proper operation of an ac circuit is that electrons be in the process of motion at any moment in time.

1-5 AMPERE—THE UNIT OF CURRENT

We must have a method for describing the amount of current that is flowing—the intensity of the electron migration. Naturally, we are not satisfied just to say that it is weak or it is strong. We need a way to measure in numbers exactly how weak or strong it is.

To measure current, it is not enough to just tell how many electrons have been moved. Rather, we must tell how many electrons have been moved in a specific amount of time. This idea is like measuring the flow of traffic on a road. We can't simply say how many cars went by. We must say how many cars went by in one hour or how many cars went by in a 24-hour day. The similarity of measuring current and measuring traffic is pointed out by Figure 1-8.

To measure current, we use the unit of **ampere,** symbolized A. Here is the definition of an ampere.

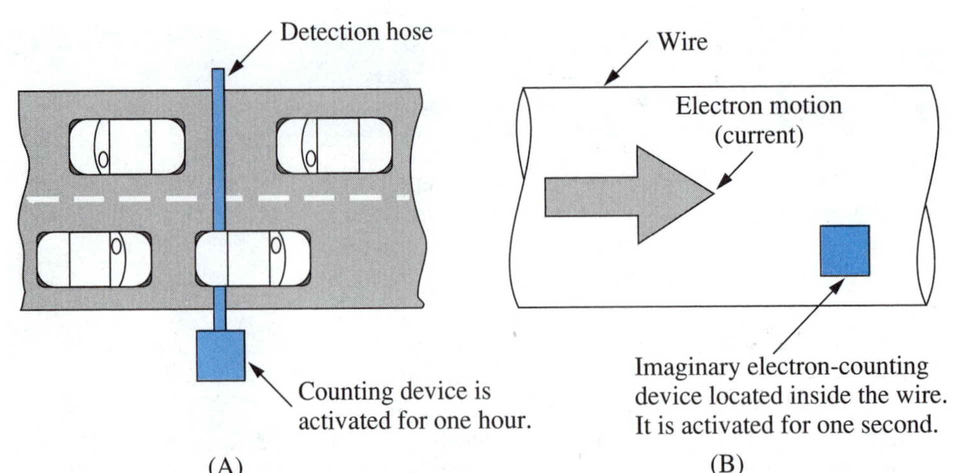

FIGURE 1-8 (A) Measuring traffic involves counting the number of cars that pass in a fixed time interval, say one hour. (B) Measuring current is like counting the number of electrons that pass in a fixed time interval of one second.

TECHNICAL FACT

One ampere is the passage of 6 240 000 000 000 000 000 electrons in an elapsed time of one second.

5

This number of electrons (6.24 billion billion) is so large that it is inconvenient to use. Therefore, we have defined the base unit of charge, the **coulomb,** as the charge contained by that many electrons. As a definition:

TECHNICAL FACT

1 coulomb (C) = the amount of charge in 6.24 billion billion electrons.

Then the ampere can be stated as:

TECHNICAL FACT

$$1 \text{ ampere} = \frac{1 \text{ coulomb}}{1 \text{ second}} \text{ or } 1 \text{ A} = 1 \text{ C / s}$$

One basic unit of current (1 A) is equal to one basic unit of charge (1 C) divided by one basic unit of time (1 s). This fundamental relationship holds true for any values, not only one basic unit. Current is always equal to charge divided by the amount of time needed to move the charge. We write

6

$$I = \frac{Q}{t}$$

EQ. 1-1

where I symbolizes the idea of current, Q symbolizes the idea of charge, and t symbolizes time.

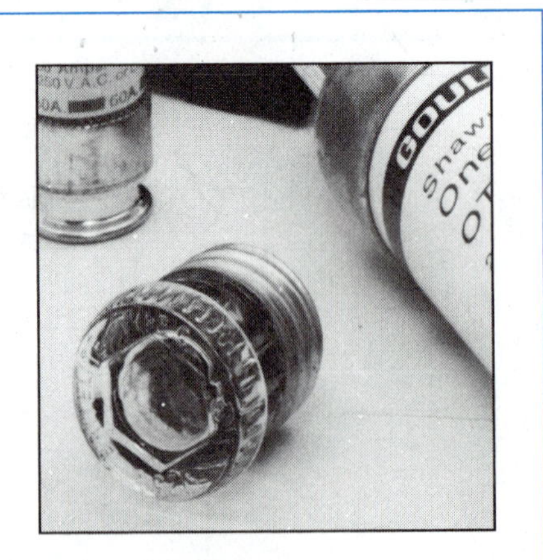

Fuses are used to prevent current from becoming dangerously large. *Courtesy of Gould, Inc., Circuit Protection Division*

EXAMPLE 1-1

Suppose that we measure 12.48 billion billion electrons passing a point in a wire in an elapsed time of 10 seconds.

a) What amount of charge has moved, in basic units?
b) What is the value of current in the wire?

SOLUTION

a) The number 12.48 billion billion electrons is 2 times as large as 6.24 billion billion electrons. Therefore, that larger number represents **2 coulombs (2 C).**
b) Inserting the known values into Equation 1-1 gives

$$I = \frac{Q}{t} = \frac{2\ \text{C}}{10\ \text{s}} = \mathbf{0.2\ A}$$

✓ SELF-CHECK FOR SECTIONS 1-4 AND 1-5

10. What caused the electrons to start jumping from atom to atom in Figure 1-7? Why didn't they jump in Figure 1-3? [2, 4]
11. Which electrons actually do the jumping in Figure 1-7, the inner-shell electrons or the outer-shell electrons? Explain why. [1]
12. If the electron motion is always in the same direction, the current is called __________ current. If the electrons reverse direction periodically, the current is called __________ current. [4]
13. Which type of current occurs in battery-operated devices like flashlights, calculators, and toys? [4]
14. Which type of current occurs in household devices like ovens, blenders, and air-conditioners? [4]
15. What is the letter symbol for the idea of current? [6]
16. What is the basic measurement unit of current? Give its letter symbol. [6]
17. What is the letter symbol for the idea of charge? [5]
18. What is the basic measurement unit of charge? Give its letter symbol. [5]
19. One coulomb passing a point in one second means a current of __________. [6]
20. How many electrons are needed to make 5 coulombs of charge? [5]
21. If this amount of charge (5 C) flows in 20 seconds, what is the current value? [6]
22. How much charge is contained by 3.12 billion billion electrons? If this amount of charge flows in a time of 0.2 second, what is the current value? [6]

FORMULAS

$$I = \frac{Q}{t}$$

EQ. 1-1

SUMMARY OF IDEAS

- Electrons and protons possess electric charge, which causes them to exert attraction and repulsion forces.
- Like charges (either both positive or both negative) repel; unlike charges attract.
- Static electricity is charge concentration resulting from the transfer of a large number of electrons from one object to another.

- On a microscopic scale, electric current is the movement of electrons.
- On a broader scale, electric current is the time flow rate of charge.
- The basic measurement unit of charge is the coulomb, symbolized C.
- The basic measurement unit of current is the ampere, symbolized A.
- 1 ampere = 1 coulomb passing in 1 second.
- Current, I, can be calculated as charge, Q, divided by time, t.
- Direct current (dc) does not change direction; alternating current (ac) changes direction periodically.

CHAPTER QUESTIONS AND PROBLEMS

1. Name the three atomic particles that make up an atom. [1]
2. Describe the electric charge on each type of atomic particle. [1]
3. In an intact atom, the number of electrons is always __________ the number of protons. [1]
4. In an intact atom, the overall net charge is __________. [1]
5. If an object has extra electrons present in its atoms, its net charge is __________. [1]
6. If an object has a shortage of electrons present in its atoms, its net charge is __________. [1]
7. If an object with excess positive charge is near an object with excess negative charge, the two objects will __________ each other. [2]
8. If an object with excess positive charge is near an object that also has excess positive charge, the two objects will__________each other. [2]
9. Give another example of two objects repelling each other, different from the example in Question 8. [2]
10. The device that removes particles of dirt from a smokestack is called an __________. [3]
11. When electrostatic attraction and repulsion forces cause massive migration of electrons, that migration is called electric __________. [4]
12. Name some of the popular metals used to carry current.
13. A lithium metal atom has __________ electron(s) in the inner shell and __________ electron(s) in the outer shell. [1]
14. T–F All of the popular current-carrying metals have at least one loosely held electron in the atom's outer shell. [1, 4]
15. T–F Electric current can pass through a solid, but not through a liquid. [4]
16. The basic measurement unit of charge is the __________ ; its symbol is __________. [5]
17. The basic measurement unit of current is the __________; its symbol is __________. [6]
18. One ampere is one coulomb per one __________. [6]
19. Find the current in amperes if a charge of 15 C flows in an elapsed time of 5 s. [6]
20. Repeat Problem 19 if a charge of 15 C flows in an elapsed time of 1 minute. [6]
21. Repeat Problem 19 if 6.24 billion billion electrons pass a point in an elapsed time of 25 s. [5, 6]
22. Suppose the current is 2 A. In an elapsed time of 30 seconds, how much charge moves? [6]
23. The mathematical symbol for the concept of charge is the letter __________. [5]
24. The mathematical symbol for the concept of current is the letter __________. [6]
25. In an ac circuit, does the current reversal happen quickly or only occasionally? [4]

CHAPTER 2

VOLTAGE AND VOLTAGE SOURCES

Some hospitals use electronic patient-monitoring systems to show cardiac waveforms and other information. The system being set up by these technicians monitors eight patients. *Courtesy of NASA*

OUTLINE

NEW TERMS TO WATCH FOR

voltage
electromotive force
volt
battery
electrode
electrolyte
recharge
battery charger
schematic symbol
generator
rectifier
dc power supply
solar cell

Voltage can be thought of as an electron-moving force. In fact, it is sometimes called EMF, which stands for electromotive force. Voltage and the common devices that produce voltage are the topics of this chapter.

After studying this chapter, you should be able to:

1. Describe the nature of voltage.
2. Specify voltage in basic units of volts.
3. Know the clear difference between voltage and current.
4. Describe the four major kinds of voltage sources: batteries, generators, electronic rectifiers, and solar cells.

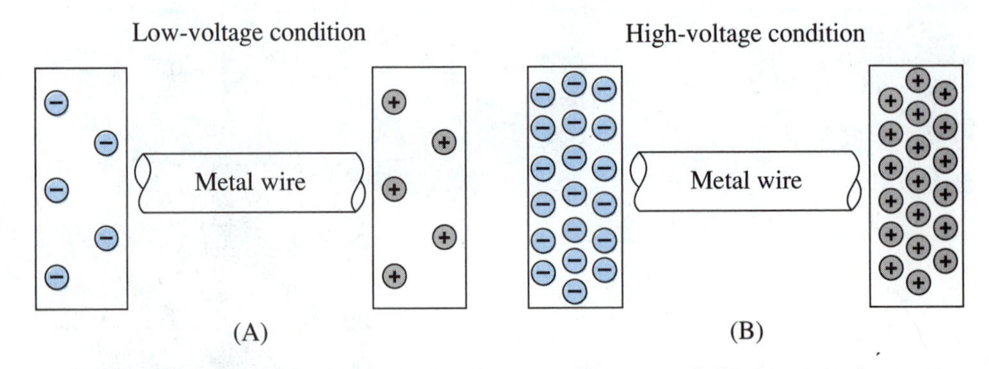

FIGURE 2-1 (A) Weaker forces on the metal's outer-shell electrons because of small net charge on the objects. (B) Stronger forces on the metal's electrons because of great net charge on the objects.

2-1 THE VOLTAGE IDEA

Look at the contrast between Figures 2-1(A) and 2-1(B). There will be some electrostatic force on the metal's electrons in Figure 2-1(A). But the force will be much greater in Figure 2-1(B). This is because the attraction and repulsion forces are greater if the net charge is more highly concentrated. We would say that the setup in part B has greater **voltage,** since the electron-moving force is greater. An electron-moving force is often called an **electromotive force.**

To gain a good understanding of voltage, it is helpful to make an analogy to a water-tank situation. The electrical difference in Figure 2-1 is like the liquid difference in Figure 2-2. The height of water in the tank in Figure 2-2(B) is greater than in Figure 2-2(A). The force tending to push water through the supply pipe depends on the total weight of water above the pipe. Therefore, the greater amount of water in Figure 2-2(B) provides more water-moving force than in Figure 2-2(A),just like the greater amount of net charge in Figure 2-1(B) provides more electron-moving force than Figure 2-1(A).

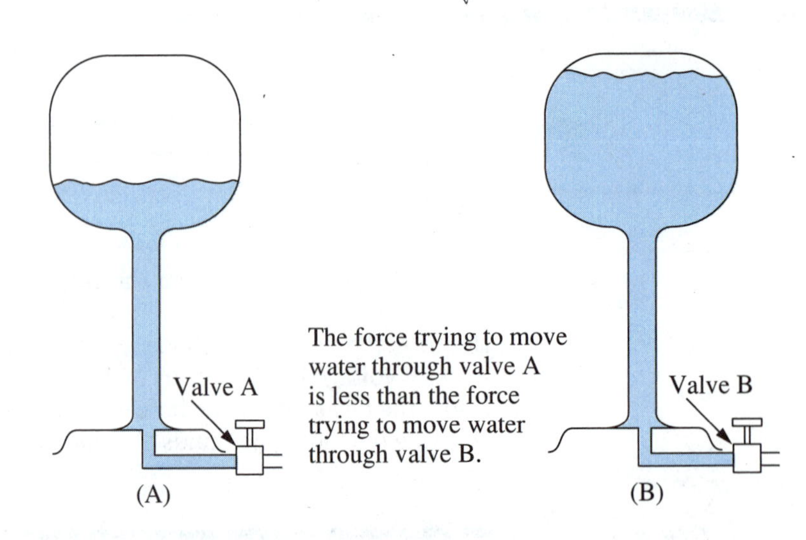

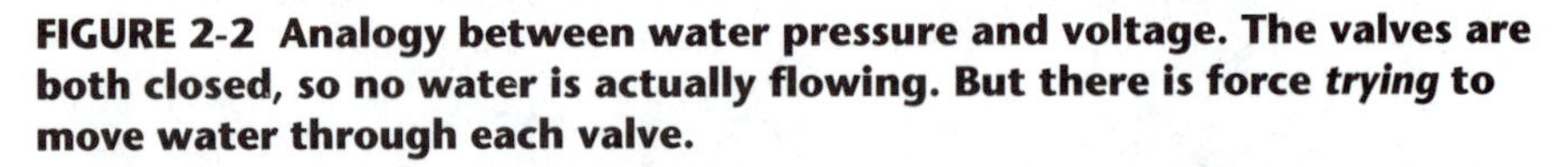
FIGURE 2-2 Analogy between water pressure and voltage. The valves are both closed, so no water is actually flowing. But there is force *trying* to move water through each valve.

2-2 MEASUREMENT UNIT—THE VOLT

As always, a new concept requires a new measurement unit. The basic measurement unit for voltage is the **volt**, symbolized V. So the words for the concept and the measurement unit are the same root word. With current, the words for the concept (current) and the measurement unit (ampere) are completely different.

Furthermore, the letter symbol for the idea of voltage is also *V*, the same letter that is used for the unit. This too is unlike the situation with current, where the letters are different—*I* for the idea of current and A for the unit.

Thus, to make the symbolic statement that "the voltage equals 6.3 volts," we write

$$V = 6.3\ \mathrm{V}$$

(In a printed book, there is a distinction between the two Vs. The *V* standing for voltage is italic—leaning to the right; the V standing for volt is roman—straight. But when people write by hand, they make no distinction.)

When we introduced the ampere unit in Chapter 1, we defined it in terms of more basic units, namely coulombs and seconds. That made it easier to understand the ampere. We cannot do the same thing for the volt at this time, because we haven't yet studied the other units that would have to be used. That will come later.

For now, just understand voltage as an electron-moving force. A weaker force is specified by a smaller number of volts; a stronger force by a larger number of volts.

EXAMPLE 2-1

In Figure 2-1, suppose that a test indicates that the voltage in part A is 4 volts, and the voltage in part B is twice as large. Write these facts using symbols.

SOLUTION

With the subscripts A and B used to distinguish between the two voltages,

$$V_A = 4\ \mathrm{V}$$

and

$$V_B = 8\ \mathrm{V}$$

An automatic printed-circuit board test setup. Every critical function of the circuit board is tested in a few seconds, and the board is assigned a pass or fail. If it fails, the program is able to locate the specific component that malfunctioned. *Courtesy of Tektronix, Inc.*

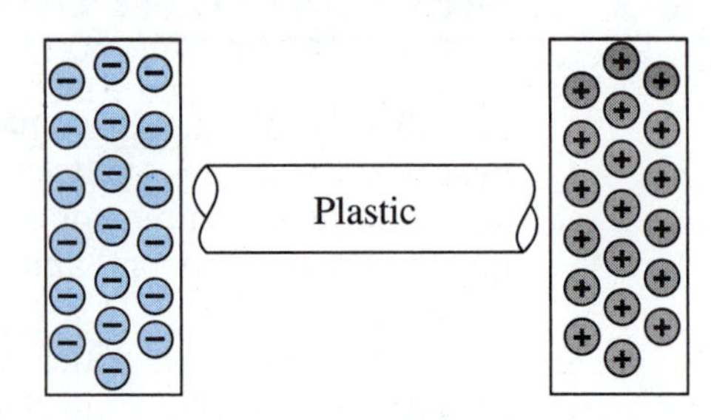

FIGURE 2-3 In plastic, the atoms have only tightly held electrons. Even though there is attraction/repulsion force (voltage) trying to make the electrons migrate, none of them are able to break free from their atoms. So there is no charge movement (no current).

2-3 VOLTAGE AND CURRENT ARE DIFFERENT THINGS

Do not confuse voltage with current. They are not the same thing. There is, however, a definite mathematical relationship between them that causes one to get larger if the other gets larger. This relationship will be described in detail in Chapter 6.

Voltage is like a force. By itself, it is not motion. Current is motion—the motion of charge. It is possible to have voltage without current, just like it is possible to have mechanical force without motion. Mechanically, this happens whenever the object receiving the force is attached so that it can't move. It happens in the fluid situation of Figure 2-2 also, where the water is feeling a force, but can't move because the valves are shut. An example of voltage without current occurs by replacing the metal wire of Figure 2-1 with a material, such as plastic, that has no loosely held electrons. This is shown in Figure 2-3.

It is not possible to have current without voltage. Current can't happen unless some voltage exists to cause it to happen. In your early study of electricity, this is a good way to think about these two concepts: voltage is the cause, current is the effect.

A further water analogy is helpful. In Figure 2-2, if we open both valves, water will begin to flow in both systems. The greater force on the water in part B causes a greater flow rate, as illustrated in Figure 2-4. Force is the cause; water flow is the effect.

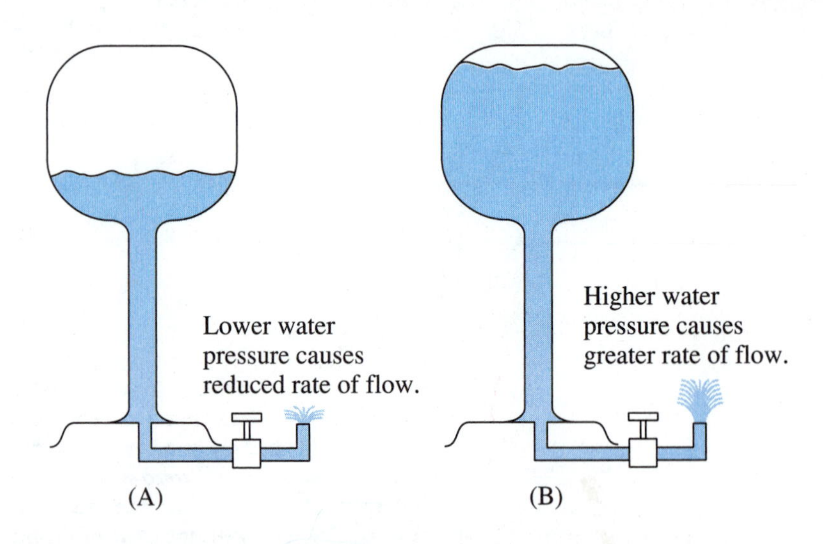

FIGURE 2-4 A water analogy will help you understand the difference between voltage and current. Voltage is like pressure; current is like flow. This analogy is helpful for initial learning. It is not rigorously true.

For electrical comparison, the high-voltage condition in Figure 2-1(B) would cause greater current flow than the lower-voltage condition in Figure 2-1(A). Again, the force-type variable (voltage) is the cause. The flow-type or motion-type variable (current) is the effect.

When speaking about current, it is proper to use the word *through*. A phrase like " the current through the wire" is all right, because current really does mean that charge is passing through something. However, when speaking about voltage, you should not say *through*. Voltage, by itself, doesn't convey motion. It doesn't necessarily mean that charge is passing through anything. Think of voltage as sitting still, trying to force charge to move from one place to another, but not necessarily succeeding. The proper word to use with voltage is *across*. This will become clearer when we learn about circuit construction in later chapters.

✓ SELF-CHECK FOR SECTIONS 2-1, 2-2, AND 2-3

1. In electrostatics, a greater amount of concentrated charge produces _________voltage. [1]
2. The letters EMF stand for_________ _________ _________. [1]
3. The basic unit of voltage is the_________. [2]
4. T-F Changing the voltage from 5 V to 10 V is equivalent to doubling the force exerted on the outer-shell electrons. [1]
5. Which concept, voltage or current, conveys the idea of motion? [3]
6. Which concept, voltage or current, conveys the idea of time? [3]

SATELLITE IMAGING

The United States National Aeronautics and Space Administration (NASA) developed the LANDSAT series of satellites for surveying the earth's surface. These satellites carry electronic sensors that detect the sun's electromagnetic radiation that is reflected back from the earth. By radio transmission, a satellite sends a "picture" of the reflected radiation pattern to a ground receiver. The pattern is processed electronically by computer to develop a color-enhanced image. From such images like the one shown here of Corpus Christi, Texas and the Gulf of Mexico, experts can get useful information.

Some of the activities that are aided by satellite imaging are measuring atmospheric conditions, detecting environmental changes, making crop estimates, locating promising areas to explore for oil or minerals, finding likely fishing spots in the ocean, and even locating people who are adrift at sea or lost in remote land areas.

In the photograph, a LANDSAT satellite is being checked prior to being launched. Both of these photos are shown in color on page 2 of the Applications color section.

Courtesy of NASA

7. Can voltage be present while current is not present? Explain. [3]
8. Can current be present while voltage is not present? Explain. [3]
9. What is wrong with this statement? "The voltage going through the copper wire is 20 V." [3]

2-4 VOLTAGE SOURCES

Dc voltage is voltage that produces dc current. It exerts an EMF that is always in the same direction. The electrostatic arrangement of Figure 2-1 is not a practical means of providing dc voltage. It very quickly loses its ability to function, for the reason explained in Section 1-4. Letting the wire ends touch the charged objects does not solve the problem. It allows the charged objects to return to the neutral state by redistributing the original charge.

4

There are four practical methods of providing usable dc voltage: batteries, generators, rectified power supplies, and solar cells.

Batteries

A **battery** produces voltage by chemical reactions. The fundamental structure of any battery is diagrammed in Figure 2-5(A).

The negative **electrode** is a metal or metal compound that has a chemical reaction with the electrolyte that causes loosely held electrons to become free. The positive electrode is a different metal or metallic compound that has an opposite chemical reaction with the electrolyte. This reaction makes free electrons become tightly held. Both reactions tend to coat the surfaces of the two electrodes with new compounds.

If a flow path exists, electrons will be forced away from the negative electrode, through the path, over to the positive electrode. Figure 2-5(B) illustrates this flow of current.

A battery cannot continue to deliver current forever. Eventually the electrodes become covered with the new chemical compounds formed by the reactions. The two reactions then slow to the point where the battery voltage becomes unusable.

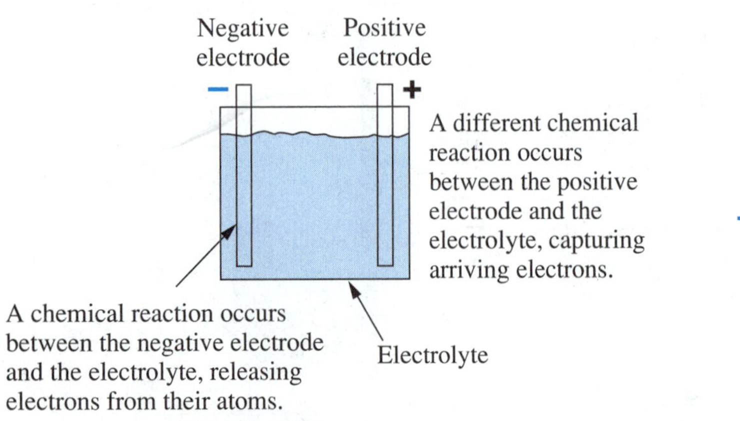

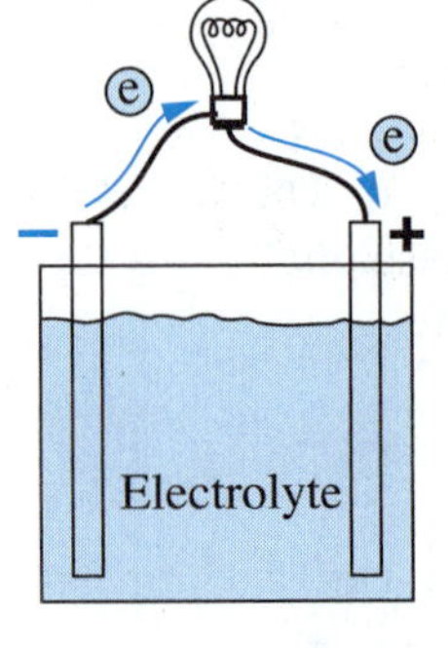

FIGURE 2-5 (A) A chemical battery contains an electrolyte. An electyrolyte can be a liquid or a porous solid soaked with liquid. (B) The negative electrode is a long-term source of electrons. The positive electrode is a long-term capturer of electrons.

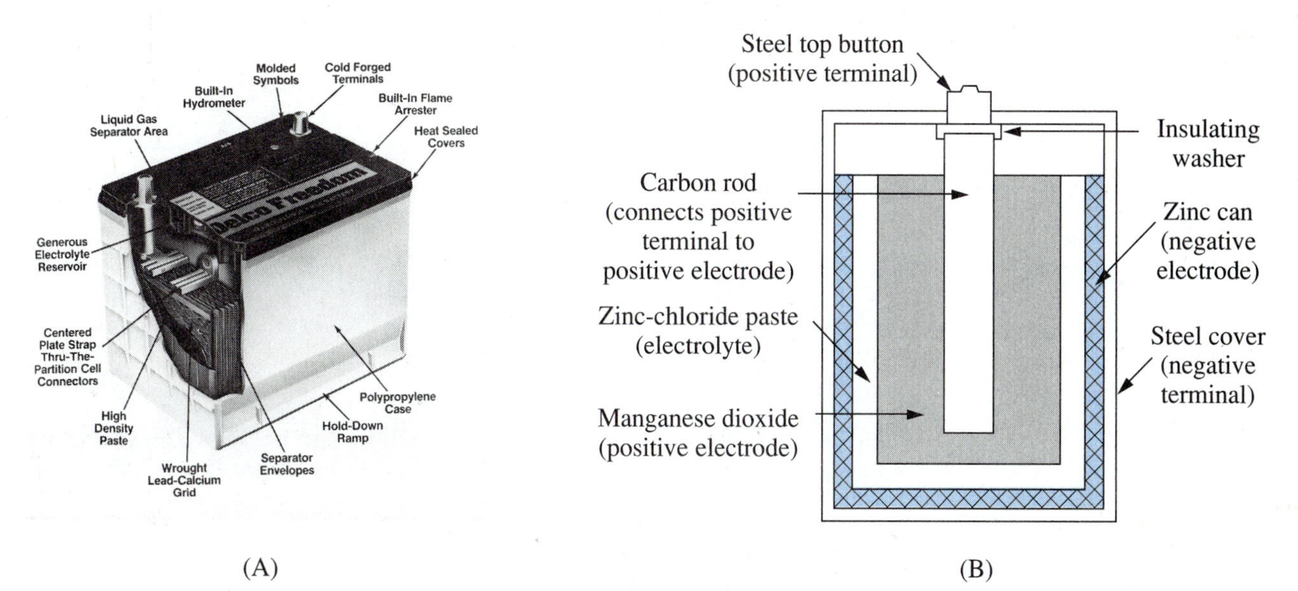

FIGURE 2-6 (A) Cutaway view of lead-acid auto battery. *Courtesy of Delco-Remy Division, General Motors Corporation.* **(B) Cross-sectional view of a carbon-zinc flashlight battery**

Automobile batteries use spongy lead for the negative electrode, and lead dioxide for the positive electrode. The electrolyte is sulfuric acid dissolved in water. An automobile battery stacks together six of the structures shown in Figure 2-5. It is called a six-cell battery. Each cell produces about 2.1 V, so the entire battery has a voltage of about 12.6 V. A cutaway view of an automobile battery is shown in Figure 2-6(A).

Standard-duty flashlight batteries often use zinc for the negative electrode and carbon surrounded by manganese oxide for the positive electrode. The electrolyte is a zinc chloride mixture. Each cell produces about 1.5 V. The structure of a carbon-zinc battery is shown in Figure 2-6(B).

Recharging a Battery. Some electrode/electrolyte combinations produce chemical reactions that can be reversed, or undone. Other combinations have reactions that cannot be reversed. If the reaction is reversible, the battery cell can be recharged after it has been in use, bringing it back to full voltage. Such cells are called secondary cells. The lead-acid automobile cell is an example. Cells that cannot be safely recharged are called primary cells. The alkaline flashlight battery is an example. To recharge a secondary cell, a higher-voltage source must be connected as shown in Figure 2-7.

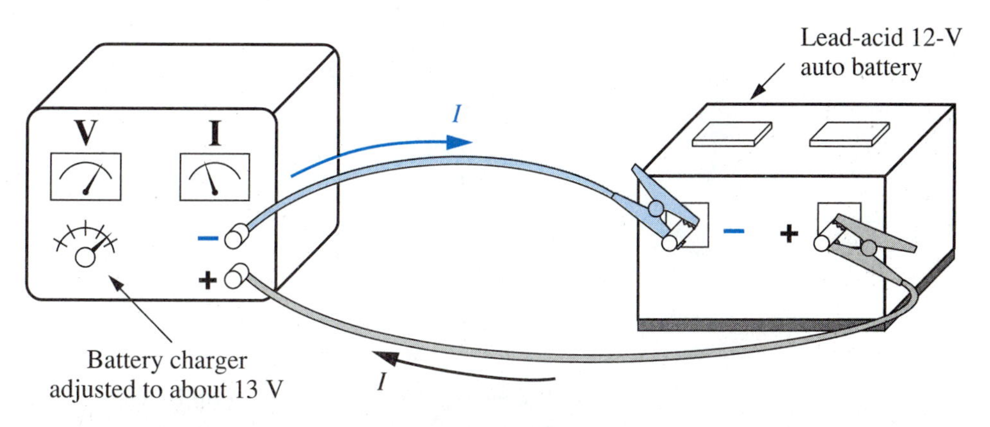

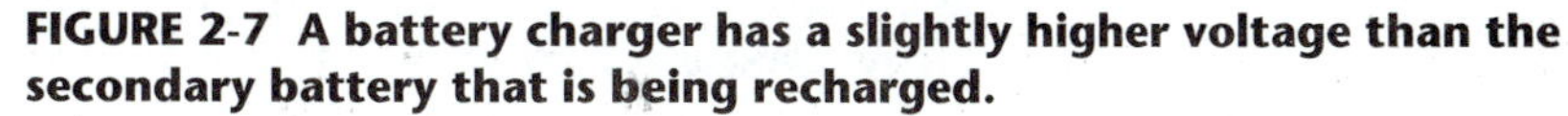
FIGURE 2-7 A battery charger has a slightly higher voltage than the secondary battery that is being recharged.

FIGURE 2-8 (A) Schematic symbol for a single-cell battery. The short line always stands for the negative terminal and the longer line for the positive terminal, even if the + and – signs are left out. The symbol *V* stands for generalized voltage of unspecified value. It may be left out. (B) Symbol for multicell battery. If the voltage value is known, it is usually specified in the drawing. Here the value is specified as 12 volts.

The negative terminal of the **battery charger** is connected to the battery's negative terminal (electrode). The battery charger's positive terminal is connected to the battery's positive terminal. Because the battery charger's voltage is slightly greater, it forces electron current to flow away from its negative terminal, through the battery itself in the reverse direction, eventually returning to the positive terminal of the battery charger.

Since the electron current flow is reversed from the battery's normal direction, its earlier chemical reactions are undone. The battery is thereby recharged to its original condition.

Figure 2-7 shows the actual physical appearances of an auto battery and a bench-top battery charger. Usually, when we draw electric circuits, it is not necessary to show actual true-to-life appearances. We can make our job easier by using symbols for the circuit devices. These **schematic symbols** are chosen for ease and quickness in drawing, not for physical resemblance to the actual devices.

The schematic symbol for a single-cell battery is shown in Figure 2-8(A). A multicell battery is symbolized in Figure 2-8(B).

The symbol in Figure 2-8(A) is also used for other dc voltage sources, besides single-cell batteries. For example, a dc generator voltage source (Section 2-4) could be symbolized as shown in Figure 2-9(A). An adjustable rectified power supply (Section 2-4) could be symbolized as shown in Figure 2-9(B).

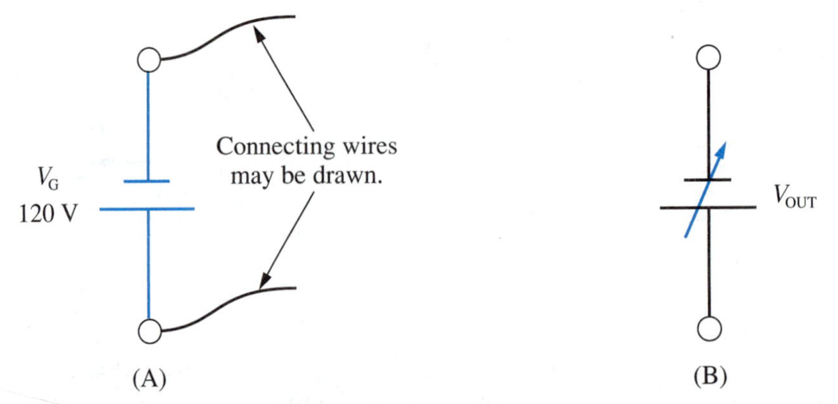

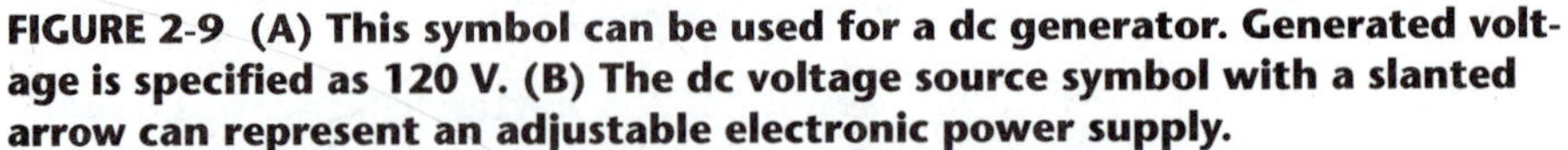

FIGURE 2-9 (A) This symbol can be used for a dc generator. Generated voltage is specified as 120 V. (B) The dc voltage source symbol with a slanted arrow can represent an adjustable electronic power supply.

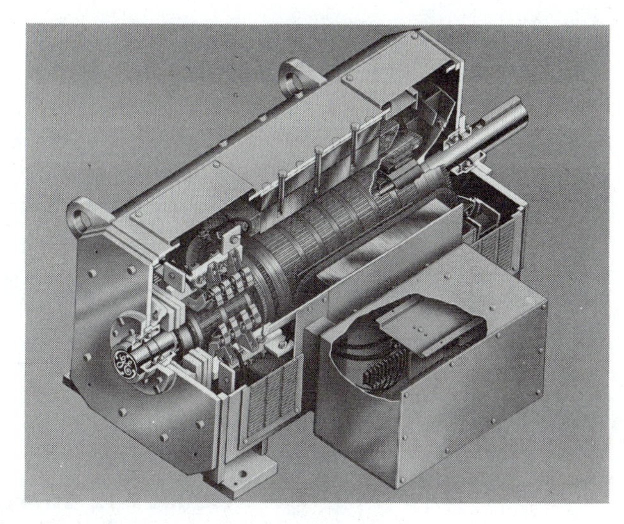

FIGURE 2-10 Cutaway view of a generator. *Courtesy of General Electric Co.*

Generators

A **generator** is a rotating machine that produces voltage by magnetic action. A strong magnet is mounted on the outside part of the machine; it stands still. The inner part of the machine contains a cylinder that rotates at high speed. This cylinder holds wire windings—that is, wire coiled many times. As the cylinder assembly rotates, the windings move past the magnets at high speed. This causes voltage to be created in the windings. This principle, called electromagnetic induction, will be explained in Chapter 14. A cutaway view of a dc generator is shown in Figure 2-10.

Generators outperform batteries in high-voltage applications. Of course, a generator must be forced to rotate by some type of mechanical engine.

Rectified Power Supplies

Dc voltage can be produced by rectifying ac voltage taken from a wall receptacle. An electronic **rectifier** converts ac to dc by allowing one voltage direction to appear but blocking the opposite voltage direction.

A **dc power supply** is a unit that combines electronic rectifiers with other electronic components to provide virtually perfect (constant) dc voltage. A dc power supply is shown in Figure 2-11.

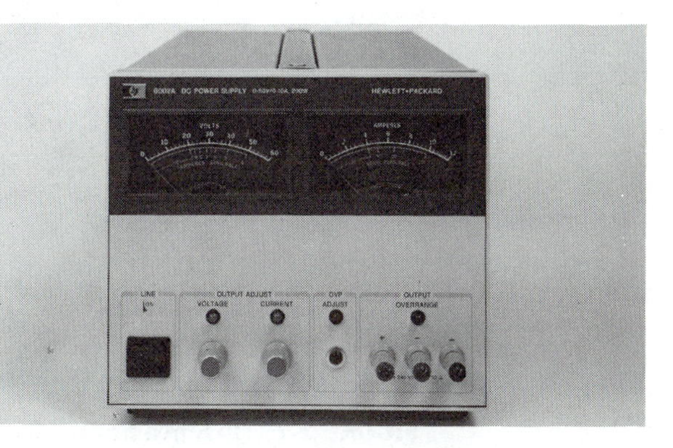

FIGURE 2-11 Adjustable dc power supply. Voltage range is 0–50 V. Maximum continuous output current is 10 A. *Courtesy of Hewlett-Packard Co.*

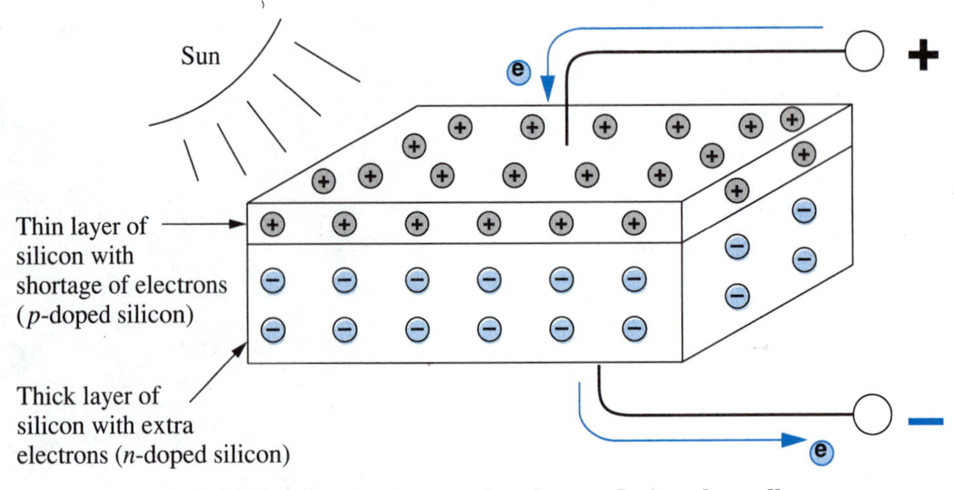

FIGURE 2-12 Structure of a photovoltaic solar cell

Solar Cells

During the last decade, rapid improvements in the technology of solar cells have boosted their importance. The basic **solar cell**, or photovoltaic cell, is diagrammed in Figure 2-12. As sunlight strikes the thin layer of *p*-doped silicon, light-energy particles called photons penetrate the layer, knocking loose more outer-shell electrons. Some of these electrons are knocked across the *p-n* junction boundary into the *n*-doped region. The excess negative charge in that region forces them out the negative lead, as Figure 2-12 shows. If a complete flow path exists, the electrons return via the positive lead to replenish the supply of electrons in the *p*-doped layer.

The research and development effort in solar cells is aimed at moving greater numbers of electrons across the *p-n* boundary. At present, sufficient numbers are knocked across the boundary to convert about 13 percent of the sunlight's energy into electricity for a particular type of solar cell called thin-film amorphous (noncrystal). This makes such solar cells practical for use on homes, as shown in Figure 2-13. They are also used in remote locations where electrical transmission wires are not available. If further improvements are able to increase that figure to about 20 percent, solar generation will become practical for both homes and large-scale commercial applications.

FIGURE 2-13 Private homes can utilize roof-mounted solar panels to supplement their energy consumption. These panels are the thin-film noncrystal type, which can be manufactured at low cost, unlike conventional crystal solar panels. *Courtesy of Energy Conversion Devices, Inc.*

In sunny climates like that of southern California, very large solar panel arrays can produce commercial quantities of electrical energy. This 160-acre solar plant near Bakersfield supplies 6.5 million watts of peak power to the distribution grid of that area's electric utility, Pacific Electric Co. *Courtesy of NASA*

✓ SELF-CHECK FOR SECTION 2-4

10. What are the three essential parts of a chemical battery? [4]
11. Chemical cells that cannot be recharged are called__________cells. Those that can be recharged are called__________cells. [4]
12. Explain why some kinds of cells can be recharged while other kinds cannot be. [4]
13. What two effects must be combined to make a generator work? [4]
14. Using electronic devices to convert ac voltage into dc voltage is called __________. [4]
15. Describe some of the natural advantages of solar cells over the other three common voltage sources. [4]

SUMMARY OF IDEAS

- Voltage is thought of as a force tending to move electrons (charge). It is symbolized *V*.
- The basic measurement unit of voltage is the volt, symbolized V.
- Voltage and current can be understood by water analogy. Voltage is like water pressure; current is like water flow.
- Voltage can cause current only if there is a complete flow path. Without such a path, voltage may be present but there can be no current.
- Voltage doesn't move. Therefore, do not use the word *through* when speaking about voltage.
- Dc voltage exerts electron-moving force always in the same direction. Ac voltage reverses force direction periodically.
- The four important sources of dc voltage are batteries, generators, electronic rectifiers, and solar cells.
- Batteries are inexpensive and easily portable, but they do not produce large values of voltage and current.
- Generators are expensive and not so easily portable, but they can produce large values of voltage and current. They must be forced to spin by a separate machine or engine.
- Electronically rectified dc voltage sources require an ac voltage source. They can produce large values of voltage and current and they are easily adjustable.
- Solar cells are completely renewable and nonpolluting voltage sources that are very useful in specialized applications.

CHAPTER QUESTIONS AND PROBLEMS

1. What letter of the alphabet is used to symbolize the concept of voltage? What letter is used to symbolize the measurement unit of voltage?
2. The pressure in a water piping system is like the__________in an electrical system. [1]
3. T-F It is possible to have voltage, but to have no current. [3]
4. T-F It is possible to have current, but to have no voltage. [3]
5. T-F Larger voltage tends to cause larger current. [1, 3]
6. Use symbols to make this statement: "The voltage across the device labeled *a* equals 25 volts." [2]
7. In a chemical cell, the reaction between the negative electrode and the electrolyte causes what to happen? [4]
8. The reaction between the positive electrode and the electrolyte causes what to happen? [4]
9. Draw a sketch of a chemical cell with an external flow path between the electrodes. Show in which direction the electron current flows. [4]
10. A modern car battery has__________individual cells. [4]
11. What is the approximate output voltage of a car battery? [4]
12. Cells that cannot be recharged are called__________cells. [4]
13. Give an example from daily life of primary cell(s). [4]
14. Cells that can be recharged are called__________cells. [4]
15. Give an example from daily life of secondary cell(s). [4]
16. Give an electrical description of how a battery charger acts on a secondary cell. [1, 4]
17. What does a battery charger do chemically to a secondary cell? [4]
18. Draw the symbol for a single-cell battery.
19. In the symbol for a single-cell battery, the shorter line is the__________terminal, and the longer line is the__________terminal. [4]

To answer Questions 20 through 25, choose one of the four major dc voltage sources.

20. Which one is most portable? [4]
21. Which is environmentally safest? [4]
22. Which is best suited to very large values of voltage and current? [4]
23. Which is best suited for use on a laboratory workbench? [4]
24. Which is used to start automobile engines? [4]
25. Which can be coupled with a farm tractor for use at distant locations on a farm? [4]

CHAPTER 3

RESISTANCE

OUTLINE

This warship is a guided-missile cruiser. Radar signals from its antenna find the location and speed of enemy aircraft. The radar information is electronically processed by two missile control systems shown in the bottom photo. These systems steer the missile toward its target. *Courtesy of Unisys Corporation*

NEW TERMS TO WATCH FOR

resistance	potentiometer
ohm	linear pot
tolerance	audio taper pot
temperature stability	conductor
nominal value	insulator
reliability factor	printed circuit (PC) board
precision resistor	semiconductor
variable resistor	incandescent lamp
rheostat	filament

There are three fundamental ideas in electricity. The first two, current and voltage, we have studied already. Resistance is the third idea.

After studying this chapter, you should be able to:

1. Describe the meaning of electrical resistance.
2. Define the ohm unit in terms of voltage and current.
3. List the four common types of manufactured resistors, and compare their advantages and disadvantages.
4. Use the resistor color code.
5. State the difference between a rheostat and a potentiometer.
6. Tell the relationship between resistance, cross-sectional area, length, and resistivity.
7. Distinguish among conductors, insulators, and semiconductors.

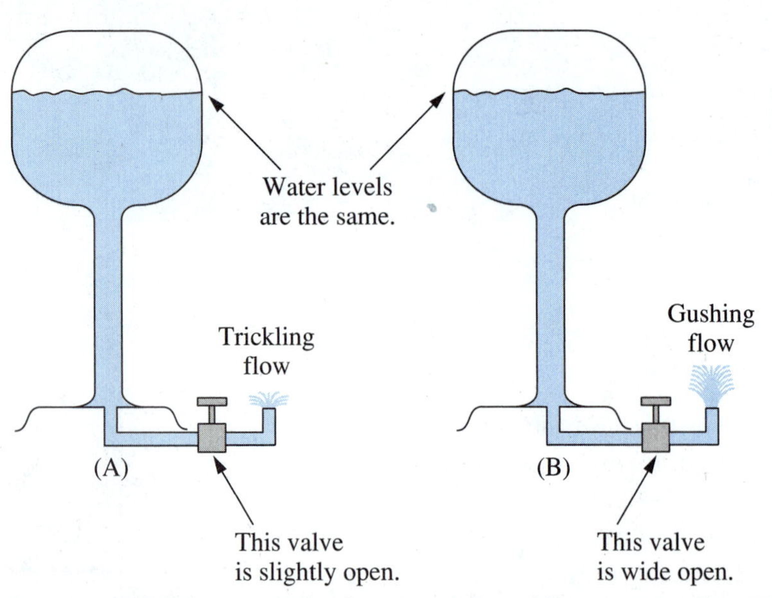

FIGURE 3-1 (A) A slightly open valve has great opposition to water flow. This is like a large resistance in an electrical circuit. (B) A wide-open valve has little opposition to water flow. This is like a small resistance in an electrical circuit.

3-1 MEANING OF RESISTANCE

1

Resistance is the ability to oppose current. If a device allows very little current to pass when voltage is applied, then the device has high resistance. If it allows a large current to flow, the device has low resistance.

Again, a water analogy is helpful for understanding the fundamental meaning. Electrical resistance is like a restrictive valve in a water system. This analogy is shown in Figure 3-1. Large resistance is like a very restrictive valve. Small electrical resistance is like a valve that has little restriction. The letter R is used to symbolize the concept of resistance.

3-2 UNIT OF RESISTANCE—OHM

The basic measurement unit of resistance is the **ohm**, symbolized Ω. Thus, to make the symbolic statement, "the resistance equals 10 ohms," we write

$$R = 10\ \Omega$$

TECHNICAL FACT

One ohm is the amount of resistance that permits one ampere of current to flow when a voltage of one volt is applied.

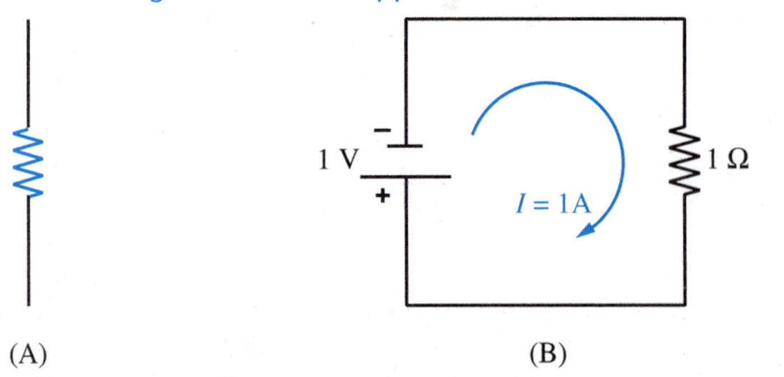

FIGURE 3-2 (A) Symbol for resistance. The letter R with a subscript is often written alongside this symbol. (B) Diagram of electric circuit with a 1-V source connected to a 1-Ω resistance. Current flow is then 1 A.

The schematic symbol for resistance is the zigzag symbol shown in Figure 3-2(A). A 1-V dc voltage source connected to a 1-Ω resistance is shown schematically in Figure 3-2(B). That diagram illustrates the fact of 1 Ω allowing 1A with 1 V applied.

✓ SELF-CHECK FOR SECTIONS 3-1 AND 3-2

1. Draw the schematic symbol for resistance.
2. The basic measurement unit for resistance is the __________.[2]
3. T-F A large resistance in an electric circuit is like a severe flow restriction in a water piping system.[1]
4. If one volt is across one ohm of resistance, what amount of current flows? [2]

3-3 FIXED RESISTORS

Resistance is present in electrical devices whether we want it or not. In many cases, we definitely do want it. A resistor is an electrical component whose specific purpose is to provide resistance. Most resistors are fixed in value. Their resistance in ohms cannot be varied. Some resistors are variable by a manual adjustment. There are four major types of fixed resistors; (1) carbon-composition, (2) deposited-film, (3) cermet, and (4) wirewound.

Carbon-Composition Resistors

A carbon-composition resistor has a piece of resistive material with embedded leads, surrounded by insulating material that is formed into a cylinder. A cutaway view of a carbon-composition resistor is shown in Figure 3-3(A).

The resistive material is a mixture of powdered carbon and powdered insulating material, held together by bonding compound. Its resistance depends on the proportion of carbon to insulating material. Carbon-composition resistors can be manufactured over the range from 2 Ω to 10 000 000 Ω.

The physical appearance of carbon-composition resistors is shown in Figure 3-3(B). The colored stripes indicate the resistance by the standard resistor color code. We will study the color code in Section 3-4.

The physical size of a resistor determines how quickly it can get rid of waste heat. The largest resistor in Figure 3-3(B) can get rid of (dissipate) heat at the rate of 2 watts; the smallest at the rate of 1/8 watt.

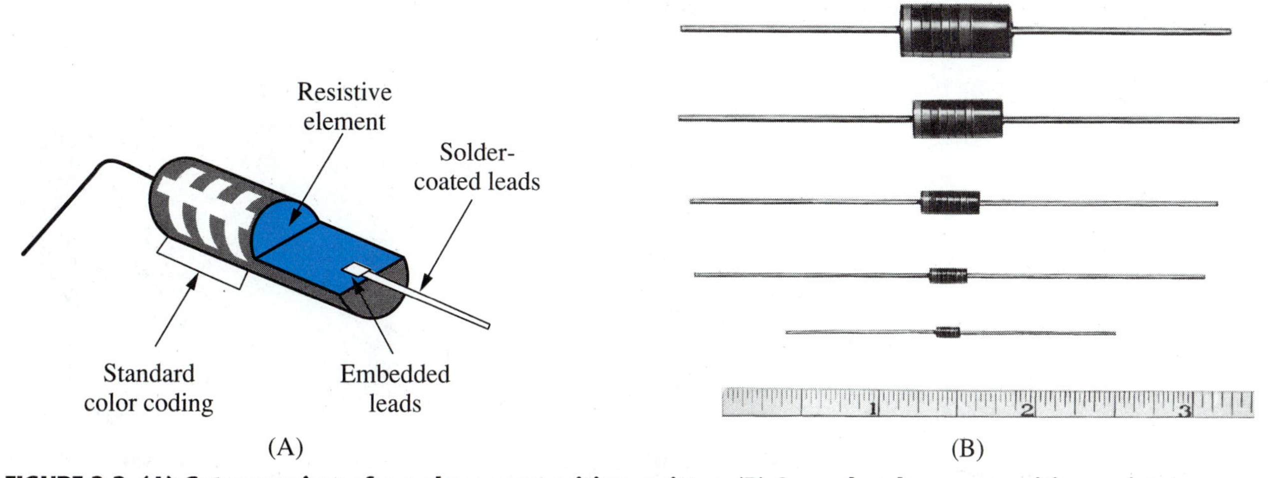

FIGURE 3-3 (A) Cutaway view of a carbon-composition resistor. (B) Several carbon composition resistors.
Courtesy of Allen-Bradley, a Rockwell International Company

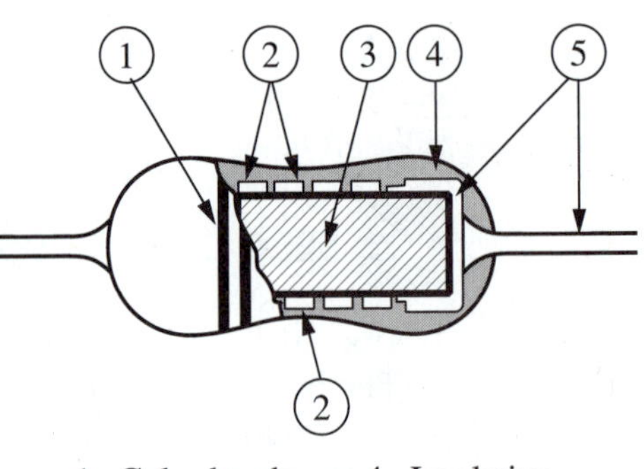

FIGURE 3-4 Cutaway drawing showing the structure of a film resistor

Carbon-composition resistors are usually manufactured to a **tolerance** of ± 5%. That means that the actual resistance is guaranteed to be no more than 5% higher and no less than 5% lower than the indicated resistance. Some carbon-composition resistors are manufactured to ± 10%.

Deposited-film Resistors

A thin film of material can be sprayed onto an insulating ceramic base. Or a film can be deposited on the base by vaporization methods. Either process can be used to make a deposited-film resistor, shown in cutaway view in Figure 3-4. The film material is either carbon or metal. It is usually cut into a spiral pattern around the ceramic substrate base, as Figure 3-4 shows. Several metal-film resistors are pictured in Figure 3-5(A). Carbon-film resistors are shown in Figure 3-5(B).

Deposited-film resistors can be made to closer tolerances than carbon-composition resistors. Carbon-films usually have tolerances of ± 5% or ± 2%. Metal-films are better yet, typically ± 1%.

Deposited-film resistors have good **temperature stability.** This means that their resistance does not change very much when they get hot. As you would expect, deposited-film resistors cost quite a bit more than carbon-composition resistors.

Cermet Resistors

Cermet is a resistive material that is a mixture of powdered ceramic and powdered metal. The cermet mixture is worked into a paste and applied in a spiral pattern around the insulating substrate, as in Figure 3-4.

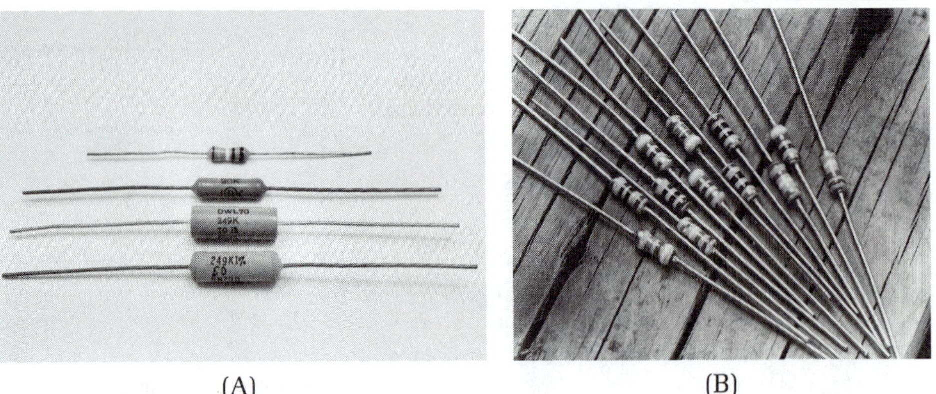

(A) (B)

FIGURE 3-5 (A) Metal-film resistors. (B) Carbon-film resistors.

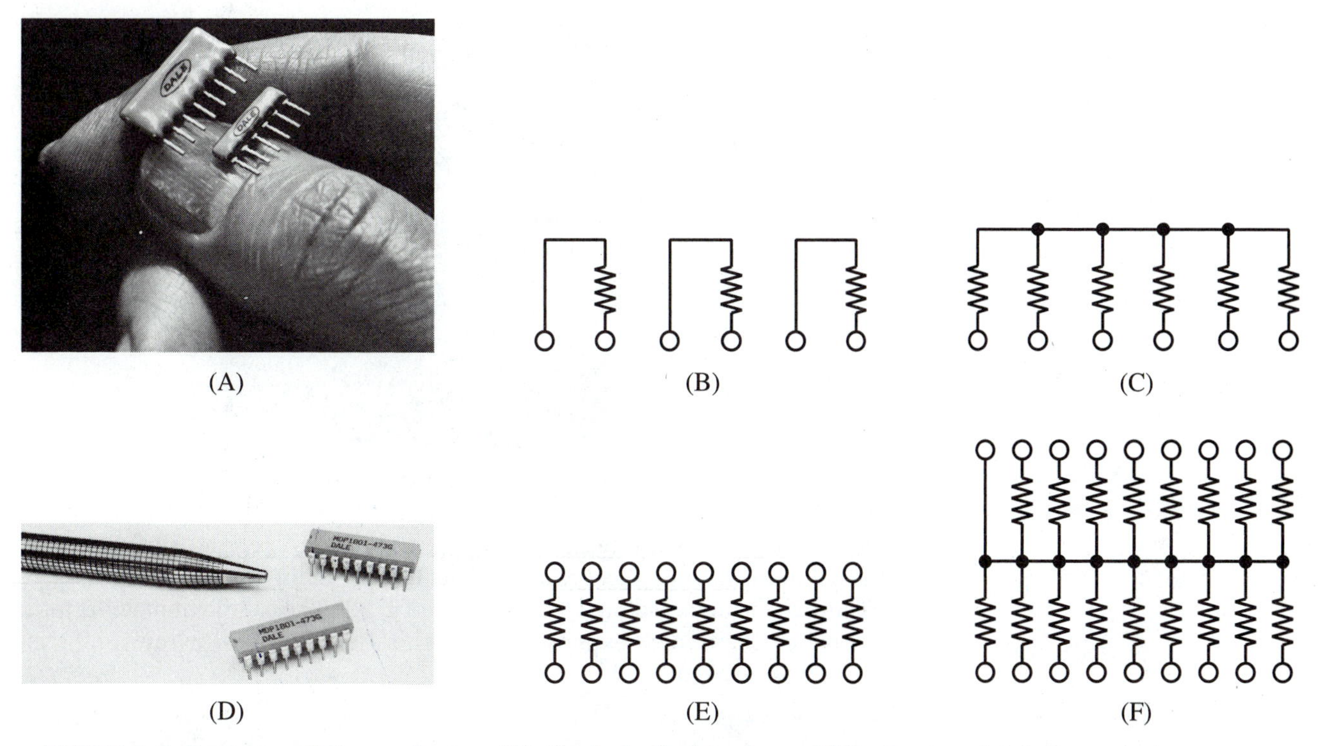

FIGURE 3-6 Cermet resistor packages. (A) Single-in-line package (SIP). *Courtesy of Dale Electronics Inc.* **(B and C) Common configurations. (D) 18-pin dual-in-line package (DIP).** *Courtesy of Dale Electronics Inc.* **(E and F) Common configurations.**

Cermet resistors typically are made to ± 2% tolerance, with ± 1% units also available. Cermet resistors are often used in networks containing several resistors in a single package. The resistors in a package may be isolated from one another or they may share common terminals.

The single-in-line package (SIP) structure is shown in Figure 3-6(A). Two possible internal resistor arrangements are shown in Figures 3-6(B) and (C).

The dual-in-line package (DIP) structure is shown in Figure 3-6(D). Two common internal resistor arrangements are shown in Figures 3-6(E) and (F).

Wirewound Resistors

Wirewound resistors have a length of wire wrapped around an insulating base. The assembly is then coated with enamel. Resistance values from less than 1 Ω to more than 100 000 Ω can be produced by controlling the thickness and length of the wire. Figure 3-7 shows a cutaway view.

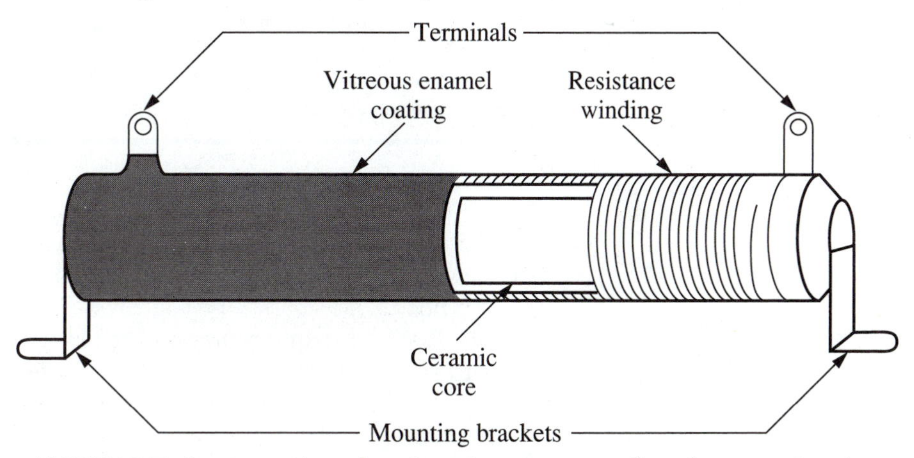

FIGURE 3-7 Cutaway view showing the structure of a wirewound resistor

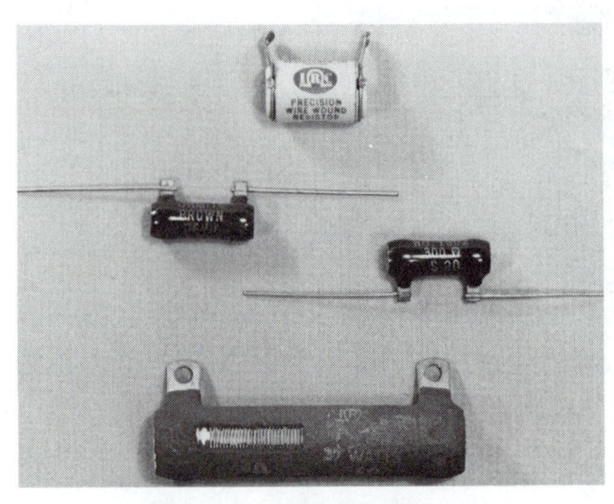

FIGURE 3-8 Wirewound resistors

Wirewound resistors tend to produce their heat near the surface, which enables them to get rid of heat rapidly. This is their main advantage.

They can be manufactured to tolerances of ± 1%. Their difficult manufacturing process makes them more expensive than other types. Several wirewound resistors are shown in Figure 3-8.

3-4 COLOR CODE

The resistance value is usually indicated by colored stripes that surround the body of a resistor. This is shown in Figures 3-3, 3-4, and 3-5. Colored stripes are better than printed numbers because the stripes can be clearly seen no matter what position the resistor is in.

The color-to-number code is given in Table 3-1. Remember that the first color in the list, black, stands for 0, not 1.

TABLE 3-1

COLOR	NUMBER (DIGIT)
Black	0
Brown	1
Red	2
Orange	3
Yellow	4
Green	5
Blue	6
Violet	7
Gray	8
White	9

Four-stripe Resistors

Most resistors that are manufactured to tolerances of ± 10% or ± 5% have four stripes, as shown in Figure 3-9. We start at the stripe that is closer to one end of the resistor's body.

The first two stripes represent the first two numbers in the resistance value. The third stripe tells how many zeros must follow the first two digits in order to express the value in ohms. (We'll discuss the fourth stripe in a moment.)

For example, if the first three stripes are Red, Violet, Orange, the resistance in ohms is 27 000 Ω, since

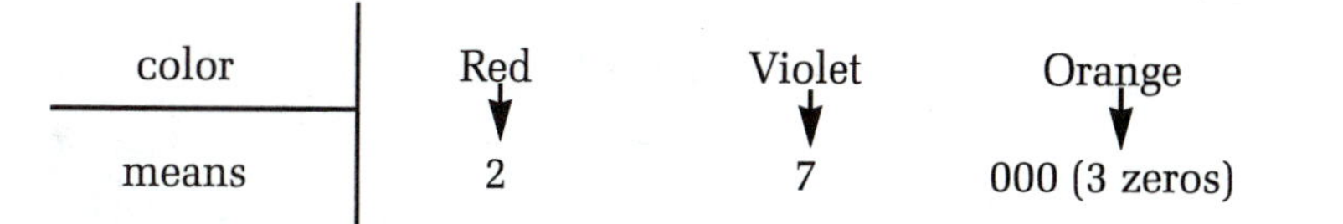

EXAMPLE 3-1

Assume that the resistor has four stripes altogether. The first three stripes are given. Write the resistance.

a) Brown, green, red
b) Blue, gray, yellow
c) Yellow, orange, brown
d) Orange, black, red
e) White, brown, black
f) Violet, green, green

SOLUTION

a) Brown, green stands for 15. Red means 2, so we attach 2 zeros, giving **1500 Ω.**
b) Blue, gray is 68. Yellow means 4, so add 4 zeros. **680 000 Ω**
c) Yellow, orange is 43. Brown means 1 so add 1 zero. **430 Ω**
d) Orange, black is 30. Don't leave out the zero. Red means 2, so add 2 more zeros. **3000 Ω**
e) White, brown is 91. Black means zero, so we add zero zeros. In other words, don't add any zeros, just leave the 2-digit number by itself. **91 Ω**
f) Violet, green is 75. The final green stripe means add 5 zeros. **7 500 000 Ω**

The first three stripes indicate the **nominal value** of resistance. Nominal value is the value that the resistor is supposed to have. It is the value that the manufacturer was trying to make. Manufacturing processes are never exact. They must always permit some amount of error. This is the meaning of manufacturing tolerance. It tells how close to the nominal value the actual value is guaranteed to be.

Tolerance is usually given as a percent. This was mentioned several times in Section 3-3. For example, a nominal 200-Ω resistor with tolerance of ± 10% is guaranteed to have an actual resistance between 180 Ω and 220 Ω. This range of values is gotten by subtracting 10 % of the nominal value, then adding 10% of the nominal value. The fourth stripe in Figure 3-9 indicates the resistor's tolerance.

TECHNICAL FACT

A silver fourth stripe means ±10% tolerance. A gold fourth stripe means ±5% tolerance.

47 x2000
2000
47
14000
8000x
94000
59

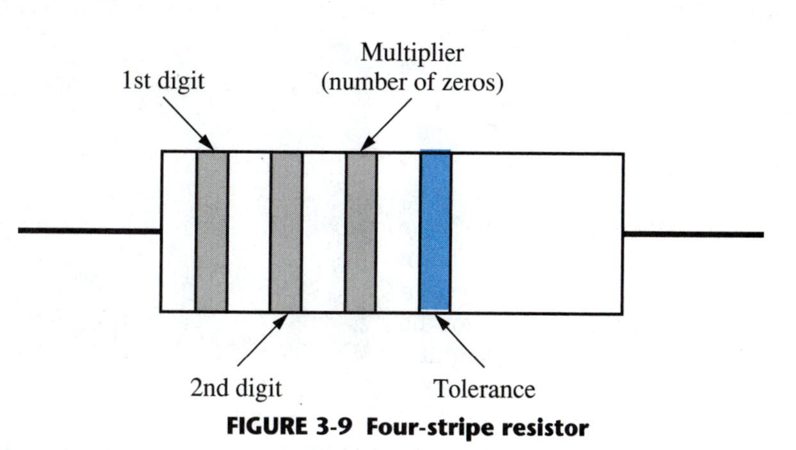

FIGURE 3-9 Four-stripe resistor

EXAMPLE 3-2

State the guaranteed range of values for the following color-coded resistors.

a) Brown, green, brown, silver
b) Green, blue, brown, gold
c) Red, red, red, gold

SOLUTION

a) Nominal value is 150 Ω. Tolerance is ± 10%. The guaranteed minimum value is given by

$$\begin{aligned} R_{min} &= 150 - (0.10)150 \\ &= 150 - 15 \\ &= \mathbf{135\ \Omega} \end{aligned}$$

Guaranteed maximum is

$$\begin{aligned} R_{max} &= 150 + (0.10)150 \\ &= 150 + 15 \\ &= \mathbf{165\ \Omega} \end{aligned}$$

b) These colors indicate 560 Ω ± 5%.

$$R_{min} = 560 - (0.05)560 = 560 - 28 = \mathbf{532\ \Omega}$$
$$R_{max} = 560 + (0.05)560 = 560 + 28 = \mathbf{588\ \Omega}$$

c) 2200 Ω ± 5%

$$R_{min} = 2200 - (0.05)2200 = 2200 - 110 = \mathbf{2090\ \Omega}$$
$$R_{max} = 2200 + (0.05)2200 = 2200 + 110 = \mathbf{2310\ \Omega}$$

Resistors Below 10 Ohms

If the third stripe is black, the largest value that can be color coded is 99 Ω (white, white, black). The first stripe is never allowed to be black, so the smallest value is 10 Ω (brown, black, black).

GLASS ART

The studio glass art movement in the U.S. has been made possible in part by the development of super-insulated electric furnaces. Such furnaces are capable of interior temperatures of 1800°F, and their temperature can be varied very precisely. Precise, slow cooling is especially necessary to maintain color separation in the glass.

Here, a pair of artists are working on a piece of blown glass artwork. Page 3 in the Applications color section shows one of the artists placing complete pieces in an electric oven for the slow cool-down to room temperature.

Courtesy of NASA

We need a way of color-coding resistance values less than 10 Ω. This is done by using gold and silver as third stripes.

TECHNICAL FACT

If gold is used for the third stripe, it means move the decimal point one place to the left. Silver as a third stripe means move the decimal point two places to the left.

These alternative meanings for gold and silver are outlined in Table 3-2.

TABLE 3-2

	AS 3RD STRIPE (MULTIPLIER)	AS 4TH STRIPE (TOLERANCE)
Gold	x 0.1	± 5%
Silver	x 0.01	± 10%

EXAMPLE 3-3

State the nominal value and tolerance for the following color-coded resistors.

a) Orange, white, gold, gold
b) Yellow, violet, silver, gold
c) Brown, black, gold, silver

SOLUTION

a) Orange, white is 39. Gold as the third stripe means move the decimal point one place to the left, which is the same as multiplying by 0.1, so the nominal value is **3.9 Ω**. Gold as the fourth stripe means the tolerance is ± **5%.**
b) Yellow, violet is 47. Silver as the third stripe means move the decimal point two places to the left, which is the same as multiplying by 0.01. The nominal resistance is **0.47 Ω**. The tolerance is ± **5%** because the fourth stripe is gold.
c) Brown, black is 10. The gold third stripe shifts the decimal point one place, for **1.0 Ω**. Tolerance is ± **10%** because of the silver fourth stripe.

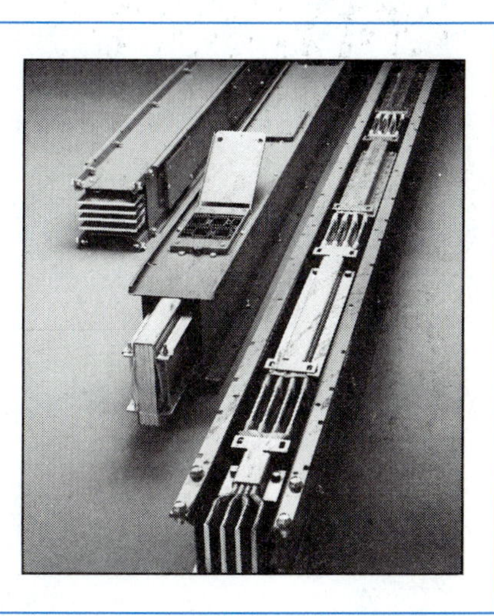

Sectionalized "busway," for carrying large currents. These busways have four solid copper conductors. The sections are bolted together and suspended from the ceiling or mounted on walls. They are the most economical and safe way to distribute electric power throughout an industrial plant.
Courtesy of Eaton Corporation, Cutler-Hammer Products

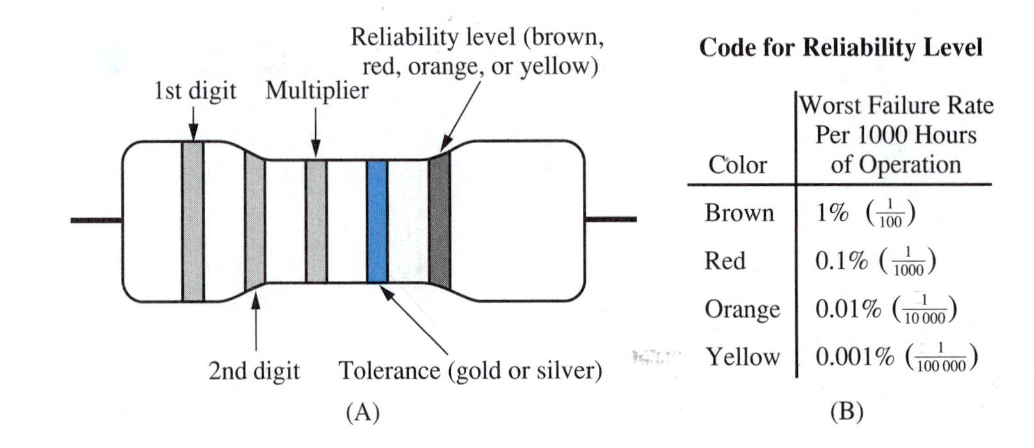

FIGURE 3-10 (A) In a five-stripe resistor, if the fourth stripe is gold or silver, the resistor is for military specification. The fifth stripe tells the reliability of the resistor under harsh operating conditions. (B) Color code for reliability ratings

Five-stripe Resistors

Some color-coded resistors have five stripes. There are two possibilities with five-stripe resistors.

1. It may be a resistor manufactured for military purposes. In that case, the fifth stripe represents the **reliability factor** of the resistor. This is shown in Figure 3-10. The reliability specification tells the maximum failure rate of a group of resistors under carefully defined operating conditions.
2. It may be a **precision resistor,** having a tolerance of ± 2% or better. Because the resistor is manufactured to such close tolerance, its nominal value can be expressed more precisely. It contains three significant digits rather than only two. The first three stripes give the significant digits, the fourth stripe is the multiplier, and the fifth stripe indicates tolerance. This is shown in Figure 3-11.

EXAMPLE 3-4

For the following color-coded resistors, tell all the information that is conveyed.

a) Orange, yellow, gray, red, brown
b) Green, brown, black, gold, orange

SOLUTION

a) The fourth stripe is not gold or silver, so the resistor is a precision unit with three significant digits. Orange, yellow, gray means 348. Red multiplier means add 2 zeros, for **34 800 Ω.** The brown fifth stripe indicates ± 2% tolerance.
b) With a gold fourth stripe, the resistor is a mil-spec unit. Green, brown, black indicates **51 Ω.** The gold fourth stripe indicates ± 5% tolerance. The orange fifth stripe shows that in a large group of these resistors, less than 0.01 % (1 out of 10 000) will fail during a 1000-hour endurance test.

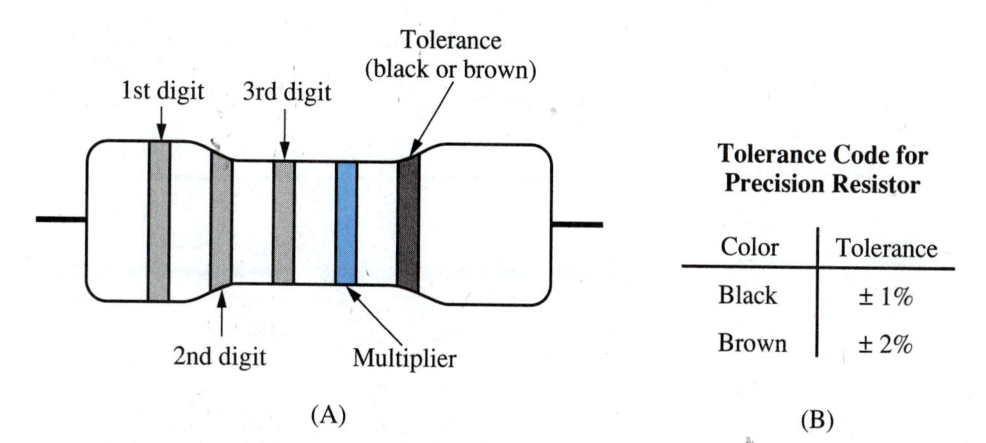

FIGURE 3-11 (A) In a five-stripe resistor, if the fourth stripe is not gold or silver, the resistor is a precision unit with very close tolerance. (B) Color code for precision tolerances

3-5 STANDARD VALUES

For stock resistors, only certain standard resistance values are available. Table 3-3 shows the standard nominal values for ± 10% and ± 5% tolerance resistors. The range shown in the table is from 10 Ω to 100 Ω. The number sequence is repeated for larger decades, up to 10 000 000 Ω or beyond.

TABLE 3-3 The Standard Resistor Values

±10% TOLERANCE	±5% TOLERANCE
10	10
	11
12	12
	13
15	15
	16
18	18
	20
22	22
	24
27	27
	30
33	33
	36
39	39
	43
47	47
	51
56	56
	62
68	68
	75
82	82
	91
100	100

The standard values for ± 2% and ± 1% tolerance resistors are given in Table 3-4. This table shows the decade from 100 Ω to 1000 Ω. The same significant figures are used for the lower decades and the higher decades.

TABLE 3-4 Precision Standard Values

±2% TOLERANCE			
100	178	316	562
105	187	332	590
110	196	348	619
115	205	365	649
121	215	383	681
127	226	407	715
133	237	422	750
140	249	442	787
147	261	464	825
154	274	487	866
162	287	511	909
169	301	536	953

±1% TOLERANCE							
100	133	178	237	316	422	562	750
102	137	182	243	324	432	576	765
105	140	187	249	332	442	590	787
107	143	191	255	340	453	604	806
110	147	196	261	348	464	619	825
113	150	200	267	357	475	634	845
115	154	205	274	365	487	649	866
118	158	210	280	374	499	665	887
121	162	215	287	383	511	681	909
124	165	221	294	392	523	698	931
127	169	226	301	407	536	715	953
130	174	232	309	412	549	732	976

EXAMPLE 3-5

Suppose you have performed a calculation that indicates that a resistance of 448 Ω would be perfect for a particular application. It is not necessary to have that exact amount of resistance, but you would like to get as close as possible.

a) If your company keeps ± 5% resistors in stock, what nominal value would you look for?

b) What would be its color-coded stripes?

SOLUTION

a) Look at Table 3-3 in the ± 5% column. The nearest nominal values in the 100–1000 decade are 430 Ω and 470 Ω. 448 Ω is only 18 Ω different from 430 Ω; it is 22 Ω different from 470 Ω. Therefore, choose **430 Ω.**

b) **Yellow, orange, brown, gold.**

EXAMPLE 3-6

Repeat Example 3-5, this time assuming that your company stocks ± 1% resistors.

SOLUTION

a) In Table 3-4, the nearest nominal values are 442 Ω and 453 Ω. The 453 Ω value is closer to 448 Ω, so choose **453 Ω.**

b) The digits 453 are coded yellow, green, orange. No zeros should be attached, so the fourth stripe is black. The tolerance ± 1% is coded by a black fifth stripe. Therefore, **yellow, green, orange, black, black.**

✓ SELF-CHECK FOR SECTIONS 3-3, 3-4, AND 3-5

5. Name the four main types of manufactured fixed resistors. [3]
6. What are the usual manufacturing tolerances for carbon-composition resistors? [3]
7. Give the nominal resistance and tolerance for the following color-coded resistors: [4]
   a) Brown, blue, red, silver
   b) Blue, gray, brown, gold
   c) Orange, white, gold, silver
   d) Yellow, violet, black, gold, red
   e) Gray, red, green, brown, brown
8. Calculate the range of actual resistance (R_{min} to R_{max}) for each of the following resistors: [4]
   a) Brown, red, black, silver
   b) Orange, white, brown, silver
   c) Gray, red, red, gold
9. Give the nominal resistance and tolerance for the color-coded resistors shown in Figure 3-12. (See page 1 in the Circuits color section for the color version of Figure 3-12.) [4]
10. Find the range of actual resistance (R_{min} to R_{max}) for each resistor in Question 9. [4]
11. Your company stocks ± 5% resistors. A calculation indicates that the target value of resistance is 790 Ω. To get as close as possible, what colors should you search for? [4]

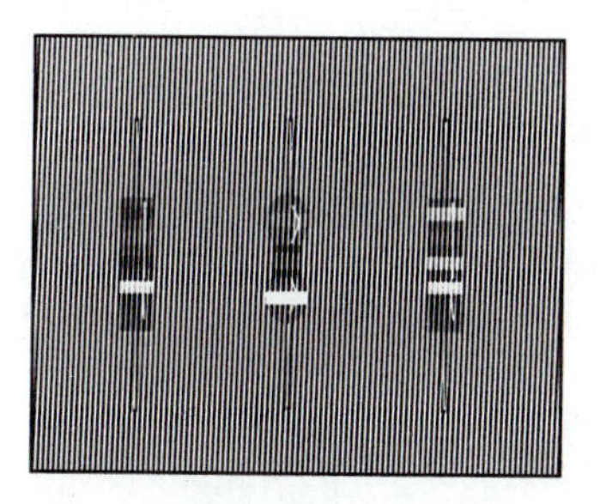

FIGURE 3-12

TROUBLESHOOTING TIP

It occasionally happens that a wrong value resistor is installed in a circuit because its color code looks similar to the color code of the correct value. For example, a technician might mistakenly install a 300-Ω resistor (Orange, Black, Brown) where the proper value is supposed to be 10 000 Ω (Brown, Black, Orange). This mistake is easily made by glancing at the stripes and reading them backward.

Here are some other mixups that occasionally happen:

11 Ω (Br, Br, Bk)	for 100 Ω (Br, Bk, Br)
200 Ω (R, Bk, Br)	for 1000 Ω (Br, Bk, R)
360 Ω (O, Bl, Br)	for 16 000 Ω (Br, Bl, O)
430 Ω (Y, O, R)	for 24 000 Ω (R, Y, O)
620 Ω (Bl, R, Br)	for 1600 Ω (Br, Bl, R)
3300 Ω (O, O, R)	for 22 000 Ω (R, R, O)

When troubleshooting a newly built circuit, be on the lookout for such resistor mixups.

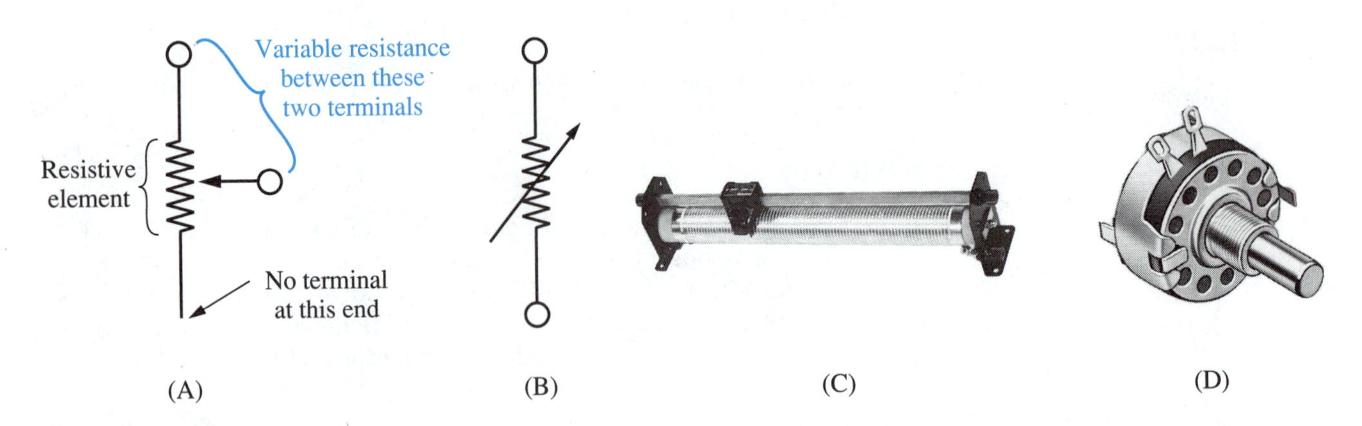

FIGURE 3-13 (A) Rheostat symbol that suggests actual construction. (B) Another symbol for a rheostat. (C) Rheostat with spiral wirewound element; movable tap slides in straight motion along the winding. *Courtesy of Biddle Instruments.* **(D) Rheostat with enclosed element; movable tap slides in circular motion along the surface of the element.** *Courtesy of Allen-Bradley, a Rockwell International Company*

3-6 VARIABLE RESISTORS

In many circuits, it is necessary to adjust the resistance. For this purpose, we use **variable resistors.** Variable resistors have a movable tap that slides over the body of the resistive material. There are two kinds of variable resistors, rheostats and potentiometers. **Rheostats** have only two terminals available to the user; one is the movable tap and the other is attached to one end of the resistive element. The schematic symbols for a rheostat are shown in Figures 3-13(A) and (B). Photographs of rheostats are shown in Figures 3-13(C) and (D).

5

Potentiometers have three terminals available: the movable tap and both ends of the resistive element. A potentiometer (pot, for short) is symbolized in Figure 3-14(A). A straight-line-motion wirewound potentiometer is shown in Figure 3-14(B). An enclosed circular-motion pot is shown in Figure 3-14(C).

A **linear pot** is designed so that the amount of resistance change per amount of tap motion is the same everywhere. The rate at which resistance changes is the same no matter which end of the element the tap is nearer.

A nonlinear pot behaves in a different way. Near one end of the element the resistance change is slight, per amount of tap movement. Near the other end of the element the resistance change is great. Nonlinear pots are often used in audio equipment. They are sometimes called **audio taper pots.**

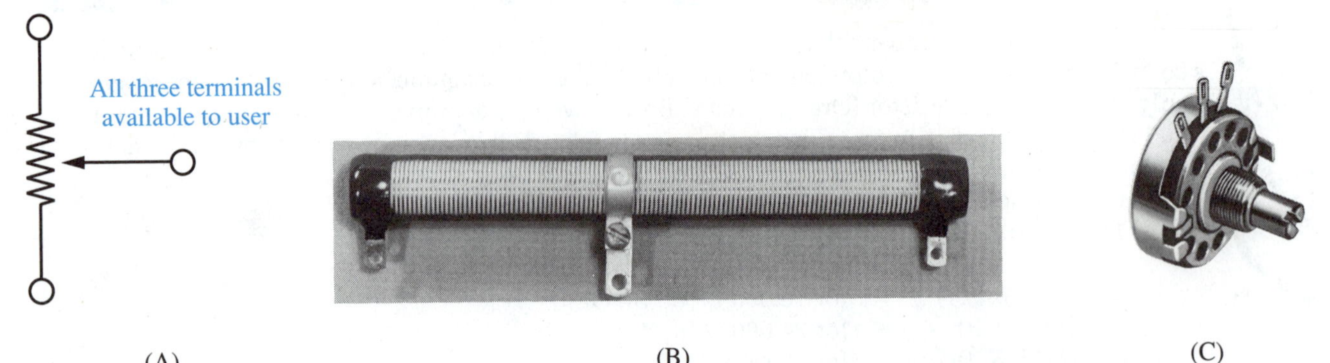

FIGURE 3-14 (A) Symbol for potentiometer. As tap is moved, resistance between one end terminal and tap increases while resistance between the other end terminal and tap decreases. (B) Straight-line pot. (C) Circular-motion pot. *Courtesy of Allen-Bradley, a Rockwell International Company*

FIGURE 3-15 Ganged pots. *Courtesy of Allen-Bradley, a Rockwell International Company*

Sometimes two or more potentiometers are controlled by the same shaft. This enables a single adjustment to control two or more circuits. Such ganged potentiometers are shown in Figure 3-15.

3-7 FACTORS THAT DETERMINE RESISTANCE

The resistance of any piece of material depends on three things:

1) Length—how long it is.
2) Cross-sectional area—how wide it is.
3) Resistivity—an atomic structural property of that specific material.

TECHNICAL FACTS

1. Longer length causes greater resistance.
2. Larger cross-sectional area causes less resistance.
3. Greater resistivity causes more resistance.

These three facts are illustrated in Figures 3-16 (A), (B), and (C).This relationship is given by the formula

$$R = \frac{rl}{A}$$

EQ. 3-1

where r stands for resistivity, l is length, and A is cross-sectional area.

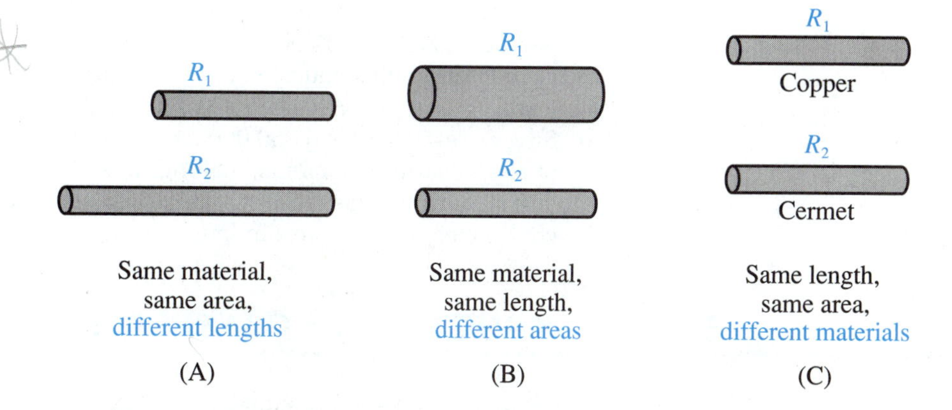

FIGURE 3-16 (A) R_2 is greater than R_1 because of longer length. (B) R_2 is greater than R_1 because of thinner cross-sectional area. (C) R_2 is greater than R_1 because cermet's resistivity is greater than copper's. Cermet has fewer loosely held electrons.

EXAMPLE 3-7

The resistivity of copper is 0.000 000 017 2, measured in basic units called ohm-meters. A certain round copper wire has cross-sectional area of 0.000 001 31 square meters and length of 125 meters. Calculate the resistance of the copper wire.

SOLUTION

From Equation 3-1

$$R = \frac{rl}{A} = \frac{(0.000\ 000\ 017\ 2)\ (125)}{(0.000\ 001\ 31)} = \mathbf{1.64\ \Omega}$$

In the United States, wire length is often expressed in feet, rather than meters. Cross-sectional area is often expressed in circular mils (cmil), rather than square meters. Resistivity must then be given in units of ohm-circular mils per foot.

✓ SELF-CHECK FOR SECTIONS 3-6 AND 3-7

12. Explain the difference between a variable resistor and a fixed resistor.
13. How can you tell a potentiometer from a rheostat? [5]

To answer Questions 14 through 16, look at the pot symbol in Figure 3-14 (A). If that pot's control shaft is turned mechanically by 10 degrees, suppose that the resistance between the top end and the movable tap increases by 50 Ω.

14. With that 10-degree turn, the resistance between the bottom end terminal and the movable tap__________by 50 Ω. (Answer increases or decreases.) [5]
15. Assume that the pot is linear. If the shaft is turned a further 10 degrees, the resistance from top end to movable tap will__________an additional__________Ω.
16. If the pot were nonlinear, describe the resistance change that would occur from that further 10 degrees of shaft rotation.
17. A certain piece of wire has resistance of 3.0 Ω. If it is cut in half, the resistance of each half will be__________Ω. [6]
18. Imagine a bar of aluminum 2 meters long, with rectangular cross section. Its resistance is 0.14 Ω. If the bar is sliced lengthwise into 2 equal pieces, each piece will have a resistance of__________Ω. [6]
19. Copper's resistivity is about 0.000 000 017 ohmmeter. The resistivity of tungsten is about 0.000 000 055 ohmmeter, roughly 3.2 times as great. There are two conducting bars with identical physical dimensions, one made of copper and the other made of tungsten. What can you say about their resistances? [6]
20. Silver's resistivity is 0.000 000 015 9 ohmmeter. This is the lowest resistivity of all the natural elements, making silver the best electrical conductor. A certain silver track has cross-sectional area of 0.000 000 02 square meters; it is 3 centimeters (0.03 meter) in length. Calculate its resistance. [6]

3-8 CONDUCTORS AND INSULATORS

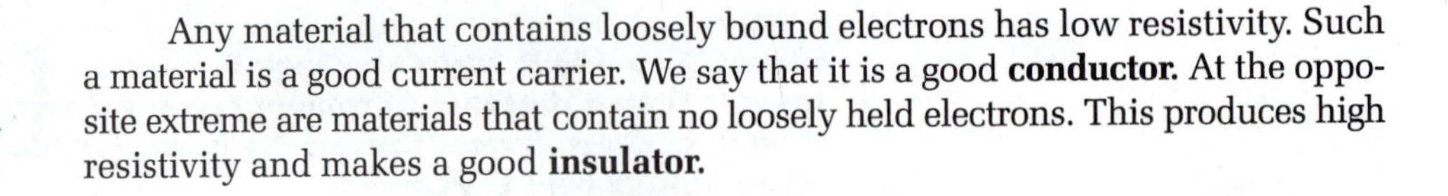
Any material that contains loosely bound electrons has low resistivity. Such a material is a good current carrier. We say that it is a good **conductor.** At the opposite extreme are materials that contain no loosely held electrons. This produces high resistivity and makes a good **insulator.**

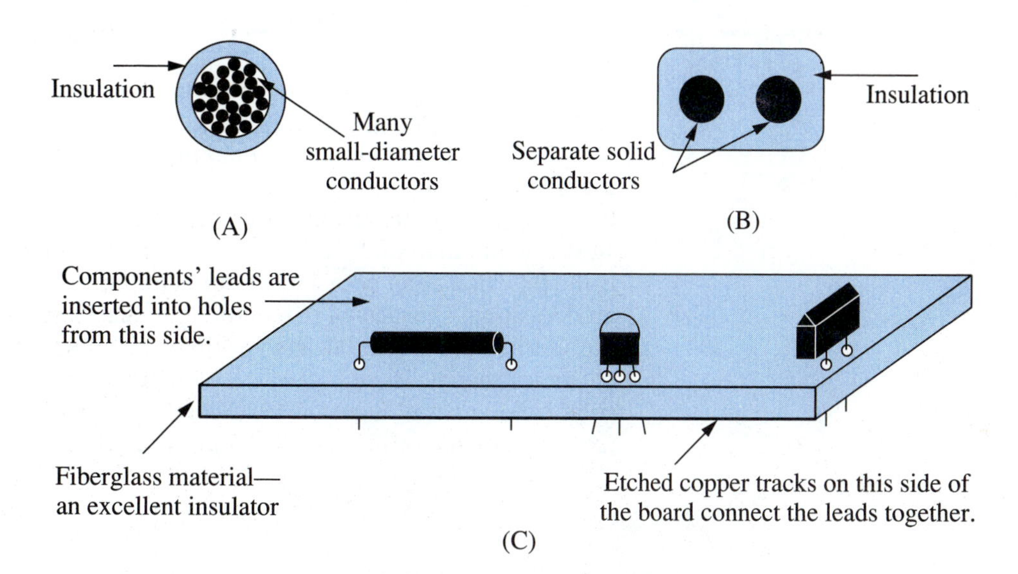

FIGURE 3-17 (A) Cross-sectional view of stranded wire with insulation. (B) Two-conductor insulated cable. (C) A printed circuit board is made of thin insulating material. The top side of the board, which the leads are inserted into, is called the component side. The bottom side, with the conductive tracks, is called the copper side, or the foil side.

Good insulators are important for preventing current where it is not wanted. Insulators are used to surround wire or to separate electrical terminals. For instance, most wire has an outside wrap of insulation. Then, if the wire touches something, there is no possibility of current accidentally leaving the wire. This is shown in Figures 3-17(A) and (B).

As another example, electrical components are often mounted on a **printed circuit (PC) board**, as shown in Figure 3-17(C). When the component leads are inserted into holes drilled through the PC board, the board's insulating ability prevents current from passing between the leads. Copper conducting tracks join those particular leads where a current flow path is desired.

Some of the popular insulating materials are plastics, rubber, fiberglass, and mica. Good insulators have resistivities that are many billion times larger than for good conductors.

Semiconductor electronic devices are manufactured and tested in *clean rooms*. The air in these rooms has all but the tiniest particles filtered out.
Courtesy of Hewlett-Packard Company

3-9 SEMICONDUCTORS

Between the two extremes are materials that have medium resistivity. These materials are called **semiconductors.**

In the past four decades, techniques have been developed for chemically doping silicon, the element with atomic number 14. The *n*-doping process is the injection of an occasional arsenic atom into the silicon crystal. This produces an occasional extra outer-shell electron within the silicon crystal. The *p*-doping process is the injection of an occasional boron atom into the silicon crystal. This produces an occasional shortage of one outer-shell electron. The solid-state electronic revolution has happened because of the many devices that have been invented by clever arrangements of layers of *n*-doped and *p*-doped silicon. Examples are the solar cell shown in Figure 2-12, discrete transistors, integrated circuits (ICs), silicon controlled rectifiers (SCRs), and numerous other devices.

3-10 LAMPS AND RESISTIVE DEVICES

The simplest lamp is the **incandescent lamp.** It has a coiled tungsten wire, called a **filament**, inside a glass bulb from which all the air has been removed. In some cases, the air is replaced with an inert gas that has no effect on the filament. A sketch of an incandescent bulb is shown in Figure 3-18(A). Its symbol is given in Figure 3-18(B).

The filament of an incandescent lamp usually has a few hundred ohms of resistance. Forcing current to flow through it creates heat. This raises its temperature drastically to several thousand degrees Fahrenheit. At that point, the tungsten begins to glow, giving off light.

An incandescent bulb is an example of a device that has resistance even though it is not a resistor per se. There are many such devices. Some of them depend on their resistance in order to operate. Electric heating elements in ovens, space heaters, and water heaters are examples.

Some resistive electrical devices have resistance only because it is unavoidable. Motors and solenoids are examples of this type.

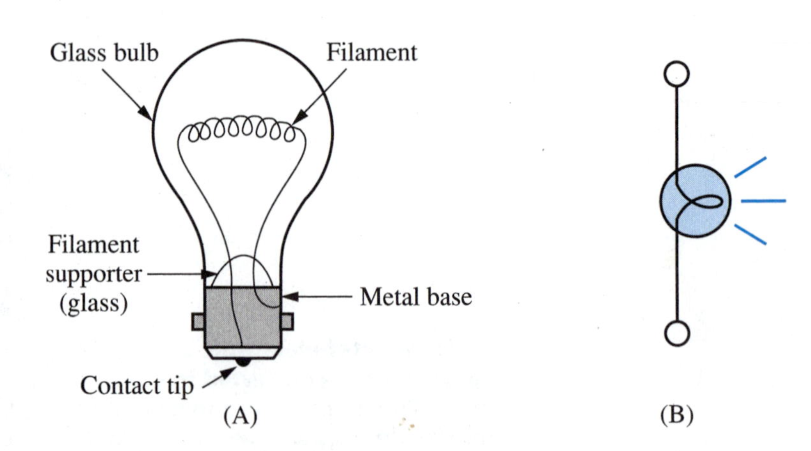

FIGURE 3-18 Incandescent lamp. (A) Structure. (B) Symbol

FORMULAS

For ± 5% tolerance resistors:

$$R_{minimum} = R_{nominal} - (0.05)R_{nominal}$$
$$= (0.95)R_{nominal}$$

$$R_{maximum} = R_{nominal} + (0.05)R_{nominal}$$
$$= (1.05)R_{nominal}$$

Similarly, for other tolerances

$$R = \frac{r\,l}{A}$$

EQ. 3-1

SUMMARY OF IDEAS

- Resistance is the ability to oppose current. It is symbolized *R*.
- The basic measurement unit of resistance is the ohm, symbolized Ω.
- 1 Ω is the resistance that allows current of 1 A if voltage of 1 V is applied.
- The four main types of fixed resistors are: carbon-composition, deposited-film, cermet, and wirewound.
- A resistor's nominal resistance is indicated by color-coded stripes. The color code must be memorized.
- Color-coded stripes are also used to indicate tolerance and reliability.
- The physical size of a resistor determines how quickly it can get rid of heat.
- Some resistors are variable. Those with two terminals are called rheostats; those with three terminals are potentiometers.
- The resistance of any piece of material is determined by three things: length, cross-sectional area, and resistivity.
- Resistivity is an atomic property of the material.
- Low resistivity makes for a good conductor; high resistivity makes a good insulator.
- Semiconductors have medium resistivity. The entire field of solid-state electronics depends on the use of doped semiconductors.

CHAPTER QUESTIONS AND PROBLEMS

1. Greater resistance tends to allow__________current. [1]
2. What is the symbol for the idea of resistance?
3. What is the basic unit of resistance? What is its symbol? [2]
4. If 1 V is applied and current of 1 A is flowing, the resistance value is__________Ω. [2]
5. During manufacture, what methods are used to control the resistance of a deposited-film resistor? [3, 6]
6. A two-terminal variable resistor is called a__________. [5]
7. A three-terminal variable resistor is called a__________. [5]
8. T-F In a wirewound resistor, thinner wire tends to produce greater resistance. [6]
9. T-F In a wirewound resistor, shorter overall wire length tends to produce greater resistance. [6]
10. Give the resistance and tolerance for each of the color-coded resistors in Figure 3-19. (See page 1 in the Circuits color section for the color version of Figure 3-19.) [4]

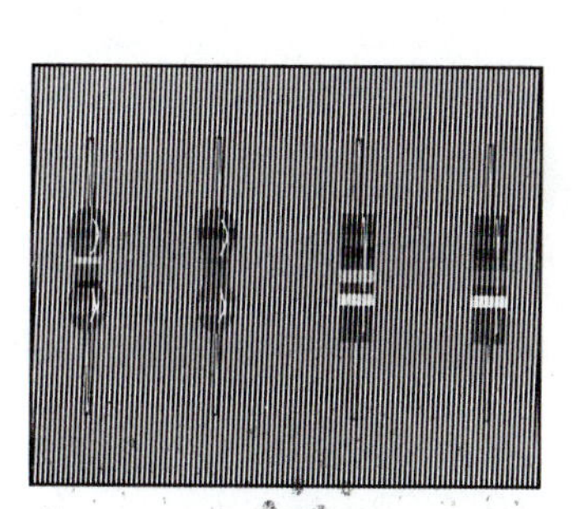

FIGURE 3-19

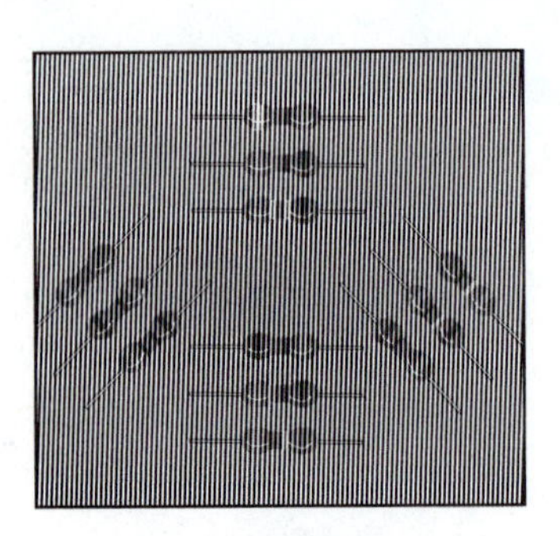

FIGURE 3-20

11. Give the minimum and maximum values for each of the resistors in Problem 10. [4]
12. When gold is used as a third stripe, what does it indicate? [4]
13. When silver is used as a third stripe, what does it indicate? [4]
14. Select the resistor from Figure 3-20 that matches or comes closest to matching each of the following values. (See page 1 in the Circuits color section for the color version of Figure 3-20.) [4]

a) 270 Ω	d) 170 Ω	g) 1 200 000 Ω
b) 63 Ω	e) 11 000 Ω	h) 205 000 Ω
c) 4450 Ω	f) 610 000 Ω	i) 1 450 Ω

To answer Questions 15 through 18, refer to the rheostat symbol of Figure 3-13 (A). The overall resistance of the entire element is 1000 Ω.

15. If the movable tap is moved all the way up to the top, what value of resistance appears between the terminals? [5]
16. If the movable tap is moved all the way down to the bottom, what value of resistance appears between the terminals? [5]
17. If the movable tap is moved to the exact center of the element, what value of resistance appears between the terminals? (Assume the rheostat is linear.) [5]
18. If the movable tap is moved three-fourths of the way down from the top, what value of resistance appears between the terminals? [5]
19. Number 24 gage copper wire has cross-sectional area of 0.000 000 205 square meter. Copper's resistivity is given in Section 3-7. What is the resistance of a 100-meter-long spool of this wire? [6]
20. Aluminum's resistivity is 0.000 000 028 3 ohmmeter. Calculate the resistance of 50 meters of No. 12 gage aluminum wire, having cross-sectional area of 0.000 003 31 square meters. [6]
21. An insulator has__________resistivity. [7]
22. A good conductor has__________resistivity. [7]
23. A semiconductor has__________resistivity. [7]
24. The actual resistive element in an incandescent lamp is called the__________. It is made of __________.
25. What causes light to be produced by an incandescent lamp?
26. Why must the air be removed from the glass enclosure of an incandescent lamp?

CHAPTER 4

SCHEMATIC DIAGRAMS

Center for monitoring and recording thunderstorm electrical activity. *Courtesy of NASA*

OUTLINE

NEW TERMS TO WATCH FOR

schematic diagram
switch
closed
open
knife switch
toggle switch
load device
control device
polarity
continuity tester

Testing and troubleshooting of electric circuits is done by referring to a schematic diagram of the circuit.

After studying this chapter, you should be able to:

1. Use schematic symbols to make a circuit diagram.
2. Understand the symbols for an open switch and a closed switch.
3. Name the five basic parts of all electric circuits and explain the function of each part.
4. Identify the polarity of load devices; relate polarity to current direction.
5. Combine two or more dc voltage sources.
6. Explain the meaning of hole current; relate it to electron current.

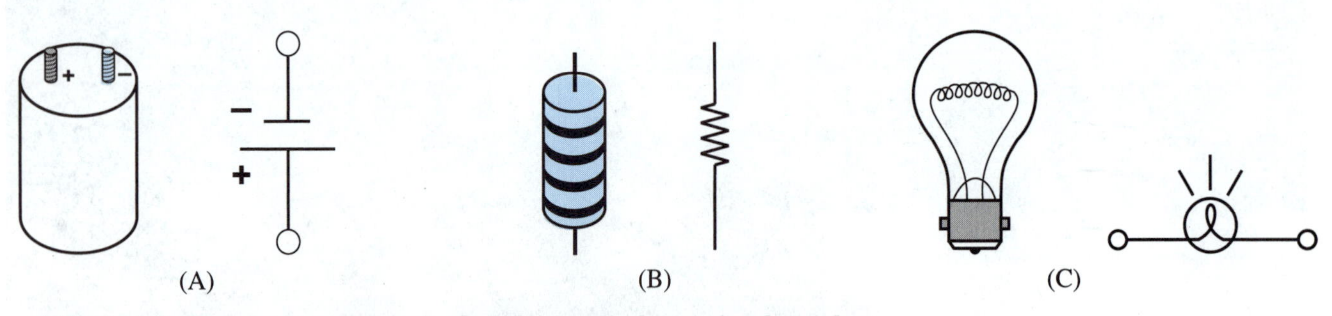

FIGURE 4-1 Electrical component schematic symbols are simple to draw.

4-1 CIRCUIT SYMBOLS

When we represent an electric circuit by a drawing on a piece of paper, it isn't necessary to show the actual physical appearance of the circuit components. Instead, we use schematic symbols to stand for the circuit components. These schematic symbols may have little or no resemblance to their actual appearance.

For instance, we saw the symbol for a voltage source in Section 2-4. It has no resemblance to a physical dry-cell battery. This is pointed out in Figure 4-1(A). The same is true for a manufactured resistor, as shown in Figure 4-1(B).

For an incandescent lamp, the symbol has some resemblance to the physical device, but it is just a slight resemblance. Compare them in Figure 4-1(C).

1

In general, we make our circuit drawings as simple as possible by the use of schematic symbols. Thus, the physical circuit of Figure 4-2(A) would be represented by the **schematic diagram** of Figure 4-2(B).

EXAMPLE 4-1

Draw the schematic diagram of the circuit in Figure 4-3(A).

SOLUTION

The two-terminal variable resistor is a rheostat. The symbols for a rheostat were presented in Section 3-6. The schematic diagram is shown in Figure 4-3(B).

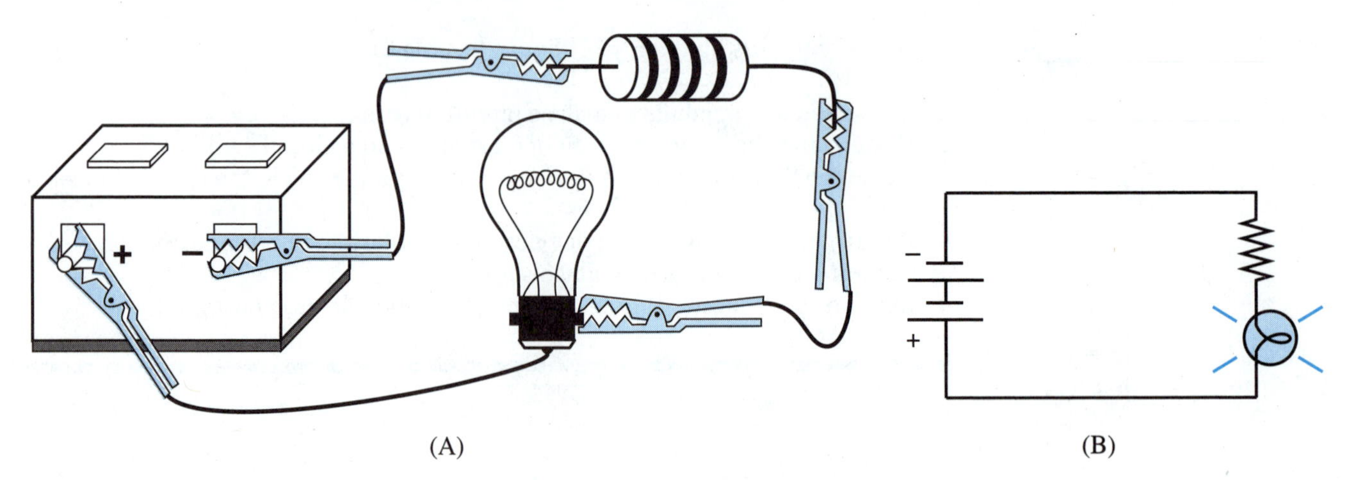

FIGURE 4-2 (A) Physical appearance of a lamp circuit. (B) Schematic diagram

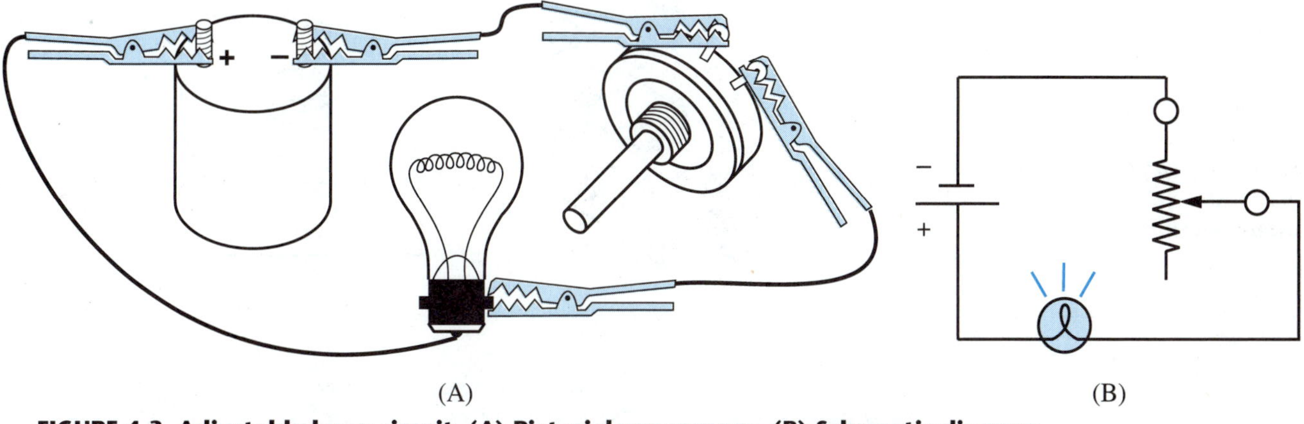

FIGURE 4-3 Adjustable lamp circuit. (A) Pictorial appearance. (B) Schematic diagram

4-2 SWITCH SYMBOLS

Many circuits are provided with a **switch.** The circuit is turned on, or energized, when the switch is **closed.** It is turned off, or deenergized, when the switch is **open.** Figure 4-4(A) shows the physical appearance of a **knife switch** in the open position.

In Figure 4-4(A), the conducting metal blade of the switch is not in contact with the spring clip on the right side. Therefore, there is no complete path for current to flow from the screw terminal on the left to the screw terminal on the right. We say that the open switch has interrupted the current path. This situation is symbolized very simply in Figure 4-4(B).

In Figure 4-5(A), the metal blade is contacting the spring clip on the right. This provides a current flow path between the screw terminals. We say that the closed switch has completed the current path. This is represented by the schematic symbol in Figure 4-5(B).

Switches have different physical structures. For example, a residential wall **toggle switch** is shown in Figure 4-6(A). A panel-mount toggle switch is shown in Figure 4-6(B). These switches would be drawn with the same schematic symbols shown in Figures 4-4(B) and 4-5(B).

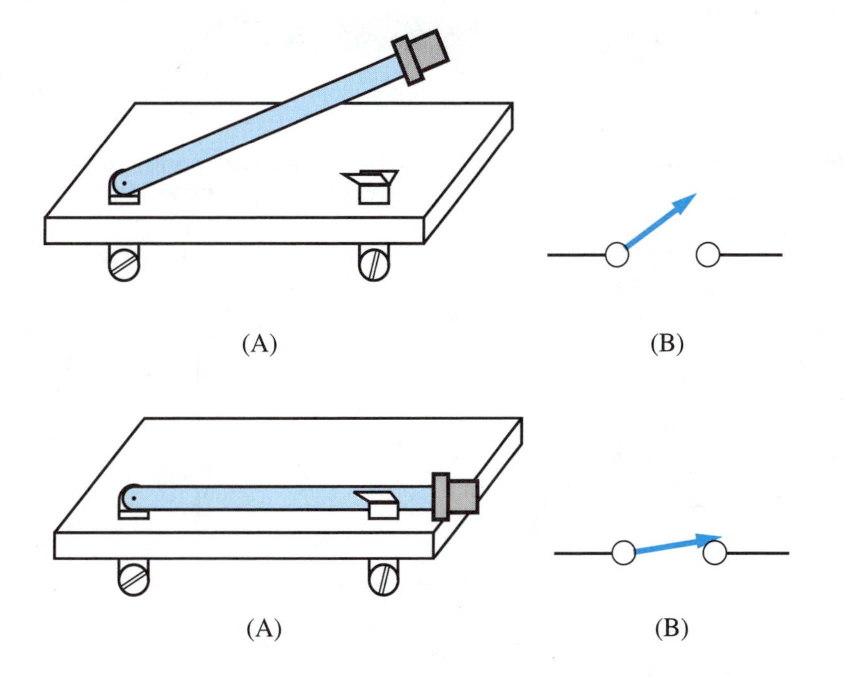

FIGURE 4-4 (A) Knife switch in the open position. (B) Schematic symbol

FIGURE 4-5 (A) Knife switch in the closed position. (B) Schematic symbol

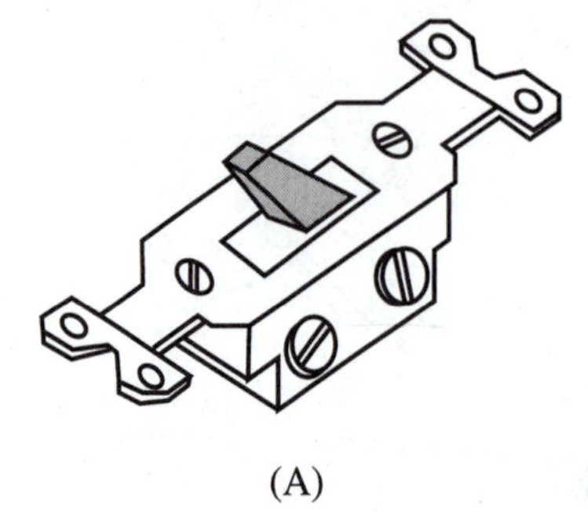

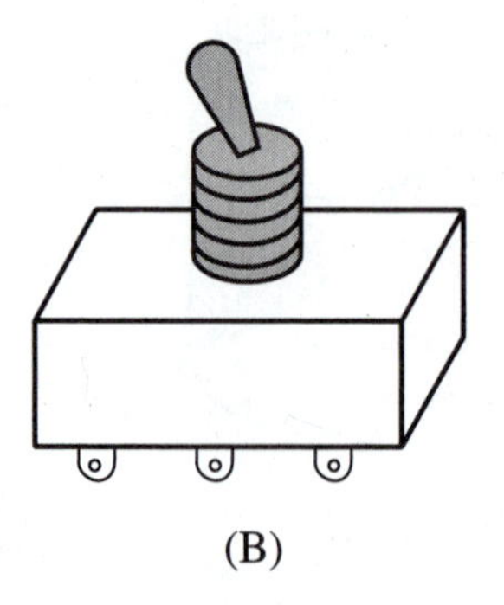

(A) (B)

FIGURE 4-6 Physical appearances of toggle switches. (A) Residential wall unit. (B) Panel-mounted unit

EXAMPLE 4-2

For the circuit of Figure 4-7:

a) Draw a schematic diagram, assuming that the switch is open.
b) Repeat, this time with the switch closed.

SOLUTION

a) The adjustable dc power supply can be symbolized with a slanted arrow through the dc source symbol. An open switch causes the lamp to produce no light, as indicated in **Figure 4-8(A).**
b) With the switch closed, the lamp glows. This is shown in the schematic of **Figure 4-8(B).**

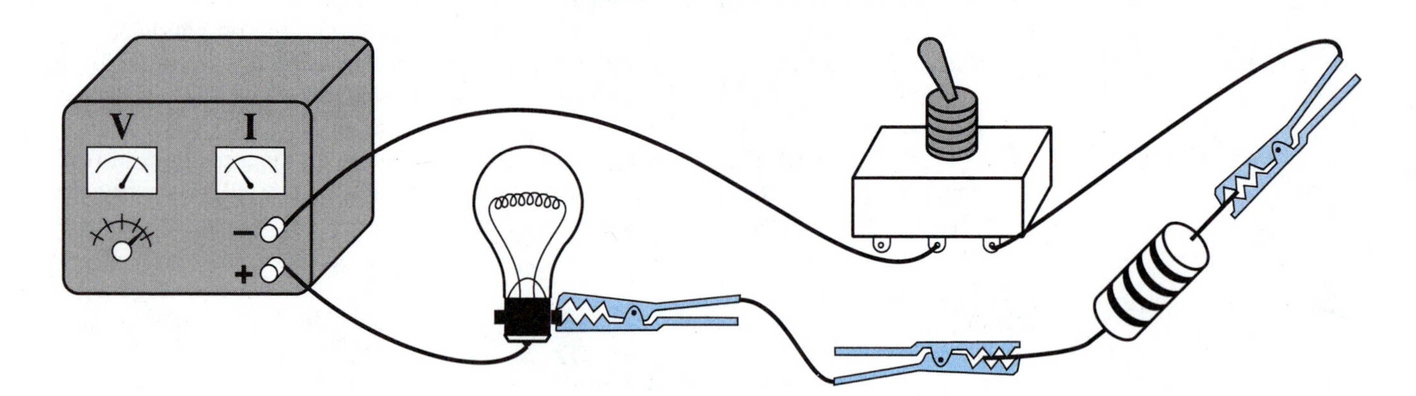

FIGURE 4-7 Pictorial diagram of an adjustable dc power supply driving a resistor-lamp combination, controlled by a toggle switch

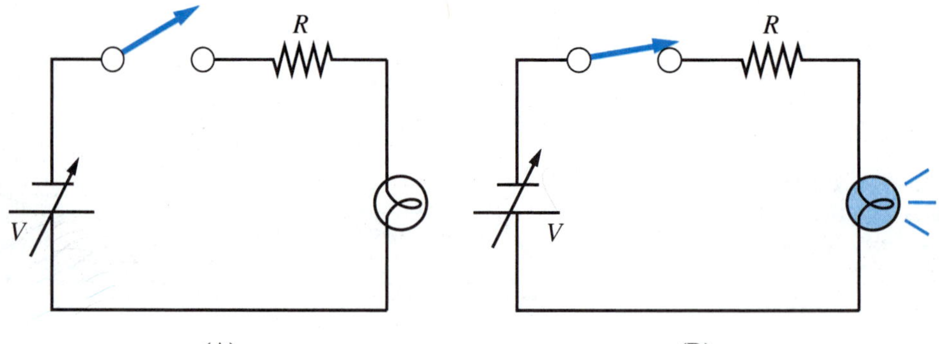

FIGURE 4-8 Schematic diagrams. (A) Switch open, lamp deenergized (not glowing). (B) Switch closed, lamp energized (glowing)

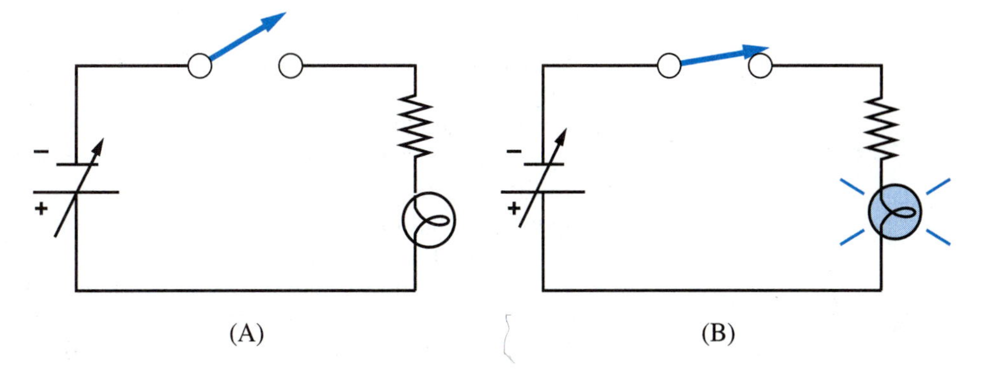

FIGURE 4-9 (A) Schematic diagram of lamp circuit deenergized, or turned off. Lamp is not glowing. (B) Energized lamp circuit with closed switch. Lamp is glowing.

Figure 4-9 shows the five basic parts of all electric circuits.

TECHNICAL FACT

All electric circuits have five basic parts:

1. Voltage source
2. Conductors
3. Insulation
4. Load device
5. Control device

3

The voltage source exerts force on the electrons to cause current to flow. It is the origin of the electric energy for the circuit. The conductors join the source to other devices in the circuit, furnishing a current flow path. Insulation prevents the conductors from touching other items outside the circuit. The conductors are usually coated with insulation. The structure of the voltage source is also bound to contain some insulation.

The **load device** produces something useful when current flows through it. The load converts the source's electrical energy into a different form of energy that is useful to us.

The **control device** in Figure 4-9 is the switch. There are other control devices that automatically turn off the circuit if it malfunctions. Fuses, circuit breakers, and similar devices perform this function. Variable resistors can also be used as control devices.

Although real circuits always contain some kind of control device, schematic diagrams are sometimes drawn without one. That is done only as a shortcut to make the diagram simpler.

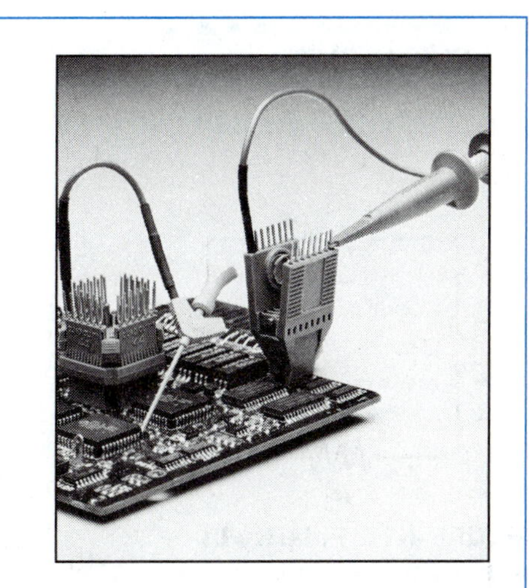

Clips for attaching to very small circuit devices. *Courtesy of Tektronix, Inc.*

✔ SELF-CHECK FOR SECTIONS 4-1 AND 4-2

1. Explain why electric circuits are represented by schematic diagrams rather than physically realistic diagrams. [1]
2. Draw a schematic diagram of a dc voltage source driving two lamps that are in line with each other. Include the on-off switch. [1]
3. In the diagram for Problem 2, show how you can relocate the switch to a different position in the circuit while still maintaining on-off control. Explain why this works. [1]
4. Figure 4-3(B) shows a circuit with a lamp that can be brightened or dimmed by a variable resistor. Referring to that diagram, tell which direction the movable tap must move (up or down) in order to make the lamp brighter. [3]
5. Draw a schematic diagram of a battery driving two lamps, with each lamp separately controlled by its own switch. [1]

4-3 POLARITY

FIGURE 4-10 An isolated load device has no polarity, but an energized load device does have polarity.

We learned in Chapter 2 that dc voltage sources have one negative terminal and one positive terminal. Electron current flows away from the negative terminal, through the conductors and load device, and returns to the positive terminal.

When this occurs, the load terminals also become polarized. That is, one load terminal becomes negative and the other becomes positive. This is shown in Figure 4-10.

In a simple circuit like Figure 4-10, it is easy to tell the **polarity** of the load terminals by this rule:

TECHNICAL FACT

The load terminal that is closer to the negative source terminal is negative itself. The load terminal closer to the positive source terminal is positive itself.

Figure 4-11 illustrates this polarity rule for a circuit with more than one load device.

Another correct way to determine load polarity is by this rule:

TECHNICAL FACT

Electron current enters a load device at its negative terminal. Current leaves a load device at its positive terminal.

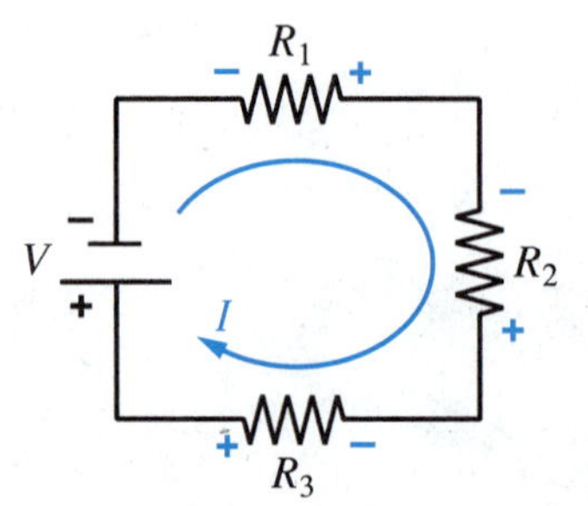

FIGURE 4-11 Polarity in a multiload circuit

This rule is also easy to verify by looking at the circuits of Figures 4-10 and 4-11.

It may be harder to tell load polarity in complex circuits. For instance, the circuit of Figure 4-12 contains four resistors and a lamp. If we consider the lamp to be the load device, it is impossible to apply the first rule. Neither lamp terminal is closer to the negative terminal of the source.

Let us try to find out the lamp polarity by applying the second rule. We've still got trouble because we can't tell the direction of current through the lamp just by looking at the diagram of Figure 4-12. Given only that diagram, there is no way to know the lamp polarity. However, if information were available about the resistance values of R_1, R_2, R_3, and R_4, it would be possible to predict the current direction through the lamp. Knowing that, the lamp's terminal polarity could be told.

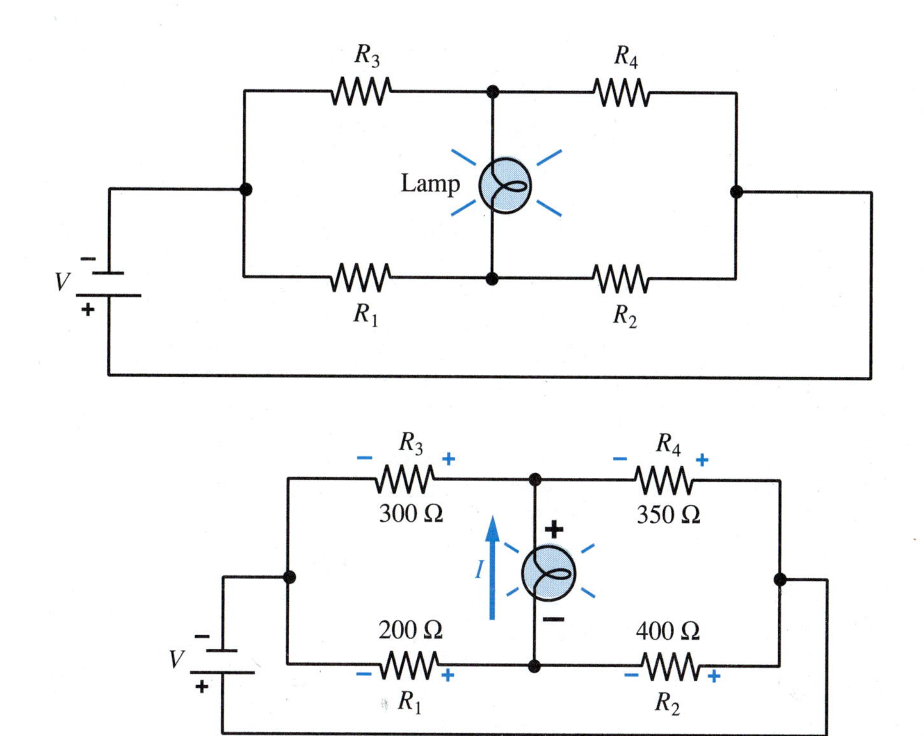

FIGURE 4-12 A more complicated circuit where we can't tell the lamp polarity simply by inspection

FIGURE 4-13 If we have proved that current, *I*, passes through the lamp in the direction shown (bottom-to-top), the lamp polarity must be as shown. Resistor polarities are then easy to tell.

EXAMPLE 4-3

In Figure 4-12, suppose $R_1 = 200\ \Omega$, $R_2 = 400\ \Omega$, $R_3 = 300\ \Omega$, and $R_4 = 350\ \Omega$. By calculation methods that we will study in Chapter 11, it can be proved that the electron current passes through the lamp from bottom to top. Tell the lamp's terminal polarity. Also mark the polarities of all four resistors.

SOLUTION

The rule is that current enters at the negative terminal and leaves at the positive terminal of a load. Therefore, since current is entering at the bottom and leaving at the top, the lamp's polarity is **negative on bottom, positive on top.** This is illustrated in Figure 4-13.

The resistor polarities are known by the first polarity rule. **All the left sides are closer to the negative source terminal, so they are negative themselves. The right-side terminals are all positive,** as shown in Figure 4-13.

TROUBLESHOOTING TIP

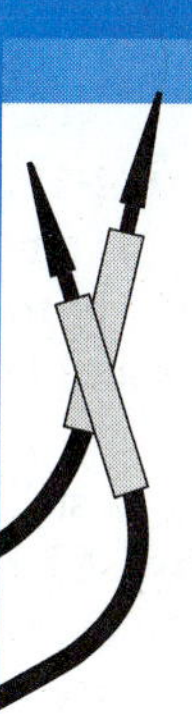

Many real-life circuits have conductors that are quite long. For example, a residential wiring circuit has its voltage source in the fuse or circuit breaker panel, which is typically located in the basement. For a second-floor bedroom circuit, the conductors must run some distance through the basement, then up two house stories, and then some additional distance on the second floor. Such a circuit is symbolized in the figure shown in (A).

If trouble occurs in a long circuit of this type, we need a convenient way to find out if the conductor paths are broken. One way to do this is with a continuity tester. A **continuity tester** is a troubleshooting tool that contains a voltage source and a lamp, and has two leads with attached probes. A continuity tester is shown schematically in the figure shown in (B).

Continued on page 50

Here is the method to test a long circuit like the one shown in (A).

1) Disconnect both ends of both conductors.
2) Twist the two conductors together at one end. This has been done at the load end in the figure shown in (C).
3) Place the continuity tester on the other end of the conductor pair. If both conductors are all right, electrical continuity exists, and the lamp will light up. If either conductor is broken, the test lamp will remain dark.

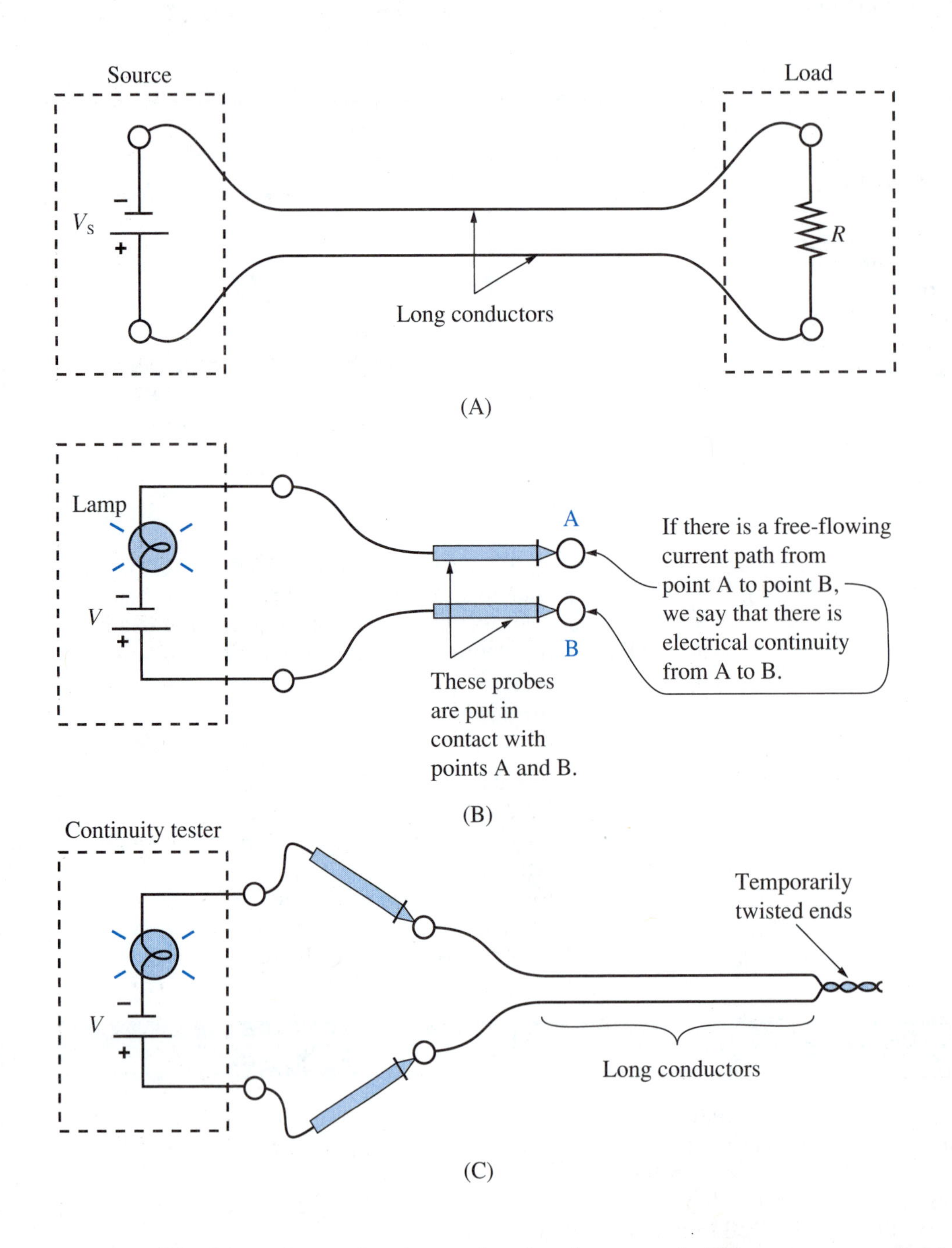

(A) In residential wiring, the conductors are insulated separately. They are surrounded by another tough plastic insulating jacket. The entire assembly is called a romex cable. (B) The lamp will light up if there is electrical continuity from point A to point B. If there is no continuity, the lamp stays dark. (C) If the tester lights up to show electrical continuity, the conductors must be OK. If it shows lack of continuity, at least one conductor is broken.

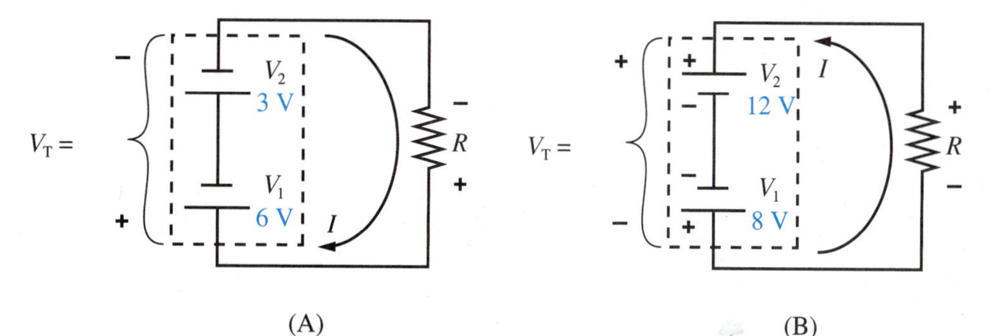

FIGURE 4-14 (A) V_1 and V_2 add to produce V_T. The polarity of V_T is negative on top and positive on bottom. (B) V_1 subtracts from V_2 to produce V_T. The polarity of V_T is positive on top and negative on bottom, the same as V_2.

4-4 COMBINING VOLTAGE SOURCES

Voltage sources are commonly connected together in line. If they are connected as shown in Figure 4-14(A), their values add to produce the total voltage, V_T. The voltages add because both sources are helping each other by trying to push current in the same direction. In other words, if V_1 were all by itself in Figure 4-14(A), it would circulate current in the clockwise (CW) direction. The same is true for V_2. Therefore,

$$V_T = V_1 + V_2 = 6\text{ V} + 3\text{ V} = 9\text{ V}$$

The overall polarity of V_T is the same as the polarity of either source alone.

In Figure 4-14(B), the voltage sources V_1 and V_2 are opposing each other. If V_1 were alone, it would circulate current CW. If V_2 were alone, it would circulate current counterclockwise (CCW). Since the two sources are trying to push current in opposite directions, the smaller voltage subtracts from the larger. That is,

$$V_T = V_2 - V_1 = 12\text{ V} - 8\text{ V} = 4\text{ V}$$

5 The overall polarity of V_T is the polarity of the larger source, V_2 in this case.

Most dc power supplies have built-in meters for displaying both voltage and current.
Courtesy of Hewlett-Packard Company

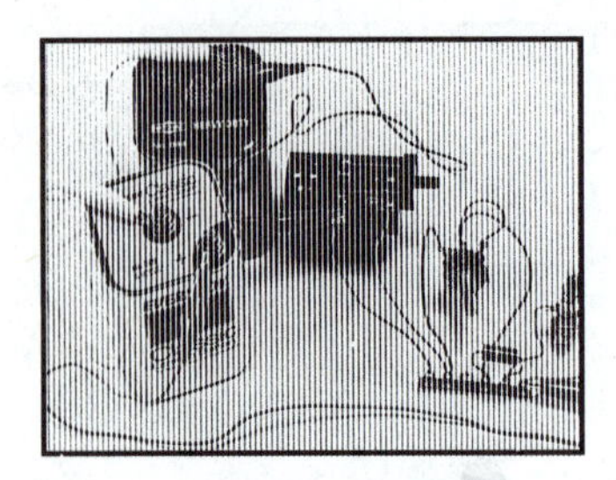

FIGURE 4-15

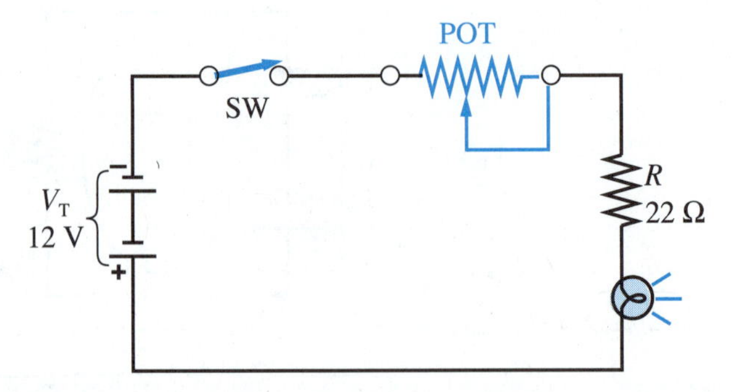

FIGURE 4-16 Schematic diagram of circuit in Figure 4-15. The right end terminal of the potentiometer has been jumpered to the tap terminal. This effectively converts the pot into a variable resistor, like a rheostat. Using a jumper is preferred to simply leaving the unused end terminal hanging in a disconnected state.

EXAMPLE 4-4

Draw the schematic diagram of the circuit in Figure 4-15. (See page 2 in the Circuits color section for the color version of Figure 4-15.)

SOLUTION

The two voltage sources are aiding each other, producing a total voltage, V_T, of 12 V. The knife switch is closed, so the lamp is energized and glowing. The fixed resistor is 22 Ω. The total pot element resistance is not visible in this picture, so it cannot be indicated on the schematic. All available information is represented in Figure 4-16.

✔ SELF-CHECK FOR SECTIONS 4-3 AND 4-4

6. For the two-resistor circuit of Figure 4-17(A), mark the polarities of R_1 and R_2. Show the direction of current through the circuit. [4]

For Questions 7 through 10, refer to the circuit of Figure 4-17(B).

7. Assume that all three switches are closed. Mark the polarity of each lamp. [4]
8. Show the direction of current through each lamp. [4]
9. Suppose switch SW2 is opened. What happens to the current through lamp 2? What can you say about the polarity of lamp 2? [4]
10. With SW2 open as in Question 9, what happens to the current direction and voltage polarity of lamp 1 and lamp 3? Explain this. [4]

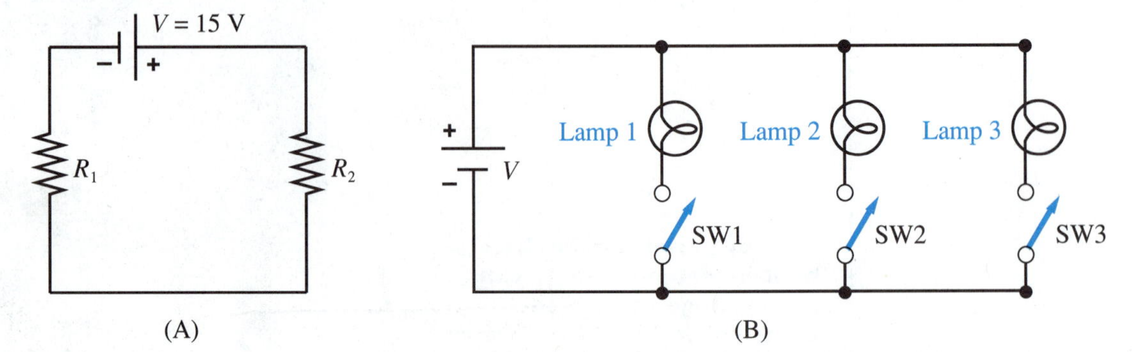

FIGURE 4-17 (A) Two-resistor circuit. (B) Three independently controlled lamps operating from the same source

SAFETY ADVICE

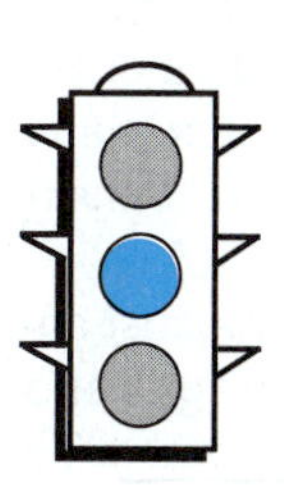

Figure 4-15 (pictured in this chapter and in the Circuits color section) has soldered connections. Becoming a good solderer takes a lot of practice. But before you even warm up your first soldering iron, there are some safety issues you must be aware of.

1. **Wear eye protection.**
2. **Use needle-nose pliers to mechanically bend and crimp wires onto their solder terminals before soldering. This assures that if the solder joint must be desoldered at a later time, the springy wire cannot suddenly spring away from the terminal and throw molten solder.**
3. **For printed circuit board soldering, insert the component leads through the holes, then bend the leads slightly on the copper side of the board. This will hold them steady when you apply the soldering iron.**
4. **Wipe your soldering iron on a damp sponge to remove solder residue. Never shake the residue off; it may fly toward your face.**

MARTIAL ARTS

In the martial arts, there is a need for an objective method of measuring the force of a kick or punch. The pads in these photos contain a very thin layer of piezoelectric material underneath the target area. A piezoelectric material produces a tiny voltage when it is compressed. The greater the compression force, the greater its output voltage. The momentary voltage burst produced by a karate blow is transmitted by cable to the clip-on module. There it is electronically amplified and converted to digital format. The force of the kick in pounds is then read by looking into the module's display window.

Courtesy of NASA

4-5 ELECTRON CURRENT VERSUS HOLE CURRENT

We have described current as the migration of negative-charged electrons from atom to atom through a conductor. You should know that many people prefer to visualize current as the migration of positive charge in the opposite direction. Figure 4-18 illustrates how this works.

Figure 4-18(A), represents the starting point in time, just after switch SW has closed. With SW completing the circuit, an electron jumps out of the atom on the far right, on its way back to the positive terminal of the source. That leaves a "hole" in the atom on the right. A hole is a place where an electron is missing. Since the nucleus still has its full amount of positive charge, the net charge on the far right atom becomes positive. Therefore, we think of a hole as possessing positive charge.

In Figure 4-18(B), which is a bit later, an electron jumps into the far right atom; that atom returns to zero net charge. Now the atom that is second from the right is missing an electron, so its net charge becomes positive. It is as if the hole moved from right to left.

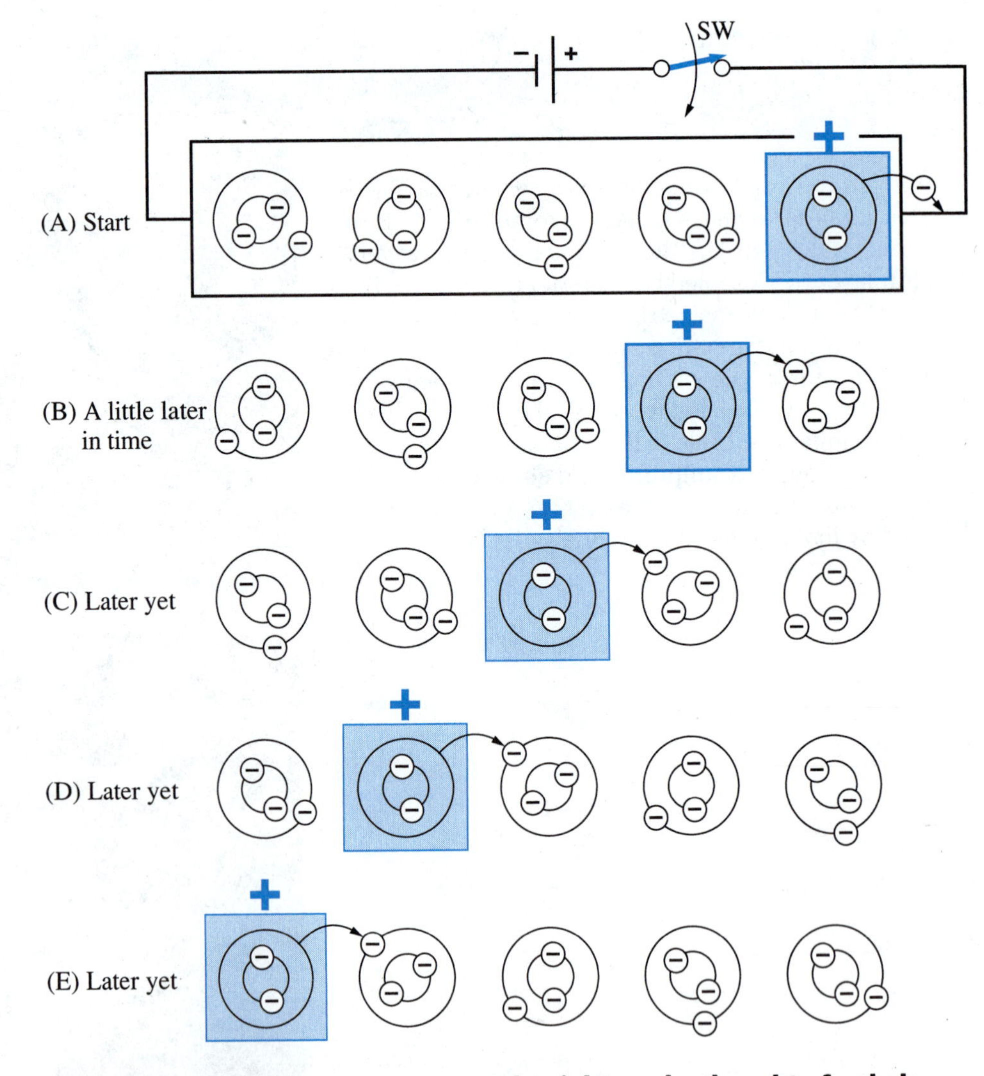

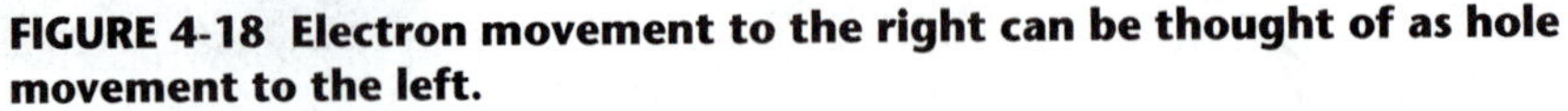

FIGURE 4-18 Electron movement to the right can be thought of as hole movement to the left.

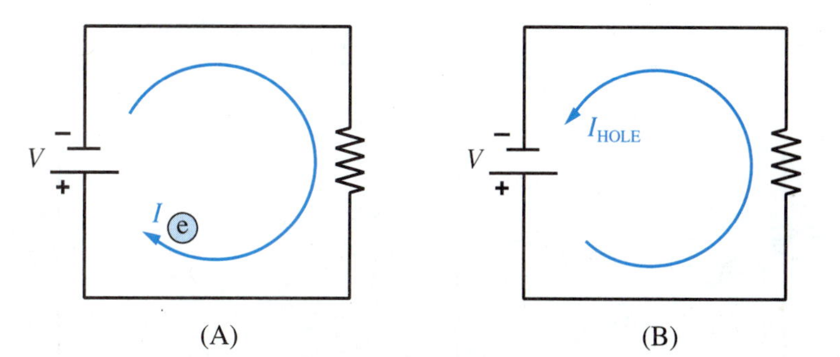

FIGURE 4-19 (A) The action in the circuit can be visualized as electron flow in the CW direction, from negative to positive. (B) Or the action can be visualized as hole flow in the CCW direction, from positive to negative. Hole-flow current is also called *conventional* current.

As time passes, this process is repeated in parts (C), (D), and (E) of Figure 4-18. This description leads to the following fact:

From an external viewpoint, electron movement in one direction is essentially the same as hole movement in the opposite direction.

Because the two views of current are equivalent, either one may be regarded as correct. In Figure 4-19, the hole-flow view in part (B) is just as valid as the electron-flow view in part (A). To be consistent, we will use the electron-flow view of current throughout the book.

FORMULAS

For voltage sources aiding each other

$$V_T = V_1 + V_2$$

For voltage sources opposing each other

$$V_T = V_1 - V_2$$

SUMMARY OF IDEAS

- Circuit devices are drawn using schematic symbols that don't necessarily resemble the actual devices.
- Switches are used to turn circuits on or off.
- A closed switch turns a circuit on by completing the path; an open switch turns a circuit off by interrupting the path.
- All circuits contain five basic parts: 1) Voltage source; 2) Conductors; 3) Insulation; 4) Load device; 5) Control device.
- Electron current enters the negative terminal of a load device and exits from the positive terminal.
- If voltage sources are pushing in the same direction, their voltage values are added.
- If voltage sources are pushing in opposite directions, their voltage values are subtracted.
- The hole-flow view of current is commonly used. Hole current leaves the positive terminal of the source and returns to the negative terminal, opposite from electron current.

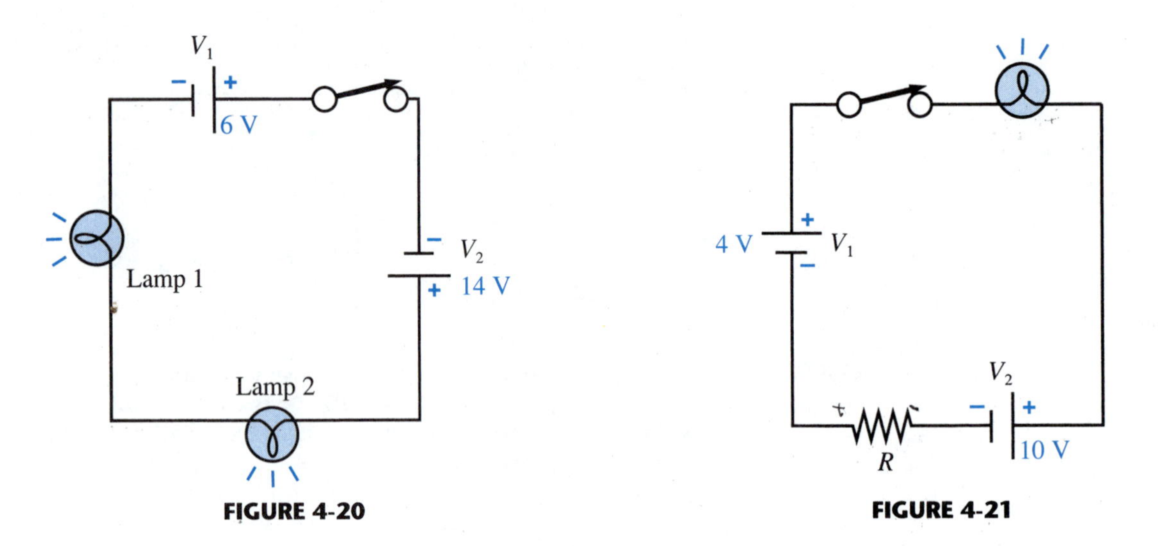

FIGURE 4-20

FIGURE 4-21

CHAPTER QUESTIONS AND PROBLEMS

1. Draw a schematic diagram of a circuit with a 24-V battery driving a 48-Ω resistor, controlled by a switch. [1, 2]
2. To turn a circuit on, its switch must be __________(open or closed). [3]
3. To turn a circuit off, its switch must be __________(open or closed). [3]
4. In the circuit drawing of Problem 1, show the electron current, labeled $I = 0.5$ A. Mark the polarity of the resistor terminals. [4]
5. The two types of switches that were described (mechanically) in this chapter are called__________switches and__________switches. [3]
6. Which type of switch has its electrical contact mechanism actually visible?
7. In an electric circuit, describe the function of the voltage source. [3]
8. Repeat Question 7 for the conductors. [3]
9. Repeat Question 7 for the insulation. [3]
10. Repeat Question 7 for the load device. [3]
11. Repeat Question 7 for the control device (switch). [3]
12. The load device that produces light is called a__________.
13. What is the name of a load device that receives electrical energy from the source and produces sound?
14. Repeat Question 13 for the device that produces mechanical rotation (turning) of a shaft.
15. Repeat Question 13 for the device that produces intense heat, causing two pieces of metal to melt and become fused together.
16. In Section 4-3, the statement was made that neither lamp terminal in Figure 4-12 is closer to the negative source terminal. Give a careful explanation of why this is true. [4]
17. In Figure 4-20, what is the value of total voltage? [5]
18. In Figure 4-20: [4]
    a) Draw the direction of electron current.
    b) Mark the polarity of lamp 1 and lamp 2.
19. In Figure 4-21, what is the value of total voltage? [5]
20. In Figure 4-21: [4]
    a) Draw the direction of electron current.
    b) Mark the polarity of the lamp and the resistor.
21. In Figure 4-22, what is the value of total voltage? [5]
22. In Figure 4-22: [4]
    a) Draw the direction of electron current.
    b) Mark the polarity of R_1, R_2, and R_3.

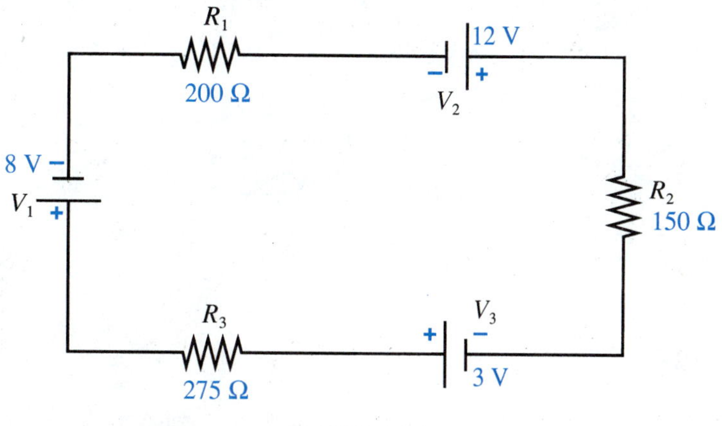

FIGURE 4-22

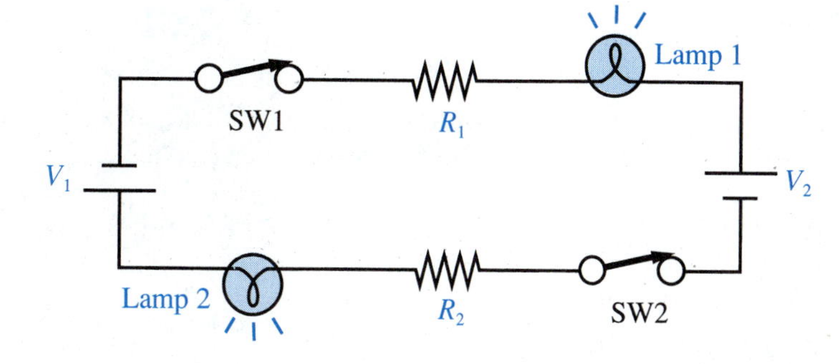

FIGURE 4-23 Controlling with more than a single switch

23. In Figure 4-23, the circuit diagram shows both switches closed. Suppose SW1 is opened while SW2 remains closed. [2, 3]
    a) Would lamp 1 remain lighted? Explain.
    b) Would lamp 2 remain lighted? Explain.
24. Hole current flows away from the voltage source's__________terminal, passes through the load, and reenters the source by its___________terminal. [6]
25. Another name for hole current is__________current. [6]

CHAPTER 5

MEASURING ELECTRIC CIRCUITS

A technician inspects the transmit signal-generating circuits for the military radar unit shown on page 8 in the Applications color section. The copper tubing carries temperature-regulated and dryed air to pressurize the housings of critical circuit sections. This entire assembly is installed inside the stationary bottom base of the radar system shown in that photo.
Courtesy of Datron Systems, Inc.

OUTLINE

NEW TERMS TO WATCH FOR

ammeter
series
analog
voltmeter
parallel
digital
ohmmeter
isolate
deenergized
multimeter
VOM
DMM

The action in an electric circuit can't be seen with our eyes, nor heard with our ears. In fact, unless something goes wrong with a circuit, none of our senses can tell us what's happening in that circuit. To find out what is taking place in a circuit, we must use meters to make measurements of the electrical variables.

After studying this chapter, you should be able to:

1. Connect an ammeter for measuring current.
2. Tell the difference between a series connection and a parallel connection.
3. Connect a voltmeter for measuring voltage.
4. Know the difference between an analog meter and a digital meter.
5. Connect an ohmmeter for measuring resistance.
6. Obtain a dc voltage measurement from the pointer position of an analog multimeter, or VOM.
7. Troubleshoot quickly a suspected voltmeter measurement error.
8. Obtain a dc current measurement or a resistance measurement from the pointer position of a VOM.
9. Obtain dc voltage, dc current, or resistance measurements from the front panel of a digital multimeter, or DMM.

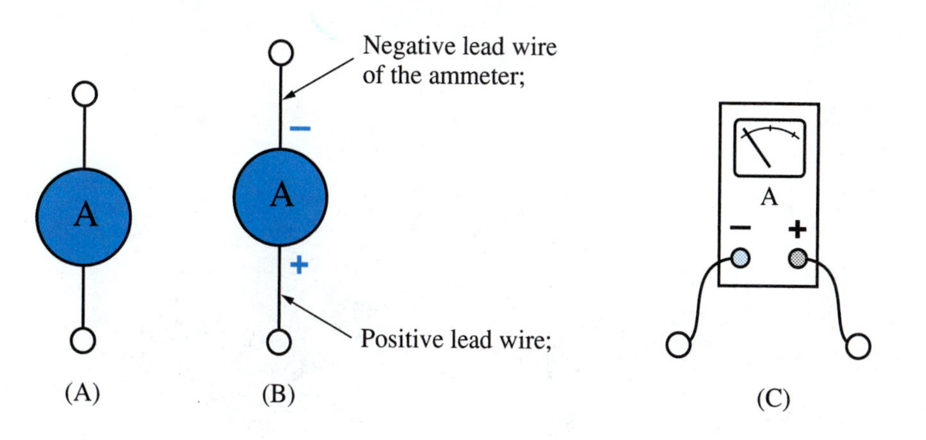

FIGURE 5-1 (A) Ammeter symbol. (B) A dc ammeter has a negative terminal with a black lead wire, and a positive terminal with a red lead wire. (C) Pictorial symbol of an ammeter

5-1 MEASURING CURRENT

The meter, or measuring instrument, that measures current is called an **ammeter**. Its symbol is a circle with the letter *A* inside, as shown in Figure 5-1(A). Ammeters used in dc circuits are usually polarized, with one negative lead and one positive lead. This is shown in Figure 5-1(B).

When first learning how to use an ammeter, we sometimes draw the ammeter pictorially, rather than schematically. This is shown in Figure 5-1(C). It resembles an actual physical ammeter with a movable pointer. To see this, glance ahead for a moment to Figure 5-3.

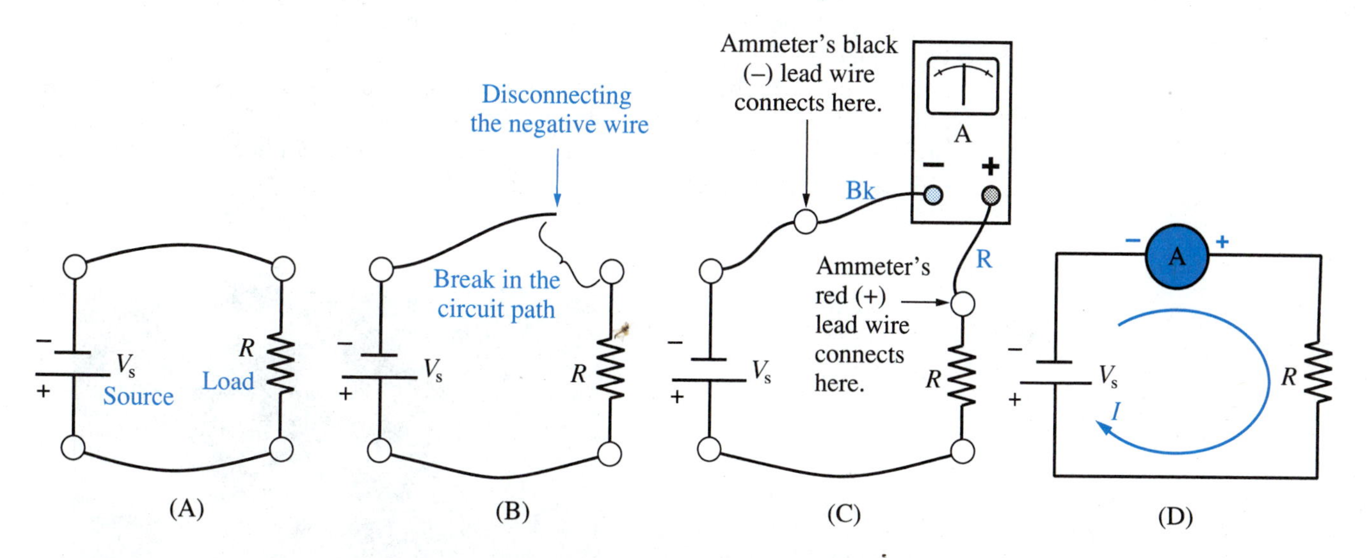

FIGURE 5-2 Procedure for making a current measurement. (A) Original circuit. (B) A break has been created on the negative side of the load. It could just as well be made on the positive side. That choice is up to you. (C) The ammeter (pictorial view) is used to "rejoin the break," or "plug the hole." The ammeter's black lead must attach to the side of the break closer to the negative source terminal. Its red lead must attach to the other side of the break, closer to the positive source terminal. (D) The current measurement can now be taken. (Schematic view of an ammeter.)

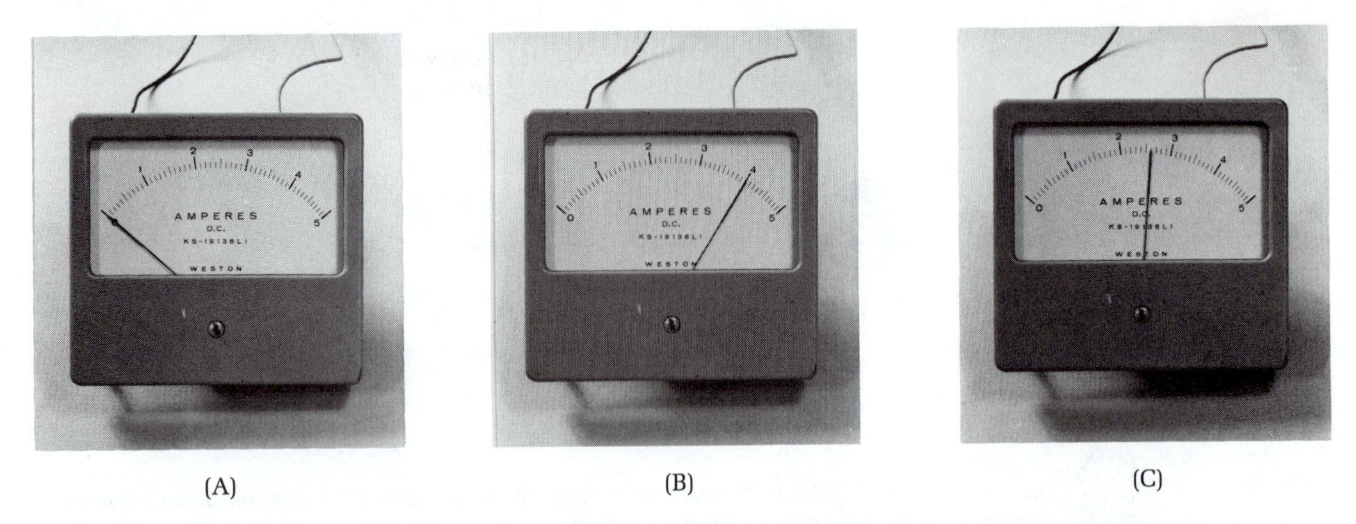

FIGURE 5-3 Dc ammeter, panel-mount type. (A) At zero. (B) Measuring 4 A. (C) Measuring 2.6 A.

TECHNICAL FACT

An ammeter must be connected into a circuit *in line with* the load.

This requires that you disconnect one of the wires to the load. In other words, you must first break open the circuit. Installing an ammeter is demonstrated in Figure 5-2.

Figure 5-2(A) is an operating circuit whose current is to be measured. In Figure 5-2(B), one of the wires joining the source to the load has been disconnected. This creates a break in the current path, which stops the current and causes the load to stop functioning.

The ammeter is connected in Figure 5-2(C), again establishing a complete path for the current. The load begins functioning again. The ammeter leads must be polarized as shown. They must not be reversed.

The circuit's final appearance is diagrammed in Figure 5-2(D). This shows clearly that the ammeter is in line with the load, also called in **series** with the load. Current is entering the ammeter by its negative (black) lead, and leaving the ammeter by its positive (red) lead. The ammeter now indicates the amount of current in the circuit.

Figure 5-3(A) shows a photograph of an **analog** ammeter. This type of ammeter indicates the current value by moving its pointer across a printed scale. In Figure 5-3(B), it reads 4 amperes (4 A). In Figure 5-3(C), the current has been reduced to 2.6 A, as the pointer indicates.

Small hand-held probes are used to detect signal voltage on densely packed integrated-circuit assemblies. *Courtesy of Hewlett-Packard Company*

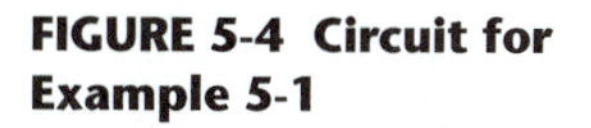

FIGURE 5-4 Circuit for Example 5-1

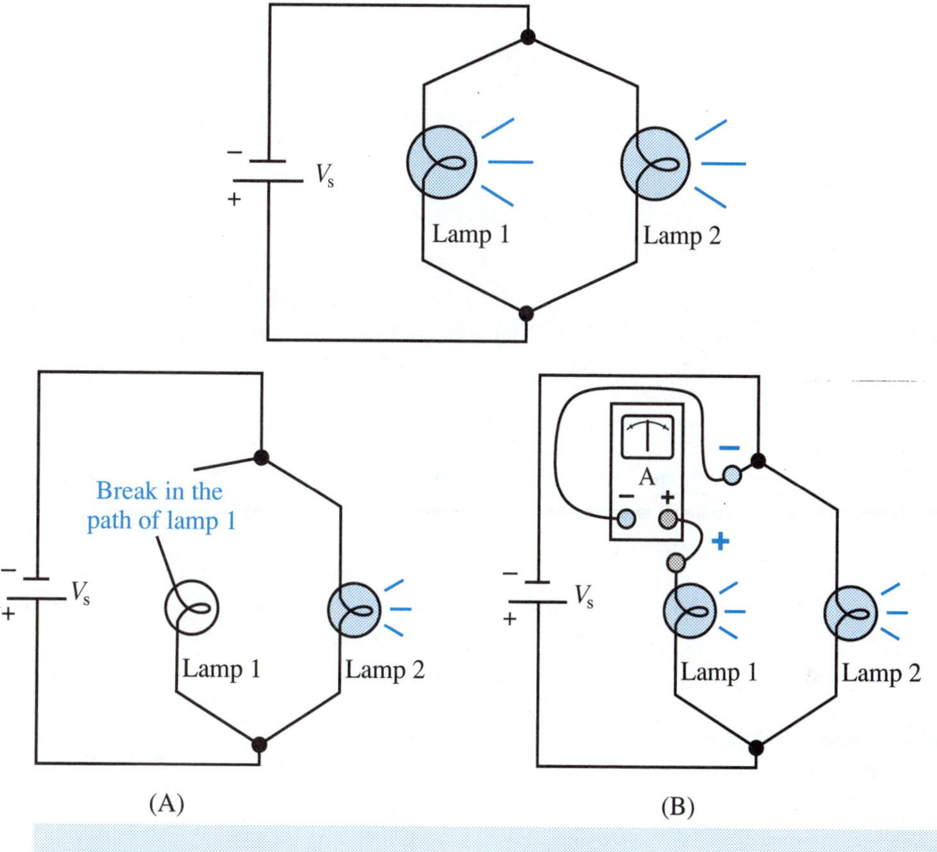

FIGURE 5-5 Measuring the current in lamp 1. (A) First break open the flow path to lamp 1. The current through lamp 1 is temporarily interrupted. (B) Connecting the ammeter across the break, with proper lead polarity. Current begins flowing again in lamp 1.

EXAMPLE 5-1

The circuit of Figure 5-4 has two lamps being lighted at the same time.

a) Show how to connect an ammeter to measure the current through lamp 1.
b) Repeat for lamp 2.

SOLUTION

a) To measure current through lamp 1, we must break open that lamp's flow path, without affecting the flow path of lamp 2. This is shown in **Figure 5-5(A)**. Then we connect the ammeter across the break, as shown in **Figure 5-5(B)**.
b) Break open the path through lamp 2, without disturbing lamp 1. The break can be on the top of lamp 2, as shown in **Figure 5-6(A)**. Then use the ammeter to bridge across the break, as shown in **Figure 5-6(B)**.

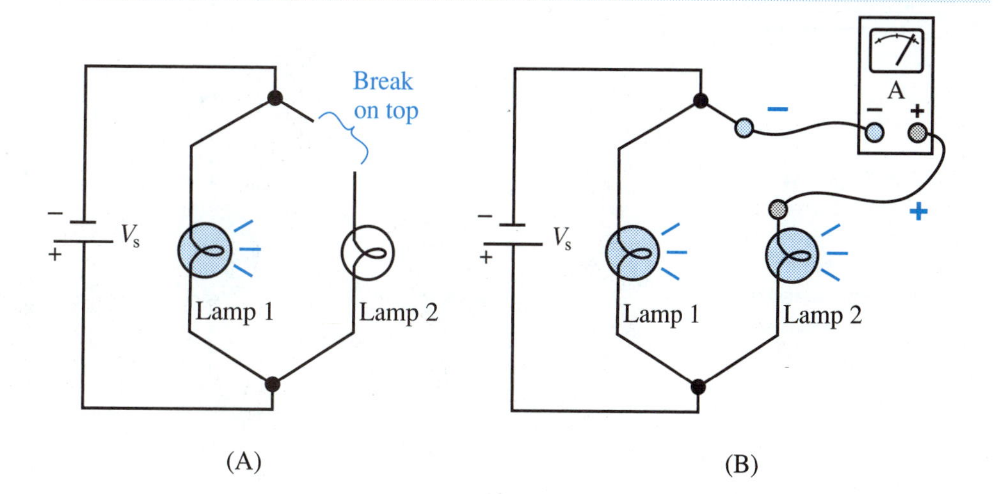

FIGURE 5-6 Measuring current in lamp 2

EXAMPLE 5-2

Show another location where an ammeter could be inserted in Figure 5-4 to measure the current in lamp 2.

SOLUTION

The flow path can be broken just as well on the bottom side of lamp 2, as shown in **Figure 5-7(A)**. The ammeter is then inserted as shown in **Figure 5-7(B)**. The same value of current will be indicated as in **Figure 5-6(B)**.

5-2 MEASURING VOLTAGE

A **voltmeter** is the instrument that measures voltage. Its symbol is shown in Figure 5-8(A). Dc voltmeters generally are polarized, as shown in Figure 5-8(B), with the usual black/red lead-wire colors.

Again, a voltmeter may be shown pictorially as we are first learning how to connect it. This is done in Figure 5-8(C).

TECHNICAL FACT 3

A voltmeter must be connected *across* the load, from one side to the other.

This type of connection is shown in Figure 5-9. It is not necessary to break the circuit open. For that reason, measuring voltage is easier than measuring current.

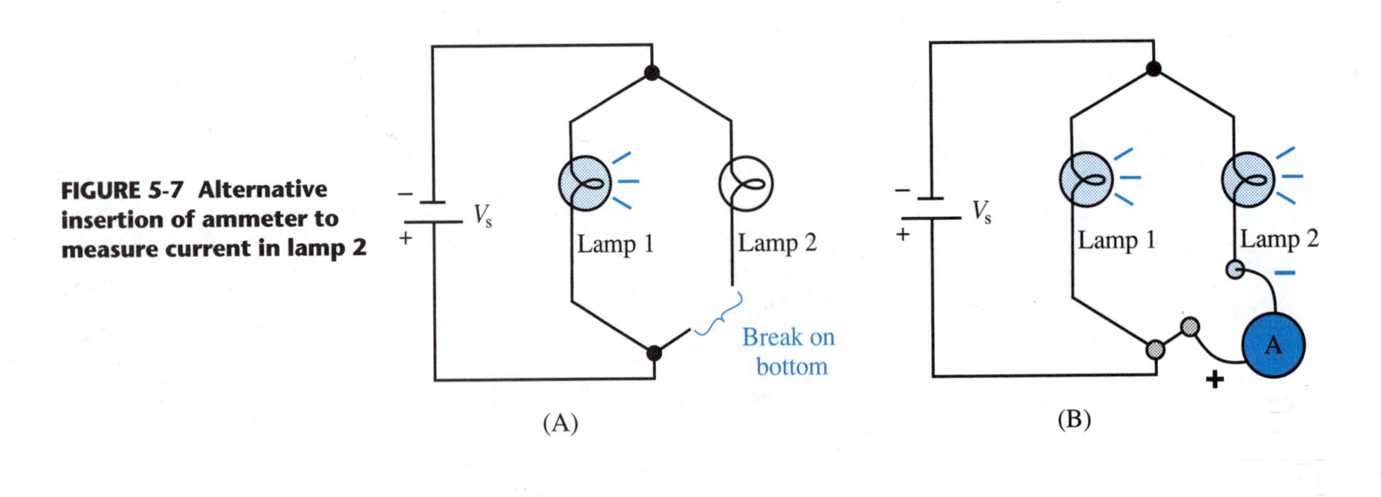

FIGURE 5-7 Alternative insertion of ammeter to measure current in lamp 2

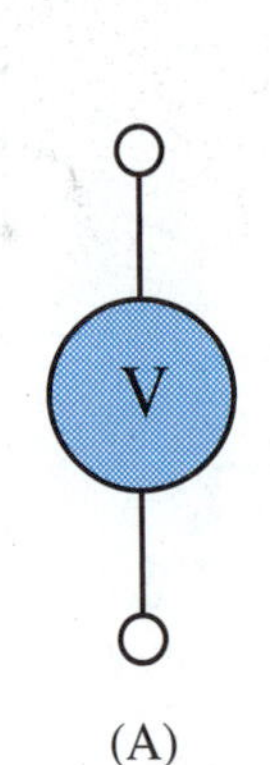

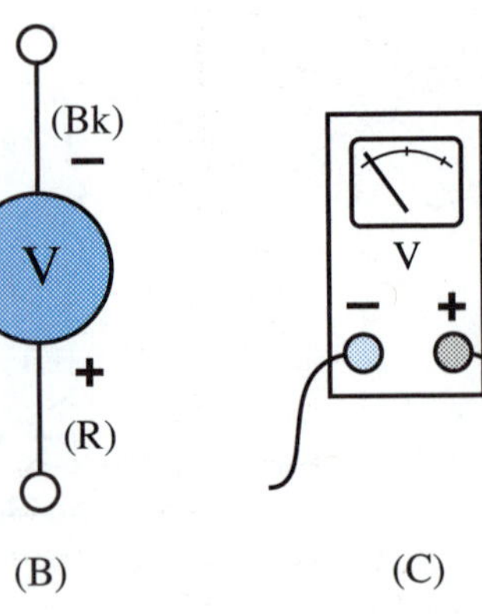

FIGURE 5-8 (A) Schematic symbol for voltmeter. (B) Black lead goes with the negative terminal, red lead with positive, just like a dc ammeter. (C) Pictorial voltmeter

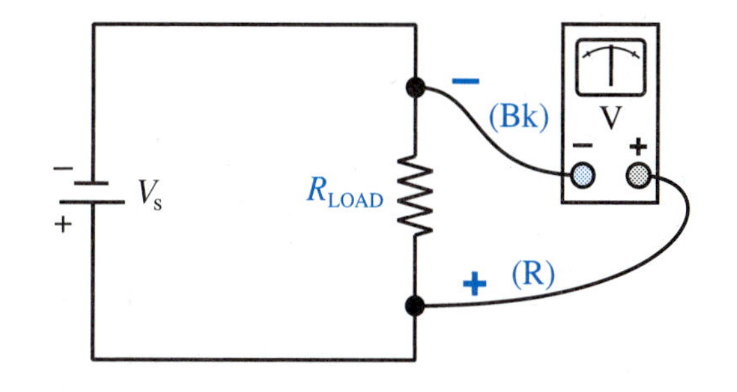

FIGURE 5-9 The basic voltmeter connection is across the load, also called in parallel with the load. Proper lead polarity must be observed.

The voltmeter's polarity is important. Its black lead must attach to the side of the load that is closer to the negative source terminal. The red lead goes closer to the positive source terminal. This is shown in Figure 5-9 for a single-load circuit.

In Figure 5-10(A), the voltmeter is connected to measure the voltage across R_1 in a two-load circuit. It has been moved to measure the voltage across R_2 in Figure 5-10(B).

An analog voltmeter positions a pointer to indicate voltage. A **digital** type lights up the proper numerals to give a direct visual readout. Figure 5-11(A) shows a digital voltmeter giving a measurement of 188.8 V. An even more precise model is shown in Figure 5-11(B). The pictorial symbol for a digital voltmeter is given in Figure 5-11(C). Digital ammeters are common too.

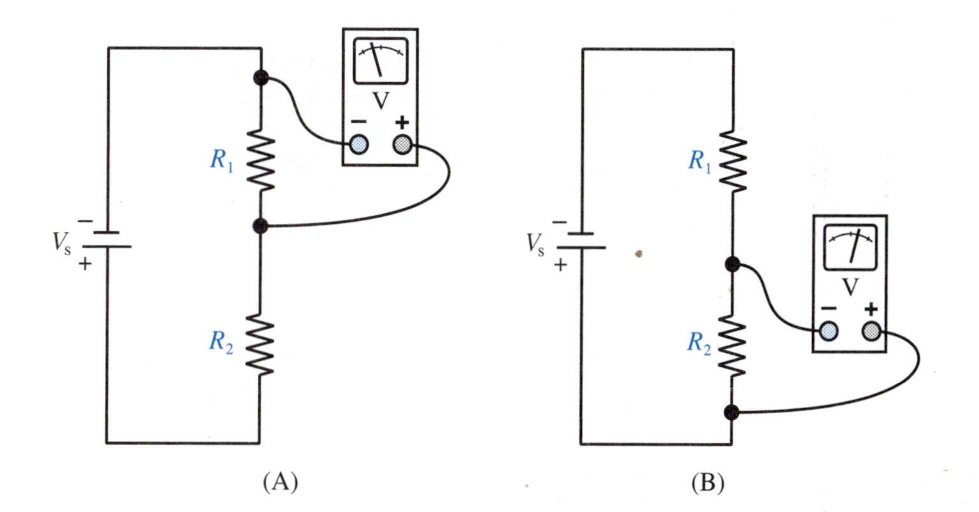

FIGURE 5-10 Measuring individual voltages in a two-resistor circuit. (A) Voltmeter in parallel with R_1. (B) In parallel with R_2

(A) (B) (C)

FIGURE 5-11 (A) Digital voltmeter. The left digit can only be 0 or 1. The right three digits can be any decimal digit, 0 through 9. This is called a 3½ digit meter. (B) 4½ digit voltmeter. The minus sign indicates that the actual voltage polarity is opposite from the expected polarity. *Courtesy of Simpson Electric Co.* **(C) Pictorial symbol**

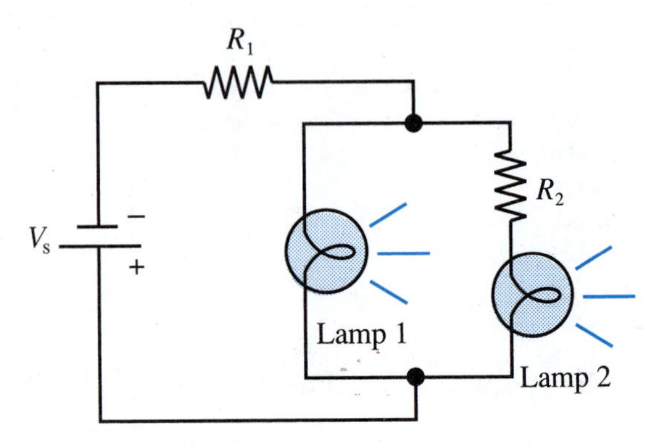

FIGURE 5-12 Circuit for Example 5-3

EXAMPLE 5-3

a) In Figure 5-12, show how to connect a voltmeter to measure the voltage across lamp 1.
b) Repeat for lamp 2.

SOLUTION

a) Connect the voltmeter from one side to the other side of lamp 1, as shown in Figure 5-13(A). It is not necessary to break open the path through lamp 1.
b) Connect the voltmeter from the top terminal to the bottom terminal of lamp 2. The proper polarity is negative on the top and positive on the bottom, as in Figure 5-13(B).

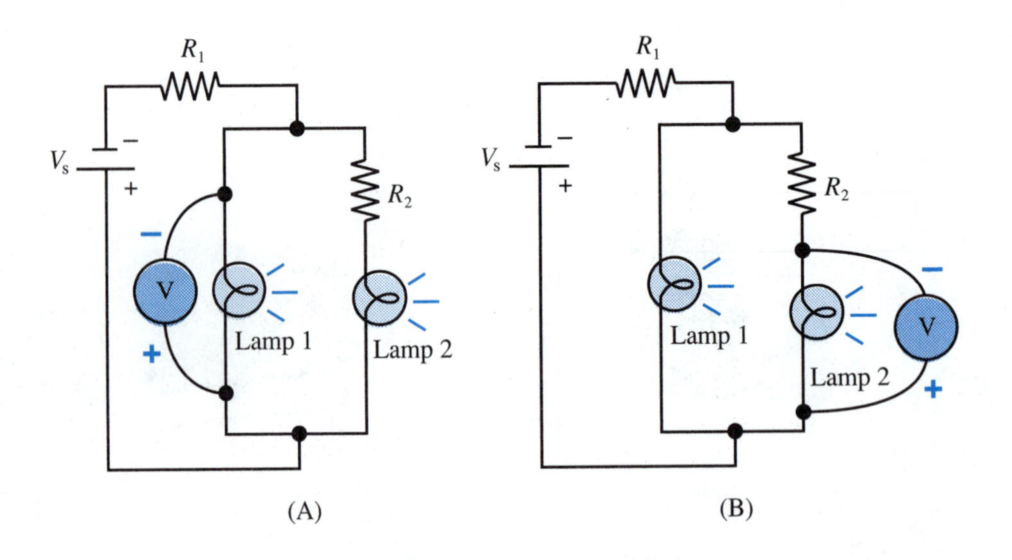

FIGURE 5-13 (A) Measuring the voltage across lamp 1. (B) Measuring the voltage across lamp 2

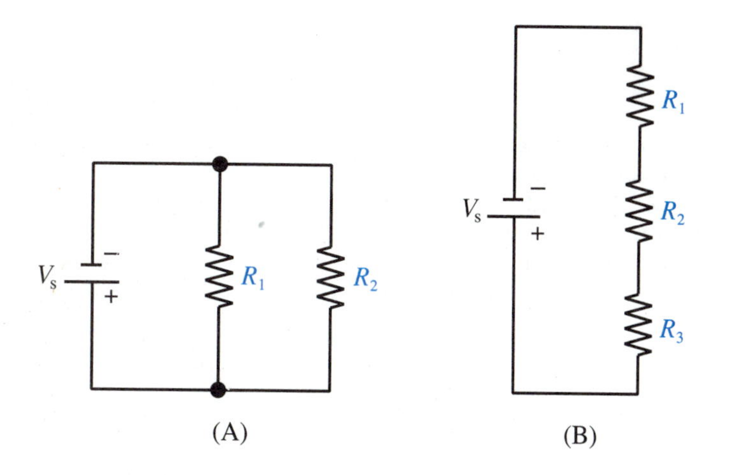

FIGURE 5-14

✔ SELF-CHECK FOR SECTIONS 5-1 AND 5-2

1. In the circuit of Figure 5-14(A), show how to connect an ammeter to measure the current through R_1. [1]
2. Repeat Question 1 for R_2. [1]
3. Show how to measure the current through the source in Figure 5-14(A). [1]
4. In the circuit of Figure 5-14(B), show two different ammeter connections for measuring the current through R_1. [1]
5. Repeat Question 4 for the current through R_2. [1]
6. From your answers to Questions 4 and 5, what can you say about the current through R_1 compared to the current through R_2? [2]
7. Repeat Question 4 for the current through R_3. What can you say comparing the current through R_3 to the current in R_1 and R_2? [1, 2]
8. In the circuit of Figure 5-14(B), show how to connect a voltmeter to measure the voltage across R_1. [3]
9. Repeat Question 8 for R_2. [3]
10. In Figure 5-14(B), show how to connect a voltmeter to measure the voltage across the *combination* of R_1 and R_2. [3]
11. Repeat Question 10 for the combination of R_2 and R_3. [3]
12. The way the circuit is constructed in Figure 5-14(B), is there any way to find the voltage across the combination of R_1 and R_3 with a single voltmeter measurement? [3]

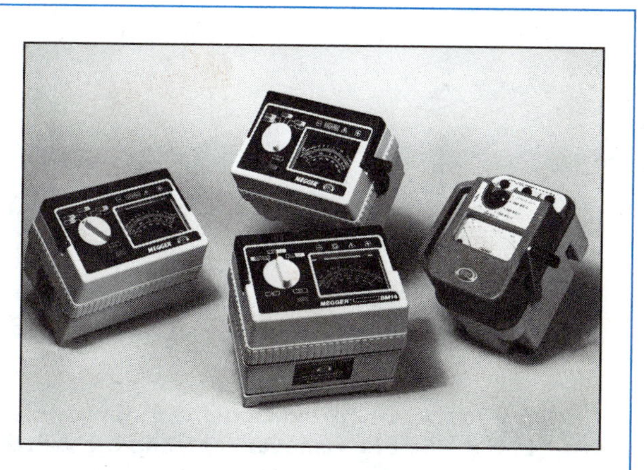

Several models of megohmmeters. They are used for measuring the resistance of wire insulation, and for detecting insulation faults that occur under high-voltage stress.
Courtesy of Biddle Instruments

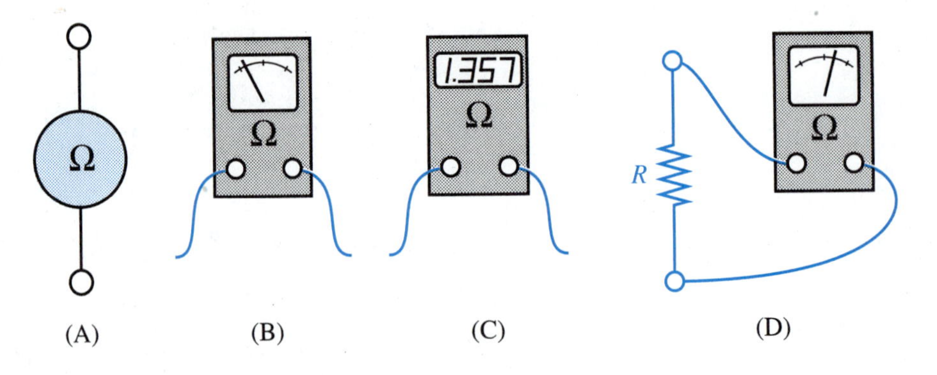

FIGURE 5-15 (A) Ohmmeter symbol. (B) Pictorial ohmmeter, analog type. (C) Digital pictorial symbol. (D) An ohmmeter connects in parallel.

5-3 MEASURING RESISTANCE

An **ohmmeter** is used to measure resistance. Its schematic symbol is shown in Figure 5-15(A). The pictorial symbols for an analog ohmmeter and a digital ohmmeter are given in Figures 5-15(B) and (C).

TECHNICAL FACT

5

An ohmmeter connects across (in parallel with) the device whose resistance is being measured.

Figure 5-15(D) shows a proper ohmmeter connection.

It is best to completely **isolate** the device before connecting the ohmmeter. Thus, to measure R_1 in the circuit of Figure 5-16(A), it is best to completely remove R_1 from the circuit before attaching the ohmmeter. This is shown in Figure 5-16(B).

Very often a device cannot be easily removed from the circuit. In that case, its resistance can still be measured if two conditions are satisfied:

1. There must be no voltage existing across the device—it must be **deenergized.**
2. There must be nothing else in parallel with the device.

For example, resistor R_1 in Figure 5-16(A) can be successfully measured simply by opening the switch before attaching the ohmmeter, as shown in Figure 5-16(C).

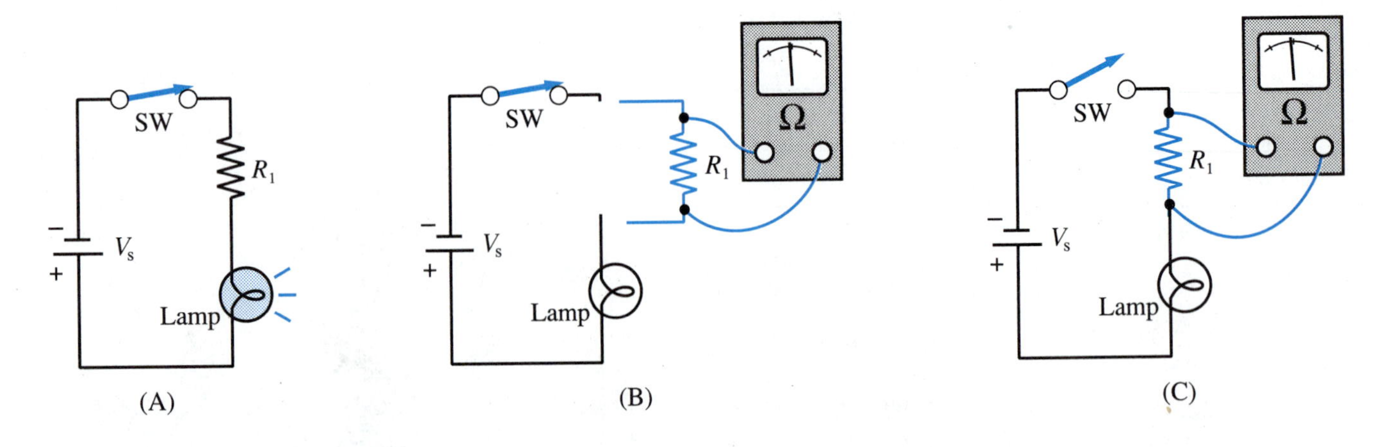

FIGURE 5-16 (A) Energized circuit. (B) If possible, completely remove R_1 before measuring it with an ohmmeter. (C) If R_1 can't be removed, opening the switch will isolate it in this circuit. The ohmmeter can then be connected.

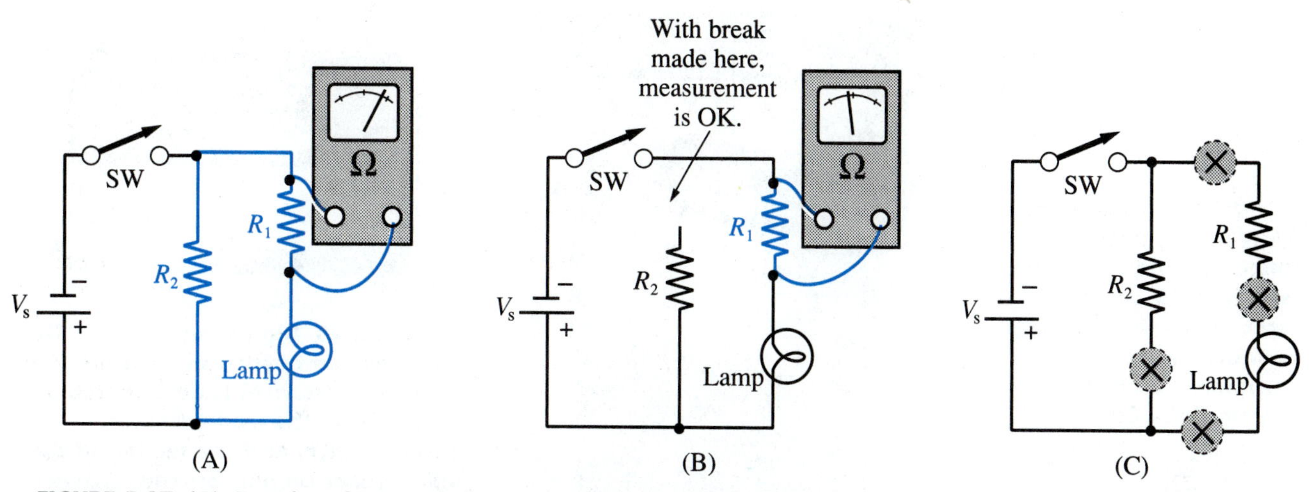

FIGURE 5-17 (A) Opening the switch deenergizes R_1, but does not isolate it. Ohmmeter measurement is not correct as shown. (B) Disconnecting the top lead of R_2 eliminates the parallel path around R_1. Measurement is now correct. (C) Disconnecting the bottom of R_2 would work just as well. So would disconnecting either lead of R_1 itself.

However, in the circuit of Figure 5-17(A), simply opening the switch does not ensure a correct resistance measurement. The ohmmeter "sees" R_1 in parallel with a path formed by R_2 and the lamp. Figure 5-17(B) shows how to get a proper measurement of R_1. The break could be made just as well at any of the locations marked by *X* in Figure 5-17(C). When measuring a standard load-type device like a resistor or a lamp, ohmmeter lead polarity doesn't matter. That is why we haven't specified red and black leads in Figures 5-15, 5-16, and 5-17.

However, there are some resistive devices that are naturally polarized. When measuring their resistance, ohmmeter lead polarity becomes important. For instance, when testing and measuring semiconductor electronic devices such as rectifiers, transistors, or light-emitting diodes (LEDs), we must be careful to connect the positive (red) ohmmeter lead to one particular terminal and the negative (black) lead to the other particular terminal. Exact instructions for performing such tests are given in a course on semiconductor electronics.

The reason for the positive and negative polarization of ohmmeter leads is because an ohmmeter contains an internal dc voltage source, as shown in Figure 5-18. This voltage source is used to push current through the load device, which is then sensed by an internal ammeter within the ohmmeter. The ohmmeter combines voltage and current to find resistance. There are both analog and digital ohmmeters.

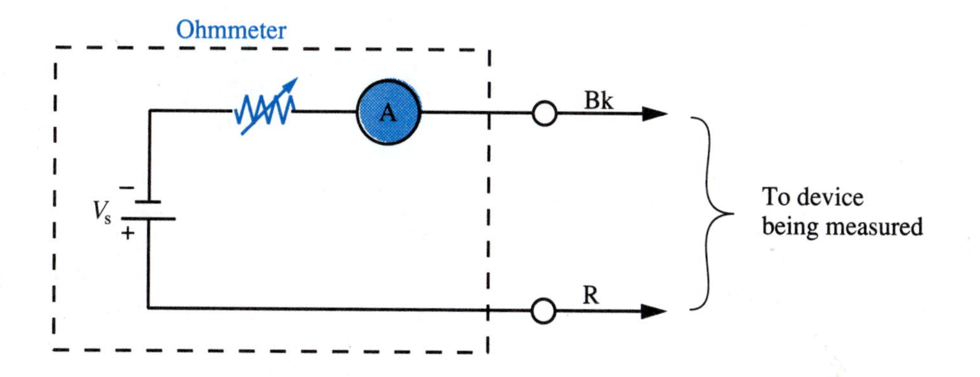

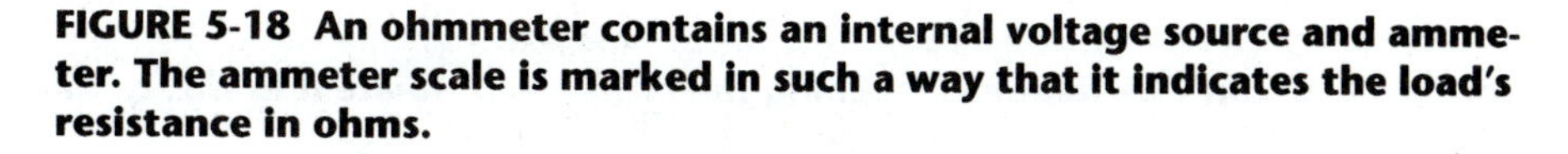

FIGURE 5-18 An ohmmeter contains an internal voltage source and ammeter. The ammeter scale is marked in such a way that it indicates the load's resistance in ohms.

FIGURE 5-19 Analog multimeter. By setting the function/range selector switch in the center, it can be made into an ammeter, voltmeter, or ohmmeter. *Courtesy of Simpson Electric Co.*

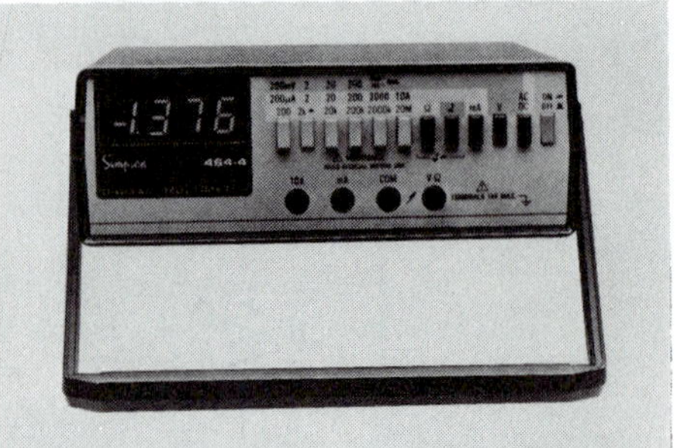

FIGURE 5-20 Digital multimeter. The measurement function (current, voltage, or resistance) is determined by which jacks the lead wires are plugged into, and the setting of the black function pushbutton switches on the right side. The maximum value that can be measured is determined by the white range pushbutton switches on the left side.
Courtesy of Simpson Electric Co.

5-4 MULTIMETERS

It is common to combine the three measurement functions into a single instrument, called a **multimeter.** An analog multimeter, sometimes called a volt-ohm-milliammeter or **VOM**, is shown in Figure 5-19. A digital multimeter, or **DMM**, is shown in Figure 5-20.

Reading an Analog Multimeter (VOM)

In a typical VOM like the one in Figure 5-19, the topmost scale is for measuring resistance. It is labeled OHMS in Figure 5-19. The second scale, labeled DC, is for measuring dc voltage and dc current. The next scale is for measuring ac voltage, and the bottom scale is for measuring relative power. Let us concentrate on using the top two scales.

Measuring Dc Voltage. To measure dc voltage, first place the small selector switch in the + dc position, on the far left of the front panel. Then you must place the main selector switch in one of its voltage positions. These are the positions that start with 2.5 V, at about 10 o'clock on the front panel ; then 10 V, 25 V, 50 V, and finally 250 V, at about 7 o'clock on the front panel.

The number that you select represents the largest voltage value that can be measured in that position. Thus, with the selector switch in the 10-V position, you can measure voltage up to 10 V, but no larger.

There are three sets of numbers below the DC scale. The first (top) set goes from 0 to 250 on the far right. The second (middle) set goes from 0 to 50; the third (bottom) set goes from 0 to 10.

You must read the set of numbers that corresponds to the position of the main selector switch. For example, with the selector switch in the 10-V position (we say, "the meter is on the 10-V scale"), you read the numbers that go from 0 to 10. The main scale marks are each worth 1 volt, though only the even-numbered marks are labeled (2, 4, 6, 8, 10). The odd-numbered marks are not labeled. There are five spaces between the main marks, so each small mark represents one-fifth of 1 volt, or 0.2 volt.

If the main selector switch is set on the 50-V scale, you must read the numbers that go from 0 to 50. Each main scale mark is worth 5 volts, but only the 10-, 20-, 30-, 40-, and 50-V marks are labeled. Each small mark represents one-fifth of 5 volts, or 1 volt per mark.

With the selector switch set to 250 V, you must read the numbers going from 0 to 250. Each main scale mark is worth 25 volts. Each small mark between the main marks represents one-fifth of 25 volts, or 5 volts per mark.

If the main selector switch is set to the 25-V scale, you read the numbers that go from 0 to 250, but you must mentally shift the decimal point one place to the left. Then the labels on the main marks actually stand for 5.0 V, 10.0 V, 15.0 V, 20.0 V, and 25.0 V. The unlabeled main marks mean 2.5 V, 7.5 V, 12.5 V, 17.5 V, and 22.5 V. Each small mark between the main marks is worth one-fifth of 2.5 V, or 0.5 V. If the main selector switch is set to 2.5 V, you again read the numbers from 0 to 250, but mentally shift the decimal point two places to the left. Each main scale mark is worth 0.25 V. Each small mark is one-fifth of 0.25 V, or 0.05 V.

EXAMPLE 5-4

What value of voltage is being measured by the VOM in Figure 5-21?

SOLUTION

The selector switch is on the 10-V scale, so read the numbers that go from 0 to 10. The main mark between the 2 and the 4 means 3 V. The pointer is two small marks to the right of 3, so the voltage is 3.0 V + 2 x 0.2 V or **3.4 V.**

EXAMPLE 5-5

What value of voltage is being measured by the VOM in Figure 5-22?

SOLUTION

The selector switch is on the 25-V scale, so read the numbers that go from 0 to 250, but mentally shift the decimal point one place to the left. The pointer is one small mark to the right of the main mark that means 22.5 V. Therefore, the voltage is 22.5 V + 1 x 0.5 V or **23.0 V.**

FIGURE 5-21 VOM setup and pointer position for Example 5-4. It is assumed that the meter is properly connected across the device being measured.

FIGURE 5-22 Pointer position for Example 5-5

TROUBLESHOOTING TIP

If you get a voltmeter reading that seems unreasonable, it is possible that the meter really is in error. (This very seldom happens.) To find out whether your suspect voltmeter is in error, you could remove it, then install a different voltmeter. However, it is faster to just place a second voltmeter right in parallel with the suspect one. This is shown in the figure.

The same idea works for a suspect ammeter. Of course, you must break the circuit open in order to insert the second ammeter in series with the original one. This slows down the process, which lessens its appeal.

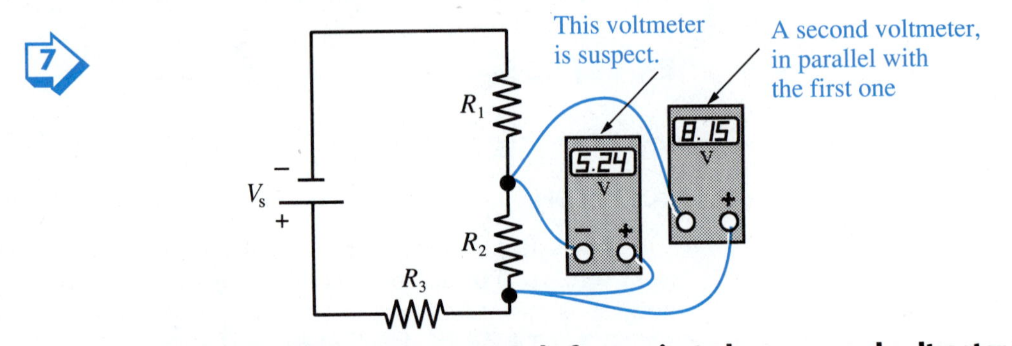

If the first voltmeter is suspected of error, just place a second voltmeter in parallel with it. If the second voltmeter gives a different reading than the first, your suspicion is probably correct.

Measuring Dc Current. To measure dc current, again place the small selector switch in the + dc position. Then place the main selector switch in one of its four current positions. These are the 500 mA, 100 mA, 10 mA, and 1 mA positions on the front panel. Just as with voltage, the scale number selected means the largest amount of current that can be measured. The symbol mA stands for milliampere, which is one-thousandth of an ampere. This notation will be explained in Chapter 6.

The current in milliamperes is taken from the pointer's position on the DC scale, the same scale used for dc voltage. For the 10-mA scale position, you read the set of numbers that goes from 0 to 10. For the 1-mA scale position, read the numbers from 0 to 10, but mentally shift the decimal point one place to the left. For the 100-mA scale position, again read the numbers from 0 to 10, but mentally place an additional zero on the numbers. This is equivalent to shifting the decimal point one place to the right.

For the 500-mA scale position, read the numbers from 0 to 50, but mentally shift the decimal point to the right by placing an additional zero on the numbers.

EXAMPLE 5-6

What value of current is being measured by the VOM in Figure 5-23?

SOLUTION

The selector switch is on the 100-mA scale position, so mentally place an additional zero on the numbers from 0 to 10. The pointer is three small marks past the 30-mA main mark. Therefore, the current is 30 mA + 3 x 2 mA = **36 mA**.

Measuring Resistance. To measure resistance, place the small selector switch in the + dc position. Then place the main selector switch in one of the three

FIGURE 5-23 VOM setup and pointer position for Example 5-6. It is assumed that the meter is properly connected in series.

FIGURE 5-24 VOM setup and pointer position for Example 5-7

resistance positions. These are the R x 1, the R x 100, and the R x 10 000 positions. They are located from 3 o'clock to 5 o'clock on the VOM's front panel.

An analog ohmmeter must be zeroed. To do this, clip the meter leads directly together. Then adjust the zero ohms knob on the right side of the front panel, until the pointer is directly on the zero mark at the right end of the OHMS scale. Disconnect the leads and you are ready to take a measurement.

The selector switch numbers are multipliers. To get the resistance, you first obtain the raw number from the pointer's location on the OHMS scale at the top. Then multiply that raw number by the selected multiplier.

Notice two things about the OHMS scale on a VOM.

1. The numbers are smaller on the right and larger on the left. This is opposite from the DC current / voltage scale.
2. The scale is not linear. That is, the numbers are physically far apart on the right, but close together on the left. Because of this, the amount that a mark is worth is not consistent.

EXAMPLE 5-7

What value of resistance is being measured by the VOM in Figure 5-24?

SOLUTION

On the OHMS scale, we have main marks labeled 2 and 5. There are two main marks between 2 and 5, so they mean 3 and 4. There are five spaces between 4 and 5, so each small mark is worth one-fifth of 1, or 0.2. The pointer is two small marks to the left of 4. Therefore, the raw number is 4 + 2 x 0.2 = 4.4. The selector switch gives a multiplier of 100. Therefore, R = 4.4 x 100 Ω = **440 Ω**.

EXAMPLE 5-8

What value of resistance is being measured by the VOM in Figure 5-25?

SOLUTION

The pointer is between main marks labeled 30 and 40. There is one small mark between 30 and 40, so it represents 35. Since the pointer is a bit to the left of the 35 mark, we could approximate its raw number as 37. The selector switch gives a multiplier of 10 000. Therefore, $R = 37 \times 10\,000\ \Omega =$ **370 000 Ω.**

Reading a Digital Multimeter (DMM)

9

A digital multimeter is quite easy to use. The black lead wire plugs into the common ground (COM) jack. The red lead plugs into the V-Ω jack if you are measuring voltage or resistance. It plugs into the mA or 10 A jack when you are measuring current.

Switch the instrument on and select DC. Choose the measurement function you want by pushing one of the function switches labeled V, mA, or Ω. The musical note switch enables you to test for electrical continuity (near zero ohms). If the resistance between the meter leads is very low, the instrument gives an audible tone. This is a quick way to check whether a circuit is complete.

Push one of the six white range switches, to select the largest value of the measured variable. For example, if you push in the third switch from the left, the largest

COOL SUIT

The cockpit of a race car gets very hot. The engine throws off considerable heat, of course. Also, the aerodynamic design of the car does not allow any cool air to flush heat away from the driver. Cockpit temperatures of 130–140 degrees Fahrenheit are common. At such temperatures, drivers can suffer dehydration or heat exhaustion.

A solution to this problem has now been developed—body vests and helmet liners with liquid circulation passages. A body vest, or "cool suit," worn by a stock-car driver is shown. The driver of a grand prix-formula car is shown here attaching his liquid-circulating helmet liner. Both systems use electrical thermal-sensing devices to monitor skin temperature. As skin temperature rises, an electronically controlled pump motor speeds up. This delivers a greater flow of refrigerated coolant through the circulation passages.

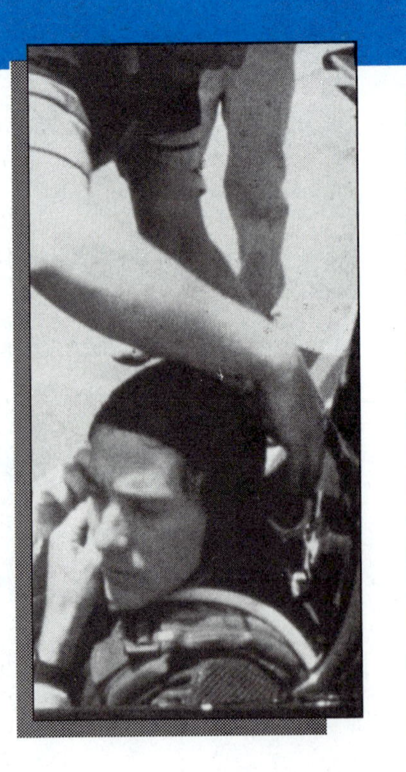

Courtesy of NASA

FIGURE 5-25 Pointer position for Example 5-8

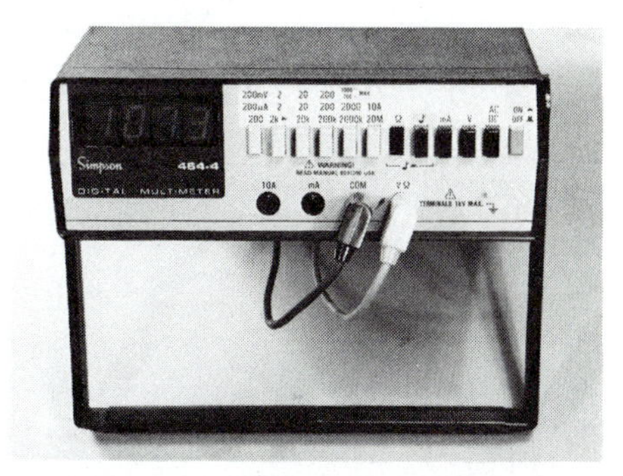

FIGURE 5-26 DMM setup for Example 5-9. It is assumed that the meter is properly connected in parallel.

voltage you will be able to measure is 20 V; the largest current will be 20 mA; and the largest resistance will be 20 kΩ (20 000 Ω, explained in Chapter 6).

In general, you should select the smallest range that is still large enough to handle the measurement. Thus, if you anticipate a voltage of about 10 to 15 V, you should select the 20-V range switch.

EXAMPLE 5-9

What measurement is given by the DMM in Figure 5-26?

SOLUTION

The red lead wire is in the V-Ω jack and the Ω function switch is pushed in. Therefore, the DMM is acting as an ohmmeter, measuring resistance. The fourth range switch from the left has been selected. This means that the maximum resistance is 200 Ω. The digital readout indicates **107.3 Ω.**

✔ SELF-CHECK FOR SECTIONS 5-3 AND 5-4

13. T-F The best way to measure the resistance of a circuit component is to completely remove it from the circuit. [5]
14. T-F When measuring the resistance of a resistor, it is all right if the resistor is energized. [5]
15. In Figure 5-17(A), it is desired to measure the resistance of R_2. If R_2 cannot be completely removed, show how to connect the ohmmeter. [5]
16. Repeat Question 15 for measuring the lamp's resistance. (This would be its cold resistance.) [5]
17. A volt-ohm-milliammeter (VOM) is a(n)____________type of meter. (Answer analog or digital.) [4]
18. The letters DMM stand for____________ ____________. [4]

SUMMARY OF IDEAS

- To measure the current through a circuit component, an ammeter must be connected in line with (in series with) the component.
- It is necessary to break the circuit open in order to insert an ammeter.
- To measure the voltage across a circuit component, a voltmeter must be connected from one side of the component to the other. This is called in parallel with the component.
- A voltmeter connection does not require that the circuit be broken open, so it is easier than making an ammeter connection.
- Many dc ammeters and voltmeters are polarized. The lead color code is black for negative, red for positive.
- To measure the resistance of a circuit component, an ohmmeter must be connected from one side to the other (in parallel). The component must be deenergized and it must be isolated from all other components.
- A multimeter is an instrument that combines the functions of ammeter, voltmeter, and ohmmeter.
- An analog meter has a pointer that moves across a printed scale.
- A digital meter gives a visual display by lighting up numerals.

CHAPTER QUESTIONS AND PROBLEMS

1. T-F A voltmeter connects in parallel with the device it is measuring. [2, 3]
2. An ammeter connects in_______with the device it is measuring. [1, 2]
3. For a polarized dc ammeter, the electron current must enter into the meter by its negative black lead. [1]
4. When connecting a polarized dc voltmeter, which terminal of the load receives the black lead? Which receives the red lead? [3]
5. T-F Connecting a voltmeter requires that the circuit be broken open. [3]
6. Any meter having a pointer moving across a printed scale is called a(n) __________meter. [4]
7. Any meter that gives a readout indication by illuminating decimal numerals is called a(n)__________meter. [4]
8. Which variable is more difficult to measure, current or voltage? Explain why. [2]
9. Describe how an ohmmeter must be connected to the device it is measuring. [5]
10. Can we measure a lamp's operating resistance by connecting an ohmmeter across it while it is lighted? Explain. [5]

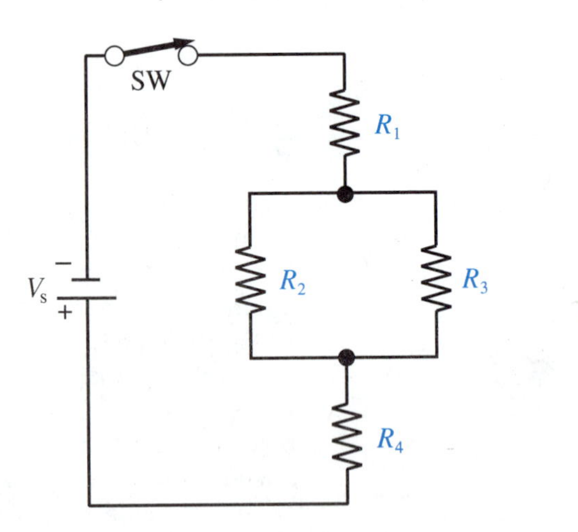

FIGURE 5-27 Circuit for Questions 17-26

11. Assume that opening the on-off switch in a circuit immediately deenergizes the entire circuit. Even then, a simple across-the-device ohmmeter connection may not necessarily be correct. Explain this. [5]
12. T-F Lead polarity is always important when connecting an ohmmeter. [5]
13. T-F An ohmmeter contains its own energy source. [5]
14. What is the name of the instrument that can be used as an ammeter or a voltmeter or an ohmmeter?
15. In Question 14, what letters are used to refer to this instrument if it is analog? If it is digital? [4]
16. Which type of meter, analog or digital, is quicker and easier to read? [4]

For Questions 17 through 26, refer to Figure 5-27.

17. Show how to connect an ammeter to measure the current through R_1. [1]
18. Show how to connect an ammeter to measure the current through R_2. [1]
19. Show two ways to connect an ammeter to measure the current through R_3. [1]
20. Show how to connect an ammeter to measure the current through R_4. [1]
21. Show how to connect a voltmeter to measure the voltage across R_1. [3]
22. Show how to connect a voltmeter to measure the voltage across R_2. [3]
23. Show how to connect a voltmeter to measure the voltage across R_4. [3]
24. What is the simplest way to get an ohmmeter measurement of the resistance of R_1? [5]
25. Repeat Question 24 for the resistance of R_4. [5]
26. Describe how to get an ohmmeter measurement of the resistance of R_2. [5]

CHAPTER 6

OHM'S LAW

Completely equipped test bench, with computer graphic display of test data. Most electronic test data is recorded in non-basic units. *Courtesy of National Instruments Co.*

OUTLINE

NEW TERMS TO WATCH FOR

Ohm's law triangle
exponent
coefficient
significant figures
engineering units
kilo
mega
milli
micro

The three most important ideas in electricity are current, voltage, and resistance. These three ideas are mathematically related to each other in a certain definite way, called Ohm's law.

After studying this chapter, you should be able to:

1. Describe the effect that voltage has on current.
2. Describe the effect that resistance has on current.
3. Calculate the current if voltage and resistance are known.
4. Calculate the resistance if current and voltage are known.
5. Calculate the voltage if current and resistance are known.
6. Use power-of-10 notation to write very large and very small numbers.
7. Perform the four arithmetic operations (multiply, divide, add, and subtract) with power-of-10 numbers.
8. Do Ohm's law calculations using engineering notation.
9. Troubleshoot a high-temperature load device by using Ohm's law to find its hot resistance.

FIGURE 6-1 Variable-voltage circuit with voltage and current being measured

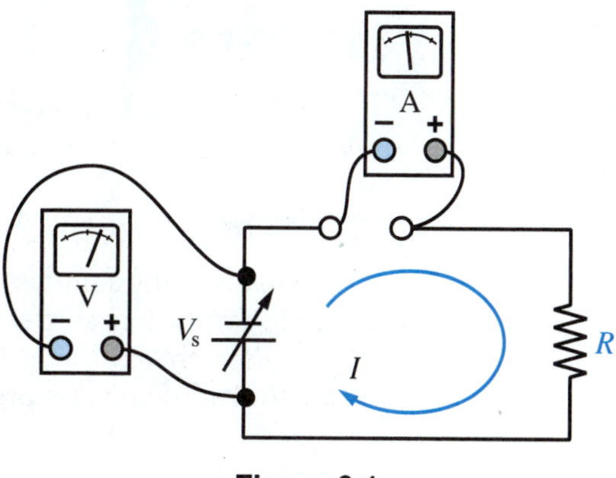

Figure 6-1

6-1 THE MEANING OF OHM'S LAW

There is a two-part experiment that is quite easy to perform. The first part uses a variable voltage driving a constant resistance with a voltmeter and an ammeter attached, as shown in Figure 6-1. If we vary the voltage, we will discover that the current changes in direct proportion. This means that if we double the voltage, the current also doubles. If we increase the voltage by 25%, the current increases by 25%. If we decrease the voltage by 10%, the current decreases by 10%.

The second part of the experiment is shown in Figure 6-2. There we have constant voltage driving a variable resistance. If we vary the resistance, we will find that the current changes by inverse proportion. This means that doubling the resistance causes the current to decrease to half the original amount. Increasing (multiplying) the resistance by any factor causes the current to be decreased (divided) by that same factor.

These two results put together are Ohm's law. In words, Ohm's law is stated as:

TECHNICAL FACT

Current is directly proportional to voltage and inversely proportional to resistance.

The mathematical formula that represents this statement is:

$$I\text{ (current)} = \frac{V\text{ (Voltage)}}{R\text{ (Resistance)}} \quad \text{or} \quad I = \frac{V}{R}$$

EQ. 6-1

FIGURE 6-2 Variable-resistance circuit. The resistance value can be measured by opening switch 1, then closing switch 2. To make the circuit operate again, reverse the switching procedure. The current can then be measured.

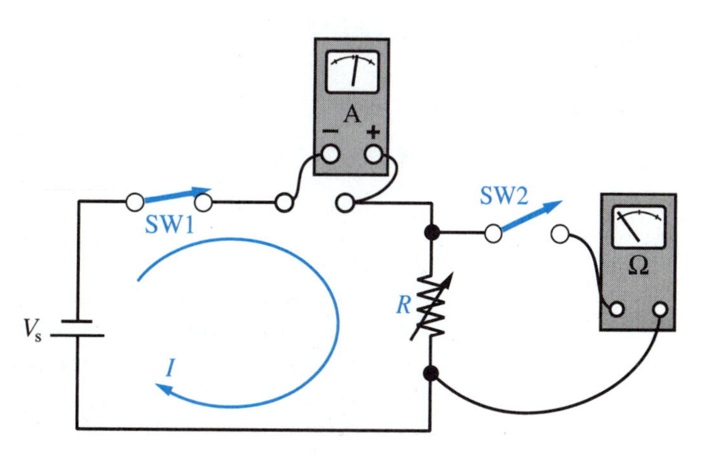

EXAMPLE 6-1

In Figure 6-1, suppose that the resistance is known to be 12 Ω and the voltmeter measures a source voltage of 24 V. Calculate the current.

SOLUTION

When solving a numerical problem, the proper procedure is to write the symbolic formula first. Second, substitute the known values of the variables on the right side of the equal sign. Last, perform the arithmetic and write the result. Following this procedure with Equation 6-1, we have

$$I = \frac{V}{R}$$

$$I = \frac{24\ \text{V}}{12\ \Omega} = \mathbf{2\ A}$$

EXAMPLE 6-2

Now let the voltage in Figure 6-1 be adjusted to 36 V. Find the new current.

SOLUTION

By the same procedure as before,

$$I = \frac{V}{R}$$

$$I = \frac{36\ \text{V}}{12\ \Omega} = \mathbf{3\ A}$$

Notice that the voltage was increased by a factor of 1.5 (36 V/24 V = 1.5). The current also increased by a factor of 1.5 (3 A/2 A = 1.5). This is the meaning of direct proportion.

EXAMPLE 6-3

In Figure 6-2, suppose the voltage source is known to be 15 V. The ohmmeter measures a resistance of 6 Ω. Calculate the current.

SOLUTION

$$I = \frac{V}{R} = \frac{15\ \text{V}}{6\ \Omega} = \mathbf{2.5\ A}$$

EXAMPLE 6-4

Now let the resistance be adjusted to 12 Ω. Calculate the new current.

SOLUTION

$$I = \frac{V}{R} = \frac{15\ \text{V}}{12\ \Omega} = \mathbf{1.25\ A}$$

As we went from Example 6-3 to Example 6-4, the resistance was increased (multiplied) by a factor of 2. That caused the current to be decreased (divided) by a factor of 2 (2.5 A/2 = 1.25 A). This is the meaning of inverse proportion.

6-2 REARRANGING OHM'S LAW

Any mathematical formula, or equation, can be rearranged to solve for a different variable. Equation 6-1 is set up to solve for current, I. It is mathematically correct to interchange the positions of I and R, giving

$$R = \frac{V}{I}$$

EQ. 6-2

This equation enables us to calculate resistance, R, if voltage, V, and current, I, are known.

EXAMPLE 6-5

A certain heater has a current of 1.5 A when driven by a 120-V source. Calculate its resistance.

SOLUTION

From Equation 6-2,

$$R = \frac{V}{I}$$

$$R = \frac{120\text{ V}}{1.5\text{ A}} = \mathbf{80\ \Omega}$$

The third rearrangement of Ohm's law enables us to solve for voltage, V, if current, I, and resistance, R, are known. Cross-multiplying I and R in Equation 6-1 or Equation 6-2 gives us

$$V = IR$$

EQ. 6-3

EXAMPLE 6-6

Suppose that the heater of Example 6-5 is connected to a new voltage source, causing the current to decrease to 1.375 A. Find the voltage of the new source.

SOLUTION

The heater is the same, so R is still 80 Ω. From Equation 6-3,

$$V = IR$$

$$V = (1.375\text{ A})(80\ \Omega) = \mathbf{110\ V}$$

6-3 THE OHM'S LAW TRIANGLE

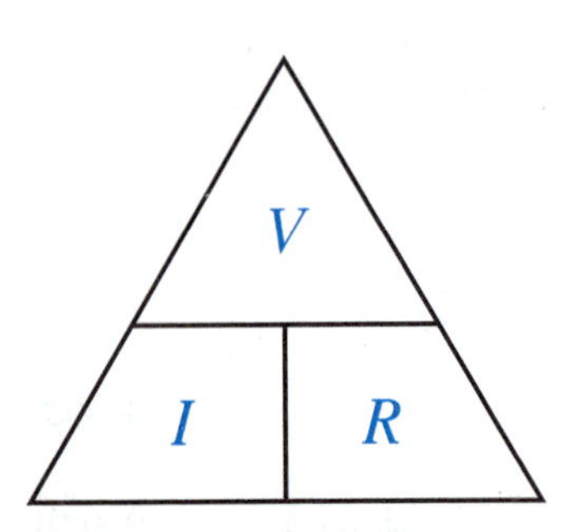

FIGURE 6-3 Ohm's law triangle. This helps you remember when you must multiply and when you must divide.

If you have trouble remembering the three forms of Ohm's law, use the memory aid in Figure 6-3. If you are solving for voltage, cover the V. I and R appear alongside each other, reminding you to multiply them. If you are solving for current, cover the I. V appears above R, reminding you to divide voltage, V, by resistance, R. When solving for resistance, cover the R. The triangle shows that V must be divided by I.

EXAMPLE 6-7

A certain lamp is lighted in a circuit. Measurements indicate that $I = 0.12$ A and $V = 16$ V. Find the lamp's resistance.

SOLUTION

Covering R in Figure 6-3 reminds us that

$$R = \frac{V}{I}$$

so

$$R = \frac{16 \text{ V}}{0.12 \text{ A}} = \mathbf{133.3\ \Omega}$$

We have rounded the result to 4 figures. In electrical calculations, results are usually rounded to 3 or 4 figures.

Up to this point, our calculations have had easy numbers and simple neat answers. Of course, in real life the numerical values usually aren't easy and neat. Example 6-7 is our first taste of this. Real-life numbers are not a serious difficulty if you perform your calculations with a hand-held calculator. From this point forward, you should use a calculator for all arithmetic operations—multiplication, division, addition, and subtraction.

✔ SELF-CHECK FOR SECTIONS 6-1, 6-2, AND 6-3

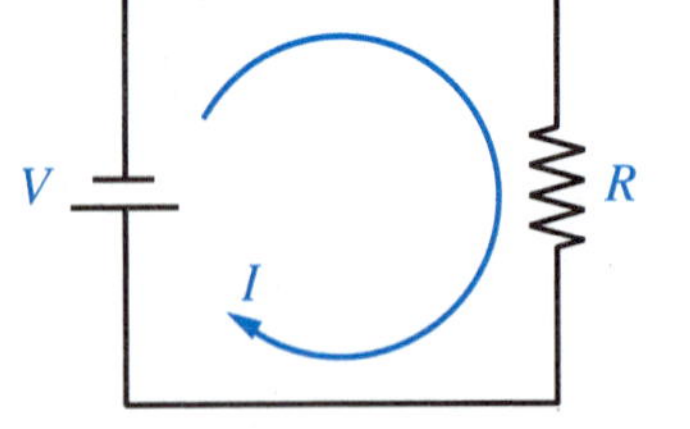

FIGURE 6-4

1. In the simple circuit of Figure 6-4, suppose source voltage $V = 15$ V and $R = 5\ \Omega$. Find current, I. [3]
2. Repeat Problem 1 for a new voltage value of 7.5 V. Compare answers. What does this mean about the relation between V and I? [1, 3]
3. In Figure 6-4, suppose $V = 8$ V and current, I, is 0.5 A. What is the value of resistance, R? [4]

In automated chemical testing applications, great amounts of data are stored quickly by fast-sampling computer programs.
Courtesy of Hewlett-Packard Company

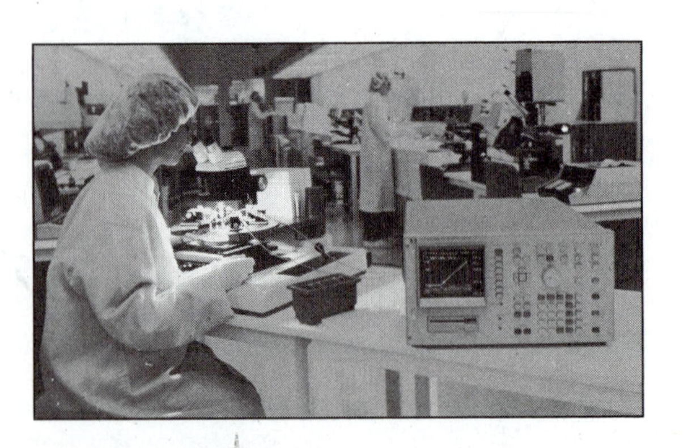

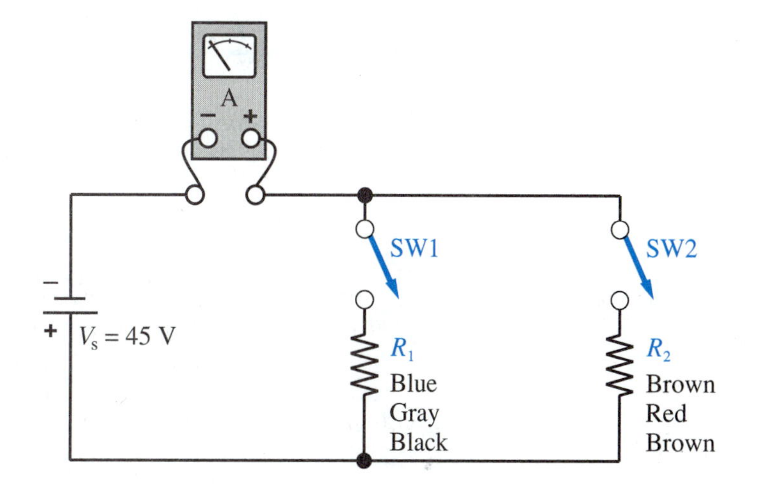

FIGURE 6-5 Circuit with two load resistors, separately switched

4. In Problem 3, suppose R is changed so that $I = 0.25$ A. What is the new value of R? Compare answers to Problems 3 and 4. What does this mean about the relation between R and I? [2, 4]
5. In Figure 6-4, suppose resistor, R, is color-coded brown-green-brown. The current is measured as 0.4 A. What is the value of source voltage, V? [5]
6. In the circuit of Figure 6-5, the two resistors are separately controlled by switches SW1 and SW2. The color codes are given for R_1 and R_2. Source voltage = 45 V. [2,3]
   a) If SW1 is closed and SW2 is open, what current value will be measured by the ammeter?
   b) Repeat part a for SW1 open and SW2 closed.
   c) For a fixed voltage, a larger resistance results in a__________current.
   d) If SW1 and SW2 are both open, what value of current will be read by the ammeter? Explain this.

6-4 POWERS OF 10

In the problems that we have solved so far, the numbers have been simple to handle. They have not been extremely large or extremely small. In reality, many of the numbers we have to deal with are very large or very small. We must have a method for handling such numbers.

Large Numbers

The number 100 is equal to 10 x 10 or 10^2. The number 1000 is equal to 10 x 10 x 10, or 10^3. We can write

$$100 = 1 \times 10^2$$
$$1000 = 1 \times 10^3$$

These two examples show that the power of 10 (the **exponent**) is equal to the number of zeros to the right of 1. The same idea holds for larger numbers.

Numbers that do not start with a 1 are handled in a similar way. For instance, the number 2400 can be written 2.4×10^3 by applying the following rule:

Place a pencil on the decimal point. (If there is no decimal point shown, it is assumed to be at the right end of the number.) Move the pencil to the left, counting places as you go, until only one digit remains to its left. The number of places that you have moved is the power of 10. The nonzero digits, with the decimal point in the new position, are multiplied by the power of 10.

EXAMPLE 6-8

Convert these large numbers to power-of-10 notation.
a) 4700 b) 91 000 c) 560 000 d) 1 200 000 e) 33 520

SOLUTION

a) 4.7.0.0. Moving the decimal point three places to the left leaves 4.7, so this number can be written $\mathbf{4.7 \times 10^3}$.
b) The decimal point moves four places to the left, so we have $\mathbf{9.1 \times 10^4}$.
c) Moving five places gives $\mathbf{5.6 \times 10^5}$.
d) Move six places for $\mathbf{1.2 \times 10^6}$.
e) Moving four places gives $\mathbf{3.352 \times 10^4}$.

The result in part e has more digits in the front part (called the **coefficient**). This happened because the original number had more leading digits that were not zeros. We say that this number has more **significant figures.**

When going the opposite way, from power-of-10 notation back to standard notation, just reverse the procedure.

EXAMPLE 6-9

Convert these numbers from power-of-10 notation back to standard notation.
a) 8.2×10^4 b) 6.753×10^2 c) 3.658×10^6 d) 8.43×10^9
e) 4.84×10^1 f) 7.95×10^0 g) 2.07×10^3

SOLUTION

a) 8.2.0.0.0. Move the decimal point four places to the right, inserting zeros as needed. The result is **82 000.**
b) Moving two places to the right gives **675.3.**
c) To move six places to the right requires three additional zeros to be inserted—**3 658 000.**
d) **8 430 000 000** (Eight billion, four hundred thirty million)
e) Moving just one place to the right gives **48.4**. This is ten times 4.84, as you would expect, since 10^1 means that we're simply multiplying by 10.
f) Move the decimal point zero places, so the number just stays as it is, **7.95.** Keeping a number the same is equivalent to multiplying by 1. This is reasonable, since $10^0 = 1$.
g) **2070**. The embedded zero is treated just like any other digit.

Small Numbers

Numbers less than 1 can be represented by negative powers of 10. This notation is shown by the following equivalent numbers.

$$0.1 = 10^{-1}$$
$$0.01 = 10^{-2}$$
$$0.001 = 10^{-3}$$

The same pattern is repeated for smaller numbers and for numbers that have digits other than 1. For instance, the number 0.000 38 can be written as 3.8×10^{-4} by applying the following rule:

Place a pencil on the decimal point. Move the pencil to the right, counting places as you go, until it moves to the right of the first nonzero digit. The number of places that you have moved is the negative power of 10. The nonzero digits, with the decimal point in the new position, are multiplied by that power of 10.

EXAMPLE 6-10

Convert the following numbers to power-of-10 notation.
a) 0.0035 b) 0.641 c) 0.000 000 782 d) 0.000 15 e) 0.049 03

SOLUTION

a) Moving three places to the right gives $\mathbf{3.5 \times 10^{-3}}$.
b) Move one place, for $\mathbf{6.41 \times 10^{-1}}$.
c) $\mathbf{7.82 \times 10^{-7}}$
d) $\mathbf{1.5 \times 10^{-4}}$
e) $\mathbf{4.903 \times 10^{-2}}$. Once the decimal point has been moved to the right of the closest nonzero digit, an embedded zero is treated like any other digit.

To go the opposite way, from power-of-10 to standard notation, just reverse this procedure.

EXAMPLE 6-11

Convert the following numbers back to standard notation:
a) 8.3×10^{-2} b) 9.65×10^{-4} c) 4.13×10^{-1} d) 6.72×10^{-9}

SOLUTION

a) Move the decimal point two places to the left, inserting a zero as needed. **0.083**
b) Move four places to the left, inserting as many zeros as needed. **0.000 965**
c) **0.413**
d) **0.000 000 006 72**

Arithmetic With Power-of-10 Numbers

Writing very large or very small numbers in power-of-10 notation makes it easier to multiply and divide them.

Multiplying. For multiplication, we multiply the coefficients, and add the powers of 10 (the exponents). For example,

$$(2.0 \times 10^3) \times (4.0 \times 10^2) = 8.0 \times 10^5$$
$$(1.5 \times 10^4) \times (3.0 \times 10^{-3}) = 4.5 \times 10^1 \text{ (since +4 added to −3 gives +1)}$$
$$(2.2 \times 10^{-2}) \times (1.5 \times 10^{-4}) = 3.3 \times 10^{-6} \text{ (since −2 added to −4 gives −6)}$$

Dividing. For division, we divide the coefficients, then subtract the exponent of the divisor (the number below the line) from the exponent of the dividend (the number above the line).
For example,

$$\frac{6.0 \times 10^4}{1.5 \times 10^1} = 4.0 \times 10^3$$

$$\frac{8.0 \times 10^2}{4.0 \times 10^5} = 2.0 \times 10^{-3} \text{ (since 5 subtracted from 2 gives −3)}$$

$$\frac{9.0 \times 10^3}{1.5 \times 10^{-2}} = 6.0 \times 10^5 \text{ (since subtracting −2 is the same as adding 2)}$$

Adding and Subtracting. For addition and subtraction of power-of-10 numbers, the two numbers must have identical exponents. Then we simply add or subtract the coefficients and leave the exponent alone. But if the two numbers don't have identical exponents, we must make them identical by moving the decimal point in one of the numbers.

For example, when the number pairs have identical exponents:

$$(3.2 \times 10^3) + (5.7 \times 10^3) = 8.9 \times 10^3$$
$$(4.1 \times 10^{-4}) + (1.9 \times 10^{-4}) = 6.0 \times 10^{-4}$$
$$(6.3 \times 10^6) - (2.2 \times 10^6) = 4.1 \times 10^6$$

Here is an example where the number pairs don't have identical exponents.

$$(3.25 \times 10^2) + (1.4 \times 10^3) = (3.25 \times 10^2) + (14.0 \times 10^2) = 17.25 \times 10^2$$

Since the original numbers don't have identical exponents, we shifted the decimal point in the second number to lower its exponent to 2. It would have been just as correct to shift the first number's decimal point in the opposite direction, giving

$$(0.325 \times 10^3) + (1.4 \times 10^3) = 1.725 \times 10^3$$

these numbers are the same

LAND SURVEY AND WEATHER FORECASTING BY SATELLITE

State and local governments throughout the U.S. are trying to develop long-range plans for land usage. One major concern is the rapid rate at which farmland is being developed for residential and commercial uses. Government agencies use satellite imaging of the earth's surface to gather reliable information about land use. Such imaging is based on the earth's reflected radiation of solar energy. This method takes less time and costs less than a land-based survey.

A soil scientist shown here is inspecting a land-use image for the state-funded Florida Mapping and Monitoring of Agricultural Lands Project (FMMALP). A colored image from FMMALP is shown on page 4 in the Applications color section.

A satellite image of a hurricane is also shown. From this image, meteorologists are able to tell the storm's size and strength. Combined with images taken at later times, its speed and direction can be measured accurately. A color-enhanced view of this image is shown on page 4 in the Applications color section.

Courtesy of NASA

EXAMPLE 6-12

Add or subtract the following pairs of numbers.

a) $(3.92 \times 10^{6}) - (1.51 \times 10^{6})$
b) $(4.05 \times 10^{-3}) - (6.82 \times 10^{-3})$
c) $(3.08 \times 10^{4}) - (1.26 \times 10^{5})$
d) $(7.92 \times 10^{-6}) + (2.88 \times 10^{-5})$

SOLUTION

a) $\mathbf{2.41 \times 10^{6}}$ b) $\mathbf{-2.77 \times 10^{-3}}$

c) Change 1.26×10^{5} to 12.6×10^{4} ; subtracting coefficients gives $\mathbf{-9.52 \times 10^{4}}$.

The alternative is to perform the subtraction with a hand-held calculator. You are then relieved from having to shift the decimal place to make the exponents equal.The keystroke sequence would be :

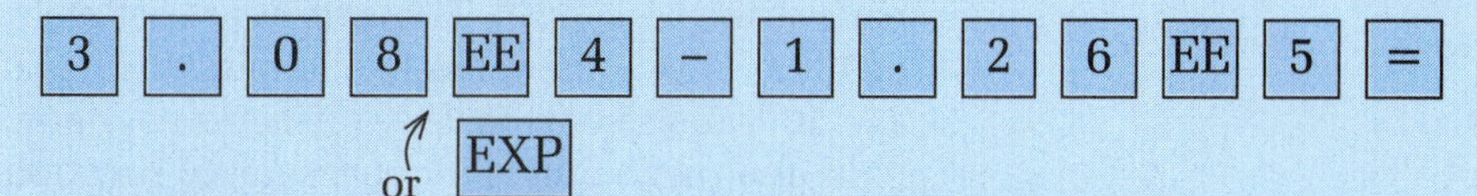

giving a displayed answer of –9.52 04 or equivalent.

d) Change 7.92×10^{-6} to 0.792×10^{-5} ; adding coefficients gives $\mathbf{3.67 \times 10^{-5}}$.
Or the addition can be performed directly with a hand-held calculator. The keystroke sequence is:

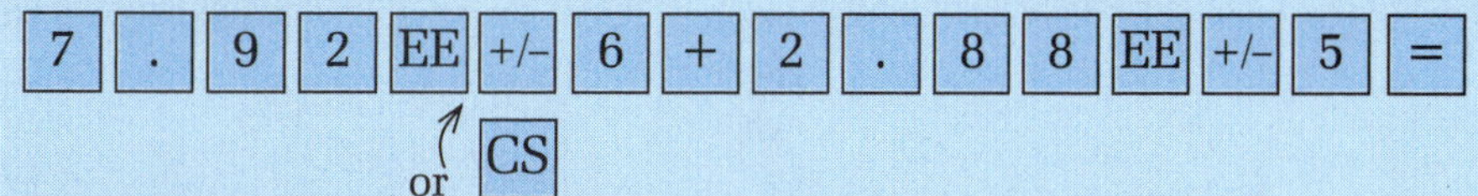

giving a displayed answer of 3.672 –05 or equivalent.

✔ SELF-CHECK FOR SECTION 6-4

7. Convert the following numbers to power-of-10 notation. [6]
   a) 4700 c) 0.0036
   b) 6 200 000 d) 0.000 015
8. Convert the following numbers to standard notation. [6]
   a) 5.72×10^{2} c) 8.3×10^{-2}
   b) 6.1×10^{4} d) 4.9×10^{-5}
9. Multiply the following numbers. [7]
   a) $(8.1 \times 10^{2}) \times (4.5 \times 10^{3})$ c) $(4.2 \times 10^{-4}) \times (1.9 \times 10^{-1})$
   b) $(9.2 \times 10^{4}) \times (6.6 \times 10^{-2})$ d) $(6.1 \times 10^{-5}) \times (4.8 \times 10^{3})$
10. Divide the following numbers. [7]
    a) $(6.8 \times 10^{4}) / (3.0 \times 10^{2})$ c) $(4.7 \times 10^{-2}) / (5.1 \times 10^{3})$
    b) $(4.1 \times 10^{3}) / (6.5 \times 10^{5})$ d) $(3.9 \times 10^{3}) / (8.2 \times 10^{-4})$
11. Add or subtract the following numbers. [7]
    a) $(6.4 \times 10^{2}) + (3.1 \times 10^{2})$ c) $(4.1 \times 10^{3}) + (5.6 \times 10^{2})$
    b) $(8.1 \times 10^{4}) - (6.0 \times 10^{4})$ d) $(7.3 \times 10^{4}) - (9.2 \times 10^{3})$

6-5 NONBASIC UNITS

As we deal with electrical variables, we sometimes find situations where the basic unit of a variable is too small for convenience. Other times, the basic unit may be too large for convenience. As an example of a too small unit, suppose that we are dealing with the resistance of wire insulation. Typical insulation resistances, expressed in basic units of ohms, fall in the range 20 000 000 Ω to 100 000 000 Ω. In power-of-10 notation, this is 20×10^6 Ω to 100×10^6 Ω.

Written the first way, using standard notation, these numbers have so many zeros that they are difficult to write or to speak. Power-of-10 notation is much easier, but if we intend to deal with many numbers in this range, writing " $\times 10^6$ " every time becomes a bother. And saying "times ten to the sixth power" every time we speak is also a bother.

To avoid such bothers, electrical technicians and engineers prefer to use large multiples of the basic unit. Such nonbasic units, or **engineering units**, rise in multiples of 1000 (10^3). They are identified by prefixes attached to the basic unit. As examples, the prefix for 10^3 is "kilo,"symbolized "k;" the prefix for 10^6 is "mega," symbolized "M." So in the insulation example above, the resistance values would be expressed as 20 MΩ to 100 MΩ.

There are also engineering units that are smaller than the basic unit (fractional parts of the basic unit). These units are gotten by dividing the basic unit by factors of 1000. This is the same as changing the negative exponent in jumps of 3 (10^{-3}, 10^{-6}, 10^{-9}, and so on). Table 6-1 lists the common nonbasic unit prefixes, their symbols, and their equivalent power-of-10 notation and standard notation.

TABLE 6-1 Engineering Units

PREFIX	SYMBOL	EXPONENT	MULTIPLIER
terra	T	$\times 10^{12}$	1 000 000 000 000
giga	G	$\times 10^{9}$	1 000 000 000
mega	M	$\times 10^{6}$	1 000 000
kilo	k	$\times 10^{3}$	1000
Basic unit	—	—	1
milli	m	$\times 10^{-3}$	0.001
micro	μ	$\times 10^{-6}$	0.000 001
nano	n	$\times 10^{-9}$	0.000 000 001
pico	p	$\times 10^{-12}$	0.000 000 000 001

Note that mega is symbolized capital M, while milli is lower case m. You must memorize the meaning of the four prefixes **kilo**, **mega**, **milli**, and **micro**, since they are seen so often.

EXAMPLE 6-13

Write the following quantities using prefixed engineering units.

a) 27 000 Ω
b) 0.0026 V
c) 0.000 035 A
d) 1 500 000 Ω
e) 0.000 000 024 A
f) 1 200 000 000 Ω

SOLUTION

Move the decimal point in jumps of three steps, until you have either one, two, or three digits to the left of the decimal point.

a) Jumping three steps to the left gives 27×10^3 Ω = **27 kΩ.**
b) Jumping three to the right gives 2.6×10^{-3} V = **2.6 mV.**
c) Jumping six to the right gives 35×10^{-6} A = **35 μA.**
d) Jumping six to the left gives 1.5×10^6 Ω = **1.5 MΩ.**
e) Jumping nine to the right gives 24×10^{-9} A = **24 nA.**
f) Jumping nine to the left gives 1.2×10^9 Ω = **1.2 GΩ.**

6-6 OHM'S LAW WITH ENGINEERING UNITS

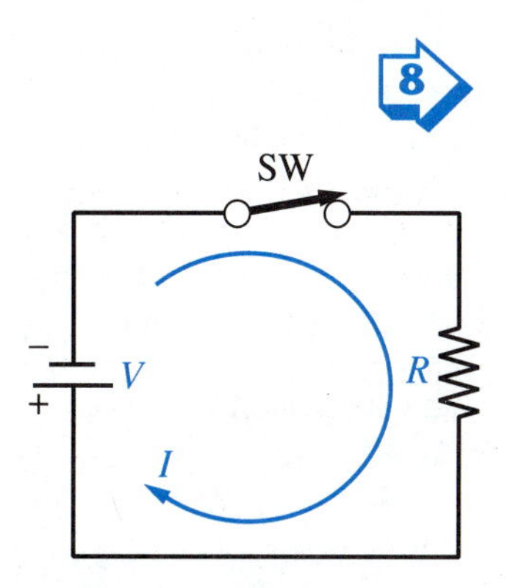

FIGURE 6-6 **Applying Ohm's law with engineering notation to a single-resistor circuit**

If electrical variables are given in engineering units, just convert them to power-of-10 notation when you put them into the Ohm's law formula. Then do the multiplication or division according to the rules in Section 6-4.

EXAMPLE 6-14

In Figure 6-6, suppose $V = 12$ V and $R = 3.0$ kΩ. Calculate current, I.

SOLUTION

From Ohm's law,

$$I = \frac{V}{R} = \frac{12 \text{ V}}{3 \times 10^{3} \Omega} = 4.0 \times 10^{-3} \text{ A} = \mathbf{4.0 \text{ mA}}$$

Performing this division with a hand-held calculator, the keystroke sequence would be: [1] [2] [÷] [3] [EE] [3] [=]

giving a displayed answer of 4. −03 or equivalent.

An ammeter calibrated in engineering units is shown in Figure 6-7.

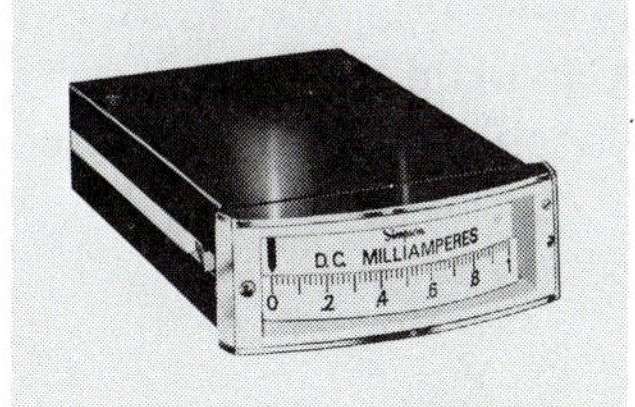

FIGURE 6-7 **Ammeter with full-scale value of 1 milliampere. Each small scale mark represents 0.02 mA.** *Courtesy of Simpson Electric Co.*

EXAMPLE 6-15

In Figure 6-6, suppose R is 1.8 MΩ and the voltage source is adjustable. What amount of source voltage will produce a current of 7 μA?

SOLUTION

$$V = IR = (7 \times 10^{-6} \text{ A}) \times (1.8 \times 10^{6} \ \Omega) = 12.6 \times 10^{0} \text{ V} = \mathbf{12.6 \text{ V}}$$

Notice that microamps and megohms "cancel" each other, since their exponents are equal but opposite in sign. The result comes out in basic units of volts.

Performing this multiplication with a hand-held calculator, the keystroke

sequence would be: [7] [EE] [+/−] [6] [x] [1] [.] [8] [EE] [6] [=]

giving a displayed answer of 12.6 or equivalent.

EXAMPLE 6-16

In Figure 6-6, suppose the voltage source is fixed at 120 V, but R is adjustable. What amount of resistance, R, will produce a current of 2.5 mA?

SOLUTION

$$R = \frac{V}{I} = \frac{120 \text{ V}}{2.5 \times 10^{-3}} = 48 \times 10^{3} \ \Omega = \mathbf{48 \text{ k}\Omega}$$

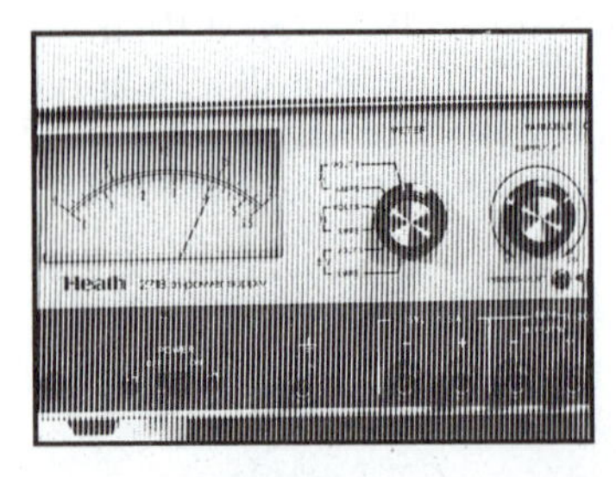

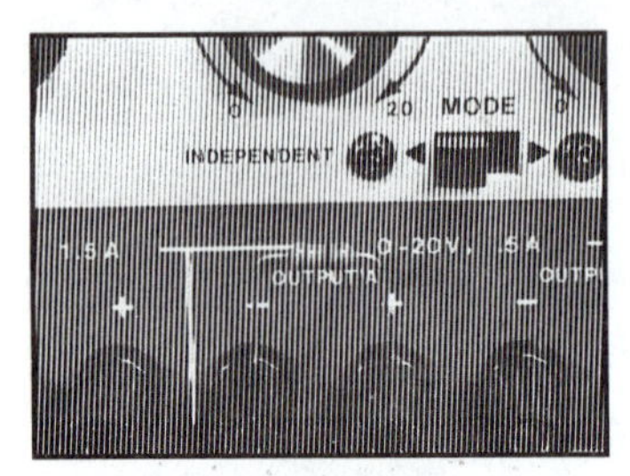

FIGURE 6-8

EXAMPLE 6-17

Figure 6-8 shows an adjustable dc power supply with a resistor connected to its terminals. The panel meter is indicating the output voltage. The resistor's stripes are visible in the close-up view. What is the value of current in the resistor? (See page 2 in the Circuits color section for the color version of Figure 6-8.)

SOLUTION

The pointer of the panel meter indicates 15 V. The stripes on the resistor represent a nominal resistance of 7500 Ω or 7.5 kΩ.

$$I = \frac{V}{R} = \frac{15\text{ V}}{7.5 \times 10^3} = 2.0 \times 10^{-3}\text{ A or } \mathbf{2.0\text{ mA}}$$

Of course, since the resistance is guaranteed only to ± 5%, the actual current could differ from 2.0 mA by ± 5% (from 1.9 mA to 2.1 mA).

✓ SELF-CHECK FOR SECTIONS 6-5 AND 6-6

12. In the circuit of Figure 6-6, let V = 200 mV and I = 50 μA. What is the value of R? [4, 8]
13. In Figure 6-6, if I = 3.7 mA and R = 43 kΩ, what is the value of V? [5, 8]
14. In Figure 6-6, suppose V = 100 μV and R = 2 kΩ. Calculate I. [3, 8]
15. The three-position selector switch in Figure 6-9 enables any one of three load resistors to be connected to the source. [3,8]
    a) Calculate the current when the switch is in position 1.
    b) Repeat for position 2.
    c) Repeat for position 3.
16. We wish to change the resistances in Figure 6-9 so that the current starts at 4 mA in position 1, then increases by a factor of 10 for each new switch position. Calculate the values of R_1, R_2, and R_3. [4, 8]

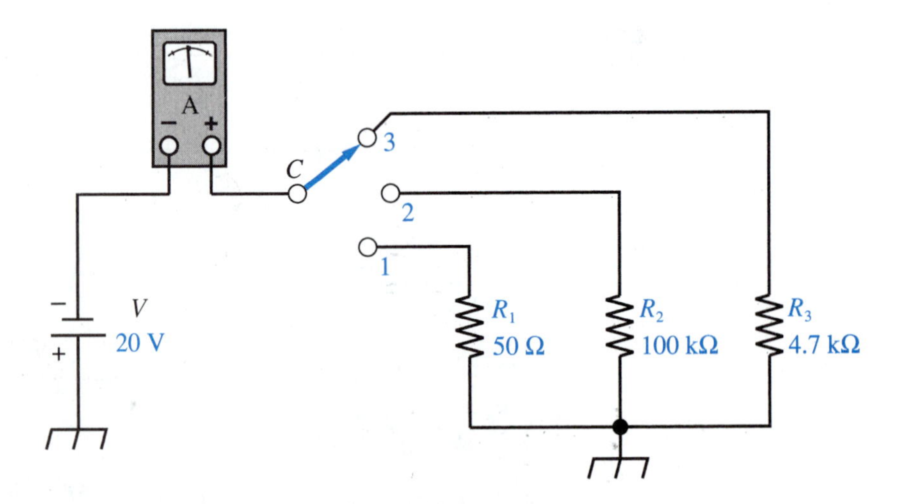

FIGURE 6-9 Three-position switch for controlling three loads

TROUBLESHOOTING TIP

Sometimes, a load's resistance cannot be found by a simple ohmmeter measurement. This occurs whenever the load operates at a very hot temperature. The problem is that the low-voltage ohmmeter battery cannot bring the load up to its normal operating temperature. Therefore, the ohmmeter measures the load's "cold resistance," which is much lower than its true operating resistance, or "hot resistance."

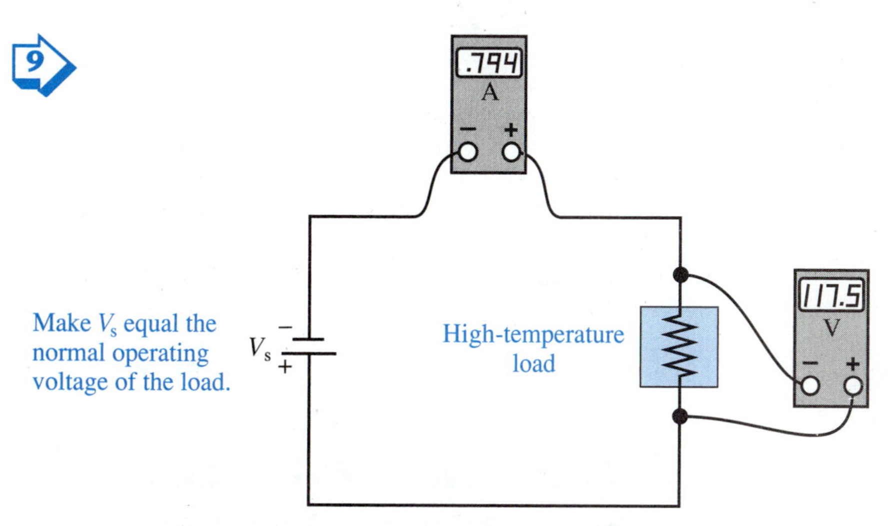

Finding hot resistance of a load

During troubleshooting, if you need to test the hot resistance of a high-temperature load, do it as shown in the figure. The load's actual operating resistance is found indirectly by applying Ohm's law. For the measurements indicated in the figure, we have

$$R_{hot} = \frac{V}{I} = \frac{117.5 \text{ V}}{0.794 \text{ A}} = \mathbf{148\ \Omega}$$

FORMULAS

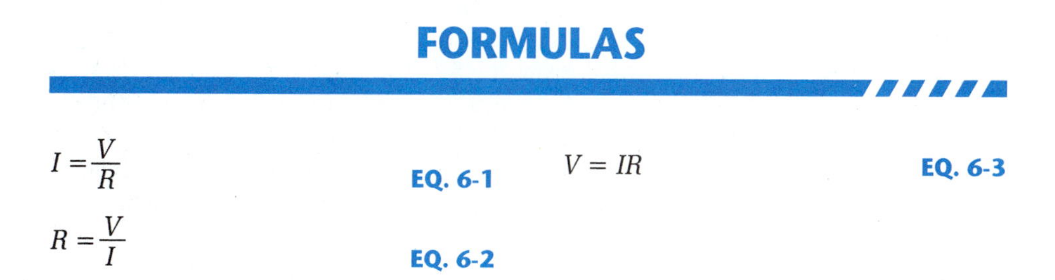

$$I = \frac{V}{R}$$ **EQ. 6-1**

$$R = \frac{V}{I}$$ **EQ. 6-2**

$$V = IR$$ **EQ. 6-3**

Electronic instruments usually have adjustable sensitivity, so they can measure very small fractions of the basic measurement unit, or multiples of the basic unit.
Courtesy of Hewlett-Packard Company

SUMMARY OF IDEAS

- Current is directly proportional to voltage.
- Current is inversely proportional to resistance.
- The Ohm's law formula, $I = V/R$, gives the mathematical relationship among current, voltage, and resistance.
- The Ohm's law formula can be rearranged to solve for any variable, if the other two variables are known.
- The Ohm's law triangle is a helpful memory aid when you are learning to use Ohm's law.
- Very large and very small numbers are confusing to write in standard notation. They should be written in power-of-10 notation.
- We add exponents when multiplying power-of-10 numbers.
- When dividing power-of-10 numbers, we subtract the exponent of the denominator from the exponent of the numerator.
- When adding or subtracting power-of-10 numbers, we must force the exponents to be the same if they aren't already the same.
- Engineering units are popular among electrical technicians and engineers. For large units, they are factors of thousands. For small units, they are factors of one-thousandth part.

CHAPTER QUESTIONS AND PROBLEMS

1. All other things being equal, higher resistance results in__________current. [2]
2. In words, Ohm's law tells us two things: (1) Current is directly proportional to __________; (2) Current is inversely proportional to __________. [1, 2]
3. A certain soldering iron has a resistance of 280 Ω when operated from a 115-V source. How much current does it take from the source? [3]
4. A lamp operated at 24 V has a hot resistance of 160 Ω. What amount of current does it carry? [3]
5. The manufacturer specifies that a space heater will carry 2.5 A of current when operated at 120 V. What is the resistance of the space heater? [4]
6. The current through a 56-Ω resistor is 0.2 A. What is the voltage across the resistor? [5]
7. A certain car horn carries 7 A when it is operated by the 12-V electrical system. What is its resistance? [4]
8. An electric welding machine must deliver 80 A of current through a metal-to-metal junction that has a hot resistance of 0.3 Ω. What is the output voltage of the welder? [5]
9. A certain speaker has a resistance of 8 Ω. To produce barely audible sound, the speaker must carry 0.075 A. What is the minimum voltage required to drive the speaker? [5]
10. The speaker in Problem 9 can carry a maximum current of 1.5 A without damage. Any current in excess of 1.5 A will cause damage. What is the maximum allowable voltage that can be applied? [5]
11. Express the following numbers in power-of-10 notation. [6]
    a) 390.0 c) 12 500 000
    b) 1600 d) 530 000
12. Convert these power-of-10 numbers to standard notation. [6]
    a) 3.9×10^2 c) 13.3×10^3
    b) 6.2×10^4 d) 27.0×10^5
13. Express the following numbers in power-of-10 notation. [6]
    a) 0.0085 c) 0.431
    b) 0.000 92 d) 0.000 006 4

14. Convert these power-of-10 numbers to standard notation. [6]
    a) 5.2×10^{-3} c) 31.5×10^{-6}
    b) 8.4×10^{-2} d) 6.3×10^{-8}
15. Perform the following multiplications and express the answers in power-of-10 notation. [7]
    a) $(8.3 \times 10^{2}) \times (2.6 \times 10^{3})$ c) $(2.7 \times 10^{2}) \times (4.1 \times 10^{-4})$
    b) $(5.1 \times 10^{4}) \times (1.9 \times 10^{-2})$ d) $(8.9 \times 10^{-3}) \times (3.4 \times 10^{-2})$
16. Perform the following divisions and express the answers in power-of-10 notation. [7]
    a) $\dfrac{6.8 \times 10^{3}}{1.7 \times 10^{2}}$ c) $\dfrac{6.65 \times 10^{-3}}{1.9 \times 10^{2}}$
    b) $\dfrac{8.1 \times 10^{5}}{5.4 \times 10^{-1}}$ d) $\dfrac{7.5 \times 10^{-4}}{3.75 \times 10^{-1}}$
17. Perform the following additions and subtractions. [7]
    a) $(6.3 \times 10^{2}) + (3.2 \times 10^{3})$ c) $(7.8 \times 10^{5}) - (3.5 \times 10^{6})$
    b) $(4.9 \times 10^{4}) - (6.8 \times 10^{3})$ d) $(4.1 \times 10^{-2}) + (5.0 \times 10^{-3})$
18. Convert these quantities to engineering notation. [6, 8]
    a) 33 000 ohms d) 0.008 volt
    b) 0.000 04 ampere e) 0.000 000 085 second
    c) 1 200 000 ohms
19. A radio receiver presents a resistance of 75 Ω to the antenna input signal. The antenna signal voltage is 500 microvolts (μV). How much current flows in the antenna downleads? [3, 8]
20. What is the resistance of a certain semiconductor device that passes 80 milliamperes (mA) of current when 600 millivolts (mV) of voltage is applied? [4, 8]
21. For the semiconductor device in Problem 20, how much voltage must be applied to produce 200 microamperes (μA) of current? Express the voltage in engineering units. [5, 8]
22. A 220-kΩ resistor has 12 V across it. Find the value of current, expressed in engineering units. [3, 8]
23. Figure 6-10(A) shows a resistor connected to the output terminals of a dc power supply. The panel voltmeter is indicating the supply voltage. (See page 2 in the Circuits color section for the color version of Figure 6-10(A). [3, 8]
    a) Calculate the nominal current.
    b) Give the range of actual current that is possible. Assume that the panel voltmeter has no error.
24. Repeat Problem 23 for the circuit in Figure 6-10(B). (See page 2 in the Circuits color section for the color version of Figure 6-10(B).)[3, 8]
25. The resistance of the human body varies greatly, depending on where the electrical contact is made on the body, moisture conditions, and so on. But roughly speaking, body resistance is about 100 kΩ. The amount of current that is harmful to internal organs is roughly 500 μA; again, this depends on many factors. Using these values, calculate the amount of voltage that should be considered dangerous to a human. [5, 8]

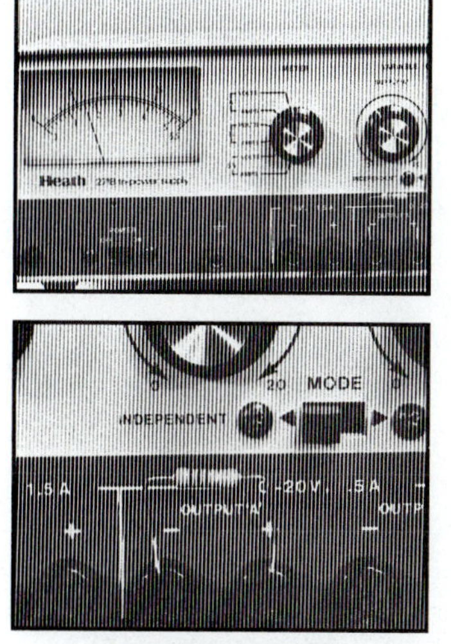

FIGURE 6-10 (A)

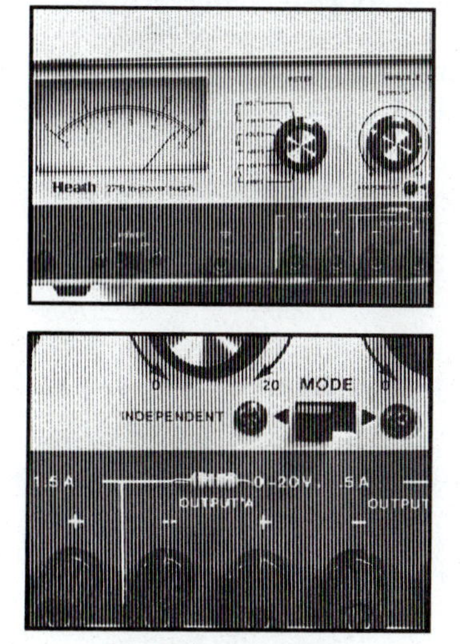

FIGURE 6-10 (B)

CHAPTER 7

POWER AND ENERGY

While a practical, rechargeable-battery electric car is projected to be on the market by the mid 1990s, this photovoltaic solar-powered car is still many years away from practicality.
Courtesy of NASA

OUTLINE

7-1 The Meaning of Energy
7-2 The Meaning of Power
7-3 Commercial Energy Measurement
7-4 Calculating Power
7-5 Efficiency
7-6 Wattmeters

NEW TERMS TO WATCH FOR

work
joule
potential
energy
energy conversion
power
watt
kilowatthour
power formula triangle
speaker
power rating
efficiency
electric motor
torque
rotational speed
wattmeter
voltage coil
current coil

In all circuits, electrical energy is taken from the voltage source and converted into a useful form by the load device. Therefore, the energy concept is important to us. Power is closely related to energy, and is also an important idea.

After studying this chapter, you should be able to:

1. Calculate electrical energy if voltage and charge are known.
2. Describe the relationship between power and energy.
3. For any particular situation, choose appropriate units of energy, joules, or kilowatthours.
4. State the relationship among power, voltage, and current. Calculate any one of these if the other two are known.
5. State and use the relationship among power, voltage, and resistance. Do the same for power, current, and resistance.
6. Troubleshoot for higher-than-normal conductor resistance.
7. Understand the efficiency of an electrical load.
8. Relate the following variables: input power, output power, waste heat power, and efficiency.
9. Use a four-terminal wattmeter properly. Do the same for a three-terminal wattmeter.

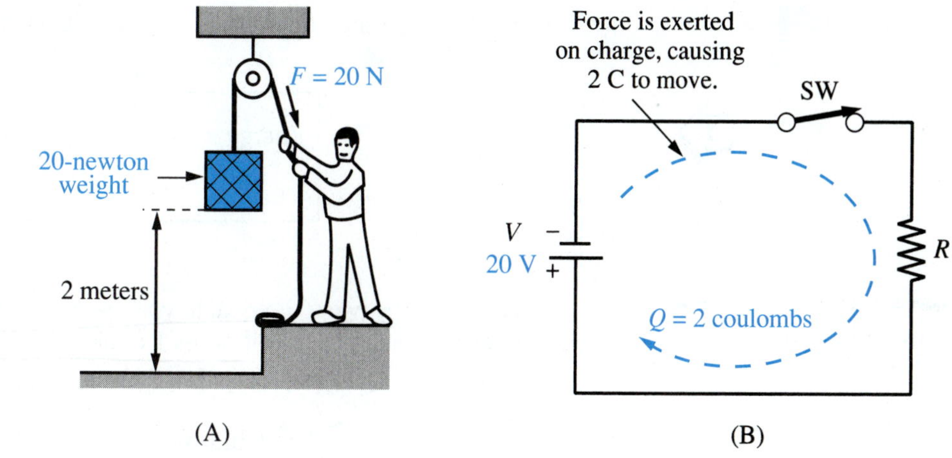

FIGURE 7-1 (A) For a mechanical force of 20 N applied through a distance of 2 m, the work done is 40 J. (B) For an electromotive force of 20 V applied for long enough to move a charge of 2 C, the work done is also 40 J.

7-1 THE MEANING OF ENERGY

In mechanics, the **work** performed in moving an object is equal to the distance moved multiplied by the force exerted. In equation form,

$$\text{Work} = \text{force} \times \text{distance}$$
$$W = F \times d$$

EQ. 7-1

The basic metric unit of force is the newton, symbolized N. One newton is about 0.225 pound. The basic unit of distance is the meter, symbolized m. One meter is about 3.28 feet. Using Equation 7-1, we can state this fact.

TECHNICAL FACT

The basic metric unit of work is the **joule,** symbolized J. One joule of work is equal to a force of 1 newton multiplied by a distance of 1 meter. (1 J = 1 N x 1 m)

In Figure 7-1(A), the person has exerted a force of 20 N through a distance of 2 m. Therefore, the work performed is calculated as

$$W = F \times d$$
$$= (20\text{ N})\,(2\text{ m}) = 40\text{ J}$$

After this amount of lifting work has been performed, the weighted object has the capability to do that same amount of work in falling. For instance, if a very soft material is placed underneath, it could be squashed flat by releasing the weight. The weight would exert the same 20-N force on the way down, and the movement would be the same 2-m distance. (Assuming the material is soft enough to be completely flattened.) Therefore, the work on the way down would be 20 N x 2 m = 40 J.

We say that the weighted object has the **potential** to do 40 J of work after it is lifted up. Another way of expressing this idea is to say that we have stored 40 J of **energy** in the object by lifting it.

TECHNICAL FACT

Stored energy is the potential to do work. Both are measured in joules.

Because energy and work are essentially the same thing, we use the same letter to symbolize them both—the letter *W*.

In an electrical situation, we have no mechanical force; instead we have voltage, V. Also, movement over a distance is not the outcome of electrical activity.

Instead, we get movement of charge, Q. Therefore, in an electric circuit

$$\text{Energy} = \text{voltage x charge}$$
$$W = VQ$$

EQ. 7-2

TECHNICAL FACT

Electrical energy (work) is equal to voltage multiplied by charge. One joule of energy is given by 1 volt times 1 coulomb. (1 J = 1 V x 1 C)

In Figure 7-1(B), if the switch remains closed long enough for the battery to move 2 C of charge, the battery has done an amount of work given by Equation 7-2 as

$$W = VQ$$
$$= (20 \text{ V})\,(2 \text{ C}) = 40 \text{ J}$$

Another way of saying this is that 40 J of stored energy has been taken from the battery. The stored **energy is converted** from electrical form to some other form by the load device. In the circuit of Figure 7-1(B), the load device is a resistor, so the energy is simply converted to heat.

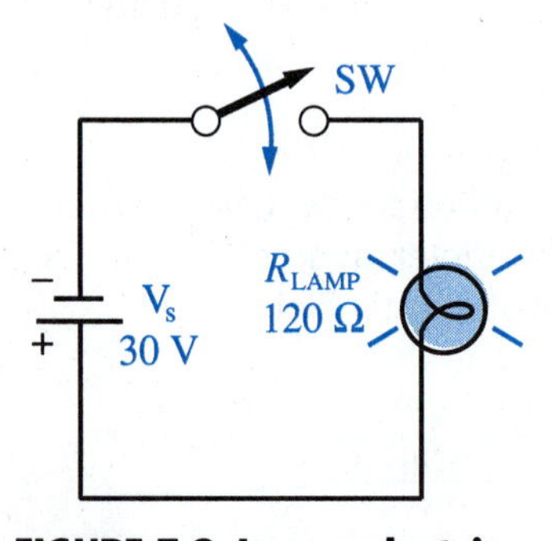

FIGURE 7-2 In any electric circuit, a source delivers energy to a load device.

EXAMPLE 7-1

In Figure 7-2, the hot resistance of the lamp filament is 120 Ω. The switch is closed for a time duration of 10 s.

a) Find the circuit's current with the switch closed.
b) Calculate the amount of charge that is moved during the time the switch remains closed.
c) How much energy is taken out of the source?
d) Describe where that energy goes.

SOLUTION

a) By Ohm's law,

$$I = \frac{V}{R_{\text{lamp}}} = \frac{30 \text{ V}}{120\ \Omega} = \mathbf{0.25\ A}$$

b) Rearranging Equation 1-2, we get

$$Q = It = (0.25 \text{ A})\,(10 \text{ s}) = \mathbf{2.5\ C}$$

c) From Equation 7-2,

$$W = VQ = (30 \text{ V})\,(2.5 \text{ C}) = \mathbf{75\ J}$$

d) The electrical energy is converted into two other forms, light and heat. It would be better if most of the energy went into light and only a small amount into heat. But incandescent lamps don't work that way. They convert most of their energy into heat.

7-2 THE MEANING OF POWER

Power is the time rate of energy. It is a measure of how quickly energy is being used. The symbol for power is *P*. As an equation,

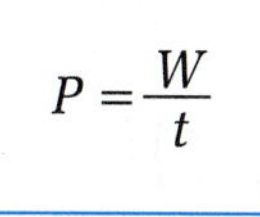

$$P = \frac{W}{t}$$

EQ. 7-3

The basic measurement unit for power is the **watt**, symbolized W.

TECHNICAL FACT

The power is 1 watt if 1 joule of energy is used in 1 second. (1 W = 1 J / s)

EXAMPLE 7-2

Go back to the lamp circuit of Figure 7-2, in Example 7-1.

a) When the switch is closed, what is the circuit's power value?
b) Describe what the dc voltage source is doing in terms of power. Also give your answers to parts (c) and (d) in terms of power.
c) Describe what the conductors are doing.
d) Describe what the load device, the lamp, is doing.

SOLUTION

a) In Example 7-1, we found the energy to be 75 J. From Equation 7-3,

$$P = \frac{W}{t} = \frac{75\ \text{J}}{10\ \text{s}} = \mathbf{7.5\ W}$$

b) The dc source is *delivering* power to the circuit.
c) The conductors are *transporting* power from the source to the load.
d) The load is *consuming* 7.5 W of power.

Also, the load is *converting* the power from electrical form to other forms. We sometimes say that the load is *burning* the power. Since the load has no ability to store energy, it must throw energy off into its surrounding environment just as fast as it receives energy from the source. To draw attention to this fact we often say that the load is *dissipating* the power.

7-3 COMMERCIAL ENERGY MEASUREMENT

The joule is the basic unit of energy in the international metric system. But it is inconveniently small to be used by electric utility companies. Instead, they measure energy in units of **kilowatthours**. A kilowatthour meter is shown in Figure 7-3.

TECHNICAL FACT

The common commercial unit of energy is the kilowatthour, symbolized kWh. It is the amount of energy conversion that occurs when the power is 1 kilowatt for a time duration of 1 hour.

FIGURE 7-3 Five-digit kilowatthour meter.
Courtesy of ABB Electric Metering Systems

AUTOMOTIVE ENGINE CONTROL

A modern automobile engine has its fuel and ignition systems controlled by a microcomputer like the one shown here. Sensors mounted on the engine provide information to the microcomputer regarding engine temperature, throttle position, intake manifold pressure, torque load on the engine, and other variables. Based on this information, the microcomputer exerts control over the fuel injection rate, and the ignition timing. With an automatic transmission, the shifting instants are also controlled. The smaller device shown above the computer board is the sensor that measures pressure in the air intake manifold.

Courtesy of Delco Electronics Corp., Subsidiary of GM Hughes Electronics

Fuel metering used to be controlled by a carburetor. It was mechanically complicated, difficult to maintain, and it was seldom able to provide the exact correct amount of fuel flow for all driving conditions.

Ignition timing used to be controlled by ignition points. The points were actuated by a mechanical distributor cam, with an elaborate arrangement of fly-weights.

With these two engine functions controlled by a microcomputer, engine manufacturers can get more power from a given engine size. Fuel efficiency and emissions are also better, and less engine maintenance is required.

Courtesy of Ford Motor Company

EXAMPLE 7-3

How much energy is consumed by an 8 kW electric clothes dryer running for 5 hours?

a) Express the answer in kilowatthours.
b) Express the answer in joules.
c) Write the conversion factor for kWh to J.

SOLUTION

a) Rearrange Equation 7-3 to solve for energy.

$$\begin{aligned} W &= Pt \\ &= (8\text{ kW})(5\text{ h}) = \mathbf{40\text{ kWh}} \end{aligned} \qquad \textbf{EQ. 7-4}$$

b) To express this energy in joules, the 5-hour time duration must be converted to basic units of seconds as follows:

$$t = 5\text{ h x} \frac{60\text{ min}}{1\text{ h}} = 300\text{ min}$$

$$t = 300\text{ min x} \frac{60\text{ s}}{1\text{ min}} = 1.8\text{ x }10^4\text{ s}$$

Substituting the basic-unit values for P and t into Equation 7-4, we get

$$W = (8\text{ x }10^3\text{ W})(1.8\text{ x }10^4\text{ s}) = 14.4\text{ x }10^7\text{ J or } \mathbf{144\text{ MJ}}$$

c) Since the answers to parts (a) and (b) are equivalent,

$$\frac{144\text{ MJ}}{40\text{ kWh}} = 1$$

$$\mathbf{3.6\text{ MJ} = 1\text{ kWh}}$$

Interior wiring of energy-metering panel.

Energy-metering panel. Current, voltage, power, and cumulative energy transfer are being measured.
Courtesy of Siemens Energy & Automation, Inc.

✔ SELF-CHECK FOR SECTIONS 7-1, 7-2, AND 7-3

1. Work and__________are essentially the same. [1, 2]
2. The basic metric unit of energy is the__________. It is symbolized_________. [1, 3]
3. If a 4.5-V battery transfers a charge of 50 C through a load, how much energy has the battery delivered to the load? [1]
4. If a battery charger then returns 35 C to the battery of Question 3, how much energy has been restored? [1]
5. Compared to its starting condition, how much energy does the battery contain after the recharge in Question 4? [1]
6. A 6-V battery lights a 40-Ω lantern for 5 minutes. How much energy does the battery deliver? [1]
7. What is the fundamental relation between energy and power? [2]
8. The basic metric unit of power is the__________. It is symbolized__________. [2]
9. What is the value of power in Question 6? [4]
10. 1 joule per second is 1__________. [2]
11. A certain air conditioning unit has a power rating of 1.0 kW when driven by a 120-V source. If it is operated for 12 hours per day, find its daily energy consumption. Express the result in both units of energy, kWh and J. [1, 2, 3]
12. A typical energy cost rate is 6.5 cents per kWh. At that rate, how much is the daily cost to operate the unit in Question 11?

7-4 CALCULATING POWER

We need a method to calculate electric power from information about the three fundamental variables—current, voltage, and resistance. Let us start by proving the relation between power, current, and voltage. Equation 7-3 tells us that power is energy divided by time.

$$P = \frac{W}{t}$$

Equation 7-2 states that energy is voltage multiplied by charge.

$$W = VQ$$

By substituting VQ for energy into Equation 7-3, we get

$$P = \frac{VQ}{t} = V\frac{Q}{t}$$

The Q / t part in the preceding equation is charge per unit of time, or simply current, I. Therefore, we can write

$$P = VI \qquad \text{EQ. 7-5}$$

which leads to this fact.

TECHNICAL FACT

Power is calculated by multiplying voltage times current. 1 watt is 1 volt times 1 ampere. (1 W = 1 V x 1 A)

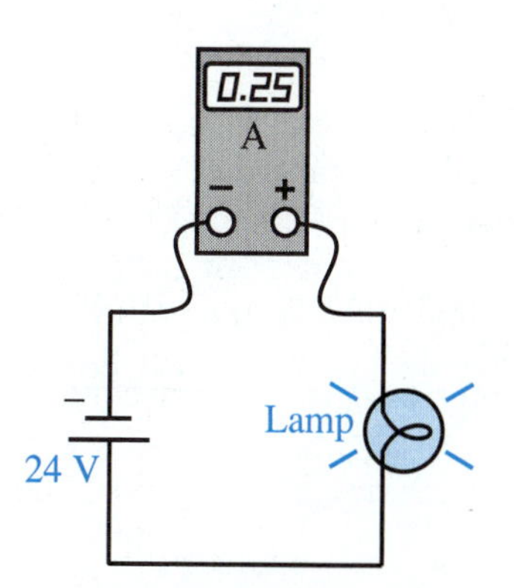

FIGURE 7-4 Power can be calculated from *V* and *I*.

EXAMPLE 7-4

In Figure 7-4, a 24-V source establishes a current of 0.25 A through the lamp. What is the circuit's power?

SOLUTION

From Equation 7-5,

$$\begin{aligned} P &= VI \\ &= (24 \text{ V})(0.25 \text{ A}) = \mathbf{6\ W} \end{aligned}$$

Equation 7-5 gives the relation among power, voltage, and current. It can be rearranged to solve for any one of those variables if the other two are known, just like Ohm's law. That is,

$$V = \frac{P}{I}$$

EQ. 7-6

and

$$I = \frac{P}{V}$$

EQ. 7-7

The **power formula triangle** shown in Figure 7-5 is helpful in remembering these relations.

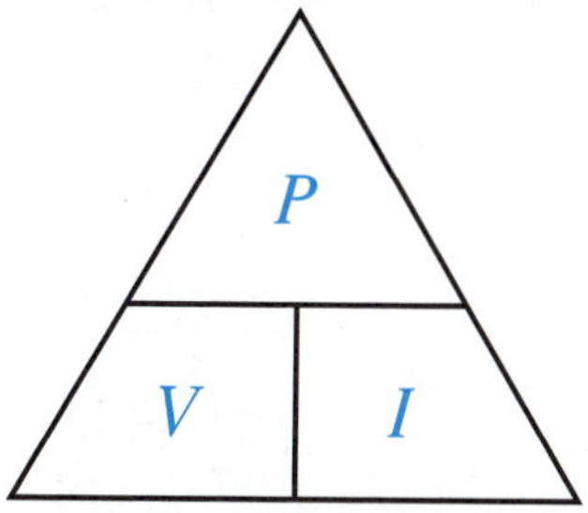

FIGURE 7-5 The power formula triangle helps us know when to multiply and when to divide.

EXAMPLE 7-5

A certain reading lamp has a rating of 150 W when driven by 120 V. How much current does it carry?

SOLUTION

From Equation 7-7, or by looking at the power formula triangle,

$$\begin{aligned} I &= \frac{P}{V} \\ &= \frac{150 \text{ W}}{120 \text{ V}} = \mathbf{1.25\ A} \end{aligned}$$

It is sometimes necessary to calculate power from knowledge of voltage and resistance. A formula for doing that is derived by substituting Ohm's law into the power formula, Equation 7-5.

$$P = VI = V\left(\frac{V}{R}\right)$$

$$P = \frac{V^2}{R}$$

EQ. 7-8

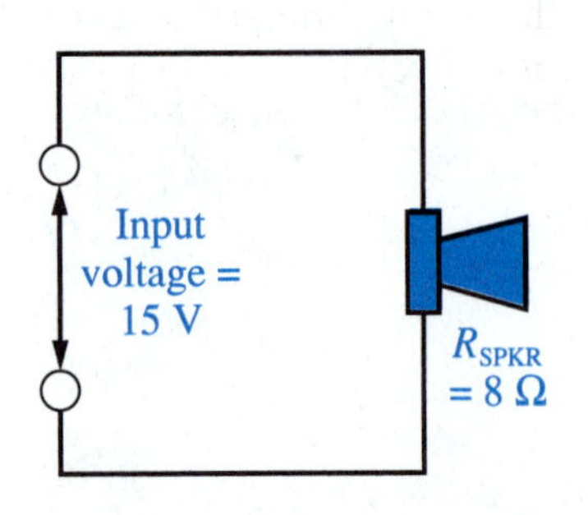

FIGURE 7-6 A speaker winding has a certain resistance, which can be symbolized R_{spkr}. 8 Ω is typical. The input voltage is ac rather than dc, but that has no effect on calculating power.

EXAMPLE 7-6

A **speaker** is a load device that converts electrical energy into sound energy and waste heat. We will examine the structure of a speaker in Chapter 13 when we study magnetism. For now, accept the idea that a speaker has a certain amount of resistance in its structure.

The speaker in Figure 7-6 has an internal resistance of 8 Ω. If a voltage of 15 V is applied, how much power does the speaker receive from the source?

SOLUTION

From Equation 7-8,

$$P = \frac{V^2}{R}$$

$$= \frac{(15\text{ V})^2}{8\ \Omega} = \mathbf{28.1\ W}$$

The squaring operation is very simple on a hand-held calculator. The keystroke sequence is: [1] [5] [x^2] [÷] [8] [=]

giving a displayed answer of [28.125]

As usual, Equation 7-8 can be rearranged to solve for either of the other two variables, R or V.

$$R = \frac{V^2}{P}$$

EQ. 7-9

$$V^2 = PR$$

$$V = \sqrt{PR}$$

EQ. 7-10

EXAMPLE 7-7

For the 8-Ω speaker in Figure 7-6, how much input voltage would be required to increase the power to 40 W?

SOLUTION

From Equation 7-10,

$$V = \sqrt{PR}$$

$$= \sqrt{(40\text{ W})\ (8\ \Omega)} = \sqrt{320} = \mathbf{17.9\ V}$$

The square root operation is usually the second function of the x^2 key. If so, the symbol $\sqrt{x}$ will be just above the x^2 key. You access the second function of a key by first pressing the second key (often labeled INV for "invert"). The keystroke sequence for this problem would be :

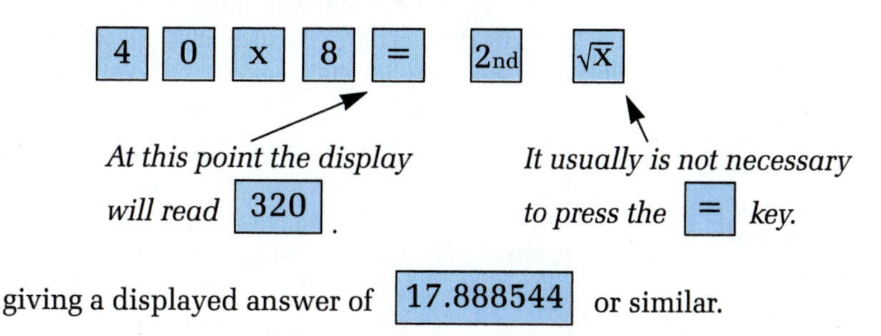

giving a displayed answer of 17.888544 or similar.

There are also occasions when it is necessary to calculate power from knowledge of current and resistance. Substituting V = IR into the power formula Equation 7-5, we get

$$P = IV = I\,(IR) = I^2R$$

$$P = I^2R \qquad \textbf{EQ. 7-11}$$

This equation can be rearranged to solve for either R or I.

$$R = \frac{P}{I^2} \qquad \textbf{EQ. 7-12}$$

$$I^2 = \frac{P}{R}$$

$$I = \sqrt{\frac{P}{R}} \qquad \textbf{EQ. 7-13}$$

TECHNICAL FACT

For a load device like a speaker, the **power rating** is the maximum power that can be delivered to the load without damaging it. If a greater than rated amount of power is delivered, the load will overheat or be harmed mechanically.

EXAMPLE 7-8

Suppose that the speaker in Figure 7-6 has a power rating of 60 W. Calculate its maximum safe current.

SOLUTION

From Equation 7-13,

$$I_{max} = \sqrt{\frac{P_{max}}{R}}$$

$$= \sqrt{\frac{60\text{ W}}{8\ \Omega}} = \sqrt{7.5} = \mathbf{2.74\ A}$$

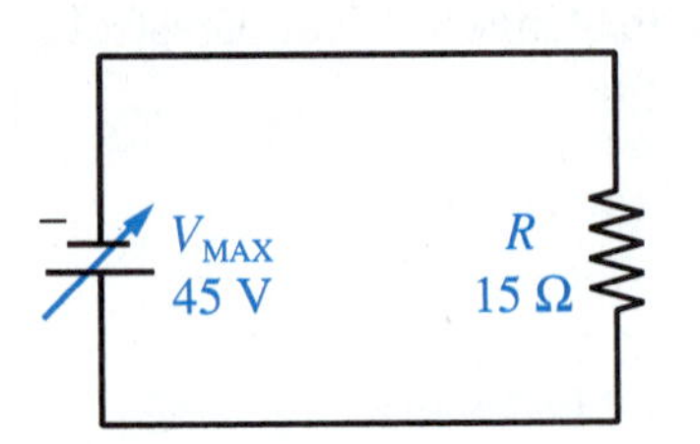

FIGURE 7-7 Load power can be varied by adjusting the source voltage.

✔ SELF-CHECK FOR SECTION 7-4

13. A heating element in a kitchen appliance carries 3.2 A when driven by a 120-V source. What is its power? [4]
14. A soldering pencil consumes 40 W while carrying a current of 0.35 A. What is its resistance? [5]
15. The circuit of Figure 7-7 contains an adjustable voltage source driving a 15-Ω resistor. Find the maximum power that can be delivered to the resistor. [5]
16. In Figure 7-7, what value of voltage would cause a power of 50 W? [5]
17. In Figure 7-7, what value of voltage would cause a power of 100 W? [5]
18. Compare the answers to Questions 16 and 17. The power was doubled, but the voltage was not nearly doubled. Explain this. [5]
19. A television set has a power consumption of 420 W when operated from a 117-V source. What amount of current does it take from the source? [4]
20. An 8-Ω loudspeaker is carrying a current of 2.1 A. What is its power consumption? What does the speaker do with that power? [5]
21. A rheostat has a 24-V source connected to its terminals. What adjusted value of resistance will cause power dissipation of 5 W? [5]
22. Deposited-film resistors are commonly manufactured with a power rating of 1/2 W. For a 10-kΩ, 1/2-W film resistor, what is the maximum safe current? [5]

TROUBLESHOOTING TIP

In some situations, a load may not work properly because its actual applied voltage is too low, even though the source voltage is the proper value. This can happen if the conductor resistances are higher than they ought to be. Look at the figure.

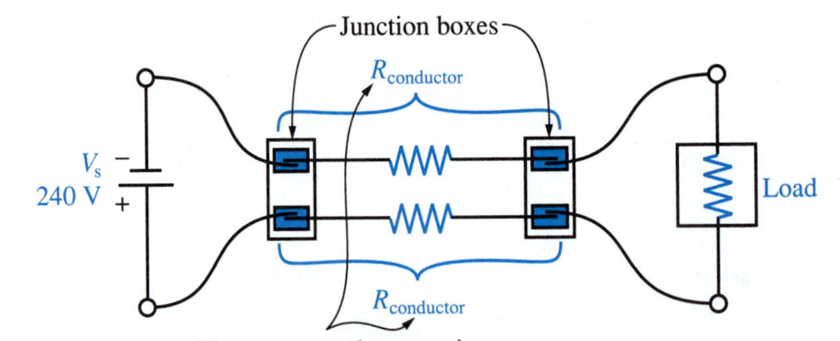

(A)

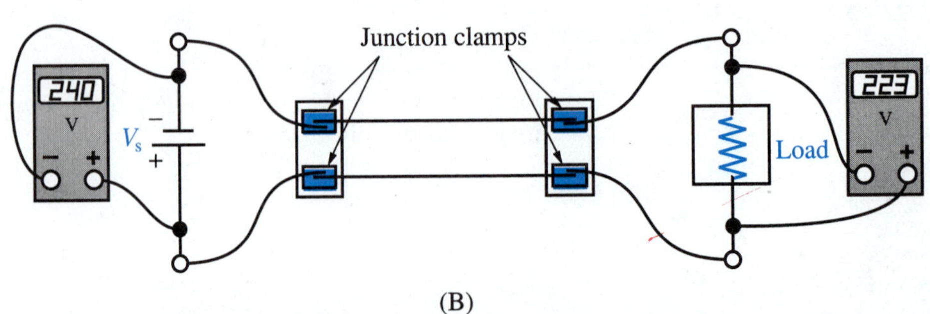

(B)

(A) Circuit conductors always have some nonzero resistance ($R_{conductor}$), but it is usually so small that it is negligible. (B) If there is a great difference between load voltage and source voltage, the conductor resistance is too high. This might even be a fire hazard.

Continued on page 103

In Figure (A), if the conductors are not properly joined together at their junction points, the resistance, $R_{conductor}$, may become large enough to cause trouble. This often occurs when copper conductors are joined to aluminum conductors. Special clamps and deoxidizing grease must be used to join these different metals.

To find out if too much voltage loss is occurring along the conductors, compare the voltage measured at the source end of the circuit to the voltage measured at the load end. Generally, the load-end voltage should be within about 1% of the source-end voltage. In Figure (B), the load voltage is about 7% lower than the source voltage, V_s. This indicates that there is a problem along the conductor run.

7-5 EFFICIENCY

Electrical load devices such as lamps, loudspeakers, and motors always produce some waste heat, besides their main output product. This waste heat is a nuisance. It causes thermal stress to the load device, and it requires the source to deliver more power than would otherwise be necessary. The situation is represented by the block diagram in Figure 7-8.

A load's **efficiency** measures its ability to provide a useful product while minimizing waste heat.

TECHNICAL FACT

Efficiency is useful output power divided by total electrical input power.

The symbol for efficiency is the Greek letter η. Therefore

7

$$\eta = \frac{P_{out}}{P_{in}}$$

EQ. 7-14

Since all the input power must be accounted for, we can write

$$P_{in} = P_{out} + P_{waste}$$

$$\text{or } P_{out} = P_{in} - P_{waste}$$

EQ. 7-15

Substituting this into Equation 7-14 gives

8

$$\eta = \frac{P_{in} - P_{waste}}{P_{in}}$$

EQ. 7-16

Equation 7-16 is an alternative way of expressing efficiency.

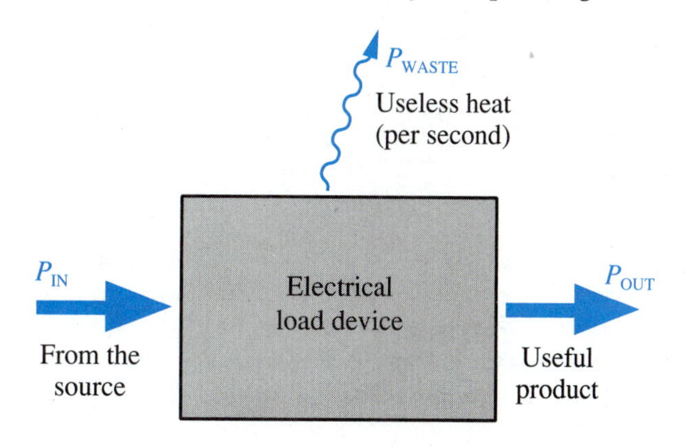

FIGURE 7-8 A load's efficiency measures its ability to provide a useful product while minimizing waste heat.

EXAMPLE 7-9

An **electric motor** is an example of a load device with rather high efficiency. In Figure 7-9, suppose $V = 120$ V and $I = 13.5$ A. At the output shaft, suppose **torque** (twisting force) equals 7.0 newton-meters ($\tau = 7.0$ N-m), and the **rotational speed** is 200 radians per second (S = 200 rad/s). These values for the mechanical variables give a mechanical output power of 1.4 kW. (See remark following the solution.)

a) Find the electrical input power, P_{in}.
b) Calculate the motor's efficiency.
c) How much power is wasted as heat?

SOLUTION

a) $P_{in} = VI = (120 \text{ V})(13.5 \text{ A}) = \mathbf{1620 \text{ W}}$
b) For an electric motor,

$$\eta = \frac{P_{out}}{P_{in}} = \frac{\text{mechanical output power}}{\text{electrical input power}} = \frac{1400 \text{ W}}{1620 \text{ W}} = \mathbf{0.864}$$

Efficiency is often expressed as a percent. Multiplying this pure decimal number by 100 gives

$$\eta = \mathbf{86.4\ \%}$$

c) From Equation 7-15

$$P_{waste} = P_{in} - P_{out}$$

$$= 1620 \text{ W} - 1400 \text{ W} = \mathbf{220 \text{ W}}$$

Another way to find wasted power is to reason this way: If 86.4% of the electric power is converted to mechanical rotation, the remaining 13.6% is thrown off as waste heat (13.6% = 100% – 86.4%). Therefore

$$P_{waste} = (0.136)\, P_{in}$$

$$= (0.136)(1620 \text{ W}) = \mathbf{220 \text{ W}}$$

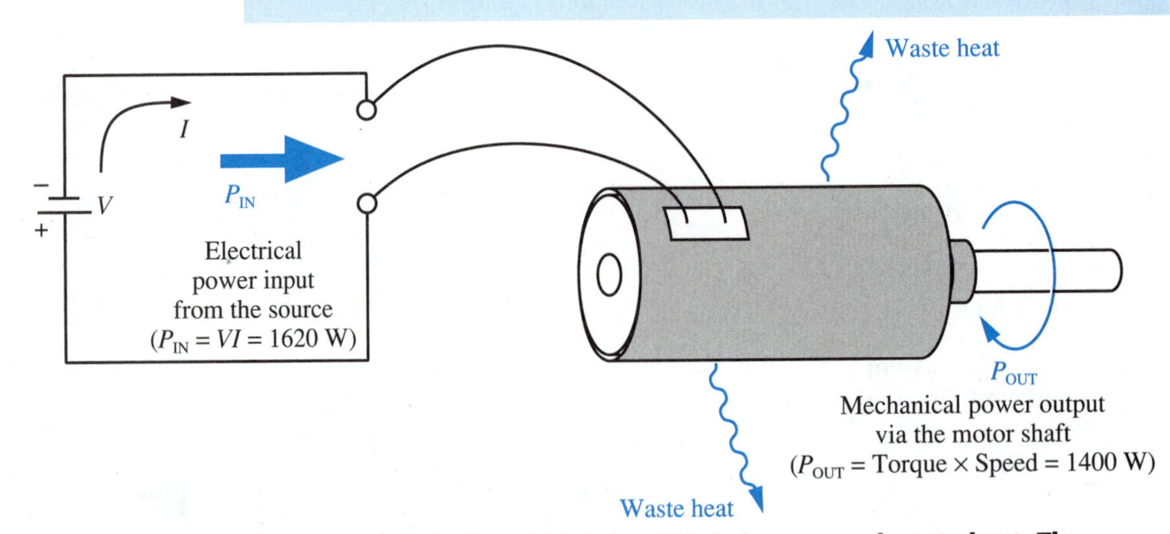

FIGURE 7-9 An electric motor converts electrical power into mechanical power and waste heat. The mechanical power is used to spin a mechanical device that is attached to the shaft.

Remark: In the United States, mechanical rotational variables are commonly measured in the following units:

Torque — pound-feet (lb-ft)
Rotational speed — revolutions per minute (rpm)
Shaft power — horsepower (hp)

These mechanical units are left over from the old-fashioned English system of measurement. We continue to use them out of habit. They will eventually be phased out as the U.S. converts to the metric system. The conversion factors to basic metric units are:

1 lb-ft = 1.356 N-m
1 rpm = 0.1047 rad / s
1 hp = 745.7 W

SAFETY ADVICE

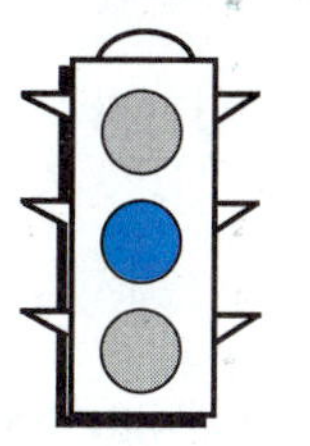

Never operate a motor without its mechanical shaft guards in place. These guards prevent you from accidentally touching or getting your clothes caught in the high-speed shaft and mechanical coupling. Motors with a pulley and belt drive are more dangerous than motors with direct-drive mechanical coupling.

Some industrial motors, called synchronous motors, spin their shafts in perfect synchronization with the rapid blinking action of fluorescent lights. In other words, at every time instant a fluorescent lamp reaches its maximum brightness, which reoccurs 120 times per second, the motor shaft is in the exact same position, mechanically. This produces a stroboscope effect that makes the motor shaft appear to be standing still. By misleading us into thinking that the motor shaft is not spinning, a missing guard becomes even more of a danger.

7-6 WATTMETERS

The instrument that measures power is a **wattmeter**. It has four terminals. Two of them connect to the internal **voltage coil**. The other two connect to the internal **current coil.** This is diagrammed in Figure 7-10(A).

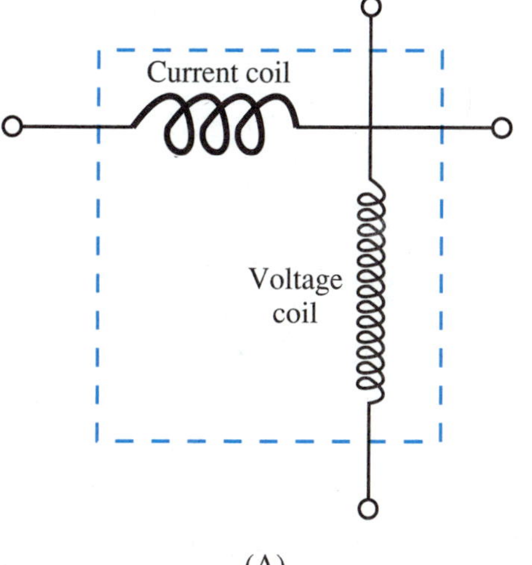

(A)

(B)

FIGURE 7-10 (A) A wattmeter contains a current coil and a voltage coil. The current coil is made of heavy-gage wire, with few turns. The voltage coil is made of thinner wire, with many turns. (B) Four-terminal wattmeter.

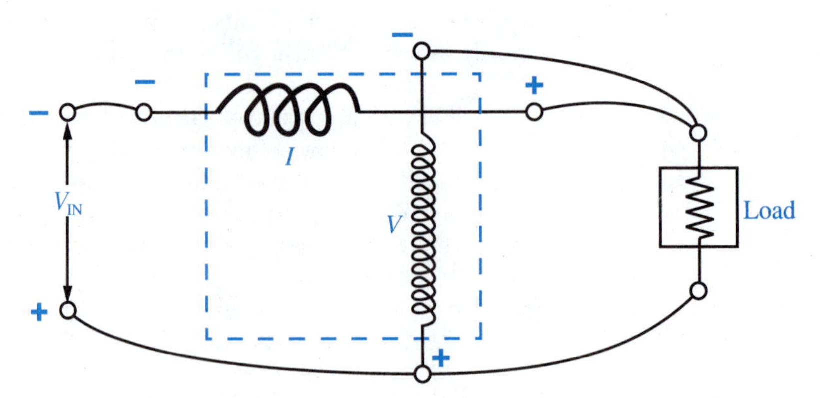

FIGURE 7-11 Proper polarization of wattmeter coils

Figure 7-10(B) is a photograph of a four-terminal wattmeter. It shows that the terminals are polarized. Here is the method for connecting a wattmeter with proper polarity.

TECHNICAL FACT

The wattmeter's current coil must be connected in line with the load; current must enter at the coil's negative terminal. The voltage coil must be connected across the load, with the coil's negative terminal connected to the load's negative terminal.

Proper wattmeter connection is shown in Figure 7-11.

Some four-terminal wattmeters have front-panel markings of LINE and LOAD. The user connects the voltage source to the LINE terminals and the load device to the LOAD terminals. The meter's internal connections guarantee proper coil polarity. This arrangement is shown in Figure 7-12(A).

Figure 7-12(A) makes it clear that one of the LINE terminals and one of the LOAD terminals are simply connected together internally. Therefore, those two terminals can be replaced by a single common terminal (COM) on the front panel. This is shown in Figure 7-12(B).

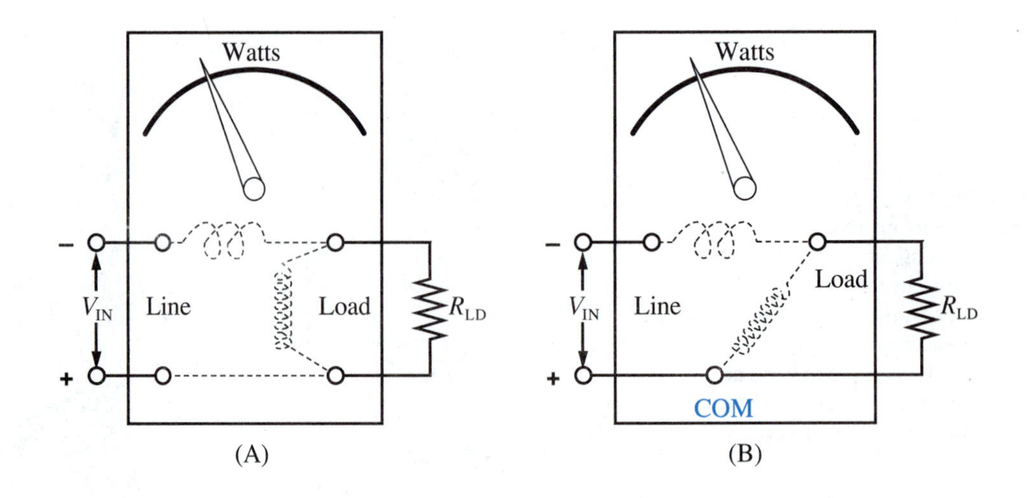

FIGURE 7-12 Alternative wattmeter panel markings. (A) Four-terminal LINE-LOAD. (B) Three-terminal LINE-LOAD.

✔ SELF-CHECK FOR SECTIONS 7-5 AND 7-6

23. All electrical loads convert part of their input power into a useful product and part of their input power into_____ _____ . [7]
24. The ratio of useful output power to total input power is called__________. [7]
25. Describe the useful output product for an electric motor. [7]
26. Is an electric motor a relatively high-efficiency device or a low-efficiency device? [7]
27. Describe the useful output product for an incandescent lamp. [7]
28. Does an incandescent lamp have relatively high efficiency or low efficiency? [7]

To answer Questions 29 through 34, refer to this statement: For an electric motor, the mechanical shaft power is given by the equation $P_{out} = \tau S$, where τ symbolizes torque and S symbolizes rotating speed.

29. The basic metric unit for τ is the_______ – _______; it is symbolized _______.
30. The basic metric unit for rotating speed is the_______ _____ _______ ; it is symbolized _______.
31. When τ in basic units is multiplied by S in basic units, P comes out in basic units of__________ .
32. The old-fashioned unit of torque that is still popular in the U.S. is the ____________–____________.
33. The unit of rotational speed that is popular in the U.S. is the ___________.
34. When using the popular U.S. units for torque and speed, the mechanical power equation is $P_{out(mech)} = 5252\ \tau S$. The unit of mechanical shaft power is then the __________, symbolized hp.
35. In a wattmeter, the __________coil is placed in line with the load so that it carries the same current as the load. [9]
36. In a wattmeter, the __________ coil is placed across the load so that it senses the load's voltage. [9]

FORMULAS

$W = F \times d$ (mechanically)	**EQ. 7-1**	$R = V^2/P$	**EQ. 7-9**
$W = V \times Q$ (electrically)	**EQ. 7-2**	$V = \sqrt{PR}$	**EQ. 7-10**
$P = W/t$	**EQ. 7-3**	$P = I^2 R$	**EQ. 7-11**
$W = Pt$	**EQ. 7-4**	$R = P/I^2$	**EQ. 7-12**
$P = VI$	**EQ. 7-5**	$I = \sqrt{P/R}$	**EQ. 7-13**
$V = P/I$	**EQ. 7-6**	$\eta = P_{out}/P_{in}$	**EQ. 7-14**
$I = P/V$	**EQ. 7-7**	$P_{out} = P_{in} - P_{waste}$	**EQ. 7-15**
$P = V^2/R$	**EQ. 7-8**	$\eta = \dfrac{P_{in} - P_{waste}}{P_{in}}$	**EQ. 7-16**

SUMMARY OF IDEAS

- Energy is the ability to do work. It is symbolized W.
- Energy can be mechanical or electrical. It can also appear as heat (thermal).
- The basic unit of energy is the joule, symbolized J.
- Electrical energy is equal to voltage times charge ($W = VQ$).
- 1 joule equals 1 volt times 1 coulomb (1 J = 1 V x 1 C).
- Power is energy per time ($P = W/t$).
- The basic unit of power is the watt, symbolized W.
- 1 watt = 1 joule per second (1 W = 1 J/s).
- A battery charger restores energy to a depleted battery.
- The common commercial unit of energy is the kilowatthour (kWh).
- Power is equal to voltage times current (1 W = 1 V x 1 A).
- Power can be calculated by $P = V^2/R$.
- Power can be calculated by $P = I^2R$.
- Efficiency is equal to useful output power divided by total electrical input power. It is symbolized η.
- Total electrical input power equals the sum of useful output power plus wasted heat power.
- The wattmeter is the instrument used for measuring power.
- Wattmeters contain a current coil and a voltage coil. These coils must be properly polarized when a wattmeter is connected into a circuit.

CHAPTER QUESTIONS AND PROBLEMS

1. T-F The idea of energy includes the idea of time. [2]
2. T-F The idea of power includes the idea of time. [2]
3. The basic metric unit of energy is the __________. It is symbolized _____. [1, 3]
4. 1 J of mechanical energy is equal to a force of__________multiplied by a distance of__________.
5. 1 J of electrical energy is equal to a voltage of__________multiplied by a charge movement of__________. [1]
6. T-F Energy and work are basically interchangeable, since energy is the ability to do work.
7. Some people prefer the letters NRG to symbolize the idea of electric energy. But we are using the letter__________. [1, 2]
8. Which variable is measured by a residential electric meter, power or energy? [2]
9. Since 1 joule is such a small amount of energy in commercial applications, electric utility companies use the __________unit. It is symbolized _______. [3]
10. In the metric system, the basic unit of power is the __________. It is symbolized _____.[2]
11. Power of 1 W is equal to energy of__________ divided by __________ of __________. [2]
12. Write the equation that expresses P in terms of I and V. [4]
13. Write the other two equations that relate P, I, and V. One of them solves for I and the other one solves for V. [4]
14. A type No. 47 bulb carries 150 mA when driven by a 14-V source. What is its power? [4]
15. A certain industrial welder operates at a voltage of 460 V and has power consumption of 50 kW. How much current does it carry? [4]
16. A certain battery charger is designed to deliver a constant current of 7 A, no matter what its output voltage is. For a power output of 100 W, what should the voltage be adjusted to? [4]
17. Write the three equations that relate P, V, and R. One solves for P, one solves for V, and one solves for R. [5]

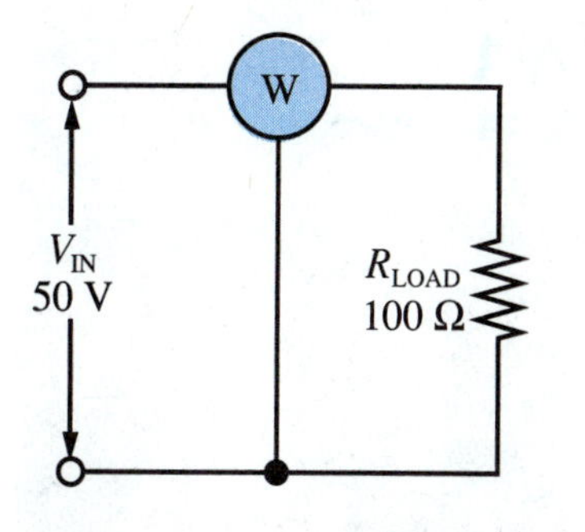

FIGURE 7-13 Schematic symbol for three-terminal wattmeter

18. A 10-Ω heating element is driven by a variable voltage source. To deliver 800 W of power, what voltage value is required? [5]
19. A three-terminal wattmeter is shown schematically in Figure 7-13. What value of power does it measure? [5, 9]
20. A rheostat is being driven by a 25-V source. What value of resistance will cause the power to equal 35 W? [5]
21. Write the three equations that relate P, I, and R. One solves for P, one for I, and one for R. [5]
22. A 75-W soldering iron has a resistance of 450 Ω. How much current does it carry? [5]
23. A welding machine is delivering 4.5 kW of power while passing a 135-A current through the weld. What is the effective resistance of the metal-to-metal contact? [5]
24. A certain incandescent bulb carries a current of 200 mA. If its filament resistance is 500 Ω, what amount of power does it consume? [5]
25. A fluorescent lamp carries 1.4 A when driven by a 120-V source. Its efficiency is 18 %. How much lighting power does it give off? [4, 7]
26. For the lamp in Problem 25, how much heat power is thrown off? [7]
27. During the winter, is the heat power in Problem 26 a total waste? Explain.[7]
28. A commercial 230-V refrigeration motor produces 2.5 kW of shaft power, with an efficiency of 0.85. Calculate its current. [4, 7]
29. A 16-Ω loudspeaker is driven by 12 V. Its sound output power is 350 mW. What is its efficiency? [5, 7]
30. A certain motor is overheating. A technician measures the motor's input voltage as 108.5 V while it is operating. At the circuit breaker box, the supply voltage measures 117.8 V. What is the probable trouble here? [6]

CHAPTER 8 SAFETY DEVICES, WIRES, SWITCHES, AND RELAYS

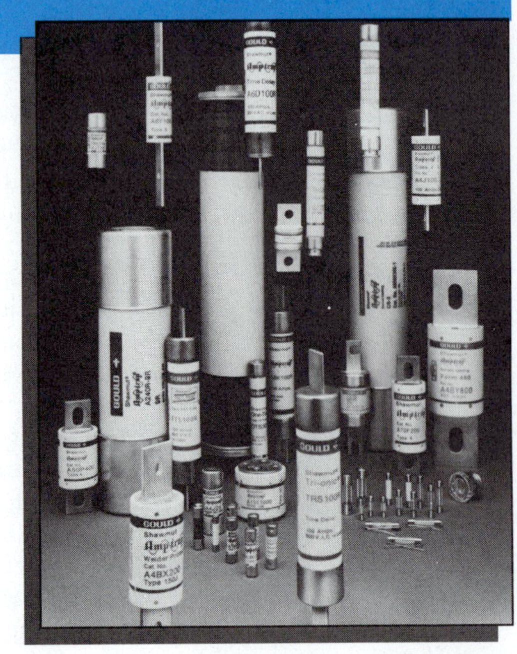

Various fuses.
Courtesy of Gould, Inc., Circuit Protection Division

OUTLINE

NEW TERMS TO WATCH FOR

overcurrent
short circuit
fuse
fast-acting fuse
slow-blow fuse
arc
circuit breaker
hot wire
common wire
grounding wire
ground-fault interrupter
leakage current
megohmmeter
shielded cable
coaxial cable
single-pole
double-pole
single-throw
double-throw
nonshorting
break-before-make
shorting
make-before-break
pushbutton (PB) switch
normally open (N.O.)
normally closed (N.C.)
momentary-contact
fixed-closure
relay
coil
contact
backup power

All circuits have five basic parts. In this chapter, we will learn about conductors and control devices.

After studying this chapter, you should be able to:

1. Describe an overcurrent and tell how one can happen.
2. Explain the function of a fuse.
3. Explain the function of a circuit breaker.
4. Explain why the grounding wire is used to provide greater safety in a three-wire grounded system.
5. Describe how a ground-fault interrupter works. Explain why a GFI provides even greater safety in bathroom and outdoor circuits.
6. Use wire-gage tables to specify the proper conductor size in a circuit.
7. Describe the electrical-control capabilities of various switches.
8. Describe the construction and method of mechanical actuation for various switches.
9. Define a normally open switch and a normally closed switch.
10. Troubleshoot a circuit to locate a defective switch or contact.
11. Describe how an electromagnetic relay works.
12. State the advantages of relay switching over direct manual switching.

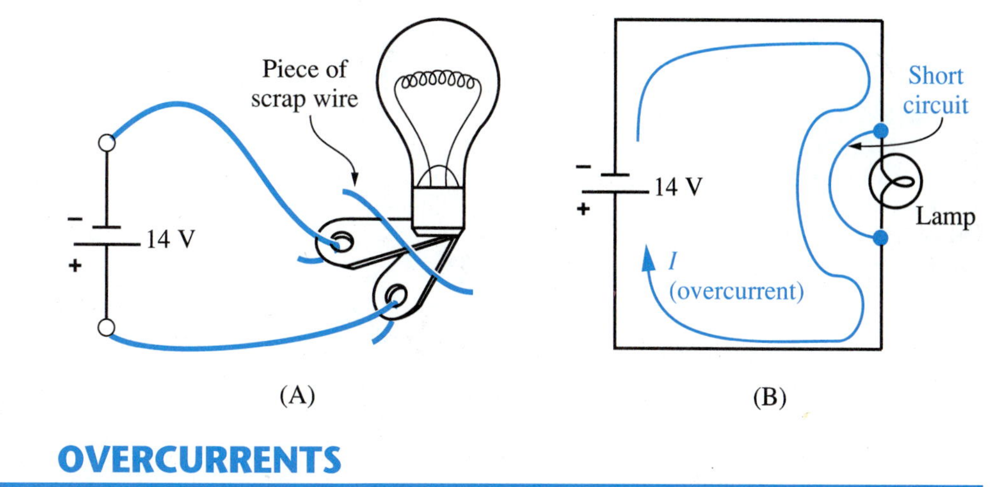

FIGURE 8-1 (A) Problems like this happen occasionally. (B) The accidental short circuit has no resistance to limit the current. Therefore, it becomes dangerously large—an overcurrent.

8-1 OVERCURRENTS

An **overcurrent** is a current too large for the wires to carry safely. Most overcurrents occur because of some malfunction in the circuit. For example, Figure 8-1(A) shows a lamp socket with exposed solder terminals. If a piece of bare wire accidentally falls across the solder terminals, a **short circuit** is created. This situation is shown schematically in Figure 8-1(B). All the current takes the zero-resistance path around the resistive lamp.

The problem shown in Figure 8-1 is called a *dead* short circuit because there is virtually zero net resistance in the circuit. A more common problem is a *partial* short circuit. This occurs when some malfunction in the load causes its resistance to become much lower than expected. Figure 8-2(A) shows a specific example.

A multiturn insulated winding has had part of its insulation damaged in Figure 8-2(A). Such damage could happen by mechanical rubbing or by overheating. The damaged turns of wire touch each other, causing the resistance of the top half to become virtually zero. In the schematic diagram of Figure 8-2(B), current bypasses the top half of the load, but still passes through the resistance of the bottom half. The resulting current is much larger than normal.

8-2 FUSES

2

A **fuse** is a safety device that melts when a dangerous overcurrent occurs. By melting, the fuse becomes an open circuit. This interrupts the current flow path and stops the overcurrent. The schematic symbol for a fuse is shown in the circuit of

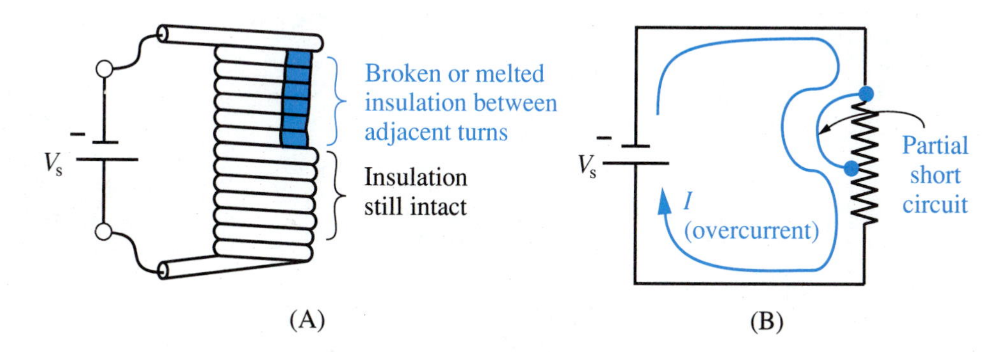

FIGURE 8-2 (A) Multiturn winding with insulated wire. If the insulation on the top half of the winding is ruined, those turns all touch each other. (B) Schematic diagram. Overcurrent occurs because net resistance is too low.

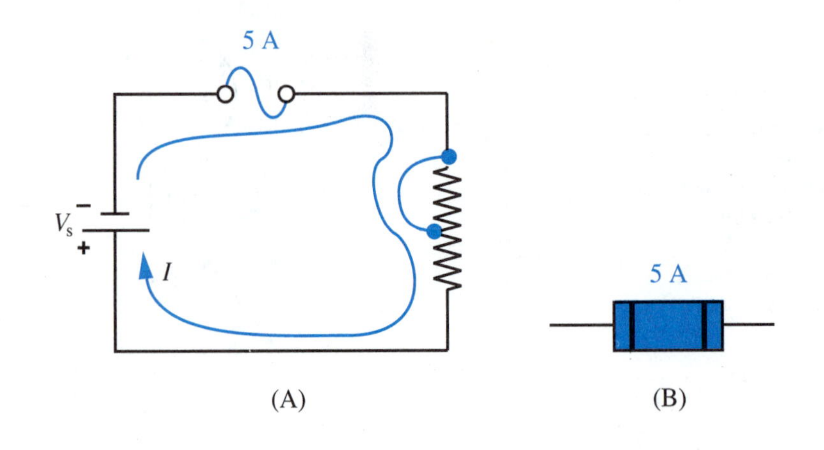

FIGURE 8-3 (A) Fused circuit. If current, *I*, exceeds 5 A due to a partial short, the fuse blows. (B) Alternative symbol for a fuse

Figure 8-3(A). The fuse's current rating is the maximum amount of current that can flow before the fuse melts, or blows. An alternative fuse symbol is shown in Figure 8-3(B).

Most fuses are cylindrical in shape. Figure 8-4(A) is a glass-body fuse that shows the meltable metal link inside. The link is often called the fuse element. A cutaway view of a larger cartridge-type fuse is shown in Figure 8-4(B). A panel-mount fuse-holder for a small glass-body fuse is shown in Figure 8-4(C). A screw-in plug fuse, the type used in residential fuse boxes, is shown in Figure 8-4(D).

Fuses differ in how quickly they blow when subjected to overcurrents. The manufacturers make **fast-acting**, normal, and **slow-blow** models. Their operating characteristics are summarized in Table 8-1.

TABLE 8-1 Approximate Time Required for Various Fuses to Blow

FUSE TYPE	SEVERITY OF OVERCURRENT		
	Slight (1.2 times rated current)	Moderate (3 to 5 times rated current)	Extreme (10 times rated current)
Fast-acting	a few minutes	1 to 10 ms	< 1 ms
Normal	a few minutes	10 to 100 ms	< 1 ms
Slow-blow	a few minutes	1 to 10 s	< 1 ms

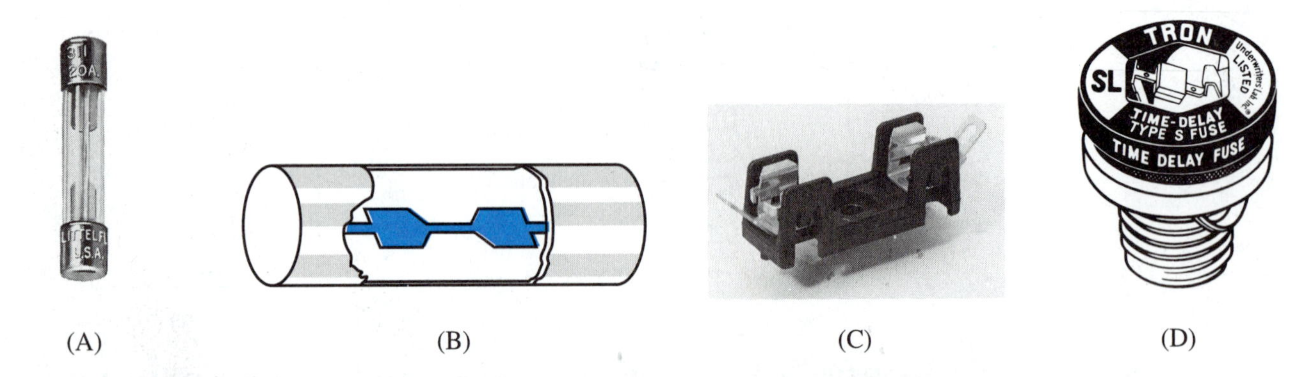

FIGURE 8-4 (A) Glass-body normal-blow fuse. *Courtesy of Littelfuse Inc.* **(B) Cutaway view of a cartridge fuse, showing its element. (C) Fuseholder.** *Courtesy of Littelfuse Inc.* **(D) Screw-in plug fuse, used in residential fuse boxes.** *Courtesy of Bussmann Div., Cooper Industries*

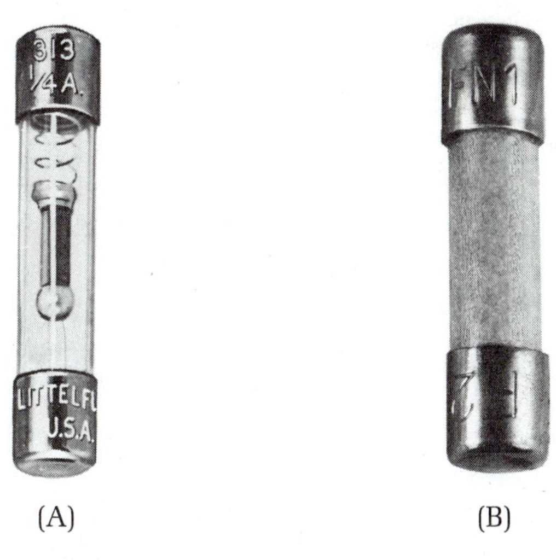

FIGURE 8-5 (A) Glass-body slow-blow fuse. (B) Ceramic-body slow-blow fuse. *Courtesy of Littelfuse Inc.*

A slow-blow glass-body fuse can be recognized by its coiled element, shown in Figure 8-5(A). A ceramic-cartridge slow-blow fuse is shown in Figure 8-5(B).

In addition to its current rating, every fuse has a voltage rating. A fuse must not be used in a circuit with a source voltage greater than this rating. This is because whenever a fuse element melts, a short-lived **arc** jumps across the air gap between the two sides, as illustrated in Figure 8-6. The greater the source voltage, the more powerful the arc tends to be. The fuse must be able to extinguish the arc quickly enough to prevent damage to the conductors leading up to the fuse body.

Any fuse will have a lower voltage rating in a dc circuit than in an ac circuit. This is because the arc produced by a dc source is harder to extinguish.

TECHNICAL FACT

To test a fuse reliably, remove it from its holder and connect an ohmmeter. An intact fuse measures nearly zero ohms. A blown fuse will show infinite ohms.

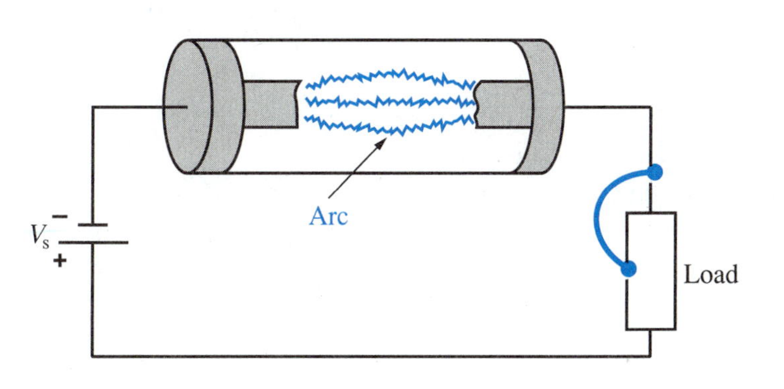

FIGURE 8-6 When a fuse blows, current arcs through the air gap in the element.

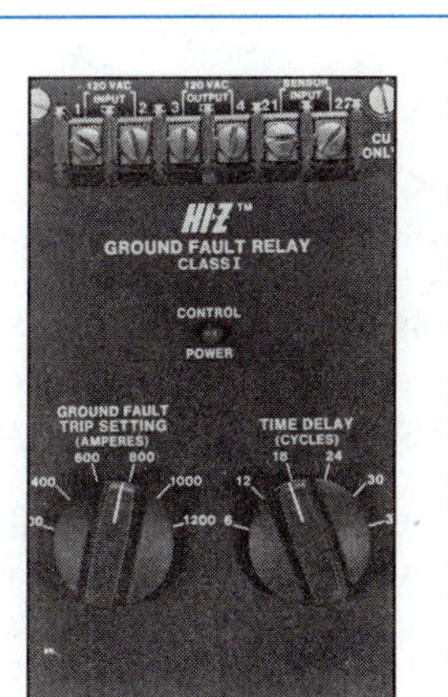

This industrial ground-fault interrupter has adjustable current sensitivity (left selector switch). It will tolerate a current imbalance for a small number of cycles of the ac line. This number is adjustable by the switch on the right. *Courtesy of Siemens Energy & Automation, Inc.*

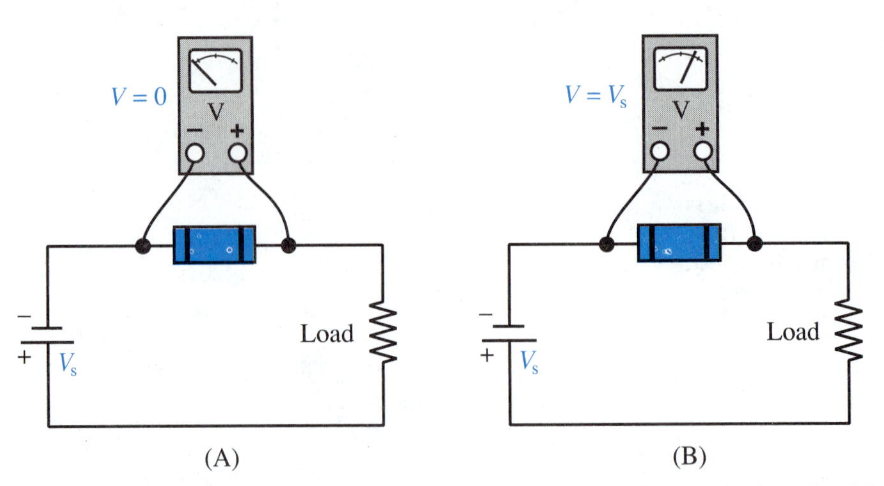

FIGURE 8-7 Testing a fuse in circuit. (A) An intact fuse will measure 0 V. (B) A blown fuse will show V_s across its terminals.

Don't do just a visual inspection to tell whether a glass-body fuse has blown. It is too difficult to see the element clearly. If it is inconvenient to remove a fuse, it can be tested in-circuit with a voltmeter as shown in Figure 8-7.

8-3 CIRCUIT BREAKERS

A **circuit breaker** has a mechanical switch contact for interrupting an overcurrent. The switch contact is controlled by a current-sensing device. The current-sensing device may be thermal, or it may be magnetically operated. Schematic symbols for the two types of circuit breaker are shown in Figure 8-8.

The advantage of a circuit breaker over a fuse is that it doesn't have to be replaced when it opens. After the basic circuit problem has been repaired, you simply press a button to reset the circuit breaker to its closed condition. A residential circuit breaker is shown in Figure 8-9.

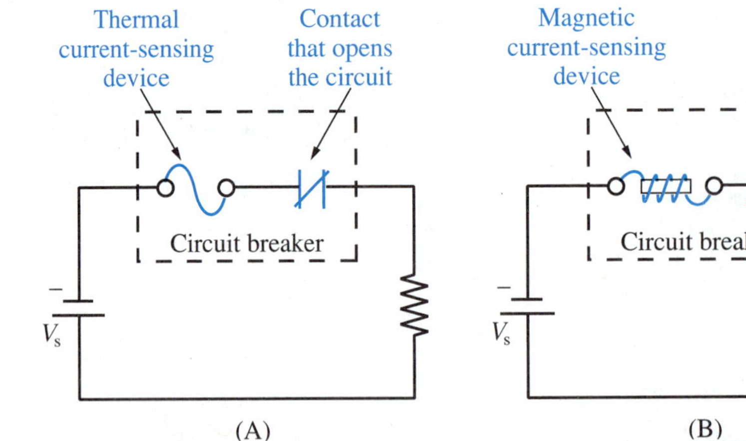

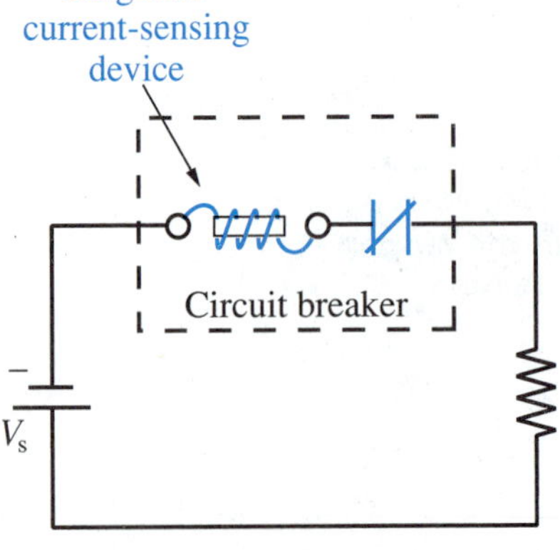

FIGURE 8-8 Circuit breaker symbols. (A) Thermal. (B) Magnetic

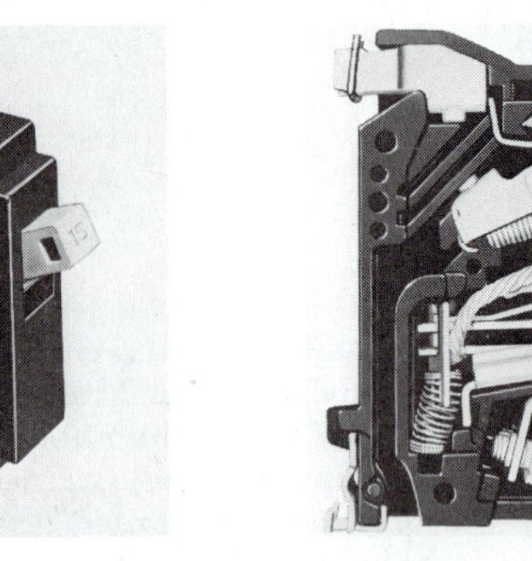

(A) (B)

FIGURE 8-9 (A) 20-A circuit breaker of the type used in residential distribution boxes. This is a thermomagnetic breaker, combining both types of current sensors. (B) Cutaway view of thermomagnetic circuit breaker. *Courtesy of Eaton Corporation, Cutler-Hammer Products*

✔ SELF-CHECK FOR SECTIONS 8-1, 8-2, AND 8-3

1. Current that is too large for a circuit to handle safely is called__________. [1]
2. Explain what a dead short circuit is. [1]
3. Explain what a partial short circuit is. [1]
4. T-F A fuse responds to an overcurrent by becoming so hot that it melts. [2]
5. T-F For a minimal overcurrent (20 % above rated value), even fast-acting fuses take a long time to melt. [2]
6. T-F For a very severe overcurrent (10 times rated value), a slow-blow fuse takes a long time to melt. [2]
7. A blown fuse measures__________ohms on an ohmmeter. [2]
8. The two current-sensing methods for circuit breakers are______________ and__________ . [3]
9. Explain the advantage of a circuit breaker over a fuse. [2, 3]
10. T-F A thermal circuit breaker contains an element that can melt. [3]

8-4 THREE-WIRE GROUNDED SYSTEMS

4

In modern electrical wiring, the metal enclosure around each home appliance is connected directly to the earth through a separate wire. This is done to prevent the possibility of a person being electrically shocked by touching the enclosure. To understand how this method works, study Figures 8-10 and 8-11.

Figure 8-10 shows an old-fashioned two-wire system. The bottom terminal of the source is deliberately connected to the earth so that the **hot** supply **wire** can never be more than 120 V away from earth potential. This is a long-standing safety practice.

When the circuit functions normally, current flows down the hot supply wire, through the load device (appliance), and back to the source by the **common** supply **wire**. If an accidental short circuit should occur between the hot wire and the metal enclosure, as shown in Figure 8-10, no overcurrent occurs. Therefore, the circuit-breaker does not open and the load continues to function. The problem is that the metal enclosure now has the same 120-V potential as the hot supply wire. If a person touches the enclosure at the same time that she contacts the earth, current will flow through her as indicated in Figure 8-10. This causes electrical shock.

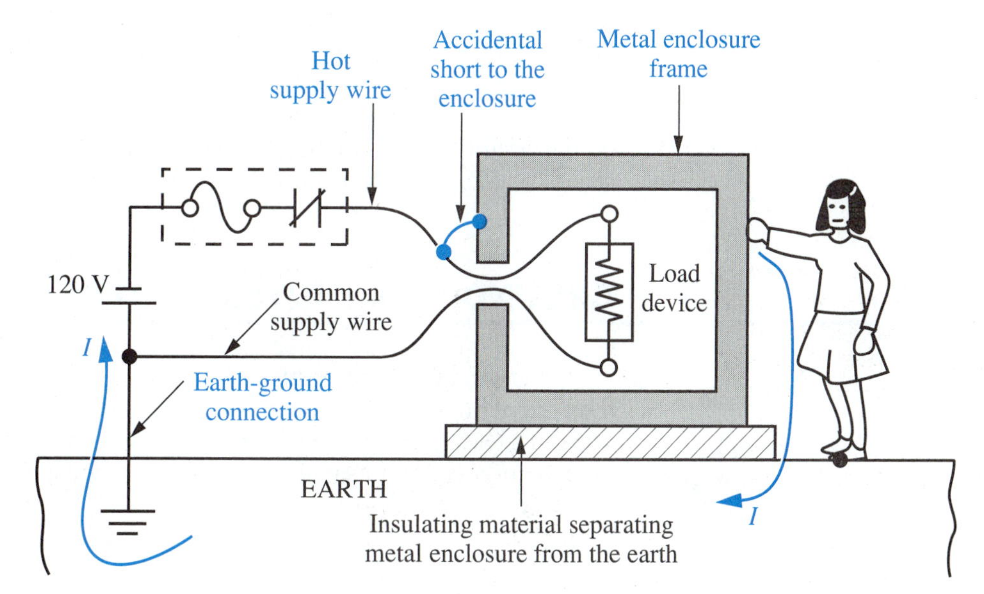

Figure 8-10 A person can be shocked in a two-wire system because the circuit breaker doesn't open if the hot wire shorts to the frame.

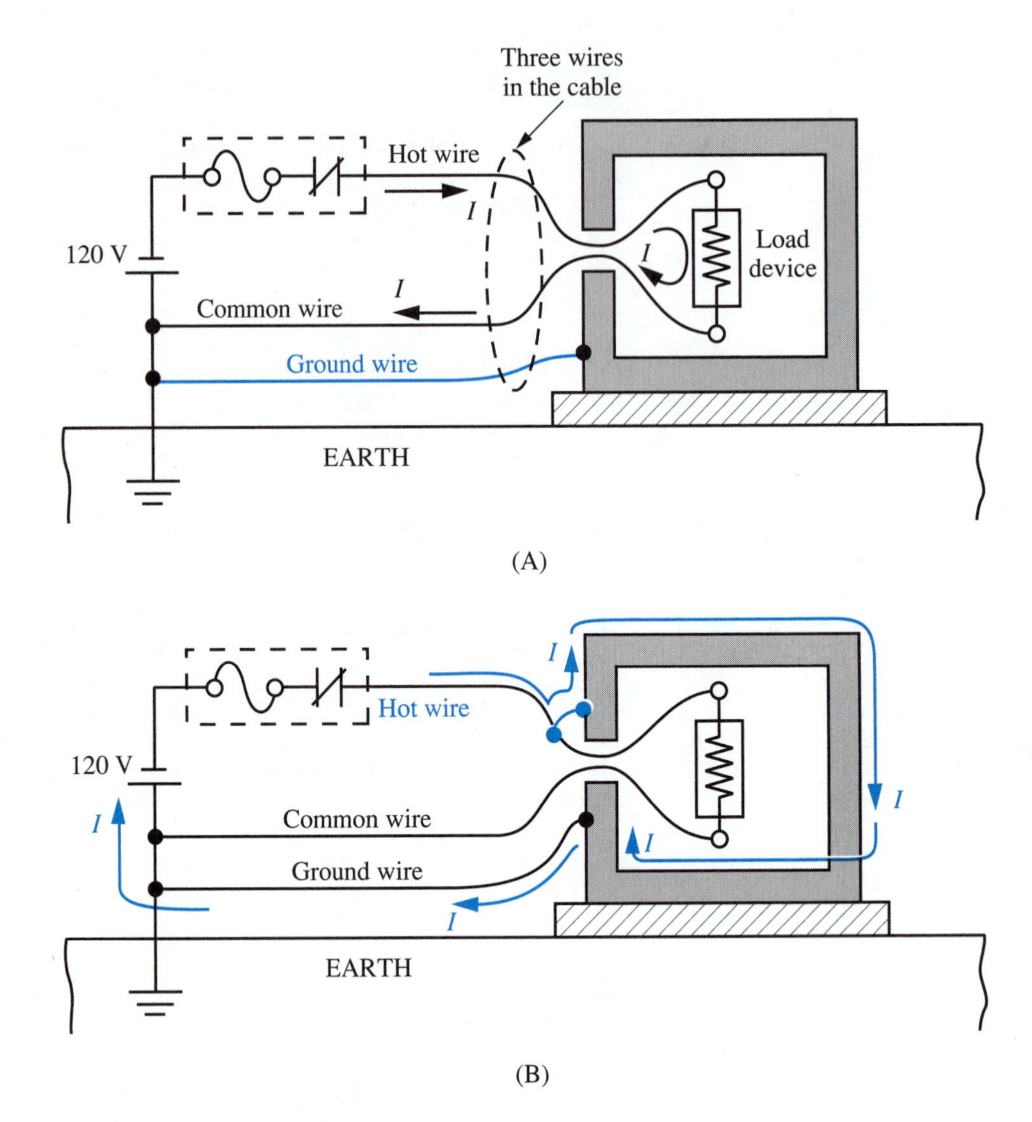

FIGURE 8-11 Grounded three-wire system. (A) Normal operation. (B) Overcurrent blows circuit breaker if hot wire shorts to frame. Actually, residential and commercial circuits are ac rather than dc, but that isn't important for understanding this safety practice.

To prevent this from happening, a third wire is brought along with the two supply wires as shown in Figure 8-11(A). The third wire has one end connected to the ground side of the source and the other end connected to the metal enclosure. When the circuit is functioning normally, no current flows in this **grounding wire.**

However, if a short circuit should happen, as indicated in Figure 8-11(B), an overcurrent occurs immediately. It flows through the metal enclosure, then through the grounding wire back to the source. The circuit breaker opens, shutting off the circuit and eliminating the dangerous condition. It is impossible to reset the circuit breaker until the short circuit has been repaired.

8-5 GROUND-FAULT INTERRUPTERS

A **ground-fault interrupter** (GFI) compares the current flowing to the load with the current flowing back from the load. Refer to Figure 8-12 to understand the operation of a GFI. A GFI is a four-terminal device, as the schematic symbol in Figure 8-12(A) indicates. It has two current-sensing coils; those two coils together control the contact.

Figure 8-12(B) shows a GFI installed in a properly functioning circuit. The two GFI coils compare I_{black} (going to the load) with I_{white} (coming back from the load).

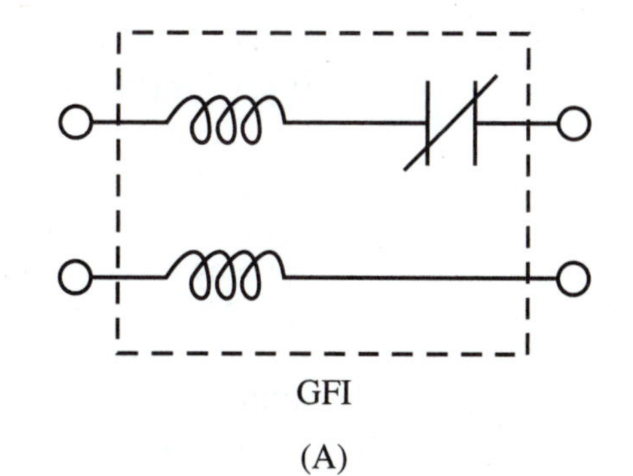

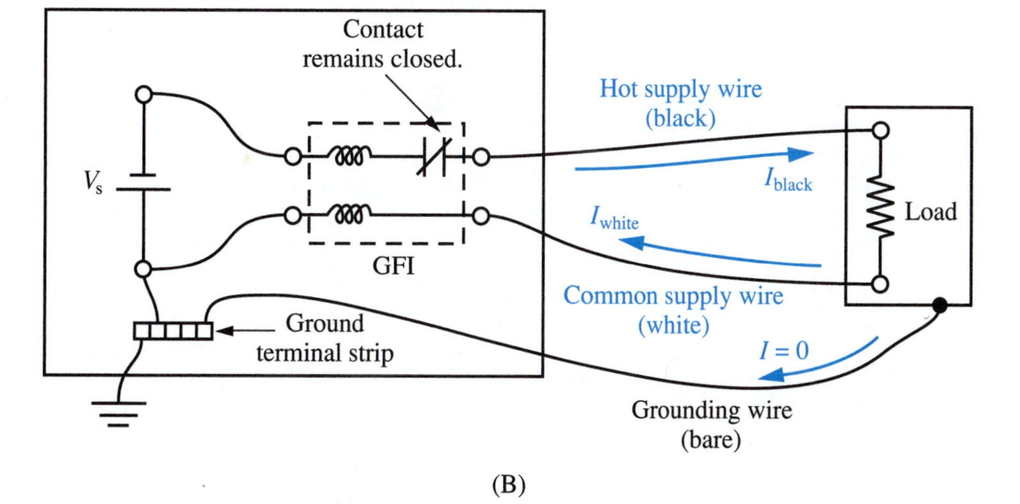

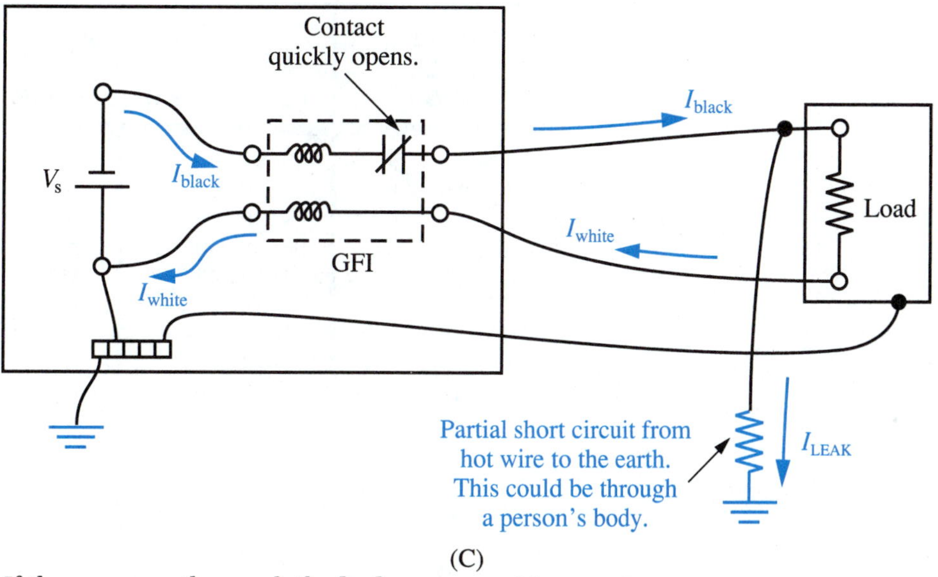

FIGURE 8-12 (A) GFI schematic symbol. (B) With circuit functioning properly, the GFI keeps the contact closed. (C) If a partial short happens, the GFI senses a difference between I_{black} and I_{white}. It opens the contact very quickly.

If they are exactly equal, the **leakage current** is zero since

$$I_{leak} = I_{black} - I_{white}$$

With a leakage, or difference, current of zero, the GFI coils balance each other. Therefore, the contact remains closed.

If a partial short circuit should occur, a portion of I_{black} will return to the source by some path other than the common white wire. This is illustrated in Figure 8-12(C) for a situation where a person accidentally touches the hot wire. When the GFI coils sense a difference between I_{black} and I_{white}, they quickly open the contact, shutting off the circuit.

The partial short circuit shown in Figure 8-12(C) is directly to the earth. However, a partial short to the metal enclosure would also cause the GFI to open.

✔ SELF-CHECK FOR SECTIONS 8-4 AND 8-5

11. Explain how a person could be shocked by touching a metal load enclosure in a two-wire system. [4]
12. Explain why a separate grounding wire is an effective safety precaution if a dead short occurs between the hot wire and the metal enclosure. [4]
13. Explain why a separate grounding wire is not necessarily an effective safety precaution if a partial short occurs between the hot wire and the metal enclosure. Make a drawing to help your explanation. [4]
14. Explain why a GFI combined with a separate grounding wire provides complete safety against partial shorts to the enclosure. Use a drawing. [5]

8-6 WIRE AND CABLE

A wire is a single conductor, either insulated or uninsulated. A cable is two or more conductors held together along their length.

AWG Number

Wires and cables are specified by their American Wire Gage Numbers (AWG#). Lower AWG numbers represent thicker wire with greater current-carrying capacity. Higher AWG numbers mean thinner wire.

TABLE 8-2 Wire Gage Specifications

WIRE GAGE AWG #	DIAMETER		CROSS-SECTIONAL AREA	
	(10^{-3} m,mm)	(mil)	(10^{-6} m^2)	cmil*
0000 (4/0)	11.68	459.8	107.2	2.116×10^5
000 (3/0)	10.40	409.5	85.01	1.678×10^5
00 (2/0)	9.265	364.8	67.42	1.331×10^5
0	8.250	324.8	53.46	1.055×10^5
2	6.543	257.6	33.62	66364
4	5.189	204.3	21.15	41749
6	4.115	162.0	13.30	26253
8	3.263	128.5	8.363	16508
10	2.588	101.9	5.259	10380
12	2.052	80.79	3.307	6528
14	1.628	64.09	2.080	4106
16	1.291	50.83	1.308	2582
18	1.024	40.32	0.8228	1624
20	0.8117	31.96	0.5175	1022
22	0.6437	25.34	0.3254	642.3
24	0.5105	20.10	0.2047	404.1
26	0.4048	15.94	0.1287	254.1
28	0.3210	12.64	0.080 95	159.8
30	0.2546	10.02	0.050 91	100.5
32	0.2019	7.949	0.032 02	63.21
34	0.1601	6.303	0.020 14	39.76
36	0.1270	5.000	0.012 66	24.99
38	0.1007	3.965	0.007 964	15.72
40	0.079 87	3.145	0.005 010	9.889

* 1 cmil = 5.066×10^{-10} m^2
1 cmil = 7.852×10^{-7} in^2

Table 8-2 lists the diameter and cross-sectional area for AWG numbers 0000 through 40. Diameter is given in basic units (10^{-3} m, or 1 mm) and also in units of mils, which are widely used in the U.S. One mil is defined as 0.001 inch. Cross-sectional area is given in basic units (10^{-6} square meters) and also in units of circular mils, popular in the U.S. One circular mil (cmil) is defined as the cross-sectional area of a round wire with 1-mil diameter. A cmil is illustrated in Figure 8-13.

The maximum current-carrying capacities for various gages of copper wire are given in Table 8-3. For aluminum wire, the current capacity is about 75 % of the same gage copper wire.

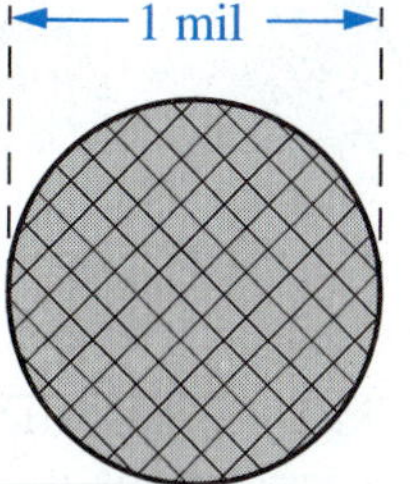

FIGURE 8-13 Area of 1 cmil. (Drawing is not to scale.)

TABLE 8-3 Current Capacity of Copper Wire

WIRE GAGE #	CURRENT CAPACITY (A)
14	15
12	20
10	30
8	45
6	65
4	85
2	115
0	150
00 (2/0)	175
000 (3/0)	200
0000 (4/0)	230

These current values are valid for standard conditions of temperature and wire packing density.

Structural Features

Wire heavier than AWG #32 may be either solid or stranded. Stranded wire is composed of several thin wires twisted together. The total cross-sectional area of all the thin wires combined is the same as the area of a single solid conductor of that same gage number. Stranded wire is more expensive than solid wire, but it is easier to work with. It can be bent into position and pulled around obstacles more readily.

The type and thickness of insulation determine the maximum operating temperature and operating voltage of a wire or cable. Some very effective insulating materials allow operation up to 250° C (480° F). Voltage ratings of several thousand volts are common.

Insulation is also graded on its toughness against mechanical abrasion and its ability to withstand water, corrosive liquids and gases, and exposure to direct sunlight.

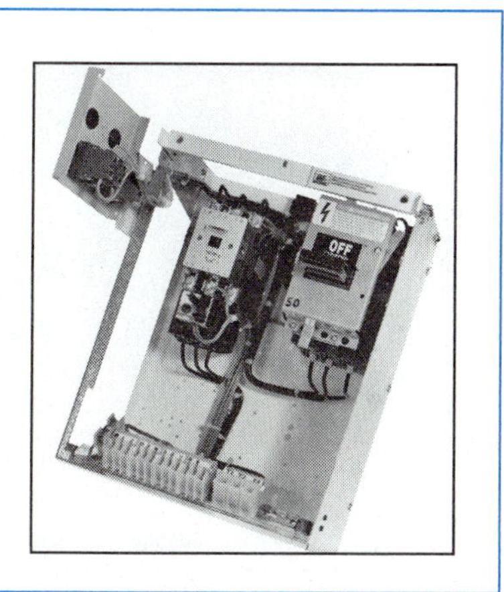

Motor-starter relay (at left) with series-connected thermal circuit breaker (at right). *Courtesy of Siemens Energy & Automation, Inc.*

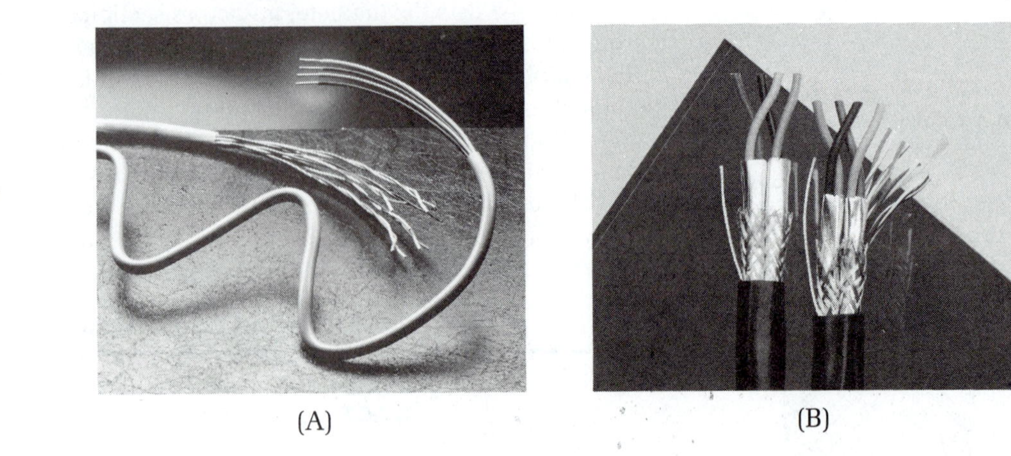

FIGURE 8-14 (A) Unshielded cable. (B) Shielded cable. *Courtesy of Belden Wire and Cable*

The condition of wire insulation can be tested by an instrument called an insulation-tester, or **megohmmeter.** This instrument measures the insulation's resistance by applying a very high voltage between a conductor and the metal enclosure frame of the equipment. An ordinary ohmmeter cannot properly test insulation because it does not apply a high enough voltage to reveal defects.

Wires and cables may be either unshielded or shielded, as shown in Figure 8-14. In a **shielded cable,** the conductors are surrounded by a copper or aluminum metal wrapping. This metal shield protects the internal wires from electrical interference called noise. This prevents distortion of the electrical signal on the wires.

A special kind of shielded cable is **coaxial cable,** Figure 8-15. It has a single solid conductor in the center, surrounded by a special insulating layer, then a braided shield, and finally the outer insulation, or jacket. The inside insulating layer is designed to have very specific characteristics for the handling of radio-frequency (RF) signals. The RF signal voltage exists between the center conductor and the braided shield.

Ribbon cables are multiconductor flat cables that usually are terminated in an edge connector. The edge connector makes a friction fit on the edge of a printed circuit board. This is shown on the left end in Figure 8-16.

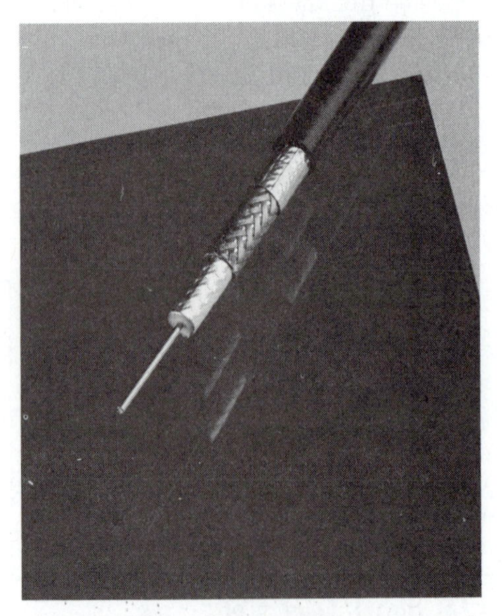

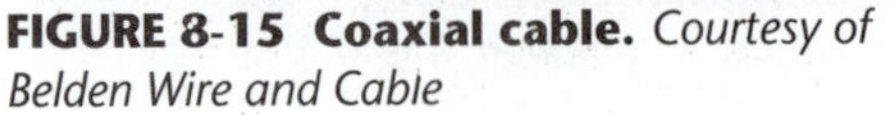

FIGURE 8-15 Coaxial cable. *Courtesy of Belden Wire and Cable*

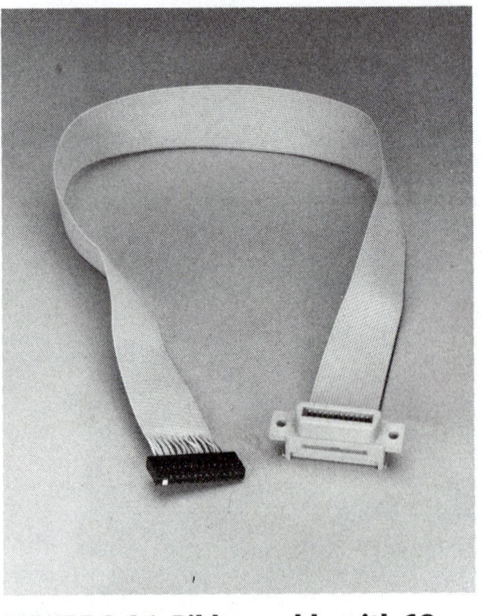

FIGURE 8-16 Ribbon cable with 12 conductors. *Courtesy of National Instruments Co.*

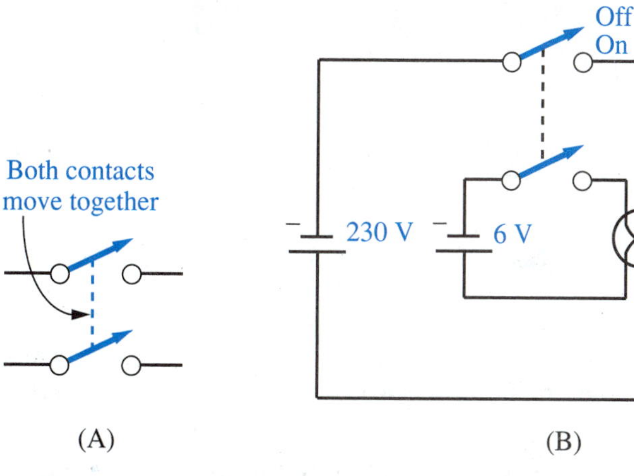

FIGURE 8-17 (A) Symbol for double-pole switch. Dashed line indicates a shared actuating mechanism. (B) In the double-pole configuration, the poles can switch completely different voltage levels.

SAFETY ADVICE

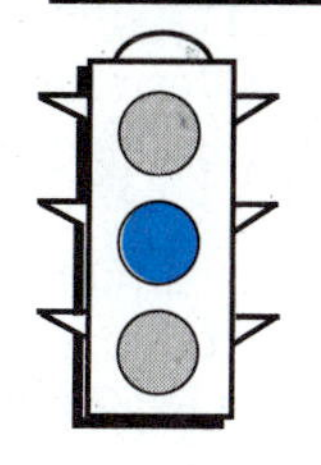

If a circuit repeatedly blows its fuse, it might not be due to a short-circuit overcurrent. It may simply be that too many load devices have been connected. In this situation, some people are tempted to replace the fuse with one having a higher current rating. Never do this.

If you permit a current that is too large for the circuit conductors to safely handle, overheating is likely to produce high-temperature spots. This is the cause of electrical fires.

8-7 SWITCHES—ELECTRICAL CONFIGURATION

7

The **single-pole** switch has just a single contact controlling a single circuit. A **double-pole** switch has two electrically separate contacts controlled by the same mechanical actuator. The schematic symbol for a double-pole switch is shown in Figure 8-17(A).

The electrical independence of the two poles is made clear in Figure 8-17(B). One pole is switching a high-power circuit, while the other pole is switching a 6-V indicator lamp.

Three poles and more are also available. Figure 8-18(A) shows a three-pole switch schematic. Four poles are shown in Figure 8-18(B).

Double-Throw Switches

All the switches in Figures 8-17 and 8-18 are called **single-throw.** This means that each pole has a single electrical contact that changes state when the

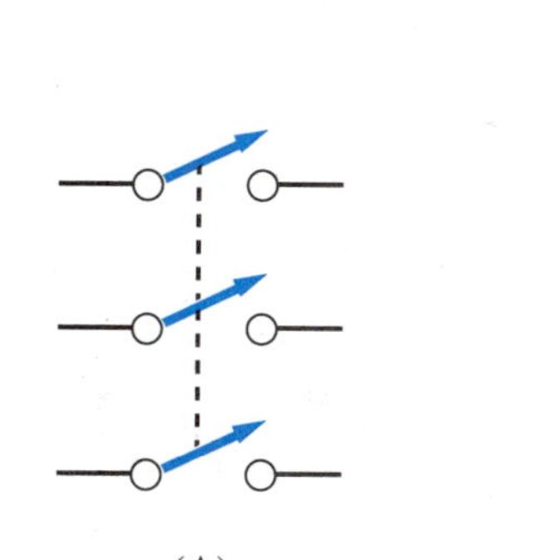

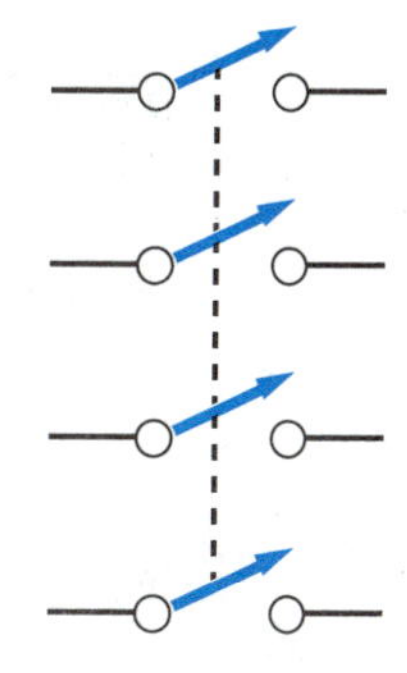

FIGURE 8-18 (A) Three-pole switch. (B) Four-pole switch

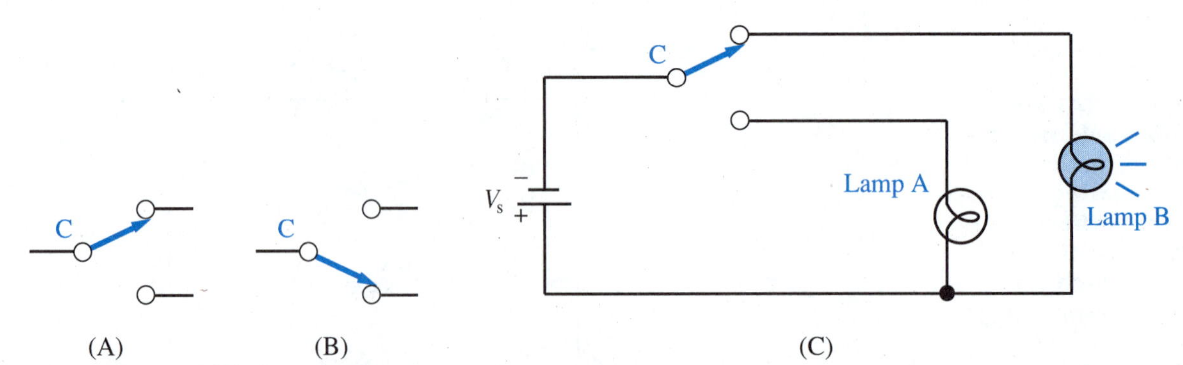

FIGURE 8-19 Single-pole double-throw switch. (A) In the up position. (B) In the down position. (C) In the up position, lamp B is lighted and lamp A is off. If the switch were actuated, lamp A would be lighted and lamp B would turn off.

switch is actuated. A **double-throw** switch is different. As shown in Figure 8-19, each pole has two electrical contacts changing states.

In Figure 8-19(A), a single-pole double-throw (SPDT) switch is in its up position. There is a closed contact between the common (C) terminal on the left and the top terminal on the right. At the same time, there is an open contact between the C terminal and the bottom terminal on the right.

When the switch is actuated, it moves to its down position, shown in Figure 8-19(B). Both contacts change state, with the C terminal to top terminal becoming open, and the C terminal to bottom terminal closing. An example circuit using a SPDT switch is shown in Figure 8-19(C).

Multiple-pole double-throw switches are readily available. Figure 8-20(A) is the schematic symbol for a double-pole double-throw (DPDT) switch. A three-pole double-throw (3PDT) unit is shown in Figure 8-20(B).

EXAMPLE 8-1

The familiar home-stairway lighting circuit uses two SPDT switches. One is mounted at the top of the stairs and one at the bottom. Draw the schematic diagram of such a stairway circuit.

SOLUTION

Either switch must be able to complete the circuit or interrupt the circuit . This is accomplished by the arrangement of **Figure 8-21.**

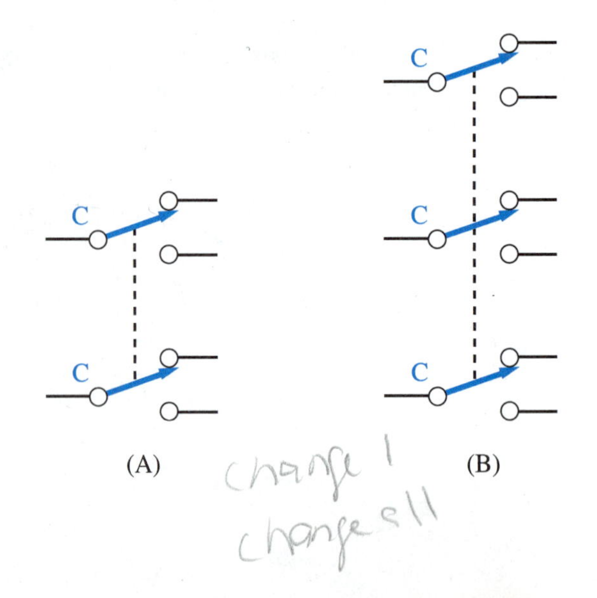

FIGURE 8-20 (A) Double-pole double-throw switch. (B) Three-pole double-throw switch

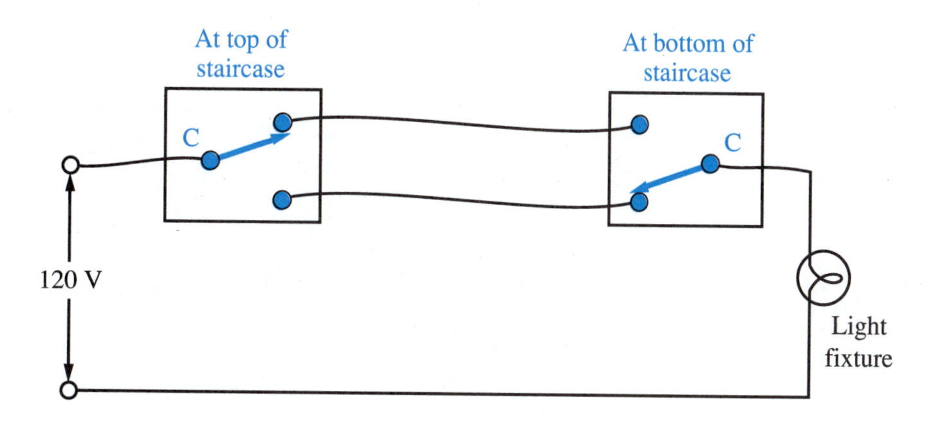

FIGURE 8-21 Lighting circuit that can be turned on or off from two locations.

ELECTRONIC SCOREBOARD

Modern stadium scoreboards do much more than give the scores of Home and Visitors. The scoreboard in this dome stadium can display high-quality images of action as it occurs on the field.

There are many video cameras stationed throughout the stadium. Every camera signal is sent by cable to the central control room in the stadium's upper deck. There they are displayed on monitor screens at the main control console, where the scoreboard operators see them simultaneously, as shown in the photos. All the while, every signal is automatically being recorded on video tape.

The operators can select any one of the real-time camera signals to appear on the scoreboard, or they can show a video tape replay of some particular action. At other times, the board serves its usual purpose of giving the current team scores or the standings of the competitors. With an operator keyboarding the action on the field into a computer, up-to-the-minute sports statistics can also be shown, as well as late-breaking news and other messages to the crowd.

Courtesy of Pontiac Silverdome.
Photos copyright 1991 by Douglas G. Ashley

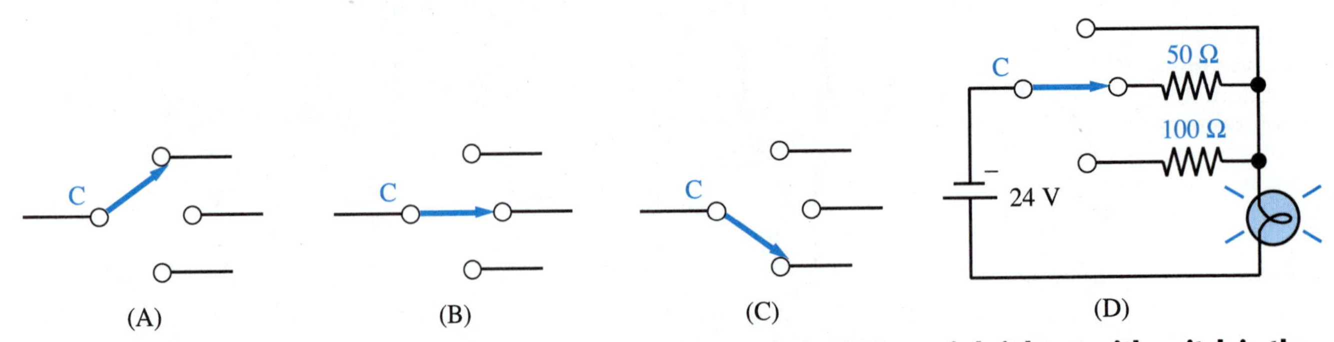

FIGURE 8-22 (A), (B), and (C) Positions of a three-position switch. (D) Lamp is brightest with switch in the top position, dimmest in the bottom position.

Multiple-Position Switches

Some switches have more than two positions. For example, a single-pole three-position switch is shown in Figure 8-22. The common terminal can be contacted with any one of the three terminals on the right, as shown in Figures 8-22(A), (B), and (C). An application is suggested in Figure 8-22(D), allowing the lamp's brightness to be changed.

EXAMPLE 8-2

A three-pole four-position switch is symbolized in Figure 8-23(A). An application is shown in Figure 8-23(B). Describe how this circuit works.

SOLUTION

The four positions are numbered from top to bottom, as shown in Figure 8-23(B).
In position No. 1, both lamp A and lamp B are off.
In position No. 2, lamp A is on alone.
In position No. 3, lamp B is on alone.
In position No. 4, both lamps A and B are on together.

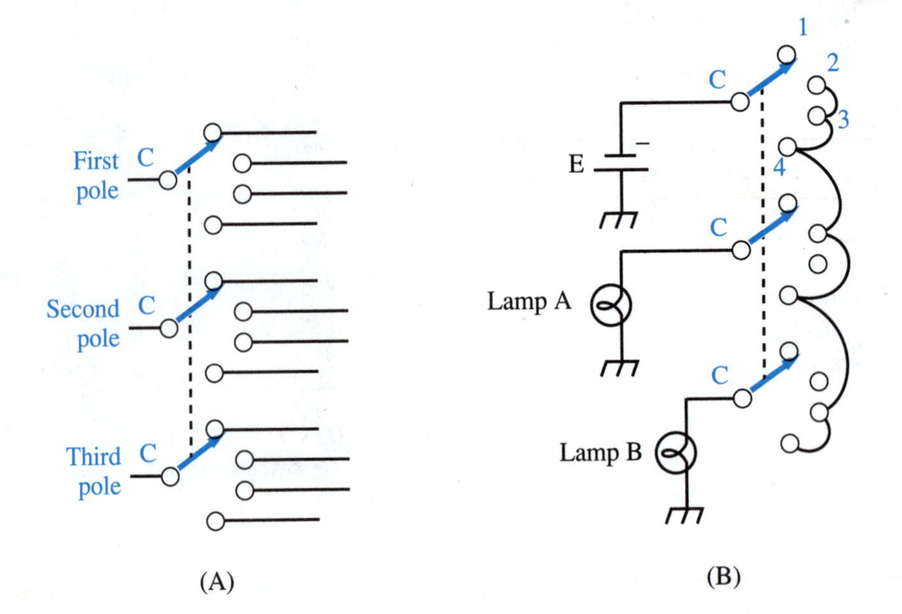

FIGURE 8-23 (A) Three-pole four-position switch. (B) Wiring the switch for Off, A, B, Both.

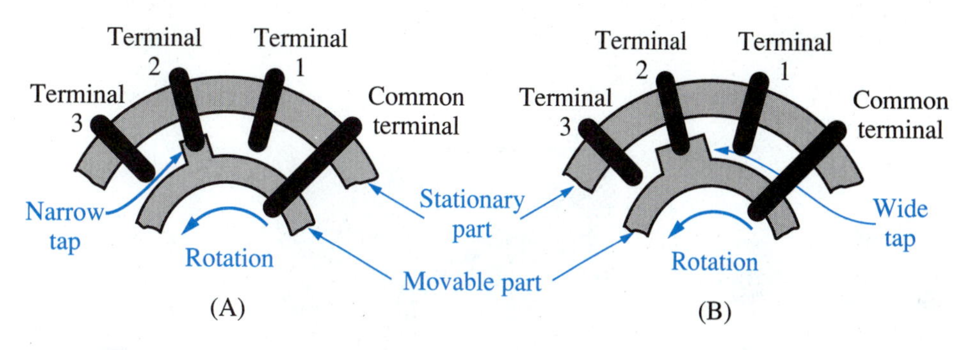

FIGURE 8-24 (A) Nonshorting switch construction. Also called break-before-make. (B) Shorting switch construction. Also called make-before-break.

Switch Ratings

Switches are rated for both current and voltage. These specify the maximum values that the switch can interrupt without damage to the contact surfaces. For example, a switch with a rating of 2 A, 120 V can safely interrupt no more than 2 A when the source voltage is 120 V.

TECHNICAL FACT

Switch ratings are generally valid only for resistive loads. If a load has inductance, its current is more difficult to interrupt.

Loads like motors, solenoid coils, and transformers contain inductance. When a switch is used with such loads, its current must be downrated. We will study the inductance idea in Chapter 14.

Most multiposition switches are the **nonshorting** variety. This means that as the switch is moved to a new position the common terminal breaks contact with the old terminal just before it closes the contact to the new terminal. Such **break-before-make** action is illustrated in Figure 8-24(A).

Some multiposition switches are the **shorting** type. With them, the common terminal makes contact with the new terminal before it breaks contact with the old terminal. In other words, the common touches both terminals temporarily. This **make-before-break** construction is illustrated in Figure 8-24(B).

Shorting switches are used whenever the voltage source requires that some load resistance be connected at all times. Certain voltage sources that contain transformers along with sensitive semiconductor devices often have this requirement.

8-8 SWITCHES—MECHANICAL ACTUATION

Knife Switches

8

A knife switch with a movable arm, or blade, is narrow enough (sharp enough) to spread apart the sides of the spring clip.

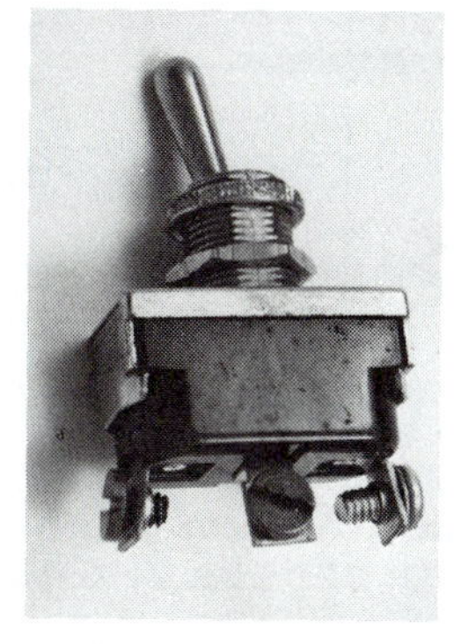

(A)

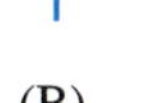

C

(B)

FIGURE 8-25 (A) Toggle switch with screw terminals. (B) Symbol for center-Off switch.

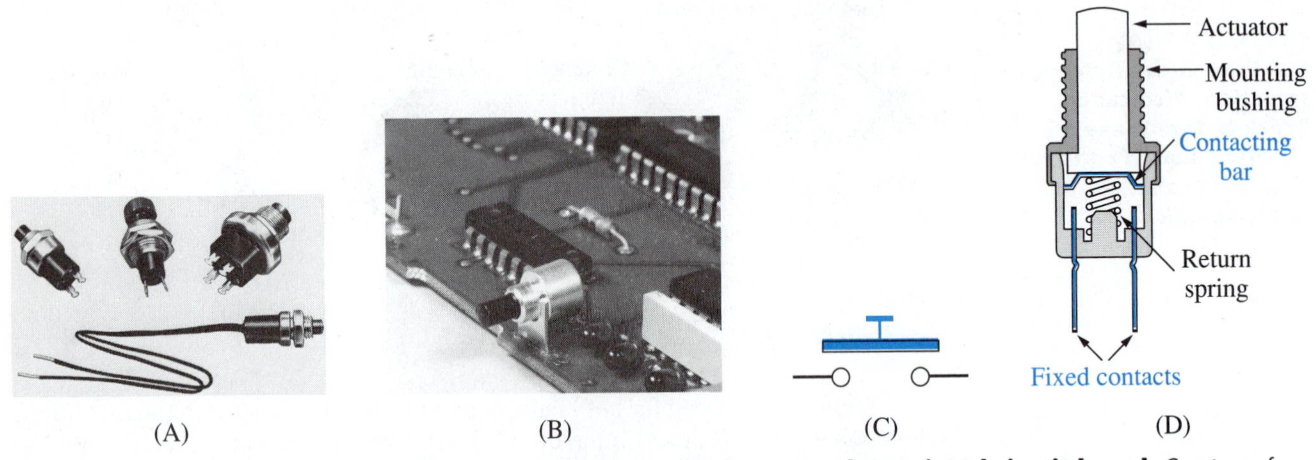

FIGURE 8-26 (A) Pushbutton switches. (B) Pushbutton switch mounted on printed circuit board. *Courtesy of Grayhill Inc.* **(C) Symbol for N.O. PB switch. (D) Cross-section of a push-button switch**

Toggle Switches

A toggle switch has an external handle for actuating the contact surfaces inside a sealed enclosure. Figure 8-25(A) is a photograph of a panel-mount toggle switch.

The symbol for a center-Off type panel-mount toggle switch is shown in Figure 8-25. With the handle in the center position, neither outside terminal contacts the common terminal. It is symbolized as shown in Figure 8-25(B).

Pushbutton Switches

Pushbutton (PB) switches are shown in Figures 8-26(A) and (B). The symbol for a SPST pushbutton switch is given in Figure 8-26(C). Visualize the horizontal contact bar moving downward as the button is pushed.

The symbol in Figure 8-26(C) is for a normally open (N.O.) PB switch. The term **normally open** means that the contact is electrically open when the switch is not actuated (in this case, when the button is not pushed). The switch closes when it is actuated by pressing the pushbutton. Figure 8-26(D) is a cross-sectional diagram showing the structure of a N.O. PB switch.

A **normally closed** (N.C.) PB switch is symbolized in Figure 8-27(A). Again, visualize the contact bar moving downward when the button is pressed. The bar is forced upward by an internal spring when the button is released. Therefore, the switch contact is electrically closed when the switch is not actuated. It opens when the switch is actuated by pressing the button.

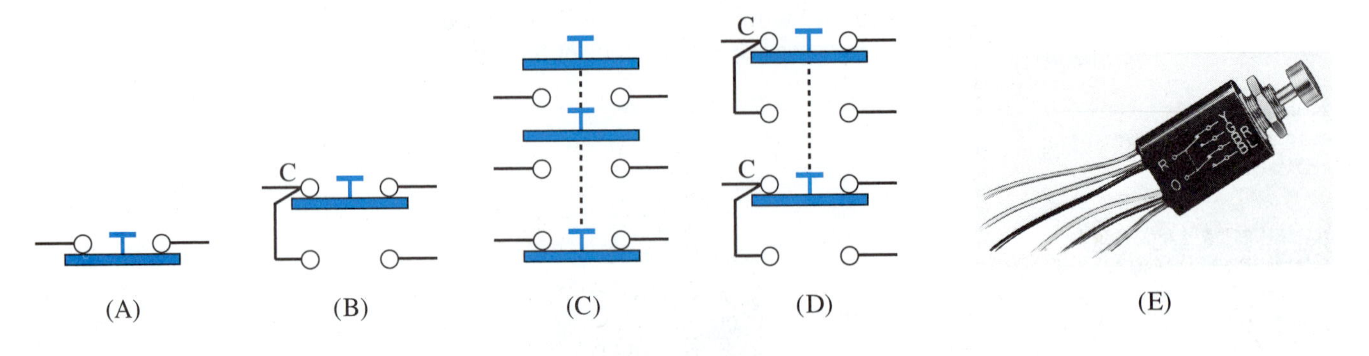

FIGURE 8-27 (A) Symbol for N.C. PB switch. (B) Single-pole double-throw PB switch. (C) Three-pole single-throw, with 2 poles N.O. and 1 pole N.C. (D) DPDT PB switch symbol. (E) Actual DPDT PB switch. *Courtesy of Switchcraft, Inc., a Raytheon Company*

FIGURE 8-28 Various rotary tap switches. *Courtesy of Grayhill Inc.*

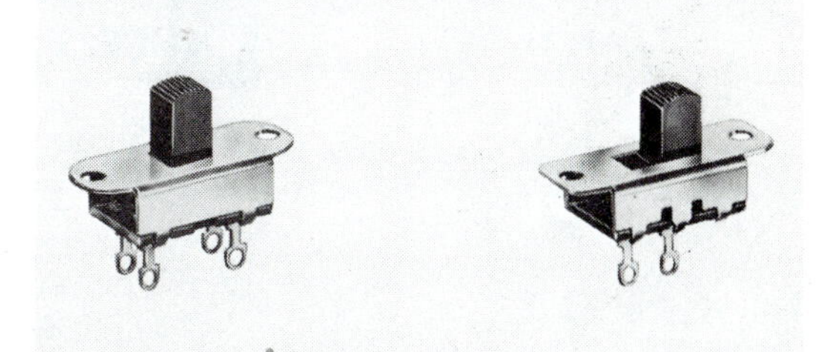

FIGURE 8-29 Slide switches. *Courtesy of Switchcraft, Inc., a Raytheon Company*

PB switches may be double-throw types, with one N.O. contact and one N.C. contact. The common terminal is usually drawn as two terminals jumpered together. This is shown in Figure 8-27(B).

Of course, PB switches may have multiple poles, as symbolized in Figures 8-27(C) and (D), and shown in Figure 8-27(E).

Most PB switches are the **momentary-contact** kind. This means that the contact(s) returns to its normal condition when the button is released. A few PB switches are the **fixed-closure** kind. With a fixed-closure PB switch, the contact(s) remains in its new state when the button is released. You must push the button a second time to get the contact(s) back into normal condition.

The distinction between momentary-contact and fixed-closure switches also applies to toggle switches, as well as certain other types of switches.

Other Switch Types

A rotary-tap switch has a center shaft that rotates like a potentiometer. This moves a common tap terminal from one position to the next, as suggested in Figure 8-24. The rotary tap structure is popular for switches with a large number of positions. Some units have twelve or more positions. Several rotary tap switches are shown in Figure 8-28.

Slide switches are actuated by moving a rectangular knob within an opening in the switch enclosure. They usually have two positions, but are available with three or more positions. Figure 8-29 shows photographs of slide switches.

A limit switch is operated by a slow-moving cam, as suggested in Figure 8-30(A). The schematic symbols are shown in Figures 8-30(B) and (C). Visualize the pivot bar as pulled downward by gravity. It lifts up when actuated by the cam.

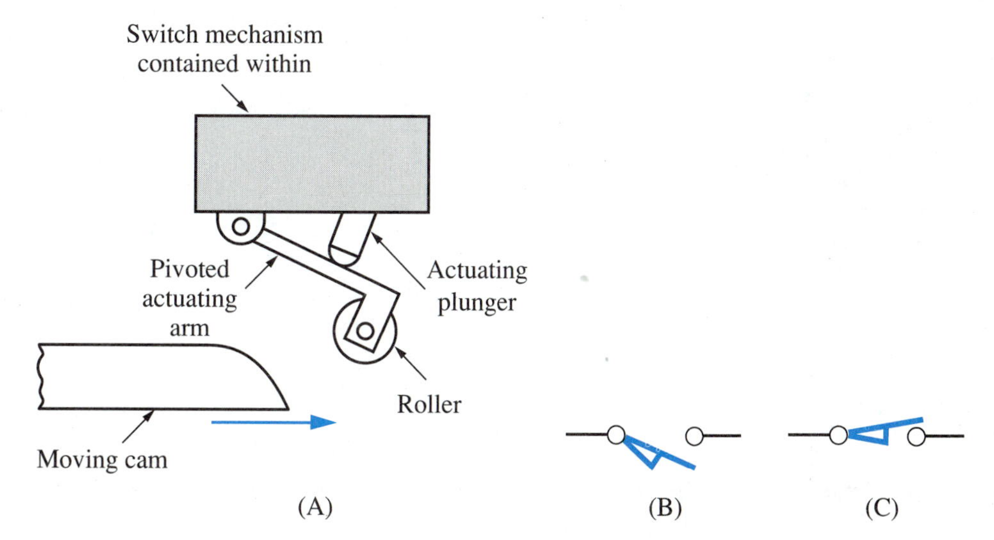

FIGURE 8-30 (A) Cam operation of a roller-arm limit switch. (B) Symbol for N.O. limit switch. (C) Symbol for N.C. limit switch.

✓ SELF-CHECK FOR SECTIONS 8-6, 8-7, AND 8-8

15. A higher AWG number means________wire. (Answer thicker or thinner.) [6]
16. For cross-sectional area of a wire, the basic metric unit is the__________ __________; a popular unit in the U.S. is the_________. [6]
17. Under normal conditions, what is the current-carrying capacity of AWG #12 copper wire? [6]
18. What is the approximate current-carrying capacity of AWG #12 aluminum wire? [6]
19. T-F If the wire insulation is thicker, the wire can be used at higher voltage levels.
20. What is the difference between a single-throw switch and a double-throw switch? [7]
21. Draw the schematic symbol for a SPDT switch. [7]
22. Draw the schematic symbol for a four-pole five-position rotary-tap switch. Then show how to wire it to perform the following switching sequence for three lamps. [7]

Position	Lamps Turned On
1	None—all Off
2	A only
3	B only
4	B and C
5	A, B, and C

Use Figure 8-23 to help you.

TROUBLESHOOTING TIP

When a switch is open, it has the job of stopping current from reaching the load. When it is closed, a switch is supposed to allow current to pass freely to the load. Occasionally, a switch may fail in one or the other of its functions. The working of a switch can be tested very easily without removing it from the circuit. Just place a voltmeter in parallel with it, as shown in the figure.

Figure (A) shows the switch closed. In the closed condition, the voltage across the switch terminals should be nearly zero. This is because there should be nearly zero resistance from one terminal to the other. By Ohm's law, any amount of current multiplied by 0 Ω gives 0 V.

Testing switch operation with a voltmeter. These voltages show proper operation. If either voltage were much different, it would indicate trouble with the switch.

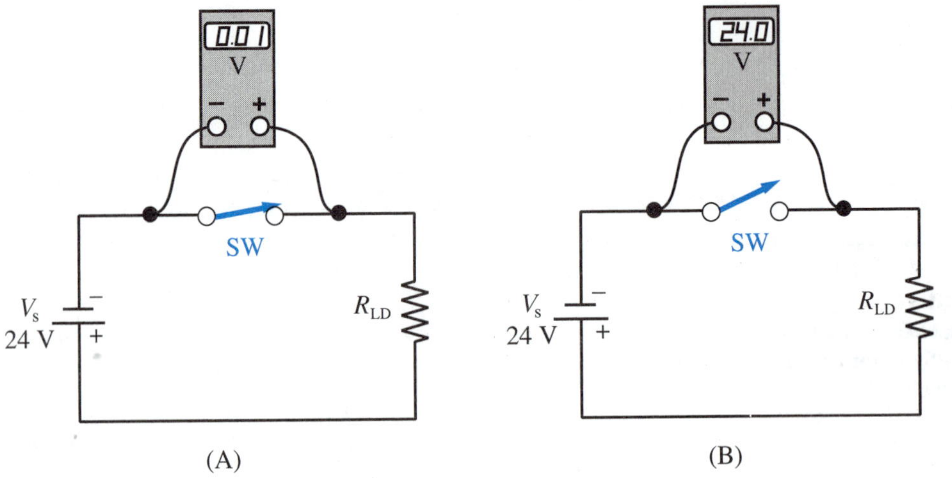

If the switch voltage measures much different from zero, the switch is defective. Most switches cannot be repaired. They are just replaced. However, some large, expensive switches have their contact surfaces accessible for cleaning with sandpaper.

Continued on page 129

Figure (B) shows the switch in the open position. Now the voltage across the switch should be the same as the source voltage, 24 V in this case. This is because the switch should allow zero current to pass. With zero current, the load voltage should be nearly zero, according to Ohm's law.

With 0 V across it, the load acts like just a plain conductor, connecting the positive terminal of the voltmeter to the positive terminal of the source. Therefore, the voltmeter is effectively in parallel with the source. If the switch voltage measures much different from 24 V, the switch is defective and must be replaced.

8-9 RELAYS

A **relay** is a magnetically operated switch. Figure 8-31(A) shows a structural diagram of a relay. A photograph of a four-pole relay (two N.O. contacts and two N.C. contacts) is shown in Figure 8-31(B).

If the relay **coil** is deenergized, no current through it, then no magnetic force is created and the tension spring holds the movable arm in the position shown in Figure 8-31(A). The contact surfaces are not touching, so the **contact** is open. This is called the normal state of the relay.

If a source energizes the coil by forcing current through it, the cylindrical iron core exerts a magnetic attraction force on the movable arm, which is also made of iron. We will study magnetic forces in Chapter 13. The magnetic force causes the movable arm to pivot, making the contact surfaces touch each other. The contact remains in this closed condition as long as the coil remains energized.

One way of symbolizing a relay is shown in Figure 8-32(A). Another schematic symbol that is popular in industrial control is shown in Figure 8-32(B). If there are

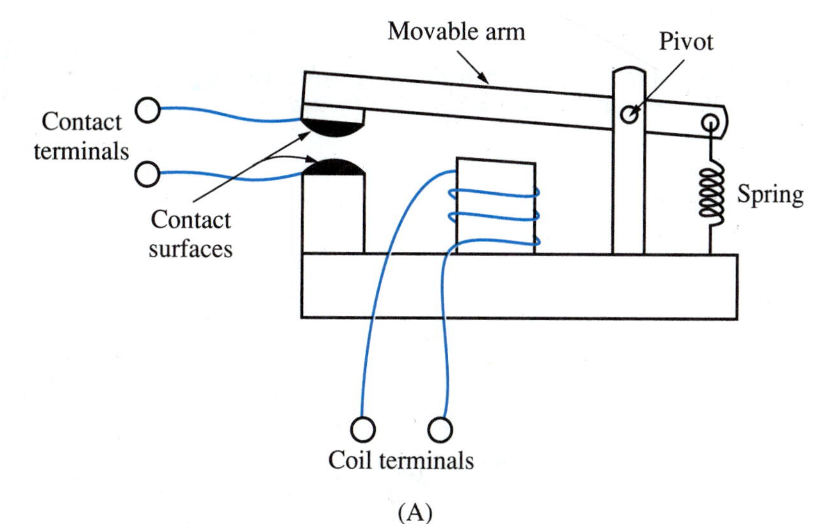

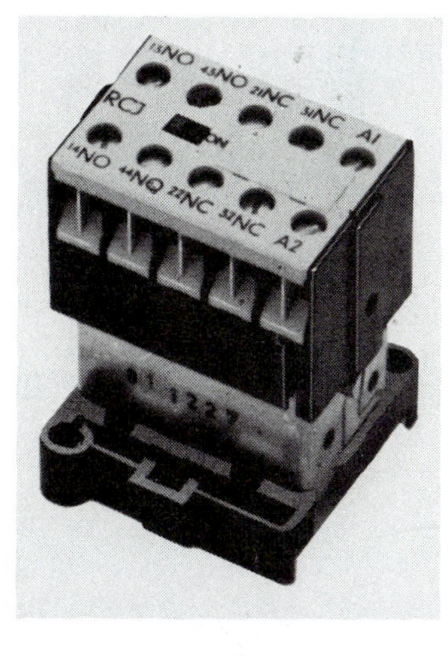

FIGURE 8-31 (A) Basic structure of an electromagnetic relay. (B) Four-pole relay with two normally open contacts and two normally closed contacts. *Courtesy of Siemens Energy & Automation, Inc.*

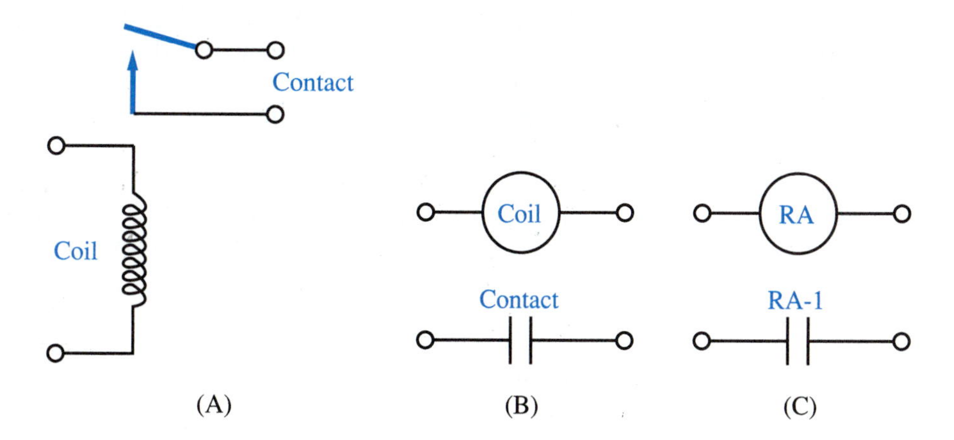

FIGURE 8-32 (A) Relay symbol. (B) Symbol used in industrial systems. (C) Labeling a relay coil and the contact that it controls. If this were a double-pole relay, the second contact would be labeled RA-2.

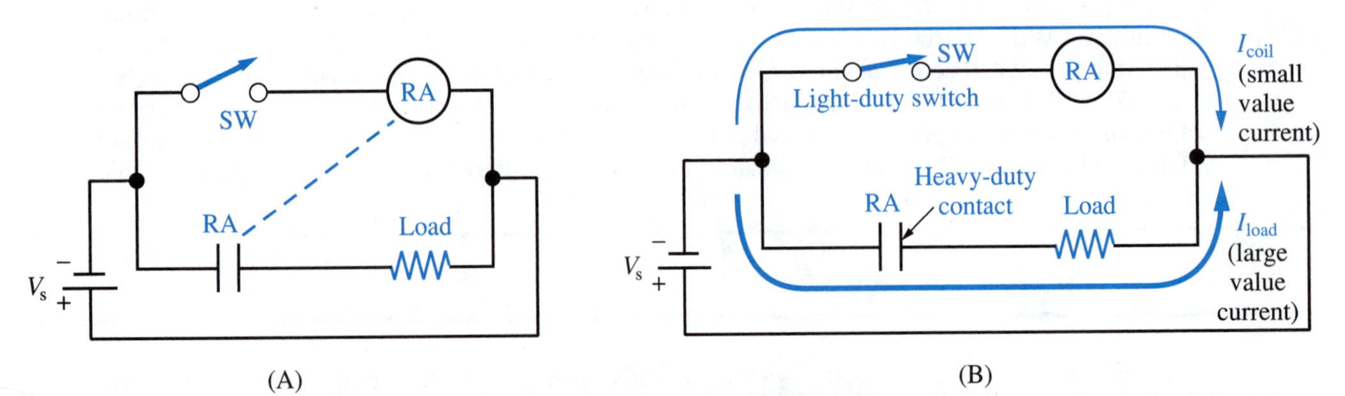

FIGURE 8-33 (A) Relay circuit. In some diagrams, the coil and its contact(s) are joined by dashed line(s). (B) Relay application of controlling a heavy-duty load with a light-duty manual switch.

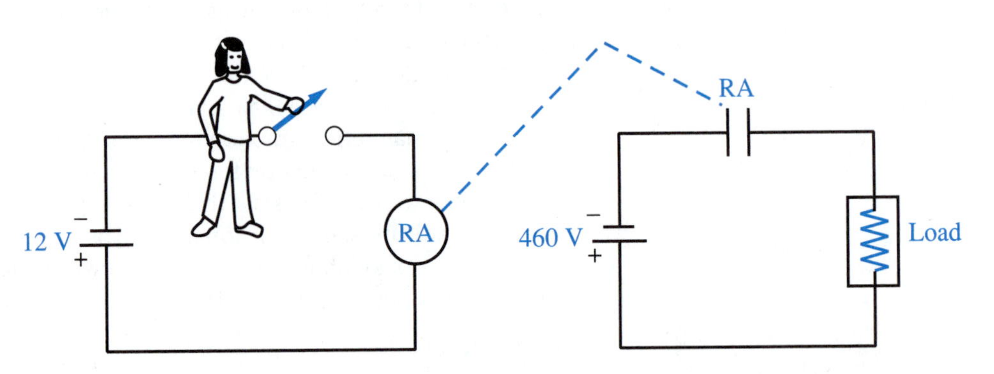

FIGURE 8-34 With a low-voltage relay coil operating a high-voltage contact, the operator avoids placing a hand in the high-voltage circuit.

several relays in a circuit, they are often labeled with letters. The first relay could be labeled RA, the second RB, and so on. Both the coil and the contact(s) are labeled, as shown in Figure 8-32(C).

Relay Uses

A simple relay circuit is shown in Figure 8-33(A). When manual switch SW is closed, coil RA energizes. That causes contact RA to close, energizing the load.

12

Relay operation of a load is better than direct manual switching of the load under certain circumstances. For example, in Figure 8-33(B) the load current is very large. If that current is handled by a heavy-duty relay contact, the human operator has an easy job. She only has to actuate a light-duty switch to energize the low-current coil.

Another important relay application is shown in Figure 8-34. The load is driven by 460 V, a very dangerous voltage. By reaching into just the low-voltage coil circuit to actuate the manual switch, the operator is safe.

Other Contact Arrangements

In Figures 8-31 through 8-34, the relay contacts are normally open (N.O.). That is, the contact is open until the coil is energized, and then it closes. Relay contacts can also be normally closed. A N.C. relay contact is symbolized in Figure 8-35.

Naturally, relay contacts can be double-throw. This is symbolized in Figure 8-36.

Many relays have two or more poles. A schematic symbol for a double-pole double-throw relay is given in Figure 8-37.

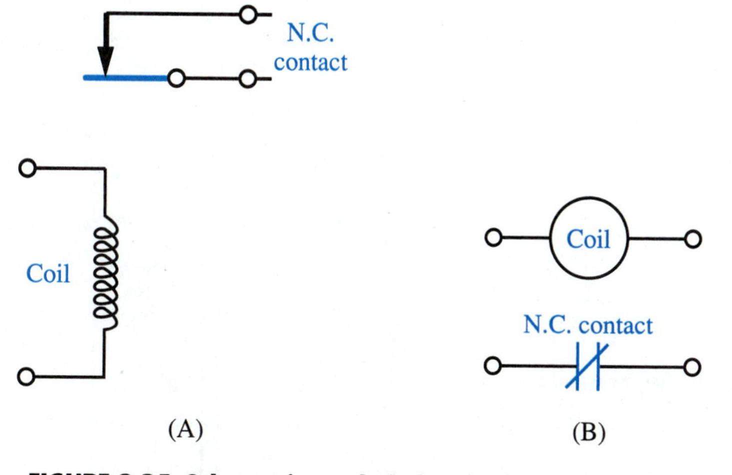

FIGURE 8-35 Schematic symbols for single-pole N.C. relay

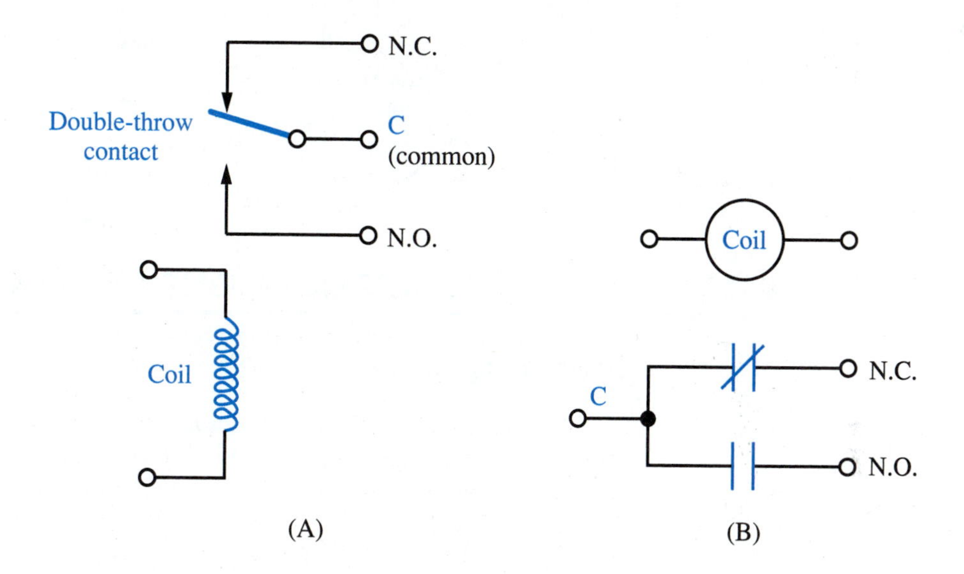

FIGURE 8-36 Schematic symbols for single-pole double-throw relay

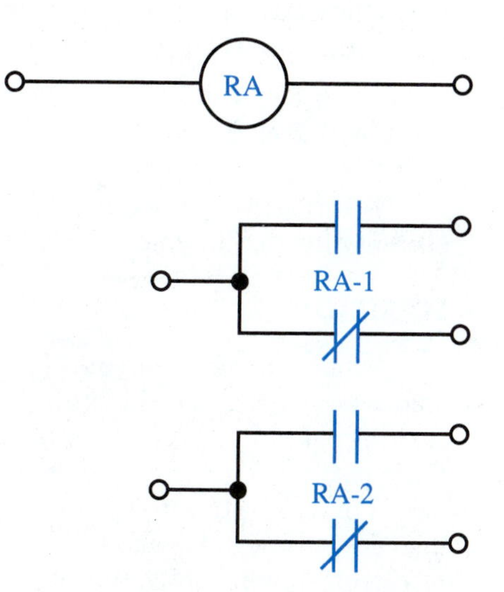

FIGURE 8-37 Symbol for DPDT relay

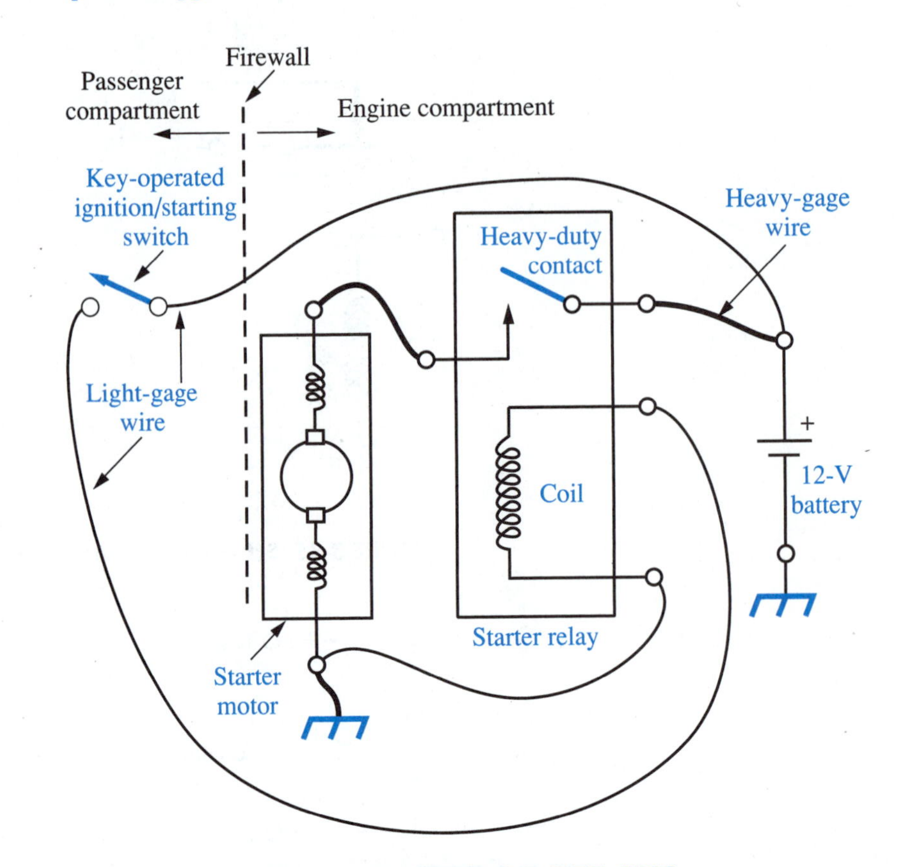

FIGURE 8-38 Automobile starting circuit

8-10 PRACTICAL APPLICATIONS OF RELAYS

Starter Relay in a Car

The starter motor of a car carries a very large current, usually over 50 A. This amount of current cannot be routed all the way from the battery to the ignition switch, then back to the engine. For one thing, the distance is too great. For another, the small key-operated ignition switch can't handle current of that magnitude. The solution to this problem is shown in Figure 8-38.

TECHNICAL FACT

In automobiles, the negative terminal of the battery is connected to the engine and chassis, which serves as a shared return path for all the electric loads on the car. Therefore, it is more convenient to think in terms of conventional current in automotive circuits.

EXAMPLE 8-3

Describe the operation of the circuit in Figure 8-38. Use the conventional current direction in your description.

SOLUTION

When the key-operated ignition / starting switch is closed, a small amount of current flows by this path: from the battery to the switch on the steering column, back into the engine compartment and through the coil of the starter relay, returning to the battery by the chassis ground.

This closes the starter relay's heavy-duty N.O. contact. A large current then flows through the short heavy-gage wire from the battery to the relay contact. From there it flows through the starter motor to chassis ground and returns to the battery.

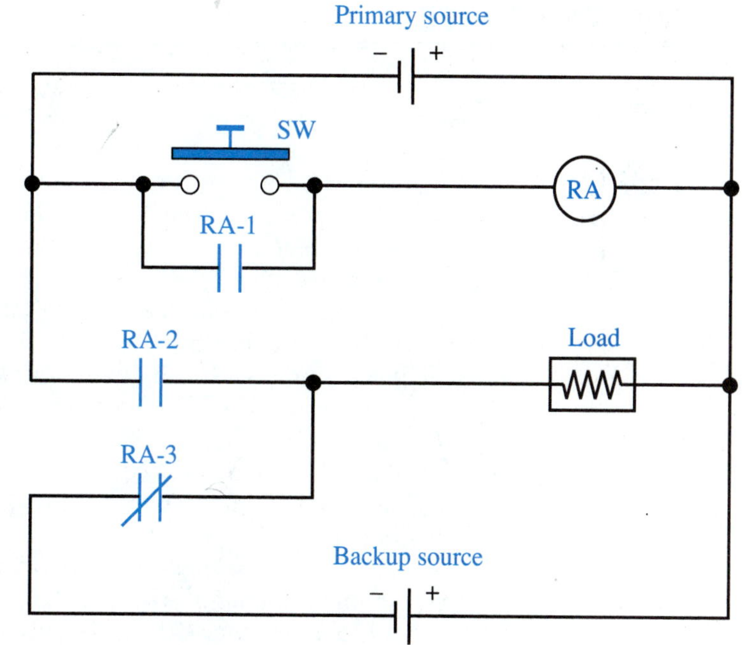

FIGURE 8-39 Automatic backup power circuit

Automatic Backup Power

Relays can be used to perform simple decision-making operations, also called logic operations. An example is shown in Figure 8-39. The operator momentarily presses pushbutton switch SW. The primary voltage source delivers current through SW into the RA coil. Energizing RA causes N.O. contact RA-1 to close, furnishing an alternate current path around SW to the coil. Therefore, the operator can release the N.O. PB switch and the RA coil remains energized.

At the same time, N.O. contact RA-2 also closes, allowing the primary voltage source to drive the load. N.C. contact RA-3 opens, disconnecting the backup source. The circuit continues to function like this as long as the primary source remains operational.

EXAMPLE 8-4

Suppose that the primary source fails in Figure 8-39. Describe how this circuit keeps the load energized.

SOLUTION

If the primary source fails, coil RA deenergizes. That causes contact RA-2 to return to its open state, which disconnects the load from the failed primary source. At the same time, RA-3 returns to its normal closed state, immediately allowing the backup source to begin driving the load. Thus, the load's power is never interrupted.

✔ SELF-CHECK FOR SECTIONS 8-9 AND 8-10

23. Explain the difference between a relay and a manual switch. [11]
24. T-F A relay allows us to control a large current by manually switching a small current. [11, 12]
25. To switch a high-voltage circuit safely, we install a relay__________in a low-voltage circuit, and the relay__________goes in the high-voltage circuit. [11, 12]
26. Explain the difference between an N.O. relay contact and an N.C. relay contact. [9]
27. Draw the industrial schematic symbols for both types of relay contact. [9]

SAFETY ADVICE

Never jam a relay in the mechanically actuated position. Sometimes, during troubleshooting, it may be necessary to artificially force a relay to actuate, even though there is no coil voltage present to energize it. If you must do this, push on the relay mechanism with an insulated tool. After you have made the observation that you need, release the mechanism, allowing the spring to return the relay to its normal deenergized position.

If your observation has to be made at some distance from the relay's location, then the only safe procedure is to have two people on the job, communicating by two-way radio.

The problem with jamming a relay in the actuated position and then wandering off some distance to make an observation, is that you might get distracted and forget to return to unjam the relay. In fact, this will inevitably happen sooner or later, if you allow yourself to get into this bad habit.

Of course, leaving a relay indefinitely jammed prevents it from performing its function. This may bring danger to personnel, to the circuit, or to machinery that is controlled by the circuit.

Inserting paper or cardboard betwen normally-open contacts to prevent electrical continuity is another relay shortcut that you might see someone using. This practice should be avoided. If you must prevent continuity during troubleshooting, unscrew the wire terminal from one side of the contact and cover it with electrical tape.

SUMMARY OF IDEAS

- An overcurrent can result from a dead short circuit or a partial short circuit.
- A fuse has an element that melts when an overcurrent occurs. The fuse is destroyed.
- A circuit breaker opens its contact when an overcurrent occurs. It is not destroyed, but can be manually reset after the overcurrent is eliminated.
- In a three-wire grounded wiring system, a short circuit to an enclosure causes the fuse or circuit breaker to blow. This protects us against electric shock.
- A GFI opens its contact if any partial short circuit to ground occurs. This provides even better protection against shock.
- Wires and cables are identified by an AWG number, which specifies their maximum current capacity.
- A single-pole switch controls just one circuit. A double-pole switch controls two separate independent circuits, and so on.
- A double-throw switch has two contacts sharing a common terminal, per pole. One contact closes when the other contact opens.
- There are many mechanical actuation methods for switches. Some of the most common are knife, toggle, pushbutton, slide, rotary, and cam-operated roller.
- A relay is a switch that is controlled by an electromagnetic coil.
- The classic relay application is to indirectly control a load with an isolated light-duty manual switch.

CHAPTER QUESTIONS AND PROBLEMS

1. Which is more expensive, a fuse or a circuit breaker? [2, 3]
2. What is the difference between a dead short circuit and a partial short circuit? Which one is more certain to trip a circuit breaker? [1, 3]

3. If a fuse is carrying an overcurrent three times as large as its rated current, is that considered a slight, moderate, or severe overcurrent? [2]
4. In Question 3, suppose the fuse is a normal type. Approximately how much time will elapse before it blows? [2]
5. Repeat Question 4 for a fast-blow type. [2]
6. Repeat Question 4 for a slow-blow type. [2]
7. Explain why fuses have voltage ratings. [2]
8. What is the difference between a thermal type of circuit breaker and a magnetic type? [3]
9. Explain why connecting a grounding wire to a load enclosure frame makes it impossible for the frame to become a shock danger. [4]
10. A ground-fault interrupter trips not because of an overcurrent but because of a __________ in the two supply-wire currents. [5]
11. T-F AWG#10 wire is thicker than AWG #14 wire. [6]
12. Based on your answer to Question 11, which wire gage has lower resistance per unit length? [6]
13. Figure 8-21 shows how a load can be controlled from either of two locations using two SPDT switches. Show how to control a load from any of three locations using two SPDT switches and one DPDT switch. [7]
14. Explain the difference between a normally open PB switch and a normally closed PB switch. Draw the schematic symbol for each. [8, 9]
15. Draw the schematic symbol for a single-pole double-throw PB switch; label the terminals. [8, 9]
16. Explain the difference between a break-before-make rotary switch and a make-before-break rotary switch. [7]
17. Draw the schematic symbol for a four-pole six-position rotary-tap switch. Then show how to wire it to perform the following switching sequence for three lamps. [7]

Position	Lamps Turned On
1	None—all Off
2	A only
3	B only
4	C only
5	A and B
6	A and C

18. Show how to wire a four-pole eight-position rotary tap switch to perform the following switching sequence for three lamps. [7]

Position	Lamps Turned On
1	None—all Off
2	A only
3	B only
4	C only
5	A and B
6	A and C
7	B and C
8	A, B, and C

19. Using a SPDT limit switch, draw the schematic diagram of a circuit that will do the following: [7, 8, 9]
    a) When the limit switch is mechanically released, lamp A is lighted and lamp B is deenergized.
    b) When the limit switch is mechanically actuated by its cam, lamp A becomes deenergized and lamp B is lighted.
20. Using a drawing, explain how a relay can be used to control a high-current load with a light-duty switch. [11, 12]

21. Using a drawing, explain how a relay can be used to control a high-voltage load with a switch located in a low-voltage circuit. [11, 12]
22. Using a drawing, explain how a relay can be used to control a high-current load with a switch located a far distance away from the load. [11, 12]
23. The normal state of a relay is with the coil__________. In this state, any N.O. contact is__________; any N.C. contact is__________. [9]
24. The nonnormal state of a relay is with the coil__________. In this state, any N.O. contact is__________; any N.C. contact is__________. [9]
25. In the backup power circuit of Figure 8-39, why does the RA coil remain energized after SW is released? [11]

CHAPTER 9

SERIES CIRCUITS

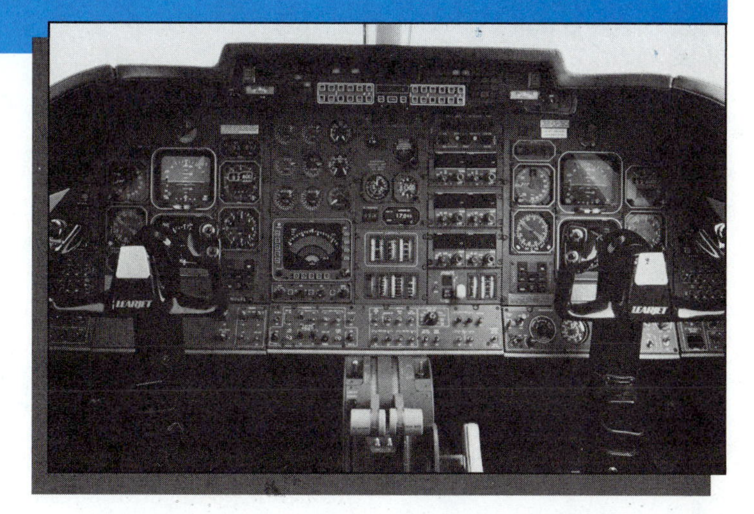

Modern airplane cockpits contain many electrical measuring instruments. *Courtesy of Learjet Inc.*

OUTLINE

9-1 Current and Voltage in Series
9-2 Ohm's Law Applied to a Series Circuit
9-3 Kirchhoff's Voltage Law
9-4 Internal Resistance of a Source
9-5 Potentiometer Voltage Dividers
9-6 Troubleshooting Series Circuits

NEW TERMS TO WATCH FOR

series circuit
total resistance
Kirchhoff's voltage law
internal resistance
no-load output voltage
open-circuit output voltage
full-load current
full-load output voltage
voltage divider
volume-control pot

When two or more load devices are connected together in-line, the circuit is called a series circuit. We will study the characteristics of series circuits in this chapter.

After studying this chapter, you should be able to:

1. Recognize a series circuit.
2. Use the rules regarding current and voltage in a series circuit.
3. Apply Ohm's law to an entire series circuit or any portion of a series circuit.
4. Apply Kirchhoff's voltage law to series circuits.
5. Explain how a real voltage source differs from an ideal voltage source.
6. Describe the use of a potentiometer as a variable voltage divider.
7. Troubleshoot a series circuit to locate a short or an open.

FIGURE 9-1 A series electrical circuit is like a series fluid circuit. The flow variable is the same everywhere within the circuit.

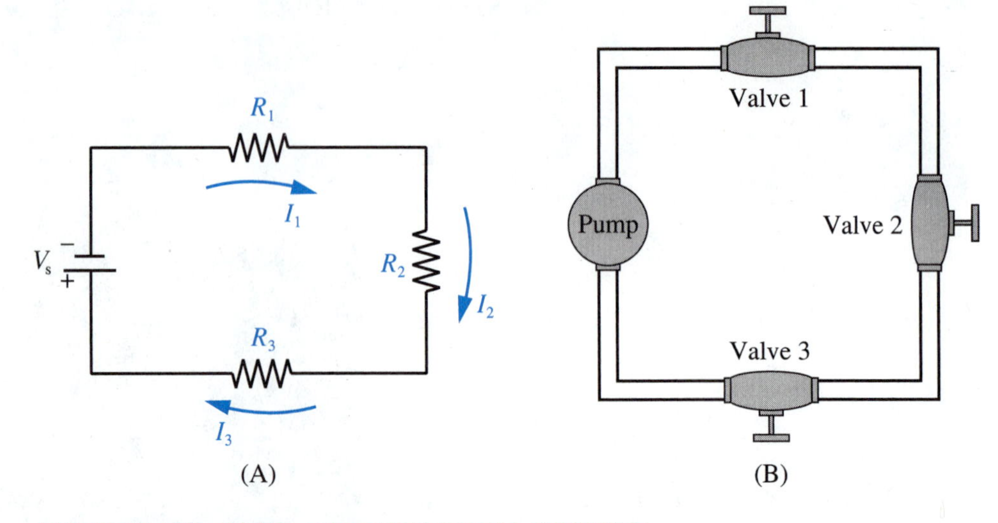

9-1 VOLTAGE AND CURRENT IN SERIES

Figure 9-1(A) shows a three-resistor **series circuit**. Compare the current through R_1 to the current through R_2. All the current passing through R_1 must continue on through R_2, since there is no wire-junction between R_1 and R_2 that would provide an alternate path back to the source. Therefore, the current through R_1 is the same as the current through R_2.

By identical reasoning, the currents through R_2 and R_3 are the same. Generalizing this to all series circuits, we can say:

TECHNICAL FACT

In a series circuit, current is the same everywhere.

In the circuit of Figure 9-1(A), there are four locations where an ammeter could be inserted. The ammeter would measure the same value of current in any one of the four locations.

To understand the equal currents in a series electrical circuit, make an analogy to a series fluid circuit like the one in Figure 9-1(B). For instance, it is easy to see that the water flow through valve 1 will always be the same as through valve 2, since there is no piping tee between those valves. The same holds for valves 2 and 3. Electric current is a "through" variable, like liquid flow, and it behaves the same way.

The individual voltages in a three-resistor series circuit are labeled in Figure 9-2. Since all resistors carry the identical current, I, we expect from Ohm's law that

FIGURE 9-2 In a series circuit, larger resistances have larger voltages and smaller resistances have smaller voltages.

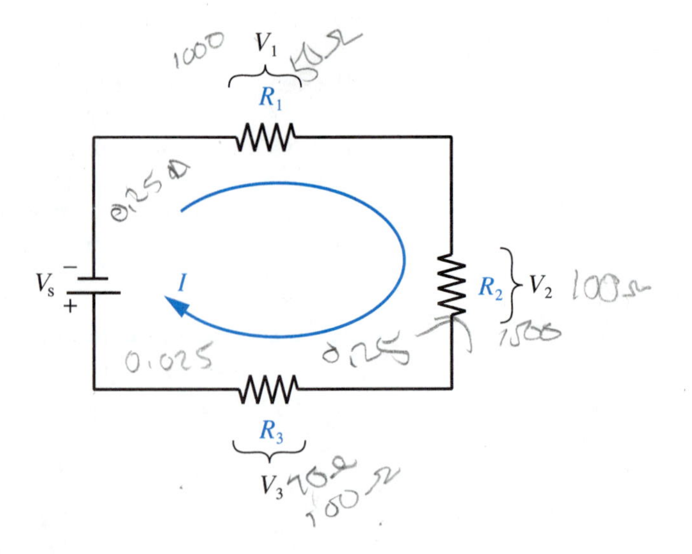

the voltages will be different. That is,

$$V_1 = IR_1$$

$$V_2 = IR_2$$

and
$$V_3 = IR_3$$

This result can be generalized for all series circuits as:

TECHNICAL FACT 2

Individual voltages are not equal in a series circuit. Voltages are proportional to resistances.

EXAMPLE 9-1

In the three-resistor circuit of Figure 9-2, suppose current is measured by inserting an ammeter between the source and R_1, giving a measurement of 0.25 A. The resistances are $R_1 = 50\ \Omega$, $R_2 = 100\ \Omega$, $R_3 = 70\ \Omega$.

a) If an ammeter were inserted between R_2 and R_3, what value would it measure?
b) If a voltmeter were connected across R_1, what amount of voltage would it measure?
c) Repeat part (b) for R_2.
d) Repeat part (b) for R_3.

SOLUTION

a) In a series circuit the current is the same everywhere, so the ammeter would measure **0.25 A.**
b) By Ohm's law,

$$V_1 = IR_1 = (0.25\text{ A})(50\ \Omega) = \mathbf{12.5\ V}$$

c)
$$V_2 = IR_2 = (0.25\text{ A})(100\ \Omega) = \mathbf{25\ V}$$

d)
$$V_3 = IR_3 = (0.25\text{ A})(70\ \Omega) = \mathbf{17.5\ V}$$

9-2 OHM'S LAW APPLIED TO A SERIES CIRCUIT

3

In Section 9-1, we saw that Ohm's law can be applied to individual resistors within a series circuit. Ohm's law can also be applied to the circuit as a whole, if the **total resistance** of the circuit is taken into account.

TECHNICAL FACT

The total resistance, R_T, of a series circuit is the sum of the individual resistances. In formula form,

$$R_T = R_1 + R_2 + R_3 + \ldots \qquad \textbf{EQ. 9-1}$$

EXAMPLE 9-2

For the resistance values given in Example 9-1, suppose that the voltage source is disconnected by opening a switch. If an ohmmeter is connected across the three-resistor combination, what value will it measure?

SOLUTION

This is a proper ohmmeter connection, since the resistors are deenergized and isolated. From Equation 9-1,

$$R_T = R_1 + R_2 + R_3$$

$$= 50\ \Omega + 100\ \Omega + 70\ \Omega = \mathbf{220\ \Omega}$$

EXAMPLE 9-3

Look at the circuit in Figure 9-3(A). By studying Figure 9-3(B), satisfy yourself that the four resistors are connected in series. (See page 3 in the Circuits color section for the color version of Figure 9-3.)

a) Calculate the circuit's total resistance, R_T.
b) What amount of current flows?
c) Labeling the resistors 1 through 4 clockwise from the left, find the individual voltages, V_1, V_2, V_3 and V_4.
d) If a voltmeter were placed across the R_2–R_3 combination, what value of voltage would it measure?

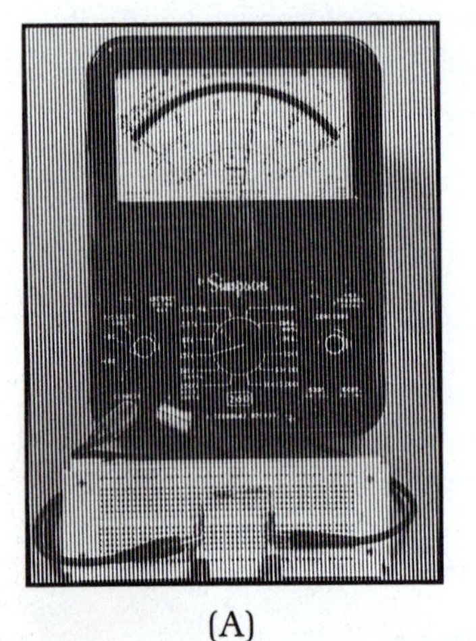

(A)

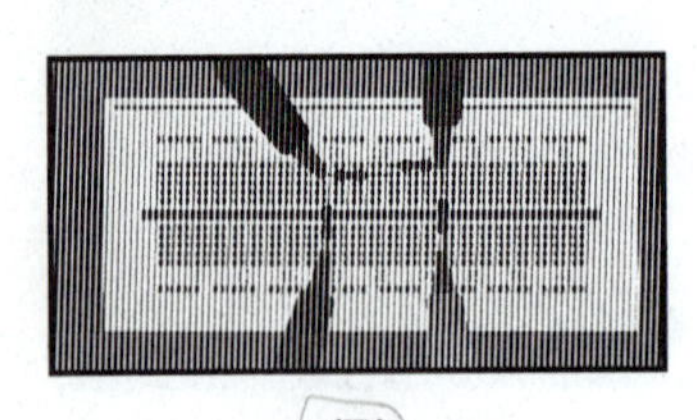

(B)

FIGURE 9-3

SOLUTION

a) The individual nominal resistance values are R_1 = 820 Ω, R_2 = 1200 Ω, R_3 = 680 Ω, and R_4 = 300 Ω. We have expressed R_2 in basic units of ohms, rather than the usual kΩ. This is because its units must be compatible with the other resistances, for addition. From Equation 9-1,

$$R_T = R_1 + R_2 + R_3 + R_4$$

$$= 820 + 1200 + 680 + 300 = \mathbf{3000\ \Omega}$$

b) The voltmeter indicates that the source voltage is 15 V. Applying Ohm's law to the circuit as a whole, we have

$$I = \frac{V_s}{R_T}$$

$$= \frac{15\ V}{3000\ \Omega} = \mathbf{0.005\ A\ or\ 5\ mA}$$

c) Applying Ohm's law to each resistor individually gives

$$V_1 = (0.005\ \text{A})(820\ \Omega) = \mathbf{4.1\ V}$$

$$V_2 = (5\ \text{mA})(1.2\ \text{k}\Omega) = (5 \times 10^{-3}\ \text{A})(1.2 \times 10^{3}\ \Omega) = \mathbf{6\ V}$$

The preceding calculation illustrates a shortcut for solving for voltage by Ohm's law : milliamperes (10^{-3}) and kilohms (10^{+3}) cancel each other, leaving an answer in basic units of volts.

Continued on page 141.

$$V_3 = (0.005\ \text{A})(680\ \Omega) = \mathbf{3.4\ V}$$

$$V_4 = (0.005\ \text{A})(300\ \Omega) = \mathbf{1.5\ V}$$

d) Because R_2 and R_3 are in series, they can be added together to find the resistance of that portion, which is symbolized R_{2-3}.

$$R_{2-3} = R_2 + R_3 = 1200\ \Omega + 680\ \Omega = 1880\ \Omega$$

Applying Ohm's law to that portion of the circuit , we get

$$V_{2-3} = I\,(R_{2-3})$$

$$= (0.005\ \text{A})(1880\ \Omega) = \mathbf{9.4\ V}$$

✔ SELF-CHECK FOR SECTIONS 9-1 AND 9-2

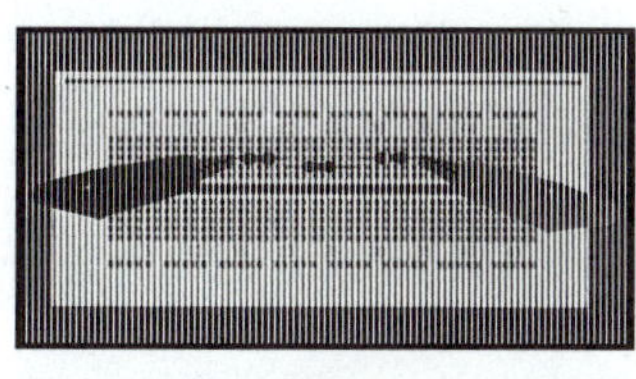

(A)

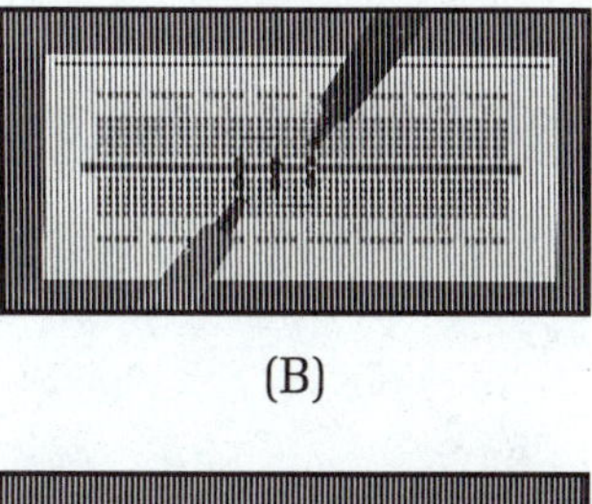

(B)

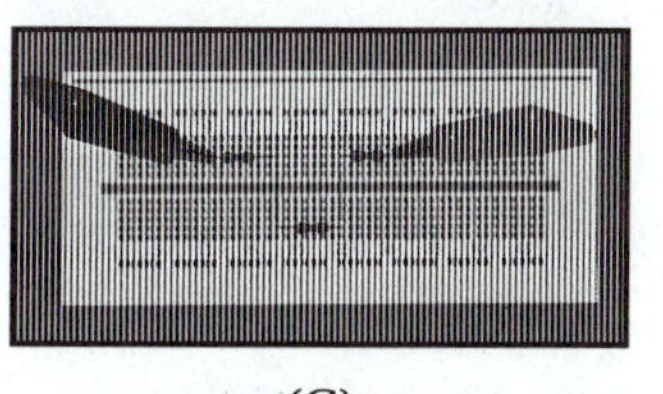

(C)

FIGURE 9-4

1. T F The smallest resistor in a series circuit has the smallest individual voltage. [3]
2. T F In a series circuit, the current that emerges from the source is reduced in value as it passes through the first resistance, and continues to get smaller as it passes through later resistances. [2]
3. In Figure 9-3, if the positions of the 820-Ω and 1200-Ω resistors were swapped, would that have any effect on their voltages? Explain. [2] NO
4. Refer to Figure 9-4 (See page 3 in the Circuits color section for the color version of Figure 9-4.) [3]
   a) What is the total resistance in Figure 9-4(A)?
   b) Repeat for Figure 9-4(B).
   c) Repeat for Figure 9-4(C).
5. Suppose the resistors in Figure 9-4(C) are energized, carrying a current of 12 mA. Find the following: [3]
   a) The source voltage, V_s.
   b) Individual voltages V_1, V_2, and V_3.
   c) The voltage across the R_1–R_2 combination.
6. In the circuit of Figure 9-4(C), suppose the resistances remain the same but source voltage V_s is reduced to 47 V. [3]
   a) Find V_1, V_2, and V_3.
   b) Find V_{1-2}.
7. For the circuit of Figure 9-5, find the following: [3]
   a) R_T
   b) R_1
   c) R_2
   d) R_4
   e) V_4

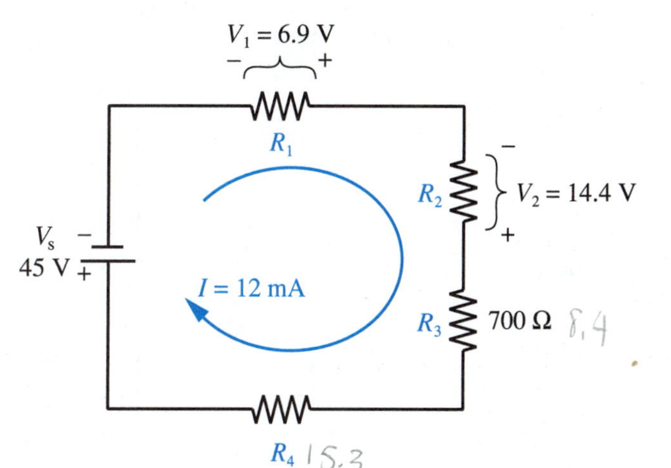

FIGURE 9-5 Four-resistor series circuit for Self-Check Problem 7

9-3 KIRCHHOFF'S VOLTAGE LAW

You may have noticed in Example 9-3 that the sum of the individual voltages equaled the source voltage (4.1 V + 6 V + 3.4 V + 1.5 V = 15 V). This isn't just a one-of-a-kind coincidence. It happens for all series circuits.

Kirchhoff's voltage law can be stated this way:

TECHNICAL FACT 4

In a series circuit, the sum of the individual voltages equals the source voltage. As a formula,

$$V_s = V_1 + V_2 + V_3 + \ldots$$

EQ. 9-2

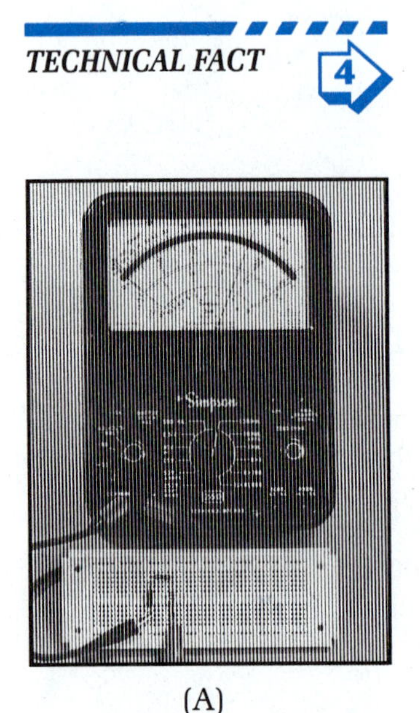

(A)

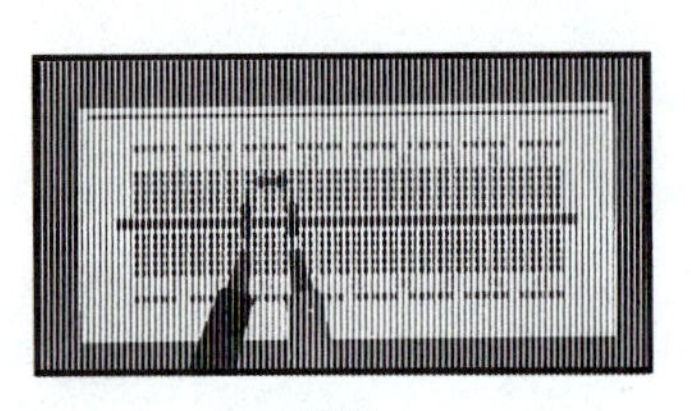

(B)

FIGURE 9-6

EXAMPLE 9-4

The circuit in Figure 9-6(A) shows that the circuit current is 70 mA or 0.07 A. What is the value of source voltage V_s? (See page 4 in the Circuits color section for the color version of Figure 9-6.)

SOLUTION

First, use Ohm's law to find the individual voltages, V_1, V_2, and V_3.

$$V_1 = IR_1 = (0.07\text{ A})(75\ \Omega) = 5.25\text{ V}$$
$$V_2 = IR_2 = (0.07\text{ A})(43\ \Omega) = 3.01\text{ V}$$
$$V_3 = IR_3 = (0.07\text{ A})(51\ \Omega) = 3.57\text{ V}$$

Applying Kirchhoff's voltage law, Equation 9-2, we get

$$V_s = V_1 + V_2 + V_3$$
$$= 5.25\text{ V} + 3.01\text{ V} + 3.57\text{ V} = \mathbf{11.83\text{ V}}$$

We could also have solved this problem by first finding R_T and then applying Ohm's law to the circuit as a whole.

$$R_T = R_1 + R_2 + R_3 = 75\ \Omega + 43\ \Omega + 51\ \Omega = 169\ \Omega$$
$$V_s = IR_T$$
$$= (0.07\text{ A})(169\ \Omega) = \mathbf{11.83\text{ V}}$$

Solving a problem by an alternative method is the best way to check an answer.

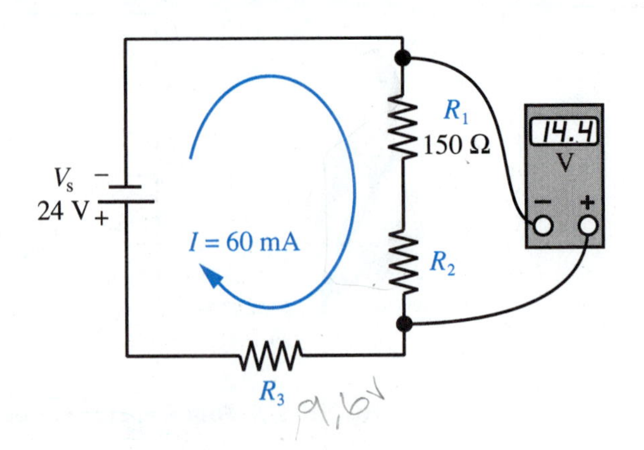

FIGURE 9-7 Series circuit for Example 9-5

EXAMPLE 9-5

Figure 9-7 shows a 24-V source driving a three-resistor series circuit. The current is known to be 60 mA and the voltage across the $R_1 - R_2$ combination is measured as 14.4 V. The only resistance that is known is $R_1 = 150\ \Omega$.

a) Find V_3.
b) Find R_3
c) Find R_2.

SOLUTION

a) The voltmeter indicates that

$$V_1 + V_2 = 14.4\ \text{V}$$

From Kirchhoff's voltage law,

$$(V_1 + V_2) + V_3 = V_s$$
$$V_3 = V_s - (V_1 + V_2)$$
$$= 24\ \text{V} - 14.4\ \text{V} = \mathbf{9.6\ V}$$

b) Applying Ohm's law to R_3, we have

$$R_3 = \frac{V_3}{I}$$

$$= \frac{9.6\ \text{V}}{60 \times 10^{-3}\ \text{A}} = \mathbf{160\ \Omega}$$

c) One approach to solving R_2 is

$$R_T = \frac{V_s}{I} = \frac{24\ \text{V}}{60 \times 10^{-3}\ \text{A}} = 400\ \Omega$$

$$R_2 = R_T - R_1 - R_3$$
$$= 400\ \Omega - 150\ \Omega - 160\ \Omega = \mathbf{90\ \Omega}$$

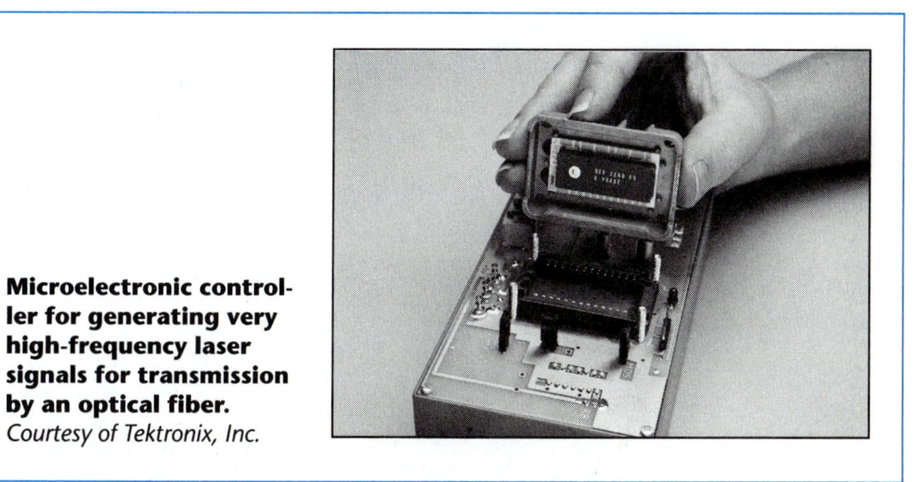

Microelectronic controller for generating very high-frequency laser signals for transmission by an optical fiber.
Courtesy of Tektronix, Inc.

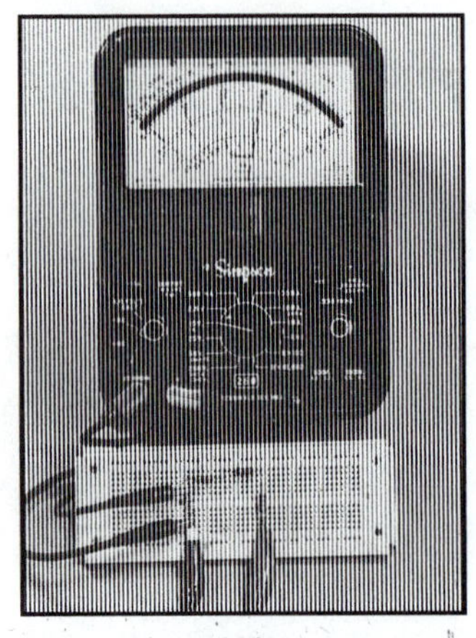

(A)

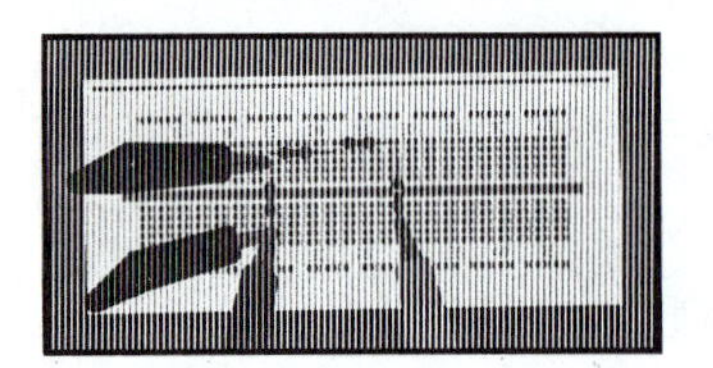

(B)

FIGURE 9-8

EXAMPLE 9-6

Refer to Figure 9-8. (See page 4 in the Circuits color section for the color version of Figure 9-8.)

a) What is the measured value of voltage across the first resistor (R_1)?
b) What is the circuit's current?
c) Solve for V_2, V_3, and V_4.
d) Find the value of the source voltage.

SOLUTION

a) The VOM is on the 10-V scale. The measured value of V_1 is **6.0 V**.
b) Applying Ohm's law to R_1, we get

$$I = \frac{V_1}{R_1} = \frac{6.0\text{ V}}{15\text{ k}\Omega} = \mathbf{0.4\text{ mA}}$$

c) From Ohm's law,

$$V_2 = (0.4\text{ mA})\,(20\text{ k}\Omega) = \mathbf{8.0\text{ V}}$$
$$V_3 = (0.4\text{ mA})\,(36\text{ k}\Omega) = \mathbf{14.4\text{ V}}$$
$$V_4 = (0.4\text{ mA})\,(10\text{ k}\Omega) = \mathbf{4.0\text{ V}}$$

d) From Kirchhoff's voltage law,

$$V_s = V_1 + V_2 + V_3 + V_4 = 6.0 + 8.0 + 14.4 + 4.0 = \mathbf{32.4\text{ V}}$$

EXAMPLE 9-7

In the three-resistor series circuit of Figure 9-9, none of the resistances are known. However, there are two voltmeters connected. Voltmeter A is measuring the voltage across the R_1–R_2 combination. Voltmeter B is measuring the $R_2 - R_3$ combination. Find the individual voltages, V_1,V_2, and V_3.

SOLUTION

Concentrate for the moment on V_B, and ignore V_A. From the V_B measurement, we know

$$V_{2\text{-}3} = V_2 + V_3 = 21\text{ V}$$

Applying Kirchhoff's voltage law, we can say

$$V_s = V_1 + (V_2 + V_3)$$
$$28\text{ V} = V_1 + 21\text{ V}$$
$$V_1 = 28\text{ V} - 21\text{ V} = \mathbf{7\text{ V}}$$

Now concentrate on V_A.

$$V_{1\text{-}2} = V_1 + V_2 = 16\text{ V}$$

From Equation 9-2,

$$V_3 = V_s - (V_1 + V_2)$$
$$V_3 = 28\text{ V} - 16\text{ V} = \mathbf{12\text{ V}}$$

V_2 can be found by comparing voltmeter V_A and voltage V_1.

Continued on page 145.

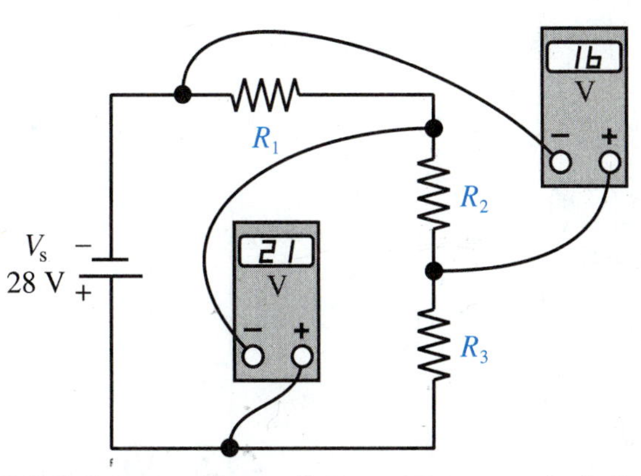

FIGURE 9-9 Interpreting voltage measurements in Example 9-7

$$V_1 + V_2 = 16\text{ V}$$
$$V_2 = 16\text{ V} - V_1 = 16\text{ V} - 7\text{ V} = \mathbf{9\text{ V}}$$

V_2 could be solved just as well by comparing voltmeter V_B and voltage V_3.

SELF-CHECK FOR SECTION 9-3

8. T-F In a series circuit, the voltage across any individual resistor is less than the source voltage. [2, 4]
9. T-F In a series circuit, the voltage across a two-resistor combination is always greater than the voltage across either individual resistor. [2, 4]
10. In a three-resistor circuit like Figure 9-2, suppose I = 25 mA, R_1 = 1 kΩ, R_2 = 1.5 kΩ, and R_3 = 700 Ω. Find the individual voltages and the source voltage. [3, 4]
11. In Figure 9-10, find V_1. From the information given, is there any way to find V_2 and V_3? [4]
12. In Figure 9-11, find V_s. [3, 4]
13. Find V_s in Figure 9-12. [4]

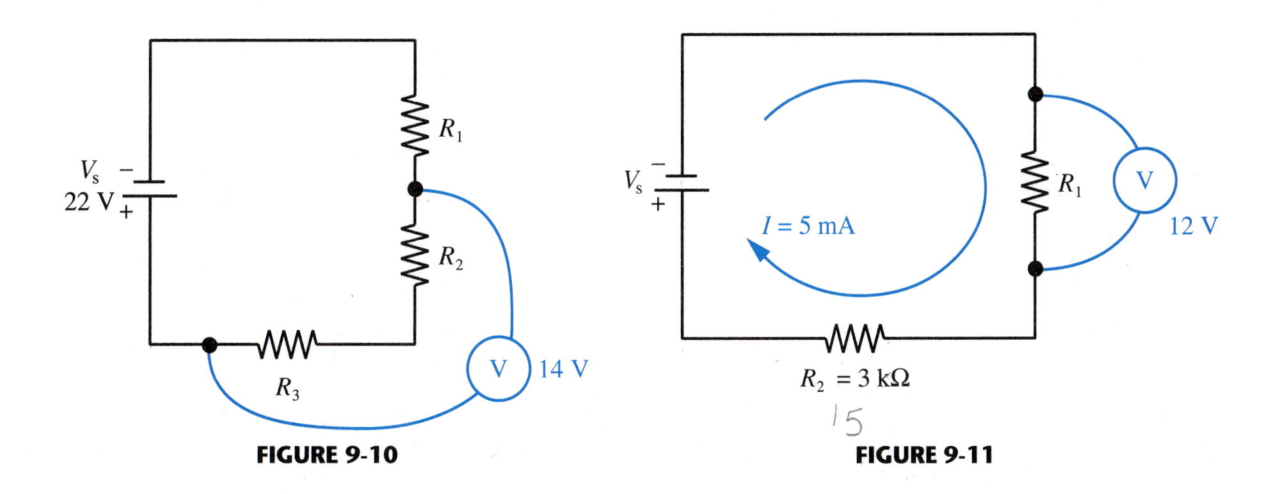

FIGURE 9-10

FIGURE 9-11

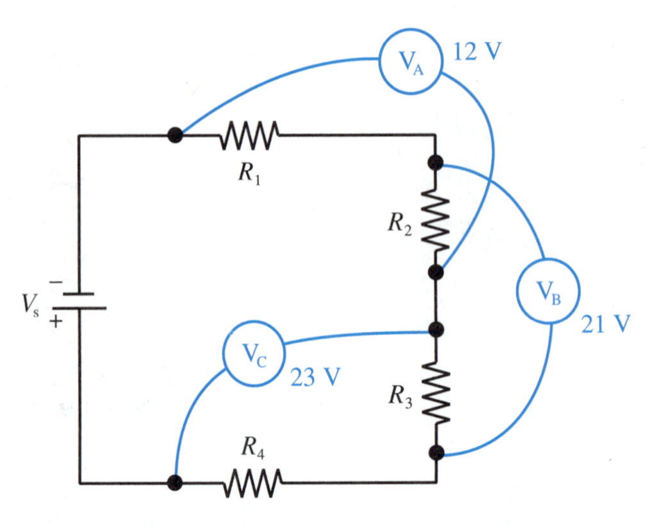

FIGURE 9-12 None of the individual voltages are known, but combination voltages are known.

SAFETY ADVICE

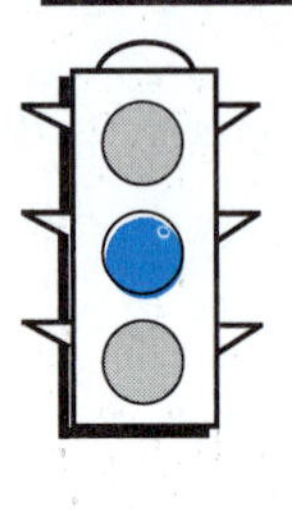

Always know exactly which switch or circuit breaker turns off the power in an area where you are working. Whenever a newcomer arrives, show him where that power-Off switch is located. This way, if an electric shock emergency ever occurs, you won't waste precious seconds looking around for a way to turn off the power.

Main On-Off switches are usually mounted so that their mechanical actuation is vertical, rather than horizontal. Vertical switches are always Down for Off, Up for On. Rehearse this fact until it becomes second nature to you. On-Off labels are sometimes hard to read.

9-4 INTERNAL RESISTANCE OF A SOURCE

All real voltage sources have a certain amount of internal resistance arising from their materials of construction. We visualize this **internal resistance** in series with the source, as shown in Figure 9-13. With R_{int} present in a circuit diagram, we treat it like any other resistance. It is added to the external component resistances to find R_T.

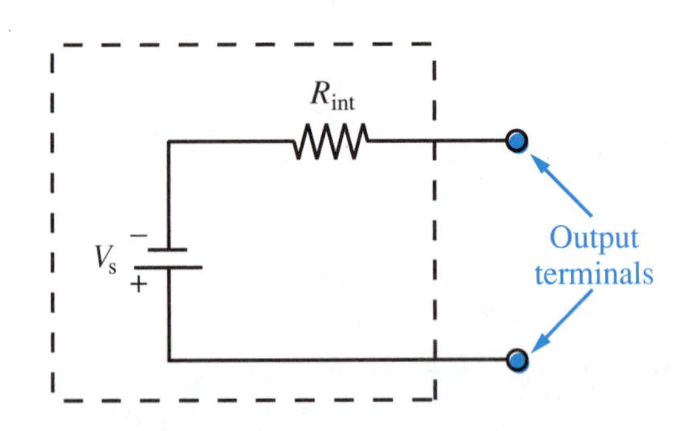

FIGURE 9-13 A real-life voltage source can be viewed as an ideal source in series with some internal resistance.

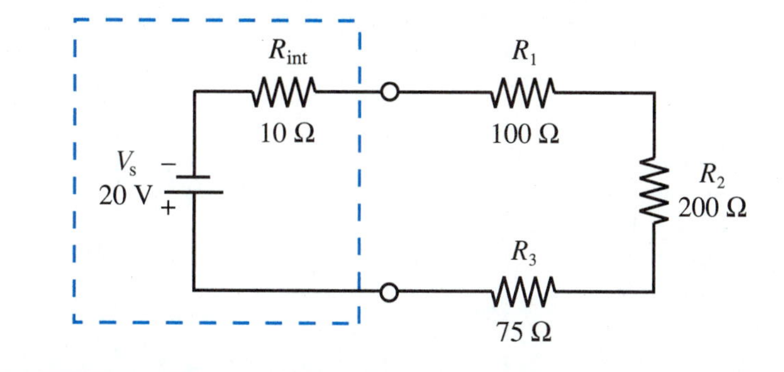

FIGURE 9-14 A real voltage source is sometimes called a nonideal voltage source. Here, a nonideal source drives a three-resistor series combination.

EXAMPLE 9-8

The voltage source in Figure 9-14 has $R_{int} = 10\ \Omega$.

a) Calculate the circuit's current.
b) Find V_1, V_2, and V_3.
c) If a voltmeter is connected across the source terminals, what value will it measure? Explain this.

SOLUTION

a) Total resistance is given by

$$R_T = R_{int} + R_1 + R_2 + R_3$$
$$= 10 + 100 + 200 + 75 = \mathbf{385\ \Omega}$$

From Ohm's law,

$$I = \frac{V_s}{R_T} = \frac{20\text{ V}}{385\ \Omega} = \mathbf{0.051\ 95\ A}$$

b)

$$V_1 = (0.051\ 95\text{ A})\ 100\ \Omega = \mathbf{5.195\ V}$$
$$V_2 = (0.051\ 95\text{ A})\ 200\ \Omega = \mathbf{10.390\ V}$$
$$V_3 = (0.051\ 95\text{ A})\ \ 75\ \Omega = \mathbf{3.896\ V}$$

c) The voltmeter will read $V_1 + V_2 + V_3$.

$$5.195\text{ V} + 10.390\text{ V} + 3.896\text{ V} = \mathbf{19.48\ V}$$

This value is different from the ideal voltage because a small amount of voltage is lost across R_{int}. The internal voltage loss can be found from Kirchhoff's voltage law or by Ohm's law. Using Kirchhoff's voltage law,

$$V_s = V_{int} + (V_1 + V_2 + V_3)$$
$$V_{int} = V_s - (V_1 + V_2 + V_3) = 20\text{ V} - 19.48\text{ V} = 0.52\text{ V}$$

By Ohm's law,

$$V_{int(loss)} = I\,R_{int} = (0.051\ 95\text{ A})(10\ \Omega) = 0.5195\text{ V or } 0.52\text{ V}$$

The source's internal voltage drop depends on the current demand by the load(s). For instance, if the circuit load has lower total resistance, current demand will rise. This causes greater internal voltage drop, reducing the output voltage available at the output terminals.

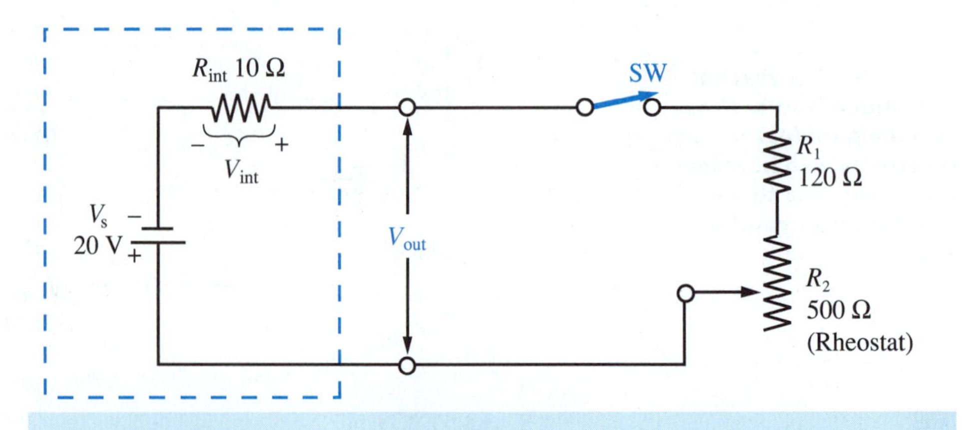

FIGURE 9-15 V_{int} **subtracts from** V_s **because it has opposing polarity. Therefore,** V_{out} **is less than** V_s **when SW is closed.**

EXAMPLE 9-9

The same nonideal voltage source from Example 9-8 is shown driving an adjustable load in Figure 9-15.

a) With SW open, what is the value of output voltage V_{out}?

b) With SW closed and R_2 adjusted to maximum resistance, what is the value of V_{out}?

c) Repeat part b with R_2 adjusted to zero ohms.

SOLUTION

a) With SW open, there is no current and therefore zero internal voltage drop.

$$V_{out} = V_s = \mathbf{20\ V}$$

This is called the **no-load output voltage** (symbolized V_{NL}), or the **open-circuit output voltage** (symbolized V_{OC}).

b) With R_2 at 500 Ω,

$$R_T = R_{int} + R_1 + 500\ \Omega$$
$$= 10\ \Omega + 120\ \Omega + 500\ \Omega = 630\ \Omega$$

$$I = \frac{V_s}{R_T} = \frac{20\ V}{630\ \Omega} = 0.031\ 75\ A$$

Applying Ohm's law to R_{int},

$$V_{int} = IR_{int} = (0.031\ 75\ A)(10\ \Omega) = 0.3175\ V$$

By Kirchhoff's voltage law,

$$V_{out} = V_s - V_{int} = 20\ V - 0.3175\ V = \mathbf{19.68\ V}$$

c) With R_2 at 0 Ω,

$$R_T = R_{int} + R_1 + 0\ \Omega$$
$$= 10\ \Omega + 120\ \Omega = 130\ \Omega$$

$$I = \frac{20\ V}{130\ \Omega} = 0.1538\ A$$

Assuming that this is the maximum current the source can deliver without overheating, it would be called the **full-load current** (symbolized I_{FL}).

$$V_{int} = IR_{int} = (0.1538\ A)\ 10\ \Omega = 1.538\ V$$
$$V_{out} = V_s - V_{int} = 20\ V - 1.538\ V = \mathbf{18.46\ V}$$

This voltage would then be called the **full-load output voltage** (symbolized V_{FL}).

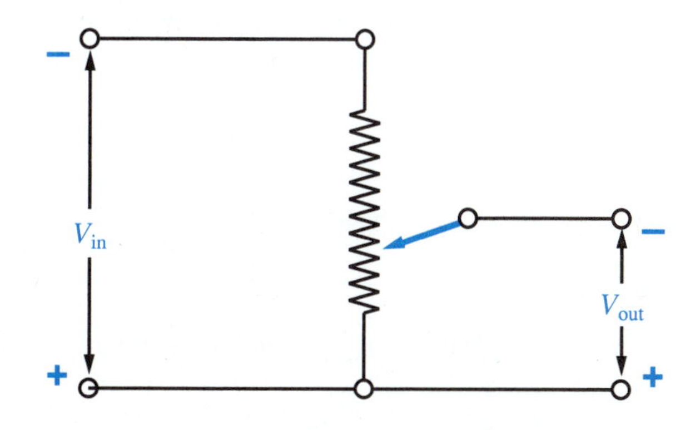

FIGURE 9-16 A pot can deliver any portion from 0 to 100% of V_{in} by adjusting its movable tap.

9-5 POTENTIOMETER VOLTAGE DIVIDERS

We saw in Chapter 3 how a pot can serve as a variable resistor. However, because it has three terminals, a pot can perform another function. This function is illustrated in Figure 9-16. Voltage V_{in} is applied to the pot's outside terminals. By adjusting the movable tap, or wiper, any portion of V_{in} can be selected. The voltage portion that is selected appears between the wiper terminal and the bottom terminal. This is considered the pot's output voltage.

For example, in Figure 9-16 suppose the applied voltage V_{in} is 50 V. If the wiper is adjusted to the exact middle of the resistive element (50% up from the bottom terminal), V_{out} will be 25 V. If the wiper is adjusted 60% of the distance up from the

SOLAR DOSIMETER

Exposure of the skin to sunlight has an effect on the ability of the body to extract vitamin D from food. The infant shown has a device called a solar dosimeter clipped to her clothing. Using a light-sensitive electronic component, it measures the cumulative amount of sunshine exposure over a period of several days or weeks. Cumulative exposure is a result of the combination of sunshine intensity along with the length of time exposed.

Medical research on the sunlight/vitamin D relationship has led to this important conclusion: breast-fed infants who receive no commercial vitamin-enriched milk supplements need one-half hour to two hours of sunshine exposure per week, in northern U.S. latitudes. With less than that amount of cumulative exposure, infants showed below-normal blood concentrations of vitamin D.

The adult subject is shown wearing a dosimeter in an ultraviolet light chamber. This is for a medical study of the relationship among ultraviolet light exposure, degree of skin pigmentation, and vitamin D blood level.

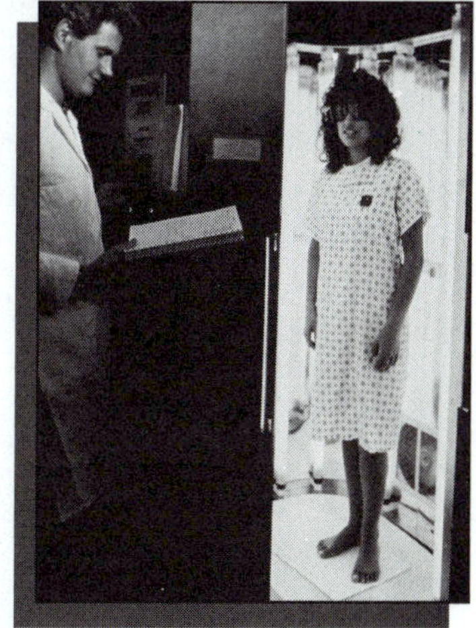

Courtesy of NASA

bottom, V_{out} will be 60% of 50 V, or 30 V.

The **voltage division** formula for a pot can be written

$$\frac{V_{out}}{V_{in}} = \frac{R_{bottom}}{R_T}$$

EQ. 9-3

where R_{bottom} symbolizes the resistance between the wiper and the bottom terminal.

EXAMPLE 9-10

Figure 9-17 shows the standard volume-control method for a radio or television. A detector circuit applies a small fixed voltage, V_{in}, across the **volume-control pot.** A portion of this voltage is taken from the pot by voltage-divider action. The V_{out} voltage is applied to the audio amplifier, which multiplies it by a certain factor, making it large enough to drive the speaker. The sound volume is varied by raising and lowering V_{out} through adjustment of the pot wiper.

As shown in Figure 9-17, $V_{in} = 1.2$ V and $R_T = 20$ kΩ. The multiplication factor of the amplifier, symbolized A_V, is 14.

a) If the wiper is adjusted so that $R_{bottom} = 15$ kΩ, what value of voltage is fed to the audio amplifier? What value of voltage is delivered to the speaker?
b) Where should the wiper be adjusted to give $V_{spkr} = 3.6$ V?

SOLUTION

a) From the voltage-divider equation (Equation 9-3),

$$\frac{V_{out}}{V_{in}} = \frac{15\text{ k}\Omega}{20\text{ k}\Omega} = 0.75$$

$$V_{out} = 1.2\text{ V }(0.75) = \mathbf{0.90\text{ V}}$$

The audio amp boosts that voltage by a factor of 14, so

$$V_{spkr} = 14\,(0.90\text{ V}) = \mathbf{12.6\text{ V}}$$

b) For V_{spkr} to be 3.6 V, the voltage fed to the amplifier from the pot must be

$$V_{out} = \frac{V_{spkr}}{A_V} = \frac{3.6\text{ V}}{14} = 0.2571\text{ V}$$

From Equation 9-3,

$$\frac{R_{bottom}}{R_T} = \frac{V_{out}}{V_{in}}$$

$$R_{bottom} = R_T\left(\frac{V_{out}}{V_{in}}\right) = 20\text{ k}\Omega\left(\frac{0.2571\text{ V}}{1.2\text{ V}}\right) = \mathbf{4.286\text{ k}\Omega}$$

This places the wiper about 21% of the way up from the bottom, since

$$\frac{4.286\text{ k}\Omega}{20\text{ k}\Omega} \cong 0.21 \text{ or } \mathbf{21\%}$$

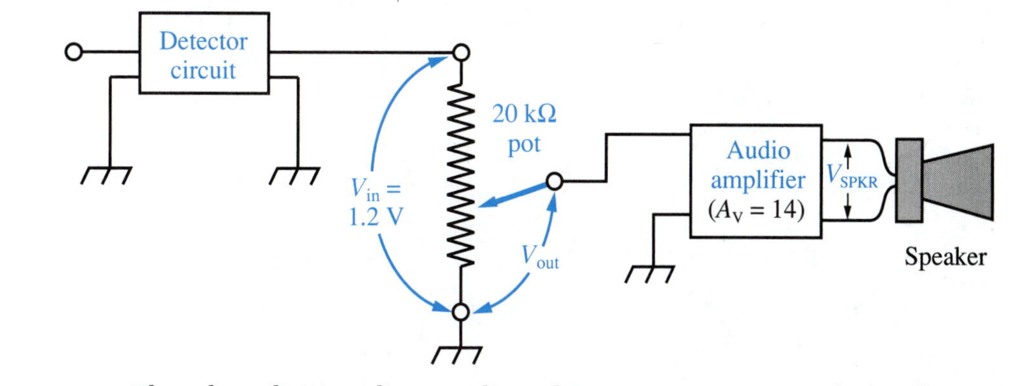

FIGURE 9-17 Controlling sound volume

The voltage division idea is not limited to potentiometers. It works for all series circuits.

TECHNICAL FACT

As a general rule, if V_{out} is taken across a certain portion of a series circuit, we can write the voltage-divider formula

$$\frac{V_{out}}{V_{in}} = \frac{R_{portion}}{R_T}$$

EQ. 9-4

EXAMPLE 9-11

Calculate V_{out} for the circuit of Figure 9-18.

SOLUTION

This is a series circuit, so Equation 9-4 can be used.

$$\frac{V_{out}}{V_{in}} = \frac{R_{portion}}{R_T} = \frac{700\ \Omega}{500\ \Omega + 700\ \Omega + 900\ \Omega} = \frac{700\ \Omega}{2100\ \Omega}$$

$$V_{out} = 24\ \text{V}\left(\frac{700\ \Omega}{2100\ \Omega}\right) = \mathbf{8.0\ V}$$

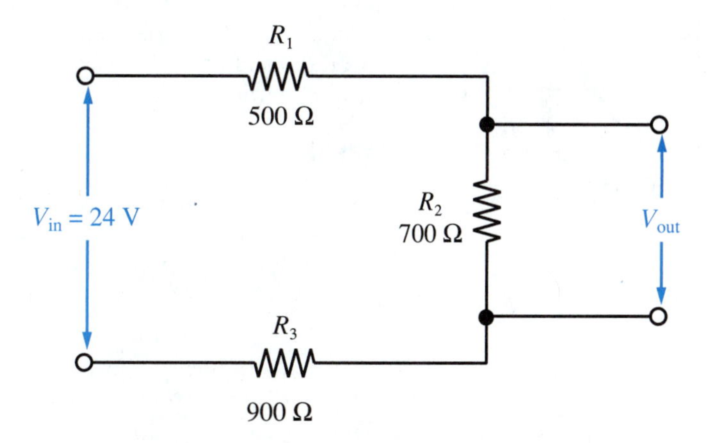

FIGURE 9-18 Voltage division

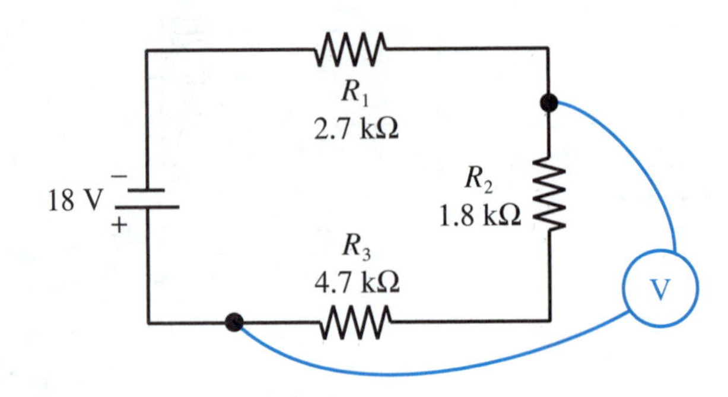

FIGURE 9-19 The voltage division method can be used for a two-resistor combination.

SELF-CHECK FOR SECTIONS 9-4 AND 9-5

14. Explain the difference between an ideal voltage source and a real voltage source. [5]
15. A certain real voltage source has a no-load output voltage of 15 V and R_{int} = 3 Ω. [3, 4, 5]
    a) What will be the output terminal voltage when the current demand is 0.2 A?
    b) What will be the output terminal voltage if a 27-Ω load resistor is connected?
16. A 1000-Ω potentiometer has V_{in} = 12 V applied across its terminals. The wiper is adjusted 300 Ω from the bottom. What is the value of V_{out}? [6]
17. For the pot in Problem 16, where should the wiper be adjusted to produce V_{out} = 9.2 V? [6]
18. Use voltage division to find the voltmeter reading in Figure 9-19.

9-6 TROUBLESHOOTING SERIES CIRCUITS

TECHNICAL FACT

In a series circuit, if one load fails in the open condition, there is no longer a complete current path. So all the other loads stop functioning.

EXAMPLE 9-12

In the circuit of Figure 9-20(A), the ammeter reads zero, even though the switch appears closed. It is suspected that one of the resistors has failed open. Find out if the source and the switch are functioning properly. If they are, find the faulty resistor without taking the circuit apart.

SOLUTION

You can verify that the source and the switch are all right by connecting a voltmeter as shown in Figure 9-20(B). The fact that the voltmeter indicates the V_s value of 10 V proves that the source and switch are both functioning properly.

Move the voltmeter to the right side of R_1. If R_1 is intact (not open), the voltmeter will still measure 10 V. This is because there will be no voltage drop across R_1, with no current in the circuit. See Figure 9-20(C).

7

Keep moving the voltmeter forward one resistor at a time. Its reading will fall to 0 V when its probe is moved to the far side of the open resistor. In this case, Figure 9-20(D) indicates that $\boldsymbol{R_2}$ **is open,** and must be replaced.

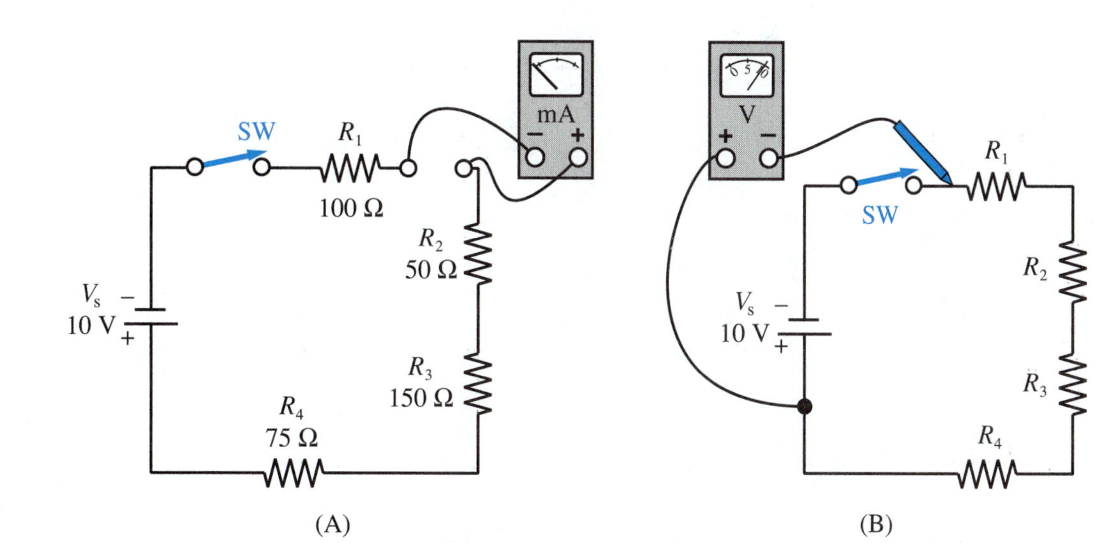

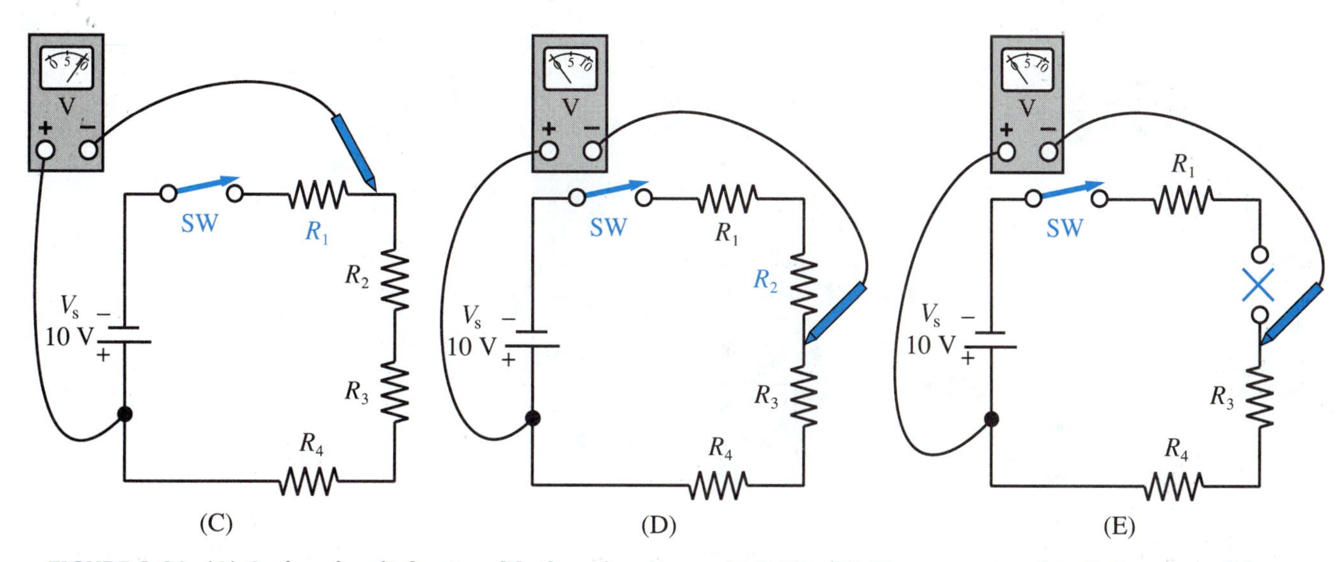

FIGURE 9-20 (A) Series circuit for troubleshooting Example 9-12. (B) The source and switch are working properly. (C) R_1 is not the fault. (D) R_2 is at fault ; it has failed open. (E) Drawing an open circuit in the place of R_2 makes it clearer why the voltmeter reads zero.

TECHNICAL FACT

In a series circuit, if one load fails in the shorted condition, a larger-than-normal current flows. This may possibly damage the other loads.

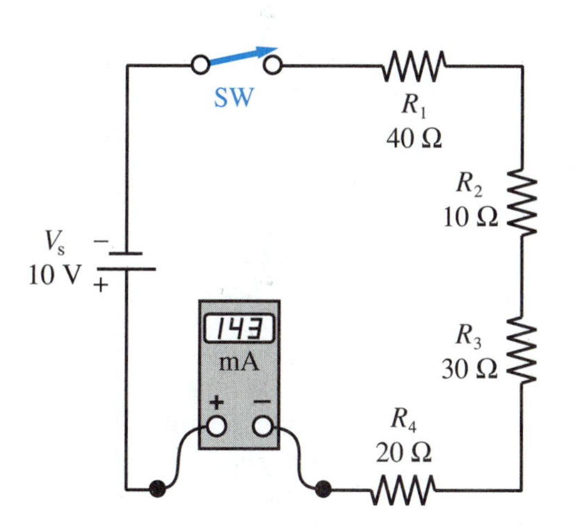

FIGURE 9-21 Series circuit with trouble

EXAMPLE 9-13

In Figure 9-21, the expected current can be found by Ohm's law.

$$I = \frac{V_s}{R_1 + R_2 + R_3 + R_4}$$

$$= \frac{10 \text{ V}}{(40 + 10 + 30 + 20)\ \Omega} = \frac{10 \text{ V}}{100\ \Omega}$$

$$= 0.10 \text{ A or } 100 \text{ mA}$$

Find the reason why the actual current is 143 mA, larger than the expected value of 100 mA.

SOLUTION

First check the source voltage with a voltmeter. If it measures all right at 10 V, then apparently one (or more) of the series resistors is shorted.

Calculate the expected voltage across each resistor. Then measure each voltage individually. The intact resistors will measure more voltage than expected. The shorted resistor will measure either zero or a less-than-expected voltage.

The actual measured voltages are indicated in Figure 9-22. Comparing expected and actual measured values, we have

	Expected	Measured
V_1	4 V	5.7 V
V_2	1 V	1.4 V
V_3	3 V	0 V
V_4	2 V	2.9 V

The conclusion is that **resistor R_3 is dead-shorted** and must be replaced.

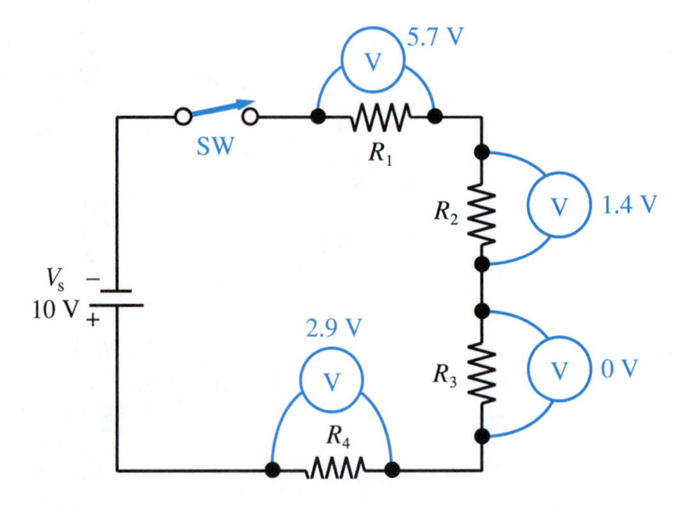

FIGURE 9-22 Results of four voltmeter measurements

FORMULAS

For series circuits:

$$R_T = R_1 + R_2 + R_3 + \ldots$$

EQ. 9-1

$$V_s = V_1 + V_2 + V_3 + \ldots$$

EQ. 9-2

For a real source with internal resistance:

$$R_T = R_{int} + R_1 + R_2 + \ldots$$

$$V_s = V_{int} + V_1 + V_2 + \ldots$$

For a pot:

$$\frac{V_{out}}{V_{in}} = \frac{R_{bottom}}{R_T}$$

EQ. 9-3

For any series circuit:

$$\frac{V_{out}}{V_{in}} = \frac{R_{portion}}{R_T}$$

EQ. 9-4

SUMMARY OF IDEAS

- In a series circuit, current is the same everywhere, but individual voltages are different from each other.
- Using total resistance, R_T, Ohm's law can be applied to a series circuit.
- Total resistance, R_T, is given by the sum of the individual resistances.
- In a series circuit, the sum of the individual resistor voltages is equal to the source voltage. This is Kirchhoff's voltage law.
- A real voltage source has some internal resistance, R_{int}. This causes some of the ideal voltage to be lost internally.
- A potentiometer can provide adjustable voltage division.
- Any series circuit can be analyzed by the voltage division method.

CHAPTER QUESTIONS AND PROBLEMS

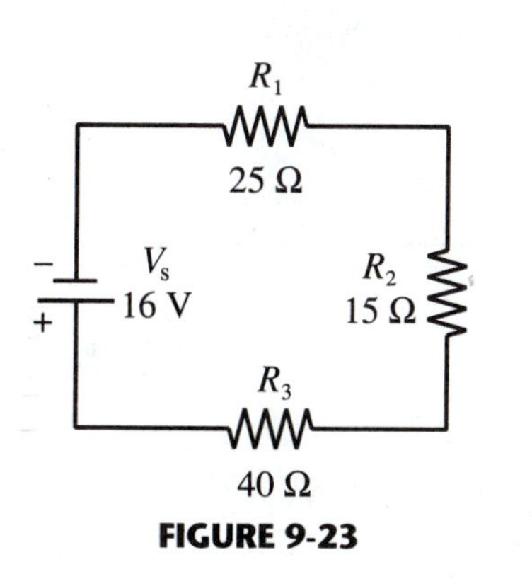

FIGURE 9-23

1. Explain why unequal voltages exist across series resistors. [2]
2. For the circuit of Figure 9-23, [3]
   a) Find the total resistance, R_T.
   b) Find the circuit's current.
   c) Calculate the individual voltages V_1, V_2, and V_3.
3. For the circuit of Problem 2, [3]
   a) Calculate the voltage across the $R_1 - R_2$ combination.
   b) Calculate the voltage across the $R_2 - R_3$ combination.
4. For the circuit of Figure 9-24, [2, 3]
   a) Find the source voltage, V_s.
   b) Which resistor has the largest voltage? Why is this reasonable?
   c) Which resistor has the smallest voltage? Why is this reasonable?

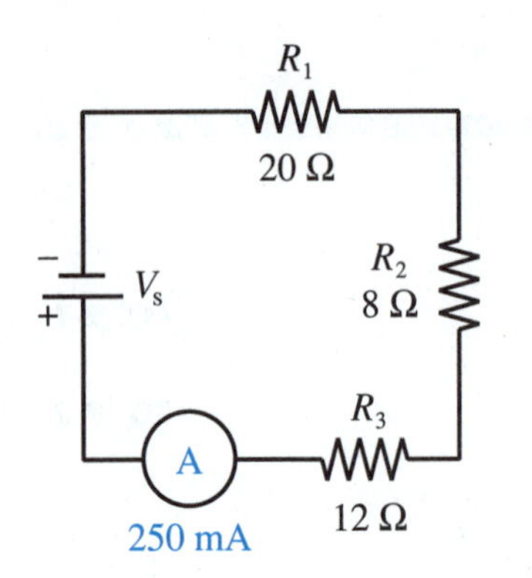

FIGURE 9-24

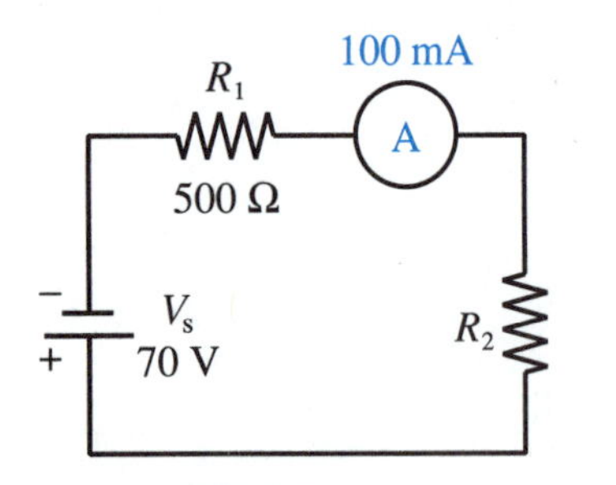

FIGURE 9-25

5. For the circuit of Figure 9-25, [3]
   a) Find the total resistance, R_T.
   b) Find R_2.
6. For the circuit of Figure 9-26, [2, 3]
   a) Find the maximum current and maximum value of V_2.
   b) Find the minimum current and minimum value of V_2.
   c) If the rheostat is adjusted to its exact center, find the current and the V_2 value.
7. For the circuit of Figure 9-27, [3, 4]
   a) Find V_1. c) Find V_3.
   b) Find V_2. d) Find R_3.
8. For the circuit of Figure 9-28, [3, 4]
   a) Find V_3.
   b) Find the source voltage, V_s.
9. For the circuit of Figure 9-29, [3, 4, 5]
   a) Find the circuit's total resistance, R_T.
   b) Calculate V_1 and V_2.
   c) Find V_{out}.
10. Repeat Problem 9 supposing R_1 is increased to 1 kΩ and R_2 is increased to 2 kΩ. Explain why V_{out} is so close to V_s. [3, 4, 5]
11. For the real voltage source in Figure 9-30, the maximum allowable current occurs when R_2 is dialed down to zero. That is the source's full-load condition. [3, 4, 5]
    a) Find the full-load output voltage.
    b) Find the full-load current.
    c) What is the no-load output voltage?

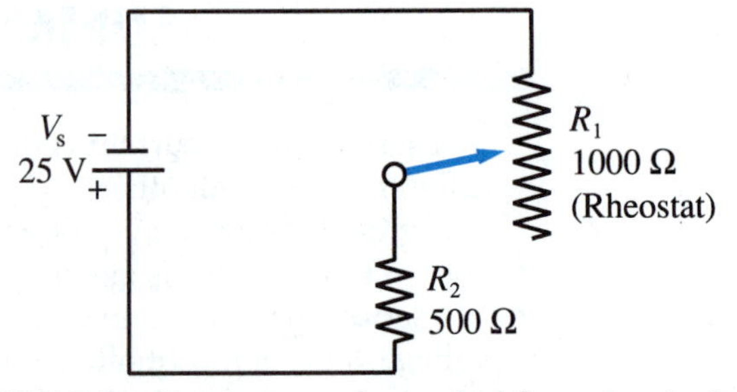

FIGURE 9-26 Current can be varied from I_{min} to I_{max}.

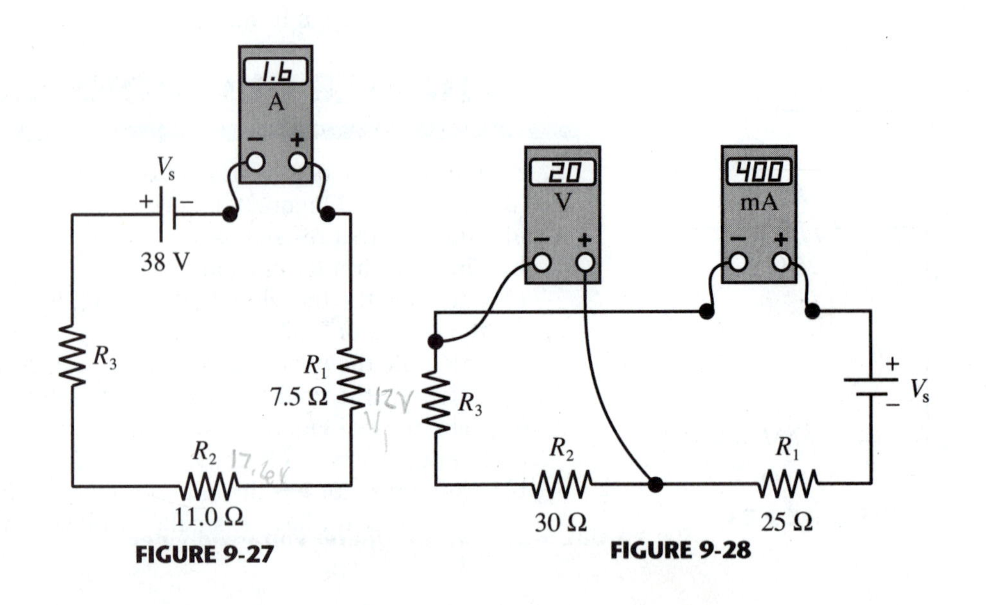

FIGURE 9-27

FIGURE 9-28

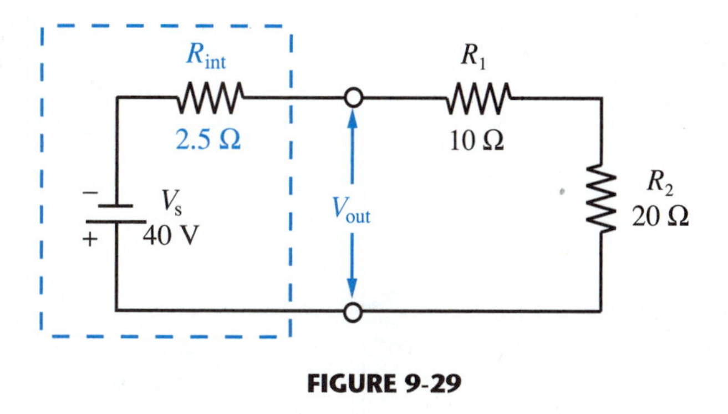

FIGURE 9-29

12. The voltage regulation of a source is defined as the difference between no-load and full-load output voltage, divided by the full-load output voltage.

$$\left(V_{reg} = \frac{V_{NL} - V_{FL}}{V_{FL}}\right)$$

Calculate the voltage regulation of the source in Problem 11. [5]
13. For the circuit of Figure 9-31, use voltage division to: [6]
    a) Find V_{out} when the wiper is adjusted 2 kΩ from the bottom.
    b) Find V_{out} when the wiper is adjusted 3.8 kΩ from the bottom.
14. Repeat Problem 13 with a 2.2 kΩ fixed resistor placed in series between the source's negative terminal and the pot. [3, 6]
15. For the circuit of Figure 9-32, [4]
    a) Find V_1 by using voltage division.
    b) Repeat for V_2 and V_3.
    c) Does the voltage division method give results that satisfy Kirchhoff's voltage law?

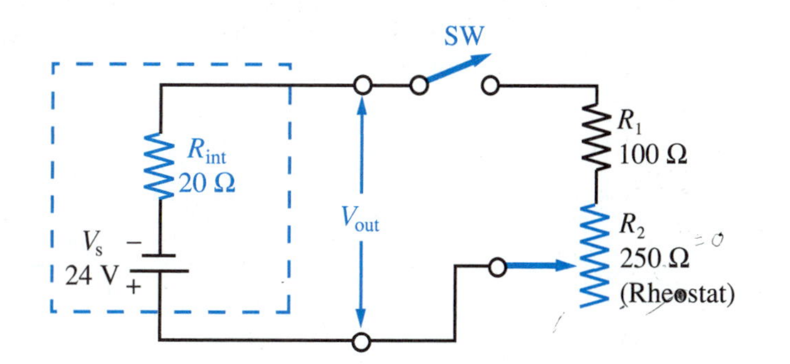

FIGURE 9-30 No-load condition occurs with SW open. Full load occurs with SW closed and R_2 dialed to 0 Ω.

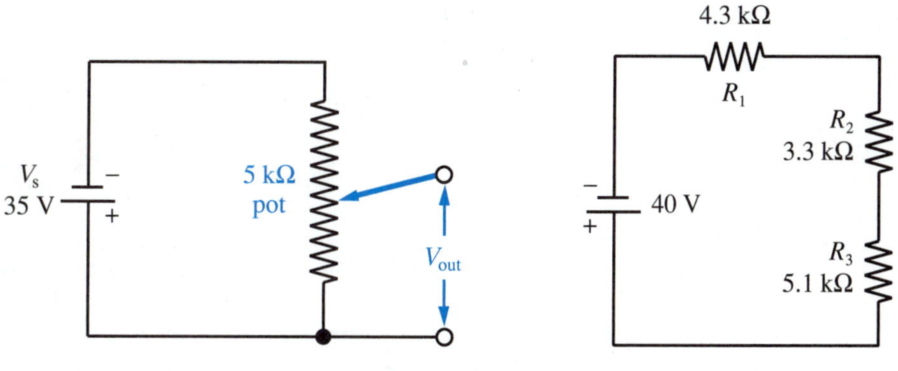

FIGURE 9-31 Potentiometer voltage-divider

FIGURE 9-32

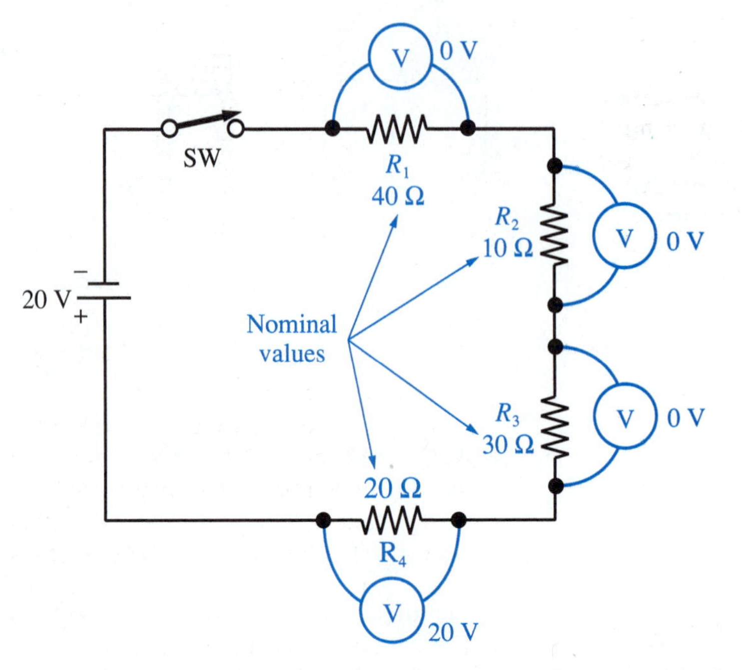

FIGURE 9-33 Results of four voltage measurements

16. In Figure 9-32, suppose a voltmeter is connected between the bottom (+) terminal of the source and the right side of R_1. Paying careful attention to the polarity of R_1, what value does the voltmeter read? [4]
17. In Figure 9-32, suppose a voltmeter is connected across the R_2–R_3 combination. What value does it measure? [4]
18. Compare your answers to Problems 16 and 17. Explain this. [4]

Troubleshooting Questions and Problems

19. The circuit of Figure 9-33 is not functioning correctly. Based on the voltmeter readings that you can see, what is the trouble? [7]
20. Is the circuit of Figure 9-34 functioning properly? If not, what is the cause of trouble? [7]

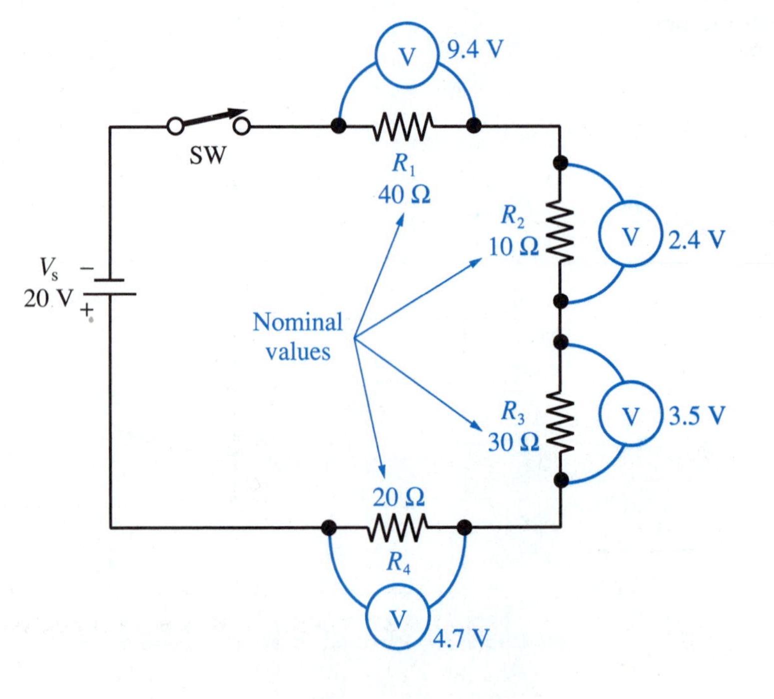

FIGURE 9-34 Results of four voltage measurements

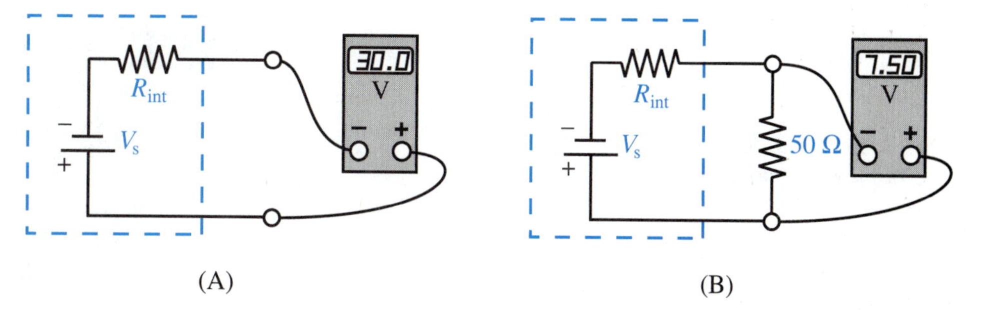

FIGURE 9-35 Real voltage source. (A) Measuring no-load voltage V_{NL}. (B) Measuring the output voltage under load.

21. A certain real voltage source is specified as having V_{NL} = 30 V, R_{int} = 50 Ω. When its open-circuit output voltage is measured in Figure 9-35(A), it reads the expected value of 30 V. When a known 50-Ω load is connected in Figure 9-35(B), the output terminal voltage is 7.5 V. What is the trouble with this circuit? [7]
22. The three-lamp circuit of Figure 9-36 is not functioning. A technician obtains four voltmeter readings as shown. What is the trouble with this circuit? [7]
23. In Figure 9-37(A), as the potentiometer wiper is turned from 0% to 100%, the voltmeter measurement varies as shown in Figure 9-37(B). What is the trouble here? [6, 7]

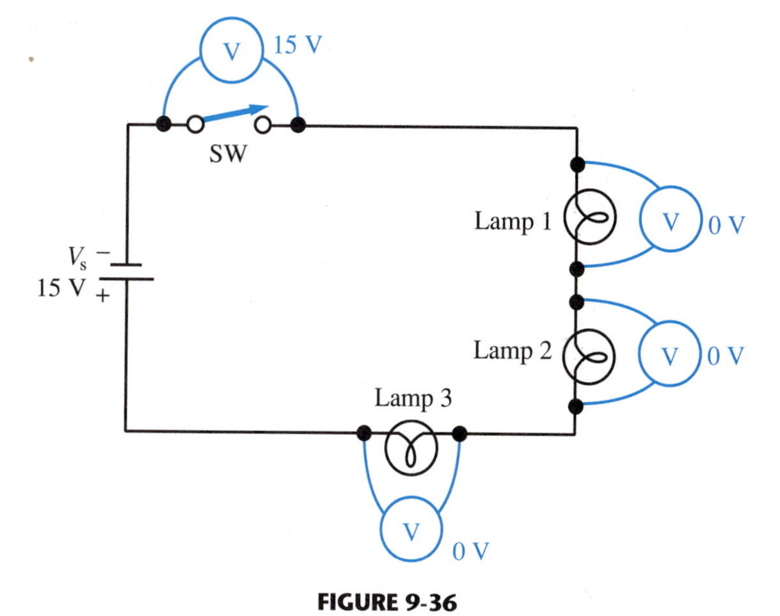

FIGURE 9-36

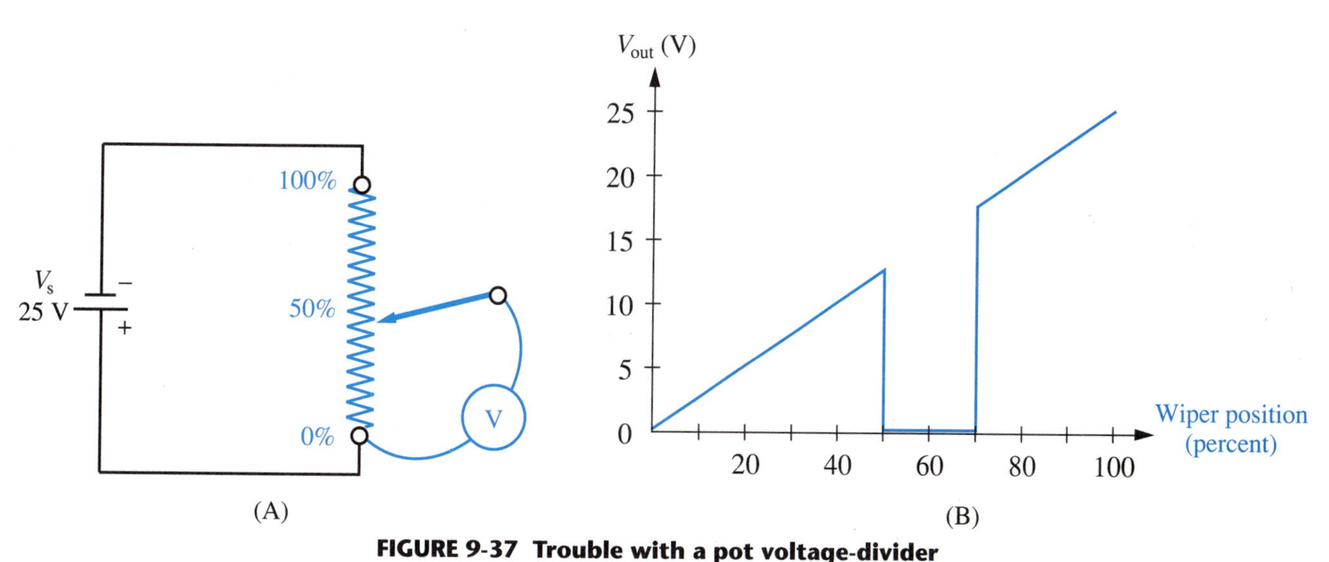

FIGURE 9-37 Trouble with a pot voltage-divider

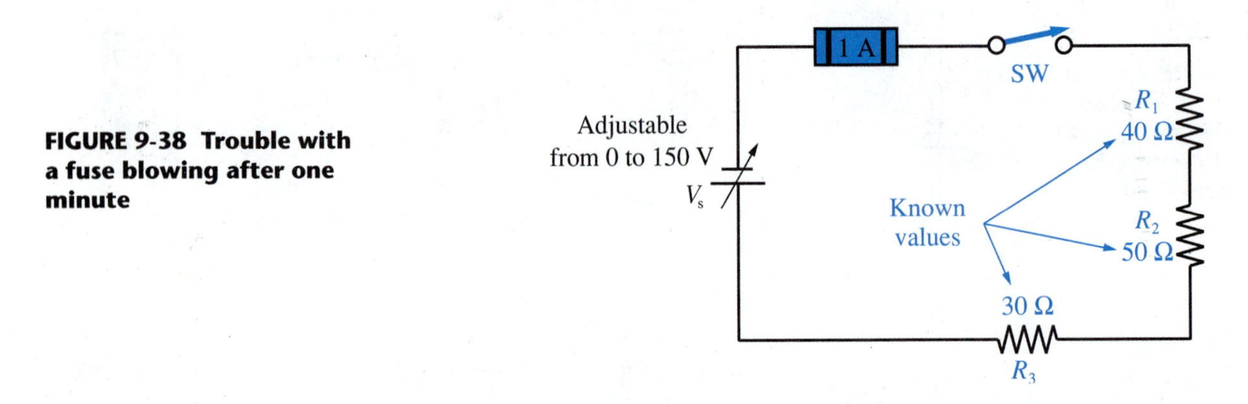

FIGURE 9-38 Trouble with a fuse blowing after one minute

24. In the volume-control circuit of Figure 9-17, suppose this happens: Starting at the bottom, the wiper tap is turned up slowly. The sound volume increases more quickly than expected, so that it reaches normal full volume with the pot at only its 50% point. Further turning of the pot beyond 50% produces no further increase in volume. What is the trouble with this circuit? [6, 7]
25. The circuit of Figure 9-38 has an adjustable dc power supply. There is no built-in meter, but the adjustment knob points to a printed scale on the front panel. It is adjusted to point to 120 V. After closing the switch, the circuit operates for about one minute, and then the fuse blows. What is the most likely trouble here? [3, 7]

C I R C U I T S

The circuits pictured in this color section correspond to certain Examples, Self-Check Questions, and Chapter Questions in the text. You will need to visually analyze these circuits in order to answer some of the questions and to follow some of the examples. Studying these real circuits will give you practice in reading the resistor color code, identifying components, and analyzing circuit problems.

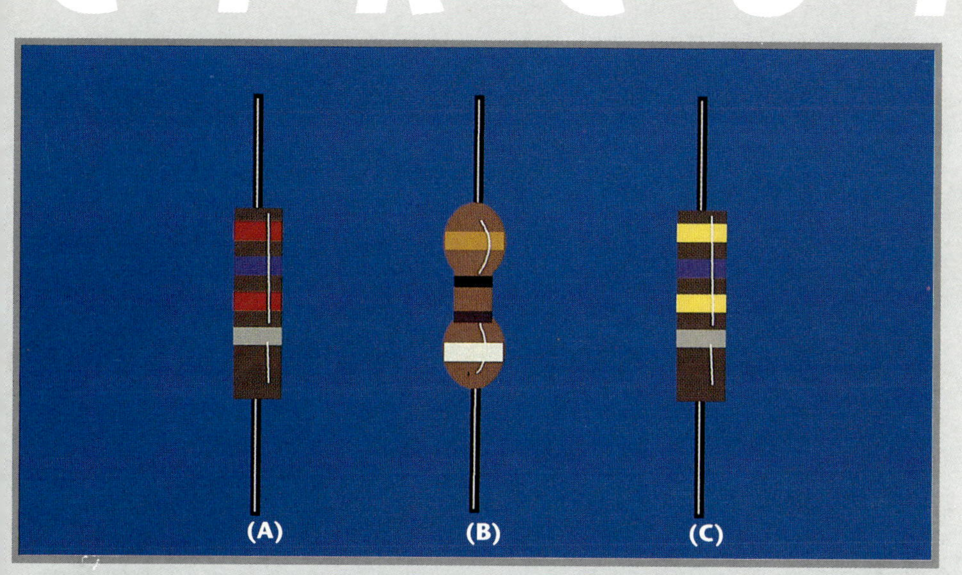

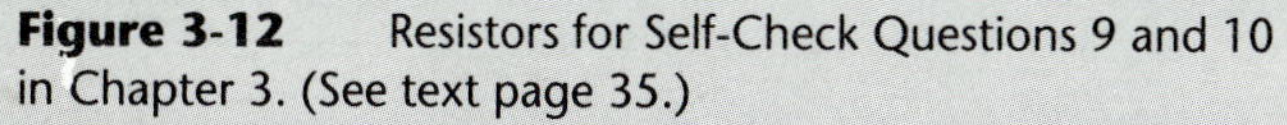

Figure 3-12 Resistors for Self-Check Questions 9 and 10 in Chapter 3. (See text page 35.)

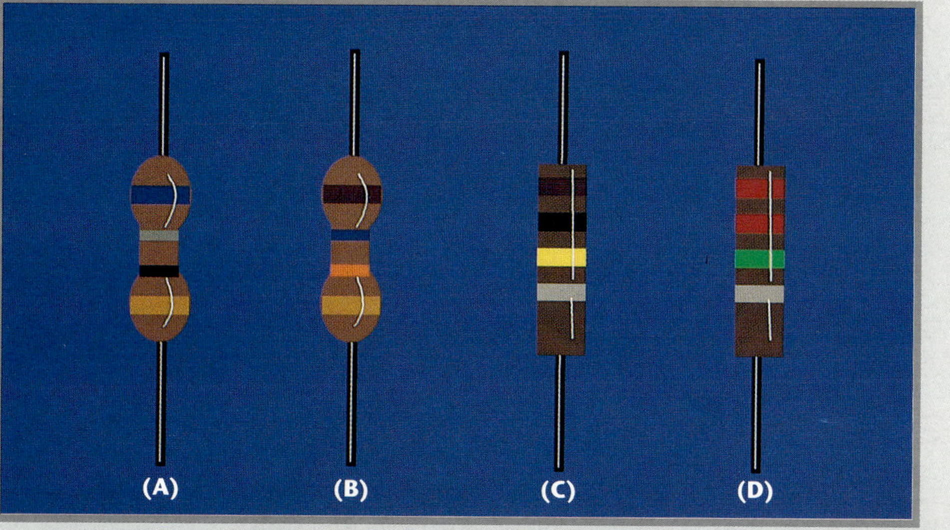

Figure 3-19 Resistors for End of Chapter Questions 10 and 11 in Chapter 3. (See text page 41.)

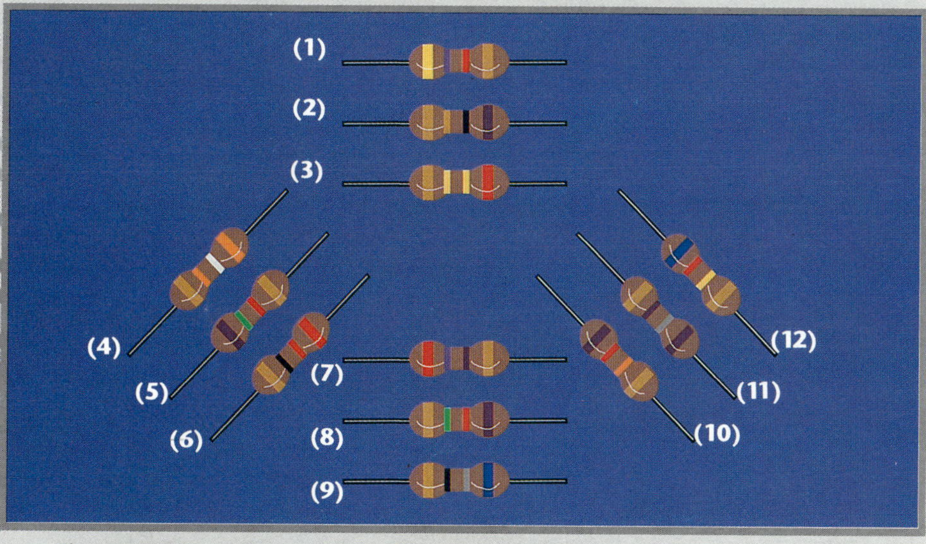

Figure 3-20 Resistors for End of Chapter Question 14 in Chapter 3. (See text page 42.)

Black = 0

Brown = 1

Red = 2

Orange = 3

Yellow = 4

Green = 5

Blue = 6

Violet = 7

Gray = 8

White = 9

Silver = ±10% tolerance

Gold = ±5% tolerance

CIRCUITS

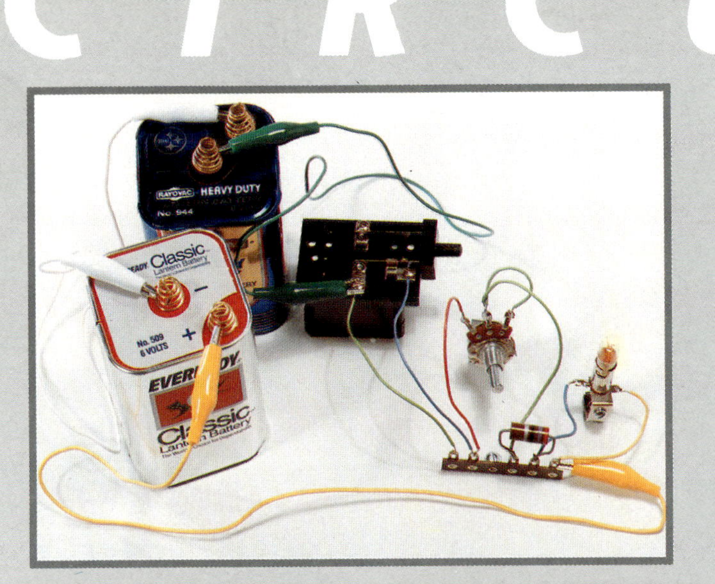

Figure 4-15
Circuit for Example 4-4 in Chapter 4.
(See text page 52.)

Figure 6-8
Setup for Example 6-17 in Chapter 6.
(See text page 88.)

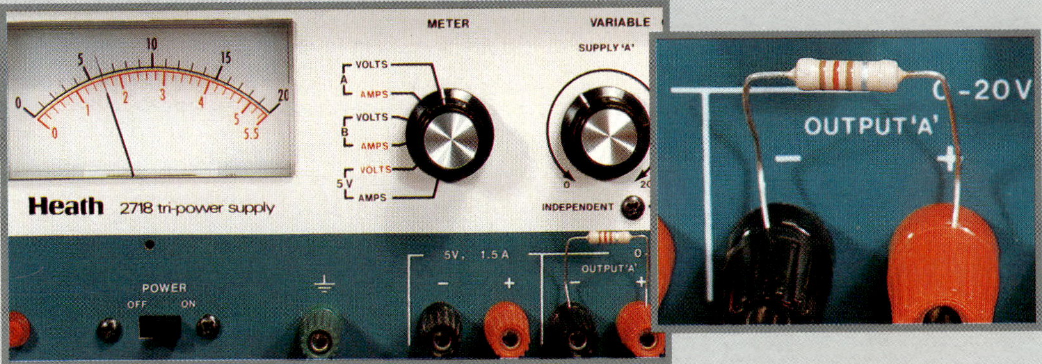

Figure 6-10(A)
Setup for End of Chapter Question 23 in Chapter 6.
(See text page 91.)

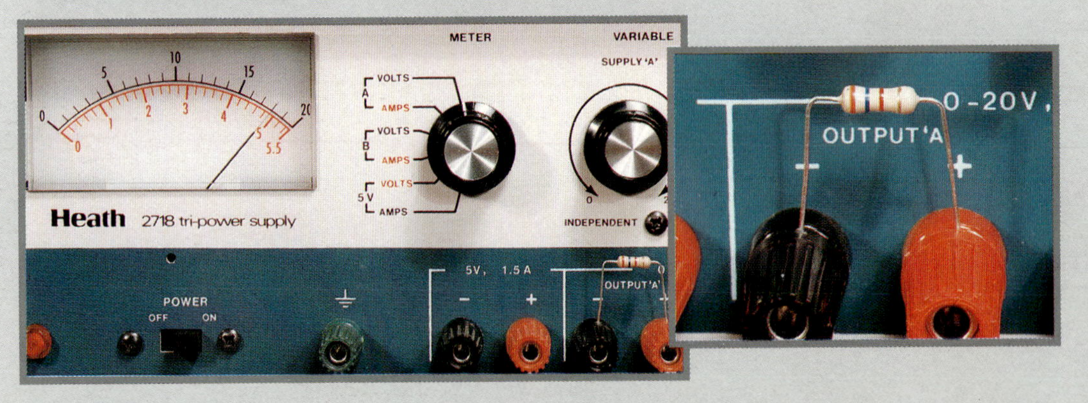

Figure 6-10(B)
Setup for End of Chapter Question 24 in Chapter 6.
(See text page 91.)

CIRCUITS

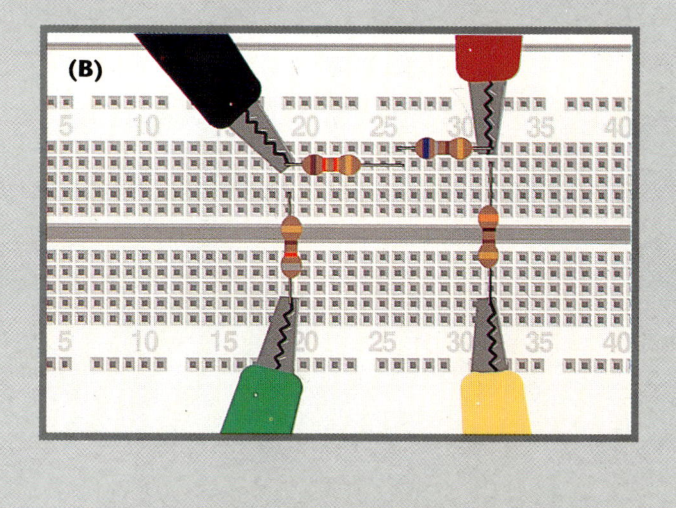

Figure 9-3
Circuit for Example 9-3 in Chapter 9 **(A)** Overall circuit showing the value of source voltage and **(B)** Close-up view of resistors. In most circuits, the dc power supply leads are black (-) and red (+). The leads from this power source are colored green (from the negative terminal) and yellow (from the positive terminal) to tell them apart from the VOM leads. (See text page 140.)

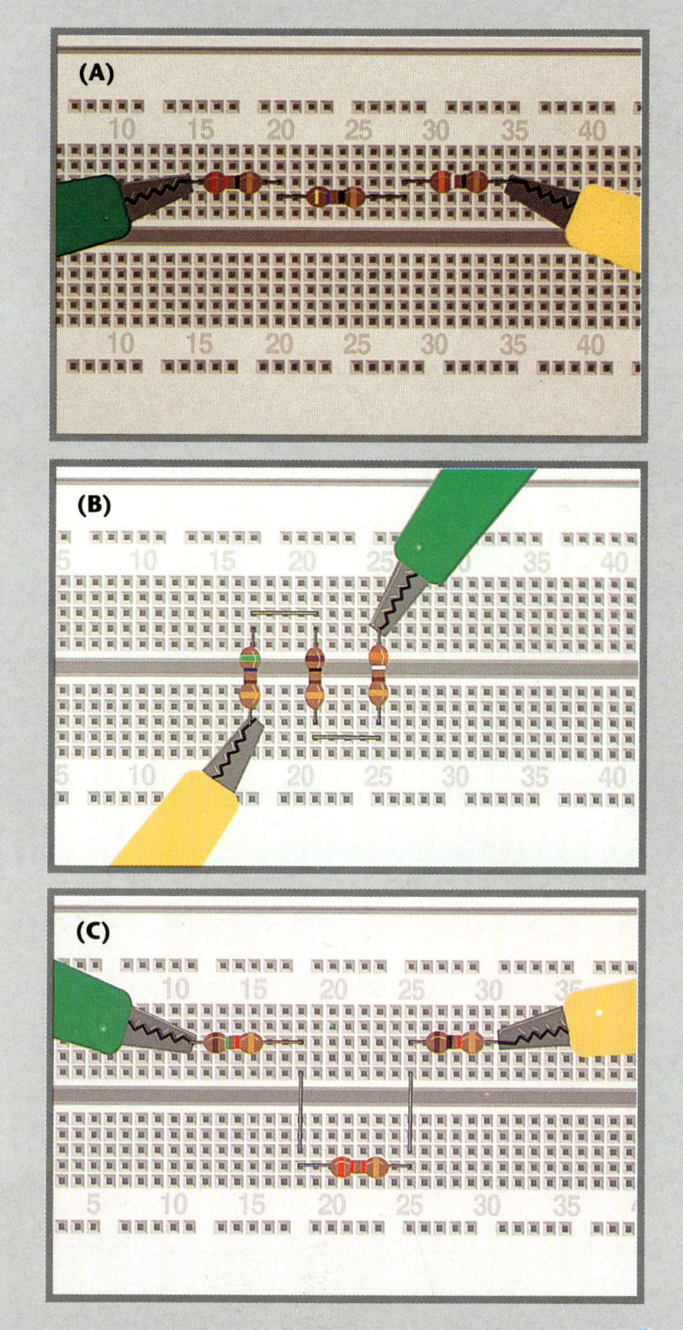

Figure 9-4
Series circuits for Self-Check Questions 4, 5, and 6 in Chapter 9. (See text page 141.)

CIRCUITS

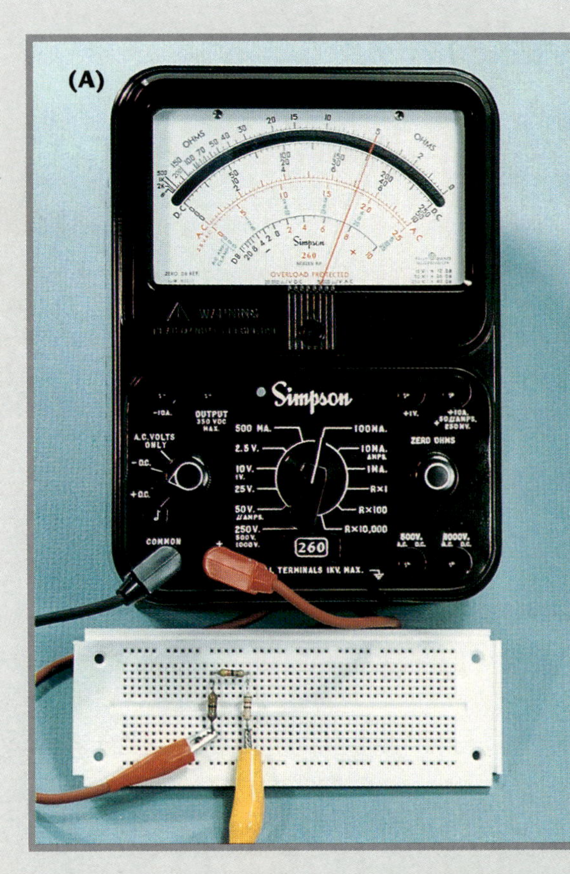

Figure 9-6
(A) Setup for Example 9-4 in Chapter 9. The VOM is acting as an ammeter. **(B)** Close-up view of resistors. (See text page 142.)

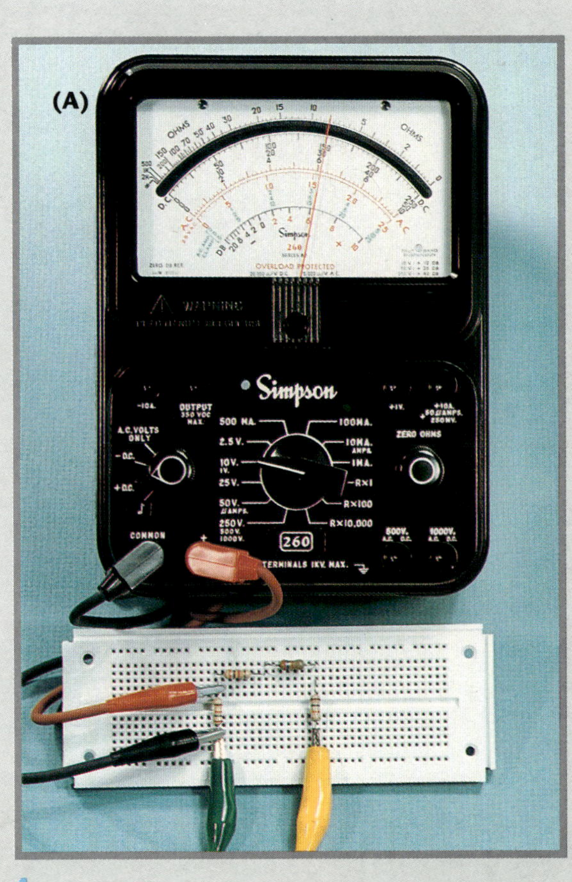

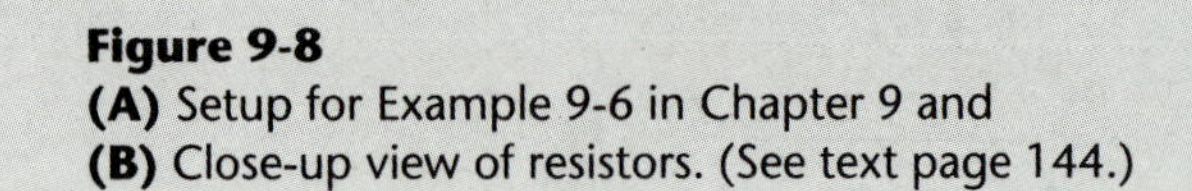

Figure 9-8
(A) Setup for Example 9-6 in Chapter 9 and **(B)** Close-up view of resistors. (See text page 144.)

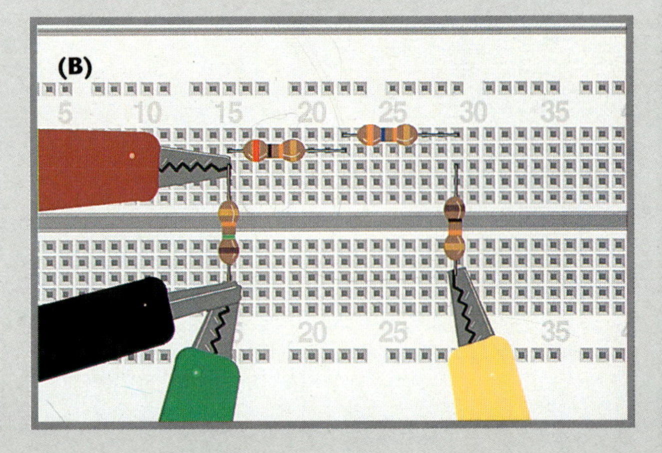

Figure 10-5
Resistor arrangements for Example 10-4 in Chapter 10.
(See text page 166.)

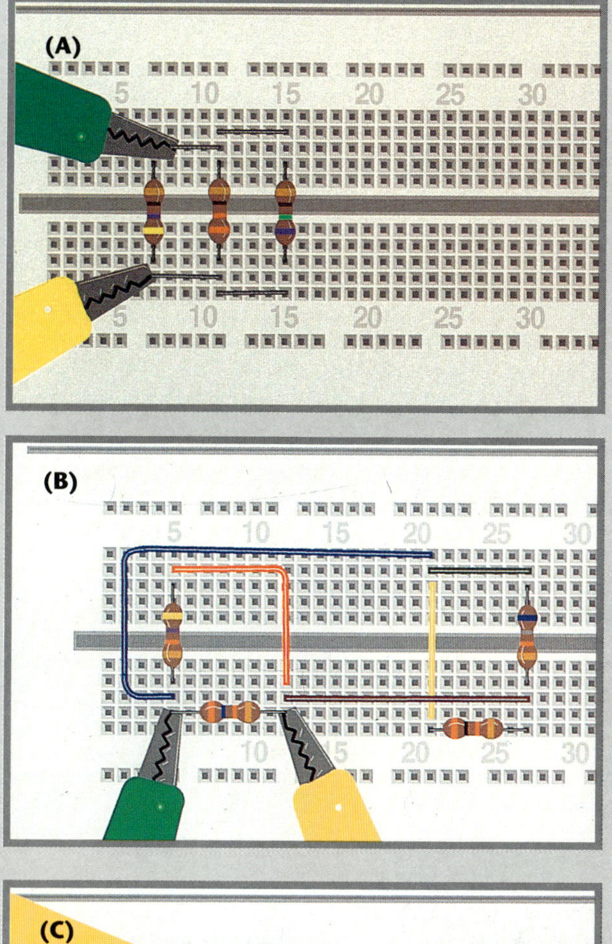

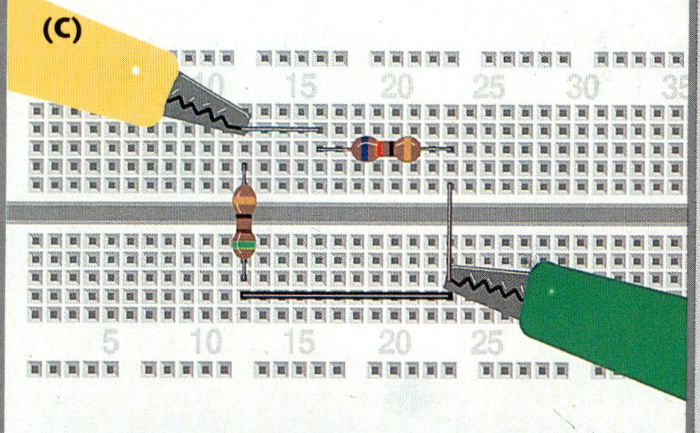

Figure 10-10 **(A)** Setup for Example 10-8 in Chapter 10 and **(B)** Close-up view of resistors.
(See text page 171.)

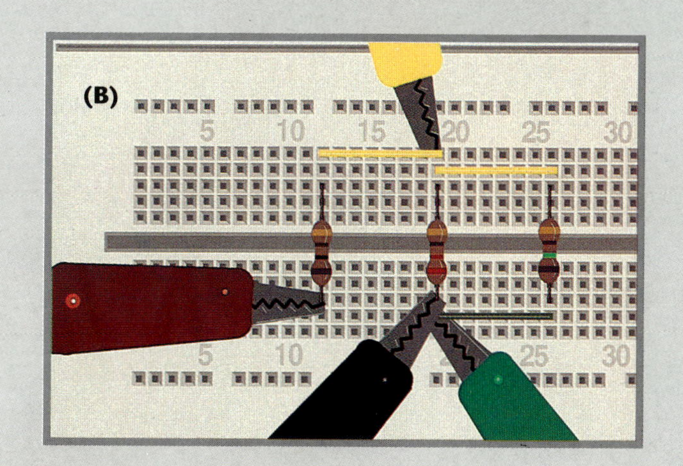

CIRCUITS

Figure 10-19 **(A)** Setup for End of Chapter Questions 11 through 16 in Chapter 10 and **(B)** Close-up view of resistors. (See text page 178.)

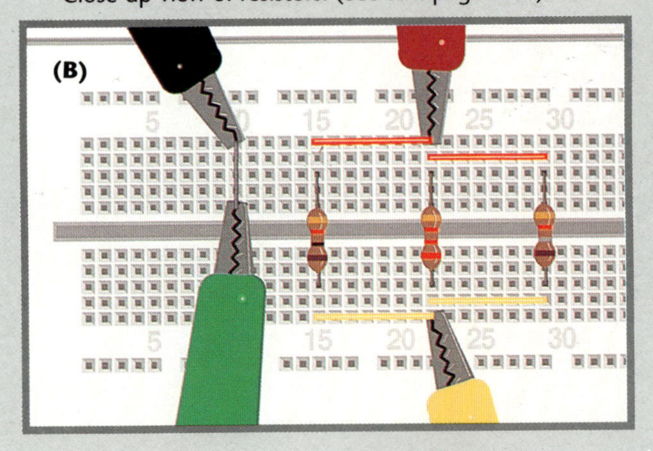

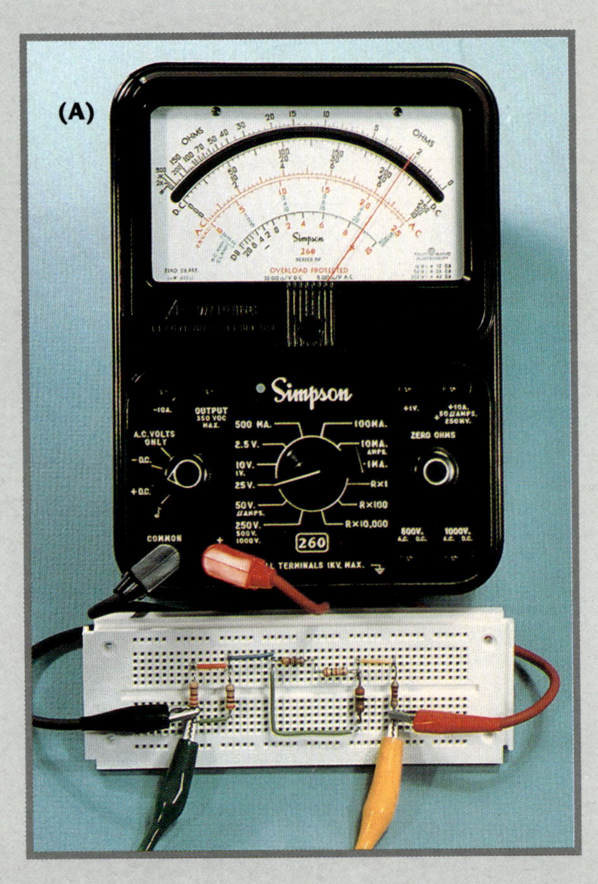

Figure 11-8 **(A)** Circuit setup for Example 11-3 in Chapter 11 and **(B)** Close-up view of circuit. (See text page 188.)

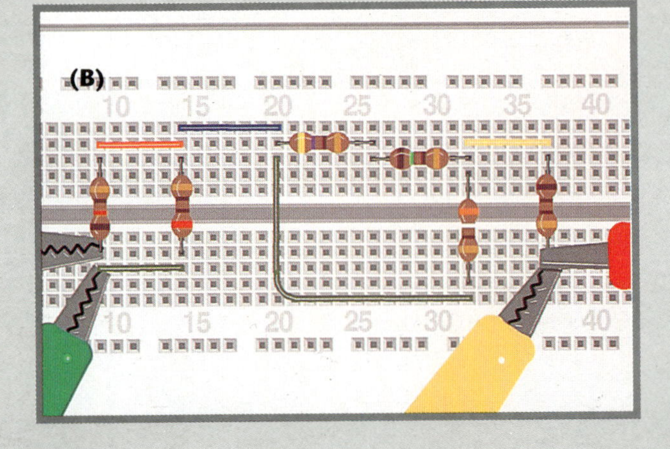

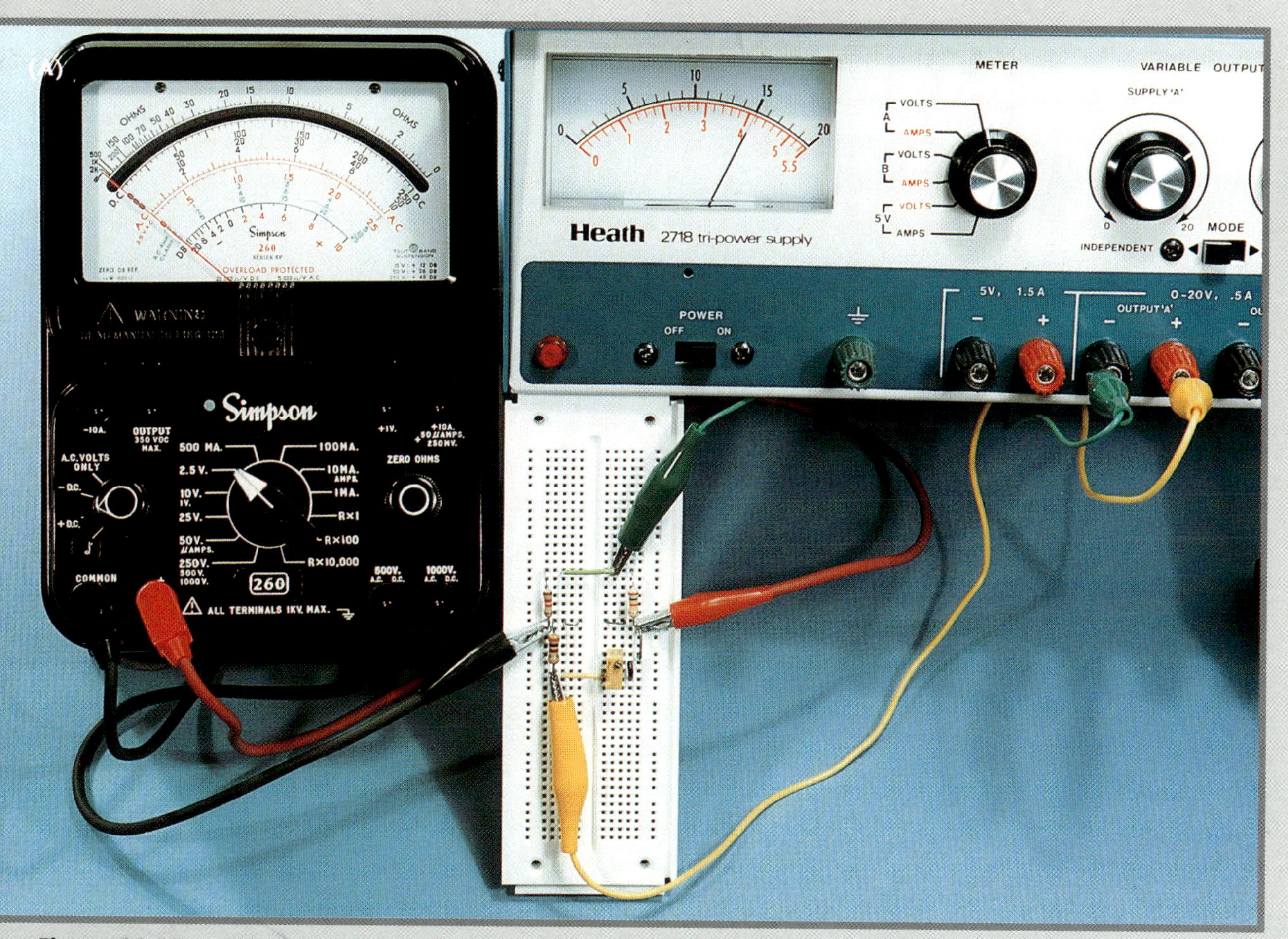

Figure 11-17 **(A)** In Chapter 11, a setup of a Wheatstone bridge. The trim pot R_4 has been adjusted to cause the voltage from A to B to equal zero. This is clear from the VOM, which is on the sensitive 2.5-V scale. **(B)** Close-up view of circuit. (See text page 193.)

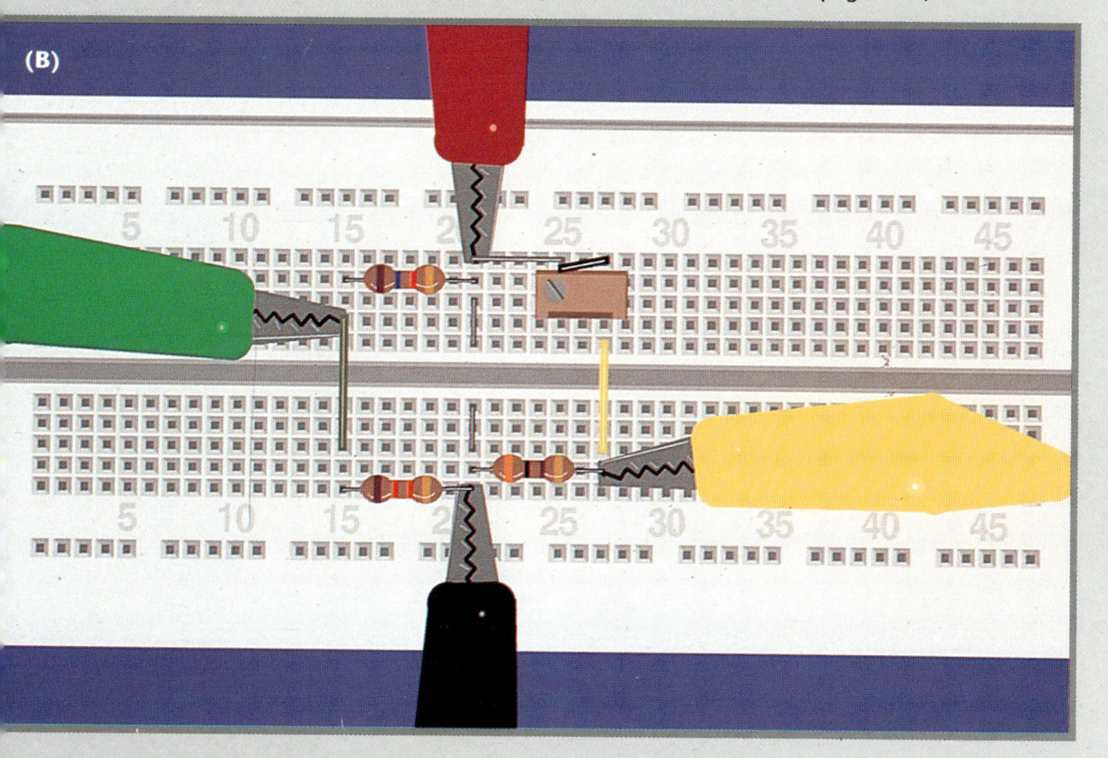

Figure 11-28 Circuit for End of Chapter Questions 7 and 8 in Chapter 11. (See text page 201.)

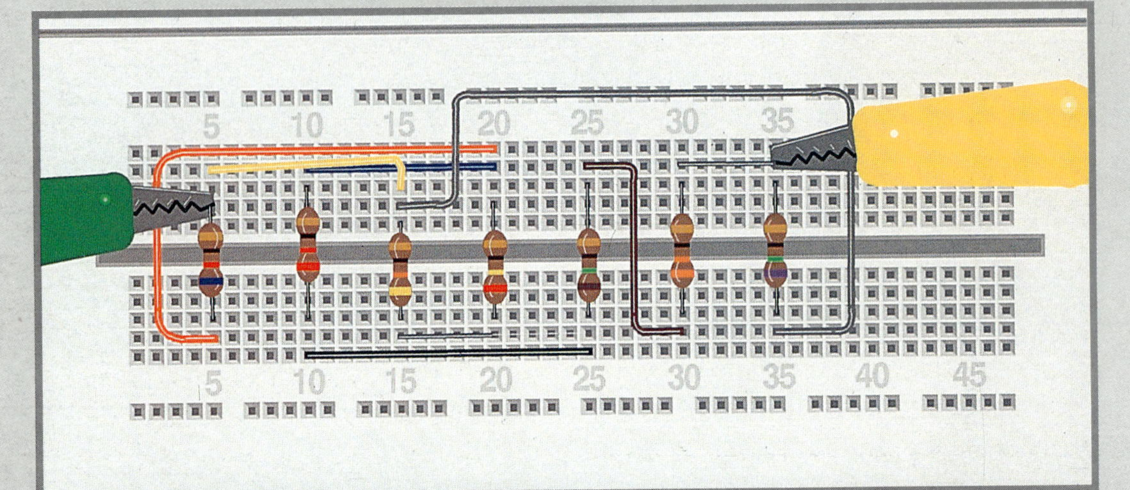

Figure 11-29 Circuit for End of Chapter Questions 9 through 12 in Chapter 11. (See text page 201.)

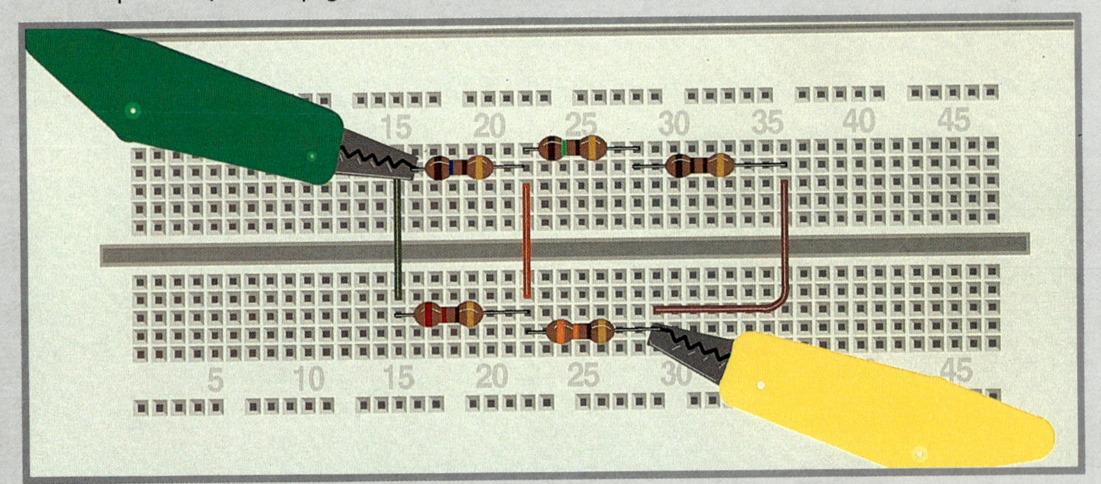

Figure 11-30 Circuit for End of Chapter Questions 13 through 16 in Chapter 11. (See text page 201.)

CHAPTER 10

PARALLEL CIRCUITS

This radar reflector dish is made of a lightweight graphite composite. It folds into a transport container that is 40 in. x 48 in. x 24 in. The dish and its container have a combined weight of 130 pounds. This radar system is used for military combat operations where easy transport and quick setup are crucial.
Courtesy of Datron Systems, Inc.

OUTLINE

NEW TERMS TO WATCH FOR

parallel circuit
branch current
reciprocal formula
1/X key
product-over-the-sum
conductance
siemens
Kirchhoff's current law
node

Some of the common electrical load devices that provide a useful output product are lights, motors, and heaters. These devices are seldom connected in series. Instead, they are connected in the other basic circuit configuration—in parallel.

After studying this chapter, you should be able to:

1. Recognize a parallel circuit.
2. Use the rules regarding current and voltage for a parallel circuit.
3. Calculate the equivalent total resistance of a parallel circuit.
4. Apply Ohm's law to an entire parallel circuit.
5. Use Kirchhoff's current law to solve for an individual branch current or the total supply current in a parallel circuit.
6. Troubleshoot a parallel circuit to locate a short circuit or an open component.

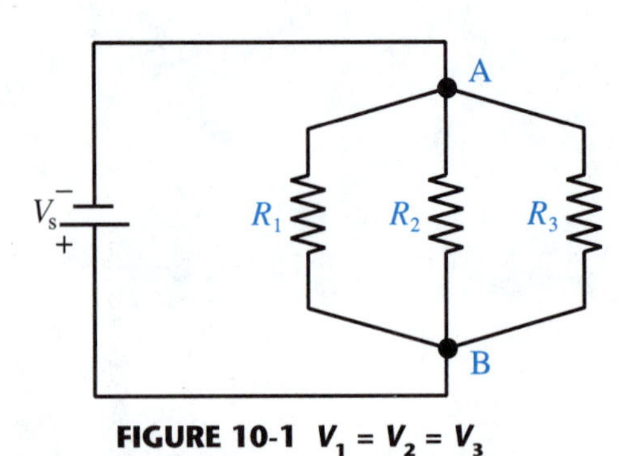

FIGURE 10-1 $V_1 = V_2 = V_3$

10-1 VOLTAGE AND CURRENT IN PARALLEL

Figure 10-1 shows one way of drawing a three-resistor **parallel circuit.** Resistors R_1, R_2, and R_3 are all connected to the same electrical points, marked A and B. Actually, point A is the same as the negative terminal of the source and point B is the same as the positive terminal. Therefore, all three resistors feel the source voltage, V_s.

Generalizing this to all strict parallel circuits, we can say:

In a parallel circuit, every individual load voltage is equal to the source voltage.

The individual **branch currents** in a three-resistor parallel circuit are labeled in Figure 10-2. Since all resistors feel the same voltage, V_s, we expect from Ohm's law that the individual currents will be different. That is,

$$I_1 = \frac{V_s}{R_1}$$

$$I_2 = \frac{V_s}{R_2}$$

and

$$I_3 = \frac{V_s}{R_3}$$

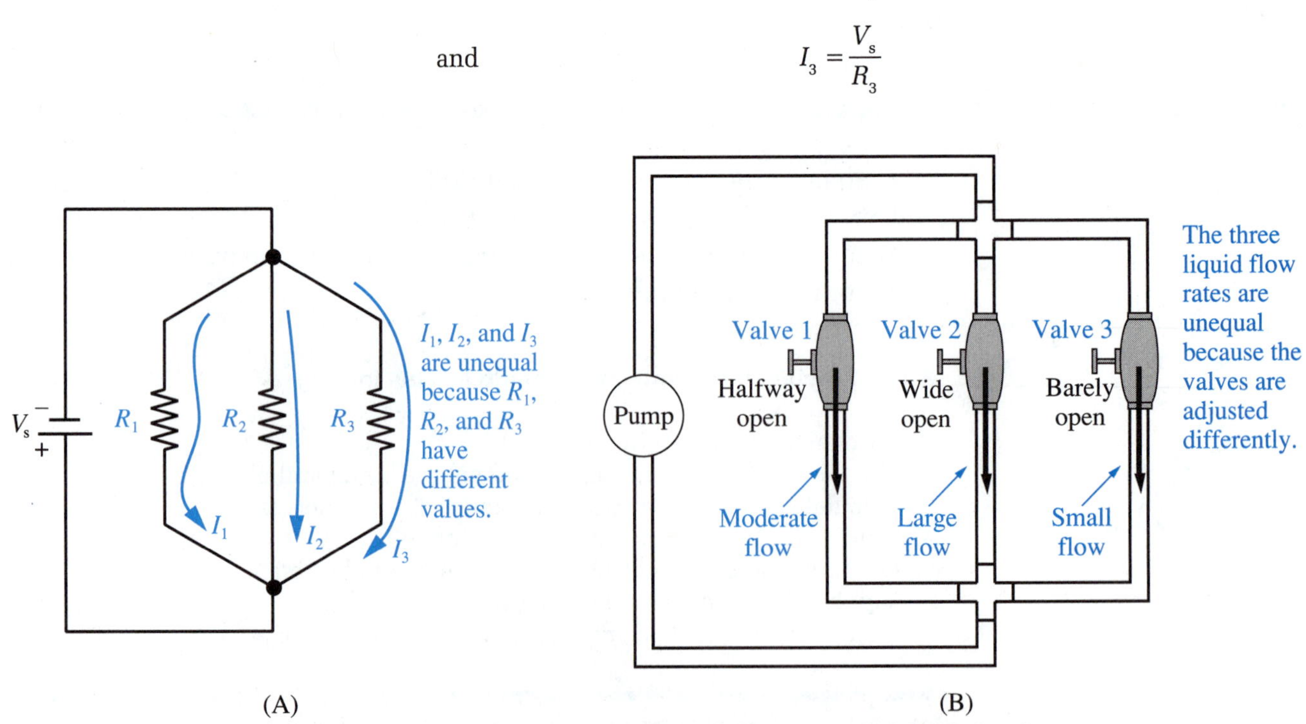

FIGURE 10-2 A parallel electric circuit is like a parallel fluid circuit.

This can be generalized to:

Individual branch currents are not equal to each other in a parallel circuit. The currents are inversely proportional to the resistances.

To understand the equal voltages and unequal currents in a parallel electrical circuit, make an analogy to a parallel fluid circuit like the one in Figure 10-2(B). It is easy to see that valve 1 has a direct piping connection to the pump. Therefore, the entire pump pressure exists across the inlet-outlet ports of valve 1. The same statement is true for valve 2 and also for valve 3. Therefore, each valve has the same pressure across its inlet-outlet ports. The voltage situation in the electric circuit is like the pressure situation in the fluid circuit.

It is also clear to see that the liquid flow rates through the valves will be different from one another, since the valves have different openings. The current situation in the electric circuit is like the flow-rate situation in the fluid circuit.

EXAMPLE 10-1

Figure 10-3 shows an easier way to draw a three-resistor parallel circuit like Figures 10-1 and 10-2(A). The source voltage is 40 V and the individual resistances are as shown.

a) If a voltmeter were connected across R_1, what value would it read?
b) If a voltmeter were connected across R_3, what value would it read?
c) If an ammeter were inserted in series with R_1, what amount of current would it measure?
d) Repeat part c for R_2.
e) Repeat part c for R_3.

SOLUTION

a) In a strict parallel circuit like this, the entire supply voltage exists across every resistor. The voltmeter would measure **40 V.**
b) Each resistor has the same voltage, **40 V.**
c) Applying Ohm's law to that individual resistor, we get

$$I_1 = \frac{V_S}{R_1} = \frac{40\ \text{V}}{80\ \Omega} = \mathbf{0.5\ A}$$

d)

$$I_2 = \frac{V_S}{R_2} = \frac{40\ \text{V}}{20\ \Omega} = \mathbf{2.0\ A}$$

e)

$$I_3 = \frac{V_S}{R_3} = \frac{40\ \text{V}}{160\ \Omega} = \mathbf{0.25\ A}$$

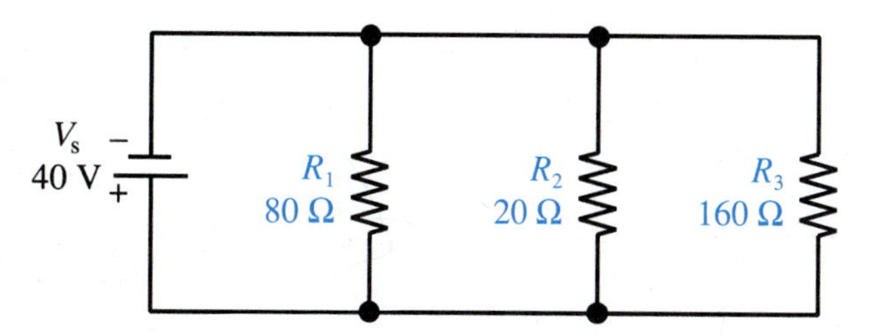

FIGURE 10-3 Three-resistor parallel circuit

Parallel Circuits Contrasted With Series Circuits

Make certain that you understand the oppositeness of parallel and series circuits. Parallel circuits have equal voltages and unequal currents. Series circuits are the other way around. This difference is summarized in Table 10-1.

TABLE 10-1

	SERIES	PARALLEL
CURRENT	Equal	Unequal
VOLTAGE	Unequal	Equal

10-2 OHM'S LAW FOR A PARALLEL CIRCUIT

In Section 10-1, we saw that Ohm's law is easy to apply to individual resistors in a parallel circuit. Ohm's law can also be applied to the circuit as a whole, if the overall total resistance of the circuit is properly calculated.

The overall total resistance of a parallel circuit is calculated by the **reciprocal formula.**

$$\frac{1}{R_T} = \frac{1}{R_1} + \frac{1}{R_2} + \frac{1}{R_3}$$

EQ. 10-1

EXAMPLE 10-2

a) In the circuit of Figure 10-3, suppose that the voltage source is disconnected by opening a switch (not shown). If an ohmmeter is placed across the three-resistor combination, what value will it measure?

b) With the circuit operating, find the total supply current, I_T. This is the current that flows in the two lines connecting to the source terminals.

SOLUTION

a) Apply the reciprocal formula, Equation 10-1. This is easily done by using the **1/X key** on a hand-held calculator.

$$\frac{1}{R_T} = \frac{1}{R_1} + \frac{1}{R_2} + \frac{1}{R_3} = \frac{1}{80} + \frac{1}{20} + \frac{1}{160}$$

$$= 0.0125 + 0.05 + 0.006\ 25$$

$$= 0.068\ 75$$

$$R_T = \frac{1}{0.068\ 75} = \mathbf{14.55\ \Omega}$$

Continued on page 165.

Using the calculator's memory to accumulate the sums, the keystroke sequence for this procedure would be:

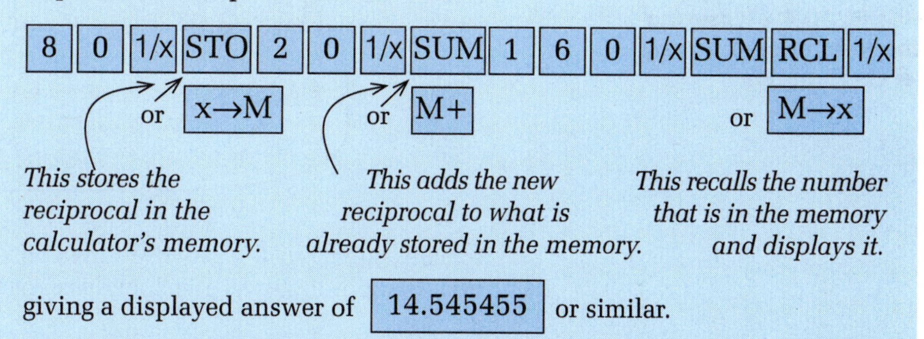

giving a displayed answer of 14.545455 or similar.

Note: Your calculator may be able to accumulate reciprocals in the display. In that case, you don't have to use the memory function. Some calculators cannot accumulate reciprocals in their display. They must use the memory.

b) Applying Ohm's law to the circuit as a whole,

$$I_T = \frac{V_s}{R_T} = \frac{40\text{ V}}{14.55\ \Omega} = \mathbf{2.75\ A}$$

When there are only two resistors in parallel, a special formula can be used to find R_T. It is called the **product-over-the-sum** formula.

$$R_T = \frac{R_1 R_2}{R_1 + R_2}$$

EQ. 10-2

EXAMPLE 10-3

Figure 10-4 shows two resistors in parallel.

a) Calculate R_T using the product-over-the-sum formula, Equation 10-2.
b) Recalculate R_T using the reciprocal formula.

SOLUTION

a) $$R_T = \frac{R_1 R_2}{R_1 + R_2} = \frac{(20)(30)}{20 + 30} = \frac{600}{50} = \mathbf{12\ \Omega}$$

On a calculator, the keystroke sequence would be:

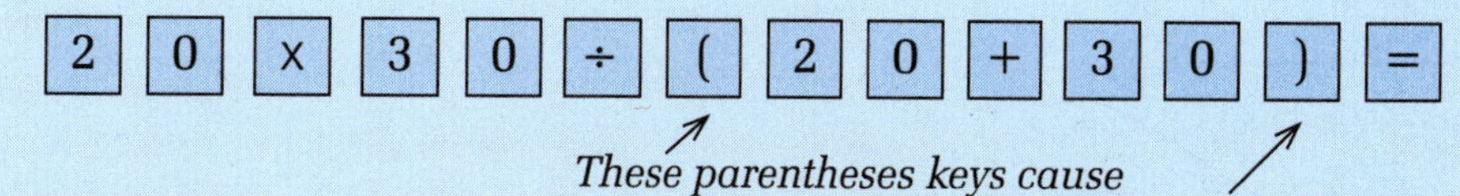

These parentheses keys cause the numbers between them to be grouped.

giving a displayed answer of 12. You must not leave out the parentheses keys in this sequence. If you do, the calculator multiplies 20 times 30, then divides that result by 20. Taking that result, it then adds 30, giving a wrong answer of 60.

b) $$\frac{1}{R_T} = \frac{1}{R_1} + \frac{1}{R_2} = \frac{1}{20} + \frac{1}{30} = 0.083\ 333$$

$$R_T = \frac{1}{0.083\ 333} = \mathbf{12\ \Omega}$$

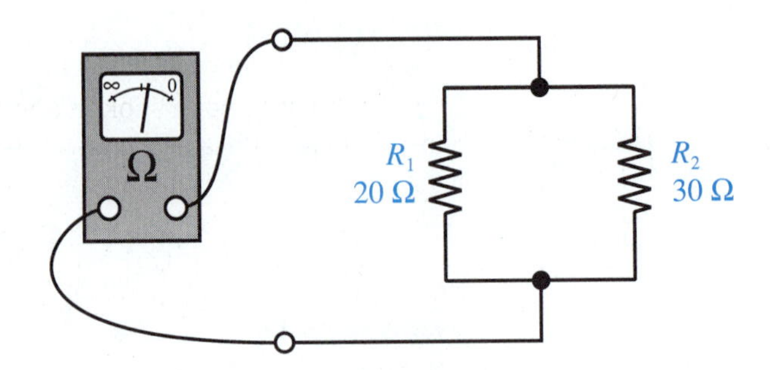

FIGURE 10-4 Measuring the resistance of a parallel combination

Actually, Equation 10-2 is derived from Equation 10-1, the reciprocal formula. Remember that product-over-the-sum does not work for three or more resistors.

(A)

(B)

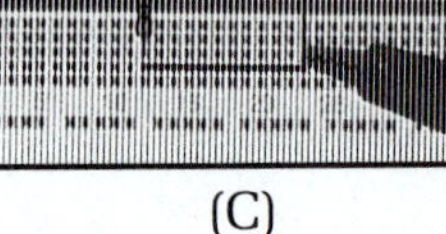

(C)

FIGURE 10-5

EXAMPLE 10-4

Refer to Figure 10-5. (See page 5 in the Circuits color section for the color version of Figure 10-5.)

a) Find the total resistance, R_T, for the resistor arrangement in Figure 10-5(A).
b) Repeat for Figure 10-5(B).
c) Repeat for Figure 10-5(C).

SOLUTION

a) It is not difficult to see that the three resistors are in parallel. From Equation 10-1,

$$\frac{1}{R_T} = \frac{1}{470} + \frac{1}{330} + \frac{1}{750} = 0.006\,491\,3$$

$$R_T = \frac{1}{0.006\,491\,3} = \mathbf{150\ \Omega}$$

b) Carefully trace the wire connections to satisfy yourself that this is a four-resistor parallel circuit. Its resistance is given by

$$\frac{1}{R_T} = \frac{1}{47\ \text{k}\Omega} + \frac{1}{36\ \text{k}\Omega} + \frac{1}{30\ \text{k}\Omega} + \frac{1}{68\ \text{k}\Omega}$$

$$R_T = \mathbf{10.3\ k\Omega}$$

c) This two-resistor parallel circuit can be solved by Equation 10-2.

$$R_T = \frac{R_1R_2}{R_1 + R_2} = \frac{62(51)}{62 + 51} = \mathbf{28.0\ \Omega}$$

Wind-tunnel flow data is automatically recorded in computer memory.
Courtesy of National Instruments Co.

Conductance

We have always spoken about a component's ability to oppose current—its resistance. It is possible to take the other point of view. We could talk about a component's ability to pass current—its **conductance.** Greater conductance permits greater current flow, other things being equal. The symbol for conductance is G. Its basic unit is the **siemens**, symbolized S. Mathematically, resistance and conductance are reciprocals of each other. That is,

$$R = \frac{1}{G} \quad \text{and} \quad G = \frac{1}{R}$$

EQ. 10-3

EXAMPLE 10-5

Refer back to Figure 10-3.

a) Using your R_T result from Example 10-2, calculate the circuit's total conductance, G_T.

b) Write the formula for total conductance of a parallel circuit.

SOLUTION

a) From Equation 10-3,

$$G_T = \frac{1}{R_T} = \frac{1}{14.55\ \Omega} = \mathbf{0.0687\ S}$$

b) Substitute the reciprocal relationship (Equation 10-3) into each term of the R_T formula
(Equation 10-1).

$$\frac{1}{R_T} = \frac{1}{R_1} + \frac{1}{R_2} + \frac{1}{R_3}$$

$$G_T = G_1 + G_2 + G_3$$

EQ. 10-4

TECHNICAL FACT

Individual branch conductances can simply be added to find the total conductance of a parallel circuit.

✔ SELF-CHECK FOR SECTIONS 10-1 AND 10-2

1. In a parallel circuit, voltages are__________and currents are__________. [2]
2. The total resistance of a parallel circuit is always______________than the __________individual resistance. [3]
3. In Figure 10-3, suppose $V_s = 60$ V, $R_1 = 10$ kΩ, $R_2 = 5.6$ kΩ and $R_3 = 18$ kΩ. [3, 4]
   a) Find R_T.
   b) Find I_T.
4. Suppose that the circuit in Problem 3 has a fourth resistor added to it, with $R_4 = 4.7$ kΩ. Calculate the new values of R_T and I_T. [3, 4]

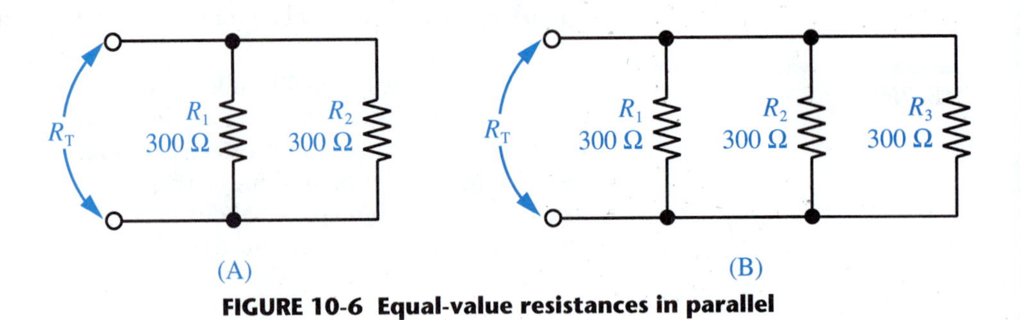

FIGURE 10-6 Equal-value resistances in parallel

5. Compare the answers for Problem 4 to the answers for Problem 3. Explain why this is reasonable. [3, 4]
6. A 150-Ω resistor is in parallel with a 100-Ω resistor. Calculate their total resistance using the product-over-the-sum formula. [3]
7. A two-resistor parallel combination has $R_1 = R_2 = 300\ \Omega$, as shown in Figure 10-6(A). Find R_T. [3]
8. A three-resistor parallel combination has $R_1 = R_2 = R_3 = 300\ \Omega$, as shown in Figure 10-6(B). Find R_T. [3]
9. Based on your answers to Problems 7 and 8, write a formula for R_T for any number of equal-value parallel resistors. [3]
10. Three parallel loads have $G_1 = 0.02$ S, $G_2 = 0.05$ S, and $G_3 = 0.10$ S. Find the circuit's total conductance, G_T.

10-3 KIRCHHOFF'S CURRENT LAW

For Strict Parallel Circuits

The three-resistor parallel circuit of Figure 10-7 can be used to explain **Kirchhoff's current law.** The total current, I_T, flows from the source to electrical junction point A. At that junction, or **node,** I_T splits into three parts, I_1, I_2, and I_3. Since overall charge flow cannot be diminished simply by entering an electrical junction, the sum of the three branch currents must equal the total current. As an equation,

$$I_T = I_1 + I_2 + I_3$$

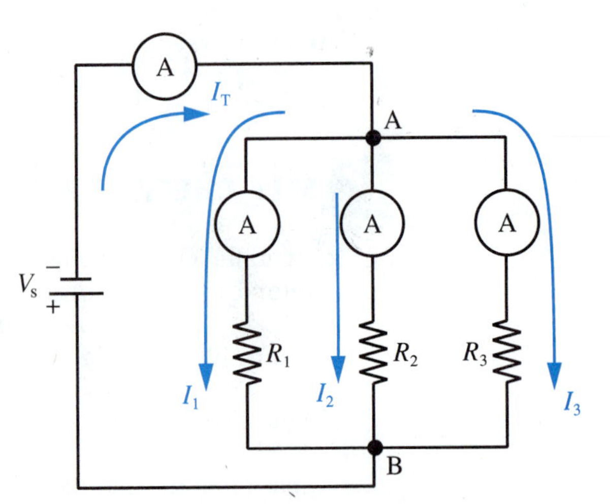

FIGURE 10-7 Individual branch currents add up to I_T in a parallel circuit.

Kirchhoff's current law for strict parallel circuits states:

TECHNICAL FACT

The sum of the branch currents equals the total current coming from the source.

5

$$I_T = I_1 + I_2 + I_3 + \ldots$$

EQ. 10-5

EXAMPLE 10-6

In Figure 10-7, the three ammeters in the branches measure $I_1 = 0.5$ A, $I_2 = 1.2$ A, $I_3 = 0.7$ A.

a) What value of current is measured by the ammeter in the negative supply line?

b) If an ammeter were placed between node B and the positive terminal of the source, what current value would it measure?

SOLUTION

a) By Kirchhoff's current law,

$$I_T = I_1 + I_2 + I_3$$
$$= 0.5 + 1.2 + 0.7 = \mathbf{2.4\ A}$$

b) Overall charge flow must be maintained in node B just as it was in node A. Charge cannot be created or destroyed simply by passing through an electrical junction point. Therefore, the current leaving point B is also given by Kirchhoff's current law.

$$I_B = I_1 + I_2 + I_3 = \mathbf{2.4\ A}$$

This result was expected since the current returning to the source's positive terminal must be the same as the amount leaving its negative terminal.

The equivalent but easier to draw schematic diagram is shown in Figure 10-8. The equivalence between the A and B nodes in both figures is clear to see.

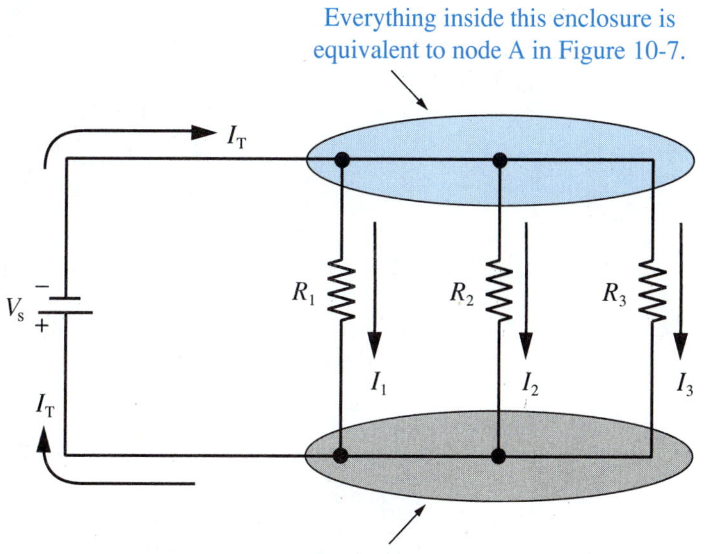

FIGURE 10-8 Identifying nodes

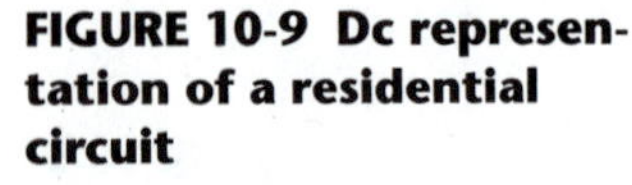

FIGURE 10-9 Dc representation of a residential circuit

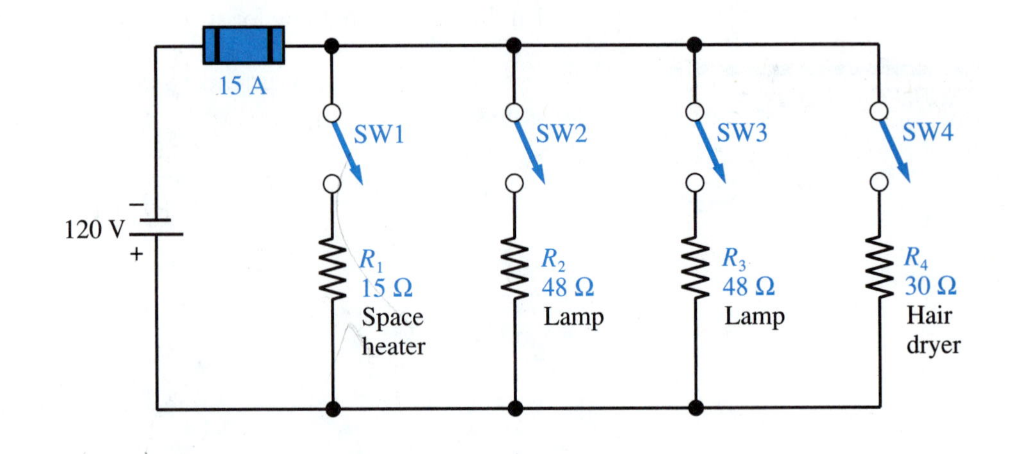

EXAMPLE 10-7

In Figure 10-9, suppose that SW1, SW2, and SW3 are all closed, connecting the space heater and both lamps in parallel across the source. SW4 is open.

a) Calculate the current through each individual branch (each individual load).

b) Find total current, I_T. Will the fuse allow this?

Now suppose SW4 is closed, connecting the hair dryer in parallel with the other three loads.

c) Calculate the new total current. Will the fuse allow this?

d) What combination of loads will the fuse allow?

SOLUTION

a) Applying Ohm's law to each resistive load gives

$$I_1 = \frac{V_s}{R_1} = \frac{120\text{ V}}{15\ \Omega} = \mathbf{8\ A}$$

$$I_2 = \frac{V_s}{R_2} = \frac{120\text{ V}}{48\ \Omega} = \mathbf{2.5\ A}$$

$I_3 = \mathbf{2.5\ A}$, the same as I_2, since the resistances happen to be the same.

b) $$I_T = I_1 + I_2 + I_3 = 8.0 + 2.5 + 2.5 = \mathbf{13\ A}$$

Yes the fuse will allow this, since 13 A is less than the fuse's current rating.

c) $$I_4 = \frac{120\text{ V}}{30\ \Omega} = 4.0\text{ A}$$

$$I_T = (I_1 + I_2 + I_3) + I_4 = 13\text{ A} + 4.0\text{ A} = \mathbf{17\ A}$$

No, the fuse will blow eventually. I_T exceeds the 15-A fuse rating.

d) If just one lamp is disconnected by opening either SW2 or SW3, the total current will decrease by 2.5 A. Then,

$$I_T = 17\text{ A} - 2.5\text{ A} = \mathbf{14.5\ A}$$

which the fuse will allow.

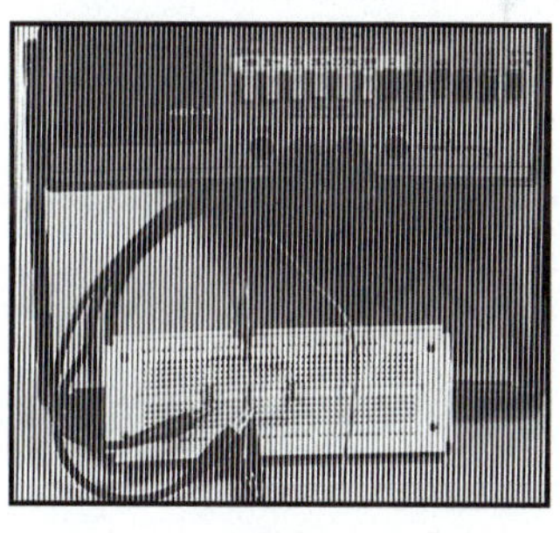

(A)

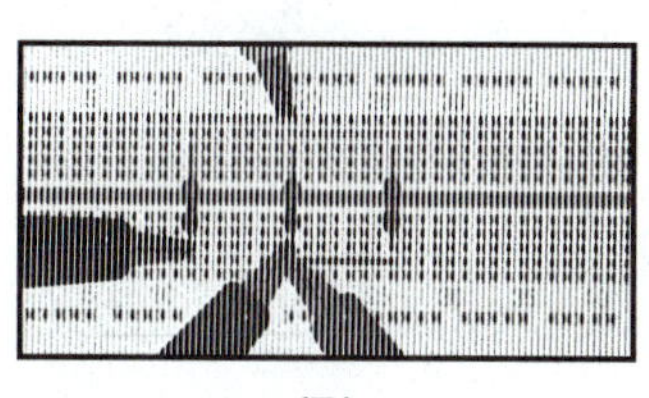

(B)

FIGURE 10-10

EXAMPLE 10-8

Refer to Figure 10-10. (See page 5 in the Circuits color section for the color version of Figure 10-10.)

a) What is the current through R_1?
b) Find I_2 , the current through R_2.
c) Find I_3 , the current through R_3.

SOLUTION

a) The DMM ammeter is in series with 100-Ω resistor R_1. Satisfy yourself that this is true. From the ammeter,

$$I_1 = \mathbf{62.6\ mA}$$

b) Applying Ohm's law to R_1 alone, we get

$$V_1 = I_1 R_1$$

$$= (62.6 \times 10^{-3}\ \text{A})\ (100\ \Omega) = 6.26\ \text{V}$$

This is a three-resistor parallel circuit. All voltages are equal in a parallel circuit, so

$$V_2 = V_3 = V_1 = 6.26\ \text{V}$$

Therefore, the source voltage, V_s, is 6.26 V.

Applying Ohm's law to R_2 alone, we get

$$I_2 = \frac{V_s}{R_2} = \frac{6.26\ \text{V}}{220\ \Omega} = \mathbf{28.5\ mA}$$

c)

$$I_3 = \frac{V_s}{R_3} = \frac{6.26\ \text{V}}{150\ \Omega} = \mathbf{41.7\ mA}$$

SAFETY ADVICE

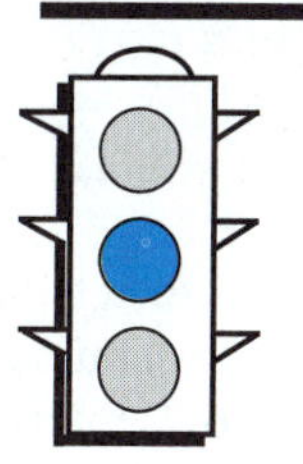

When installing an ammeter in a circuit, it is necessary to break the circuit open, as you know very well. Always turn the power off before you make your disconnection. Then install the ammeter. Then turn the power back on, to obtain the current measurement.

Never attempt to disconnect a circuit that is energized. The sudden stopping of the circuit's current can produce dangerous arcing and sparking. Of course, the On-Off switch itself must suddenly stop the circuit's current. But these switches are specially built to suppress arcing. The circuit designer will have chosen a proper switch model to safely stop this current.

Kirchhoff's Current Law for Circuit Locations in General

Kirchhoff's current law can also be stated in a more general way.

TECHNICAL FACT

For any electrical junction point, the sum of all the currents entering the junction point is equal to the sum of all the currents leaving the junction point.

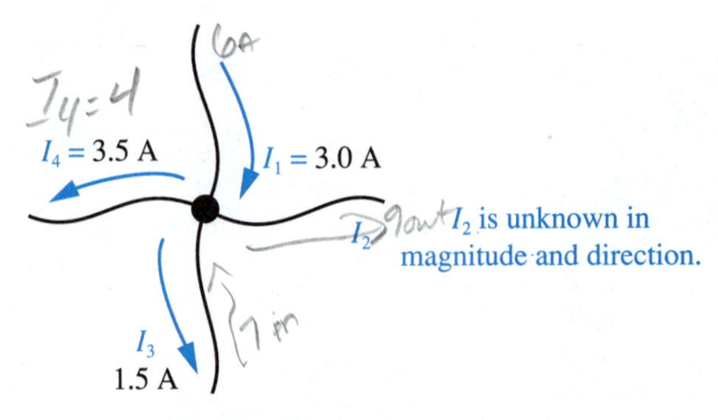

FIGURE 10-11 Four-wire node

EXAMPLE 10-9

Figure 10-11 shows a four-wire junction point. The currents in three wires are known. Their directions and magnitudes are shown in the figure. Find the current direction and magnitude in the fourth wire.

SOLUTION

Of the three currents we know, one is entering and two are leaving the node. The totals so far are:

$$I_{entering} = I_1 = 3.0 \text{ A}$$

$$I_{leaving} = I_3 + I_4 = 1.5 \text{ A} + 3.5 \text{ A} = 5.0 \text{ A}$$

Therefore, I_2 must equal the difference between these values.

$$I_2 = I_{leaving} - I_{entering} = 5.0 \text{ A} - 3.0 \text{ A} = \mathbf{2.0\ A,\ entering\ the\ node.}$$

✓ SELF-CHECK FOR SECTION 10-3

11. Kirchhoff's voltage law is applied to series circuits. Kirchhoff's ______ law is applied to parallel circuits. [5]
12. In a certain two-branch parallel circuit, the branch currents are 2.5 A and 4.25 A. What is the total current in the supply line? [5]
13. In Figure 10-7, suppose V_s = 24 V, R_1 = 48 Ω, R_2 = 96 Ω, and R_3 = 80 Ω. Calculate I_T using Kirchhoff's current law. [5]
14. In Problem 13, suppose a fourth load resistance, R_4, is connected in parallel, causing the new total current to become 1.45 A. Find the resistance of R_4.[3, 4]
15. In Figure 10-11, suppose the wire on top carries I_1 = 6 A entering the node. On the right, I_2 = 9 A leaving the node. On the bottom, I_3 = 7 A entering the node. Find the magnitude and direction of I_4, the amount on the left. [5]

Three-dimensional visualization is possible with today's powerful computers.
Courtesy of Tektronix, Inc.

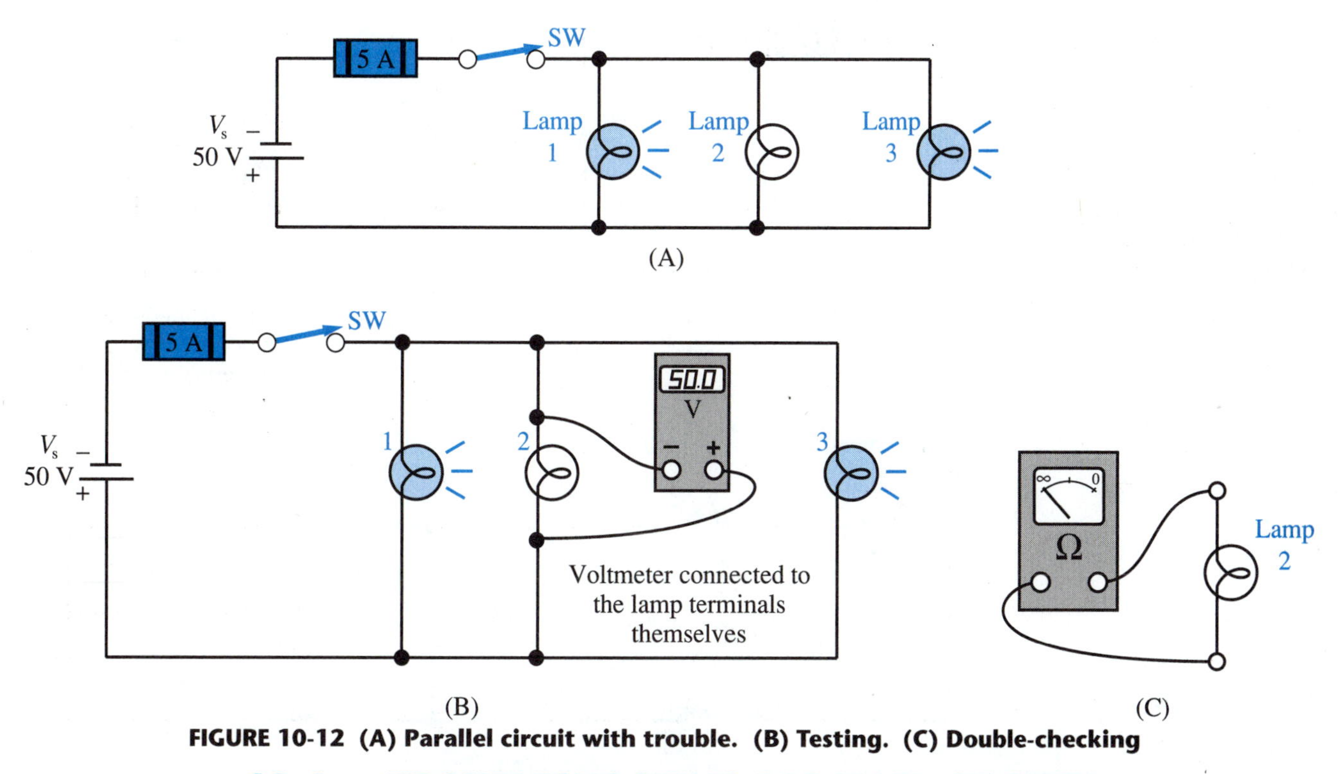

FIGURE 10-12 (A) Parallel circuit with trouble. (B) Testing. (C) Double-checking

10-4 TROUBLESHOOTING PARALLEL CIRCUITS

TECHNICAL FACT

In a parallel circuit, if one load fails in the open condition, all the other loads continue to work. Only the failed load stops working.

EXAMPLE 10-10

In the parallel circuit of Figure 10-12(A), lamp 2 has gone out. Troubleshoot the circuit to tell whether the lamp filament itself has opened or whether there is some disconnection in the circuit leading up to the lamp. Assume that the lamp terminals themselves are accessible.

SOLUTION

Connect a voltmeter across the lamp terminals as shown in Figure 10-12(B). In that figure, the voltmeter reads the V_s value of 50 V. Therefore, the wiring up to the lamp is intact and the lamp filament must be open.

To double-check this conclusion, open the switch to deenergize the entire circuit. Remove lamp 2 and measure it with an ohmmeter, as shown in Figure 10-12(C). An infinite-ohm indication confirms that the lamp filament is open.

6

On the other hand, it might happen that a voltmeter across the lamp 2 terminals measures 0 V, as shown in Figure 10-13. This would indicate that the trouble is in the wiring from the main supply lines to the lamp. For example, the connection of the lamp's bottom terminal to the bottom supply line might be faulty because of a broken wire, a cold solder joint, or a poor screw connection.

TECHNICAL FACT

In a parallel circuit, if just one load fails in the shorted condition, it causes an overcurrent that will probably blow the fuse or circuit breaker. Then all the loads are deenergized.

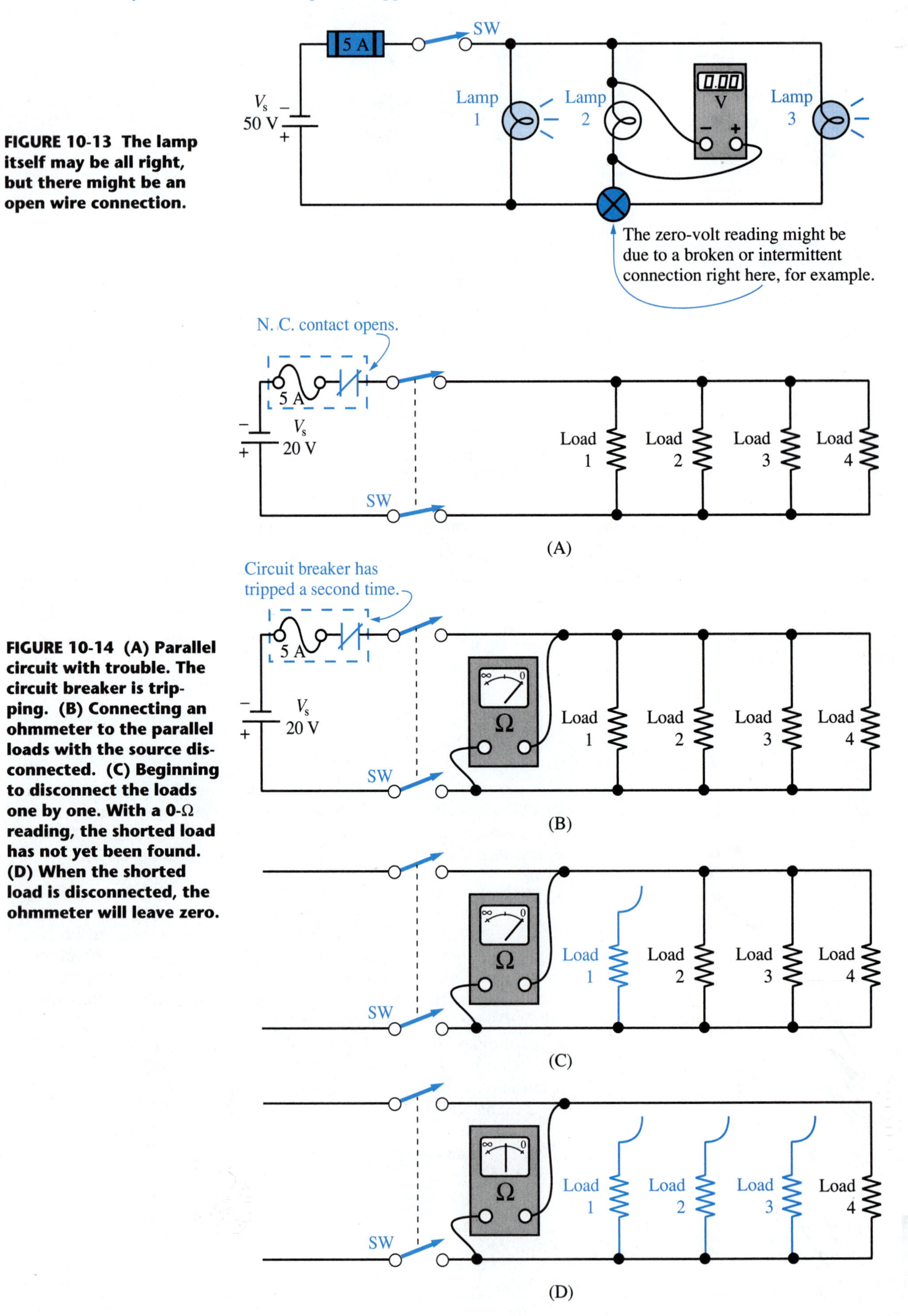

FIGURE 10-13 The lamp itself may be all right, but there might be an open wire connection.

FIGURE 10-14 (A) Parallel circuit with trouble. The circuit breaker is tripping. (B) Connecting an ohmmeter to the parallel loads with the source disconnected. (C) Beginning to disconnect the loads one by one. With a 0-Ω reading, the shorted load has not yet been found. (D) When the shorted load is disconnected, the ohmmeter will leave zero.

EXAMPLE 10-11

Suppose the circuit breaker in Figure 10-14(A) trips. How would you troubleshoot the circuit?

SOLUTION

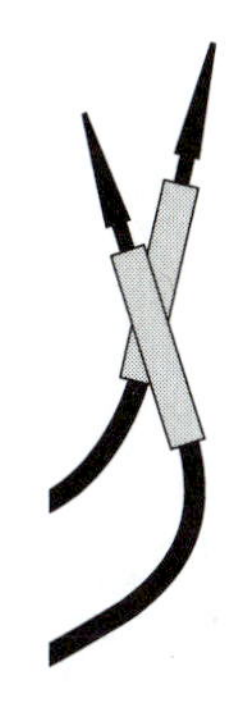

You would try one time to reclose the circuit breaker contact. If the circuit shutdown is due to a nuisance trip, then reclosing the circuit breaker will get the circuit working again. Nuisance shutdowns can occur because of a temporary surge from the voltage source or for other reasons involving outside interference.

If the circuit breaker trips a second time, then there is trouble in the parallel circuit itself. One way to proceed is to open the double-pole switch to completely isolate the loads from the source. Then place an ohmmeter across the parallel loads, as shown in Figure 10-14(B). The ohmmeter will read zero ohms, or an unusually low resistance, depending on whether the trouble is a dead-shorted load or a partially shorted load.

Disconnect one end of load 1, as shown in Figure 10-14(C). If load 1 had been the shorted load, the ohmmeter would have changed its indication from zero ohms. Continue disconnecting the loads, one at a time. This may be difficult to do, depending on how the circuit is constructed. (It may be soldered to terminal strips, soldered to copper tracks on a PC board, held together by wire nuts, crimped together, or screwed together.)

When you reach the shorted load, its disconnection will cause the ohmmeter to return to a normal reading. In Figure 10-14(D), the technician had to reach load 3 before the ohmmeter left the 0-Ω reading. Therefore, the trouble was load 3 shorted.

A final check is to remove load 3 entirely and measure it with an ohmmeter. It will have an unusually low resistance.

PHYSICAL THERAPY

This computerized physical therapy machine is used to evaluate the condition of hip, knee, and ankle joints and their associated muscles. It is also an effective rehabilitation tool. The computer receives information about position, speed, and force from electronic sensors on the machine's movable parts.

Using this information, along with preprogrammed information about the individual patient, the machine provides the optimum amount of mechanical resistance. In this way, the patient's rehabilitation program is continually monitored and varied to match the present physical condition. It even varies its mechanical resistance during a single exercise session, gradually building up as the repetitions continue, then tapering off as those muscles tire. Amount of exertion and range of motion are graphed on the video screen, allowing the patient to observe his or her own performance.

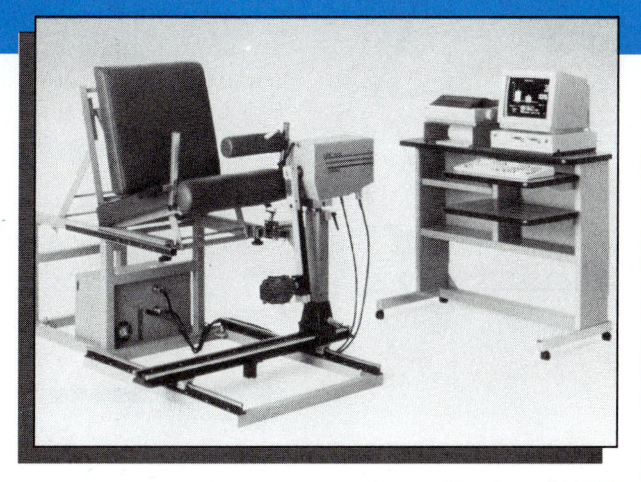

Courtesy of NASA

On page 1 in the Applications color insert, a patient is using the machine's attachments for wrist, elbow, shoulder, and back exercise therapy.

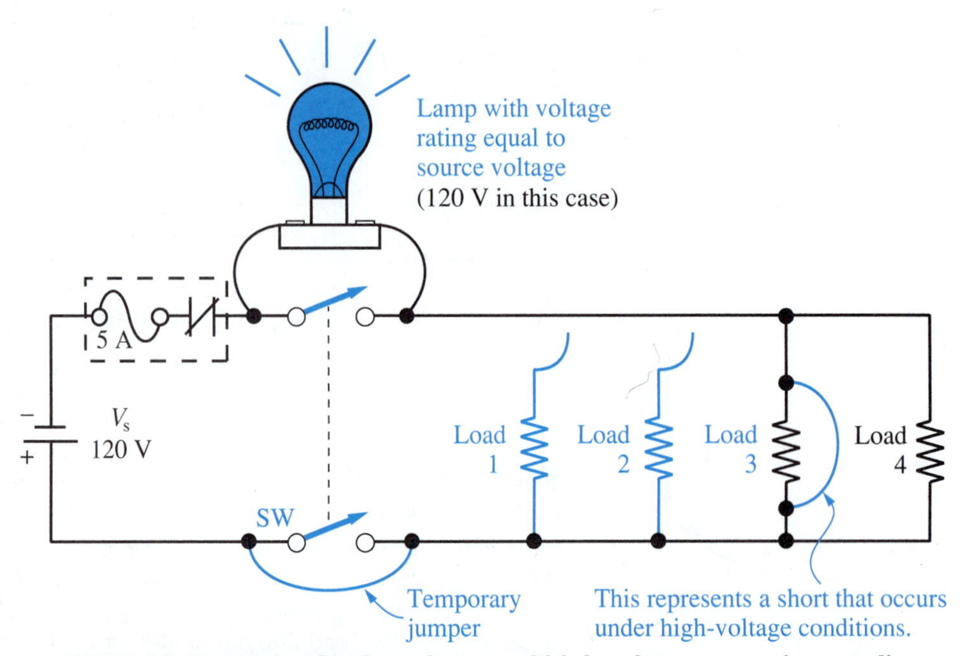

FIGURE 10-15 Testing loads under actual high-voltage operating conditions. Before beginning to disconnect another load, deenergize the circuit each time by flipping the breaker to Off.

TECHNICAL FACT

Sometimes a short circuit appears only when a load is subjected to its normal operating voltage. The short may not appear when tested by a low-voltage ohmmeter.

Damaged or weakened insulation is the usual cause of high-voltage shorts. To troubleshoot for this kind of short, one approach is to use a special high-voltage ohmmeter, or a megohmmeter.

Another approach is shown in Figure 10-15. In that figure, a lamp is temporarily wired in series with the multiple-load parallel combination. If the trouble is a high-voltage short in one of the loads, say load 3, the lamp will light at nearly full brilliance. The short circuit around load 3 makes a complete current flow path for the lamp.

As before, disconnect the loads one at a time. When load 3 is eventually disconnected, the lamp will go out, or become much less brilliant.

FORMULAS

$$\frac{1}{R_T} = \frac{1}{R_1} + \frac{1}{R_2} + \frac{1}{R_3} \ldots$$ **EQ. 10-1**

$$R_T = \frac{R_1 R_2}{R_1 + R_2}$$ **EQ. 10-2**

$$R = \frac{1}{G} \text{ and } G = \frac{1}{R}$$ **EQ. 10-3**

$$G_T = G_1 + G_2 + G_3 \ldots$$ **EQ. 10-4**

$$I_T = I_1 + I_2 + I_3 \ldots$$ **EQ. 10-5**

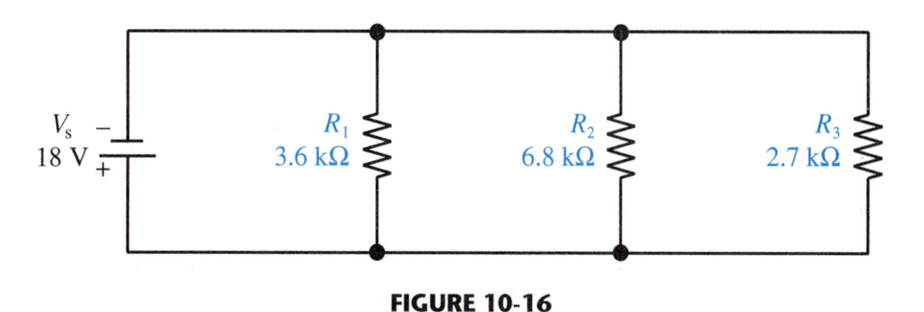

FIGURE 10-16

SUMMARY OF IDEAS

- In a parallel circuit, voltages are equal and currents are unequal.
- In a parallel circuit with three or more resistors, total resistance, R_T, is found by the reciprocal formula. The reciprocal formula is easy to use with a hand-held calculator having a 1/X key.
- If only two resistors are in parallel, R_T can be found by the product-over-the-sum formula.
- Ohm's law can be applied to a parallel circuit as a whole to find I_T.
- Conductance, G, is the tendency of a load device to allow current.
- Conductance is the reciprocal of resistance ($1/R$). It is measured in siemens.
- Kirchhoff's current law states that in a parallel circuit the sum of the individual branch currents is equal to the total supply current, I_T.
- Another way of stating Kirchhoff's current law is that the sum of the currents entering an electrical junction point is equal to the sum of the currents leaving that electrical point.

CHAPTER QUESTIONS AND PROBLEMS

1. For the circuit of Figure 10-16, [3, 4]
   a) Find the total resistance, R_T.
   b) Solve for total current, I_T, using Ohm's law.
   c) Show the two places where an ammeter could be inserted to measure I_T.
2. Suppose a fourth resistor, R_4 = 1.0 kΩ, is connected in parallel with the circuit of Figure 10-16. [3]
   a) Find the new total resistance, R_T. Compare this resistance to the total resistance from Problem 1. Explain why this is reasonable.
   b) Find the new total current, I_T. Compare this current to the total current from Problem 1 and explain.
3. In Figure 10-17, it is desired to choose a value of R_3 that will cause the total current, I_T, to be 0.2 A, as shown. What value of R_3 is needed? Solve this problem using Ohm's law and the reciprocal formula. [3, 4]

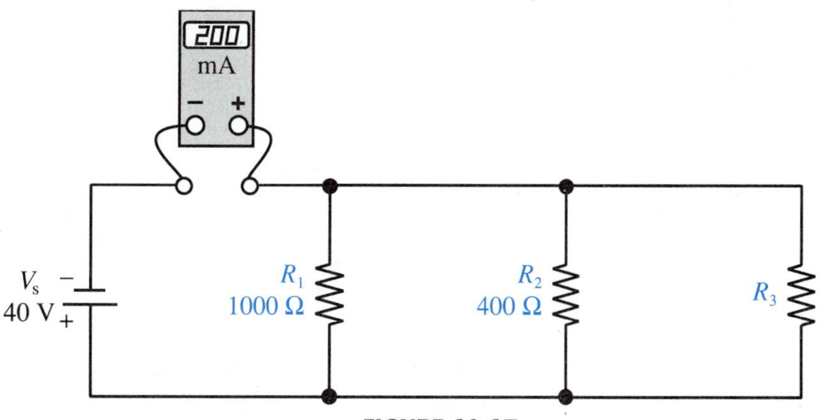

FIGURE 10-17

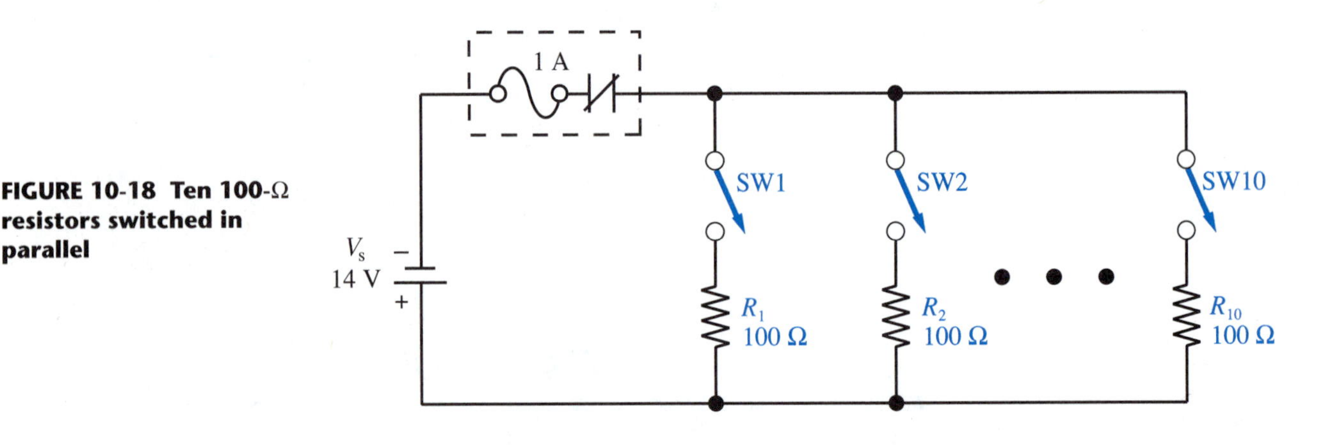

FIGURE 10-18 Ten 100-Ω resistors switched in parallel

4. The circuit of Figure 10-18 has ten switched loads, each having a resistance of 100 Ω. How many of the loads can be switched on before the circuit breaker trips? [3, 4]
5. A two-resistor parallel circuit has R_1 = 50 Ω and R_2 = 75 Ω. Find R_T by product-over-the-sum. Check your result by using the reciprocal formula. [3]
6. Find the individual branch conductances G_1, G_2, and G_3 in Figure 10-16.
7. Find the total conductance, G_T, in Figure 10-16. Compare to R_T from Problem 1, part a.
8. For a parallel circuit, $I_T = V_s G_T$. Use this with the G_T value from Problem 7 to calculate I_T . Compare to the result from Problem 1, part b.
9. In Figure 10-7, suppose I_T = 75 mA, I_1 = 18 mA, and I_3 = 36 mA. What must be the value of I_2? [5]
10. In Figure 10-17, solve for individual currents I_1 and I_2. Find I_3 from Kirchhoff's current law. What is the value of R_3? Compare to the result from Problem 3. [5]

$R_3 = 0.06$

11. Refer to Figure 10-19. (See page 6 in the Circuits color section for the color version of Figure 10-19.) For this circuit, find the following: [3, 4]
    a) Total resistance, R_T.
    b) Source voltage, V_s.
    c) Branch current I_3.
12. In Problem 11, Figure 10-19, solve for individual branch currents I_1 and I_2. Then add all three branch currents ($I_1 + I_2 + I_3$) to obtain total current, I_T. Check against the ammeter measurement in Figure 10-19. [5]
13. In Problem 11, solve for each resistor's power. Then find the circuit's total power, P_T.

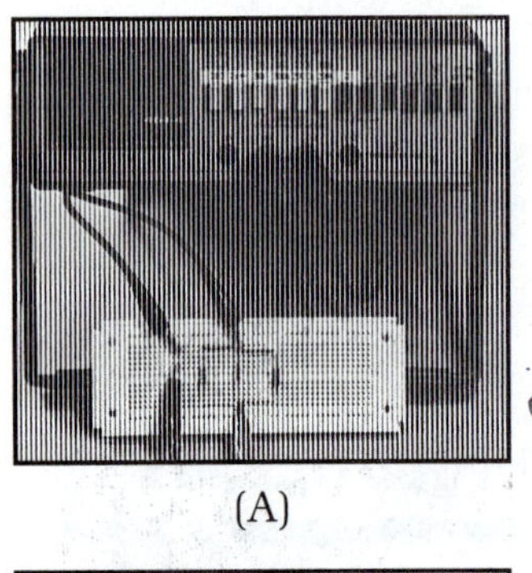

(A)

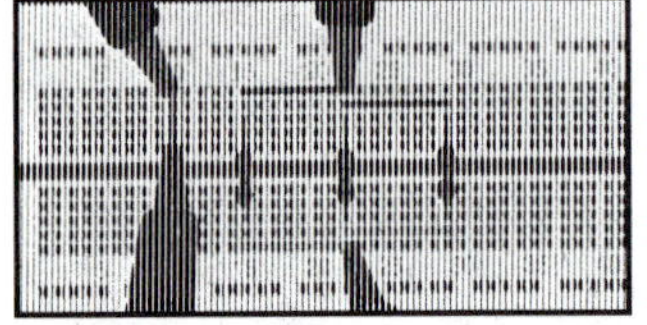

(B)

FIGURE 10-19

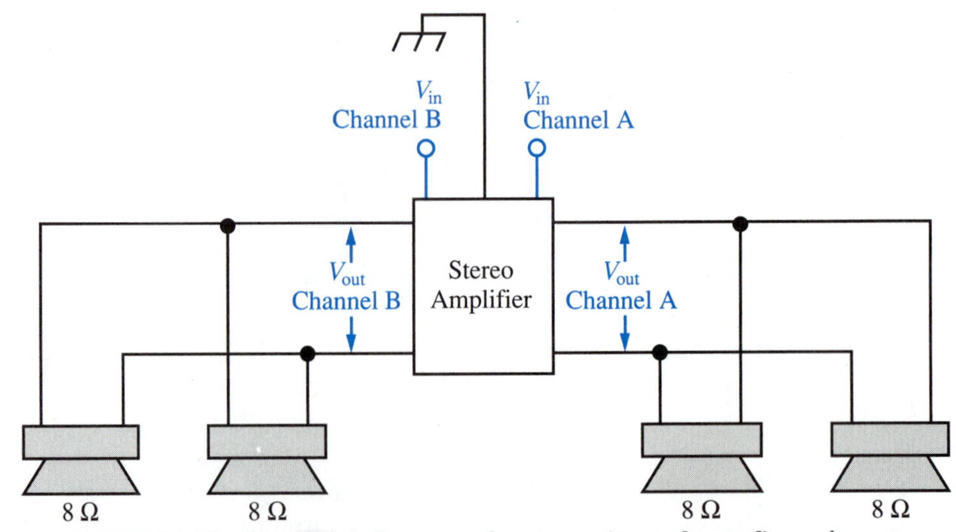

FIGURE 10-20 Simplified diagram of a stereo (two-channel) music system, with two 8-Ω speakers per channel

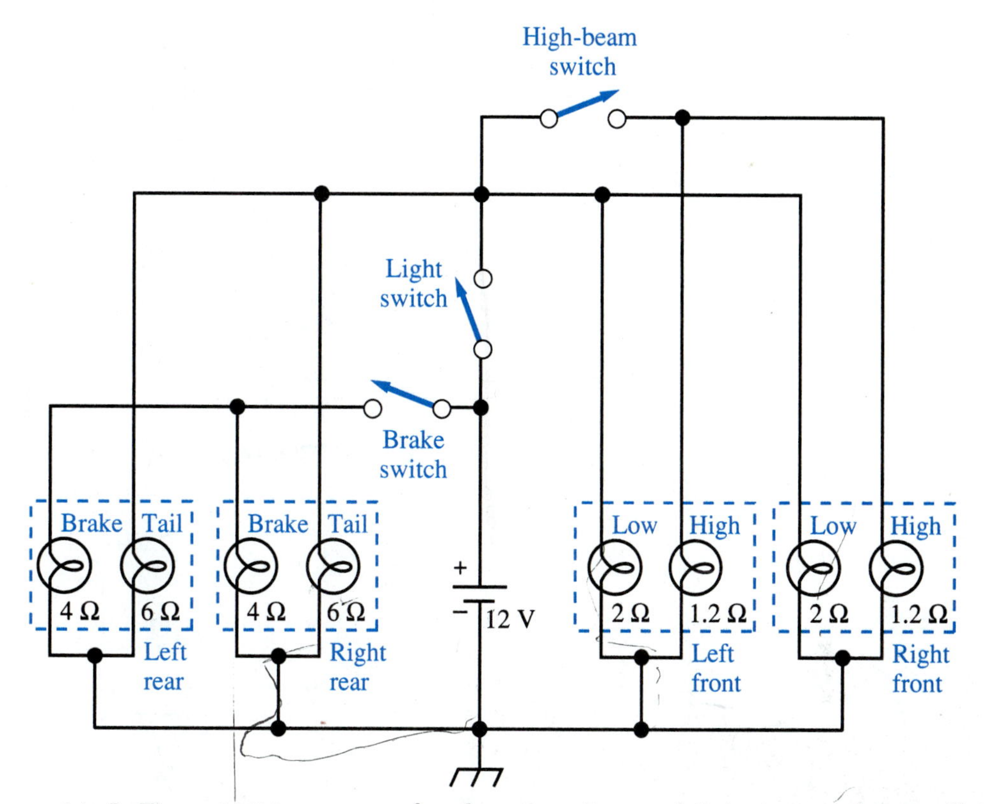

FIGURE 10-21 Light circuits on an automobile. The fuses are not shown.

14. In Figure 10-19, suppose a fourth resistor, R_4 = 1.8 kΩ, is connected in parallel with the three resistors already present. Tell which of the following variables would increase, which would decrease, and which would stay the same. [2, 3, 4]
   a) I_2 b) I_T c) V_s d) R_T e) V_1 f) P_3 g) P_T
15. Return to Figure 10-19 and suppose that R_3 was removed from the circuit. Comparing the original results from Problem 11, tell which of the following variables would increase, which would decrease, and which would stay the same. [2, 3, 4]
   a) I_2 b) I_T c) V_s d) R_T e) V_1 f) P_2 g) P_T
16. Calculate the new value for each variable that changed in Problem 15. [2, 3, 4]
17. Figure 10-20 represents a stereo (two separate music signals) amplifier with two speakers per channel. The maximum output voltage is 12 V on both channels at the same time. What is the maximum power transferred by the amplifier? [3, 4]

Use Figure 10-21 for Questions 18 through 20.

18. Figure 10-21 represents a car's light circuits. The eight bulbs have the following operating resistances (each bulb): [1, 2]

Tail light	6 Ω
Brake light	4 Ω
Low-beam headlight	2 Ω
High-beam headlight	1.2 Ω

   a) Which bulbs are turned On if the Brake switch is closed but the Light switch is open?
   b) Which bulbs are turned On if the Light switch is closed, but the other two switches are open?
   c) Which bulbs are turned On if the Light switch is closed, and the High-beam switch is closed, but the Brake switch is open?
19. Referring to Question 18, find the following: [3, 4]
   a) The battery's current and its total power for the conditions in part a .
   b) Repeat for the conditions in part b .
   c) Repeat for the conditions in part c .

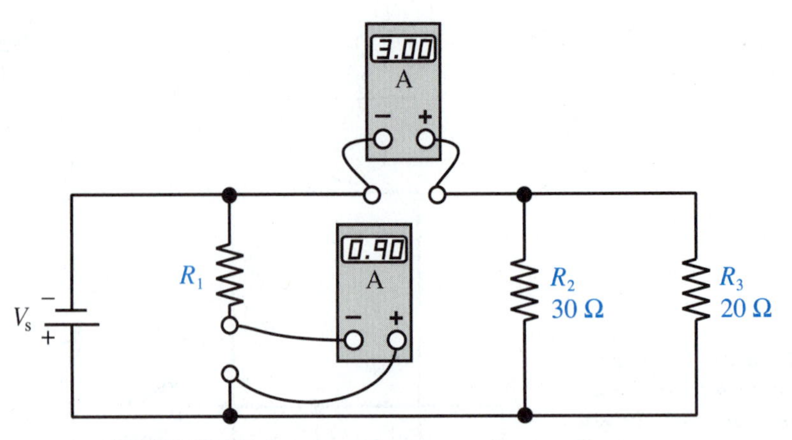

FIGURE 10-22 Interpreting ammeter readings

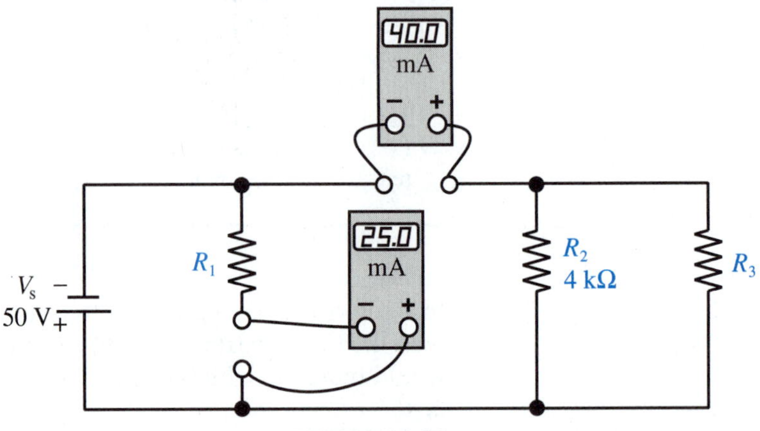

FIGURE 10-23

20. If any single bulb burns out (open-circuited), what effect does that have on the other bulbs? Explain. [6] Same
21. Referring to Question 20, explain the way in which parallel circuits are different from series circuits in regard to a single load burn-out. [6]
22. In the circuit of Figure 10-22, [2, 3, 4, 5]
    a) Find total current, I_T.
    b) Solve for source voltage, V_s.
    c) Find R_1 and R_T.
23. In the circuit of Figure 10-23, [2, 3, 4, 5]
    a) Find I_2, I_3, and I_T.
    b) Solve for R_1, R_3, and R_T.
24. In the circuit of Figure 10-24, find all the branch currents, I_1, I_2, I_3, and I_4. Also find the source voltage, V_s. [2, 3, 4, 5]
25. In the circuit of Figure 10-25, find all the branch currents, I_1, I_2, I_3, and I_4. [2, 3, 4, 5]

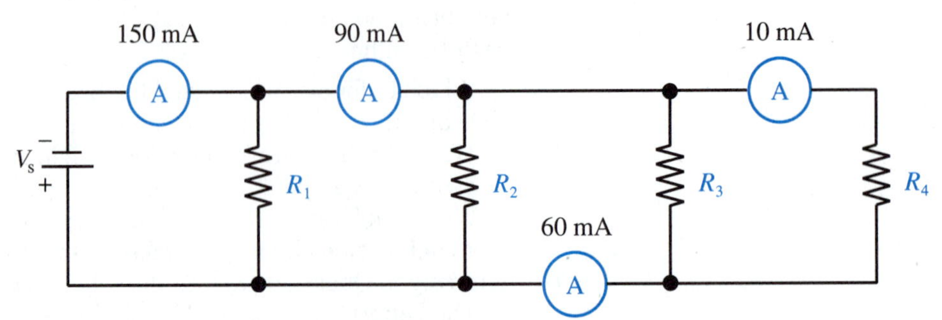

FIGURE 10-24 Several ammeters in a parallel circuit

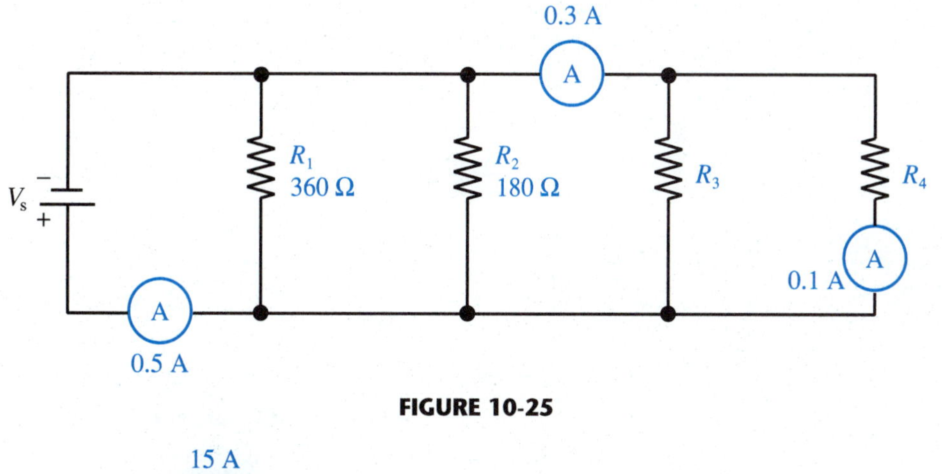

FIGURE 10-25

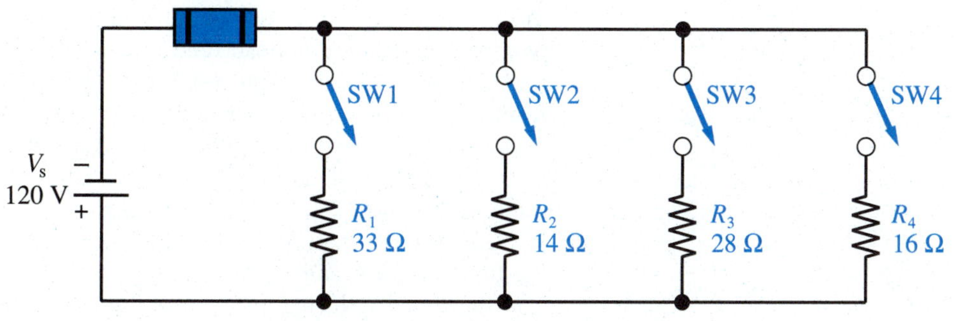

FIGURE 10-26 Switched parallel loads

26. In Figure 10-26, what combination of switches should be closed to obtain the maximum I_T without blowing the fuse? [5]
27. In Figure 10-27, solve for the values of R_1 and R_2. [5]

Troubleshooting Questions and Problems

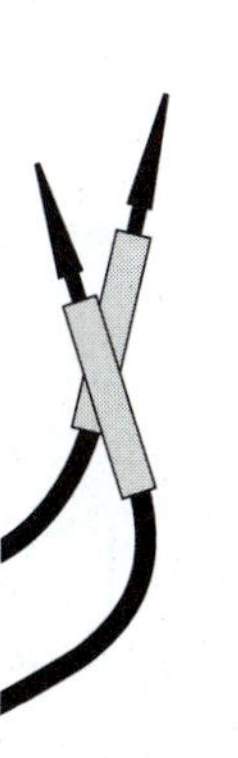

28. In a parallel circuit protected by a fuse, if one load fails open, will the fuse blow? [6]
29. For the conditions in Question 28, will the circuit's total current change? Explain. [5, 6]
30. The circuit in Figure 10-28 has four parallel resistors with the nominal values shown. The ammeter is known to be accurate. Is this circuit functioning properly? Explain. [2, 4, 6]
31. Calculate the expected branch currents in Figure 10-28 and apply Kirchhoff's current law. With the ammeter known to be correct, can you tell which resistor is open-circuited? [5, 6]

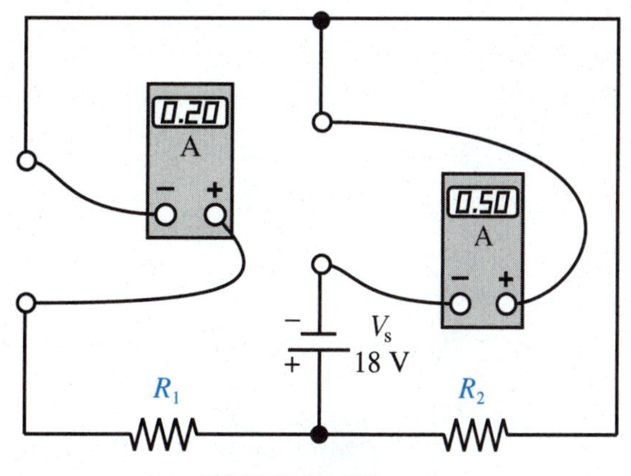

FIGURE 10-27

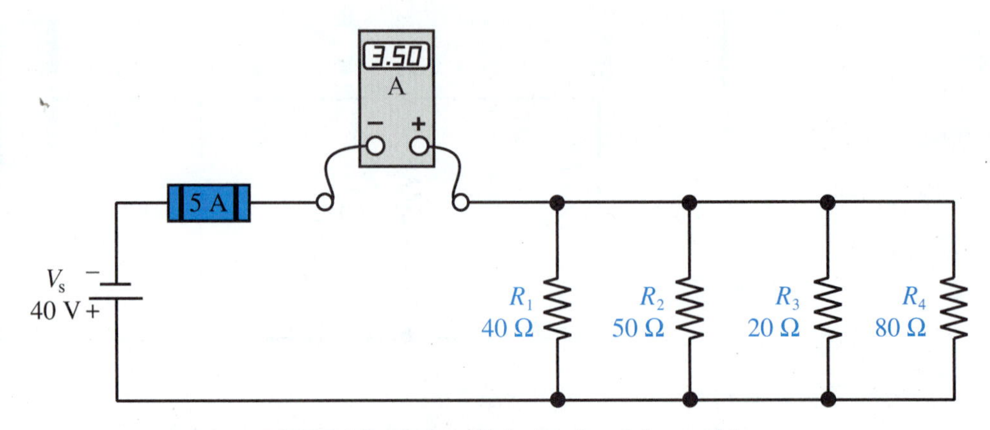

FIGURE 10-28 Parallel circuit with trouble

32. In Figure 10-29(A), the fuse blows immediately when SW is closed, indicating a short circuit. A technician deenergizes the eight-resistor circuit and connects an ohmmeter as described in Section 10-4. Then, rather than disconnecting the top lead of one resistor at a time, she separates the $R_1 - R_4$ group from the $R_5 - R_8$ group. This is done as shown in Figure 10-29(B). (This may or may not be possible, depending on the actual physical construction of the circuit). If the ohmmeter now reads 0 Ω, what has the technician learned about the location of the short? Explain. [2, 6]
33. If the ohmmeter reads a normal resistance value in Figure 10-29(B), what has the technician learned about the location of the short? Explain. [2, 6]
34. Continuing with the technique used in Questions 32 and 33, describe how you can find the shorted resistor in the fewest possible steps. [2, 6]
35. After the fuse has blown twice in Figure 10-29(A), suppose that you open the switch and connect an ohmmeter that has a 1.5-V internal battery. Suppose that the indicated resistance is not zero ohms. Instead, it is the normally expected value for a properly working eight-load circuit. What is the probable trouble in this circuit? [6]
36. Referring to the situation in Question 35, describe how you would go about locating the shorted resistor. [6]

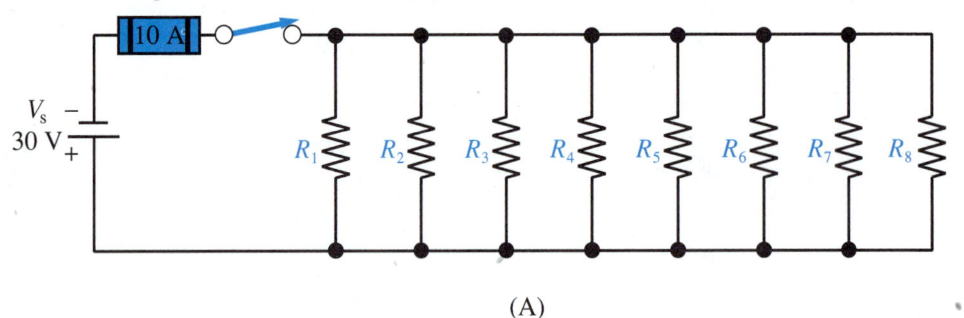

(A)

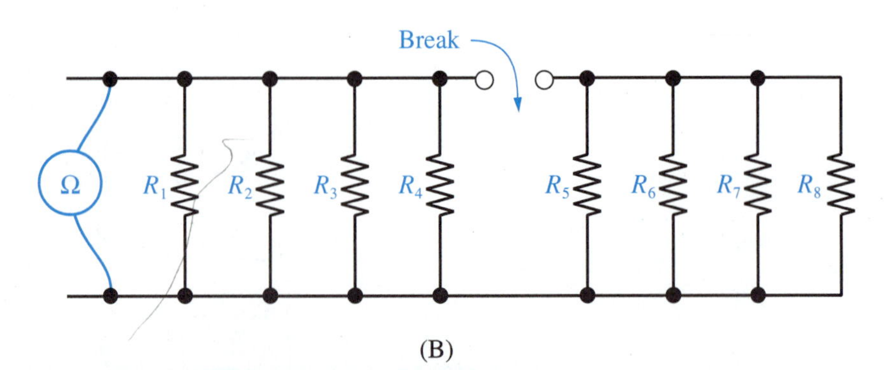

(B)

FIGURE 10-29 (A) Multiple-load parallel circuit with one load shorted. (B) Finding the shorted resistor quickly

CHAPTER 11

SERIES-PARALLEL CIRCUITS

Modern steelmaking processes use computers to monitor and control the process variables all the way through the ingot pour shown here. The computer system can then inventory each steel ingot and later track its progress through the shape-rolling operations. *Courtesy of NASA*

OUTLINE

NEW TERMS TO WATCH FOR

series-parallel circuit
equivalent resistance
simplify and reconstruct
Wheatstone bridge
balanced
nulled
resistance ratio
temperature-sensitive resistor
amplifier

Some circuits are neither strict series nor strict parallel. They are a combination of series and parallel.

After studying this chapter, you should be able to:

1. Reduce a series-parallel combination to an equivalent total resistance.
2. Reconstruct the original series-parallel circuit to specify all voltages and currents.
3. Calculate the resistance values necessary to balance a Wheatstone bridge.
4. Troubleshoot a series-parallel circuit to locate a shorted or open component.

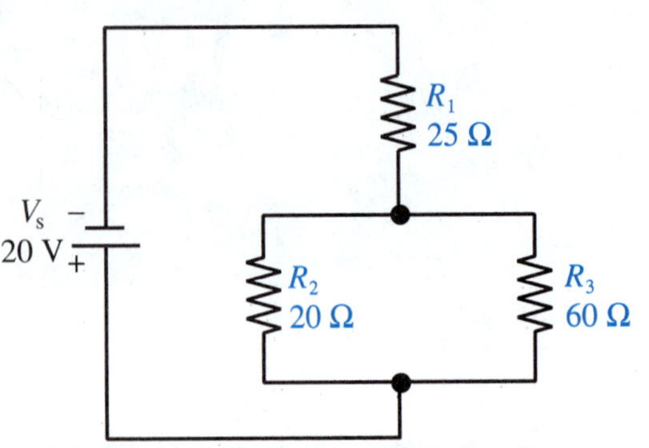

FIGURE 11-1 Series-parallel circuit

11-1 SIMPLIFYING SERIES-PARALLEL CIRCUITS

The circuit in Figure 11-1 is an example of a **series-parallel circuit.** Resistors R_2 and R_3 are in parallel with each other. The parallel combination is symbolized R_{2-3}.

The **equivalent resistance** of the R_{2-3} combination can be calculated by the product-over-the-sum formula as

$$R_{2-3} = \frac{R_2 R_3}{R_2 + R_3}$$

$$= \frac{(20)(60)}{20 + 60} = \frac{1200}{80} = 15\ \Omega$$

With R_2 and R_3 combined into a single resistance, the original circuit can be redrawn as shown in Figure 11-2(A). This is a series circuit. Total resistance, R_T, is found by adding the resistance values, giving

$$R_T = R_1 + R_{2-3} = 25\ \Omega + 15\ \Omega = 40\ \Omega$$

Thus, the original circuit has a net total resistance of 40 Ω, shown in Figure 11-2(B).

The total current from the supply can be calculated by applying Ohm's law to this total resistance.

$$I_T = \frac{V_S}{R_T} = \frac{20\ \text{V}}{40\ \Omega} = 0.5\ \text{A}$$

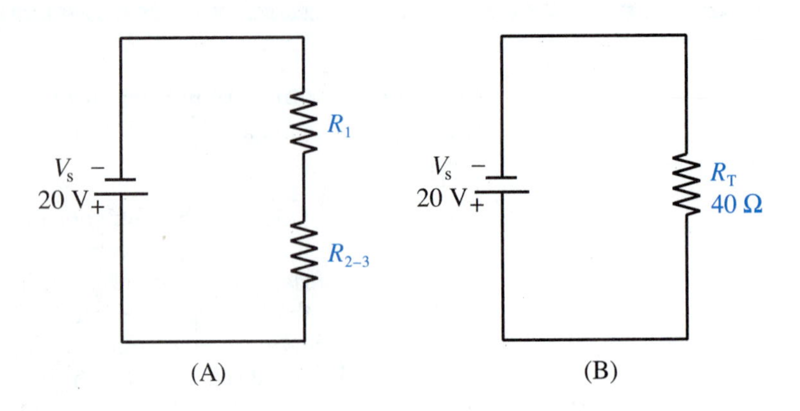

FIGURE 11-2 (A) Figure 11-1 partially simplified. (B) Simplified down to one total resistance

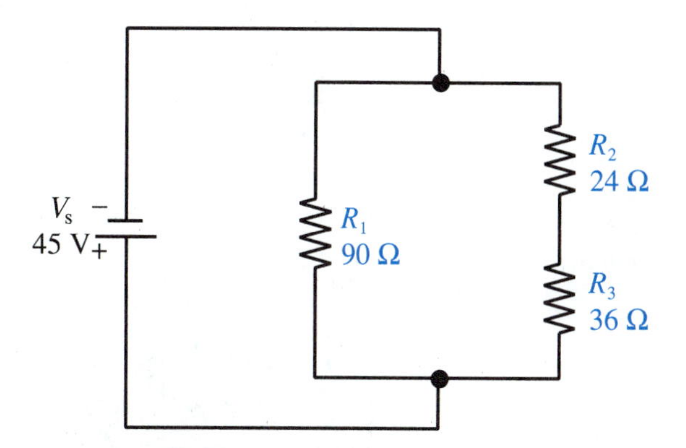

FIGURE 11-3 Another series-parallel circuit

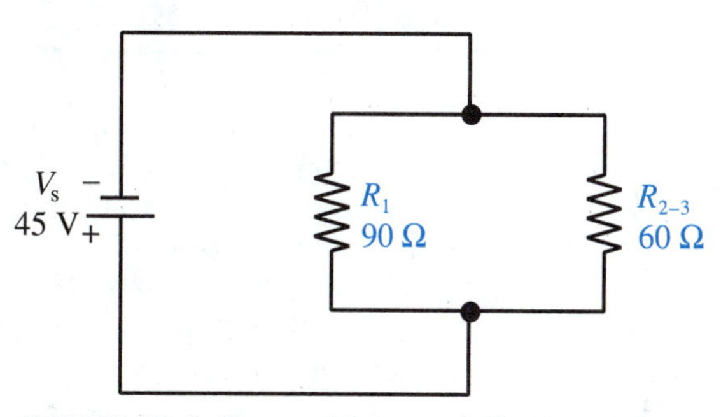

FIGURE 11-4 Figure 11-3 partially simplified

EXAMPLE 11-1

For the circuit in Figure 11-3,

a) Simplify the circuit in steps to find its net total resistance.
b) Calculate the total current from the supply.

SOLUTION

a) The two resistors, R_2 and R_3, are in series with each other. Therefore, their equivalent resistance is given by

$$R_{2-3} = R_2 + R_3 = 24\ \Omega + 36\ \Omega = 60\ \Omega$$

At this point, the original circuit can be redrawn as shown in Figure 11-4. It is clear from Figure 11-4 that R_{2-3} is in parallel with R_1. The reciprocal formula gives

$$\frac{1}{R_T} = \frac{1}{R_1} + \frac{1}{R_{2-3}} = \frac{1}{90} + \frac{1}{60}$$

$$R_T = \mathbf{36\ \Omega}$$

The simplified circuit is shown in Figure 11-5.

b) Applying Ohm's law to the net total resistance yields

$$I_T = \frac{V_s}{R_T} = \frac{45\ \text{V}}{36\ \Omega} = \mathbf{1.25\ A}$$

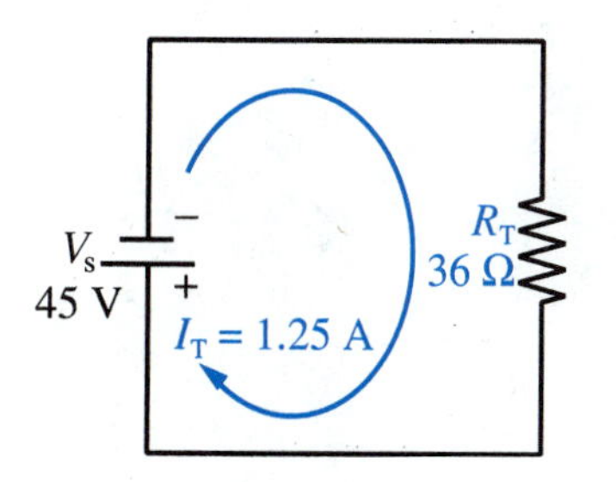

FIGURE 11-5 Figure 11-3 simplified down to R_T.

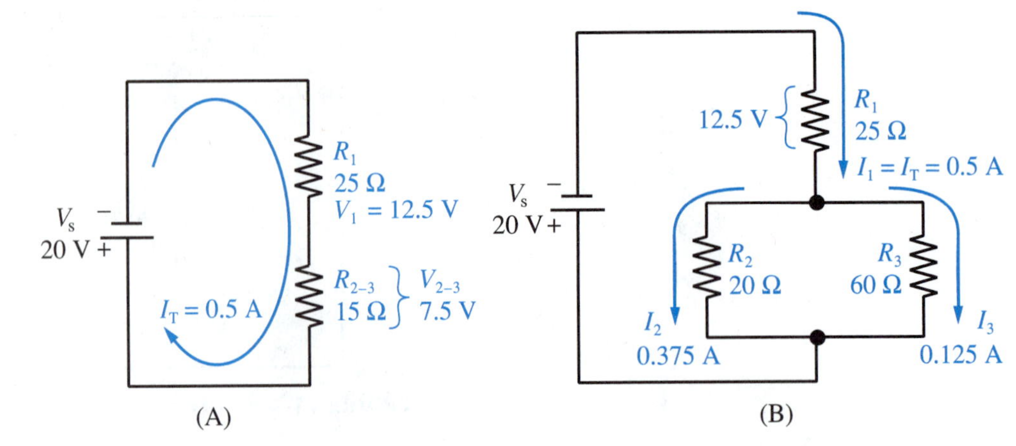

FIGURE 11-6 (A) Reconstructing the Figure 11-2 circuit. (B) Reconstructing further, back to the circuit of Figure 11-1

11-2 SIMPLIFY AND RECONSTRUCT METHOD

Once a series-parallel circuit has been reduced to a single resistance and its total current has been calculated, we can reverse the process. The circuit is now reconstructed one step at a time, using the knowledge that was obtained during the reduction process.

Return to the circuit in Figures 11-1 and 11-2. We know that the total supply current is I_T = 0.5 A. Let us specify this value of current in the partially reconstructed circuit of Figure 11-6(A).

Ohm's law for R_1 gives

$$V_1 = I_T R_1 = (0.5\ \text{A})\ 25\ \Omega = 12.5\ \text{V}$$

Ohm's law applied to the R_{2-3} combination gives

$$V_{2-3} = I_T R_{2-3} = (0.5\ \text{A})\ 15\ \Omega = 7.5\ \text{V}$$

An alternate method is Kirchhoff's voltage law, which gives

$$V_{2-3} = V_s - V_1 = 20\ \text{V} - 12.5\ \text{V} = 7.5\ \text{V}$$

Resistors R_2 and R_3 are actually in parallel, so 7.5 V appears across each one of them equally. (Remember, voltages are equal in parallel.) Applying Ohm's law to each resistor individually, we get

$$I_2 = \frac{V_{2-3}}{R_2} = \frac{7.5\ \text{V}}{20\ \Omega} = 0.375\ \text{A}$$

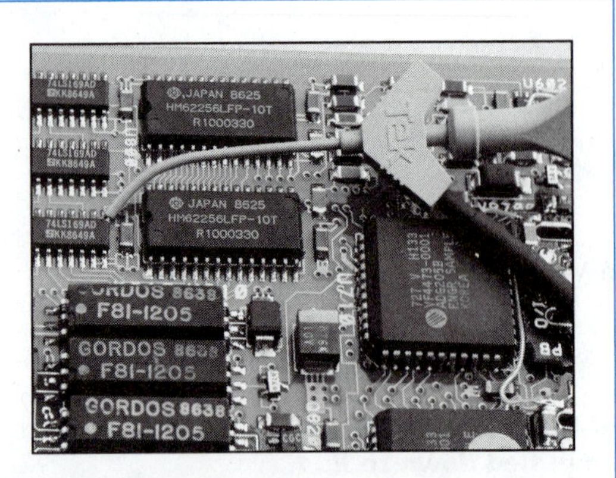

Special clips are available for attaching to densely packed printed circuit boards.
Courtesy of Tektronix, Inc.

$$I_3 = \frac{V_{2-3}}{R_3} = \frac{7.5\text{ V}}{60\ \Omega} = 0.125\text{ A}$$

These values are indicated in Figure 11-6(B).

During the reconstruction process we have found every voltage and current existing in the original circuit.

EXAMPLE 11-2

Reconstruct the circuit in Figures 11-3 and 11-4. Find every voltage and current in the original circuit.

SOLUTION

The total resistance resulted from the parallel combination of R_1 and R_{2-3}, as Figure 11-4 makes clear. Therefore, the entire 45-V supply voltage exists across R_1, as well as across R_{2-3}. Applying Ohm's law to R_1, we obtain

$$I_1 = \frac{V_s}{R_1} = \frac{45\text{ V}}{90\ \Omega} = \mathbf{0.5\text{ A}}$$

Repeating for R_{2-3} gives

$$I_{2-3} = \frac{V_s}{R_{2-3}} = \frac{45\text{ V}}{60\ \Omega} = \mathbf{0.75\text{ A}}$$

An alternative method would be to apply Kirchhoff's current law to the top node of Figure 11-4, knowing that $I_T = 1.25$ A.

$$I_{2-3} = I_T - I_1 = 1.25\text{ A} - 0.5\text{ A} = \mathbf{0.75\text{ A}}$$

These current values are shown in the partially reconstructed circuit of Figure 11-7(A).

Individual voltages V_2 and V_3 can now be found. R_2 and R_3 are in series, so each one of them carries 0.75 A, the I_{2-3} amount. From Ohm's law,

$$V_2 = I_{2-3} R_2 = (0.75\text{ A})(24\ \Omega) = \mathbf{18\text{ V}}$$
$$V_3 = I_{2-3} R_3 = (0.75\text{ A})(36\ \Omega) = \mathbf{27\text{ V}}$$

These voltage values are indicated in the fully reconstructed circuit of Figure 11-7(B).

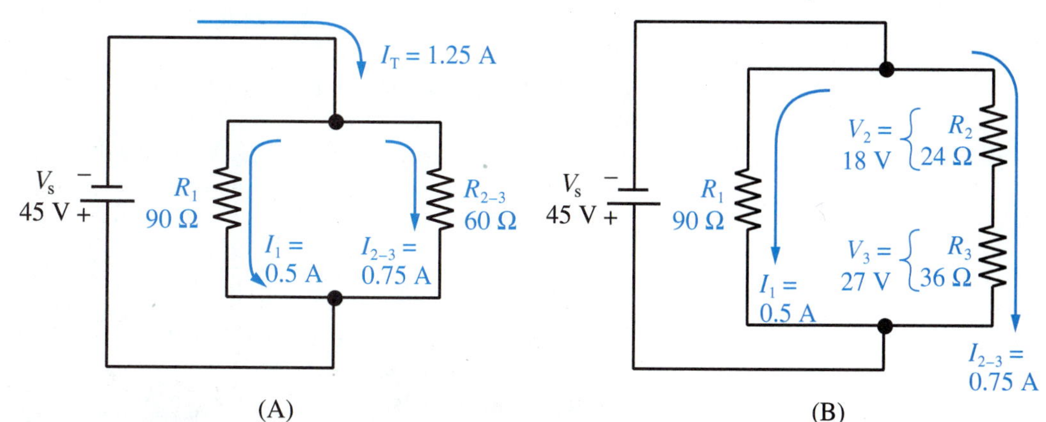

FIGURE 11-7 (A) Reconstructing the Figure 11-4 circuit. (B) Reconstructing further, back to the original circuit of Figure 11-3

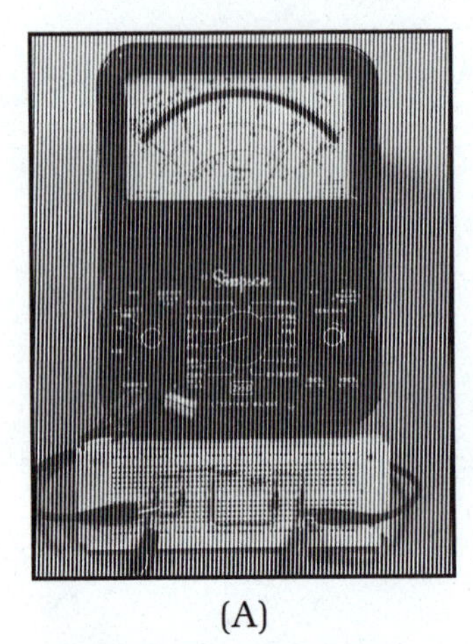

(A)

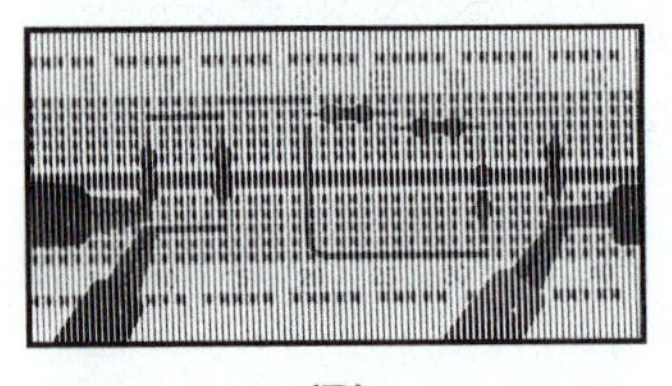

(B)

FIGURE 11-8

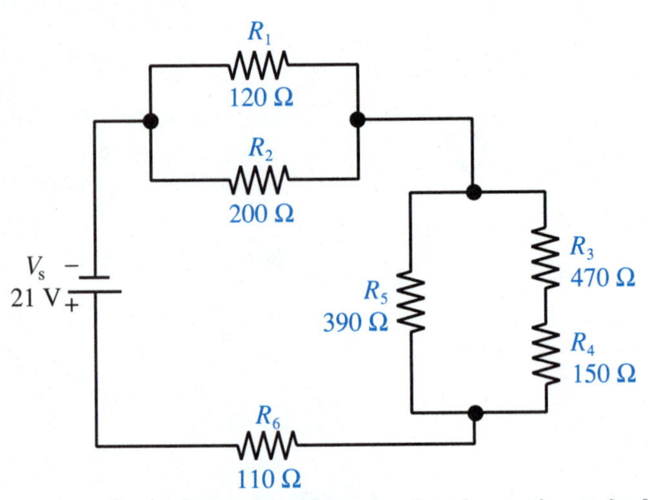

FIGURE 11-9 Complicated circuit to be solved. Schematic equivalent of Figure 11-8

EXAMPLE 11-3

Simplify and reconstruct the circuit of Figure 11-8. (See page 6 in the Circuits color section for the color version of Figure 11-8.) Find the voltage across every resistance and the current through every branch.

SOLUTION

It is easy to see that the 120-Ω resistor (R_1) and the 200-Ω resistor (R_2) are in parallel with each other. Also, the 470-Ω resistor (R_3) is in series with the 150-Ω resistor (R_4). The R_3–R_4 series combination is in parallel with the 390-Ω resistor (R_5). The R_3–R_4–R_5 combination is in series with the R_1–R_2 combination.

Finally, the 110-Ω resistor (R_6) is in series with all the preceding resistors. Drawing the circuit schematically, we get the diagram of Figure 11-9.

We can begin by combining R_1 and R_2. Using the reciprocal formula,

$$\frac{1}{R_{1-2}} = \frac{1}{R_1} + \frac{1}{R_2} = \frac{1}{120\ \Omega} + \frac{1}{200\ \Omega}$$

$$R_{1-2} = 75\ \Omega$$

Resistors R_3 and R_4 are in series, so

$$R_{3-4} = R_3 + R_4 = 470\ \Omega + 150\ \Omega = 620\ \Omega$$

This is a good point to redraw the circuit in partly simplified form, as shown in Figure 11-10(A).

This version of the circuit makes it clear that R_5 is in parallel with the R_{3-4} combination. Therefore, the resistance R_{3-4-5} can be found by the reciprocal formula.

$$\frac{1}{R_{3-4-5}} = \frac{1}{R_{3-4}} + \frac{1}{R_5} = \frac{1}{620\ \Omega} + \frac{1}{390\ \Omega}$$

$$R_{3-4-5} = 239.4\ \Omega$$

Continued on page 189.

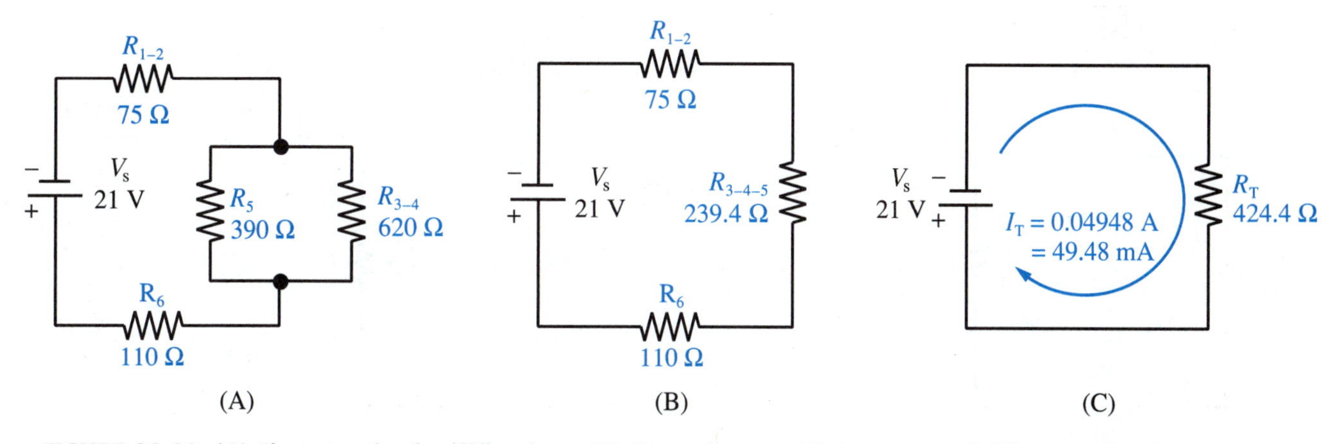

FIGURE 11-10 (A) First step in simplification. (B) Second step. (C) Last step, yielding total resistance, R_T.

Redrawing the circuit again, we get Figure 11-10(B). This equivalent circuit is a series circuit, so

$$R_T = R_{1-2} + R_{3-4-5} + R_6$$

$$= 75\ \Omega + 239.4\ \Omega + 110\ \Omega = 424.4\ \Omega$$

With $R_T = 424.4\ \Omega$, Ohm's law predicts total current of

$$I_T = \frac{V_s}{R_T} = \frac{21\ \text{V}}{424.4\ \Omega} = 0.049\ 48\ \text{A}\ \text{or}\ 49.48\ \text{mA}$$

which is indicated in Figure 11-10(C). This completes the simplification process.

We begin the reconstruction process by calculating the voltages across the three series resistances of Figure 11-10(B). Applying Ohm's law three times,

$$V_{1-2} = I_T R_{1-2} = (0.049\ 48\ \text{A})\ 75\ \Omega = \mathbf{3.71\ V}$$

$$V_{3-4-5} = I_T R_{3-4-5} = (0.049\ 48\ \text{A})\ 239.4\ \Omega = \mathbf{11.85\ V}$$

$$V_6 = I_T R_6 = (0.049\ 48\ \text{A})\ 110\ \Omega = \mathbf{5.44\ V}$$

As a check, verify that Kirchhoff's voltage law is satisfied.

$$V_{1-2} + V_{3-4-5} + V_6 = 3.71\ \text{V} + 11.85\ \text{V} + 5.44\ \text{V} = 21.0\ \text{V} = V_s$$

This voltage information is recorded in Figure 11-11(A).

Continue reconstructing by separating R_{3-4-5} into two parallel paths, R_{3-4} and R_5. Each of these paths has 11.85 V, so

$$I_{3-4} = \frac{V_{3-4}}{R_{3-4}} = \frac{11.85\ \text{V}}{620\ \Omega} \doteq \mathbf{19.11\ mA}$$

$$I_5 = \frac{V_5}{R_5} = \frac{11.85\ \text{V}}{390\ \Omega} = \mathbf{30.38\ mA}$$

Continued on page 190.

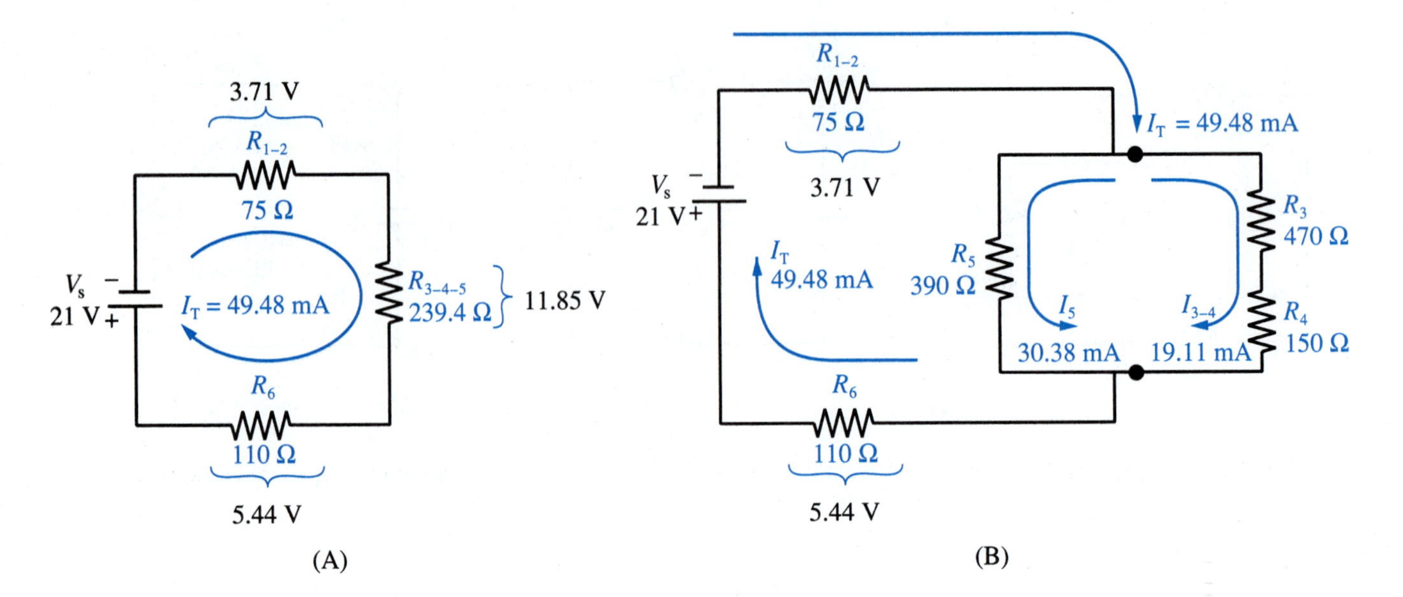

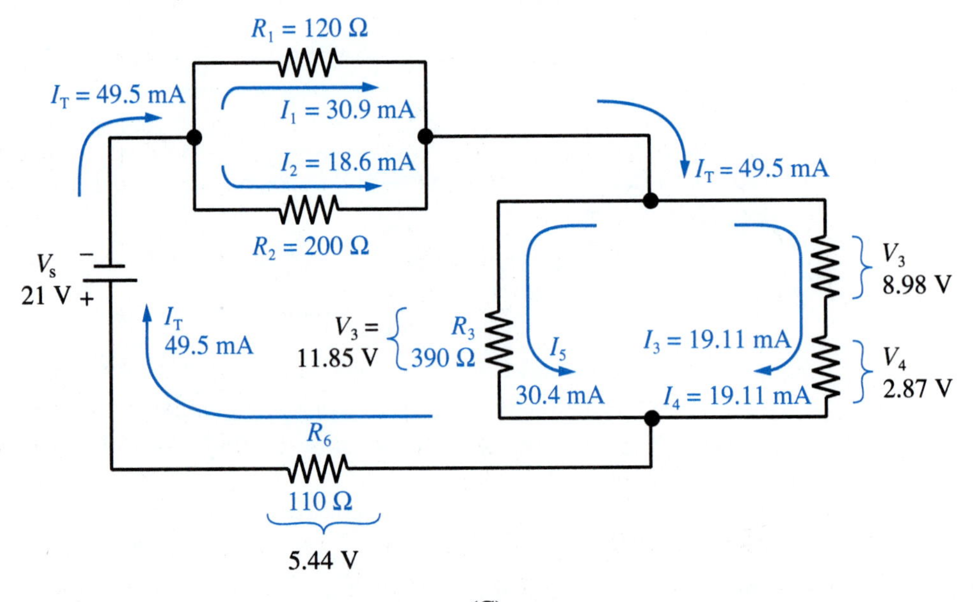

(C)

FIGURE 11-11 (A) First step in reconstructing the circuit. (B) Second step. (C) Final step, getting us back to the original circuit.

As a check, verify that Kirchhoff's current law is satisfied at the junction of R_3 and R_5.

$$I_{3-4} + I_5 = 19.11 \text{ mA} + 30.38 \text{ mA} = \mathbf{49.49\ mA}$$

which agrees with I_T to 3 significant figures. These current values are shown in Figure 11-11(B), which makes it clear that

$$I_6 = \mathbf{49.48\ mA}$$

R_3 and R_4 both carry 19.11 mA, so their voltages are

$$V_3 = I_{3-4} R_3 = (19.11 \text{ mA})\ 470\ \Omega = \mathbf{8.98\ V}$$

$$V_4 = I_{3-4} R_4 = (19.11 \text{ mA})\ 150\ \Omega = \mathbf{2.87\ V}$$

Continued on page 191.

These voltages are shown in Figure 11-11(C).

The final step in reconstructing the original circuit is to break apart R_{1-2} into two parallel paths, R_1 and R_2. Both parallel resistors have 3.71 V, so their currents are given by Ohm's law as

$$I_1 = \frac{V_1}{R_1} = \frac{3.71\text{ V}}{120\ \Omega} = \mathbf{30.92\ mA}$$

$$I_2 = \frac{V_2}{R_2} = \frac{3.71\text{ V}}{200\ \Omega} = \mathbf{18.55\ mA}$$

The fully reconstructed circuit is shown in Figure 11-11(C), with all values rounded to 2 digits to the right of the decimal point. Currents I_1 and I_2 are specified, and so are voltages V_3 and V_4. Every voltage and current in the circuit has been found.

✔ SELF-CHECK FOR SECTIONS 11-1 AND 11-2

1. Solve the circuit of Figure 11-12 for all currents and voltages. [1, 2]
2. Solve the circuit of Figure 11-13 for all currents and voltages. [1, 2]
3. Use the results of Problem 2 to solve for the power in each resistor in Figure 11-13. [1, 2]
4. Solve the circuit of Figure 11-14 for all currents and voltages. [1, 2]
5. Solve the circuit of Figure 11-15. [1, 2]

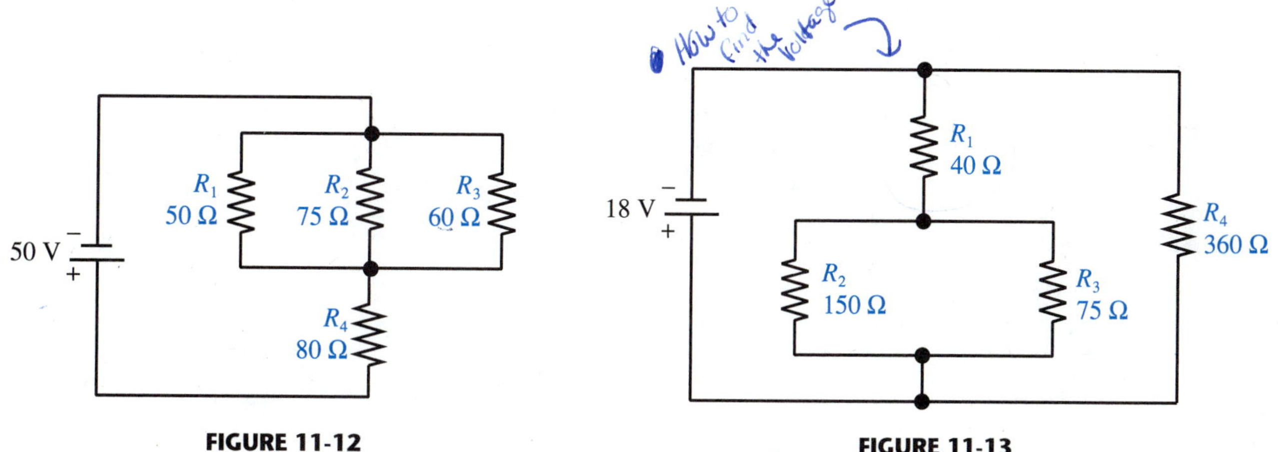

FIGURE 11-12

FIGURE 11-13

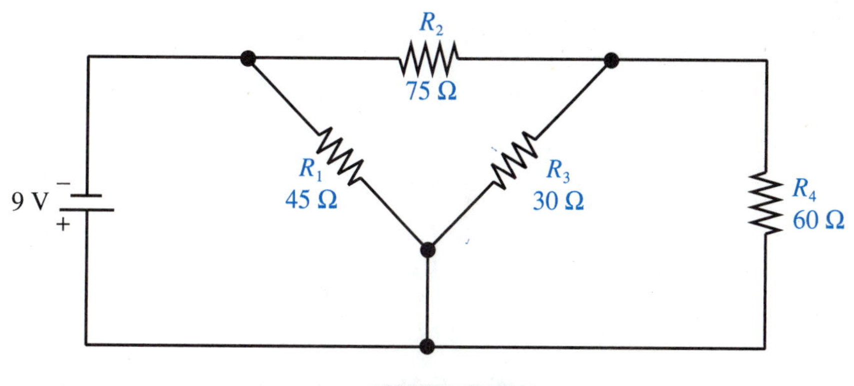

FIGURE 11-14

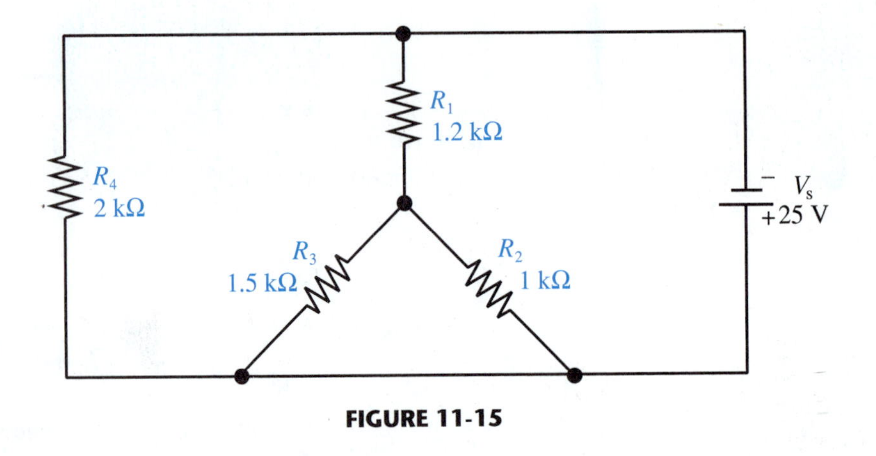

FIGURE 11-15

11-3 WHEATSTONE BRIDGES

The circuit arrangement in Figure 11-16(A) is called a bridge circuit. This is because a meter usually is connected between points A and B, thereby "bridging" from one side of the circuit to the other. The diamond shape in Figure 11-16(A) is the classic way of drawing a bridge, but the right-angled drawing in Figure 11-16(B) is simpler.

If all four elements of the bridge are resistors, as in Figure 11-16, it is called a **Wheatstone bridge,** named for the man who discovered its usefulness. Basically, the Wheatstone bridge consists of two series combinations that are in parallel with each other. The R_1–R_2 series combination on the left is in parallel with the R_3–R_4 series combination on the right. In general, at least one of the four resistors is variable.

TECHNICAL FACT

The operational purpose of a Wheatstone bridge is to adjust the variable resistor so that the voltage between points A and B becomes zero. The bridge is then said to be **balanced**, or **nulled.**

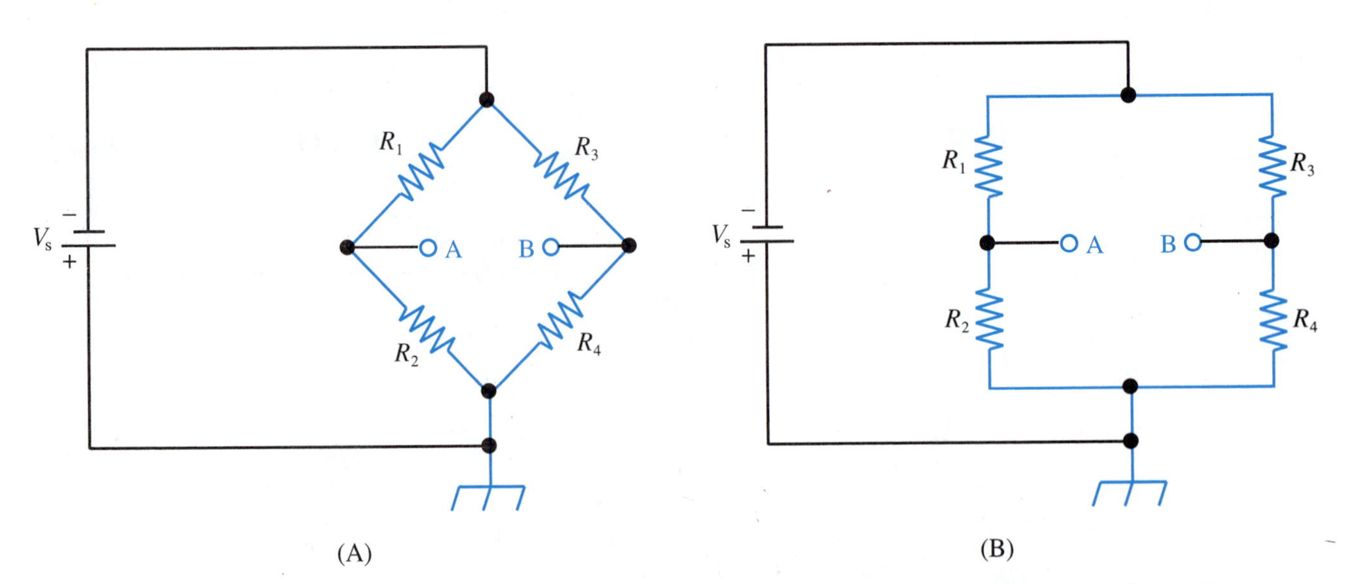

FIGURE 11-16 Wheatstone bridge. (A) Diamond-shaped schematic. (B) Simpler schematic

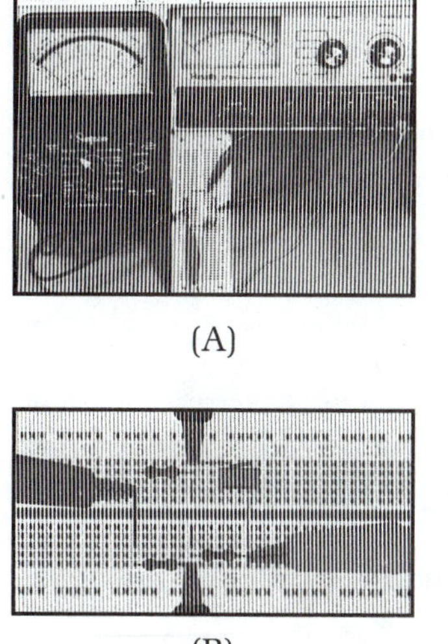

(A)

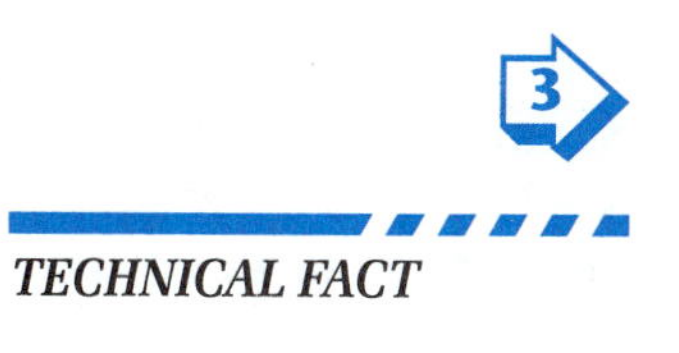

(B)

FIGURE 11-17

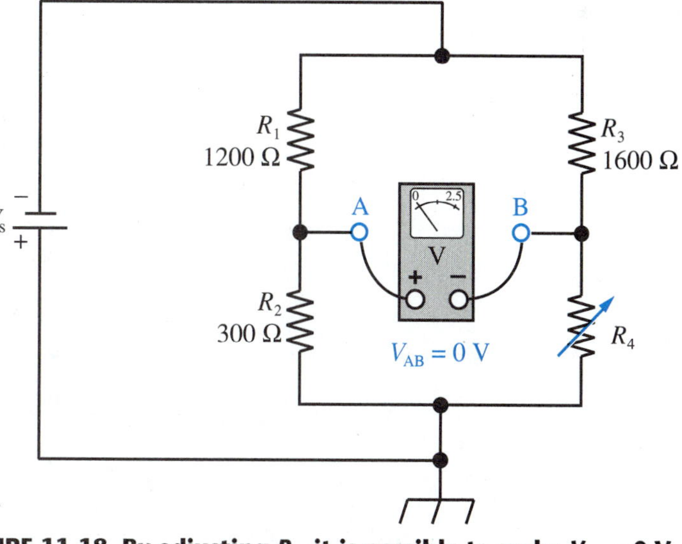

FIGURE 11-18 By adjusting R_4, it is possible to make $V_{AB} = 0$ V.

Usually R_4 is the adjustable resistor. Then the balanced condition is indicated in Figure 11-17. (See page 7 in the Circuits color section for the color version of Figure 11-17.)

The schematic drawing of this circuit is shown in Figure 11-18. In Figure 11-18, in order to get $V_{AB} = 0$ V, this must happen: The voltage division on the left side between R_1 and R_2 must be the same as the voltage division on the right side between R_3 and R_4. This will occur when the **resistance ratios** are the same on both sides. As a formula,

$$\frac{R_1}{R_2} = \frac{R_3}{R_4} \quad \text{(for balance)}$$

which is usually rearranged as

TECHNICAL FACT

$$\frac{R_4}{R_3} = \frac{R_2}{R_1} \quad \text{(for balance)} \qquad \textbf{EQ. 11-1}$$

EXAMPLE 11-4

In Figures 11-17 and 11-18, we see that $V_s = 20$ V, $R_1 = 1200\ \Omega$, $R_2 = 300\ \Omega$, and $R_3 = 1600\ \Omega$.

a) Solve for the value of R_4 required to balance the bridge.
b) Using the voltage-divider formula (Equation 9-4), calculate V_1 and V_2.
c) Again using voltage division, calculate V_3 and V_4.
d) Compare your answers to parts b and c. Explain why $V_{AB} = 0$ V.

SOLUTION

a) From Equation 11-1,

$$\frac{R_4}{R_3} = \frac{R_2}{R_1}$$

Continued on page 194.

$$\frac{R_4}{1600\ \Omega} = \frac{300\ \Omega}{1200\ \Omega}$$

$$R_4 = 1600\ \Omega\left(\frac{300\ \Omega}{1200\ \Omega}\right) = \mathbf{400\ \Omega}$$

b) The voltage-divider formula for a series circuit is

$$\frac{V_{\text{portion}}}{V_T} = \frac{R_{\text{portion}}}{R_T}$$

Applying this formula to the series circuit on the left side of the bridge, we get

$$V_1 = V_s\left(\frac{R_1}{R_1 + R_2}\right) = 15\ \text{V}\left(\frac{1200\ \Omega}{1200\ \Omega + 300\ \Omega}\right) = 15\ \text{V}\ (0.8) = \mathbf{12\ V}$$

Applying Kirchhoff's voltage law to that same series circuit on the left gives

$$V_2 = V_s - V_1 = 15\ \text{V} - 12\ \text{V} = \mathbf{3\ V}$$

c) On the right side,

$$V_3 = V_s\left(\frac{R_3}{R_3 + R_4}\right) = 15\ \text{V}\left(\frac{1600\ \Omega}{1600\ \Omega + 400\ \Omega}\right)$$

$$= 15\ \text{V}\ (0.8) = \mathbf{12\ V}$$

$$V_4 = V_s - V_3 = 15\ \text{V} - 12\ \text{V} = \mathbf{3\ V}$$

d) The left side of the bridge is dividing the 15-V source voltage into 12-V and 3-V portions. The right side of the bridge has been adjusted to do exactly the same. Therefore, **there is no difference between the two sides, causing V_{AB} to equal 0 V.**

Many industrial measurement and control instruments use a Wheatstone bridge. An example from the area of high-temperature measurement is shown in Figure 11-19. R_2 is a **temperature-sensitive resistor** capable of handling very hot temperatures. It is placed in position to feel the temperature that is being measured.

R_4 is a potentiometer with a jumper wire from the wiper terminal to the bottom terminal. It acts as an adjustable resistor. The pot shaft is turned automatically by a slow-moving motor. This is symbolized by the dashed line between the motor shaft and the pot shaft in Figure 11-19. The motor shaft is also mechanically linked to the temperature-pointer shaft. Thus, as the motor adjusts the pot position, it also adjusts the temperature-indicating pointer.

Here is how the instrument works: If the temperature changes, it makes R_2 change. This causes the bridge to become unbalanced, with V_{AB} not equal to zero. As shown in Figure 11-19, the V_{AB} voltage is wired to the input of an **amplifier.** The amplifier boosts this small voltage to a larger value, V_{out}, for running the motor. As the motor slowly turns, it adjusts the R_4 pot resistance and repositions the pointer.

At some point, the R_4 resistance becomes the exact value to rebalance the bridge. When that happens, V_{in} to the amplifier returns to zero, so V_{out} also becomes zero. With zero voltage applied, the motor stops running, freezing the pot in the balance position. At the same time, the pointer has been moved to the proper scale position to indicate the temperature being sensed by R_2.

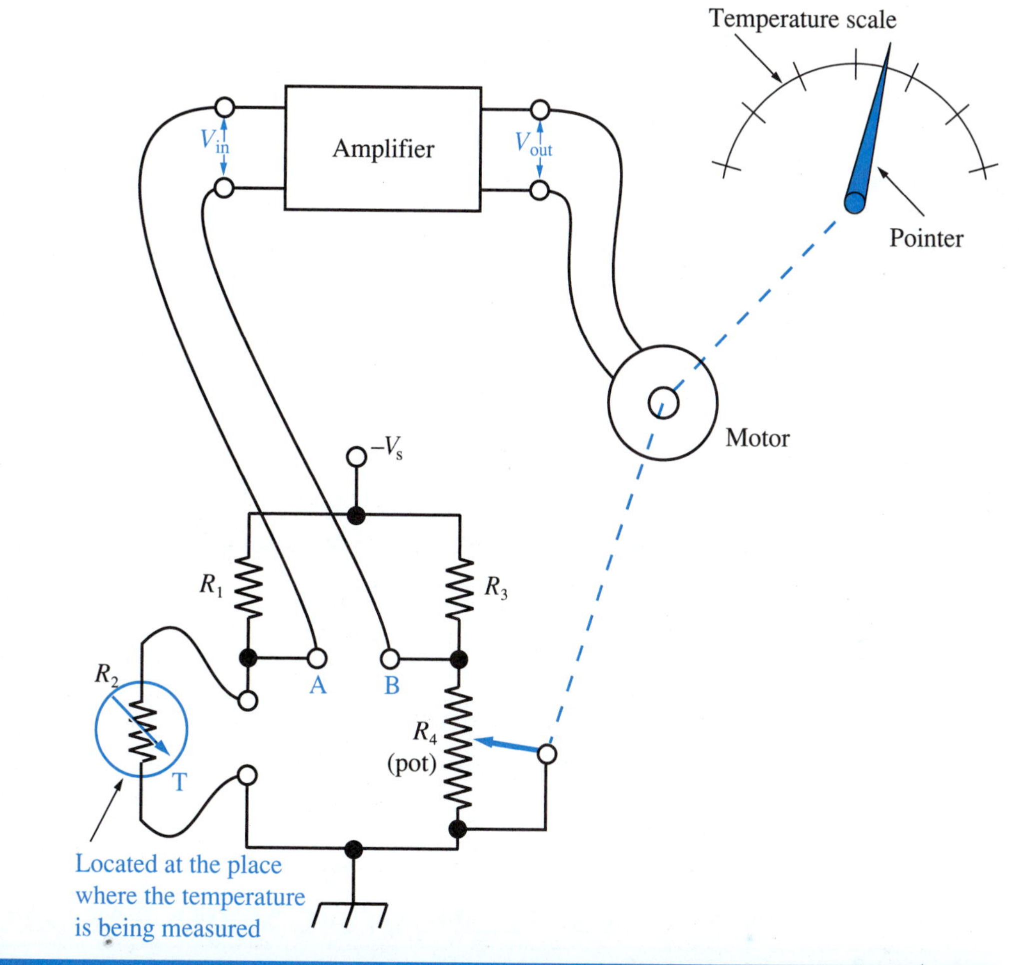

FIGURE 11-19 Layout of an industrial-type temperature instrument

WETLANDS SURVEY

The lower 48 states still have about 150 000 square miles of wetlands. This waterfowl habitat is continually watched by many of the states and several private environmental organizations.The chief private organization is Ducks Unlimited, Inc., with over 500 000 members.

The Ducks Unlimited experts shown here are inspecting displays that are generated by computer processing of satellite information. The corporation has an agreement with NASA to receive the earth-radiation data obtained by a particular satellite, called LANDSAT 5, which orbits the earth at more than 400 miles into space. This habitat information can be updated every 16 days. That is the amount of time it takes LANDSAT 5 to completely scan the entire surface of North America.

A photo image of an earth surface area is also shown. An electronically color-enhanced version is shown on page 6 in the Applications color section. Such images reveal valuable information about the ability of an area to support waterfowl.

Courtesy of NASA

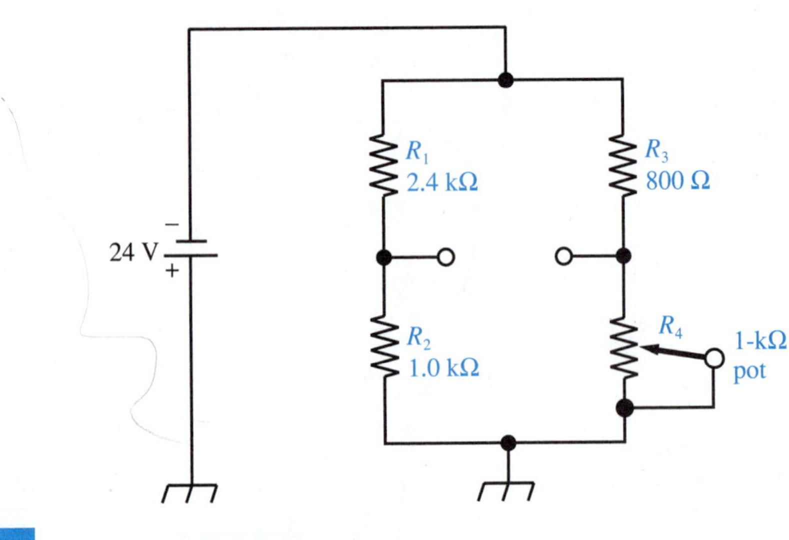

FIGURE 11-20 Wheatstone bridge for Self-Check Questions 6 through 10

✓ SELF-CHECK FOR SECTION 11-3

For Questions 6 through 10, refer to the Wheatstone bridge in Figure 11-20.

6. What value of R_4 will balance the bridge? [3]
7. With the bridge balanced, what is the value of V_2? What is the value of V_4?
8. Does the value of V_2 depend on whether or not the bridge is balanced? Explain. [3]
9. Repeat Question 8 for V_4. [3]
10. If the source voltage decreases to 23 V, will the answer to Problem 6 change? Explain this. [3]

11-4 TROUBLESHOOTING SERIES-PARALLEL CIRCUITS

As a general rule, electrical troubleshooting is a process of deciding what you expect *should* happen, and checking to see if it *does* happen. Repeat this until you

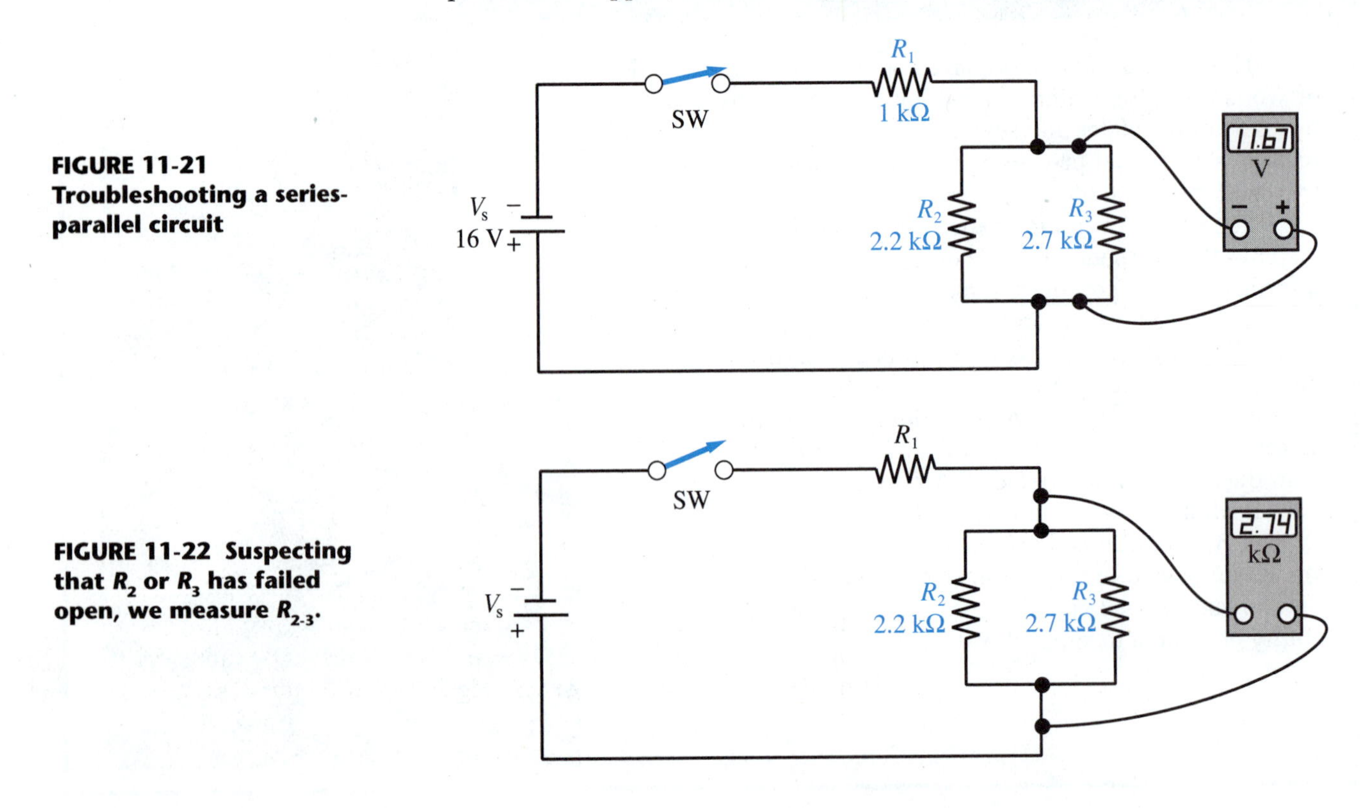

FIGURE 11-21 Troubleshooting a series-parallel circuit

FIGURE 11-22 Suspecting that R_2 or R_3 has failed open, we measure $R_{2\text{-}3}$.

discover something that is not happening the way you expected. Then use your electrical test instruments to find out the cause. Let us practice this troubleshooting approach for locating shorted or open resistors in series-parallel circuits.

EXAMPLE 11-5

Is the circuit of Figure 11-21 functioning properly? If not, locate the probable trouble.

SOLUTION

First, find out if a voltage of 11.67 V should be expected across the R_2–R_3 parallel combination. By product-over-the-sum,

$$R_{2-3} = \frac{R_2R_3}{R_2 + R_3} = \frac{(2.2\ \text{k}\Omega)(2.7\ \text{k}\Omega)}{(2.2 + 2.7)\ \text{k}\Omega} = 1.212\ \text{k}\Omega$$

Then, using voltage division,

$$V_{2-3} = V_s\left(\frac{R_{2-3}}{R_1 + R_{2-3}}\right) = 16\ \text{V}\left(\frac{1.212\ \text{k}\Omega}{(1 + 1.212)\ \text{k}\Omega}\right)$$

$$= 8.77\ \text{V (expected)}$$

So the measured value of 11.67 V is not the correct voltage that should appear here.

One possible explanation for this higher-than-expected voltage is that resistance R_{2-3} is larger than it should be. This would occur if either R_2 or R_3 has failed open. Therefore, let us deenergize the circuit and use an ohmmeter to measure R_{2-3}. This has been done in Figure 11-22. With the ohmmeter indicating 2.74 kΩ, which is the value of R_3 alone, it is clear that **R_2 has failed open.**

EXAMPLE 11-6

Is the circuit of Figure 11-23 functioning properly? If not, what is the trouble?

SOLUTION

We expect the voltage from the right side of R_1 to circuit ground to be less than 18 V. This is because there should be some voltage drop across R_1, which subtracts from the V_s value.

The voltmeter is telling us that there is zero voltage drop across R_1. How could this happen? One explanation is that R_1 is shorted. Opening the switch to deenergize and isolate R_1, we can take an ohmmeter measurement as shown in Figure 11-23(B). This ohmmeter measurement shows an actual resistance of 1.02 kΩ, so R_1 is not shorted.

Another possible explanation for zero voltage drop across R_1 in Figure 11-23(A) is that the remainder of the circuit is open somewhere. If that were true, no current would flow from the source through R_1, resulting in zero voltage drop across R_1. One possibility for an overall open condition is for R_4 to fail open. Moving the ohmmeter to R_4 in **Figure 11-23(C)** indicates that is the trouble.

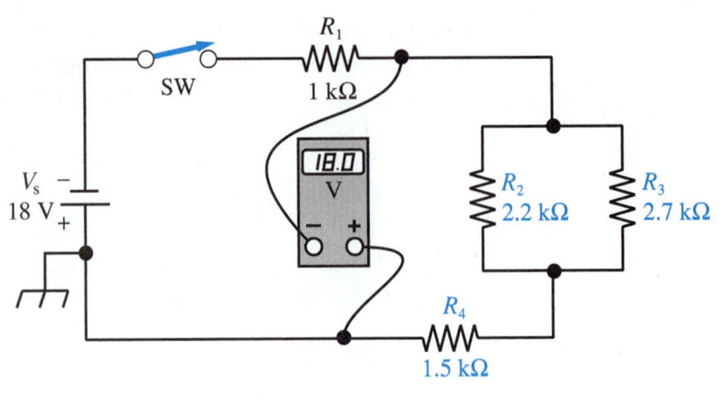
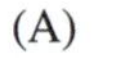

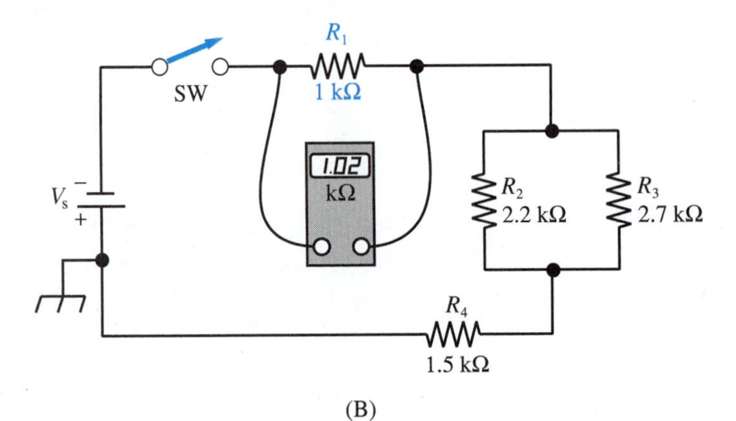

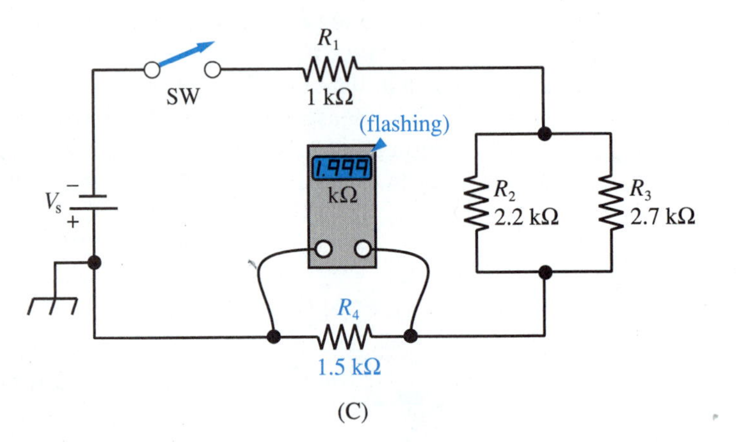

FIGURE 11-23 (A) Circuit for troubleshooting Example 11-6. (B) Testing to see if R_1 is shorted. (C) Testing whether R_4 is open. On a digital ohmmeter, an over-range, or infinite resistance, is indicated by the left digit becoming 1 and all other digits becoming 9. The display then flashes on and off.

EXAMPLE 11-7

If the ohmmeter in Figure 11-23(C) had shown a normal R_4 resistance close to 1.5 kΩ, the trouble would have to be something other than R_4. How would we locate it?

SOLUTION

We would search for the open elsewhere in the circuit. This could be done by connecting the positive lead of a voltmeter to circuit ground, then probing around the circuit with the negative lead. This is illustrated in **Figure 11-24.**

Eventually, we will find a point where the voltmeter reads 0 V. The open is between there and the last point where the meter read 18.0 V. For example, if the reading was 18.0 V at point 2 but 0 V at point 3, there must be a break between points 2 and 3.

FIGURE 11-24 Searching for the location of an open. As we move the probe clockwise, we will continue to read 18.0 V until we pass the open place. Then the reading will fall to 0 V. This reveals the physical location of the open circuit.

SAFETY ADVICE

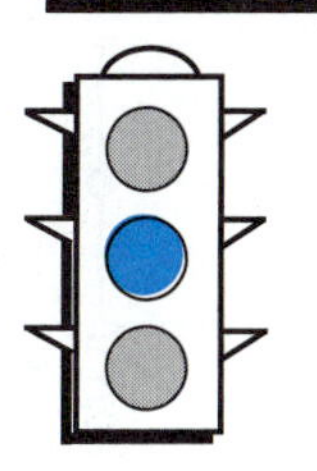

Some troubleshooting procedures require that the circuit's power remains on. Often, one lead of a voltmeter is connected to a reference point in the circuit, called the ground point. Then the other lead is moved about to measure voltages at various points, relative to the ground point. This is shown in Figure 11-24.

When making voltage measurements in a live circuit, it is absolutely necessary that you observe the precautions regarding electric shock that were listed and explained in the Working Safely section in the front of the text.

1. Remove metal rings and wristwatches.
2. Put only one of your hands into the circuit, never both hands.
3. Keep all parts of your body, especially your feet, electrically insulated from the surroundings.
4. Be certain that the plastic insulating handle of the voltmeter probe is in perfect condition.

FORMULAS

$$\frac{R_4}{R_3} = \frac{R_2}{R_1} \quad \text{(for balance)} \qquad \textbf{EQ. 11-1}$$

SUMMARY OF IDEAS

- A complicated circuit usually can be simplified to an equivalent total resistance.
- The simplification is accomplished one step at a time, by combining resistors that are in series or in parallel.
- After R_T and I_T have been found, the original circuit can be reconstructed one step at a time. Individual voltages and currents can then be specified.
- A Wheatstone bridge is a useful circuit for obtaining accurate measurements of temperature and other physical variables.
- A Wheatstone bridge becomes balanced when both sides have the same ratio of resistances ($R_4 / R_3 = R_2 / R_1$).

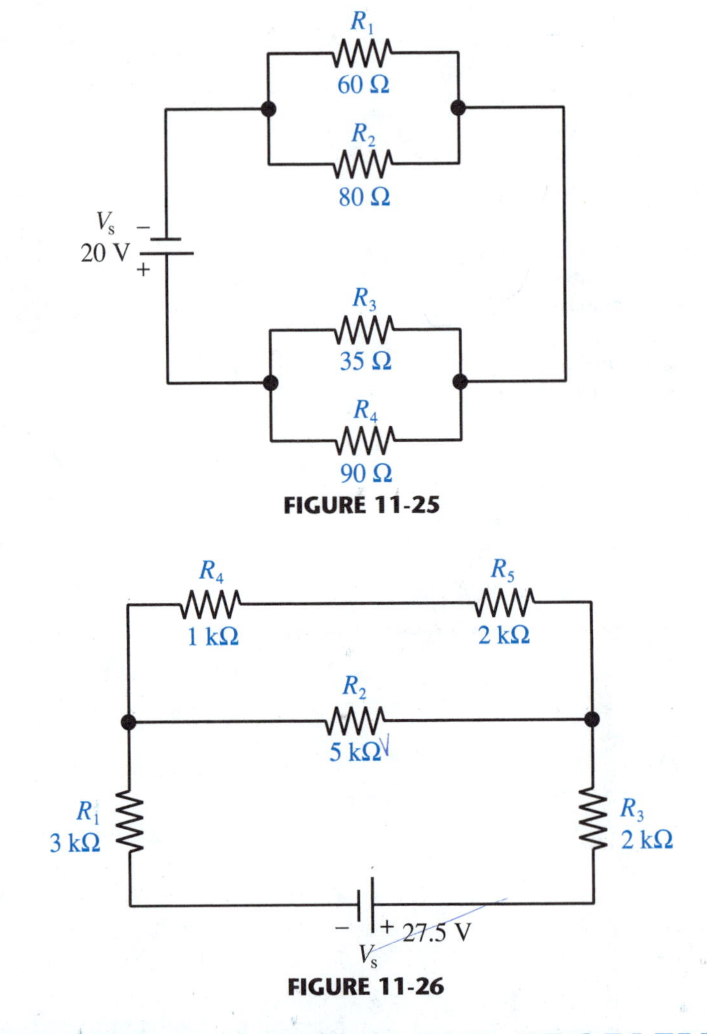

FIGURE 11-25

FIGURE 11-26

CHAPTER QUESTIONS AND PROBLEMS

1. What is the very first step in analyzing a series-parallel circuit? [1]
2. Solve the circuit of Figure 11-25. That is, find every voltage and current in the circuit. [1, 2]
3. Solve the circuit of Figure 11-26. [1, 2]
4. In the circuit of Figure 11-26, suppose a sixth resistor, $R_6 = 2.5\ \text{k}\Omega$, is connected in parallel with the ideal source. Is there a change in any of the currents or voltages for R_1 through R_5? [1, 2]
5. In Problem 4, what would change in the circuit? [1, 2]
6. Solve the circuit of Figure 11-27. [1, 2]
7. Refer to Figure 11-28. (See page 8 in the Circuits color section for the color version of Figure 11-28.) Assume that the source voltage between the green and yellow leads is 10 V. Draw the schematic diagram of this circuit.
8. Using your schematic diagram from Problem 7, analyze the circuit. Find every resistor voltage and every branch current. [1, 2]
9. Refer to Figure 11-29. (See page 8 in the Circuits color section for the color version of Figure 11-29.) Assume that the source voltage between the green and yellow leads is 16 V. Draw the schematic diagram of this circuit.
10. Using your schematic diagram from Problem 9, analyze the circuit. Find every resistor voltage and every branch current. [1, 2]
11. In Figure 11-29, what is the value of current in the blue wire? [2]
12. In Figure 11-29, what is the value of current in the orange wire? [2]

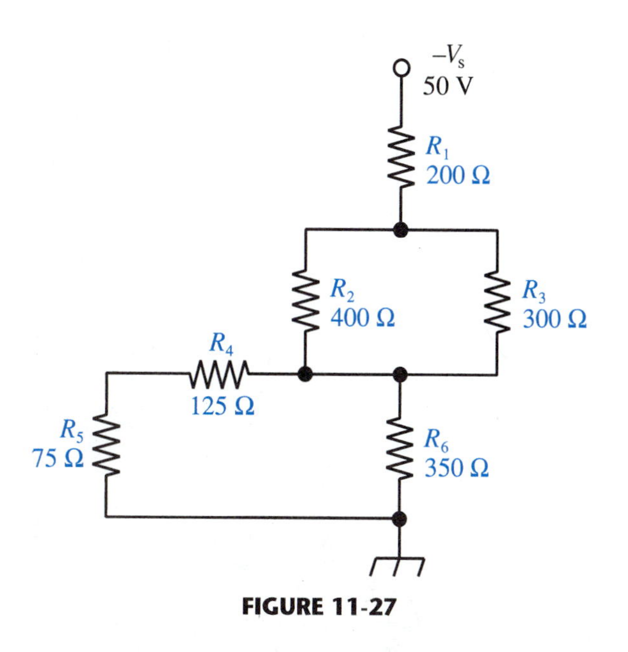

FIGURE 11-27

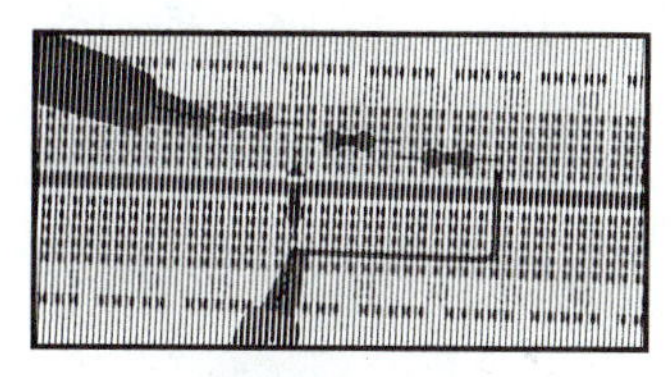

FIGURE 11-28

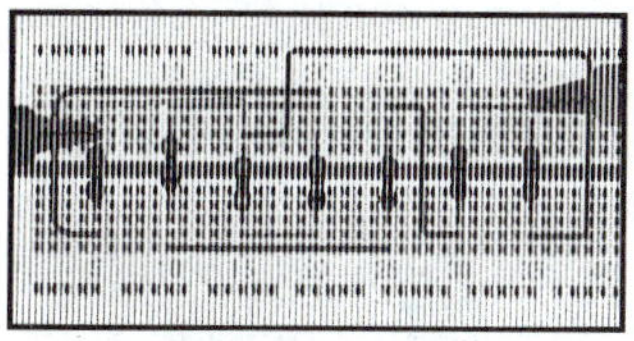

13. In the circuit of Figure 11-30, assume that the source voltage between the green and yellow leads is 20 V. (See page 8 in the Circuits color section for the color version of Figure 11-30.) Draw the schematic diagram of this circuit.
14. Using your schematic diagram from Problem 13, analyze the circuit. Find every resistor voltage and every branch current. [1, 2]
15. In Figure 11-30, what is the value of current in the green wire? [2]
16. In Figure 11-30, what is the value of current in the orange wire? [2]

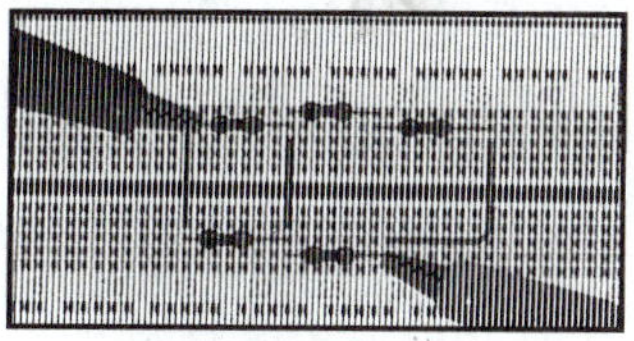

FIGURE 11-29

For Problems 17 through 20, refer to Figure 11-31, which is a dc version of a residential wiring system. Current I_1 can be found by applying Ohm's law to V_{s1} and R_1. Current I_2 can be found by applying Ohm's law to V_{s2} and R_2.

17. Suppose $R_1 = 12\ \Omega$ and $R_2 = 8\ \Omega$. Find I_1 and I_2. Then find the magnitude and direction of $I_{neutral}$ by applying Kirchhoff's current law to point N.
18. Repeat Problem 17 for $R_1 = 6\ \Omega$ and $R_2 = 7.5\ \Omega$.
19. Is it ever possible for $I_{neutral}$ to be larger than I_1? Larger than I_2?
20. If the outside wires (the black and the red) are AWG #12, what gage should the white neutral wire be?

FIGURE 11-30

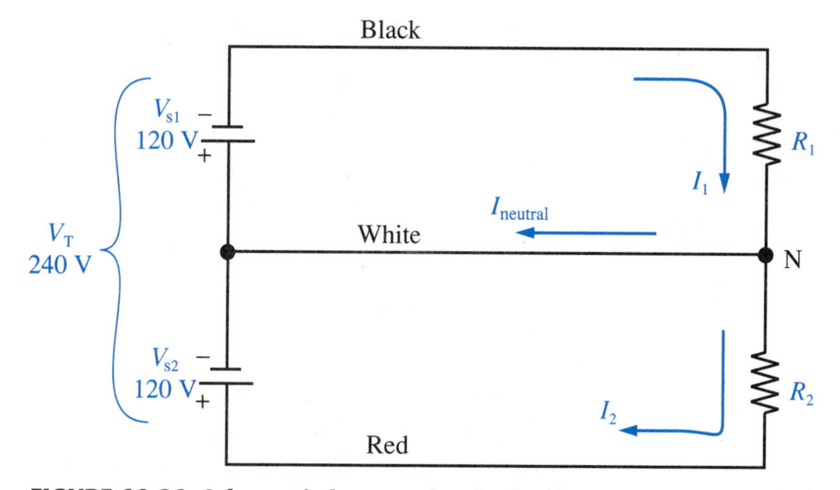

FIGURE 11-31 Schematic layout of a single-phase center-grounded 240 V/120 V residential wiring system

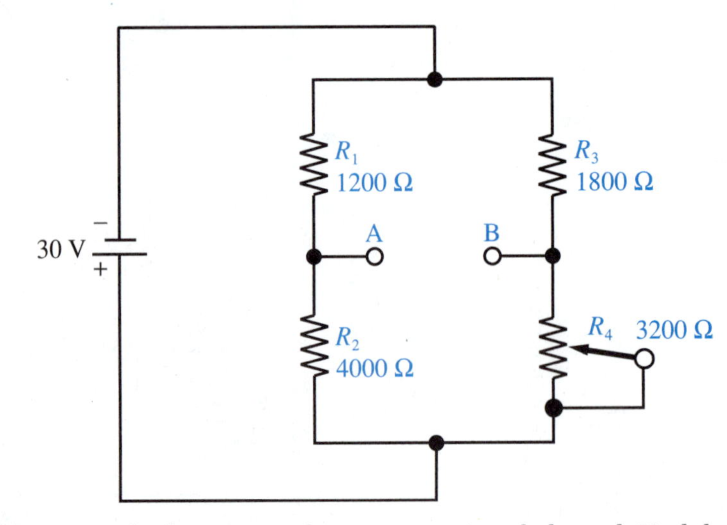

FIGURE 11-32
Wheatstone bridge for Problems 21 through 25

21. The Wheatstone bridge circuit of Figure 11-32 is unbalanced. Find the voltage across all four resistors, using voltage division and Kirchhoff's voltage law.
22. In Problem 21, solve for V_{AB}.
23. In Figure 11-32, what new value of R_4 would balance the bridge? [3]
24. Do V_1 and V_2 change as the bridge goes from unbalanced to balanced? Explain. [3]
25. Do V_3 and V_4 change as the bridge goes from unbalanced to balanced? Explain. [3]

Troubleshooting Questions and Problems

26. In the circuit of Figure 11-12, suppose that a voltmeter across the R_1–R_2–R_3 combination measures 12.7 V. Is the circuit working properly? Explain. [4]
27. Make a guess about the cause of the trouble in Problem 26. Based on your guess, predict what the voltage would become across R_1–R_2–R_3. Keep guessing until you find the specific problem that caused $V_{1\text{–}2\text{–}3}$ to equal 12.7 V. [4]
28. In Figure 11-13, suppose that an ammeter in the negative supply line measures 207 mA. Is the circuit working properly? Explain. [4]
29. Make a guess about the cause of the trouble in Problem 28. Based on your guess, predict what the current would become in the supply line. Keep guessing until you find the specific problem that caused I_T to equal 207 mA. [4]
30. In Figure 11-14, suppose that a 1-A fuse (not shown) blows immediately when the on-off switch (not shown) is closed. What is the most likely cause of the trouble? [4]
31. In Figure 11-15, suppose that a voltmeter across R_1 measures 13.6 V. Is the circuit working properly? Explain. [4]
32. For the trouble you see in Problem 31, repeat the on-paper troubleshooting analysis that you did in Problems 27 and 29. [4]
33. In Figure 11-26, suppose that a voltmeter across R_2 measures 3.93 V. Is the circuit working properly? Explain. [4]
34. For the trouble you see in Problem 33, repeat the on-paper troubleshooting analysis that you did in Problems 27, 29, and 32. [4]
35. In Figure 11-27, suppose that a voltmeter across R_1 measures 17.5 V and a voltmeter connected between the bottom of R_2 and ground also measures 17.5 V. Is the circuit working properly? Explain. [4]
36. Make a guess about the cause of the trouble in Problem 35. Does your guess explain why those two voltage measurements should both equal 17.5 V? If not, keep trying until you identify the trouble. [4]

CHAPTER 12

CAPACITANCE

Various capacitors. *Courtesy of Sprague Electric Co.*

OUTLINE

NEW TERMS TO WATCH FOR

capacitor
farad
dielectric constant
temperature-stable
disc
monolithic
Electronics Industries Association (EIA)
electrolytic
leakage resistance
tantalum
capacitance meter
LCR meter
impedance bridge
stray capacitance
chassis
electrical noise
rectify
breaker points
ignition coil

Capacitors are the second most common electrical component, behind resistors. They can store charge temporarily, which enables them to serve as temporary voltage sources. This is their function in a rectified dc power-supply smoothing circuit. They can delay the appearance of voltage until charge has been moved from one place to another. This enables them to slow down circuit action. This is their function in a standard automobile ignition circuit. They can discriminate one frequency from another in an ac circuit. This is their function in a radio tuner.

The first two abilities can be understood in a dc circuit context. We will study them in this chapter. The third ability is used in ac circuits. We will explore that ability later.

After studying this chapter, you should be able to:

1. Describe the structure of a capacitor and explain the basic charging process.
2. Relate the three variables—capacitance, voltage, and charge. Solve for any one of them if the other two are known.
3. Describe how capacitance is affected by the physical factors of (1) plate area, (2) plate spacing, and (3) dielectric constant.
4. Name the common types of capacitors. List the characteristics and advantages of each type.
5. Interpret the EIA code for ceramic capacitors.
6. Handle capacitors that are connected in series or in parallel.
7. Troubleshoot a capacitive circuit to locate a malfunctioning capacitor.
8. Explain the voltage-stabilizing function of a capacitor in various applications.

12-1 BEHAVIOR OF CAPACITORS

A **capacitor** has a very simple structure, shown in Figure 12-1. It has two metal plates separated by insulation, with leads attached to the plates.

Figure 12-2 shows how a capacitor functions in a circuit. Negatively charged electrons flow down one of the leads, say the top lead. They enter the metal atoms of the top plate, but are unable to pass through the dielectric insulating layer. Therefore, the top plate becomes a negatively charged object.

Due to charge repulsion, loosely held electrons in the metal of the bottom plate are forced away from the top plate. They flow away from the capacitor through the bottom lead, as shown in Figure 12-2(A). Therefore, the bottom plate develops a net positive charge. Of course, this charging process occurs because the external circuit forces it to occur.

If the capacitor is then disconnected from the rest of the circuit, its stored charge enables it to act like a voltage source all on its own. This is illustrated in Figure 12-2(B).

Capacitance is symbolized C. The schematic symbols for a capacitor are shown in Figure 12-3(A). The curved-line symbol is more widespread, but the symbol with both straight lines is popular in computer-aided drafting, or CAD.

If there are several capacitors in a circuit, we identify them by subscripts, the same as resistors. Figure 12-3(B) shows this.

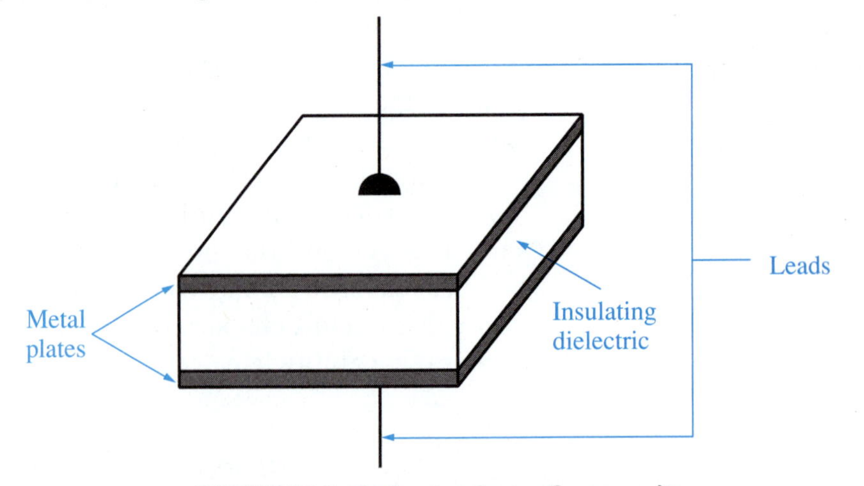

FIGURE 12-1 Basic structure of a capacitor

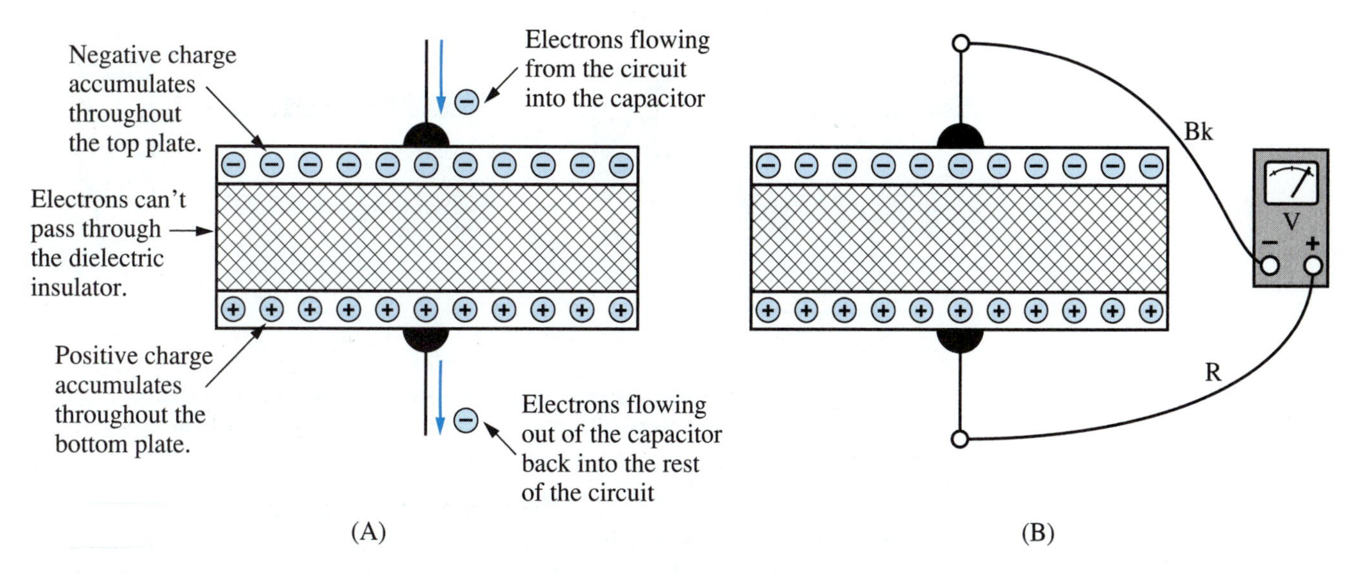

FIGURE 12-2 (A) During the charging process, the plates become oppositely charged. (B) After the charging process is finished, the capacitor possesses voltage.

12-2 MEASUREMENT UNIT—THE FARAD

The basic measurement unit of capacitance is the **farad**—symbolized F.

TECHNICAL FACT

One farad is the amount of capacitance that has 1 volt across the terminals if 1 coulomb of charge is on the plates. (1 F = 1 C / 1 V)

To understand this, look again at Figure 12-2(B). The amount of voltage being measured across that charged capacitor depends on two things:

1. The amount of charge, in coulombs, that is collected on the plates. (The amount of negative charge on the top plate is always equal to the amount of positive charge on the bottom plate.)
2. The amount of capacitance possessed by the capacitor.

Regarding item number 1, the greater the charge, Q, the greater are the attraction/repulsion forces tending to return the charge to its original neutral position. This means greater voltage, V. The relationship between charge and voltage is illustrated in Figure 12-4.

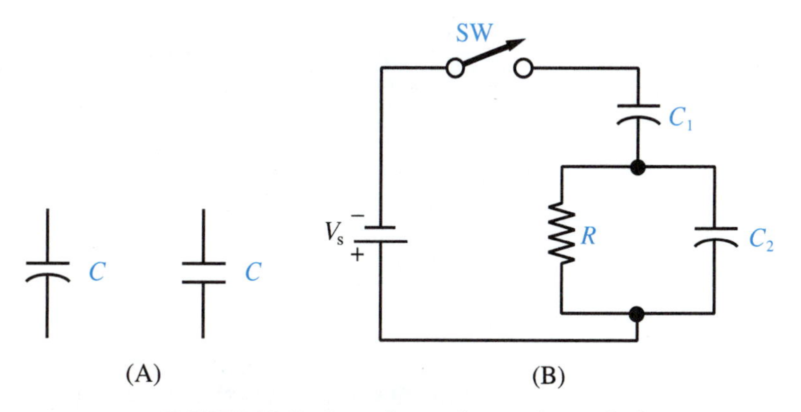

FIGURE 12-3 Capacitor schematic symbols

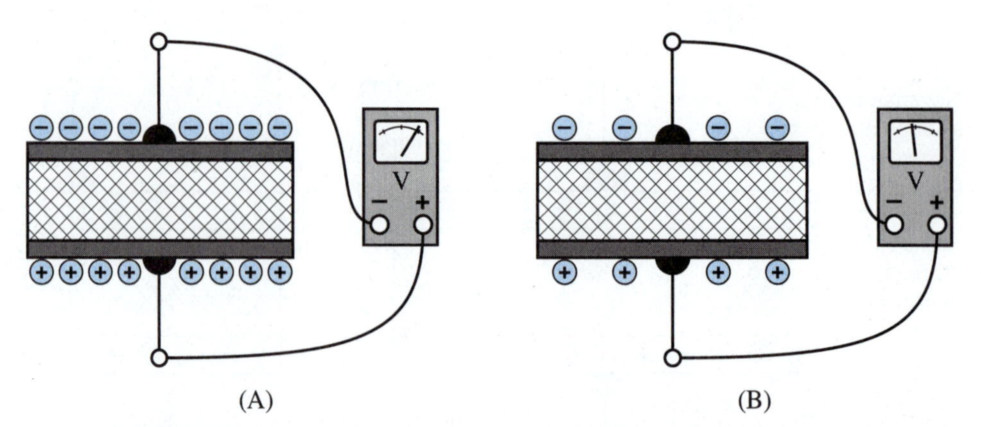

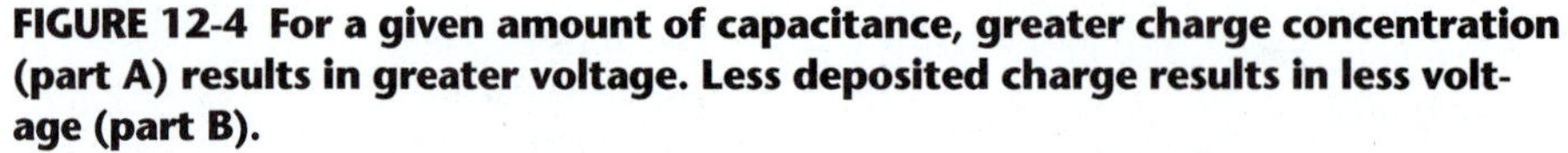

FIGURE 12-4 For a given amount of capacitance, greater charge concentration (part A) results in greater voltage. Less deposited charge results in less voltage (part B).

Regarding item number 2, a greater capacitance, *C*, causes the charge to be spread over a larger plate area. Its effect is then diluted, producing less voltage. This is illustrated in Figure 12-5.

Summarizing these relations, we have

TECHNICAL FACT

Voltage is proportional to charge, *Q*; voltage is inversely proportional to capacitance, *C*.

$$V = \frac{Q}{C}$$

EQ. 12-1

EXAMPLE 12-1

A 0.05-F capacitor has 1.5 C of charge on its plates. What amount of voltage will be measured across it?

SOLUTION

From Equation 12-1,

$$V = \frac{Q}{C} = \frac{1.5\text{ C}}{0.05\text{ F}} = \mathbf{30\text{ V}}$$

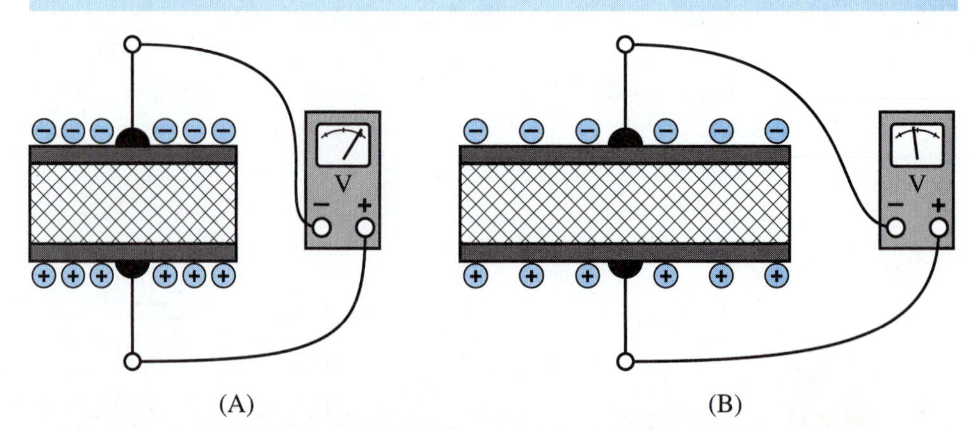

FIGURE 12-5 For a given amount of charge, smaller plate area (smaller capacitance) causes the charge to be more concentrated (part A), resulting in greater voltage. Larger plate area (larger capacitance) gives diluted charge concentration, resulting in less voltage (part B).

EXAMPLE 12-2

We wish to establish a voltage of 25 V across a capacitor by transferring 3.12×10^{18} electrons between its plates. What amount of capacitance is needed?

SOLUTION

Rearranging Equation 12-1 gives

$$C = \frac{Q}{V}$$

EQ. 12-2

Q must be converted to basic units of coulombs for use in this equation.

$$Q = 3.12 \times 10^{18} \text{ electrons} \times \frac{1 \text{ C}}{6.24 \times 10^{18} \text{ electrons}} = 0.5 \text{ C}$$

$$C = \frac{Q}{V} = \frac{0.5 \text{ C}}{25 \text{ V}} = \mathbf{0.02\ F}$$

Even though the farad is the basic unit of capacitance, for most practical capacitors it is far too large a unit. Most real capacitors are tiny fractions of a farad. Therefore, capacitor values are almost always expressed in microfarads (μF), nanofarads (nF), or picofarads (pF).

EXAMPLE 12-3

A 3-μF capacitor is charged to 45 V. What amount of charge is on its plates?

SOLUTION

Rearranging Equation 12-1 or 12-2 gives

$$Q = CV = (3 \times 10^{-6} \text{ F})\, 45 \text{ V} = 1.35 \times 10^{-4} \text{ C or } \mathbf{0.135\ mC}$$

SELF-CHECK FOR SECTIONS 12-1 AND 12-2

1. To establish voltage across a capacitor, what must happen? [2]
2. One farad equals one__________ per one__________. [2]
3. What is the voltage across a 20-μF capacitor that contains 4 mC of charge? [2]
4. What size capacitor will have 125 V when 50 μC of charge is stored? [2]
5. What are the more convenient units for expressing practical capacitor sizes?
6. Once a capacitor has been charged up, it is able to temporarily function like a__________ __________. [1, 8]
7. Express 0.002 μF in nanofarads. Express it in picofarads.
8. Express 4700 pF in microfarads.

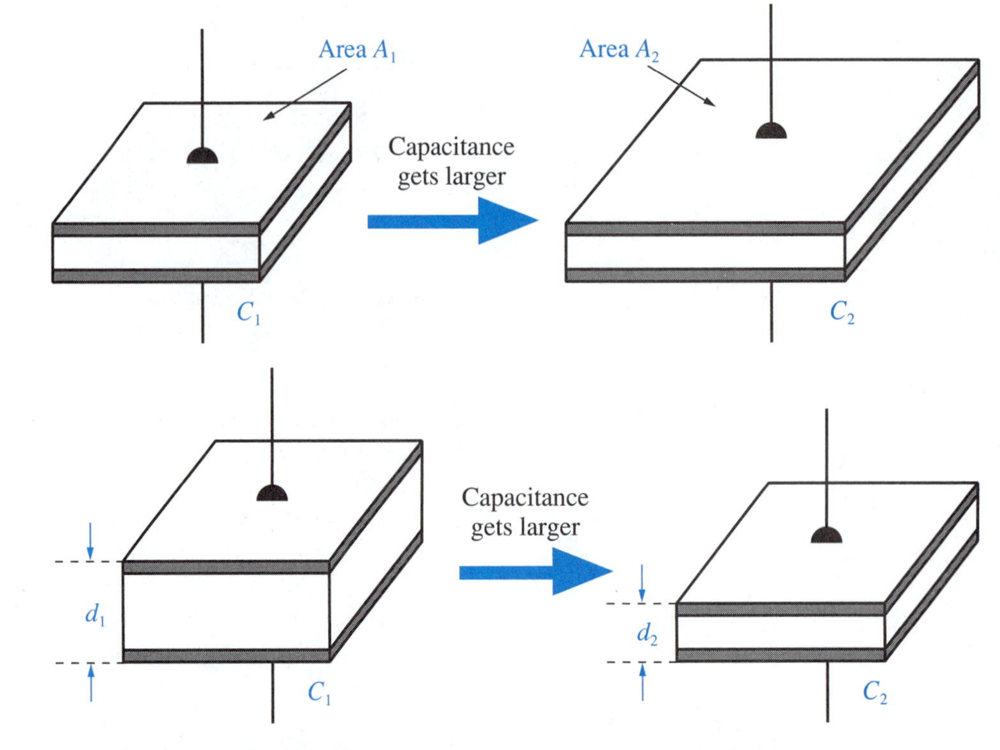

FIGURE 12-6 Larger plate area gives greater capacitance. If $A_2 = 2\ A_1$, then $C_2 = 2\ C_1$, assuming equal dielectric material and thickness.

FIGURE 12-7 Closer spacing gives greater capacitance. If $d_2 = 1/2\ d_1$, then $C_2 = 2\ C_1$.

12-3 FACTORS THAT DETERMINE CAPACITANCE

Three physical factors determine the capacitance of a capacitor. They are:

1. The area of the plates.
2. The distance between the plates.
3. The type of dielectric material used.

Other things being equal, capacitance is proportional to plate area. This relationship is illustrated in Figure 12-6.

Capacitance is inversely proportional to the distance between the plates. The thinner the dielectric separation, the greater the capacitance. This is illustrated in Figure 12-7.

Finally, capacitance is affected by the type of dielectric material that insulates the plates from each other. Some dielectric materials contain molecules that are electrically polarized. This means that although the entire molecule is electrically neutral, its charge is not evenly distributed throughout its body. One side of the molecule has a bit more negative charge and the other side has a bit more positive charge. When the capacitor plates become charged, the dielectric molecules orient themselves in opposition to the plate polarity, as shown in Figure 12-8. This is in accordance with the laws of charge attraction. The molecule's negative side is attracted to the positive capacitor plate, and vice-versa.

Electrical devices designed for critical applications have their electrical specifications and performance checked by automated testing procedures.
Courtesy of Hewlett-Packard Company

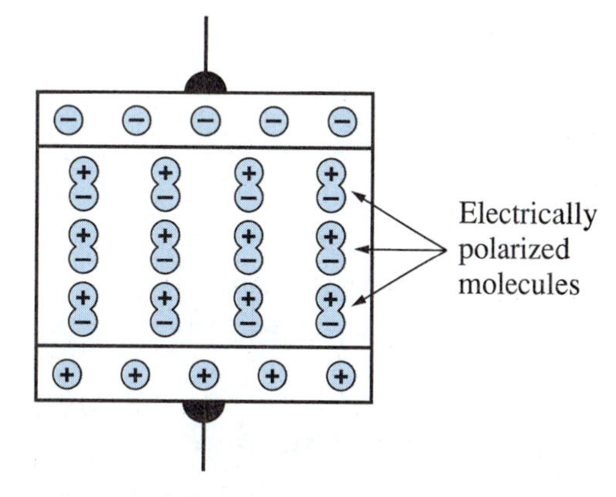

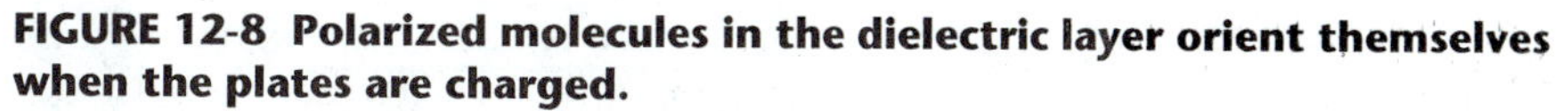

FIGURE 12-8 Polarized molecules in the dielectric layer orient themselves when the plates are charged.

The molecular orientation in Figure 12-8 reduces the effect of charge concentration on the plates. That is, with the positive end of the dielectric layer closer to the negatively charged plate, the concentration of negative charge is partially diluted. This happens also on the bottom end, near the positive plate.

Because charge is diluted, the voltage across the plates is less than it otherwise would be. Recall Equation 12-2,

$$C = \frac{Q}{V}$$

If the voltage is lessened, Equation 12-2 makes it clear that capacitance is increased. This explains how the type of dielectric affects capacitance.

Different dielectric materials have differing degrees of molecular polarization. The degree of polarization is represented by a number called the **dielectric constant**, symbolized k. Table 12-1 lists the dielectric constants of several common capacitor materials.

TABLE 12-1 Dielectric Constants of Some Capacitor Dielectric Materials

MATERIAL	DIELECTRIC CONSTANT
Vacuum	1.0
Air	1.0006
Teflon	2.0
Mineral oil	2.2
Polyethylene	2.3
Paper, parafinned	2.5 (2.0–6.0)
Polystyrene	2.6
Rubber	3.0 (3–5)
Fused quartz	3.9 (3.8–4.4)
Polyester	4.0
Polycarbonate	4.5
Mica	5.0 (5.0–8.5)
Ceramic (low k)	6 typically (varies widely)
Porcelain	6.5 (5–7)
Neoprene	6.9
Bakelite	7
Glass	7.5 (4.8–8.0)
Aluminum oxide electrolytic	8.4
Tantalum oxide electrolytic	26
Water	78
BST ceramic (high k)	7500 (varies widely)

The formula for calculating capacitance in terms of dielectric constant, plate area, and plate spacing is

$$C = (8.85 \times 10^{-12})\frac{kA}{d}$$

EQ. 12-3

Area, A, must be expressed in basic metric units of square meters, and distance, d, must be in meters. C comes out in basic units of farads.

EXAMPLE 12-4

A certain capacitor has plates that are 80 cm long by 2 cm wide. The dielectric material is mica, 0.18 mm thick.

a) Calculate the capacitance.

b) How thin would the dielectric layer have to be in order to produce a capacitance of 0.015 μF (15 nF) ?

c) What would be the capacitance in part b if parafinned paper were used as the dielectric instead of mica?

SOLUTION

a) To use Equation 12-3, area, A, and plate spacing, d, must be expressed in basic units.

$$80 \text{ cm} \times \frac{1 \text{ m}}{100 \text{ cm}} = 0.8 \text{ m}$$

$$2 \text{ cm} = 0.02 \text{ m}$$

$$A = (0.8 \text{ m}) \times (0.02 \text{ m}) = 1.6 \times 10^{-2} \text{ square meters } (\text{m}^2)$$

$$d = 0.18 \text{ mm} \times \frac{1 \text{ m}}{1 \times 10^{3} \text{ mm}} = 1.8 \times 10^{-4} \text{ m}$$

From Table 12-1, the dielectric constant for mica is 5.0.

Plugging these values into Equation 12-3 gives

$$C = (8.85 \times 10^{-12})\frac{kA}{d} = (8.85 \times 10^{-12})\frac{(5.0)\,(1.6 \times 10^{-2})}{1.8 \times 10^{-4}} = 3.93 \times 10^{-9} \text{ F or } \mathbf{3.93 \text{ nF}}$$

b) Rearranging Equation 12-3, we get

$$d = (8.85 \times 10^{-12})\frac{kA}{C} = (8.85 \times 10^{-12})\frac{(5.0)\,(1.6 \times 10^{-2})}{15 \times 10^{-9}}$$

$$= 0.0472 \text{ mm or } \mathbf{47.2\ \mu m}$$

c) This problem could be solved by plowing through Equation 12-3 again. However, an easier solution is to recognize that k for parafinned paper is one-half as large as for mica (2.5 / 5.0 = 0.5). Therefore, the capacitance will also be half as large, since C is proportional to k .

$$C = (0.5)\,(3.93 \text{ nF}) = \mathbf{1.97 \text{ nF}}$$

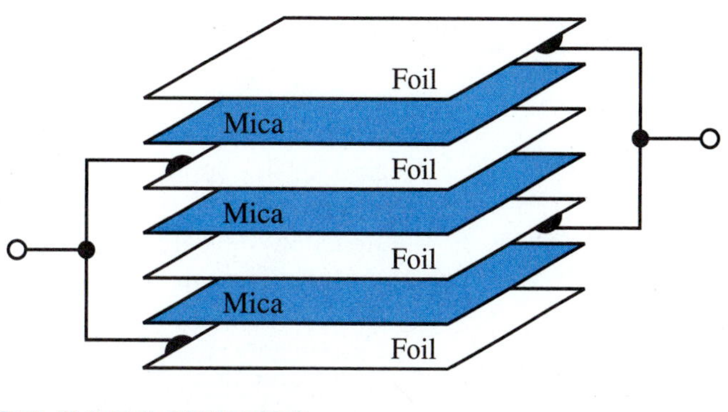

FIGURE 12-9 Multilayer structure. This structure is used for mica dielectrics and other dielectric materials as well.

12-4 TYPES OF CAPACITORS

4

The best way to classify capacitors is by the type of dielectric used. Let us concentrate on four categories of dielectric materials: (1) mica, (2) ceramic, (3) plastic, and (4) electrolytic.

Mica

Mica is a naturally occurring mineral based on the element silicon. The very pure mica used in modern capacitors is manufactured synthetically, however.

Figure 12-9 shows the structure of a mica capacitor. Several metal plates are separated by thin layers of mica. Every second metal plate is connected to the next, forming one overall plate with an attached lead wire. The remaining half of the metal plates are also jumpered together, attached to the capacitor's other lead wire. Using multiple plates in this manner increases the overall plate area, boosting the capacitance value.

The whole assembly is sealed with a protective coating. If the protective coating is neatly formed by molding equipment, the capacitor is called a molded mica unit. If the coating comes from simply dipping the assembly into a molten bath and letting it harden, the capacitor is called a dipped unit. Molded mica capacitors are shown in Figure 12-10(A). The dipped variety is shown in Figure 12-10(B).

Characteristics of Mica Capacitors

Range of Values and Manufacturing Tolerance. Mica capacitors have somewhat small capacitances, spanning the range from 1 pF to about 0.1 μF. They can be manufactured to rather close tolerance, ± 5% being typical. Units of ± 1% tolerance are also available.

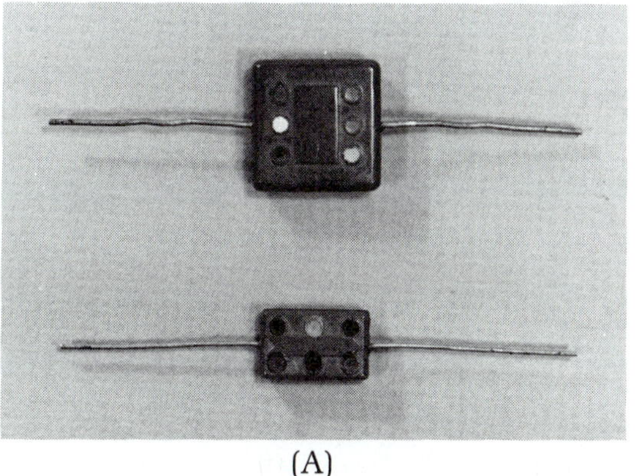

(A)

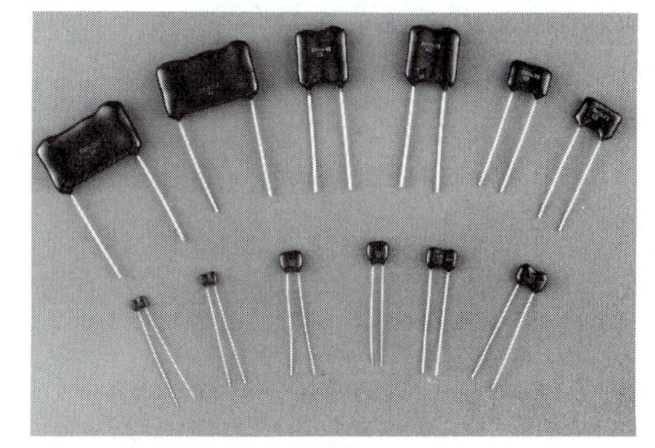

(B)

FIGURE 12-10 (A) Molded coating. (B) Dipped coating

Temperature Characteristics. All capacitors are subject to changes in capacitance resulting from temperature variation, no matter what their dielectric. A capacitor is considered higher quality if its temperature-related capacitance variation is small. It is then described as **temperature-stable.** Temperature stability is commonly expressed in percent per degree Celsius (%/°C), or in parts per million per degree Celsius (ppm/°C). Mica capacitors typically have temperature stability of about 0.01%/°C, or 100 ppm/°C, which is quite good compared to other types.

Capacitors have limits on the coldest and hottest temperatures at which they can operate. In fact, all electrical and electronic components have such temperature limits. Mica capacitors have a very wide usable temperature range. Typically, they can withstand temperature extremes from –55 to +125°C. This is equivalent to –67 to +257°F. Some mica units can operate as hot as 150°C.

Voltage Rating. Every capacitor has a certain maximum voltage that it can withstand. If a greater voltage is applied to the plates, the capacitor's dielectric layer may be destroyed. Sometimes the voltage rating is indicated by a colored stripe or dot, and sometimes it is written directly on the body. It may be symbolized plain V, or WV, which stands for continuous working volts.

A capacitor's voltage rating is determined by the type of dielectric material and its thickness. Of all the common dielectric materials, mica has the highest voltage rating per unit of thickness. Ratings of several thousand volts are possible.

Ceramic Capacitors

The characteristics of ceramic capacitors can vary widely depending on the type of ceramic used. As Table 12-1 shows, different types of ceramic materi-

SKIN CARE TESTING

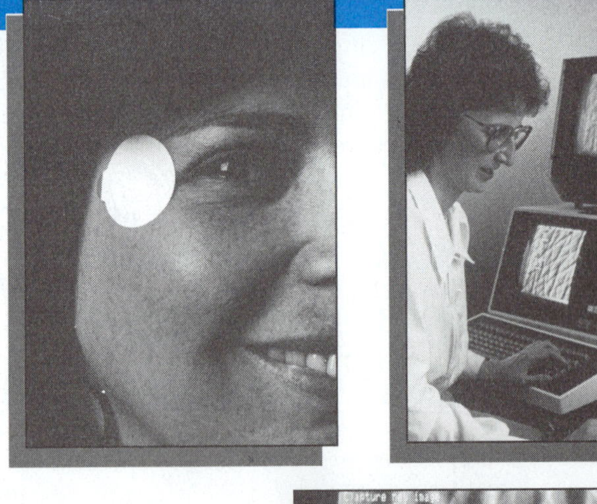

In the last few years, skin-smoothing products have been brought to the market by several major drug research companies. The most widely publicized product is Retin-A cream, from Ortho Pharmaceutical Co. Independent testing of the Retin-A effect on facial skin is shown here. A volunteer is having an impression made of the "crows-feet" wrinkles at the side of her eye. The silicon rubber impression is photographed by a magnifying camera. A closeup view of the camera image can be displayed on the video screen, with an electronically generated cross-sectional view of the peaks and valleys of the skin's wrinkles. These displays are shown in color on page 5 of the Applications color section.

By digital computer processing of many photographic frames, a three-dimensional color-enhanced image of the skin can be made. These images can then be compared, before and after treatment with the cream. You can see a before-and-after comparison for retin-A on page 5 of the Applications color section.

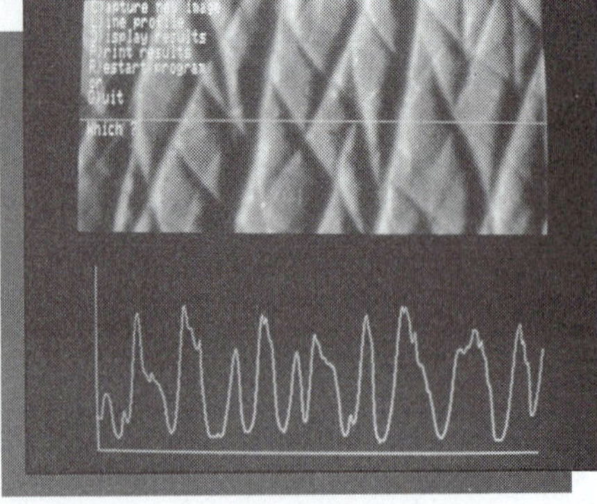

Courtesy of NASA

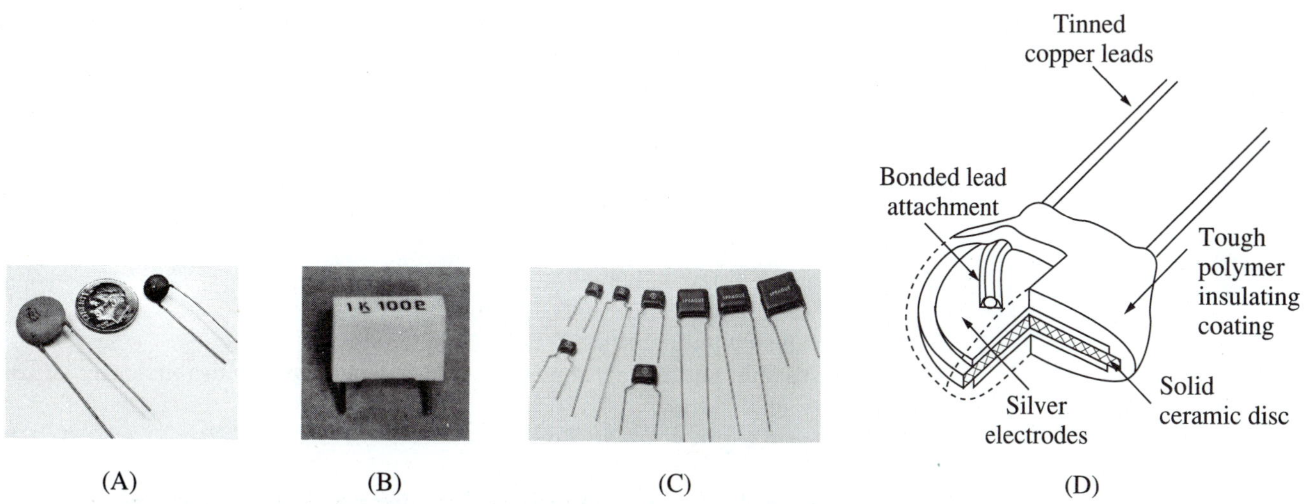

FIGURE 12-11 (A) Disc packages. (B) Molded ceramic capacitor. (C) Dipped ceramic capacitors. (D) Internal structure of ceramic disc. *Courtesy of Sprague Electric Co.*

als have tremendously different dielectric constants. The general rule is that ceramics with high *k* provide greater capacitance for a given physical size, but have worse temperature stability. Ceramics with low *k* provide less capacitance density, but have better temperature stabililty.

In the past, high-*k* ceramic capacitors came in the **disc** package shown in Figure 12-11(A). Low-*k* ceramic capacitors came in the molded package shown in Figure 12-11(B) or the dipped package shown in Figure 12-11(C). Nowadays, however, this package distinction is not always true.

The internal structure of a ceramic disc is shown in Figure 12-11(D). Because their structure is so simple, discs are the cheapest, most durable capacitor package.

The internal structure of a molded or dipped ceramic capacitor is multilayered, also called **monolithic**, the same as Figure 12-9. Multilayered ceramic capacitors are often called MLCs. In recent years, they have captured an ever-larger share of the capacitor market.

Characteristics of High-k Ceramic Capacitors. High-*k* ceramic discs tend to have rather low capacitance because their structure does not provide much plate area. Most units are below 0.1 μF, although large-diameter discs can reach 1 μF. High-*k* MLCs can reach about 10 μF.

Manufacturing tolerance cannot be held close with high-*k* ceramic capacitors. Typical tolerances are ±10% and ±20%. Larger discs tend to be worse, often with tolerance as wide as +80%, –20%.

In general, high-*k* ceramic capacitors are not as temperature-stable as mica capacitors. Nor do they have as wide a usable temperature range.

By making the dielectric layer quite thick, high-*k* ceramic discs can attain voltage ratings of several thousand volts.

Characteristics of Low-k Ceramic Capacitors. For low-*k* ceramic capacitors, the maximum capacitance value is about 0.01 μF in disc form. In multilayer form, the maximum is about 3 μF. Manufacturing tolerances can be held as close as ±0.5%. Typically, though, it is ±5% or ±10%.

The temperature stability of low-*k* ceramic capacitors is very good. Certain capacitors referred to as NP0 units are designed to have no change whatsoever over their entire usable temperature range. Actually, the best the manufacturers can guarantee is for the change to be no greater than 30 ppm/°C. Most low-*k* ceramic capacitors have a usable temperature range of –55 to +85°C, but some units can operate to +125°C.

It is not feasible to greatly increase the dielectric thickness of a low-*k* MLC. Therefore, they tend to have moderate voltage ratings, no greater than a few hundred volts.

Standard EIA Markings for Ceramic Capacitors. The **Electronics Industries Association (EIA)** has developed a code for specifying information about a ceramic capacitor in the following order:

1. Nominal capacitance. A three-digit number. If the first digit is zero, the capacitance is given in units of μF. Since a decimal point is hard to see, it may be signified by a letter. If the first digit is not zero, the capacitance is given in units of pF. Then the first two digits are the coefficient and the third digit is the multiplier.
2. Tolerance. A single letter.
3. Lowest usable temperature. A single letter.
4. Highest usable temperature. A single digit.
5. Worst possible percentage change in capacitance over the usable temperature range. A single letter.

The code is presented in Table 12-2, for items 2 through 5.

TABLE 12-2 EIA Standard Code for Ceramic Capacitors

TOLERANCE (%)	TEMPERATURE (°C) Lowest	Highest	CAPACITANCE CHANGE (%)
F ±1	X −55	5 +85	A ± 1
G ±2	Y −30	7 +125	B ± 1.5
H ±3	Z +10		C ± 2.2
J ±5			D ± 3.3
K ±10			E ± 4.7
M ±20			F ± 7.5
Z +80, −20			P ± 10
P +100, −0 (See note below)			R ± 15
			S ± 22

Note: Manufacturing tolerance of +100%, −0% is also referred to as Guaranteed Minimum Value (GMV).

EXAMPLE 12-5

A low-*k* MLC is marked 0.27 M Y7F. Give all the information you can.

SOLUTION

Nominal capacitance: 0.27 μF
Manufacturing tolerance: ± 20%
Usable temperature range: −30 to +125°C
Temperature stability: Capacitance will change by no more than ± 7.5% over the entire temperature range.

Plastic-Film Capacitors

Plastic-film capacitors are constructed as shown in Figure 12-12(A). Two thin metal-foil strips are alternated with two plastic-film dielectric strips. The most popular plastic compounds appear in Table 12-1. They are polyethylene, polystyrene, polyester, and polycarbonate. All four strips are wrapped in a tight spiral, forming a cylindrical shape. The entire assembly is then wrapped and sealed by machine, or dipped and allowed to harden.

Two wrapped plastic-film capacitors are shown in Figure 12-12(B). The dipped version is shown in Figure 12-12(C).

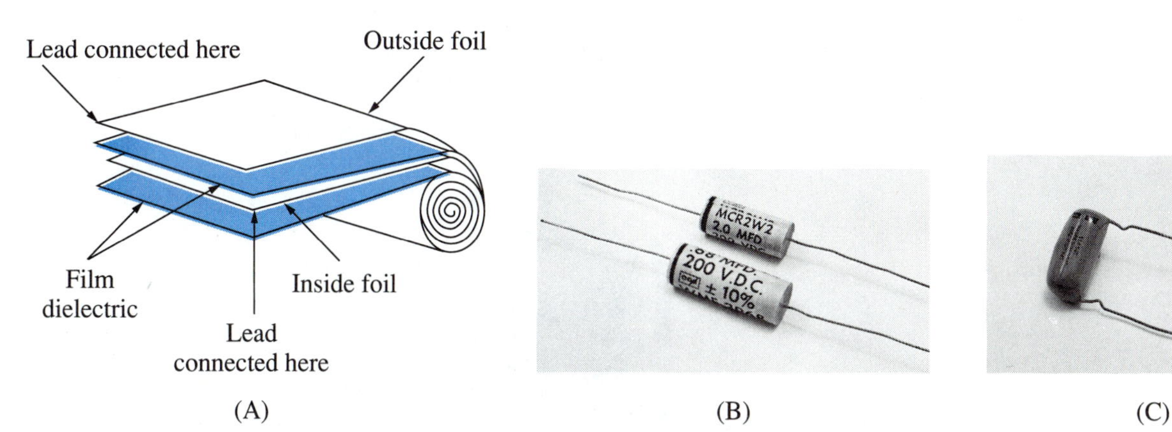

FIGURE 12-12 (A) Tubular construction of plastic-film capacitors. (B) Outer-wrapped (molded) plastic capacitors. (C) Dipped plastic capacitor.

Characteristics of Plastic-Film Capacitors. The spiral-wrap tubular construction of Figure 12-12(A) allows a large plate area to fit in a small volume. Therefore, plastic-film capacitors can attain fairly high values, up to about 20 μF. It is possible to hold their manufacturing tolerance to ± 1%. More typical tolerances are ± 5% and ± 10%.

Temperature stability of plastic-film capacitors is good. Depending on the particular type of plastic used, temperature stability can be from about 50 ppm /°C to 300 ppm /°C. Polycarbonate has the widest usable temperature range, usually −55 to +125°C, with no voltage derating. The other plastics tend to have more limited temperature ranges.

All the plastic-film dielectrics can achieve voltage ratings of several hundred volts. Specialized units may have voltage ratings as high as 15 kV.

Electrolytic Capacitors

Electrolytic capacitors have the same basic tubular construction shown in Figure 12-12(A). The final package appearance can have any of the forms shown in Figure 12-13. Figure 12-13(A) shows the lead wires coming out opposite ends of the cylinder. This is called the axial-lead package. In Figure 12-13(B), the lead wires emerge from the same end. This is called single-ended or printed-circuit mount. In Figure 12-13(C), the entire outer jacket is made of metal. This is often called the metal-can package. In some units, the case connects to one of the interior metal-foil plates. Several aluminum-plate electrolytic capacitors are shown in Figure 12-13(D).

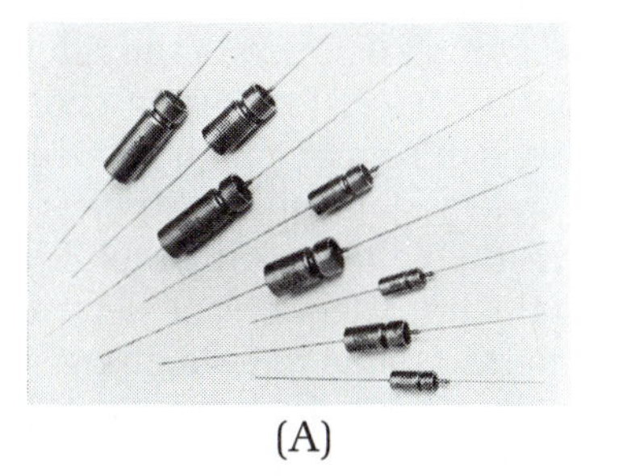
(A)

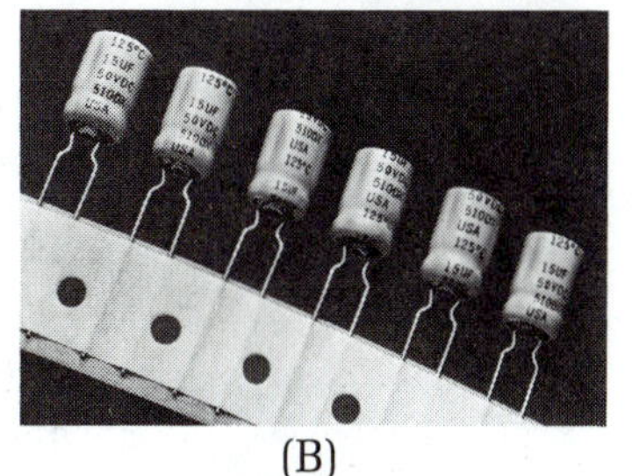
(B)

(C)

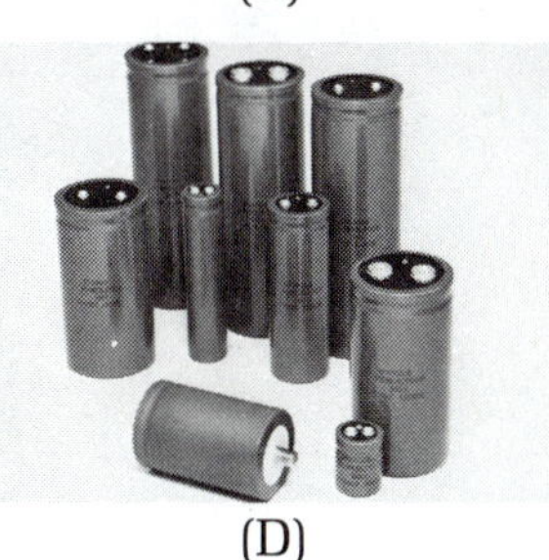
(D)

FIGURE 12-13 (A) Electrolytic capacitors in axial-lead package.
Courtesy of Sprague Electric Co.
(B) Single-ended package. This is sometimes called the radial-lead package, although that is not correct terminology.
Courtesy of Sprague Electric Co.
(C) Metal-can package. The negative plate is connected to the can. The positive plate(s) connect to the end terminal(s). (D) Collection of aluminum electrolytics.
Courtesy of Sprague Electric Co.

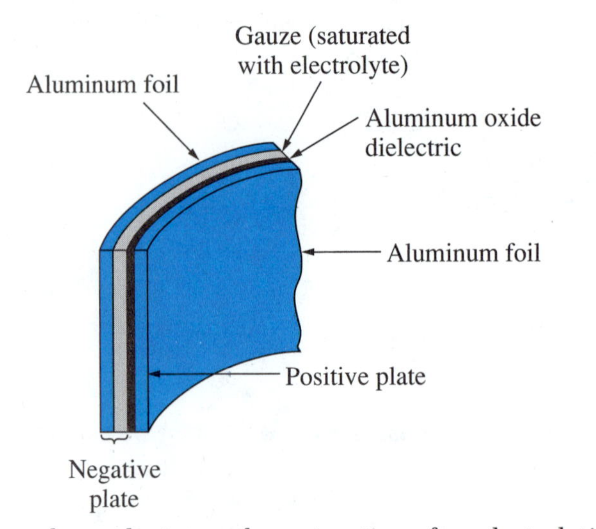

FIGURE 12-14 Internal structure of an aluminum electrolytic capacitor

Figure 12-14 shows the internal construction of an electrolytic capacitor. A liquid-soaked gauze separates the two aluminum foils. The liquid is an electrolyte chemical. When voltage is applied during the manufacturing process, the electrolyte reacts with the aluminum of the inside foil. This causes a very thin layer of aluminum oxide to form on the surface of the inside foil. The aluminum oxide becomes the dielectric.

The dielectric always forms on the inside foil during manufacturing because the dc voltage is always applied positive on the inside foil, negative on the outside foil. Therefore, when an electrolytic capacitor is used in a circuit, it must always be polarized positive on the inside, negative on the outside. It must never have reverse polarity.

The manufacturer clearly marks the lead polarity on the body of the capacitor. Caution is required when connecting an electrolytic capacitor into a circuit. If it is polarized backward, it may explode. Also for this reason, an electrolytic capacitor cannot be used in an ac circuit. When an electrolytic capacitor is drawn in a schematic diagram, its polarity should be marked.

Characteristics of Aluminum Electrolytic Capacitors. The great advantage of electrolytic capacitors is their very high capacitance density. This is because the dielectric layer is so thin, only a few molecules deep on the inside foil surface. In the range above 1 μF, electrolytic capacitors have much smaller volume and lower cost than any other type. Capacitance values to 100 000 μF and higher are available.

Because the dielectric forming process is somewhat irregular, it is difficult to hold close manufacturing tolerance. A typical tolerance range is ± 20%, with many units having tolerance in the range of +50%, –10%.

Aluminum-plate electrolytic capacitors are not very temperature-stable. However, they can be made with a usable temperature range from –40 to +150°C.

They are available with any voltage rating desired, from 3 V to several hundred volts. As the voltage rating increases, so does physical size and price.

One of the imperfections of electrolytic capacitors is their leakage. Ideally, a capacitor should be able to hold charge forever, because its dielectric is a perfect insulator. We say that its **leakage resistance** is virtually infinite. This is nearly true for some ceramic and plastic capacitors.

But the leakage resistance of aluminum electrolytic capacitors is rather low, usually 1 to 10 MΩ. Leakage resistance can be visualized in parallel with the capacitor itself, as shown in Figure 12-15. By Ohm's law,

$$I_{\text{leakage}} = \frac{V}{R_{\text{leakage}}} = \frac{40\text{ V}}{5\text{ M}\Omega} = 8\ \mu\text{A}$$

In actuality, the 8 μA current would pass through the dielectric from one plate to the other.

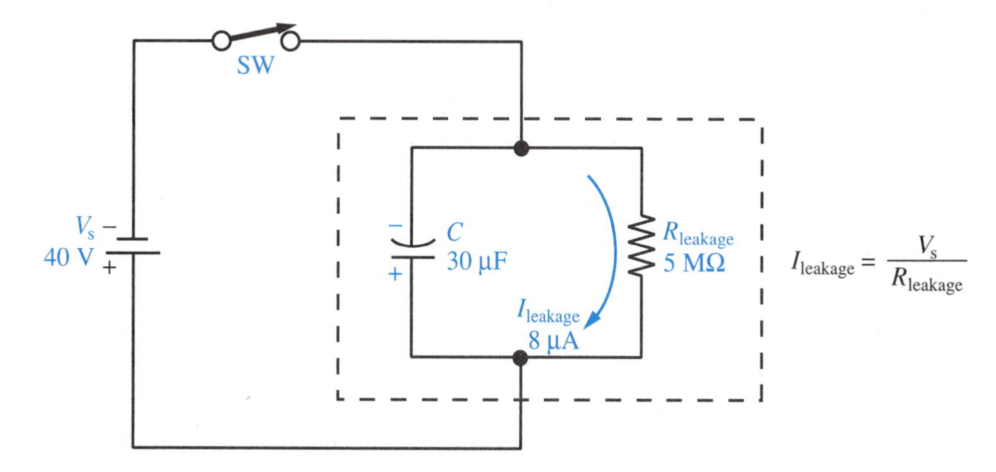

FIGURE 12-15 Real (non-ideal) capacitor model

In most circuit applications, electrolytic capacitor leakage is not a serious problem. But in some situations, it may be unacceptable.

Tantalum Electrolytic Capacitors. Instead of aluminum, the metal **tantalum** can be used for the plates of an electrolytic capacitor. This provides certain advantages.

1. Temperature stability improves considerably. Also, the usable temperature range is wider.
2. Leakage resistance is higher, usually greater than 20 MΩ.
3. Capacitance density is greater. For a given capacitance and voltage rating, a tantalum capacitor has about half the volume of aluminum.
4. Life expectancy is longer.

Tantalum electrolytics are much more expensive than aluminum. For equivalent capacitance and voltage rating, a tantalum unit costs about three to five times as much as aluminum.

✔ SELF-CHECK FOR SECTIONS 12-3 AND 12-4

9. Name the three physical factors that determine capacitance. [3]
10. If plate area is doubled, capacitance is__________. [3]
11. If the spacing between plates is doubled, capacitance is__________. [3]
12. Which dielectric, teflon or neoprene, tends to produce greater capacitance? Why? [3]
13. A capacitor has a plate area of 0.025 m^2 , and polystyrene dielectric 0.5 mm thick. What is its capacitance? [3]
14. What does working voltage mean for a capacitor?
15. What is the basic advantage of the spiral-wrapped tubular construction method? [4]
16. Speaking approximately, what is the usable temperature range for most capacitors? [4]
17. Name some of the disadvantages of an aluminum electrolytic capacitor. [4]
18. A low-*k* MLC is marked 153 J X5D. Give all the information you can about it. [5]

SAFETY ADVICE

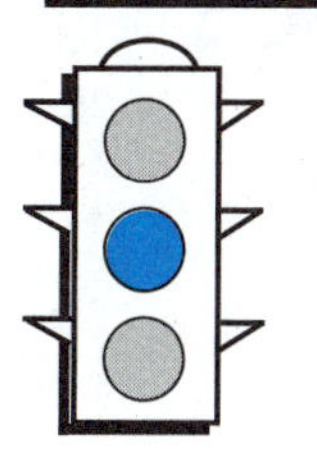

When selecting a capacitor for use in a circuit, be very careful to select a unit that has an adequate voltage rating. If a capacitor's voltage rating is not equal to or greater than the actual applied voltage, it will almost certainly be destroyed. In some cases, the destruction is in the form of a dangerous explosion. This is especially likely for electrolytic capacitors.

Also, electrolytic capacitors are liable to explode if they are polarized backwards. Pay very careful attention to the terminal polarity when installing an electrolytic capacitor into a circuit.

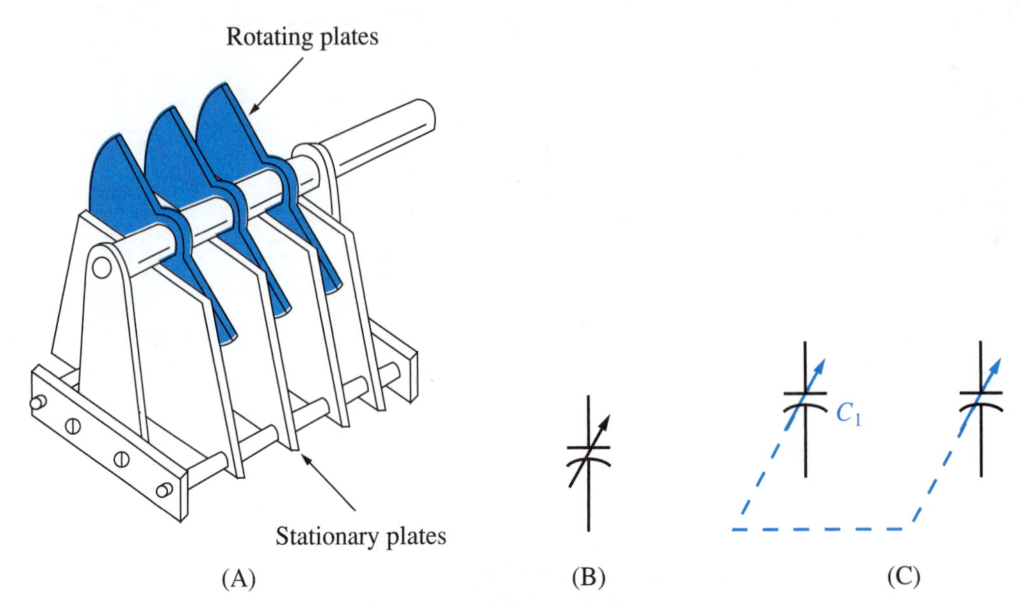

FIGURE 12-16 (A) Interleaved-plates type of variable capacitor. (B) Schematic symbol. (C) Symbolizing ganged variable capacitors.

12-5 ADJUSTABLE CAPACITORS

Variable capacitors are commonly used for tuning radio circuits, adjusting the frequency response of amplifiers, and other purposes. They are available in several styles. Figure 12-16(A) shows the structure of an intermeshing-plate air-dielectric capacitor. A portion of the rotating plates is in position across from the stationary plates. If the shaft is turned so that the portion is increased, effective plate area is increased and capacitance rises.

A variable capacitor is symbolized in Figure 12-16(B). Frequently, two or more interleaved capacitors are ganged on the same shaft. They are symbolized as shown in Figure 12-16(C).

A compression-type variable capacitor is shown in Figure 12-17. The top plate is springy and naturally lifts up from the dielectric layer, producing an air gap. Tightening the adjusting screw forces the top plate down. This reduces the overall plate spacing, causing capacitance to rise.

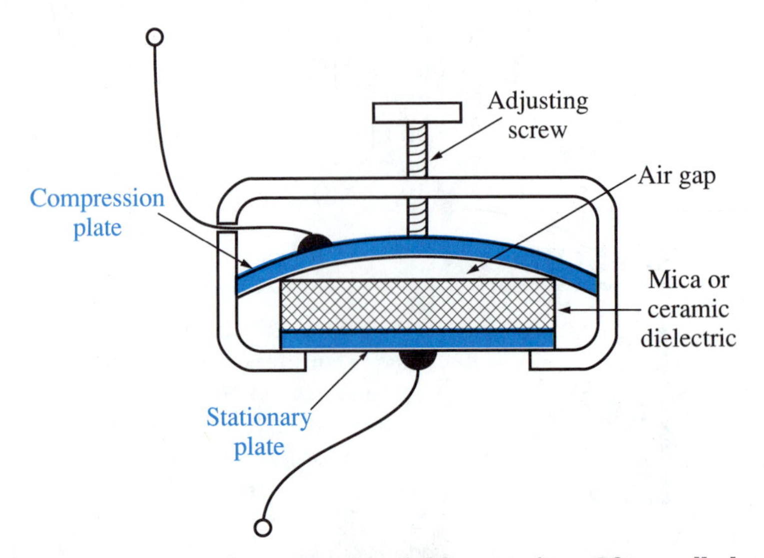

FIGURE 12-17 Compression type of variable capacitor. Often called a trimmer capacitor

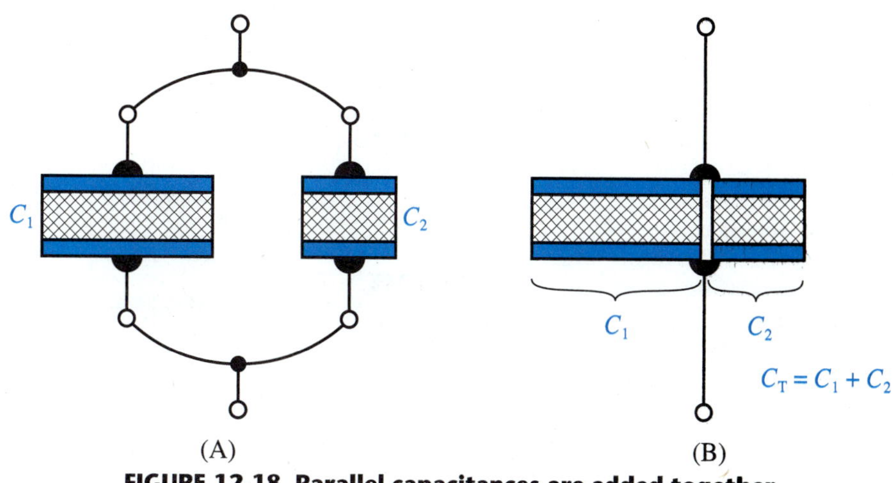

FIGURE 12-18 Parallel capacitances are added together.

12-6 CAPACITORS IN PARALLEL AND SERIES

TECHNICAL FACT

When two or more capacitors are connected in parallel, the total capacitance is equal to their sum. As a formula,

$$C_T = C_1 + C_2 + \ldots$$

EQ. 12-4

6

This can be understood by looking at Figure 12-18. Capacitances C_1 and C_2 are wired in parallel in part A. But that is no different from simply pushing the two capacitors together until their top and bottom plates touch each other, as shown in part B. This makes the total plate area equal to the sum of area A_1 plus A_2. Since capacitance is proportional to area, total capacitance is also equal to the sum of C_1 plus C_2.

The voltage rating of a parallel capacitor combination is given by the lowest rated capacitor in the group.

EXAMPLE 12-6

Three plastic-film capacitors are wired in parallel. Their ratings are : C_1 = 10 μF, 15 WV ; C_2 = 8 μF, 50 WV ; C3 = 2 μF, 25 WV . Find the total capacitance, C_T, and the effective voltage rating of the combination.

SOLUTION

From Equation 12-4,

$$C_T = C_1 + C_2 + C_3 = 10\ \mu\text{F} + 8\ \mu\text{F} + 2\ \mu\text{F} = \mathbf{20\ \mu F}$$

The voltage rating of C_1 is the lowest at **15 V.**

TECHNICAL FACT

When two or more capacitors are connected in series, the total capacitance is given by the reciprocal formula.

$$\frac{1}{C_T} = \frac{1}{C_1} + \frac{1}{C_2} + \ldots$$

EQ. 12-5

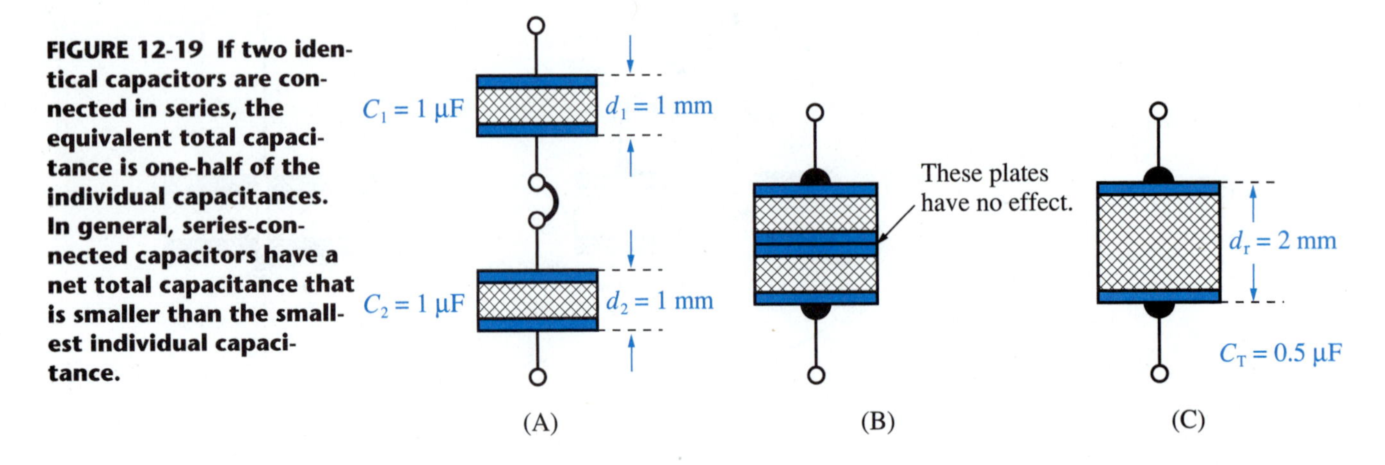

FIGURE 12-19 If two identical capacitors are connected in series, the equivalent total capacitance is one-half of the individual capacitances. In general, series-connected capacitors have a net total capacitance that is smaller than the smallest individual capacitance.

Figure 12-19 will help you understand this. Capacitances C_1 and C_2 are wired in series in part A. This is no different than allowing the bottom plate of C_1 to touch the top plate of C_2, as shown in part B. The center plate thus formed is completely isolated from the rest of the circuit. It cannot accumulate charge, so it has no effect on the overall capacitive performance. Therefore, we might as well remove it, as shown in part C.

The total spacing between plates is now 2 mm, compared to the 1-mm spacing for C_1 and C_2. Since capacitance is inversely proportional to plate spacing, C_T is half as large as C_1 or C_2.

If identical capacitors are connected in series, the overall voltage rating is the sum of the individual voltage ratings. This rule only applies to identical capacitors.

EXAMPLE 12-7

Two capacitors are wired in series. They are: $C_1 = 4$ nF, 50 WV; $C_2 = 6$ nF, 100 WV.

a) Find the total capacitance, C_T.

b) Can you tell easily the maximum voltage rating of the series combination?

SOLUTION

a) From Equation 12-5,

$$\frac{1}{C_T} = \frac{1}{4\text{ nF}} + \frac{1}{6\text{ nF}}$$

$$C_T = \mathbf{2.4\ nF}$$

b) No, it isn't easy to tell the net voltage rating, since the capacitances are not equal.

✔ SELF-CHECK FOR SECTIONS 12-5 AND 12-6

19. Name and describe the two common types of adjustable capacitors. [4]
20. Capacitors in parallel combine like resistors in__________. [6]
21. Suppose $C_1 = 700$ pF, $C_2 = 1.5$ nF, and $C_3 = 500$ pF. If all three capacitors are connected in parallel, find C_T. [6]
22. Capacitors in series combine like resistors in__________. [6]
23. A 1-μF capacitor is wired in series with a 330-nF capacitor. Find C_T. [6]

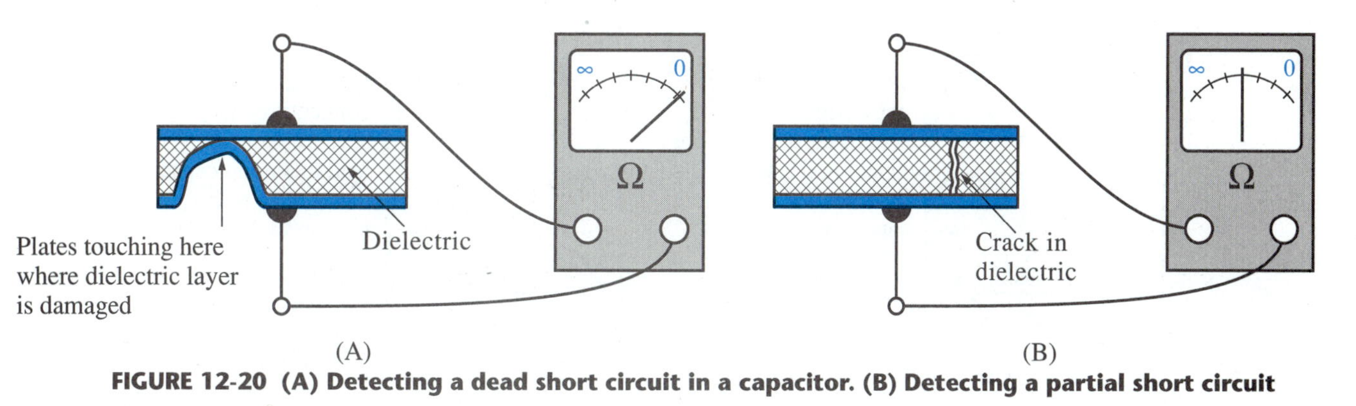

FIGURE 12-20 (A) Detecting a dead short circuit in a capacitor. (B) Detecting a partial short circuit

12-7 TROUBLESHOOTING AND MEASURING CAPACITORS

Capacitors can cause circuit trouble in two ways.

1. They can fail altogether, becoming either short-circuited or open-circuited.
2. Their capacitance value can change due to temperature or aging.

To do circuit troubleshooting, you should know how to test for capacitor failure, and also how to measure the capacitance value of a good capacitor.

Testing for Failure

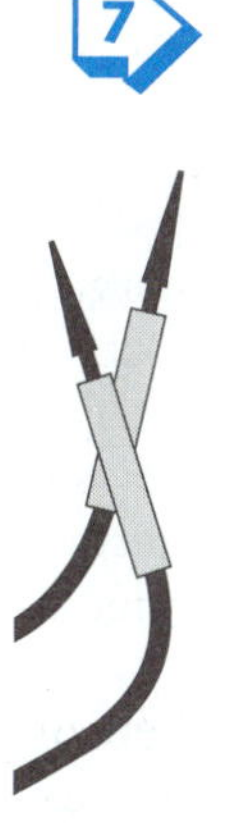

Dead Short Circuit. Sometimes a capacitor can develop a dead short circuit between plates because of damage to the dielectric layer. This is easily detected by an ohmmeter measurement, as shown in Figure 12-20(A).

A short circuit may be intermittent, occurring some of the time but not all the time. To detect this, press on the capacitor's body from different directions with an insulated tool.

Partial Short Circuit. A partial short circuit can occur if a crack develops in the dielectric layer. With a nonelectrolytic capacitor, a partial short circuit is revealed by an ohmmeter measuring a medium-value resistance, as shown in Figure 12-20(B).

Testing Large-Value Capacitors. For capacitors larger than 1 μF, a revealing test can be performed with an analog ohmmeter. It is done as follows:

1. Make sure the capacitor is initially discharged by momentarily connecting a jumper wire across its terminals.
2. Set the ohmmeter to its highest multiplier factor (usually x 10 kΩ or x 100 kΩ, but see caution given with Figure 12-21) and connect one ohmmeter lead to the capacitor. Observe proper polarity if the capacitor is electrolytic.
3. Touch the other ohmmeter lead to the capacitor while watching the pointer. If the capacitor is good, the pointer will surge to the right (towards zero ohms), then return to the left (high value of resistance). This sequence of motions is illustrated in Figure 12-21.

These "chip" ceramic capacitors will be bonded directly to the copper tracks on a printed circuit board using surface-mount techniques. The large reel feeds the capacitors to automated assembly machines during PC-board manufacture. *Courtesy of Sprague Electric Co.*

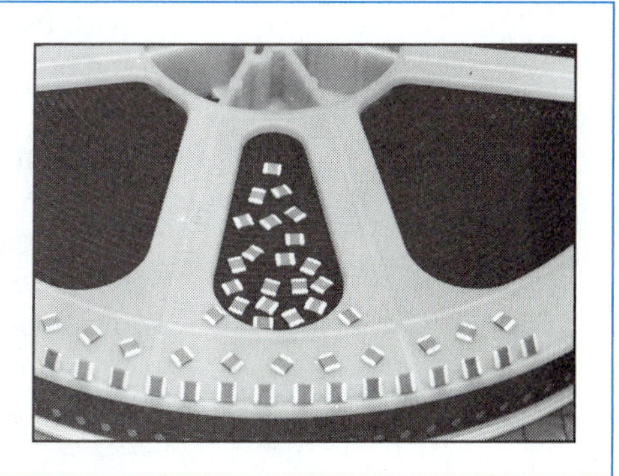

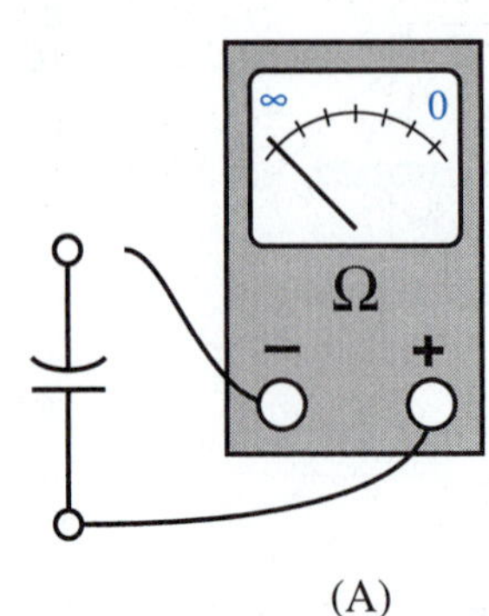

(A)

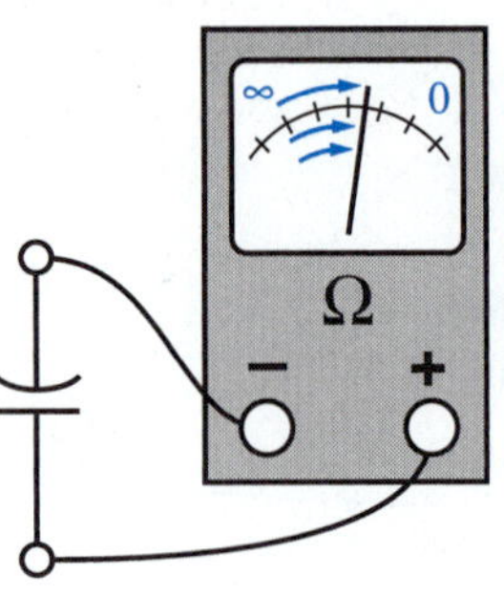

(B)

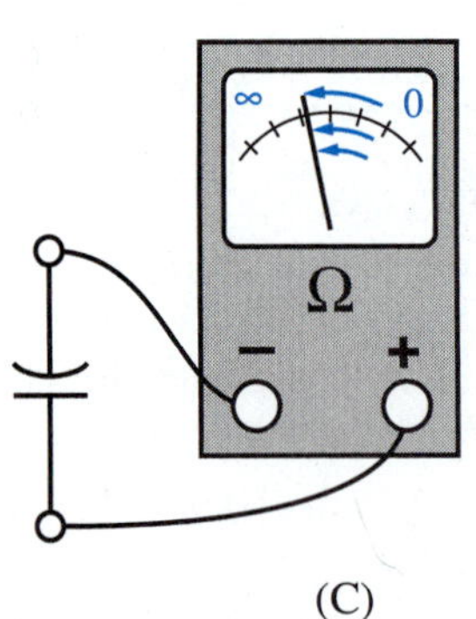

(C)

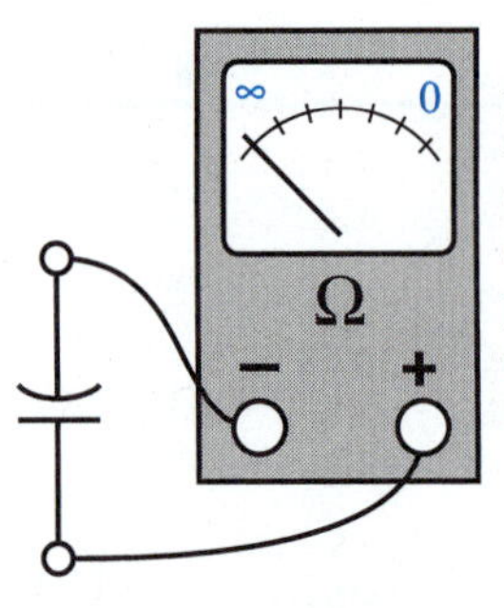

(D)

FIGURE 12-21 The ohmmeter test for capacitors in the range above 1 μF. When using this test, check to see that the ohmmeter's battery voltage does not exceed the capacitor's voltage rating.

The initial surge to the right occurs because the capacitor is being charged by the ohmmeter battery. As the capacitor voltage rises, charging current decreases, causing the pointer to return to the high-resistance end of the scale.

If the capacitor is dead-shorted, the pointer will move to the far right and remain on zero ohms, as described earlier. If the capacitor is partially shorted, the pointer will return partway to the left and stop at an intermediate value. As long as the final reading is greater than about 1 MΩ for an electrolytic capacitor, the capacitor is probably functioning properly.

If the capacitor is open-circuited, the pointer will not surge to the right at all. An open circuit occurs because one of the lead wires has separated from its plate.

High-Voltage Testing. Some capacitor problems show up only at high voltages. They cannot be detected at the low test voltages used by ohmmeters. The circuit in Figure 12-22 can be used for high-voltage testing. The procedure is the same as for the ohmmeter test.

Temperature Effects. Some capacitor troubles occur only when the unit becomes quite hot or cold. The capacitor can be heated up with a hot-air gun of the type used with heat-shrink tubing. Temperature can be lowered by spraying the unit with freeze mist. These temperature techniques are also useful for troubleshooting other kinds of electrical and electronic components.

Measuring Capacitance

Besides capacitors failing open or shorted, circuit trouble also can occur because of a change in capacitance value due to aging. Therefore, circuit troubleshooting sometimes requires the measurement of a capacitor's value. There are three instruments that are commonly used for measuring the value of a capacitor.

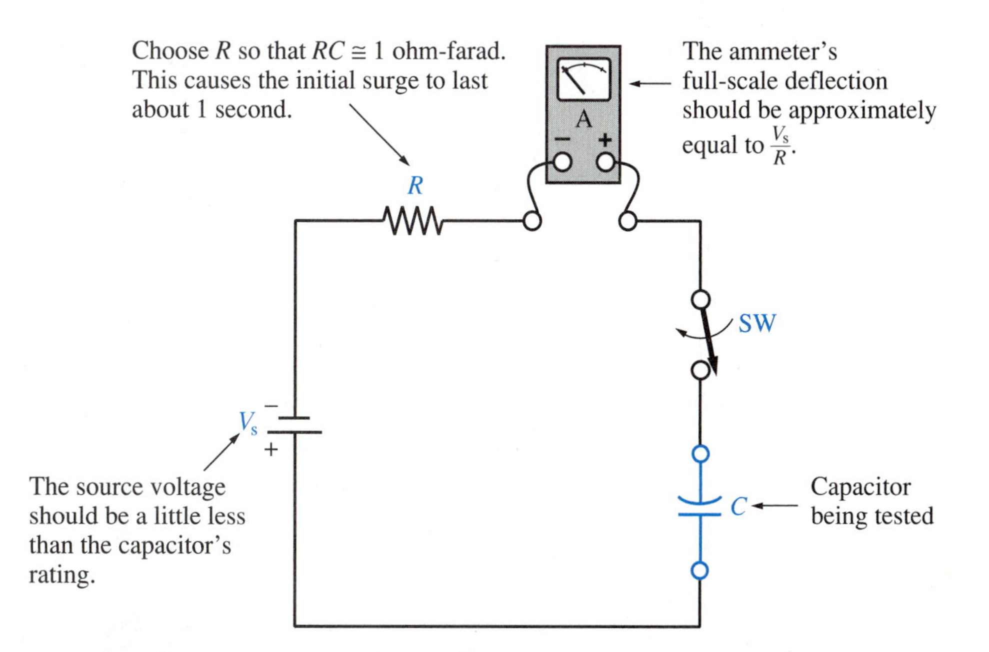

FIGURE 12-22 Testing a capacitor under high-voltage conditions

They are:

1. **Capacitance meter.** Analog and digital readout models are available.
2. **LCR meter.** This is a multipurpose instrument that can measure capacitance, resistance, or inductance. A popular model is shown in Figure 12-23(A). An advanced model is shown in Figure 12-23(B). An LCR meter can also measure the leakage resistance of a capacitor. Electrolytic capacitors are the most likely type to cause circuit trouble due to excessive leakage.
3. **Impedance bridge.** This instrument is capable of measuring capacitance, inductance, and their combinations with resistance (including leakage resistance). Figure 12-23(C) shows a model.

Small-value capacitors less than about 1 μF can be tested for a shorted condition by using the ohmmeter method described earlier. But small capacitors cannot be tested for an open-circuit failure using the ohmmeter method. This is because they charge so quickly that the ohmmeter pointer doesn't have enough time to move to the right. Instead, to test a small capacitor for an open, it must be measured

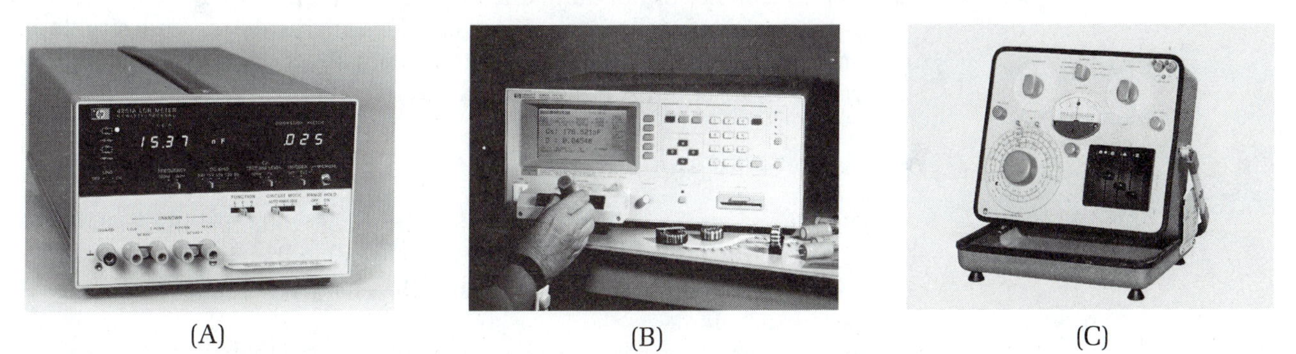

FIGURE 12-23 (A) Auto-ranging LCR meter, capable of measuring capacitors in the range from 0.1 pF to 19 990 μF (19.99 mF). The capacitor's leakage resistance is measured indirectly by the D-factor indication. With an auto-ranging meter, the operator does not have to select the proper range with a front-panel control. That task is handled internally by the instrument. *Courtesy of Hewlett-Packard Company* **(B) Auto-ranging LCR meter, capable of measuring capacitance from 0.000 01 pF to 9.9999 F (9 999 900 μF). The grounding strap on the operator's wrist prevents her body from accumulating static charge, which could damage sensitive components.** *Courtesy of Hewlett-Packard Company* **(C) Impedance bridge, capable of measuring capacitors in the range from 10 pF to 1100 μF. The capacitor's leakage is indicated by the D (dissipation factor) measurement.** *Courtesy of GenRad, Inc.*

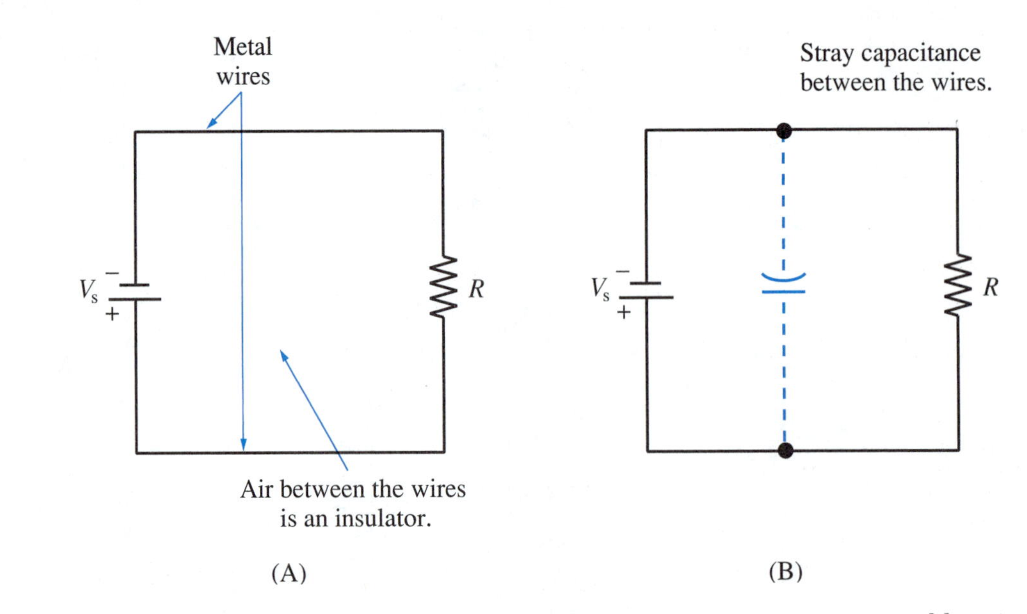

FIGURE 12-24 Wire-to-wire stray capacitance

on one of the instruments shown in Figure 12-23. An open-circuit problem is revealed by a very low capacitance reading.

12-8 STRAY CAPACITANCE

To this point, we have discussed only component capacitors. But capacitance can also be present in a circuit as a stray effect resulting from the circuit's construction.

Wire-to-Wire Capacitance

The basic elements of capacitance are two pieces of metal separated by an insulator. Therefore, every pair of wires in a circuit has a small amount of associated capacitance. This idea is illustrated in Figure 12-24.

The amount of **stray capacitance** between wires is usually very small, less than 1 picofarad. The exact amount depends on the distance between the wires, their length, whether or not they run alongside each other, and other factors.

Stray capacitance between wires can be considered an undesirable but unavoidable side effect. In most cases, it does not affect circuit operation. But in high-frequency ac circuits, it may have a significant effect.

Wire-to-Chassis Capacitance

Circuitry is often mounted on a metal surface or contained in a metal enclosure. When that is the case, the metal surface is called the **chassis.** Stray capacitance exists between every circuit wire and the chassis. It is unavoidable for the same reason as before.

Because the chassis has much greater surface area than a single wire, stray capacitance from wire to chassis is greater than from wire to wire. Therefore, it tends to be a more serious problem in high-frequency circuit behavior.

Component Stray Capacitance

All electrical components have at least two wires. The material between the wires is partially insulating. Therefore, stray capacitance exists across the component. This is illustrated in Figure 12-25(A) for a resistor. A device with three leads

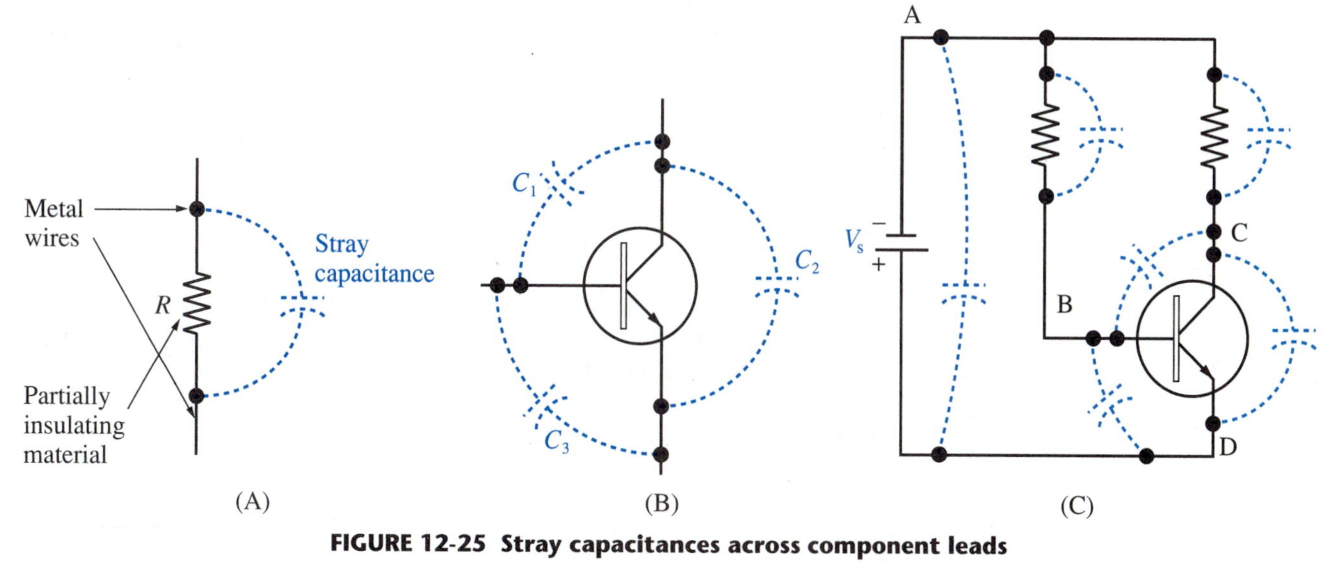

FIGURE 12-25 Stray capacitances across component leads

has three stray capacitances, as shown for the transistor in Figure 12-25(B).

There are a great many stray capacitances in a complete circuit. Figure 12-25(C) shows the wire-to-wire and component stray capacitances for a fairly simple circuit. This does not count the wire-to-chassis stray capacitances.

Correct Use of Tubular Capacitors

As described in Section 12-4, a capacitor in a tubular package has one outside foil and one inside foil. The outside foil lead is usually identified by a black band. Because of the stray capacitance problem, there are some situations where it is important to connect the outside and inside foils into the circuit in a particular way.

Figure 12-26 illustrates one of these situations. A time-varying signal voltage exists on wire X. Mixed in with that signal are some unpredictable minor voltage variations, called **electrical noise**. Suppose that it is important to keep wire Y as noise-free as possible, because wire Y feeds another circuit.

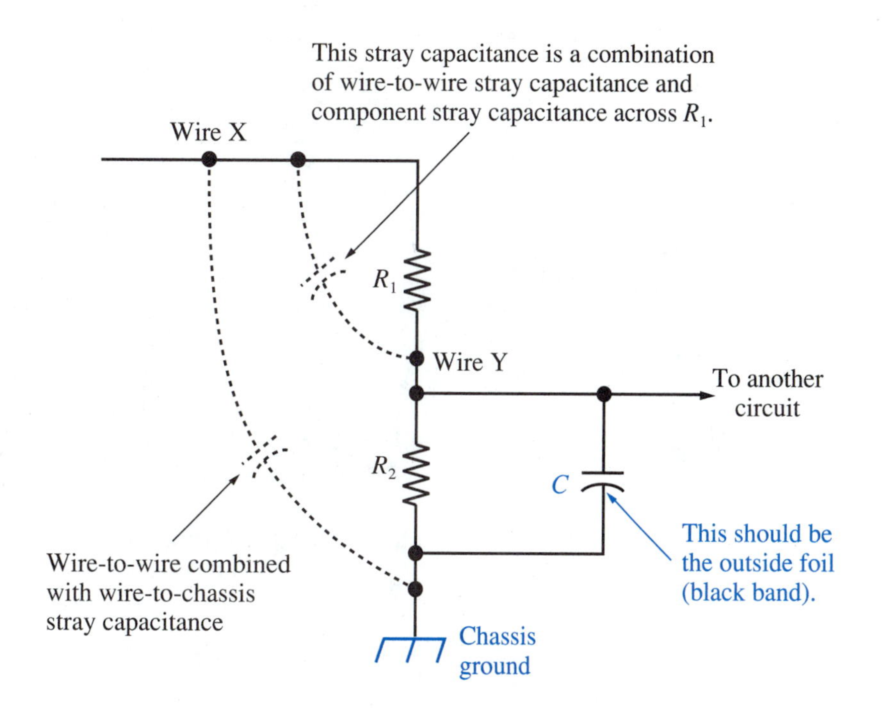

FIGURE 12-26 A tubular capacitor should have its outside foil, not its inside foil, connected to chassis ground.

A portion of the noise on wire X is bound to appear on wire Y, for two reasons. First, voltage division is occurring between R_1 and R_2. Second, there is some stray capacitance between wire X and wire Y, as described in Figure 12-26.

To minimize the portion of the noise appearing on wire Y, it is necessary to minimize the stray capacitance between wire X and wire Y. If capacitor *C* were installed with its outside foil connected to wire Y, the effective area of wire Y would be increased by the foil area. This would increase the X-to-Y stray capacitance and worsen the noise coupling.

But with the inside foil connected to wire Y and the outside foil connected to chassis ground, radiated noise from wire X is guided directly to ground and eventually back to the noise source. The noise never hits the inside foil of capacitor *C* because the outside foil intercepts it and shunts it away to ground. We say that the outside foil shields the inside foil.

As a general rule:

TECHNICAL FACT

Whenever a tubular capacitor is placed in a circuit, the outside foil should be connected to chassis ground, or as close to chassis ground as possible.

12-9 CAPACITOR APPLICATION EXAMPLES

Dc Power Supply Filter Capacitor

A capacitor can convert pulsating dc into relatively smooth dc. The **rectifying** circuit for changing ac into pulsating dc is shown in Figure 12-27(A). We are not interested now in how it operates. We are only interested in how a capacitor can smooth out the voltage pulsations that it produces. These pulsations are sketched in Figure 12-27(B).

A capacitor has been added to the basic rectifying circuit in Figure 12-27(C). As the voltage pulsations approach their peak points, they deposit charge on capacitor *C*. The proper amount of charge is deposited to make the capacitor into a temporary voltage source of value V_{peak} — the maximum (peak) voltage value of the pulsation. The amount of deposited charge is given by Equation 12-2, $Q = C(V_{peak})$.

When the natural pulsation starts heading downward, the capacitor takes over the job of maintaining a fairly steady voltage across the load, as shown in Figure 12-27(D). It cannot maintain a perfectly steady voltage because some of its deposited charge drains off its plates in order to deliver current to the load, as required by Ohm's law.

During the next pulsation, the upward-climbing rectifier voltage meets the slowly declining capacitor voltage. At that moment, the rectifier begins replenishing the charge on the capacitor. The replenishment is completed at the very top of the pulsation. The entire cycle then repeats itself.

SAFETY ADVICE

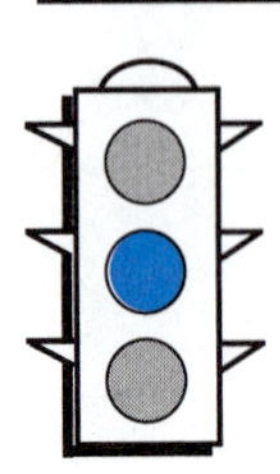

A capacitor does not necessarily discharge immediately when the circuit's power is turned off. If there is no discharge path for the negative charge to move back to the positive plate, a capacitor may store charge for many hours or even days. In such cases, there is danger of electric shock even with the power turned off.

Always check the circuit's schematic diagram for large capacitors with no clear discharge path. Such capacitors should be safely discharged by connecting a temporary short circuit directly across their terminals. A manufacturer will sometimes even give you instructions on where to place an insulated-handle screwdriver to provide a discharge path. Television receivers are notorious for this problem.

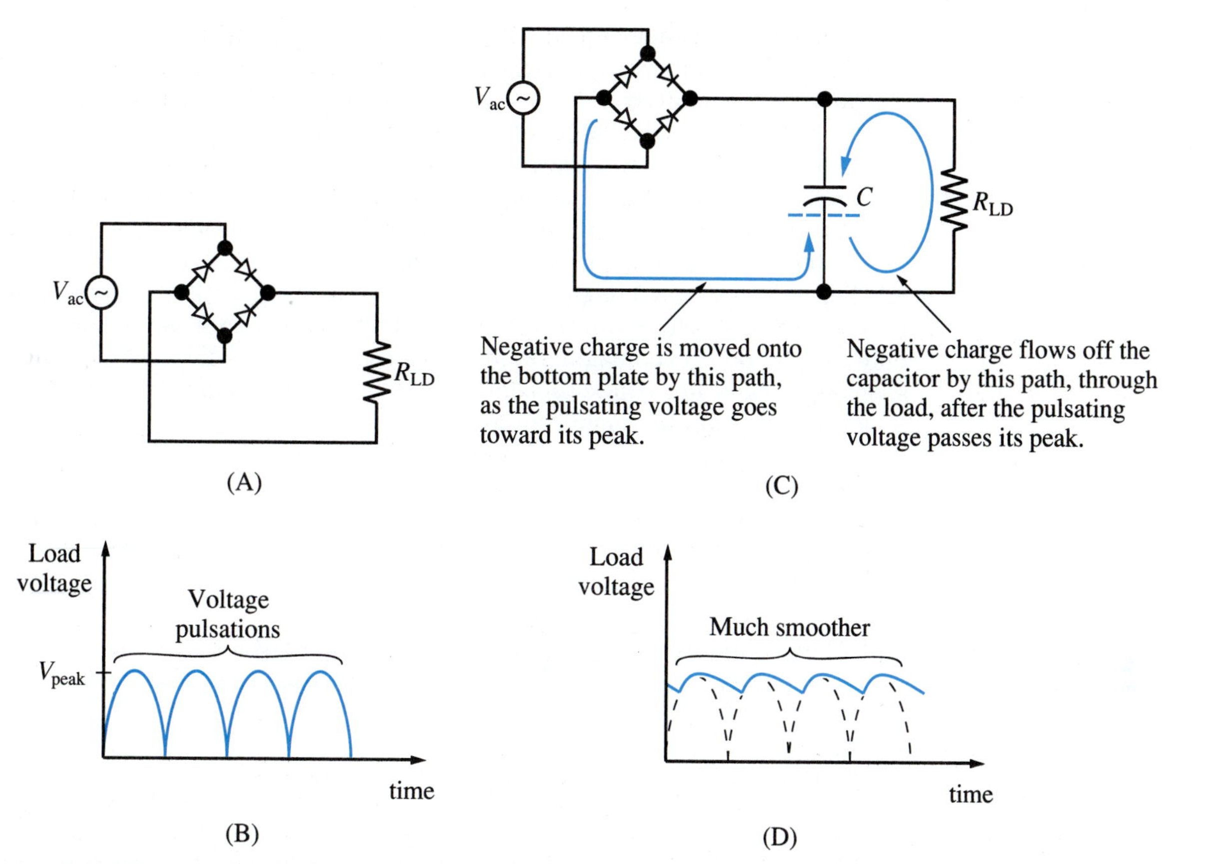

FIGURE 12-27 (A) Ac-to-dc rectifying circuit. (B) Pulsations of voltage produced by bridge rectifying circuit. (C) Smoothing capacitor installed in rectifying circuit. (D) Smoother voltage resulting from capacitor charge-storage.

Arc-suppression Capacitor in a Standard Ignition System

A standard gasoline-engine ignition system is drawn in Figure 12-28. It works like this:

1. Just before the ignition instant, the distributor cam (not shown) closes the breaker-point switch. This completes the circuit from the car battery through the primary winding of the **ignition coil.** Conventional (hole) current begins flowing as shown in Figure 12-28.
2. At the ignition instant, the distributor cam forces the breaker-points apart, opening the switch. The current shown in Figure 12-28 stops.

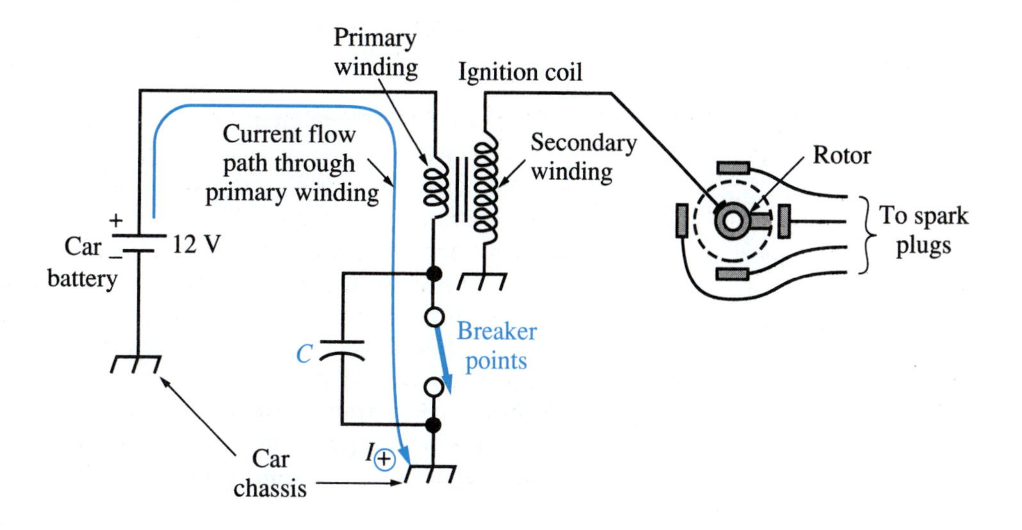

FIGURE 12-28 A standard ignition system contains an arc-suppression capacitor. The capacitor also has a second purpose of prolonging the spark plug arc.

3. Because of the abrupt stoppage of current, the primary winding generates a large voltage that attempts to keep the current flowing. We will study this concept carefully in Chapter 14.
4. The large voltage (several hundred volts) generated by the primary winding is increased to a much larger voltage (many thousands of volts) by the secondary winding. The secondary voltage arcs across the rotor gap and across the spark plug electrodes. The arc across the spark plug ignites the gasoline.

In Figure 12-28, the purpose of parallel capacitor C is to delay the primary winding voltage so that it doesn't appear across the breaker points at the very instant they open. If the voltage did appear instantly, it would arc across the narrow gap between the points, damaging their contact surfaces. By delaying the voltage slightly, the capacitor allows time for the breaker points to move far enough apart to prevent or minimize contact arcing.

The capacitor delays the voltage by requiring that some charge be transferred between its plates before any voltage can exist ($V = Q / C$). The time required for the primary winding to transfer the charge is just long enough for the points to move far enough apart for arc protection. This action by the capacitor can be summarized by the following rule.

TECHNICAL FACT

The voltage across a capacitor cannot change instantly. Voltage can change only after charge has been transferred between plates, which requires some elapsed time.

The capacitor also has a second purpose in the ignition system. It oscillates with the primary winding, prolonging the spark within the engine cylinder.

Modern automobile ignition systems do not use breaker-point switches. They use magnetic pickup and solid-state electronic switching instead.

✓ SELF-CHECK FOR SECTIONS 12-7, 12-8, AND 12-9

24. The ohmmeter test can be used for capacitors larger than about______μF. [7]
25. In the ohmmeter capacitor test, if the pointer doesn't move at all, the capacitor is__________. [7]
26. If the ohmmeter pointer moves all the way to the right and fails to return to the left, the capacitor is__________. [7]
27. T-F It is possible to completely eliminate stray capacitance by careful construction of a circuit.
28. T-F Stray capacitance is more of a problem for high-frequency ac circuits than for dc circuits.
29. When a nonelectrolytic tubular capacitor is placed in a circuit, the ______________foil should be closer to chassis ground.

Questions 30 through 33 refer to Figure 12-27.

30. During what part of the voltage pulsation does the rectifier circuit bring the capacitor back to full charge? [8]
31. During what part of the voltage pulsation does the capacitor become the source of voltage to the load? [8]
32. Why can't the filter capacitor keep the load voltage absolutely steady? [8]
33. T-F If the filter capacitor is made larger, the load voltage will become smoother. [8]
34. In Figure 12-28, when the breaker points fly apart, why doesn't a large amount of primary winding voltage appear across them immediately? [8]

FORMULAS

$$V = \frac{Q}{C}$$ **EQ. 12-1**

$$C = \frac{Q}{V}$$ **EQ. 12-2**

$$Q = CV$$

$$C = (8.85 \times 10^{-12})\frac{kA}{d}$$ **EQ. 12-3**

For parallel capacitors:

$$C_T = C_1 + C_2 + \ldots$$ **EQ. 12-4**

For series capacitors:

$$\frac{1}{C_T} = \frac{1}{C_1} + \frac{1}{C_2} + \ldots$$ **EQ. 12-5**

SUMMARY OF IDEAS

- A capacitor is a circuit element that has two metal plates separated by an insulating dielectric layer.
- When a capacitor is operating in a circuit, charge is transferred from one plate to the other plate.
- Capacitance, C, is the amount of transferred charge per unit of voltage ($C = Q / V$).
- The basic measurement unit of capacitance is the farad, symbolized F. (1 farad =1 coulomb / 1 volt)
- From the formula $C = Q / V$, any one of those three variables can be calculated if the other two are known.
- Practical capacitors have capacitances that are a small fraction of a farad. They are usually measured in microfarads, nanofarads, or picofarads (μF, nF, or pF).
- The capacitance of a parallel-plate capacitor is given by the formula

$$C = (8.85 \times 10^{-12})\frac{kA}{d}$$

- In the capacitance formula, k is a property of the dielectric material, called the dielectric constant; A is the area of the plates in square meters, and d is the distance between the plates (thickness of the dielectric layer) in meters.
- Four major categories of dielectric material are (1) mica ; (2) ceramic ; (3) plastic-film ; and (4) electrolytic. Different dielectric materials produce different characteristics in capacitors.
- Some of the important characteristics of capacitors are : (1) available range of capacitance values; (2) manufacturing tolerance achievable at reasonable cost; (3) temperature stability; (4) maximum allowable voltage.
- Electrolytic capacitors are the most economical, but they permit only one polarity of applied voltage.
- Capacitors in parallel add directly, like resistors in series.
- Capacitors in series combine by the reciprocal formula, like resistors in parallel.
- There are various instruments available for measuring and troubleshooting capacitors.
- Stray capacitance is inevitable in all circuit construction. It may be a problem in circuit operation.
- Some common applications of capacitors are: (1) to act as temporary voltage sources (power-supply filtering, for example); (2) to prevent quick change in voltage, or delay the appearance of voltage (ignition system arc-suppression, for example).

CHAPTER QUESTIONS AND PROBLEMS

1. A 4-μF capacitor has 1×10^{-5} C of charge deposited on its plates. What is the voltage across the capacitor? [2]
2. How much charge must be moved in order to produce 25 V across a 50-μF capacitor? [2]
3. We wish to establish 15 V across a capacitor by transferring 7.5 mC of charge. What size capacitor is required? [2]
4. One plate of a capacitor has a net negative charge of –100 mC. What is the net charge on the other plate? [1]
5. Other things being equal, if the plate area is increased, capacitance is__________.[3]
6. Other things being equal, if the plate spacing (dielectric thickness) is increased, capacitance is__________. [3]
7. T-F A material with highly polarized molecules tends to have a high-value dielectric constant, k. [3]
8. A certain capacitor has plate area of 0.0018 m^2 (square meter). It has polyethylene dielectric of 4.0 mm thickness. What is its capacitance? [3]
9. A capacitor with the same physical dimensions as in Problem 8 has teflon dielectric. Find its capacitance. [3]
10. How thick should a parafinned-paper dielectric be in order to make 0.005 μF with a plate area of 0.25 m^2? [3]
11. An aluminum electrolytic capacitor must always be polarized negative on the__________ foil, and positive on the__________foil. [4]
12. Which would be better for mounting on a printed circuit board, an axial-lead or single-ended capacitor? [4]
13. Explain why a disconnected electrolytic capacitor can't hold its charge forever. [4]
14. It is a fact that a capacitor can't change its voltage instantly. Therefore, at the very instant when a capacitor begins to charge, it acts like a momentary________circuit. [8]
15. When capacitors are connected in__________, their values add directly. [6]
16. T-F When two capacitors are connected in parallel, the effective voltage rating is the lower of the two voltage ratings. [6]
17. In Figure 12-29, which capacitor combination presents the greater total capacitance? Explain. [6]
18. When a capacitor is tested with an analog ohmmeter, what happens if the capacitor is very leaky? [7]

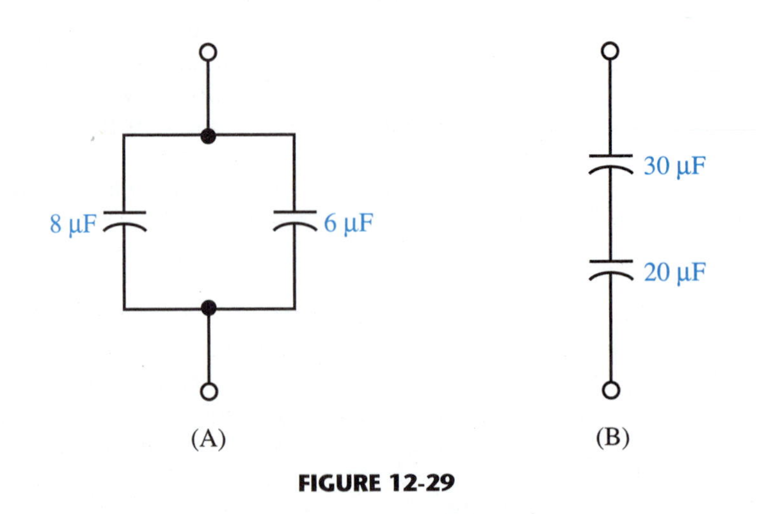

FIGURE 12-29

19. What are the shortcomings of the ohmmeter test for capacitors? [7]
20. An electronic device with three leads has__________stray capacitances.

Troubleshooting Questions and Problems

21. List some of the ways in which capacitors can cause trouble in circuit operation. [7]
22. When a good capacitor larger than about 1 μF is tested with an analog ohmmeter, describe how the pointer moves. Explain why this happens. [7]
23. In the ohmmeter test of Question 22, how does the pointer respond if the capacitor is very leaky? [7]
24. What are some capacitor problems that cannot be found by the ohmmeter test? [7]
25. A certain ceramic capacitor is suspected of being the cause of trouble in a circuit. It is marked 0.15 J Y5B, using the EIA code. When it is connected to an LCR meter, the meter reads 195 nF, with a leakage resistance of 8 MΩ. Based on this information, should the capacitor be replaced? [5,7]
26. A certain electrolytic capacitor is marked 50 μF, 10 V. When tested with an ohmmeter containing a 9-V battery, the pointer moves all the way to the right , and stays there. What is the problem? [7]
27. In Figure 12-29(A), suppose a capacitance meter measures a total capacitance of 9.2 μF. What is the most likely trouble? [6,7]

CHAPTER 13

MAGNETISM

The easiest way to handle scrap steel is with a powerful electromagnet moved by a crane. *Courtesy of O.S. Walker Company, Inc.*

OUTLINE

NEW TERMS TO WATCH FOR

magnetic field
flux lines
north pole
south pole
permanent magnet
electromagnet
left-hand rule
permeability
air-core
saturation
retentivity
residual magnetic field
magnetic shield
solenoid valve
speaker cone
weber
flux density
tesla

You probably have some familiarity with magnetism. You probably know, for instance, that a permanent magnet attracts and holds iron objects. Magnets have many other more important uses.

After studying this chapter, you should be able to:

1. Describe the strength and direction of the magnetic effect, given a magnetic field drawing.
2. Use the laws of magnetic attraction and repulsion.
3. Distinguish between a permanent magnet and an electromagnet.
4. Tell the north and south poles of an electromagnet coil by applying the left-hand rule.
5. Understand the ideas of magnetic permeability, magnetic saturation, and magnetic retention.
6. Explain the operation of a solenoid-operated valve and a moving-coil speaker.
7. Define magnetic flux and magnetic flux density.
8. Calculate magnetizing force; relate it to magnetic flux density.

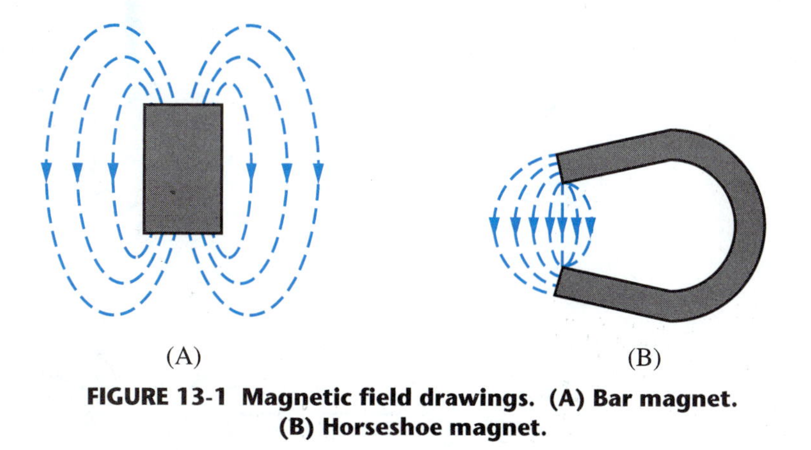

FIGURE 13-1 Magnetic field drawings. (A) Bar magnet. (B) Horseshoe magnet.

13-1 MAGNETIC FIELDS

A magnet is a device that is surrounded by a magnetic force field. This **magnetic field** is represented by magnetic **flux lines** that come out of one location on the magnet, then return to a different place on the magnet. Figure 13-1 shows two examples.

The magnetic field around a bar magnet is drawn in Figure 13-1(A). The flux lines (which aren't really visible, of course) emerge from the top end of the bar, then blossom out in three-dimensional space like a flower. The lines make a complete turn and eventually reenter the bar at the bottom end.

The familiar hardware-store horseshoe magnet has a field as shown in Figure 13-1(B). Remember that a drawing can show only the two dimensions of the paper, but the magnetic field actually has depth in three dimensions.

1

The force associated with a magnetic field is strongest where the flux lines are closest together. It is weakest where the flux lines are farthest apart. For instance, in the bar magnet of Figure 13-1(A), magnetic attraction and repulsion forces are strongest right at the ends of the bar.

13-2 NORTH AND SOUTH

Magnetic fields produce real attraction and repulsion forces. To help us describe these forces, we need a method to distinguish one end, or pole, of a magnet from the other. Here is the definition.

TECHNICAL FACT

The **north pole** of a magnet is the end that flux lines come out of. The **south pole** is the end that flux lines go back into.

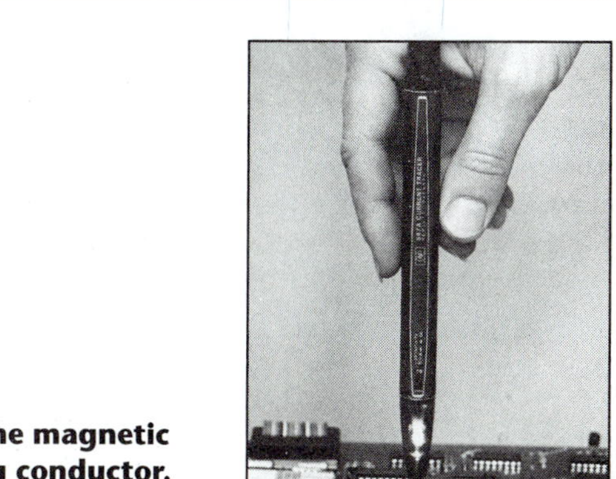

Digital current-tracers sense the magnetic field around a current-carrying conductor.
Courtesy of Hewlett-Packard Company

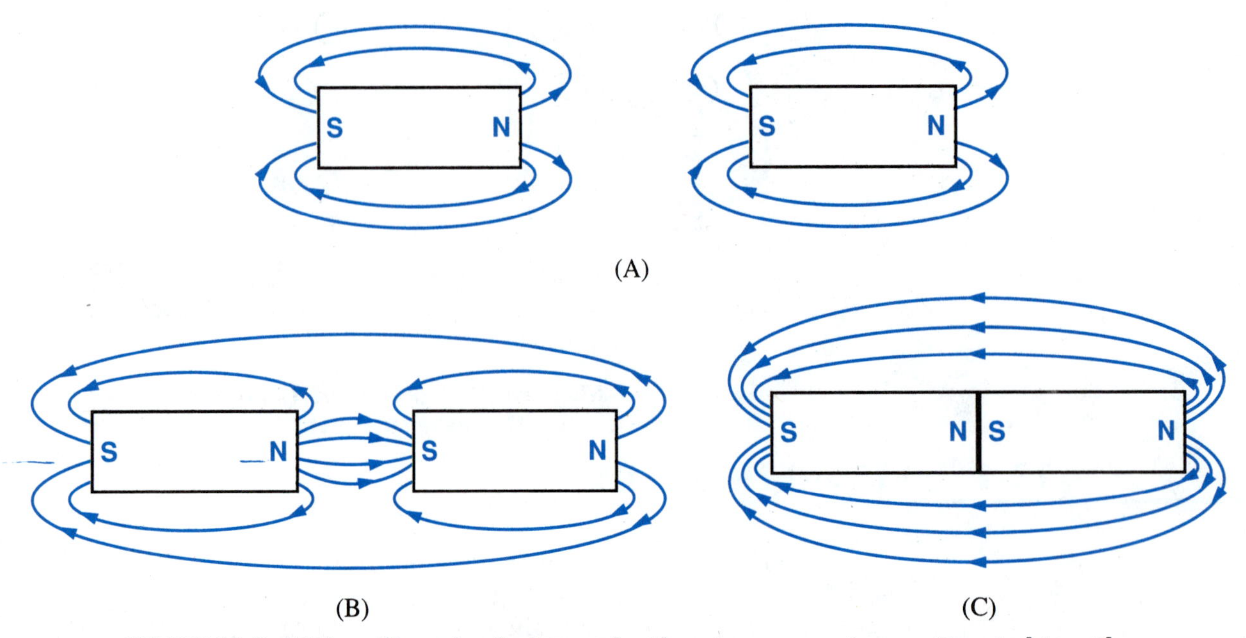

FIGURE 13-2 With unlike poles facing each other, two magnets are attracted together.

In Figure 13-1(A), the top end is north and would be labeled N. The bottom end is south. It would be labeled S. In Figure 13-1(B), the top prong is N; the bottom prong is S.

The laws of magnetic attraction and repulsion are:

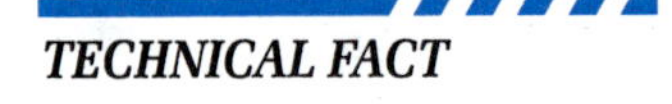

TECHNICAL FACT

Unlike magnetic poles attract each other. Like magnetic poles (both N or both S) repel each other.

The attraction of unlike poles is illustrated in Figure 13-2. In part A, the two magnets are a distance apart, but close enough to exert force on one another. Because the poles that are facing each other are unlike, one north and one south, the force is an attraction, rather than a repulsion.

If the magnets are free to move, they will start coming together as shown in Figure 13-2(B). When this happens, they begin to share flux lines, as that drawing illustrates. The closer the poles get to each other, the stronger the interaction force becomes. This is because each magnet is entering the part of the other magnet's field where its flux lines are closer together. Finally the magnets touch each other in Figure 13-2(C). This effectively creates a single larger magnet.

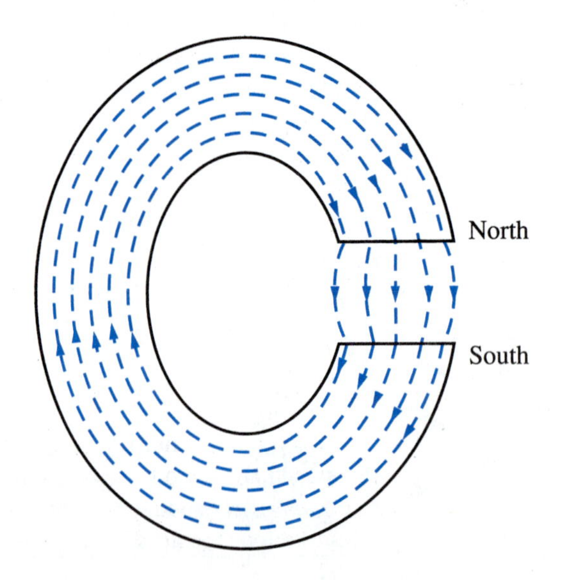

FIGURE 13-3 Flux lines are complete loops; they do not have a definite starting point and a definite stopping point.

FIGURE 13-4 (A) Arrowhead symbols represent magnetic flux lines perpendicular to the page, pointing out of the page toward the viewer. (B) Tailfeather symbols represent magnetic flux lines perpendicular to the page, pointing into the page away from the viewer.

If the magnet on the left were turned around in Figure 13-2(A), there would be two south poles facing each other. If the one on the right were turned, there would be two north poles facing each other. In either case, the like poles would repel each other, causing the two magnets to move apart.

Magnetic flux lines do not simply start at a north pole and stop at a south pole. They are continuous through the body of the magnet material. This is illustrated in the C-shaped magnet of Figure 13-3.

Sometimes we need to take a head-on view of a magnetic pole. Figures 13-1, 13-2, and 13-3 are all side views. A head-on view of a north pole would have flux lines pointing directly at the viewer, like an arrow pointing toward the viewer. This is represented on paper as a group of arrowheads, shown in Figure 13-4(A).

A head-on view of a south pole has flux pointing away from the viewer, like the back ends of arrows pointing away. This is represented as a group of tailfeathers, shown in Figure 13-4(B).

✔ SELF-CHECK FOR SECTIONS 13-1 AND 13-2

1. The invisible lines in a magnetic field are called__________lines. [7]
2. Define the north pole of a magnet. What is the difference between the north pole and the south pole? [1]
3. In a magnetic field drawing, what does the closeness (density) of flux lines represent? [1, 7]
4. T-F Magnetic flux lines exist in a continuous loop, with no definite starting and stopping points. [1]
5. A magnetic north pole is attracted by a magnetic__________pole. [2]
6. A magnetic north pole is repelled by a magnetic__________pole. [2]

13-3 PERMANENT MAGNETS

The magnets in Sections 13-1 and 13-2 are **permanent magnets**. They maintain their magnetization without any external help. They do this because they consist mostly of iron. Iron atoms are different from other elements in this way: Each iron atom has an electron organization that makes it a tiny weak magnet all by itself.

Initially, a bar of iron will have its tiny atomic magnets pointing in all different directions, as shown in Figure 13-5(A). But if an external magnetic field is impressed on the iron bar, all the individual atomic magnets become aligned. Suppose the external field is impressed in the left-to-right direction, as suggested in Figure 13-5(B). The magnetic attraction/repulsion force causes the iron atoms to align with their south ends to the left and their north ends to the right. This is shown in Figure 13-5(C).

That alignment is then maintained permanently, because the bar as a whole has been converted into a magnet. It's a vicious circle: The fact that the atomic magnets are all aligned creates a strong net magnetic field, and the strong net magnetic field keeps the atomic magnets aligned. A permanent magnet has been created.

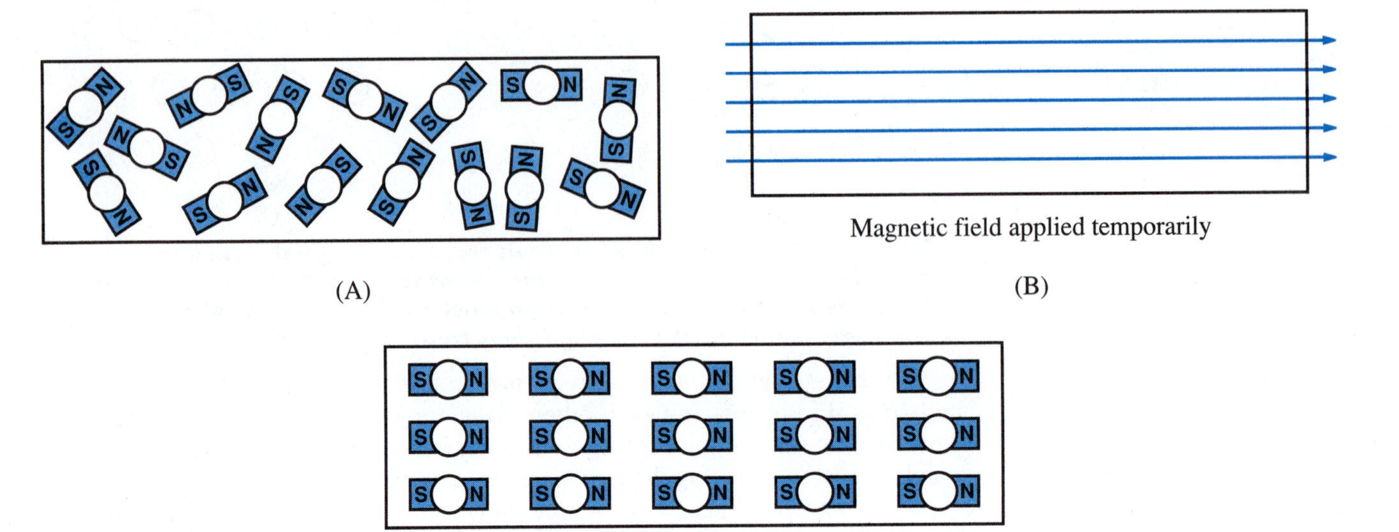

FIGURE 13-5 (A) Iron bar in the unmagnetized state. (B) Subjecting the bar to a temporary external magnetic field. (C) Iron atoms become permanently aligned.

13-4 ELECTROMAGNETISM

Permanent magnets are convenient to use, but they have two disadvantages:

1. In a moderate size, they cannot provide the very strong magnetic force fields needed for certain applications.
2. They can't be switched on and off.

To solve these disadvantages, we use **electromagnets.** An electromagnet produces a magnetic field when current flows through its winding.

To understand electromagnetism, start by looking at the straight current-carrying wire in Figure 13-6. In this case, the arrowhead represents a current flowing out of the page, toward us. An important thing happens as a result of that current. A magnetic field that is circular around the wire is created. The flux lines are circular in the clockwise direction, as Figure 13-6 shows.

Close to the wire, the magnetic field is stronger. Farther away from the wire, the field is weaker. Figure 13-6 conveys this by the densely packed flux lines close to the wire, and the spread-apart flux lines farther from the wire.

The direction of circulation of magnetic flux is given by the **left-hand rule** for a straight wire. The left-hand rule is demonstrated in three dimensions by

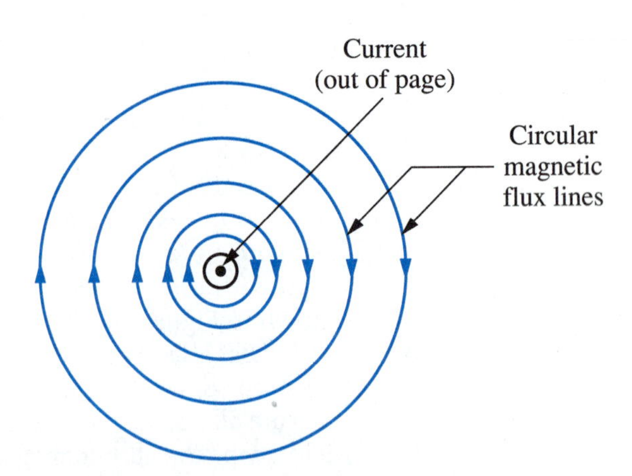

FIGURE 13-6 The magnetic field is circular around a straight current-carrying wire. Its strength decreases with distance from the wire.

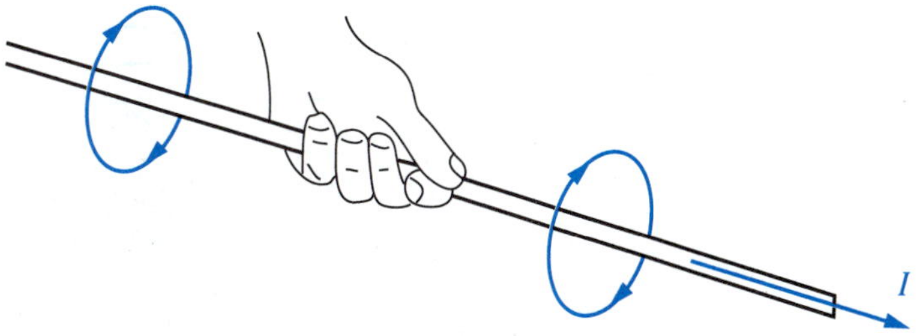

FIGURE 13-7 Left-hand rule for finding the direction of circulation of magnetic flux

Figure 13-7. The fingers curl in a clockwise direction (as opposed to counterclockwise). Therefore the flux is clockwise around the wire.

TECHNICAL FACT

Grasp the straight wire with your left hand; let your thumb point in the direction of the electron current. Your fingers will curl in the direction of the magnetic flux.

Figure 13-6 is a two-dimensional drawing. By pointing your left thumb out of the page, verify that the left-hand rule predicts clockwise flux around the perpendicular wire.

If the current-carrying wire is bent into a round loop, the magnetic field takes on the shape shown in Figure 13-8. By applying the left-hand rule to small segments of the loop (each small segment is almost straight), you can see that the overall tendency is that the flux passes through the loop in the left-to-right direction. Prove this to yourself.

The loop has an identifiable south side (left) and an identifiable north side (right). In this respect, it is like the permanent magnets of Figures 13-1 through 13-3. It is an elementary electromagnet.

A realistic functioning electromagnet is made by extending the idea of Figure 13-8. When the wire is coiled many times, each short segment of the coil contributes a little bit to the overall flux. This produces a magnetic field that points straight down the center of the hollow coil. As illustrated in Figure 13-9, the flux emerges at one end of the coil, curves back around the outside, and reenters at the opposite end.

This electromagnet has clearly identifiable north and south poles. The magnetic field is strongest right at the poles, before the flux lines begin to bend and spread apart.

If the coils are very tightly wrapped, the overall magnetic field has the identical shape as the permanent magnet of Figure 13-1(A). A tightly wrapped coil may be called by the name *solenoid* or sometimes by the name *winding*.

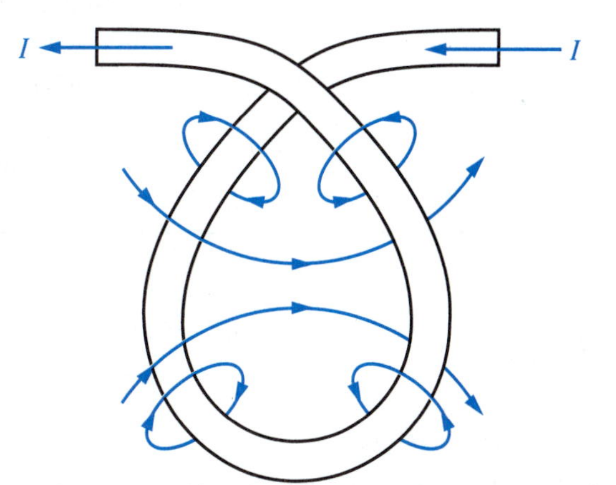

FIGURE 13-8 Magnetic flux resulting from looping the wire once

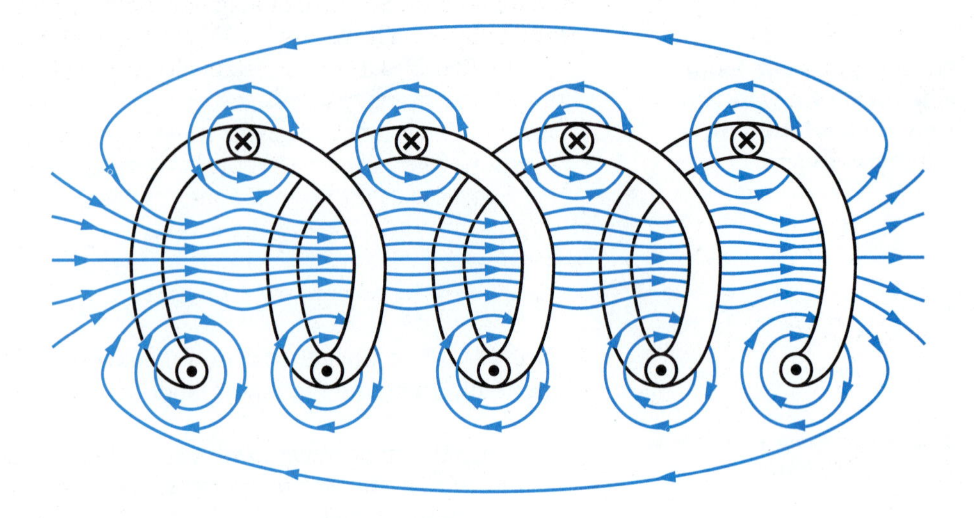

FIGURE 13-9 Magnetic field flux produced by a loosely wrapped coil

13-5 MAGNETIC CORE MATERIALS

As explained in Section 13-3, iron has the characteristic that each atom is a tiny weak magnet. Certain other metals and metallic materials also have this characteristic, in some cases to an even greater degree than iron. Such materials can be used as the core of an electromagnet. In other words, the winding can be wrapped around a piece of magnetic material, instead of a hollow form. The initial magnetic field from the electromagnet is reinforced by the internal alignment of the atomic magnets in the core material. The resulting magnetic field is then much stronger than the initial field by itself.

5

The ability of a material to align and reinforce the initial field is known as its relative magnetic **permeability**, symbolized μ_r. For example, if a particular material is able to produce a magnetic field that is three times as strong as the air-core field, that material has $\mu_r = 3$. The effect of differing permeabilities is shown graphically in Figure 13-10. In those graphs, the strength of the magnetic field is plotted on the vertical axis, and the current in the electromagnet winding is plotted on the

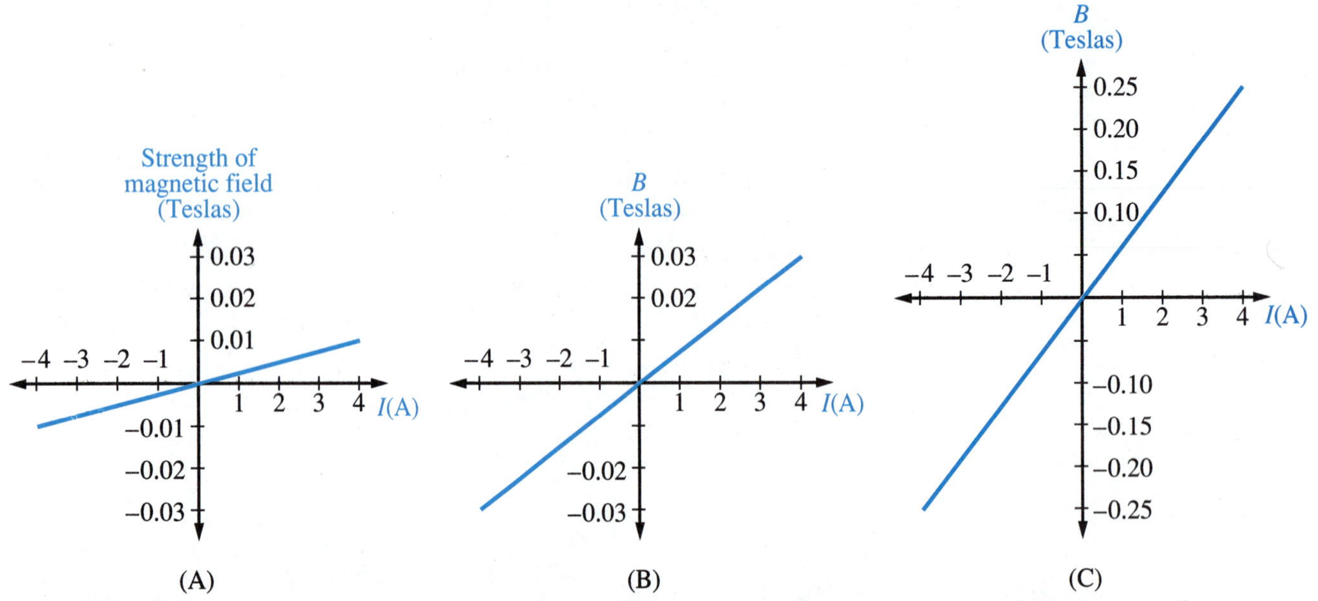

FIGURE 13-10 Effect of different core materials with the same electromagnet winding construction. (A) Air-core ($\mu_r = 1$). (B) $\mu_r = 3$. (C) $\mu_r = 25$.

horizontal axis. Strength of magnetic field is symbolized B ; it is measured in basic units of teslas. These units will be explained more fully in Section 13-8.

In Figure 13-10(A), a certain electromagnet is wrapped around a nonmagnetic core. The core may actually be hollow, or it may be a solid material that is nonmagnetic. This is called the **air-core** situation. For this particular electromagnet, a 4-A current produces a magnetic field with a strength of 0.01 tesla.

Figure 13-10(B) is for the same electromagnet wrapped around a core of magnetic material. For $I = 4$ A, the B value is 0.03 tesla, three times as large as the air-core situation. Therefore, this core material has permeability $\mu_r = 3$.

The magnetic core material for Figure 13-10(C) has $\mu_r = 25$.

EXAMPLE 13-1

For the core material of Figure 13-10(B),

a) What is the strength of the magnetic field when $I = 2$ A?
b) Describe the magnetic field when $I = -2$ A.
c) Assuming that the proportional (straight-line) relationship continues to hold, what would be the value of B for $I = 8$ A?

SOLUTION

a) Projecting up vertically from the 2-A mark, we intercept the graph at $B =$ **0.015 tesla.**
b) A negative value of current means that its direction has been reversed. By applying the left-hand rule to the electromagnets in Figures 13-8 and 13-9, it is clear to see that reversing the current direction also reverses the magnetic flux direction. The magnetic north and south poles are switched.

On the graph of Figure 13-10(B), projecting down from −2 A hits the line at $B =$ **−0.015 tesla.** If the original positive flux was left-to-right as shown in Figure 13-9, this new flux is right-to-left inside the core.

c) If the proportionality holds, doubling current from 4 A to 8 A will double B from 0.03 tesla to **0.06 tesla.**

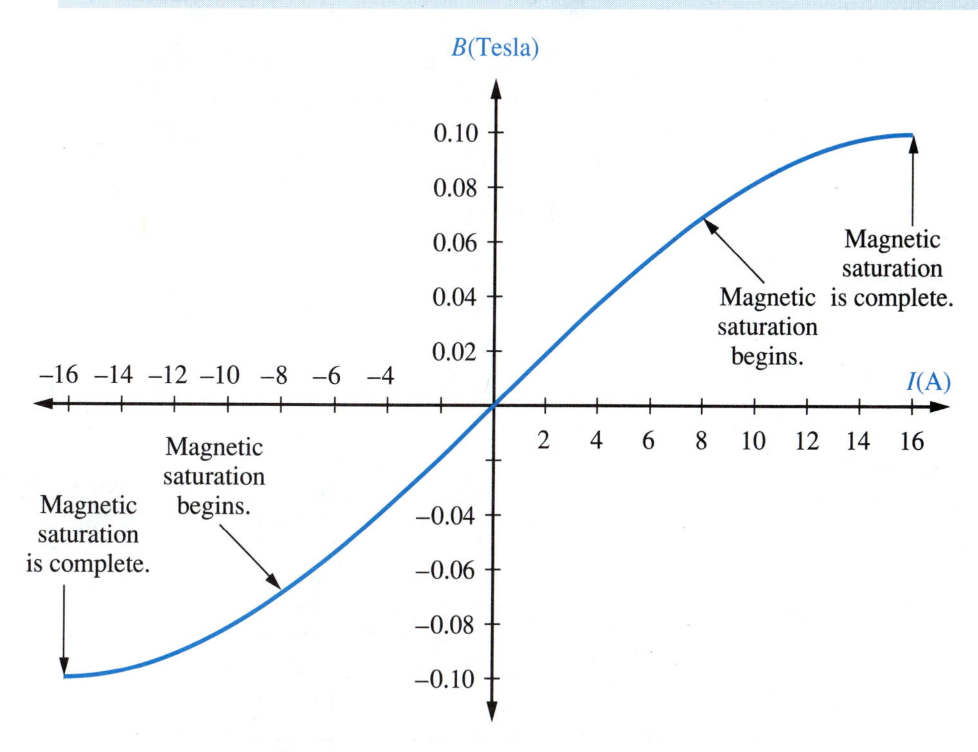

FIGURE 13-11 Magnetic core saturation

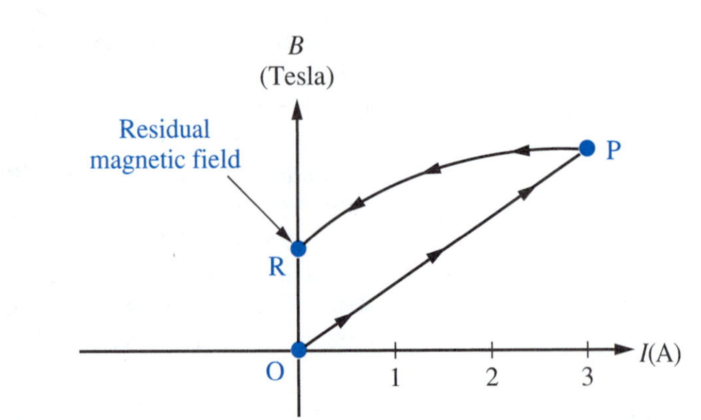

FIGURE 13-12 The residual magnetic field remains after the electromagnet's current returns to zero. This is due to retentivity of the core material.

In Example 13-1, we were careful to say *if* the proportionality holds. At some point the proportionality is bound to stop, because all the tiny atomic magnets become aligned. This occurrence is called magnetic **saturation.** The graph in Figure 13-11 illustrates the magnetic saturation effect.

An electromagnet with permeable core material will have at least some tendency to behave like a permanent magnet when the winding current is stopped. This is because of the self-perpetuating effect of the atomic magnets that was described in Section 13-3. Figure 13-12 illustrates this tendency of a core material to retain part of its recent magnetic field after the electromagnet is deenergized. In that graph, initially increasing the current from zero to point P changes the magnetic field from zero to some value. But decreasing the current from point P back to zero doesn't bring the magnetic field back to zero. Instead, the field declines to point R, its residual value.

The ability of a core material to do this is called its magnetic retention, or **retentivity.** A large amount of retentivity is desirable in certain applications, such as snap-action magnetic proximity detectors. However, in most applications, retentivity is harmful. Manufacturers of motors and transformers make every effort to minimize it.

✔ SELF-CHECK FOR SECTIONS 13-3, 13-4, AND 13-5

7. What is it about iron that makes it useful in permanent magnets? [5]
8. T-F Electromagnets can be built with stronger magnetic fields than permanent magnets. [3]
9. In Figure 13-13, electron current is entering the winding on the top lead. Using the left-hand rule, identify which electromagnet pole is north and which is south. [4]

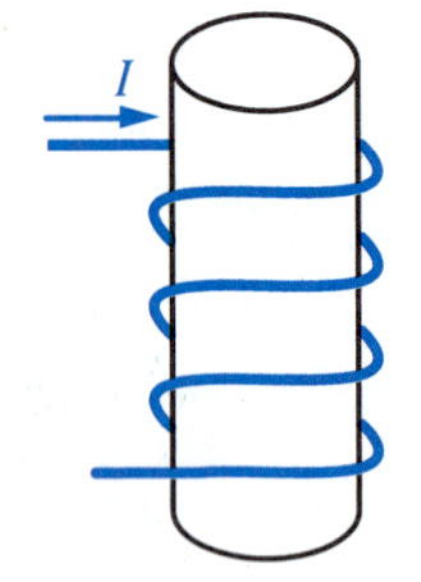

FIGURE 13-13 Electromagnet with cylindrical core

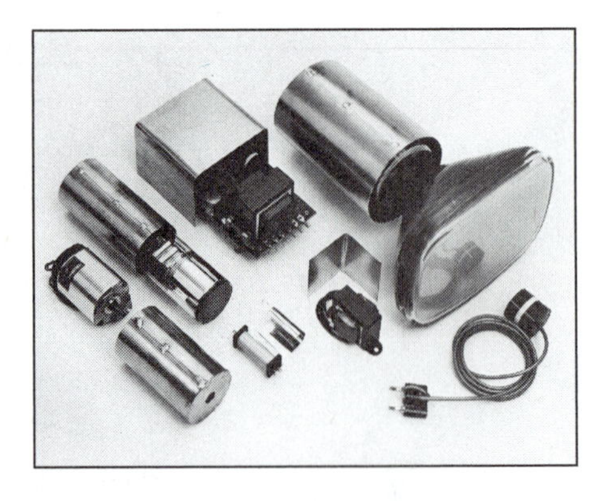

Metal with high magnetic permeability is often used to surround magnetically sensitive devices such as cathode-ray tubes, transformers, reed-relays, and stepping motors, as shown here.
Courtesy of Magnetic Shield Division

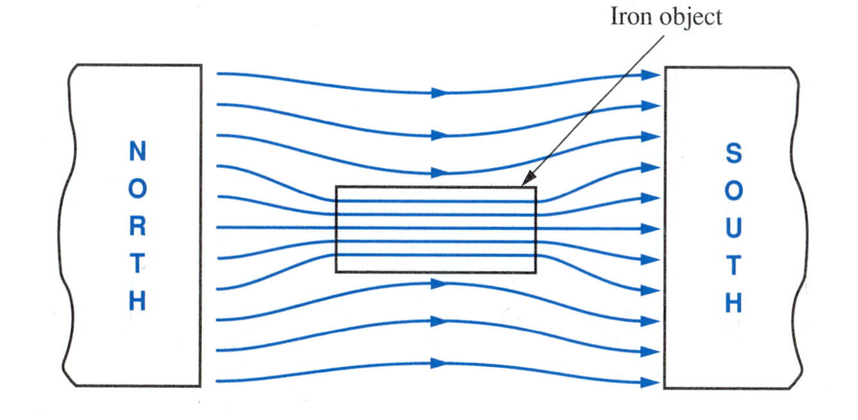

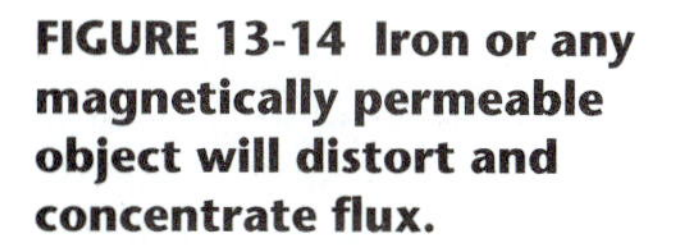
FIGURE 13-14 Iron or any magnetically permeable object will distort and concentrate flux.

10. In Question 9, suppose the current reverses, so that it enters on the bottom lead. Apply the left-hand rule again, and comment on the result. [4]
11. Two electromagnets have identical windings, carrying identical currents. One is wrapped on an air core and the second is wrapped on a metal core with $\mu_r = 8$. The air-core magnet has $B = 0.02$ tesla. What is the strength of the magnetic field in the metal core? [5]
12. Does magnetic core saturation occur abruptly or gradually? [5]
13. Refer to Figure 13-12. Draw the *B*-versus-*I* response for a different material that has greater retentivity. [5]

13-6 MAGNETIC SHIELDING

Magnetic flux can be distorted from its natural path by a piece of permeable material. For example, in Figure 13-14, the flux tends to concentrate in the iron object, rather than remain straight beween the poles.

We often take advantage of this effect in order to prevent magnetic flux from entering certain spaces. Whenever a device or circuit must be protected from magnetic effects, we surround it with a **magnetic shield.** This is illustrated in Figure 13-15.

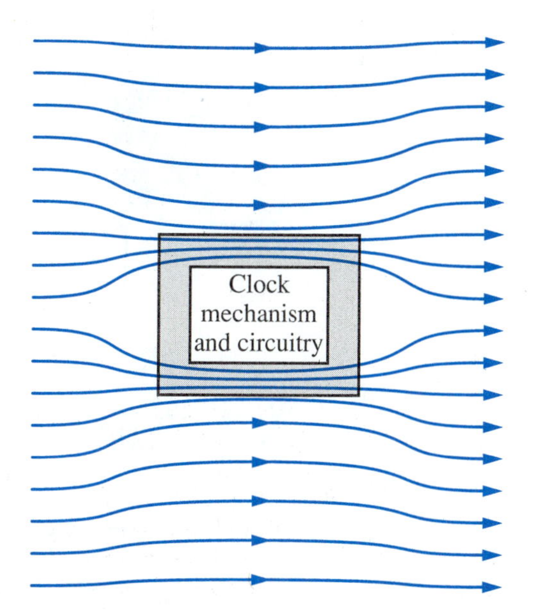

FIGURE 13-15 Magnetic shielding. The sensitive device is surrounded by an enclosure made of permeable material.

13-7 PRACTICAL APPLICATIONS OF ELECTROMAGNETS

Solenoid-operated Valve

A very common use of solenoid electromagnets is for opening and closing valves. This enables us to control natural gas valves on furnaces, water valves on washers, hydraulic oil valves on industrial machinery, and numerous other applications.

Look at Figure 13-16 to understand the operation of a **solenoid valve.** When the electromagnet is deenergized, with no current flowing through its winding, it produces no magnetic field. Therefore, it exerts no attraction force on the movable iron core. The compressed spring at the top of Figure 13-16(A) pushes the iron core downward, which causes the valve plug to press firmly against the valve seat. This seals off the fluid passage, closing the valve.

When the electromagnet is energized by an external voltage source, it creates a strong magnetic field. With current entering the top lead in Figure 13-16(B), the magnetic flux will emerge from the bottom of the coil. Verify this with the left-hand rule. The bottom of the coil is therefore a north pole.

Since the coil flux points downward, the iron core material aligns with it and also has downward flux. This flux emerges from the bottom and reenters at the top of the core. Therefore, the top of the iron core is a south pole.

The north pole of the coil attracts the south pole of the iron core. Since the iron core is free to slide, it is pulled upward into the center of the coil. This lifts the valve plug away from the valve seat, opening the valve.

Actually, the iron core is always partway inside the electromagnet coil, as shown in Figure 13-16(A). It is drawn outside in Figure 13-16(B) only to make it easier to show the flux lines.

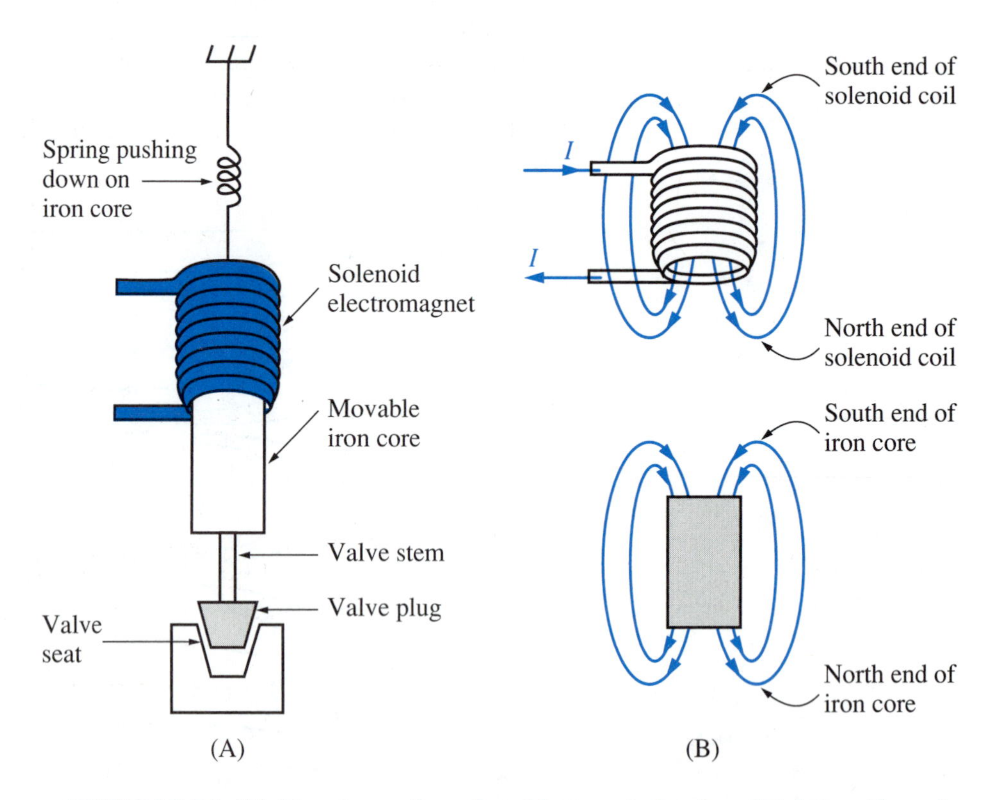

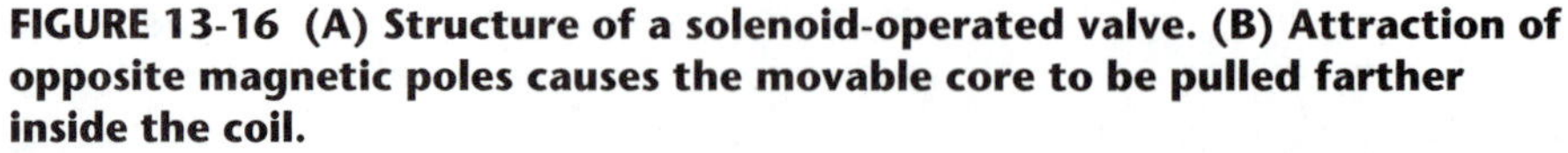
FIGURE 13-16 (A) Structure of a solenoid-operated valve. (B) Attraction of opposite magnetic poles causes the movable core to be pulled farther inside the coil.

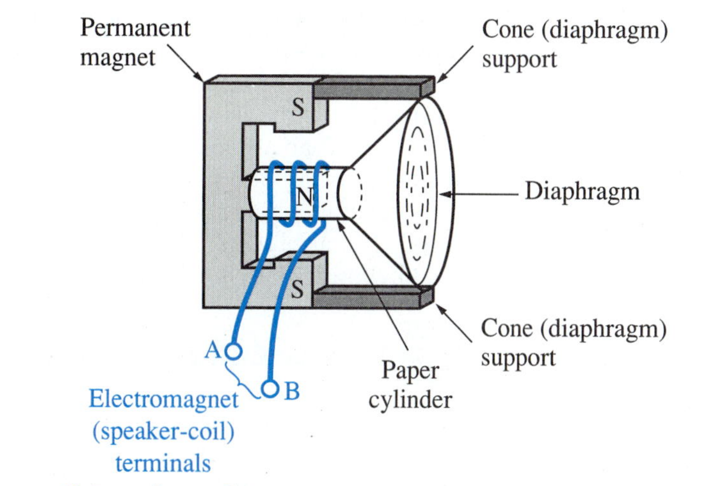

FIGURE 13-17 Rapidly reversing the current through the speaker coil causes the magnetic force to reverse, making the diaphragm vibrate.

A moving-coil speaker uses the attraction and repulsion forces between magnetic poles to create sound. The construction of a moving-coil speaker is shown in Figure 13-17. The **speaker cone** is a stiff paper-like diaphragm that tapers into a paper cylinder. The paper cylinder has an electromagnet coil glued to its outside. A permanent magnet is inserted partway into the paper cylinder.

If current enters the speaker coil on wire B, the coil produces flux pointing from right to left. This makes the left end of the coil a north pole. There is a repulsion

SUPERCONDUCTIVITY

A superconductor has virtually zero resistance to current flow. Until recently, extremely cold temperatures of about 270 degrees below zero Celsius (about −450°F) were required to make a superconductor work. Recent research has led to new materials that become superconductive at about −120°C. Researchers are trying to improve this performance even further, bringing it closer to normal environmental temperatures. If they succeed, there may be great benefits to society.

For example. superconducting electrical transmission wires could carry current from the generating plant to your neighborhood utility pole with no energy waste. At present, the U.S. electrical transmission network wastes about 8% of the total generated energy in overcoming wire resistance. Eliminating this waste would make everybody's electric bills lower. It would also reduce the amount of fossil fuel (coal, oil, and natural gas) that we must burn to generate electricity.

A more spectacular possibility is magnetically levitated transportation systems. Magnetic levitation (MagLev) can lift a vehicle up from the earth's surface and suspend it in air, between guide rails. This reduces the surface friction to zero, making fuel consumption very low even at speeds in excess of 200 miles per hour.

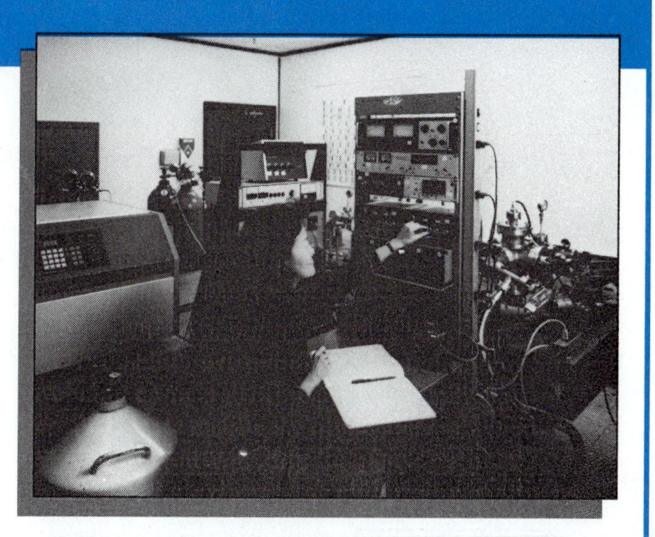

Materials scientist working in a superconductor research laboratory.
Courtesy of Energy Conversion Devices, Inc.

At this time, it is quite difficult and expensive to get enough current through our electromagnets to obtain the very great magnetic forces required to levitate a train car loaded with goods, or a mass-transit vehicle loaded with people. The MagLev vehicle must carry its own chilling equipment to lower the temperature of its electromagnets below −120°C. If the superconductive temperature can be raised significantly, this difficulty will ease, and MagLev transportation will become practical.

force between this north pole and the permanent magnet's north pole, so the speaker cone is pushed to the right.

But if current enters the speaker coil on wire A, the coil's magnetism is reversed. Now the left end of the coil is south. Check this out with the left-hand rule. There is an attraction force between the coil's south pole and the permanent magnet's north pole, so the speaker cone is pulled to the left.

In actual operation, the current through the speaker coil reverses direction rapidly (ac current). The rapid back-and-forth motion of the diaphragm produces sound vibrations in the surrounding air.

✔ SELF-CHECK FOR SECTIONS 13-6 AND 13-7

14. Explain the practice of magnetic shielding. What does it accomplish and how does it work? [1, 2]
15. Would the solenoid-operated valve in Figure 13-16 still work if the current direction were reversed? Explain. [6]
16. A stronger upward pulling force on the valve mechanism could be produced by__________the current through the solenoid coil. [6]
17. Explain your answer to Question 16. [6]

13-8 MAGNETIC VARIABLES AND MEASUREMENT UNITS

To this point, we have discussed magnetic principles without much reference to actual measurement units. Let us now focus on the important magnetic quantities and their measurement units.

Flux

7

Flux refers to the number of magnetic field lines that exist. Flux is symbolized by the Greek letter Φ. Its basic measurement unit is the **weber**, symbolized Wb.

Flux is completely independent of the size of the space that contains the field lines. Thus, in Figure 13-18(A), the flux is 4 webers. In Figure 13-18(B) the flux is still 4 Wb, even though the lines occupy a greater cross-sectional area.

In actual practice, 1 Wb is quite a large amount of flux. For instance, an average horseshoe pickup magnet sold in hardware stores has a flux of around 0.0001 Wb. A small electric drill motor emits a pole flux of about 0.001 Wb, or 1 milliweber. It would take a very large electric motor of several hundred horsepower to emit a per-pole flux of 1 Wb.

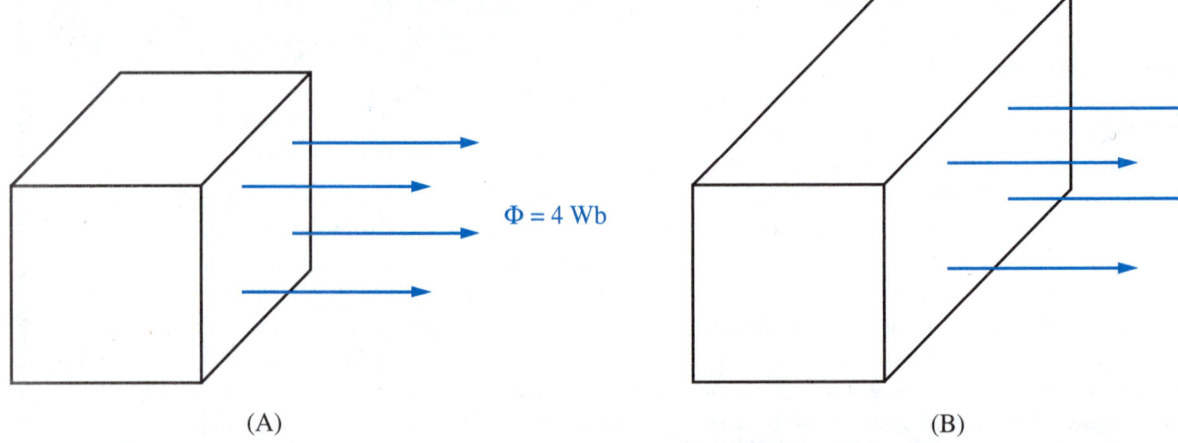

FIGURE 13-18 Magnetic flux, Φ, is simply the number of field lines that exist, without regard to how dense they are. A flux of 4 webers is far too large to be realistic. It is used here for conceptual purposes only. Therefore, the results in Example 13-2 are also unrealistic.

Flux Density

Flux density is just what its name implies. It is a measure of how closely packed the flux is. For instance, the flux density in Figure 13-18(A) is greater than in Figure 13-18(B) because the same amount of flux is confined to less area. The flux lines are therefore closer together, or denser.

Flux density is symbolized by the letter B. Its basic measurement unit is the weber per square meter, more commonly called a **tesla**, symbolized T.

The relationship between flux, Φ, and flux density, B, is given by

$$B = \frac{\Phi}{A}$$

EQ. 13-1

in which A stands for area in basic units of square meters.

EXAMPLE 13-2

In Figure 13-18, suppose the dimensions in part A are 1 meter by 1 meter and the dimensions in part B are 1 meter by 2 meters. Calculate the flux density in both cases.

SOLUTION

In part A , area is given by

$$A = (1\text{ m})(1\text{ m}) = 1\text{ m}^2$$

From Equation 13-1,

$$B = \frac{\Phi}{A} = \frac{4\text{ Wb}}{1\text{ m}^2} = 4\text{ T}$$

Repeating for part B gives

$$B = \frac{\Phi}{A} = \frac{4\text{ Wb}}{2\text{ m}^2} = \mathbf{2\text{ T}}$$

Everyday magnetic devices like solenoid valves, loudspeakers, and electric motors have flux densities much lower than the values in this example problem. Flux density is the variable that represents the strength of a magnetic field. Other things being equal, the field in Figure 13-18(A) will produce magnetic forces twice as large as the field in Figure 13-18(B).

Magnetizing Force

Magnetizing force is the ability of an electromagnet coil to align the atomic magnets in its core. The actual flux density within the core depends on the magnetizing force as well as the permeability of the core material itself. Magnetizing force depends on three variables:

1. Current through the coil.
2. The number of turns in the coil.
3. The core length over which those turns are spread.

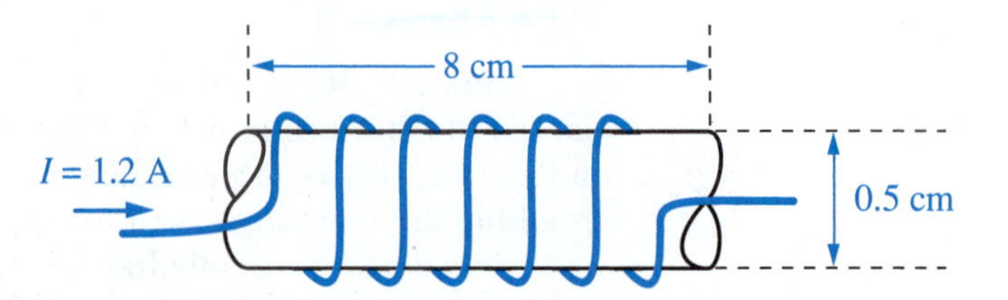

FIGURE 13-19 Cylindrical solenoid coil. Equation 13-2 is correct whenever the coil length is at least 10 times as great as the diameter.

The symbol for magnetizing force is H. Its basic unit is the ampere-turn per meter (A•t / m). In formula form,

$$H = \frac{NI}{l}$$

EQ. 13-2

where N stands for the number of coil turns and l is length in meters.

EXAMPLE 13-3

a) Calculate the magnetizing force of the electromagnet in Figure 13-19.
b) From this information, can you tell the strength of the magnetic field (flux density) inside the core?

SOLUTION

a) The coil has 6 turns. They are spread over 8 cm, or 0.08 m. From Equation 13-2,

$$H = \frac{NI}{l} = \frac{(6\text{ turns})(1.2\text{ A})}{0.08\text{ m}} = \mathbf{90\ A{\bullet}t/m}$$

b) No, you can't tell flux density just by knowing magnetizing force. A given amount of magnetizing force will produce greater or lesser flux density as the core material has greater or lesser permeability.

The relationship between flux density and magnetizing force is given by

$$B = \mu_r\,(1.26 \times 10^{-6})\,H$$

EQ. 13-3

where μ_r is the relative permeability of the core material. Air has a relative permeability of virtually 1.0.

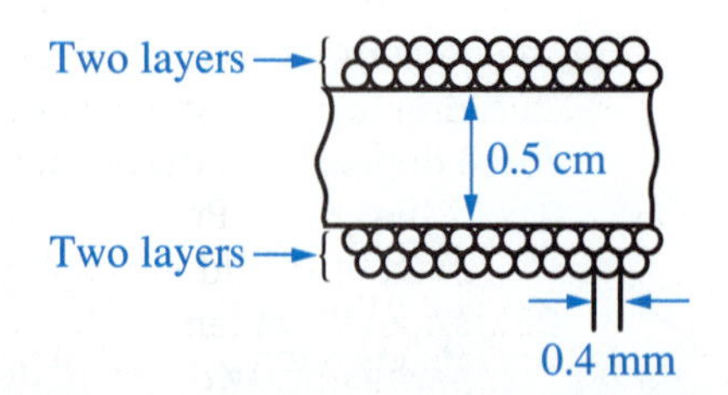

FIGURE 13-20 Closely wrapped, two-layer winding

EXAMPLE 13-4

A solenoid coil is 8 cm long with a diameter of 0.5 cm, the same dimensions as in Figure 13-19. It is wrapped with two layers of insulated wire, each wire having a thickness of 0.4 mm. Adjacent wires touch each other, as shown in Figure 13-20. The two layers of wire also touch each other.

a) If the coil carries a current of 0.6 A, calculate its magnetizing force.
b) The core material has relative permeability $\mu_r = 32$. Calculate the flux density, B.
c) Suppose a new core having $\mu_r = 45$ is substituted. Calculate the new flux density.

SOLUTION

a) The number of wire turns per layer is given by

$$8 \text{ cm} \times \frac{1 \text{ turn}}{0.4 \text{ mm}} = 8 \times 10^{-2} \text{ m} \times \frac{1 \text{ turn}}{0.4 \times 10^{-3} \text{ m}} = 200 \text{ turns / layer}$$

Since there are two layers, the total number of turns is 2 x 200 = 400 turns. Real solenoids are almost always wrapped with more than one layer.

Applying Equation 13-2,

$$H = \frac{NI}{l} = \frac{(400 \text{ turns})(0.6 \text{ A})}{8 \times 10^{-2} \text{ m}} = \mathbf{3000\ A \cdot t / m}$$

b) From Equation 13-3,

$$B = \mu_r (1.26 \times 10^{-6}) H = 32(1.26 \times 10^{-6}) 3000 = \mathbf{0.121\ T}$$

c) $$B = 45(1.26 \times 10^{-6}) 3000 = \mathbf{0.170\ T}$$

The flux density has increased in proportion to the core's permeability.

Unfortunately, magnetizing force is commonly called by two other names. These names are "field intensity" and "magnetic field strength." Both names are very misleading and should be avoided.

Magnetomotive Force

The numerator is sometimes separated out from Equation 13-2. With length ignored, the product of current and turns is called magnetomotive force, symbolized MMF. Naturally, the basic unit is simply the ampere-turn (A•t).

✔ SELF-CHECK FOR SECTION 13-8

18. An area with dimensions of 15 cm x 60 cm has 5×10^{-4} weber of magnetic flux passing through it. What is the flux density of the magnetic field? [7]
19. In Problem 18, assume that the flux is evenly distributed throughout the area. How much flux is contained in a 15 cm x 15 cm section? [7]
20. A solenoid coil is 3 cm long, contains 1200 turns, and carries a 1.5-A current. What amount of magnetizing force does it produce? [8]
21. If the coil of Problem 20 is wrapped on an air core, what is the strength of the magnetic field emerging from the pole? [8]
22. Repeat Problem 21 for a core made of iron-nickel alloy material, with relative permeability of 50. [5, 8]
23. What is the value of magnetomotive force (MMF) for the coil in Problem 20?

FORMULAS

$$B = \frac{\Phi}{A}$$ **EQ. 13-1**

$$H = \frac{NI}{l}$$ **EQ. 13-2**

$$B = \mu_r (1.26 \times 10^{-6}) H$$ **EQ. 13-3**

SUMMARY OF IDEAS

- A magnetic field drawing conveys the direction and strength of the magnetic effect in a space.
- Greater magnetic strength is indicated by denser (more closely packed) field lines.
- Magnetic flux is the number of field lines that exist in a certain space. It is symbolized Φ.
- The end of a magnet that flux lines emerge from is called its north pole. The end of a magnet that flux lines reenter is called its south pole.
- Unlike magnetic poles attract each other; like magnetic poles repel each other.
- A permanent magnet emits magnetic flux at all times. It requires no electrical input.
- An electromagnet produces flux only when it is carrying electric current in its coil.
- Magnetic flux is produced whenever current flows in a conductor.
- The left-hand rule can be used to relate current direction with magnetic flux direction.
- Wrapping wire into a spiral coil accentuates the magnetic flux density inside the coil and at the end-faces of the coil.
- In electromagnets, special core materials are commonly used to increase the magnetic flux density. This ability of the core material is represented by its relative permeability, symbolized μ_r.
- Magnetic core saturation is the fact of further increase in electromagnet current producing very little further increase in B. It occurs at large values of current.
- Retentivity is the ability of a core material to retain some of its magnetic flux even after the electromagnet coil has been deenergized.
- A portion of space can be shielded from magnetic field lines by enclosing it with an iron-based metal.
- The basic measurement unit for magnetic flux is the weber, symbolized Wb.
- Magnetic flux density is the concept of flux per unit area. It is the variable that measures the strength of a magnetic field. It is symbolized B.
- The basic measurement unit for magnetic flux density is the weber per square meter, also called the tesla, symbolized T.
- The magnetic flux density of a long solenoid-type or toroid-type electromagnet is given by $B = \mu_r (1.26 \times 10^{-6}) N I/l$.
- The expression $N I/l$ is usually factored out and called magnetizing force. It is symbolized H, measured in units of ampere-turns per meter.

CHAPTER QUESTIONS AND PROBLEMS

1. T-F In a magnetic field drawing, the strength is represented by the density of the lines. [1]
2. What symbol is used to show magnetic flux pointing into the page, away from the viewer? [1]

3. What symbol is used to show magnetic flux pointing out of the page, toward the viewer? [1]
4. Distinguish between magnetic flux and magnetic flux density. Give the symbol for each concept. [1, 7]
5. What are the basic measurement units for the magnetic variables in Question 4? [1, 7]
6. The pole face of a certain motor has dimensions of 6 x 3 cm. It emits a magnetic flux of 2.5×10^{-3} Wb. What is the field's flux density? [7]
7. In problem 6, if Φ decreases to 1×10^{-3} Wb, find the new value of B. [7]
8. Iron and alloys of iron, nickel, and cobalt are used to construct__________ magnets. [3]
9. T-F In an unmagnetized bar of iron, individual atomic magnets are randomly oriented. [3]
10. When wire is looped many times to form a solenoid coil, the resulting magnetic field is much stronger than the field from just a straight piece of wire. Explain why this happens. [3]
11. Define the term *relative magnetic permeability*. [5]
12. A certain magnetic core produces a magnetic flux density of 1.0 T when subjected to a magnetizing force of 5500 A•t / m. Calculate its relative permeability, μ_r. [5, 8]
13. Assuming that the permeability stays constant, how great a magnetizing force is required to produce a flux density of 1.5 T in the core described in Question 12? [5, 8]
14. A certain magnetic core material has a residual flux density of 0.25 T when its electromagnet cycles from 0 to 3000 A•t / m of magnetizing force H, then back to $H=0$. Sketch its curve of B versus H (like the curve of B versus I in Figure 13-12.) [5, 8]
15. Repeat Problem 14 for a different core material that retains a residual flux density of 0.4 T. [5, 8]
16. Which material, the one in Problem 14 or the one in Problem 15, would be better suited for use as a snap-action magnetic sensing device? [5]
17. Which material, Problem 14 or Problem 15, would be better suited for use in an ac motor or transformer? [5]
18. A certain single-layer air-core solenoid is 5 cm long and has a diameter of 3 mm. It is close-wrapped (adjacent turns of wire touch each other) with AWG #28 enameled wire (wire diameter = 0.3210 mm, from Table 8-2). [8]
    a) How many turns does it possess?
    b) What will be the magnetizing force, H, for a current of 75 mA?
    c) How strong will the B field be at the poles, with I = 75 mA?
19. For the solenoid in Problem 18, calculate magnetizing force H and flux density B if the current is raised to 150 mA. [8]
20. Suppose that the solenoid of Problem 19 is provided with a ferromagnetic (iron-based) core instead of air. If the core has a relative permeability of $\mu_r = 120$ and has a linear B–H curve, calculate the flux density for a current of 150 mA. [8]
21. For the solenoid of Problem 20, what value of current will produce a magnetic flux density of 0.094 T? [8]
22. The inability of a magnetic core to increase its flux density, B, in proportion to its magnetizing force, H, at large values of H, is called__________. [5]
23. T-F For most magnetic materials, the saturation point is easily identifiable because it occurs abruptly. [5]
24. All other things being equal, which would produce a greater pulling force on the valve plug in Figure 13-16, an iron core with $\mu_r = 90$ or an iron core with $\mu_r = 120$? Why? [5, 6]
25. All other things being equal, which would produce a greater pulling force on the valve plug, a solenoid coil with 3000 turns or one with 4000 turns? Why? [5,6]

CHAPTER 14

INDUCTANCE

OUTLINE

14-1 Inductors Oppose Change in Current
14-2 Unit of Measurement—Henry
14-3 Factors that Determine Inductance
14-4 Types of Inductors
14-5 Inductors in Series and Parallel
14-6 Mutual Inductance
14-7 Troubleshooting and Measuring Inductors
14-8 Applications of Inductors

Modern radiographic instruments, combined with up-to-the-moment electronic processing of the radiographic signal, enable a medical team to view their microsurgical actions inside a patient's heart or other organ. *Courtesy of General Electric Medical Systems*

NEW TERMS TO WATCH FOR

inductance	Lenz's law	E-I frame	mutual inductance	tachometer
induced voltage	henry	toroid	suppression (of electrical noise)	

An inductor is essentially an electromagnet. It has a multiturn winding wrapped around a core. When current flows through the winding, it creates a magnetic field in the core material.

An inductor is different from an electromagnet only in its application. Electromagnets are used to produce mechanical force. The solenoid-operated valve and the moving-coil loudspeaker described in Section 13-7 are typical examples. However, inductors are not used to produce mechanical force. Their action is strictly electrical.

After studying this chapter, you should be able to:

1. Describe the action of an inductor when its current is changed by an external circuit.
2. Use Lenz's law to find the polarity of induced voltage.
3. Relate the three variables of induced voltage, inductance, and rate of change of current.
4. Find the inductance of a cylindrical or toroidal inductor, knowing its length, cross-sectional area, number of winding turns, and core permeability.
5. Distinguish between iron-core, ferrite-core, and air-core inductors.
6. Calculate the total inductance of two or more inductors connected in series or in parallel.
7. Describe how mutual inductance can be reduced or entirely eliminated.
8. Troubleshoot an inductive circuit to locate a shorted or open inductor.
9. Explain how an inductor can be used to suppress undesirable voltage bursts that are coupled by stray capacitance.
10. Explain the principle of operation of an inductive-pickup tachometer.

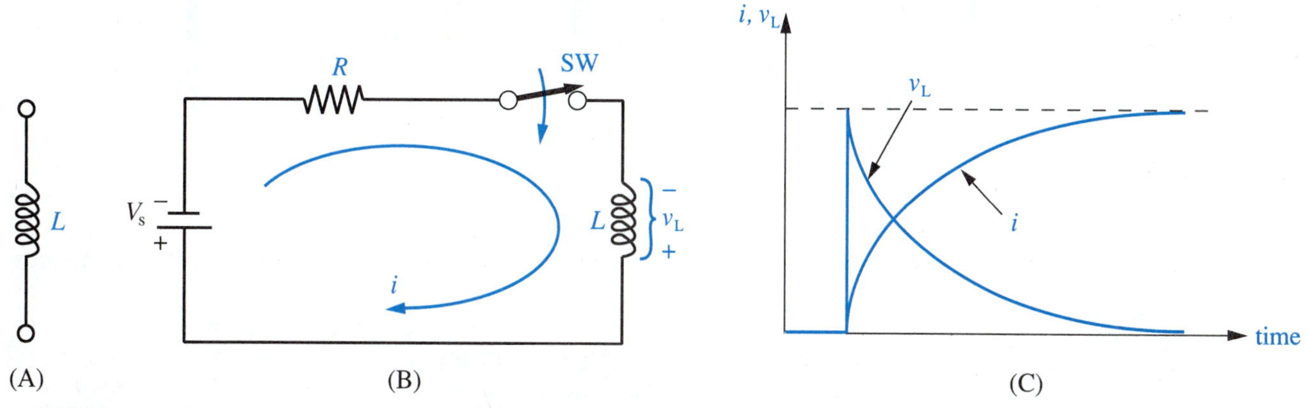

FIGURE 14-1 (A) Schematic symbol for inductor. The terms *coil* and *choke* are sometimes used to mean inductor. (B) Forcing current to begin flowing through an inductor. (C) Waveform graphs of induced voltage, v_L, and current, *i*. Voltage that is created by an inductor is said to be induced voltage. When we speak about an instantaneously varying voltage, as opposed to a dc voltage, we symbolize it with lower-case *v*, rather than capital *V*. The same is true for instantaneous current. We use lower-case *i*, rather than capital *I*.

14-1 INDUCTORS OPPOSE CHANGE IN CURRENT

An inductor has the property of inductance, which is symbolized *L*. The inductor schematic symbol in Figure 14-1(A) shows this. **Inductance** is a quality that tends to prevent current from changing. As an example of this, look at the circuit of Figure 14-1(B).

Initially, SW is in the open position and zero current is flowing through the inductor. If SW is suddenly closed, the source attempts to establish clockwise current in the circuit loop. The inductor reacts to this attempt by automatically creating a voltage, v_L, that opposes the source voltage. In this case inductor voltage v_L is negative on top and positive on bottom, as shown in Figure 14-1(B).

The creation of voltage v_L is only temporary, however. As time passes, voltage v_L gradually declines, eventually dropping to zero. The sudden appearance of v_L and then its decline are shown in the waveform graph of Figure 14-1(C). As induced voltage v_L declines, the source is able to overcome the inductor's opposition. The resulting current build-up is also graphed in Figure 14-1(C). In summary, the inductor is able to delay the establishment of steady current, but cannot permanently prevent it.

Do not think that an inductor opposes only increase in current. An inductor also opposes any decrease in its current that occurs as a result of external circuit action. An example is shown in Figure 14-2. In that figure, SPDT switch SW is initially in the up position. Therefore, a certain steady current is flowing through the inductor. This value of current is determined simply by Ohm's law (V_s / R), since the inductor's induced voltage is zero if the switch has been up for a long while.

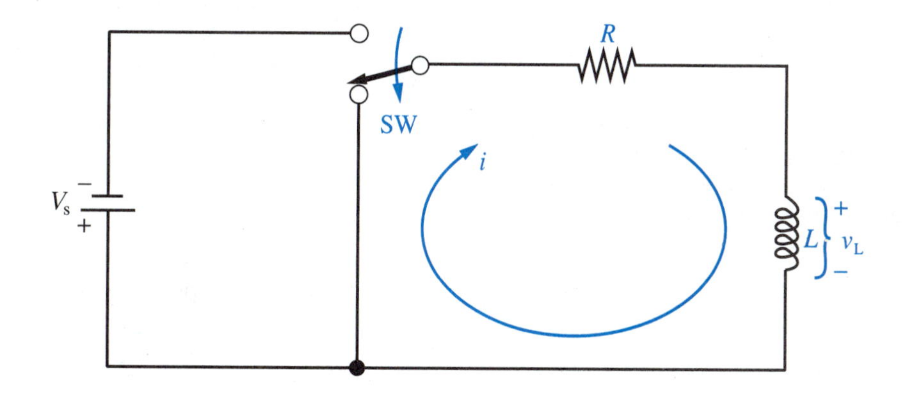

FIGURE 14-2 Circuit switching action tending to decrease inductor current. Induced voltage v_L has a polarity that helps the source, rather than opposing it.

If SW is suddenly thrown to the down position, the voltage source becomes disconnected and the inductor current tends to stop. The inductor reacts to this removal of its current by temporarily inducing a voltage with the polarity shown in Figure 14-2. In this way, the inductor is able to keep the current temporarily flowing in the original clockwise direction. As time passes, v_L declines to zero and current i does the same. Again, the inductor cannot permanently prevent the current change from occurring, but it can prevent the change from happening immediately.

Inductor action in a dc switching circuit can be summarized by the statement:

TECHNICAL FACT

An inductor prevents current from changing instantly. It induces voltage v_L with the proper polarity to maintain the current at a constant value for just an instant after a switching action.

The portion of this statement that pertains to the polarity of v_L is called **Lenz's law.**

14-2 UNIT OF MEASUREMENT—HENRY

The basic measurement unit of inductance is the **henry**, symbolized H. The idea of inductance deals with the creation of voltage in response to current change. As an equation,

$$v_L = L\left(\frac{\Delta i}{\Delta t}\right)$$

EQ. 14-1

in which the Δ symbol stands for "change in."

Thus, Equation 14-1 tells us that induced voltage v_L depends on the time rate of change of current ($\Delta i/\Delta t$ is change in current per unit of time). The time rate of change of current, in basic units of amperes per second, multiplied by inductance, in basic units of henrys, equals induced voltage in volts.

From this, we can define 1 henry as:

TECHNICAL FACT

One henry is the amount of inductance that causes an induced voltage of 1 volt when the current is changing at a rate of 1 ampere per second.

EXAMPLE 14-1

The resistor-inductor (*RL*) circuit of Figure 14-3 has a variable source voltage.

a) Suppose that the source voltage is being varied in such a way that the circuit current is increasing at a rate of 0.2 A/s. Find the magnitude of induced voltage v_L and state its polarity.

b) Suppose that the source voltage is being reduced in such a way that the current is decreasing at a rate of 0.4 A/s. Find the magnitude of v_L and state its polarity.

Continued on page 253.

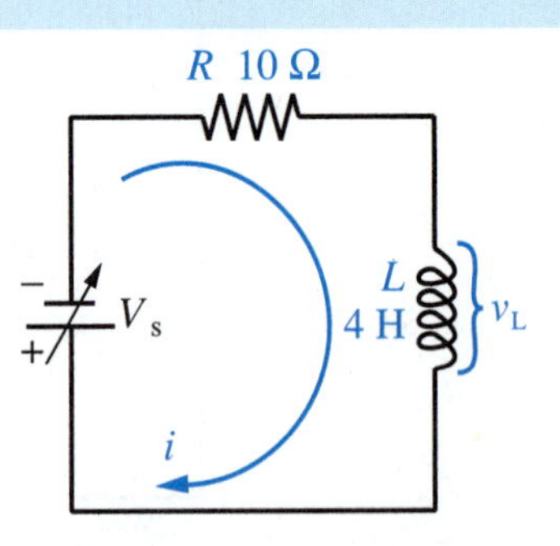

FIGURE 14-3 The inductor generates voltage v_L in response to any change in current *i*.

SOLUTION

a) With current increasing in the CW direction, the v_L polarity must be negative on top, positive on bottom. This opposes the voltage source and tends to retard the increase in current.

From Equation 14-1,

$$v_L = L\left(\frac{\Delta i}{\Delta t}\right)$$

$$= (4\ \text{H})\left(\frac{0.2\ \text{A}}{\text{s}}\right) = \mathbf{0.8\ V}$$

b) With current still flowing in the CW direction but decreasing, the v_L polarity will be negative on bottom, positive on top. This aids the voltage source and tends to keep the current flowing in the CW direction, trying to prevent the change.

$$v_L = L\left(\frac{\Delta i}{\Delta t}\right) = 4\ \text{H}\left(\frac{0.4\ \text{A}}{\text{s}}\right) = \mathbf{1.6\ V}$$

✔ SELF-CHECK FOR SECTIONS 14-1 AND 14-2

1. __________ is the electrical property that tends to oppose change in current. [1]
2. For the circuit of Figure 14-1(B), carefully explain why the polarity of v_L is negative on top when the switch suddenly closes. [2]
3. For the circuit of Figure 14-1(B), carefully explain why the circuit current, i, builds up gradually to its final steady value, as shown in the waveform of part C. [2, 3]
4. T-F Lenz's law states that the polarity of v_L opposes the source voltage when the circuit action is tending to increase current and aids the source voltage when the circuit action is tending to decrease current. [2]
5. In Figure 14-4, suppose SW has been in position 1 for a while. [2]
   a) If SW is moved to position 2, describe the polarity of v_L.
   b) SW remains in position 2 long enough for the current to stabilize. Then it is moved to position 3. Describe the polarity of v_L.
6. The basic measurement unit of inductance is the________; it is symbolized__________. [3]
7. A certain inductor induces 5 V when its current is changing at a rate of 8 A/s. Find its inductance. [3]

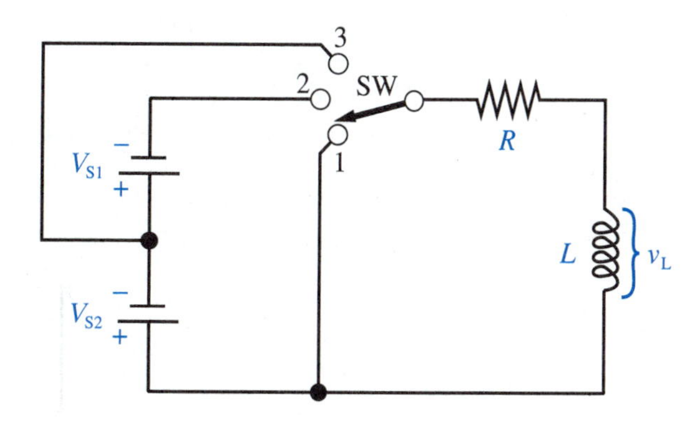

FIGURE 14-4 Every time the switch is moved to a new position, causing a change in the inductor's current, the inductor temporarily induces a voltage.

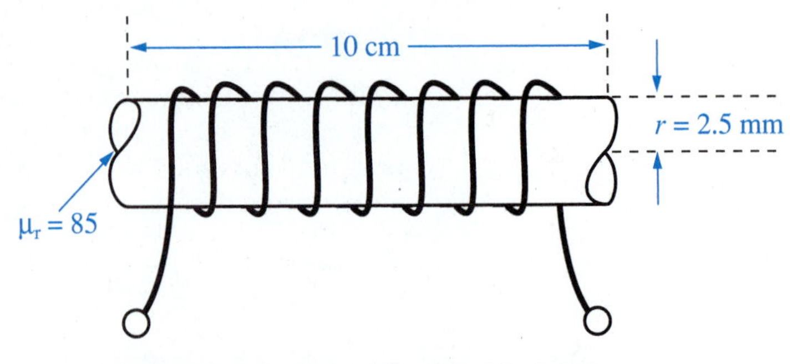

FIGURE 14-5 Long cylindrical inductor

14-3 FACTORS THAT DETERMINE INDUCTANCE

There are four physical features that affect the inductance of an inductor. They are:

1. Number of turns in the winding, N.
2. Cross-sectional area of the core, A.
3. Length of the core, l.
4. Magnetic permeability of the core material, μ_r.

For long cylindrical-shaped and toroid-shaped inductors, the inductance is given by

$$L = \mu_r(1.26 \times 10^{-6})\frac{N^2A}{l}$$

EQ. 14-2

Note that inductance varies as the square of the number of turns, N. Thus, for given physical dimensions of area and length, doubling the number of turns increases inductance, L, by four times.

EXAMPLE 14-2

The cylindrical inductor of Figure 14-5 has length of 10 cm and cross-sectional radius of 2.5 mm, as shown. The core material has relative permeability of 85.

a) If 500 turns of wire are wrapped around the core, what will be the inductance?
b) How many turns would have to be placed around the core to produce inductance $L = 100$ mH?

SOLUTION

First calculate the cross-sectional area in basic units of square meters. For a circular cross section,

$$A = \pi r^2 = 3.1416\ (2.5 \times 10^{-3}\ \text{m})^2 = 1.964 \times 10^{-5}\ \text{m}^2$$

Length, l, is 0.01 m, in basic units.

a) Applying Equation 14-2 gives

$$L = \mu_r\ (1.26 \times 10^{-6})\frac{N^2 A}{l}$$

Continued on page 255.

$$= 85\ (1.26 \times 10^{-6}) \frac{(500)^2\ (1.964 \times 10^{-5}\ \text{m}^2)}{0.01\ \text{m}}$$

$$= \mathbf{52.6 \times 10^{-3}\ H} \text{ or } \mathbf{52.6\ mH}$$

b) To raise the inductance to 100 mH, the number of winding turns must be increased. Rearranging Equation 14-2 gives

$$N^2 = \frac{Ll}{\mu_r\ (1.26 \times 10^{-6})\ A}$$

$$N = \sqrt{\frac{Ll}{\mu_r\ (1.26 \times 10^{-6})\ A}}$$

$$= \sqrt{\frac{(100 \times 10^{-3}\ \text{H})\ (0.01\ \text{m})}{85\ (1.26 \times 10^{-6})\ (1.964 \times 10^{-5}\ \text{m}^2)}} = \sqrt{4.756 \times 10^5}$$

$$= \mathbf{690\ turns}$$

L increased by a factor of almost 2, but the number of turns, *N*, had to be increased by a factor of only about 1.4 (690 / 500 = 1.38). This is because inductance goes up as the square of *N*.

14-4 TYPES OF INDUCTORS

Inductors can be classified on the basis of their type of core material. If the core is a ferromagnetic (containing iron) material with high permeability, the inductor is referred to as an iron-core unit.

Many inductors are wrapped on a core of iron oxide mixed with inert material. Such an inductor is called a ferrite-core unit. Ferrite cores tend to yield lower inductance due to their lower magnetic permeability. However, the ferrite core gives certain advantages. Inductance tends to be more nearly constant because of better *B*-versus-*H* linearity (less of the saturation problem shown in Figure 13-11). There is also less heating of the core generated by a certain nonideal effect—the eddy-current effect. The eddy-current idea will be explained in Section 17-8.

Some inductors are wrapped around a hollow form. They are referred to as air-core units. They have very low inductance due to air's low magnetic permeability. An inductor winding is sometimes wrapped on a solid material with no magnetic permeability ($\mu_r = 1$). Such an inductor is also called an air-core unit.

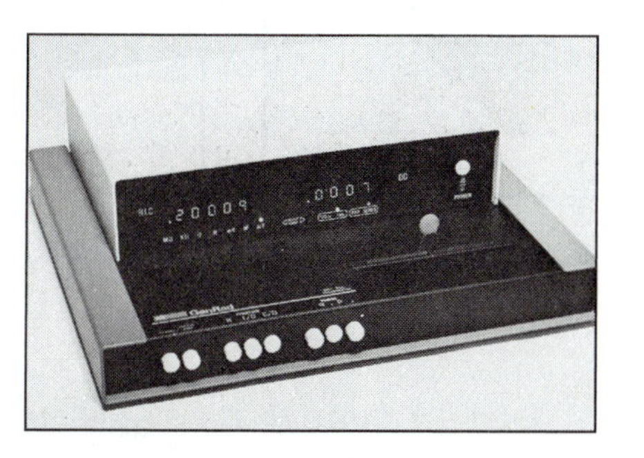

LCR digital meter for very accurate measurement of inductance and Q-factor. This instrument can also measure capacitance and associated leakage resistance, or D-factor. It is auto-ranging.
Courtesy of GenRad, Inc.

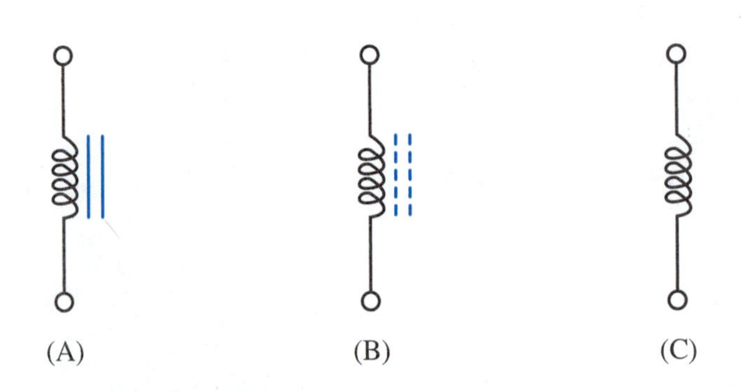

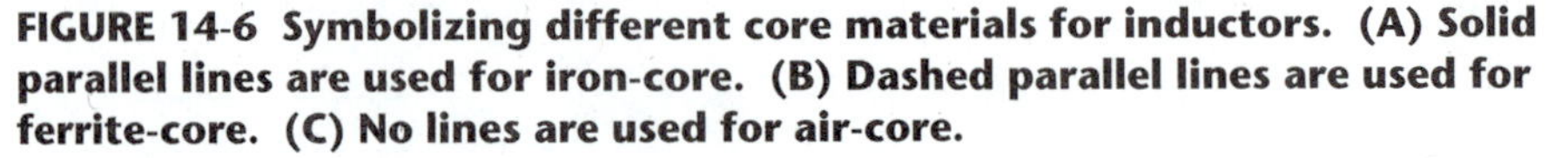
FIGURE 14-6 Symbolizing different core materials for inductors. (A) Solid parallel lines are used for iron-core. (B) Dashed parallel lines are used for ferrite-core. (C) No lines are used for air-core.

The three core types are symbolized schematically as shown in Figure 14-6.

Most iron-core inductors are constructed on an **E-I frame,** as shown in Figure 14-7(A). The winding turns are wrapped around the center leg of the E piece. Then the frame assembly is completed by attaching the I piece.

Actually, an E-I frame consists of many thin E and I pieces separated by thin layers of insulating material. This is made clear by the edge view in Figure 14-7(B). Such laminated construction reduces heat generated in the core by the undesirable eddy-current effect.

The purpose of the I piece is to provide a complete path for the magnetic flux lines. If it were not present, flux emerging from the center leg of the E piece would have to pass through air as part of its path. A flux path containing an air gap is said to have higher magnetic reluctance, compared to a path that has permeable material all the way. Higher magnetic reluctance tends to weaken the flux density in a magnetic path, just like higher resistance tends to weaken the current in an electric circuit path. With the I piece in place, magnetic reluctance is lowered, flux density is strengthened, and inductance, L, is increased.

Figure 14-7(C) shows a photograph of a 10-henry E-I iron-core inductor.

A **toroid** inductor is one with a doughnut-shaped core, as shown in Figure 14-8(A). Most toroid cores are ferrite material. A toroid core tends to have very low magnetic reluctance because there is no air gap and the flux path is continually

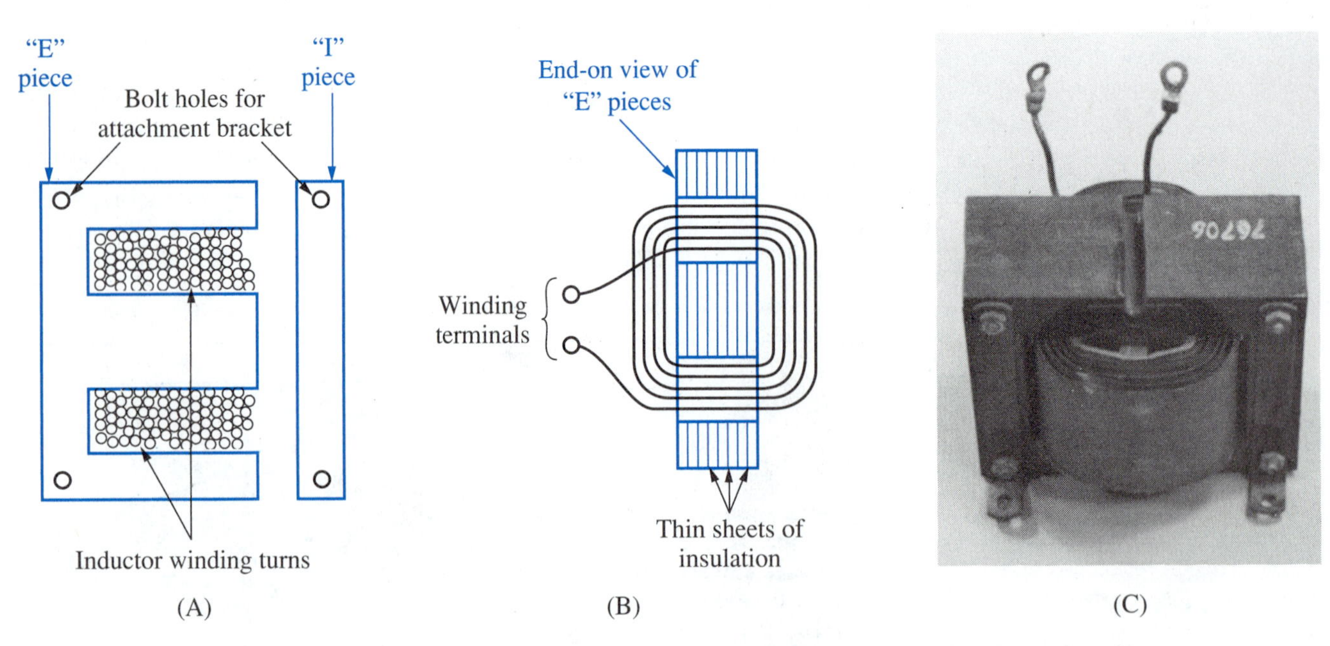

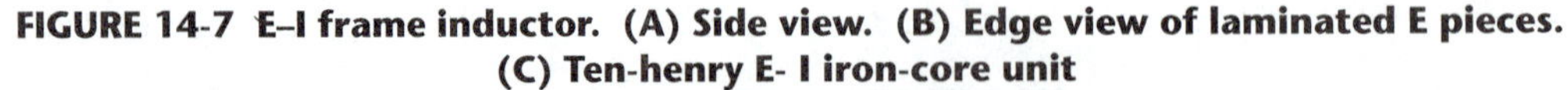
FIGURE 14-7 E–I frame inductor. (A) Side view. (B) Edge view of laminated E pieces. (C) Ten-henry E- I iron-core unit

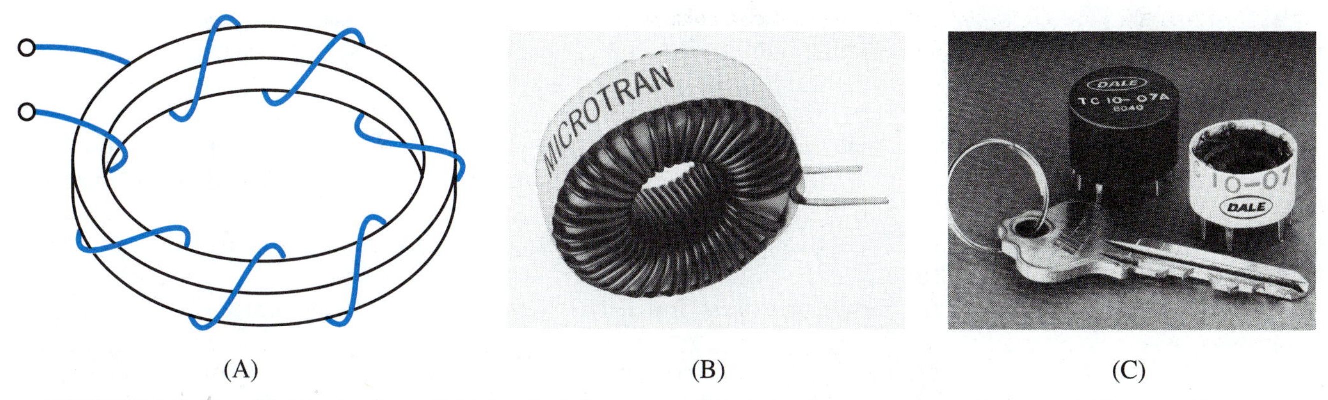

FIGURE 14-8 Toroidal inductor. (A) Physical diagram. (B) Actual appearance. *Courtesy of Microtran Company, Inc.* **(C) Tapped toroid with protective casing and solder tabs.** *Courtesy of Dale Electronics, Inc.*

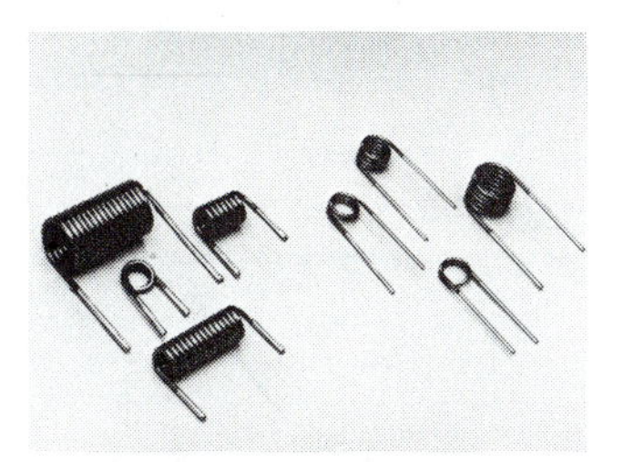

(A)

(B)

FIGURE 14-9 (A) Exposed-winding cylindrical air-core inductors.
(B) Cylindrical solid-core inductors. *Courtesy of Dale Electronics, Inc.*

curved, instead of having sharp turns. Therefore, a toroid structure provides large inductance for its size.

Figure 14-8(B) shows an open-winding loosely wrapped toroid inductor. An encased toroid is shown in Figure 14-8(C).

Cylindrical-core inductors often have exposed windings, like the air-core units shown in Figure 14-9(A). Solid-core devices are usually covered with an outer coating, as shown in Figure 14-9(B). Some cylindrical-core inductors have their windings sealed and protected in a molded enclosure. Figure 14-10 shows two molded inductors.

Cylindrical inductors like those in Figures 14-9 and 14-10 often are magnetically shielded. Then they can't be affected by stray magnetic flux from outside sources. Also, a shield prevents their own flux from leaking out to affect nearby magnetically sensitive devices.

A variable inductor usually has a movable ferrite piece, called a slug. Changing the position of the slug within the winding causes the effective core permeability to change, thereby varying the inductance. A cross-sectional view of such an inductor is shown in Figure 14-11.

✔ SELF-CHECK FOR SECTIONS 14-3 AND 14-4

8. Name the physical characteristics that determine the inductance of an inductor. [4]
9. For the inductor in Example 14-2, suppose that the number of winding turns is changed to 1200. Everything else stays the same. Calculate the inductance. [4]

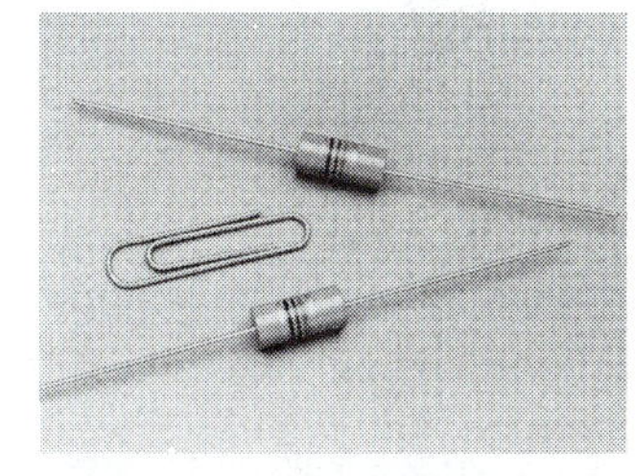

FIGURE 14-10 Molded-package inductors. The color-coded stripes indicate the nominal inductance in units of microhenrys. *Courtesy of Dale Electronics, Inc.*

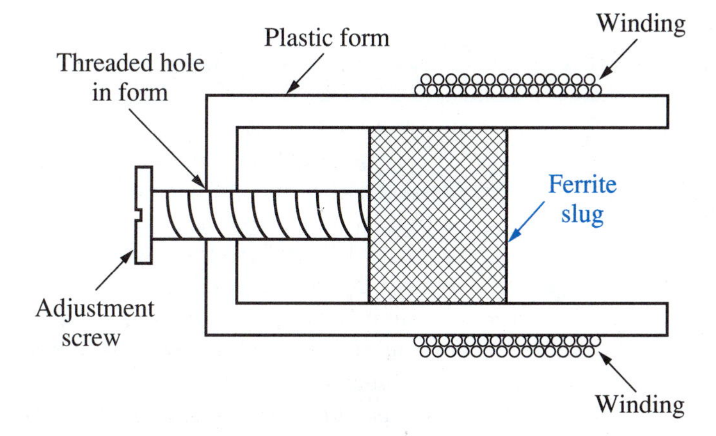

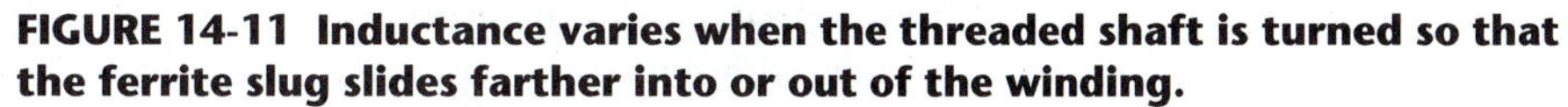

FIGURE 14-11 Inductance varies when the threaded shaft is turned so that the ferrite slug slides farther into or out of the winding.

10. A certain inductor has 200 winding turns and possesses 40 mH of inductance. Another inductor is identical except that it contains 600 winding turns. What is the inductance of the second inductor? [4]
11. Name the three basic core types for inductors. [5]
12. Draw the schematic symbol for each of the inductors in Question 11. [5]
13. Of the three basic core types, which one is most likely to experience the problem of inductance change due to magnetic saturation? [5]
14. T-F Any inductor with a nonmagnetic core is called an air-core unit. [5]
15. An inductor with a doughnut-like core is called a__________inductor. [4]
16. Inductor cores that furnish a complete closed path with no air gap have__________ magnetic reluctance.
17. In Question 16, such an inductor tends to have__________inductance, other things being equal.
18. In Figure 14-11, draw where the ferrite slug should be positioned to produce maximum inductance. [4, 5]

14-5 INDUCTORS IN SERIES AND PARALLEL

When inductors are connected in series, the total inductance is given by

$$L_T = L_1 + L_2 + L_3 + \ldots$$

EQ. 14-3

Equation 14-3 is valid only if the inductors don't interact with each other. This means that the magnetic field flux produced by one inductor must not pass through the coils of any other inductor.

Inductors in parallel combine like resistors in parallel, by the reciprocal formula.

$$\frac{1}{L_T} = \frac{1}{L_1} + \frac{1}{L_2} + \frac{1}{L_3} + \ldots$$

EQ. 14-4

Again, the inductors must not interact in order for Equation 14-4 to hold true.

14-6 MUTUAL INDUCTANCE

Two inductors are sometimes oriented so that the flux produced by one passes through the winding coils of the other—they share magnetic flux. This tends to happen when their axes are aligned in the same direction, as shown in Figure 14-12(A). When flux sharing takes place, we say that there is **mutual inductance** between L_1 and L_2. Such mutual inductance would be symbolized M_{1-2}.

In many circumstances, mutual inductance is undesirable. It can be eliminated or reduced by spreading the inductors farther apart physically. Another method is to rotate the axis of one inductor by 90 degrees, as shown in Figure 14-12(B).

Also, inductors can be magnetically shielded, as pictured in Figure 13-15. This completely eliminates mutual inductance. The photograph in Section 13-5 shows typical shields.

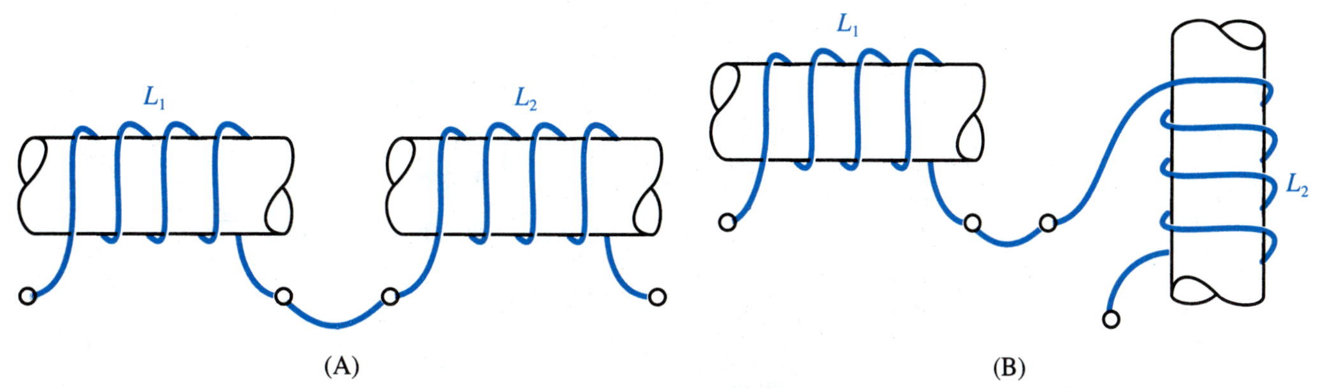

FIGURE 14-12 L_1 **and** L_2 **are connected in series. (A) Mutual inductance,** M_{1-2} **will exist. Equation 14-3 will not hold. (B) Rotating one of the inductors tends to reduce** M_{1-2}**.**

14-7 TROUBLESHOOTING AND MEASURING INDUCTORS

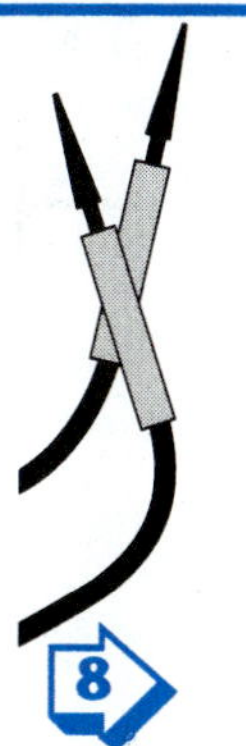

A good inductor has a certain amount of wire resistance in its multiturn winding. This resistance will be quite important to us later, in Chapter 18. If the winding is made with relatively few turns of heavy wire, the wire resistance tends to be rather low, perhaps only a few ohms. If the winding is made with many turns of thinner wire, its resistance will be higher, often in the range of several hundred ohms.

A good inductor will measure this wire resistance when it is tested by an ohmmeter. This is pictured in Figure 14-13(A). If an inductor fails open, it can be detected by an ohmmeter test. In Figure 14-13(B), the broken-open winding gives an infinity measurement on an ohmmeter.

Inductors seldom fail dead-shorted. It is more likely that some of the winding turns will short together due to damaged insulation.

Testing a partially shorted inductor with an ohmmeter gives a lower-than-normal resistance reading. This lower-than-normal ohmmeter measurement reveals the trouble only *if* you happen to know the normal resistance value for that inductor. In most troubleshooting situations, you will not know the normal winding resistance value. This is because inductor manufacturers don't mark the winding's resistance on the body. Only its inductance is marked.

Of course, if you have an identical inductor that is known to be good, then you can compare resistances. If the suspected inductor measures lower resistance than the inductor that is known to be good, the suspected inductor is partially shorted.

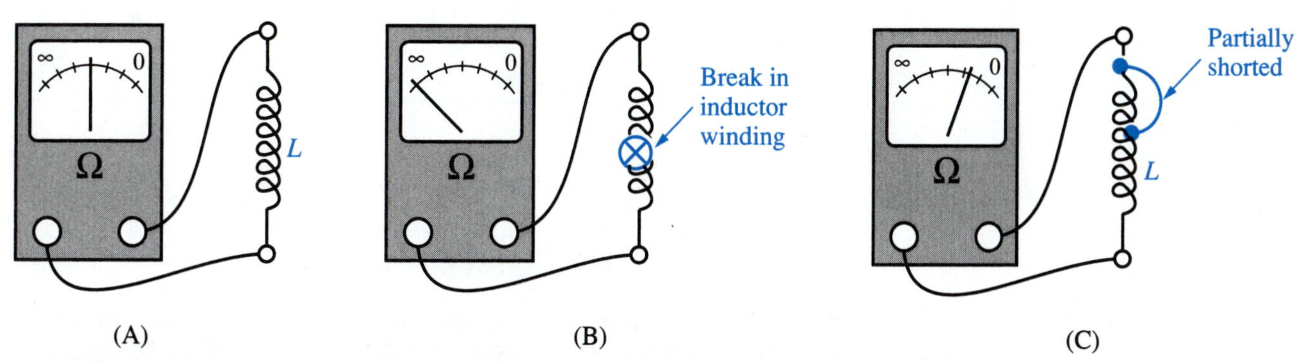

FIGURE 14-13 (A) A good inductor measures some medium amount of resistance. (B) An inductor that has failed open measures ∞ Ω. (C) A partially shorted inductor measures lower-than-normal resistance.

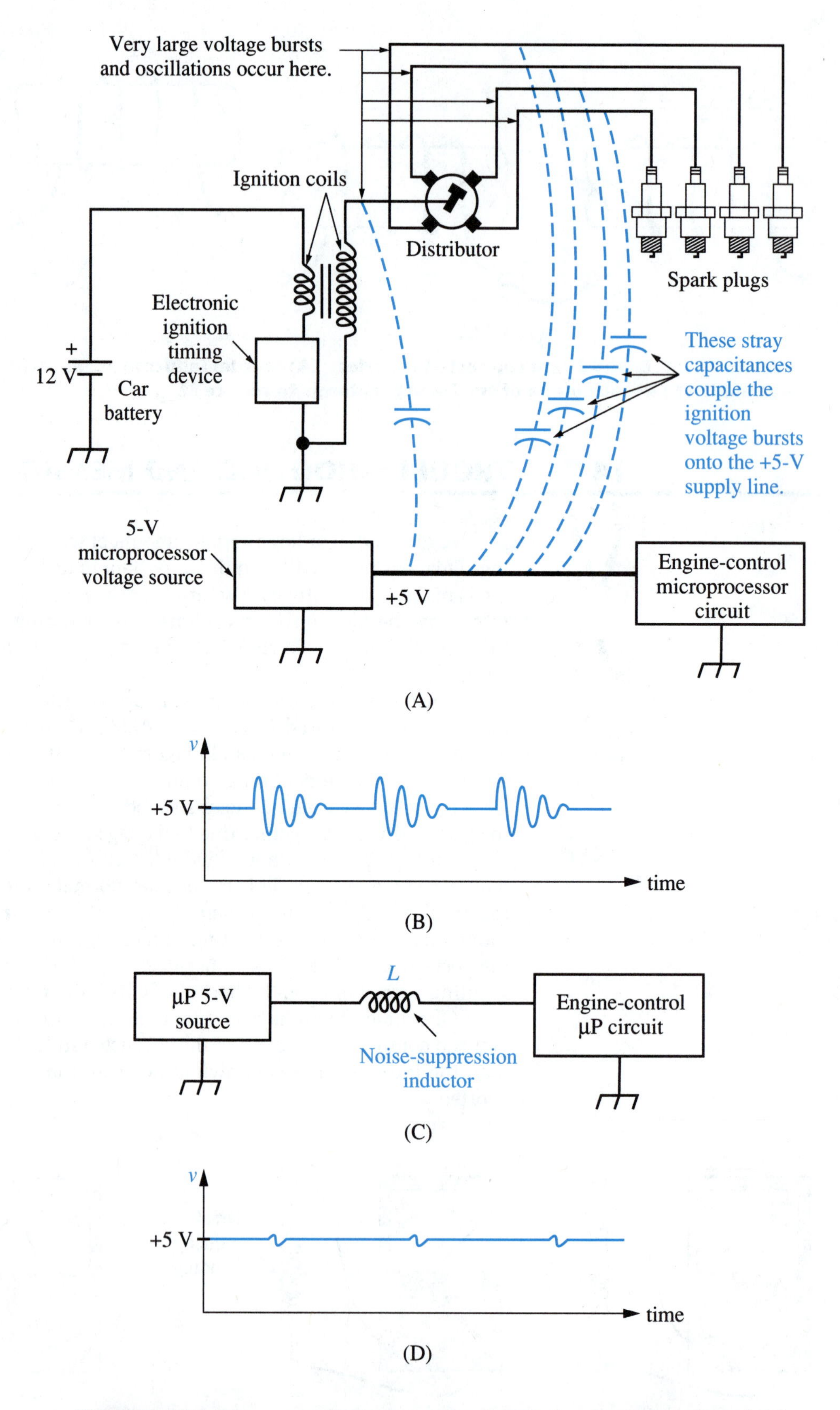

FIGURE 14-14 Stray capacitances couple electrical noise bursts from ignition wires to the μP. A series inductor can dramatically reduce the noise bursts. A capacitor can also be placed in parallel with the μP to increase the inductor's effectiveness.

A more common way to detect partial shorts is to measure the inductance, L, on an LCR meter (Section 12-7; also photo in Section 14-2) or impedance bridge. A partially shorted inductor will measure an L value much lower than the nominal inductance.

14-8 APPLICATIONS OF INDUCTORS

Automotive Noise Suppression

The dc voltage source for an automobile engine-control microprocessor (μP) must be very steady. However, the voltage source is subject to severe bursts of voltage from the ignition system. This problem is illustrated in Figure 14-14.

As illustrated in Figure 14-14(A), large voltage surges are routed from the ignition coil, through the distributor, to the spark plugs. The spark plug connecting wires are only a short distance from the microprocessor's 5-V supply line. Therefore, there is a fairly large amount of stray capacitance between the ignition wires and the +5-V line. These stray capacitances tend to couple voltage bursts and oscillations onto the supply line, as shown in Figure 14-14(B).

Such voltage bursts can be nearly eliminated by placing an inductor in the supply line, as shown in Figure 14-14(C). When a fast ignition-voltage burst hits the supply line, it tends to cause a fast current change. The inductor reacts to the fast current change by inducing a counter-voltage of opposing polarity, which nearly smothers the burst. We say that the inductor has **suppressed the electrical noise** burst. The result is an almost steady supply-line voltage, as shown in Figure 14-14(D).

Tachometer Pickup Inductor

10

A **tachometer** measures rotational speed, often in units of revolutions per minute (rpm). One way to construct a tachometer is shown in Figure 14-15(A). As the permanent magnet rotates past the inductor, the flux in the core changes. The changing flux causes the inductor to create voltage . (The inductor reacts the same way as it would to a current change in its winding.)

Each rotation of the shaft creates one cycle of output voltage. Figure 14-15(B) shows three cycles, corresponding to three mechanical rotations. Rotational speed is found by measuring the frequency (number of cycles per second) of the output voltage.

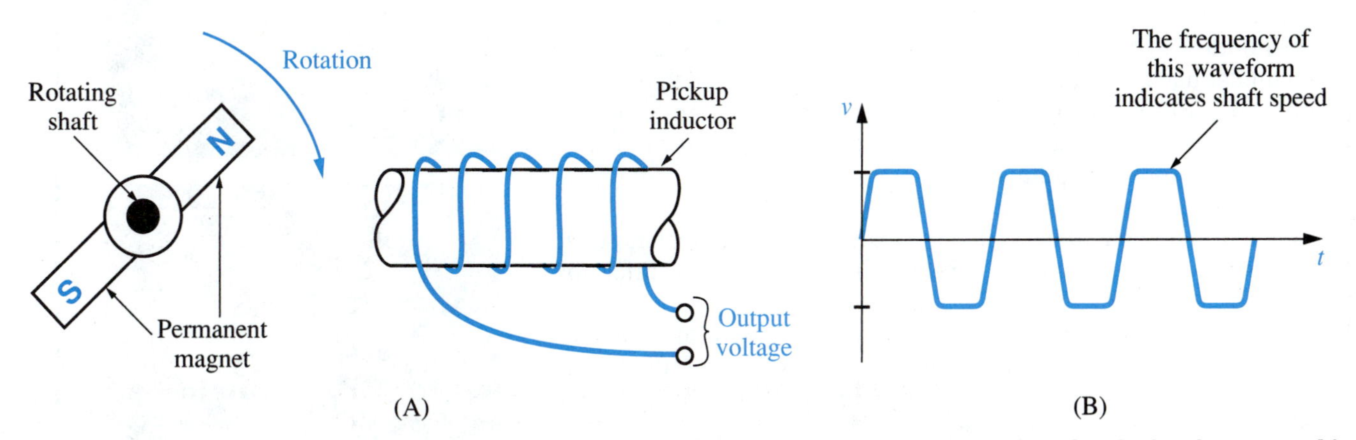

FIGURE 14-15 (A) In this tachometer design, a permanent magnet is attached to the shaft whose speed is being measured. (B) Output voltage waveform

✓ SELF-CHECK FOR SECTIONS 14-5 THROUGH 14-8

19. A 0.4-H inductor is wired in series with a 0.6-H inductor. There is no magnetic interaction between them. What is the total inductance, L_T? [6]
20. If the two inductors in Question 19 were connected in parallel, find L_T. [6]
21. If it was found that L_1 and L_2 in Questions 19 and 20 were interacting magnetically, which of the following steps would tend to reduce or eliminate the interaction? [7]
    a) Positioning L_2 so that its axis is at a 90-degree angle to L_1.
    b) Moving L_1 and L_2 closer together.
    c) Surrounding L_1 with a ferromagnetic enclosure.
    d) Surrounding L_2 with a ferromagnetic enclosure.
22. A certain inductor is suspected of causing trouble. It is marked with nominal inductance, $L = 50$ mH. An ohmmeter test indicates that its winding resistance is 25 Ω. Does that indicate that the inductor is bad? [8]
23. Describe the operation of an inductor that suppresses electrical noise. [9]
24. In Question 23, how is the inductor connected to the load, in series or in parallel? [9]
25. In Figure 14-15, if an electrical measurement device senses the tachometer output voltage making 30 cycles per second, what is the rotational speed in rpm units? [10]

INSULIN INJECTION

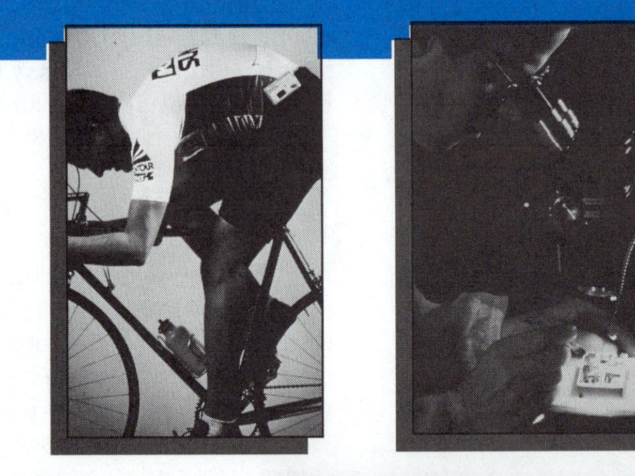

Diabetes is a disease in which a person's pancreas gland is not able to make enough natural insulin to keep the blood's sugar concentration at a normal level. More than one million diabetic patients in the United States must give themselves daily hypodermic injections of manufactured insulin. For many diabetics, the injections must be carefully timed depending on their meals and physical activities.

Insulin pumps that deliver a continuous flow of insulin at a slow rate have been developed recently. The external version of the pump has a thin plastic tube about 30 inches long, with a hypodermic needle at the end. It weighs less than 4 ounces. The wearer places the needle under the skin on the abdomen, then clips the credit-card-sized pump to a piece of clothing. The insulin flow rate is set by pushbutton switches that enter a program into an electronic microcomputer housed inside the pump. Flow rate can be varied throughout the day or from day to day, depending on the patient's needs.

In the photo at upper right the pump's tiny parts are being assembled under magnification.

The internal version of the pump is surgically implanted in the patient's abdomen. It holds enough insulin to last several months. It is refilled through a small catheter tube just below the surface of the skin, using a hypodermic needle.

The 3-inch diameter, 1/4-inch thick unit, with an attached catheter, is shown in the bottom photo, along with its radio transmitter. The unit is programmed by holding the small radio transmitter on the skin directly above it.

Courtesy of NASA

FORMULAS

$$v_L = L\left(\frac{\Delta i}{\Delta t}\right)$$ **EQ. 14-1**

$$L = \mu_r\,(1.26 \times 10^{-6})\frac{N^2A}{l}$$ **EQ. 14-2**

In series: $$L_T = L_1 + L_2 + L_3 + \ldots$$ **EQ. 14-3**

In parallel: $$\frac{1}{L_T} = \frac{1}{L_1} + \frac{1}{L_2} + \frac{1}{L_3} + \ldots$$ **EQ. 14-4**

SUMMARY OF IDEAS

- An inductor is a circuit element that has the electrical quality of inductance, which is symbolized *L*.
- Inductance is a current-stabilizing quality, in a similar way that capacitance is a voltage-stabilizing quality.
- If a circuit-switching action tends to change an inductor's current, the inductor generates (induces) a voltage, v_L, which tends to oppose the change (to maintain the same value of current).
- Induced voltage v_L declines to zero as time passes. The inductor then has no further effect on the circuit's behavior (ideally).
- After v_L has disappeared, an ideal inductor acts like a zero-resistance short circuit to a steady dc current.
- The voltage induced by an inductor depends on the time rate of change of its current.
- The physical structure of an inductor is the same as that of an electromagnet.
- The basic measurement unit of inductance is the henry, symbolized H.
- For a toroidal or long cylindrical inductor, inductance is given by

$$L = \mu_r\,(1.26 \times 10^{-6})\frac{N^2A}{l}$$

- Inductors in series add directly, like resistors. This assumes that the inductors are noninteracting.
- Noninteracting inductors in parallel combine by the reciprocal formula.
- Two common applications of inductors are: (1) to suppress voltage bursts and current surges in series with a protected load; (2) to induce voltage pulses in response to mechanical motion.

CHAPTER QUESTIONS AND PROBLEMS

1. The basic unit of inductance is the__________; it is symbolized________. [3]
2. In a dc circuit subjected to a switching action, describe the effect that an inductor has on the circuit. [1]

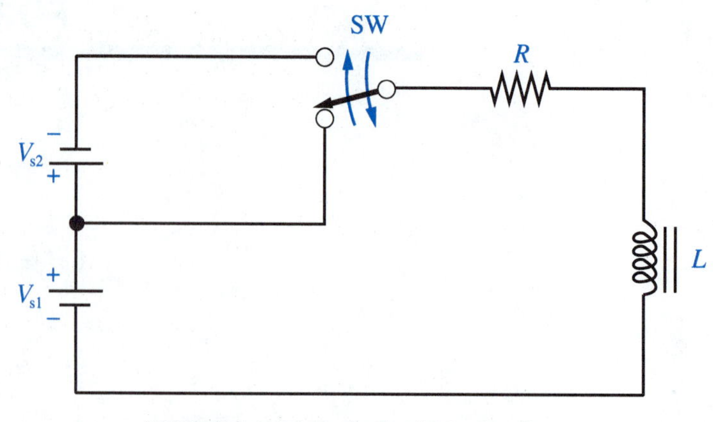

FIGURE 14-16 Switched RL circuit

3. In Figure 14-16, suppose switch SW is thrown into the up position after being in the down position for a while. Describe the reaction of the inductor, including polarity. [1, 2]
4. Suppose SW in Figure 14-16 is thrown back into the down position after being up for a while. Describe the reaction of the inductor, including polarity. [1, 2]
5. Suppose that $L = 3$ H in Figure 14-16. If the initial rate of change of current is 4 A / s when the switch is thrown into the up position, what amount of voltage v_L is initially induced? [3]
6. The toroid core in Figure 14-17 has a center-axis length, l, of 7.0 cm. It has a circular cross-section with diameter $d = 0.5$ cm.The relative permeability of the core material is 6. [4]
   a) Calculate the cross-sectional area from $A = \pi\, d^2 / 4$ ($\pi \cong 3.14$). Express in basic units of square meters.
   b) If it is desired to build an inductor with $L = 1$ mH, how many turns of wire should be wrapped around the toroid?
7. If it is desired to raise the inductance to 3 mH using the toroid core of Problem 6, how many turns, N, must the winding have? [4]
8. All other things being equal, increasing the core's permeability causes the inductance, L, to be__________. [4]
9. All other things being equal, increasing the core's length causes the inductance, L, to be__________. [4]
10. Besides permeability and length, what other physical factors have an effect on inductance? [4]
11. What type of core material would a 20-H inductor have? [4, 5]
12. In Figure 14-7, the center leg of the E piece is twice as wide as the outside legs. Explain why this is necessary. [5]
13. An inductor with only 3 μH of inductance probably has which type of core? [5]

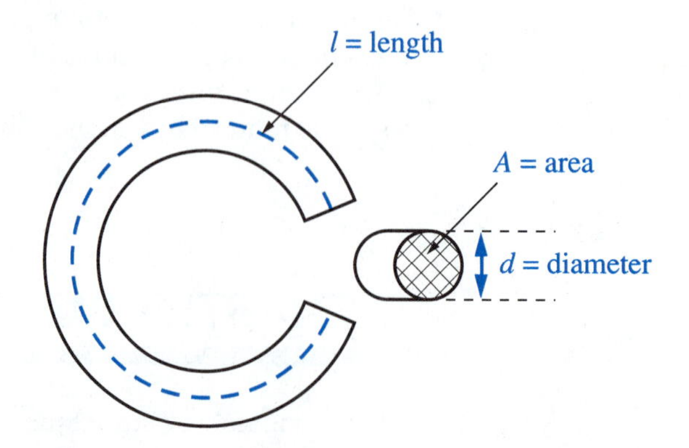

FIGURE 14-17 Toroidal core with circular cross section

14. T-F To produce maximum inductance from a variable inductor, the movable ferrite slug should be exactly centered in the winding. [4,5]
15. What is the disadvantage of an air-core inductor compared to ferrite- and iron-core inductors? [5]
16. Name an advantage of an air-core inductor compared to ferrite- and iron-core inductors. [5]
17. T-F The essential advantage of a toroid core over a cylindrical core is that the toroid core provides a lower reluctance path for the magnetic flux. [5]
18. What requirement must be satisfied so that series and parallel inductors can be combined by the normal (resistor-like) methods? [6, 7]
19. What are the three methods that can be used to reduce or eliminate mutual inductance? [7]
20. In the tachometer of Figure 14-15, does the strength of the permanent magnets have any effect on the frequency of the generated voltage? Why? [10]
21. In Figure 14-15, what is affected on the voltage waveform if the permanent magnet is made stronger? [10]

Troubleshooting Questions

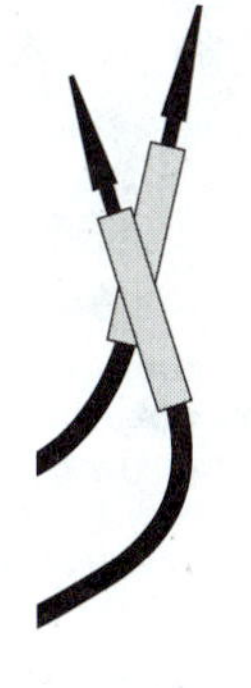

22. What will be the ohmmeter indication for an open inductor winding? [8]
23. Repeat Question 22 for a dead-shorted inductor winding. [8]
24. Repeat Question 22 for a partially shorted inductor winding. [8]
25. A certain 20-mH inductor is suspected of causing trouble in a circuit. Suppose that you isolate the inductor and connect an LCR meter. This is shown in Figure 14-18, with the LCR meter measuring 21.3 mH. What is your conclusion? [8]
26. A certain 10-mH inductor is suspected of causing trouble in a circuit. Suppose that you do not have an LCR meter, but you measure its winding resistance with an ohmmeter as 85 Ω. There is an identical 10-mH inductor elsewhere in the circuit that is known to be working properly. Its winding resistance measures 250 Ω. What is your conclusion? [8]

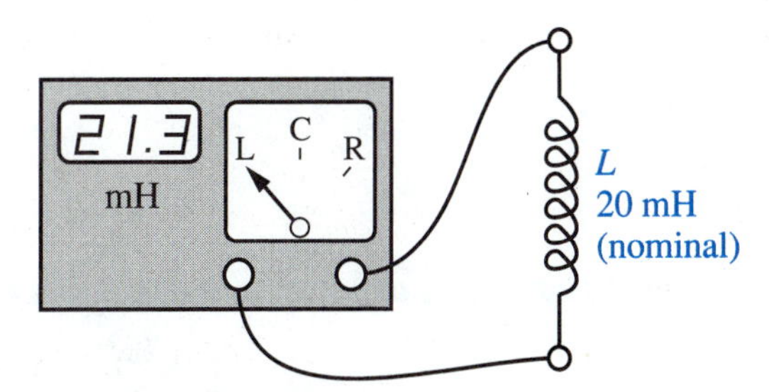

FIGURE 14-18 Measuring a suspect inductor with an LCR meter

CHAPTER 15

TIME CONSTANTS

These technicians are testing the operation of a controller that eases windmill generators onto the electric utility grid, rather than switching them on abruptly. This is necessary in order to minimize the initial transient current surge between the generator and the power lines. *Courtesy of NASA*

OUTLINE

15-1 Transient Waveforms for *C* and *L*
15-2 Time-Constant Facts
15-3 *RC* Time Constants
15-4 *RL* Time Constants
15-5 Inductive Kickback

NEW TERMS TO WATCH FOR

transient
time-constant function
final capacitor voltage
initial capacitor current
final inductor current
initial inductor voltage
inductive kickback

In electric circuits, a transient is a short-lived voltage or current waveform. The transients produced by capacitors and inductors following a dc switching action are of special interest to us.

After studying this chapter, you should be able to:

1. Define a dc transient and explain what causes dc transients.
2. Recognize a rising transient waveform and a falling transient waveform when seen on graph paper or an oscilloscope.
3. Calculate the time constant, τ, for a series *RC* or *RL* circuit.
4. Use the time-constant idea to describe the behavior of rising and falling transients.

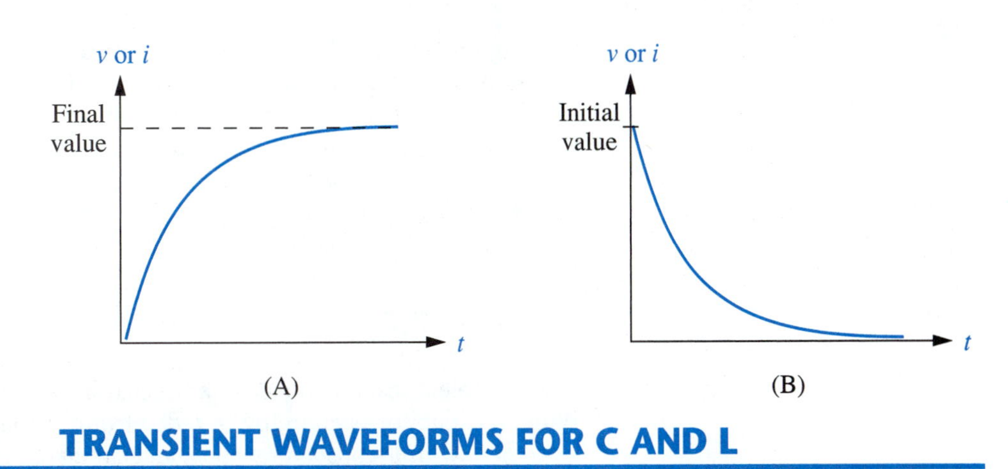

FIGURE 15-1 Time-constant transient waveforms of voltage or current versus time. (A) Rising. (B) Falling.

15-1 TRANSIENT WAVEFORMS FOR C AND L

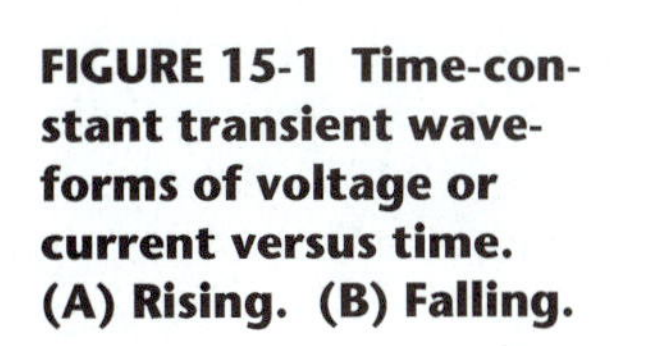

The sudden application or removal of a dc signal always causes a capacitor or inductor to produce a **transient.** A waveform graph describes the instantaneous behavior of these transients. All transient waveforms of either voltage or current have one feature in common: they all follow the **time-constant functions**. The rising and falling transient time-constant functions are shown in Figure 15-1.

These waveforms are not completely new to us. We have seen them previously in Figure 14-1(C), with the rising function describing the build-up of inductor current and the falling function describing the appearance and decline of inductor voltage.

They also apply to a capacitor-charging circuit. In Figure 15-2(A), for example, sudden closure of SW would produce a v_C-versus-time waveform that would have the characteristics of the rising function. This is shown in part B, with the final value of v_C equal to the source voltage, V_s. The *i*-versus-*t* waveform would track the falling function. This is shown in part C, with the initial value of current i_C determined by Ohm's law, since the capacitor has initial voltage of zero.

All rising and falling time-constant waveforms have the characteristic that they start fast and finish slow. For example, early in Figure 15-2(B), the capacitor voltage climbs rapidly. As time passes, v_C continues to increase, but at a slower rate. The reduced steepness of the curve shows this. Finally, near the completion of the transient, the voltage just creeps slowly toward its final steady value.

The same basic description applies to the falling i_C waveform in Figure 15-2(C). The i_C value decreases rapidly at first. Later, it is still descending, but not so steeply. At the finish, i_C creeps slowly down to zero.

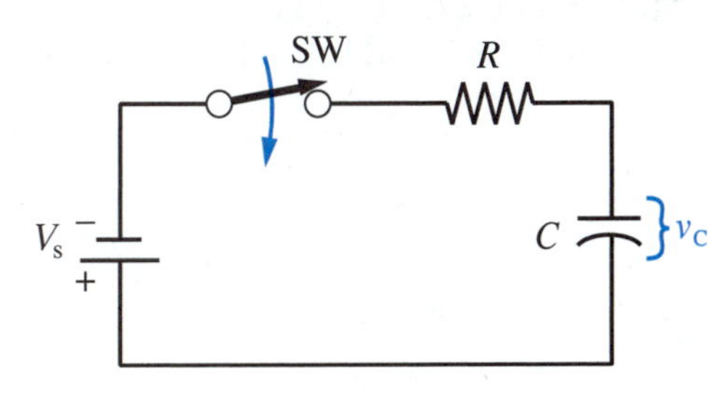

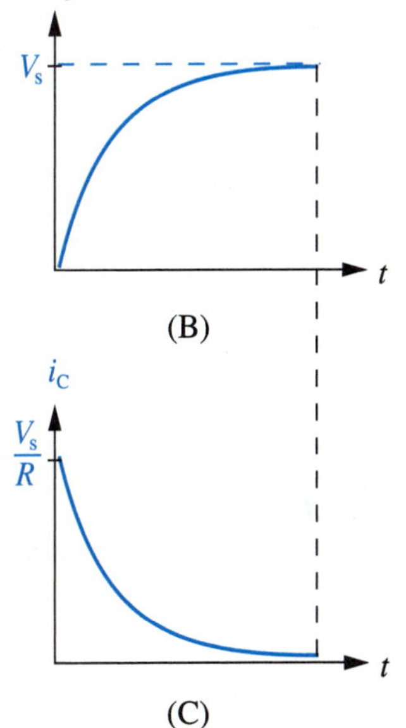

FIGURE 15-2 Resistor-capacitor charging circuit produces time-constant transient waveforms.

In Figure 15-2, the exact amount of time required for these waveforms to rise and fall depends on the specific values of capacitance and resistance. This will be explained in Section 15-2. In an inductive circuit, the exact transient time depends on the values of inductance and resistance. Section 15-3 will explain this.

The word *fluxing* means that an inductor is in the process of creating magnetic flux, because its current is building up after a switch has closed. It is like charging for a capacitor. Also, defluxing an inductor is like discharging a capacitor.

✓ SELF-CHECK FOR SECTION 15-1

1. What two electrical components produce transient time-constant waveforms? [1]
2. When a capacitor is charging in a dc circuit, capacitor voltage v_C follows the __________ time-constant function, and circuit current i follows the __________ time-constant function. (Answer rising or falling.) [1, 2]
3. When an inductor is fluxing in a dc circuit, inductor voltage v_L follows the ______________ time-constant function, and circuit current i follows the ______________ time-constant function. [1, 2]
4. Time-constant waveforms start__________ and finish.__________ [2]

15-2 TIME-CONSTANT FACTS

Time constant is symbolized by the Greek letter τ. Here is the definition of one time constant for any transient waveform.

TECHNICAL FACT

One time constant (1 τ) is the amount of time required for a transient waveform to go through 63% of its change.

Figure 15-3 shows the vertical (voltage or current) axis scaled in percent, and the horizontal time axis scaled in τ. In the rising waveform of Figure 15-3(A), the electrical variable starts at zero and reaches 63% of its final value after an amount of time equal to 1 τ. In the falling waveform of Figure 15-3(B), the electrical variable starts at 100%. It then declines by 63% (to 37%) in an elapsed time of 1 τ, following the switching action. The 63% figure is universal. This means that it applies to all circuits, no matter what the specific value of the time constant happens to be.

More detailed time-constant curves are shown in Figure 15-4. They display the universal facts listed in Table 15-1.

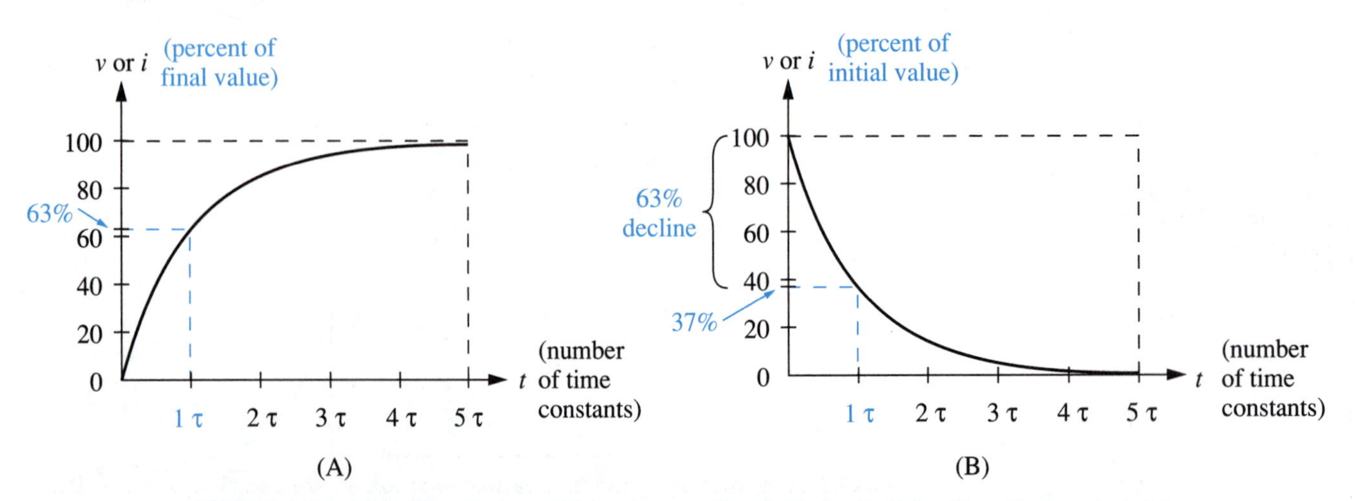

FIGURE 15-3 The electrical variable changes by 63% during the first time constant.

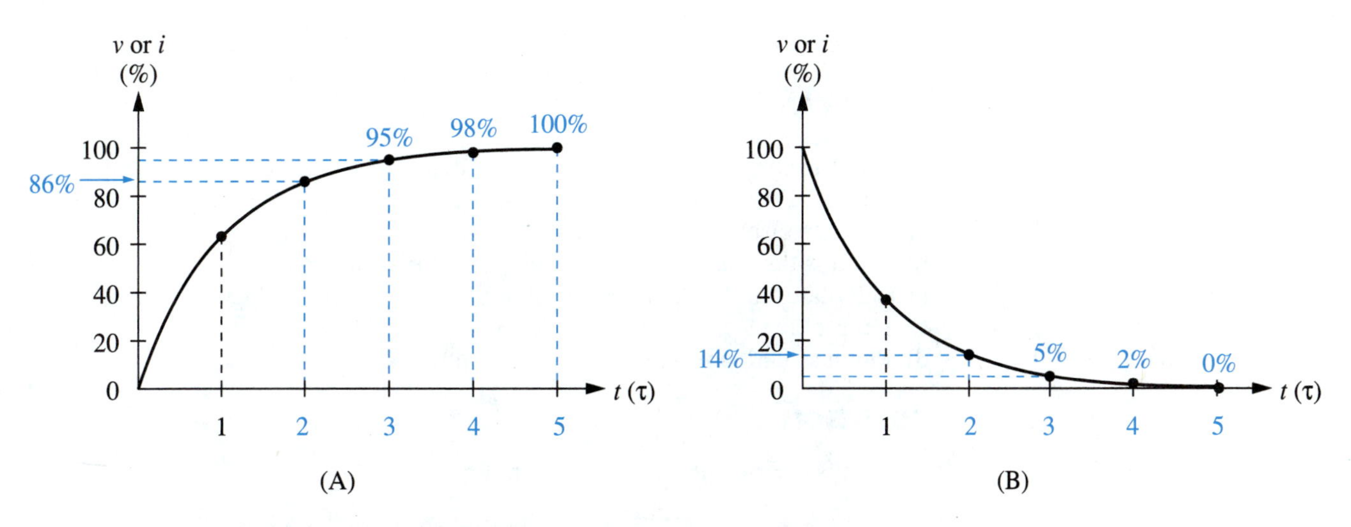

FIGURE 15-4 Detailed time-constant curves. These percentages are universal.

TABLE 15-1 Facts Regarding Universal Time-Constant Curves

AFTER THIS NUMBER OF τ	THE ELECTRICAL VARIABLE CHANGES BY THIS PERCENT	FALLING CURVE'S VALUE (100% MINUS PERCENT CHANGE)
1	63%	37%
2	86%	14%
3	95%	5%
4	98%	2%
5	virtually 100%	virtually 0%

15-3 RC TIME CONSTANTS

When a capacitor is charged or discharged through a series resistor, the value of one time constant is given by

TECHNICAL FACT

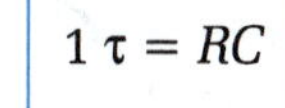

$$1\ \tau = RC \qquad \text{EQ. 15-1}$$

R must be expressed in basic units of ohms and *C* in basic units of farads; τ will be in basic units of seconds.

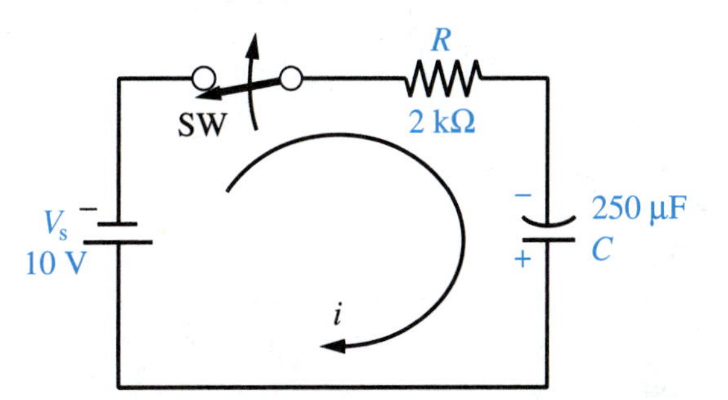

FIGURE 15-5 Knowing specific values of *R* and *C*, we can find out specific charging times.

EXAMPLE 15-1

In the circuit of Figure 15-5, the switch closes at time $t = 0$. The capacitor is initially completely discharged.

a) Find the circuit's time-constant value.
b) What is the final steady value of capacitor voltage?
c) Calculate the instantaneous value of v_C for the following time instants: $1\,\tau$, $2\,\tau$, $3\,\tau$, $5\,\tau$.
d) Plot a scaled graph of v_C versus actual time.

SOLUTION

a) This is a series RC circuit, so from Equation 15-1,

$$1\,\tau = RC = (2 \times 10^3\ \Omega)\ (250 \times 10^{-6}\ \text{F}) = \mathbf{0.5\ s}$$

b) The capacitor will charge to the source voltage value, **10 V**.

TECHNICAL FACT

In a series RC charging circuit, the capacitor will eventually reach a final voltage given by

$$v_{C\,(\text{final})} = V_s$$

c) Referring to Table 15-1, we have

4

At $1\,\tau$ (0.5 s), $v_C = 0.63(v_{C\,\text{final}}) = 0.63(10\ \text{V}) = \mathbf{6.3\ V.}$
At $2\,\tau$ (1.0 s), $v_C = 0.86(10\ \text{V}) = \mathbf{8.6\ V.}$
At $3\,\tau$ (1.5 s), $v_C = 0.95(10\ \text{V}) = \mathbf{9.5\ V.}$
At $5\,\tau$ (2.5 s), $v_C = 1.00(10\ \text{V}) = \mathbf{10.0\ V.}$

d) Plotting the four points from part c and the zero point produces the transient waveform graph of **Figure 15-6.**

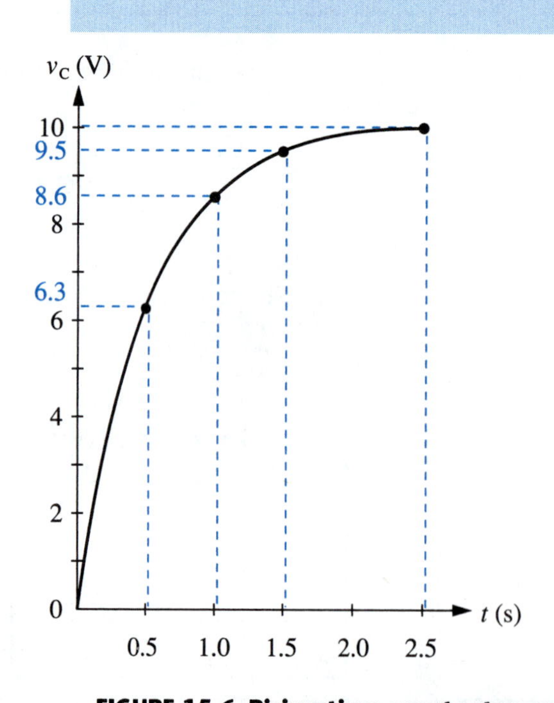

FIGURE 15-6 Rising time-constant waveform of v_C versus t. Note that the final capacitor voltage equals the source voltage.

EXAMPLE 15-2

For the same circuit, Figure 15-5, describe the current transient.

a) Find the initial value of circuit current, $i_{(initial)}$.
b) Calculate the instantaneous actual value of i for the following time instants: 1 τ, 2 τ, 3 τ, 5 τ .
c) Plot a scaled graph of i versus time, t.

SOLUTION

a) By Kirchhoff's voltage law, $V_s = v_R + v_C$. At $t = 0$, the capacitor has $v_C = 0$. Therefore, $V_s = v_R + 0$, which means that the entire source voltage, V_s, appears across resistor R at $t = 0$.

Applying Ohm's law to the resistor,

$$i_{\text{initial}} = \frac{V_s}{R} = \frac{10\text{ V}}{2\text{ k}\Omega} = \mathbf{5\text{ mA}}$$

TECHNICAL FACT

When charging a series RC circuit, the initial value of current can be found from

$$i_{(\text{initial})} = \frac{V_s}{R} \quad \text{since } v_{C\,(t\,=\,0)} = 0$$

b) Referring to the decreasing percentages in Table 15-1,

At 1 τ (0.5 s), $i = 0.37(i_{(\text{initial})}) = 0.37(5\text{ mA}) =$ **1.85 mA.**
At 2 τ (1.0 s), $i = 0.14(5\text{ mA}) =$ **0.70 mA.**
At 3 τ (1.5 s), $i = 0.05(5\text{ mA}) =$ **0.25 mA.**
At 5 τ (2.5 s), $i =$ **0.**

c) Plotting the four data points from part b and the initial point produces the graph of **Figure 15-7.**

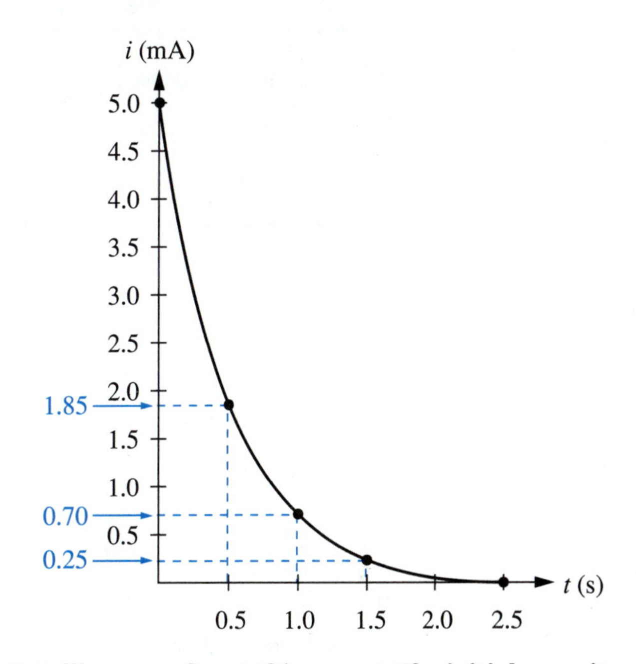

FIGURE 15-7 Falling waveform of *i* versus *t*. The initial capacitor current is equal to V_s / *R*.

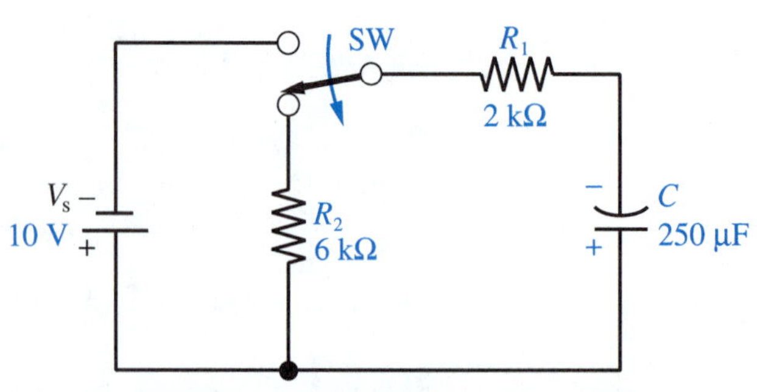

FIGURE 15-8 RC discharging circuit

EXAMPLE 15-3

Let us add another resistor, R_2, to the circuit of Figure 15-5. The new circuit, containing a double-throw switch, is shown in Figure 15-8. With the capacitor fully charged, suppose that SW is thrown into the down position at $t = 0$.

a) Find the circuit's time constant.
b) Calculate the instantaneous values of v_C and i for the following time instants: 1 τ, 2 τ, 3 τ, 5 τ.
c) Plot a waveform graph of v_C versus t.

SOLUTION

a) The capacitor must discharge through the series combination of $R_1 + R_2$. From Equation 15-1,

$$1\,\tau = (R_1 + R_2)\,C = (2\text{ k}\Omega + 6\text{ k}\Omega)\;250\;\mu\text{F}$$
$$= (8\text{ k}\Omega)\;250\;\mu\text{F} = \mathbf{2.0\ s}$$

b) $v_{C\,(\text{initial})} = 10\text{ V}$. From Ohm's law,

$$i_{(\text{initial})} = \frac{V_{C\,(\text{initial})}}{R_T} = \frac{10\text{ V}}{8\text{ k}\Omega} = 1.25\text{ mA}$$

Referring to the decreasing percentage column in Table 15-1, we get

t	v_C	i
1 τ (2.0 s)	0.37(10 V) = 3.7 V	0.37(1.25 mA) = 0.463 mA
2 τ (4.0 s)	0.14(10 V) = 1.4 V	0.14(1.25 mA) = 0.175 mA
3 τ (6.0 s)	0.05(10 V) = 0.5 V	0.05(1.25 mA) = 0.0625 mA
5 τ (10.0 s)	0	0

c) Plotting the initial point and the four points in part b produces the falling time-constant curve of **Figure 15-9.**

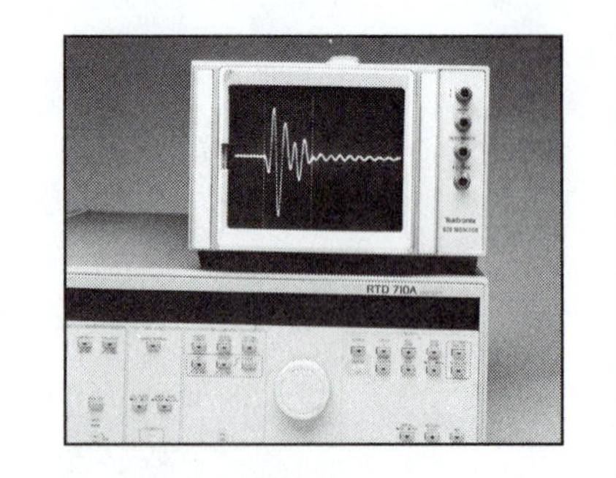

Waveform display.
Courtesy of Tektronix, Inc.

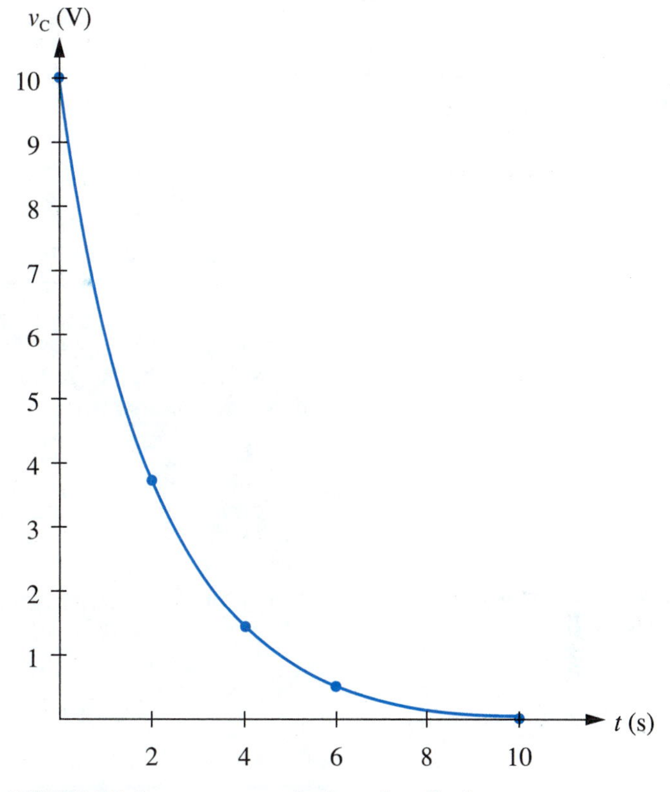

FIGURE 15-9 v_C **versus** t **during the discharge**

15-4 RL TIME CONSTANTS

When an inductor is fluxed (has current built up) or defluxed (has current taken away) through a series resistor, the time constant is given by

TECHNICAL FACT

$$1\ \tau = \frac{L}{R}$$

EQ. 15-2

L must be expressed in henrys, R in ohms, and τ will be in basic units of seconds.

EXAMPLE 15-4

In Figure 15-10, the ideal 7-H inductor begins fluxing when SW is closed.

a) What is the value of the time constant?
b) What is the value of the final steady current?
c) Plot a graph of actual current i versus time, using $t = 0$, $1\ \tau$, $2\ \tau$, $3\ \tau$, and $5\ \tau$.

SOLUTION

a) For a series RL circuit, Equation 15-2 gives

$$1\ \tau = \frac{L}{R} = \frac{7\ \text{H}}{70\ \Omega} = \mathbf{0.1\ s}$$

Continued on page 274.

FIGURE 15-10 Knowing *R* and *L*, we can find the circuit's time-constant value and fluxing times.

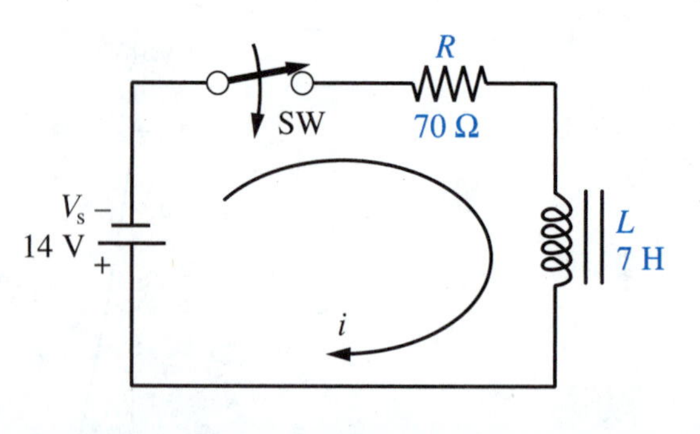

b) The ideal inductor induces a voltage that eventually declines to zero. After that, it has no ability to oppose current. Resistor R becomes the only current-limiting component in the circuit. From Ohm's law,

$$I_{\text{(final)}} = \frac{V_s}{R} = \frac{14\ \text{V}}{70\ \Omega} = \mathbf{0.2\ A}$$

TECHNICAL FACT

When fluxing a series RL circuit with an ideal inductor, the final value of dc current can be found from

$$I_{\text{(final)}} = \frac{V_s}{R} \quad \text{since } v_{L\ \text{(final)}} = 0\ \text{V (ideally)}$$

c) From the increasing percentages in Table 15-1,

t	*i*
1 τ = 0.1 s	0.63(0.2 A) = 0.126 A
2 τ = 0.2 s	0.86(0.2 A) = 0.172 A
3 τ = 0.3 s	0.95(0.2 A) = 0.190 A
5 τ = 0.5 s	1.00(0.2 A) = 0.2 A

These data points produce the rising waveform plot of **Figure 15-11.**

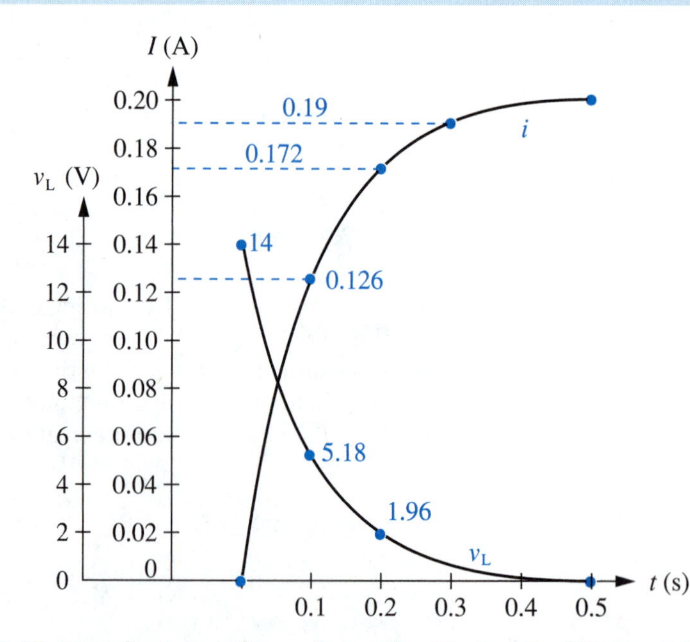

FIGURE 15-11 Transient waveforms for series RL circuit of Figure 15-10. Note that the final inductor current is equal to V_s / *R*. Also, the initial inductor voltage is equal to V_s.

EXAMPLE 15-5

Describe the inductor voltage transient for the circuit of Figure 15-10 from Example 15-4.

a) What is the initial value of v_L at the instant of switch closure?

b) Plot a graph of v_L versus actual time t, using $t = 0$, 1 τ, 2 τ, 3 τ, and 5 τ.

SOLUTION

a) At the instant of switch closure, the current equals zero. This is because the inductor does not permit current to change in zero elapsed time. With zero current, resistor voltage $v_R = 0$, by Ohm's law.

From Kirchhoff's voltage law,

$$V_s = v_R + v_L$$
$$14\text{ V} = \cancelto{0}{v_R} + v_L$$
$$v_{L\,(\text{initial})} = \mathbf{14\text{ V}}$$

TECHNICAL FACT

When fluxing a series RL circuit, the initial value of inductor voltage v_L is given by

$$v_{L\,(\text{initial})} = V_s$$

b) From the decreasing percentages in Table 15-1,

t	v_L
0	14 V
1 τ = 0.1 s	0.37(14 V) = 5.18 V
2 τ = 0.2 s	0.14(14 V) = 1.96 V
3 τ = 0.3 s	0.05(14 V) = 0.7 V
5 τ = 0.5 s	0 V

These points produce the falling transient waveform in **Figure 15-11.**

✓ SELF-CHECK FOR SECTIONS 15-2, 15-3, AND 15-4

5. A transient makes_____% of its overall change during the first time constant. [4]
6. A transient makes_____% of its overall change during the first two time constants. [4]
7. The total duration (virtual) of a transient is_____time constants. [4]
8. A 20-μF capacitor is charged through a 1.5-kΩ resistor. Find τ. [3]
9. What would be the time constant value if the same 20-μF capacitor were discharged through the 1.5-kΩ resistor? [3]
10. For the component values in Problems 8 and 9, the total time duration of the transient would be_____ms. [4]
11. In a series *RL* circuit with $R = 5$ kΩ and $L = 100$ mH, what is the value of the time constant? [3]
12. For the *L* and *R* values in Problem 11, what percentage of total current buildup would occur in the first 60 μs? [4]

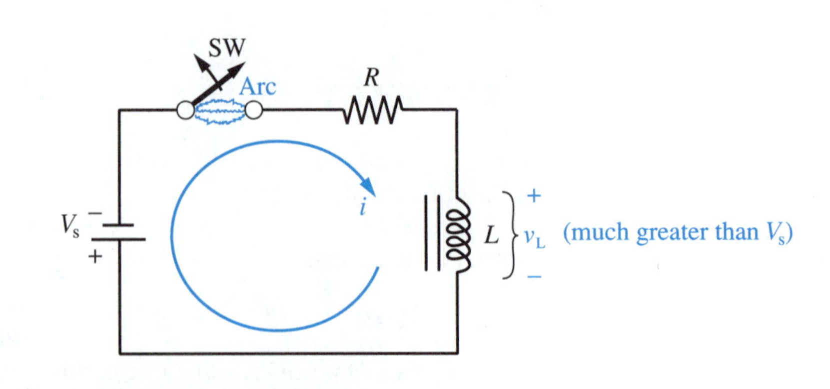

FIGURE 15-12 Inductive kickback occurs when an inductor's current is suddenly interrupted. The momentary induced voltage, v_L, may be several thousand volts.

15-5 INDUCTIVE KICKBACK

In Figure 15-10, if the switch is suddenly opened after the inductor has been fluxed, here is what happens. The inductor induces a very large voltage, which causes an arc across the switch contacts. This is indicated in Figure 15-12.

The inductor does this in order to maintain an instantaneously constant value of current at the moment the switch opens. The total resistance of the circuit is determined by the resistance of the air between the switch contact surfaces. This resistance is very large, many megohms. To ionize the air molecules so that they will conduct current, the inductor's initial voltage must rise to a very high value, as indicated in Figure 15-12.

The induced polarity is negative on the bottom, positive on the top. This is because the inductor is acting as a temporary source, maintaining the same direction of current that existed just prior to the switch opening.

The arc is short-lived, lasting less than 1 second. Even so, repeated occurrence damages the contact surfaces.

SAFETY ADVICE

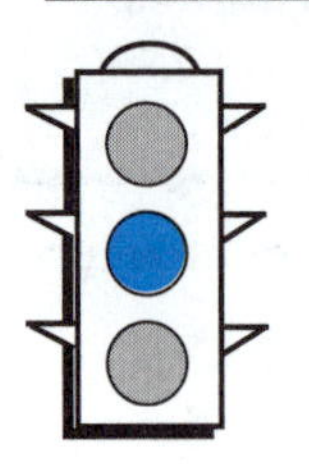

Most real-life electrical loads contain some inductance. This is because most loads have their conductors wound, rather than straight. The most common example of this is the electric motor. Every standard motor contains at least two multiturn windings.

Because of the inductive kickback problem, it is more difficult to interrupt the current in an inductive load like a motor, than to interrupt current in a resistive load. As the interrupting switch is opened, you must be careful not to expose yourself to the arc that occurs across the separating contact surfaces. The arc can severely burn your skin, and the intensely bright flash can harm your eyes.

A manual switch that has the job of carrying current to a motor (or other inductive load), and interrupting that current when the load is shut off, is often called a disconnect switch. See Figure 15-13. Motor disconnect switches are contained in a locked steel enclosure, shown in Figure 15-14. The enclosure is designed so that its door must be closed and locked to allow the disconnect switch handle to be moved. This is called a mechanical interlock.

Never try to defeat this safety feature by jamming apart the interlock mechanism. Furthermore, always stand off to the side of the enclosure, not directly in front of it, when you move the handle to Off. On very rare occasions, combustible gases may collect inside the enclosure. The interrupting arc can ignite these gases, causing an explosion that blasts the door open. This dangerous explosive situation cannot occur if the system has been designed and installed in compliance with the National Electrical Code, and properly maintained afterwards.

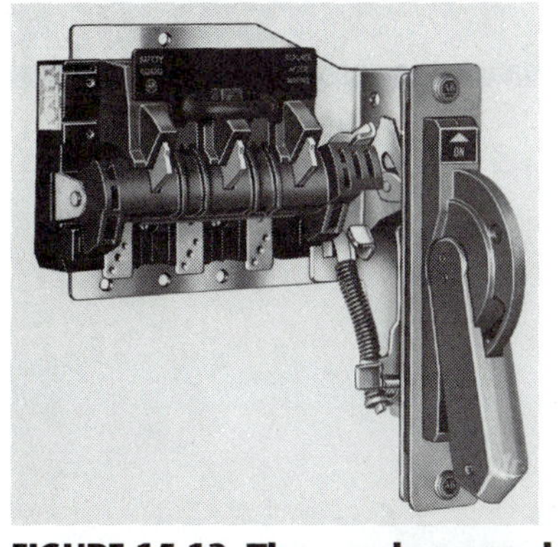

FIGURE 15-13 Three-pole manual disconnect switch, used in industrial and commercial wiring systems. It is shown in the open or "OFF" position. *Courtesy of Allen-Bradley, a Rockwell International Company*

FIGURE 15-14 Six-pole disconnect switch, mounted in its steel enclosure. Fuseholders are also mounted but no fuses are installed. *Courtesy of Eaton Corporation, Cutler-Hammer Products*

AUTOMATED ASSEMBLY

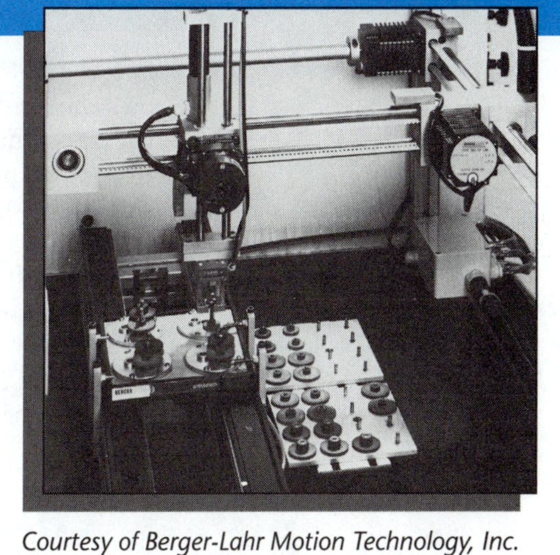

Courtesy of Berger-Lahr Motion Technology, Inc.

This automatic assembly machine is assembling gear boxes that will be used with small electric motors. The overall assembly process is performed by several machines similar to this one, stationed one next to the other. The combined group of machines is called a work cell.

This is a three-dimensional, or three-degree-of-freedom machine, plus gripper. The three capabilities for motion are: 1) horizontally left and right; 2) horizontally forward and back; and 3) vertically up and down. Rack-and-pinion mechanisms accomplish the two horizontal motions. The left-and-right rack is clearly visible across the center of this photo, a few inches below the left-and-right guide bar. The forward-and-back rack is visible on the far right of the photo, below the forward-and-back guide bar.

This particular machine has the job of placing four gears and their center shafts into each gear box partial assembly. The individual gears are picked up by the gripper from the holding table to the right of the dual slide-rails. First, a gear from one of the rearmost eight pegs is picked up and placed in the assembly. There is a downward-facing optical sensor alongside the machine's gripper base. It detects whether or not a peg still has a gear available. That way, the gripper never grabs air.

After installing the first gear, the machine runs forward to another table that holds the shafts. That table is out of view in this photo. It picks up the correct kind of shaft, then runs back and places the shaft in its proper location in the assembly. That action is what we are seeing at this moment.

Then a different kind of gear is assembled, from one of the eight pegs in the next-forward group on the holding table. Then its matching shaft is fetched and installed. And so on, until all four different gears and shafts are assembled.

The machine then begins work on the next partial assembly. At the moment shown here, three partial assemblies are finished and the machine is working on the fourth. When this batch of four is finished, it will be automatically taken away on the slide-rails. A new batch of base-plates will then be brought forward and the machine will begin working on them.

Of course, other automated equipment in the work cell keeps the gear-holding table and the shaft-holding table loaded with parts.

Some assembly machines can have their set of motions changed without any rewiring. Instead, a human simply enters a new set of motion instructions through a keyboard programming device. Such reprogrammable machines are called assembly robots.

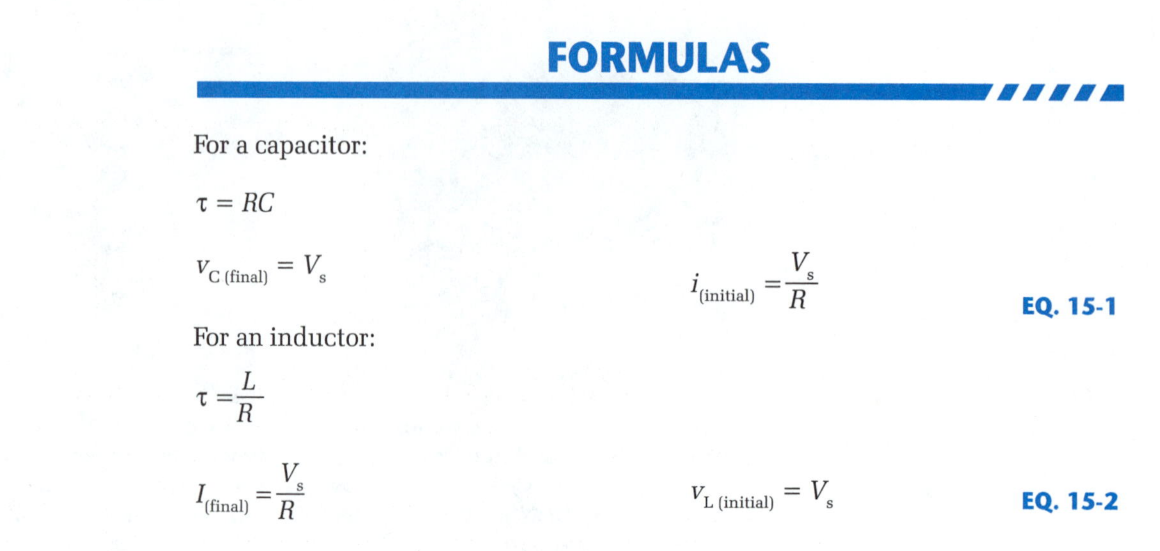

FORMULAS

For a capacitor:

$\tau = RC$

$v_{C\,(\text{final})} = V_s \qquad i_{(\text{initial})} = \dfrac{V_s}{R}$ **EQ. 15-1**

For an inductor:

$\tau = \dfrac{L}{R}$

$I_{(\text{final})} = \dfrac{V_s}{R} \qquad v_{L\,(\text{initial})} = V_s$ **EQ. 15-2**

SUMMARY OF IDEAS

- A transient is the reaction of a capacitor or inductor after a dc source is switched on or off. Both voltage and current are involved in a transient.
- Dc transients follow either the rising time-constant function or the falling time-constant function.
- Time constant is symbolized τ.
- In the first 1 τ, 63% of the total change occurs. In 2 τ, 86%. In 3 τ, 95%.
- Virtually all the total change occurs in 5 τ.
- For a series *RC* circuit, $\tau = RC$.
- A capacitor will not allow voltage to change instantaneously.
- For a series *RL* circuit, $\tau = \dfrac{L}{R}$.
- An inductor will not allow current to change instantaneously.

CHAPTER QUESTIONS AND PROBLEMS

1. T-F More voltage change occurs during the first time constant than occurs during the second time constant. [2]
2. For a rising time-constant curve, instantaneous voltage equals 63% of final voltage after__________. [4]
3. For a falling time-constant curve, instantaneous voltage equals 14% of initial voltage after__________. [4]
4. T-F In a series *RC* circuit, the fully charged value of capacitor voltage, $v_{C\,(\text{final})}$, is equal to the source voltage, V_s. [4]
5. T-F In a series *RC* charging circuit, with the capacitor initially discharged, the initial current at $t = 0$ is given by V_s / R. [4]

For Problems 6 and 7, suppose that the switch in Figure 15-15 is thrown into the down position after being up for a while. [3, 4]

6. a) Find the charging time constant.
   b) How long will it take to fully charge the capacitor?
   c) What is the value of v_C at $t = 1$ ms? At $t = 3$ ms?
7. a) What is the initial value of current i at $t = 0$?
   b) What is the value of i at $t = 1$ ms? At $t = 3$ ms?
   c) At what time will $i = 0$?

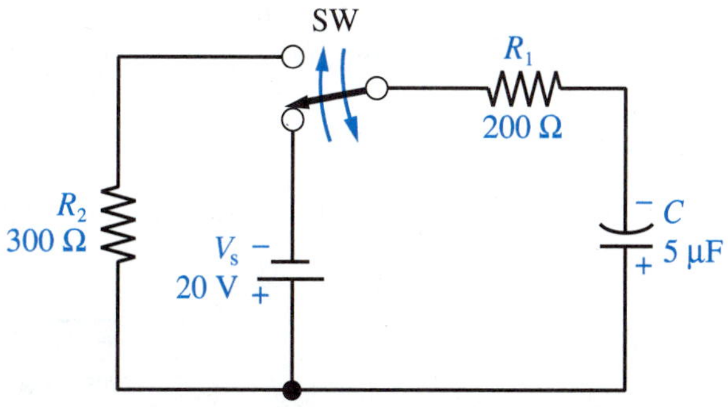

FIGURE 15-15 Capacitor charging and discharging circuit

For Problems 8 and 9, refer to Figure 15-15 again and suppose that the switch is thrown into the up position after being down for quite a while. [3, 4]

8. a) Find the value of the discharging time constant.
   b) What is the value of v_C at $t = 2.5$ ms? At $t = 7.5$ ms?
9. a) What is the value of current i at $t = 2.5$ ms? At $t = 7.5$ ms?
   b) At what time will $i = 0$?
10. T-F In a series *RL* fluxing circuit with the inductor initially deenergized, the initial induced voltage at $t = 0$ is equal to the source voltage, V_s. [4]
11. T-F In a series *RL* fluxing circuit with an ideal inductor, the final steady value of current, I_{final}, is equal to V_s / R. [4]
12. Immediately after a switch has been thrown, a capacitor's__________and an inductor's__________must maintain the same values that existed just before the switch was thrown. [1, 4]

For Problems 13 and 14, suppose that the switch closes in the circuit of Figure 15-16. Both inductors are ideal, with no internal resistance. [3, 4]

13. a) What is the circuit's total inductance, L_T?
    b) Find the circuit's time constant.
    c) What is the final value of current?
    d) Calculate current i at $t = 20$ ms. Repeat for 40 ms, 60 ms, and 100 ms.
    e) Plot a waveform graph of i versus t. Use $t = 0$ and the time instants from part d.
14. a) What is the initial voltage across the $L_1 - L_2$ combination (v_{LT})?
    b) Calculate v_{LT} at $t = 20$ ms. Repeat for 40 ms, 60 ms, and 100 ms.
    c) Plot a waveform graph of v_{LT} versus t. Use $t = 0$ and the time instants from part b.

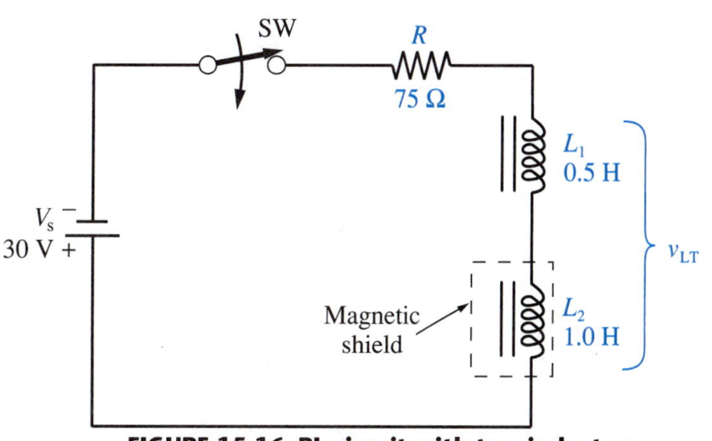

FIGURE 15-16 RL circuit with two inductors

15. When a switch opens on an energized inductor, the kickback voltage works to maintain the current in the__________direction as prior to opening. [1, 4]
16. Explain why inductor kickback voltage can be much greater than the source voltage. [1, 4]
17. If the switch is opened in Figure 15-17, which pole will experience more serious arcing across the contacts? Explain why. [1]

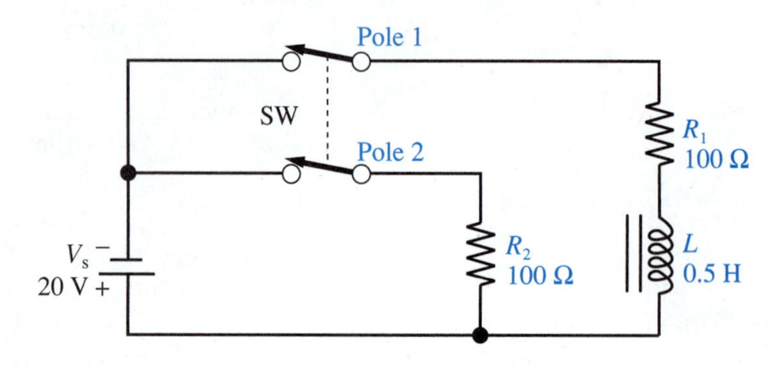

FIGURE 15-17

CHAPTER 16 ALTERNATING CURRENT AND VOLTAGE

Besides the use of radar for enforcing traffic laws, permanent radar systems can be used to encourage public cooperation in driving zones that deserve special carefulness.
Courtesy of CMI/MPH subsidiaries of MPD, Inc.

OUTLINE

NEW TERMS TO WATCH FOR

sine wave
alternating current (ac)
ac voltage
instantaneous value
peak voltage (or current)
peak-to-peak voltage
effective voltage
rms voltage
average ac power
cycle
period
frequency
hertz
ac alternator
electrical degrees
rotor
armature winding
in phase
out of phase
lead
lag
power factor

So far, we have studied only dc circuits. In homes, businesses, and factories, ac circuits are far more common than dc circuits.

After studying this chapter, you should be able to:

1. Describe the operation of a sine-wave ac circuit and tell how it is different from a dc circuit.
2. Identify the peak value of an ac current or voltage waveform. Also, identify the peak-to-peak value.
3. Relate the peak value to the effective (rms) value.
4. Relate average ac power to effective voltage and current.
5. Identify the period of an ac waveform.
6. Relate period to frequency.
7. Describe the basic structure and operation of an ac alternator.
8. Recognize when current and voltage are out of phase.

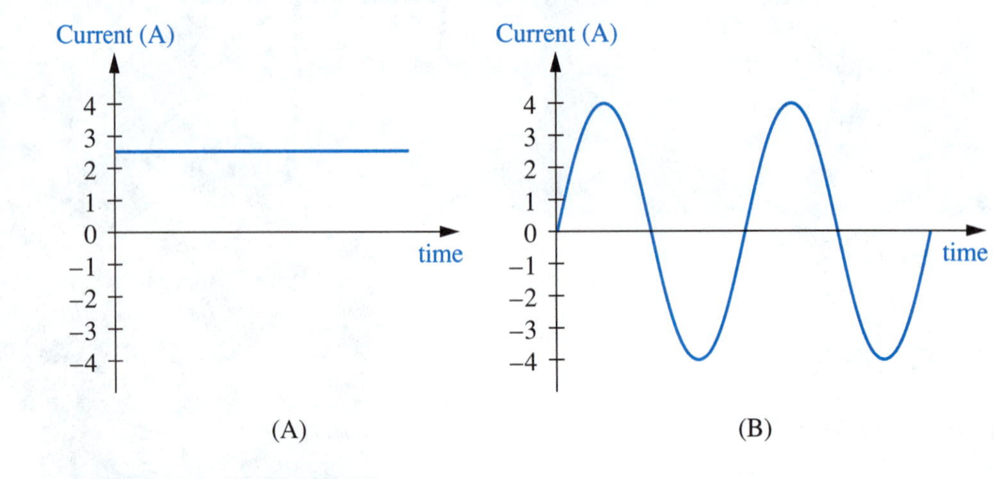

FIGURE 16-1 (A) Dc current waveform. Magnitude is constant and the direction never changes. (B) Ac current waveform. Magnitude continually changes, and the direction reverses. This is a sine wave ac, which means that the detailed shape of the waveform follows the mathematical sine relationship.

16-1 SINE WAVEFORMS

As we know from Chapters 14 and 15, a waveform is a graph of current or voltage versus time. For example, Figure 16-1(A) shows a waveform of direct current (dc). The amount, or magnitude, of the current is constant. In this case, it is 2.5 A. The waveform is always positive; it is never negative. This means that the current always flows in one particular direction, which is considered the positive direction. The current never reverses direction.

1

By contrast, look at the **alternating current (ac)** waveform in Figure 16-1(B). In that waveform, the magnitude of current is continuously changing. As time passes, the magnitude changes from 0 to 4 amperes, then back again to 0 A. Then the waveform becomes negative. This means that the current reverses direction in the circuit.

The same relationship holds between dc and ac voltage. This is shown in Figures 16-2(A) and (B).

The schematic diagrams of Figure 16-3 show the symbol for an ac voltage source. In Figure 16-3(A), the polarity of the source voltage, v_s, is positive on the top terminal, negative on the bottom terminal. This is considered to be the positive polarity in the voltage waveform graph of Figure 16-2(B). This voltage polarity causes the instantaneous current, i, to flow in the counterclockwise direction. Counterclockwise is considered to be the positive current direction in the ac current waveform of Figure 16-1(B).

The negative voltage polarity produces the negative current direction, as shown in Figure 16-3(B). This instantaneous circuit action is represented by the negative halves of the waveforms in Figures 16-1(B) and 16-2(B).

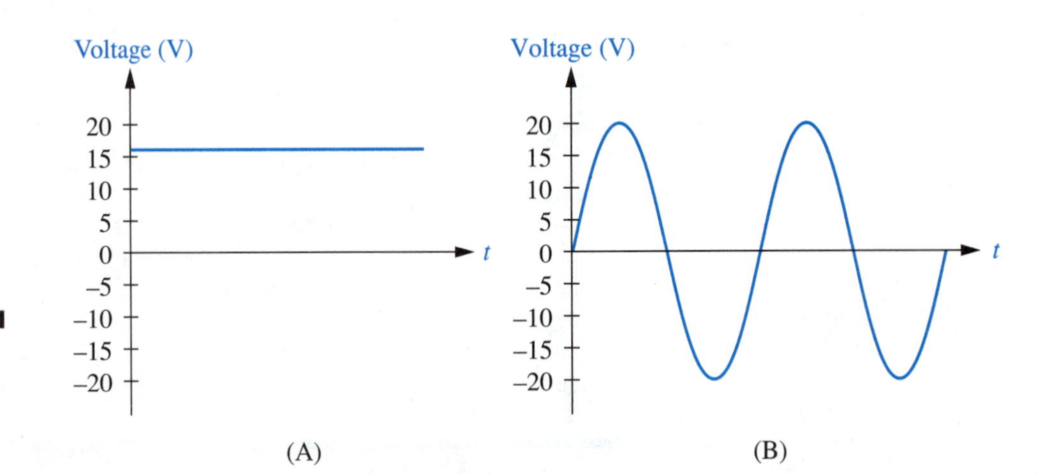

FIGURE 16-2 (A) Dc voltage waveform. Magnitude is constant and the polarity never changes. (B) Ac voltage waveform. Magnitude continually changes and the polarity switches. Again, the wave's shape follows the mathematical sine relationship.

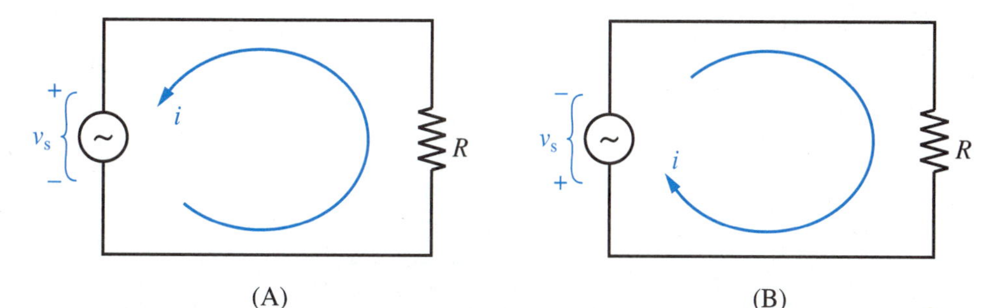

FIGURE 16-3 The instantaneous value of voltage is symbolized by the letter *v* (lower-case *v*, not capital *V*). Instantaneous value of ac current is symbolized *i*, rather than capital *I*. (A) Positive polarity of source voltage, v_s, produces the positive direction of current, *i*. (B) Negative polarity of v_s produces the opposite (negative) direction of current, *i*.

EXAMPLE 16-1

For the single-resistor circuit of Figure 16-3, suppose that the ac source voltage varies between +20 V and –20 V, as sketched in the waveform of Figure 16-2(B). If the resistance is 5 Ω, draw the waveform of current, *i*, to scale.

SOLUTION

Ohm's law applies to instantaneous ac voltage and current, just like dc. When $v_s = +20$ V,

$$i = \frac{v_s}{R} = \frac{+20\ \text{V}}{5\ \Omega} = +4\ \text{A}$$

When $v_s = -20$ V, $i = -4$ A. Therefore, the current waveform varies between +4 A and – 4 A. This is drawn to scale in Figure 16-1(B).

INFRARED SCANNING

There are electronic devices that detect infrared rays. These rays are electromagnetic waves like light rays, but they have lower frequencies than the light that is visible to human eyes. Infrared rays are emitted by all warm objects. The warmer an object is, the greater the intensity of its infrared radiation. The ability to detect warm bodies by their infrared radiation has obvious military applications. These photos show an important civilian application.

A low-flying airplane has an infrared scanner mounted beneath its nose or wing. The scanner detects and records on tape the infrared pattern below. By flying slowly over buildings or residential neighborhoods, the scanner makes an image on video tape of the exact locations on the buildings where heat energy is being lost. A typical image is shown in the bottom photo.

From this information, maintenance and repair crews know just where a building's roof is damaged or leaking. Homeowners can tell whether their ceiling insulation is adequate, and whether their homes have any moisture damage from roof leaks.

Courtesy of NASA

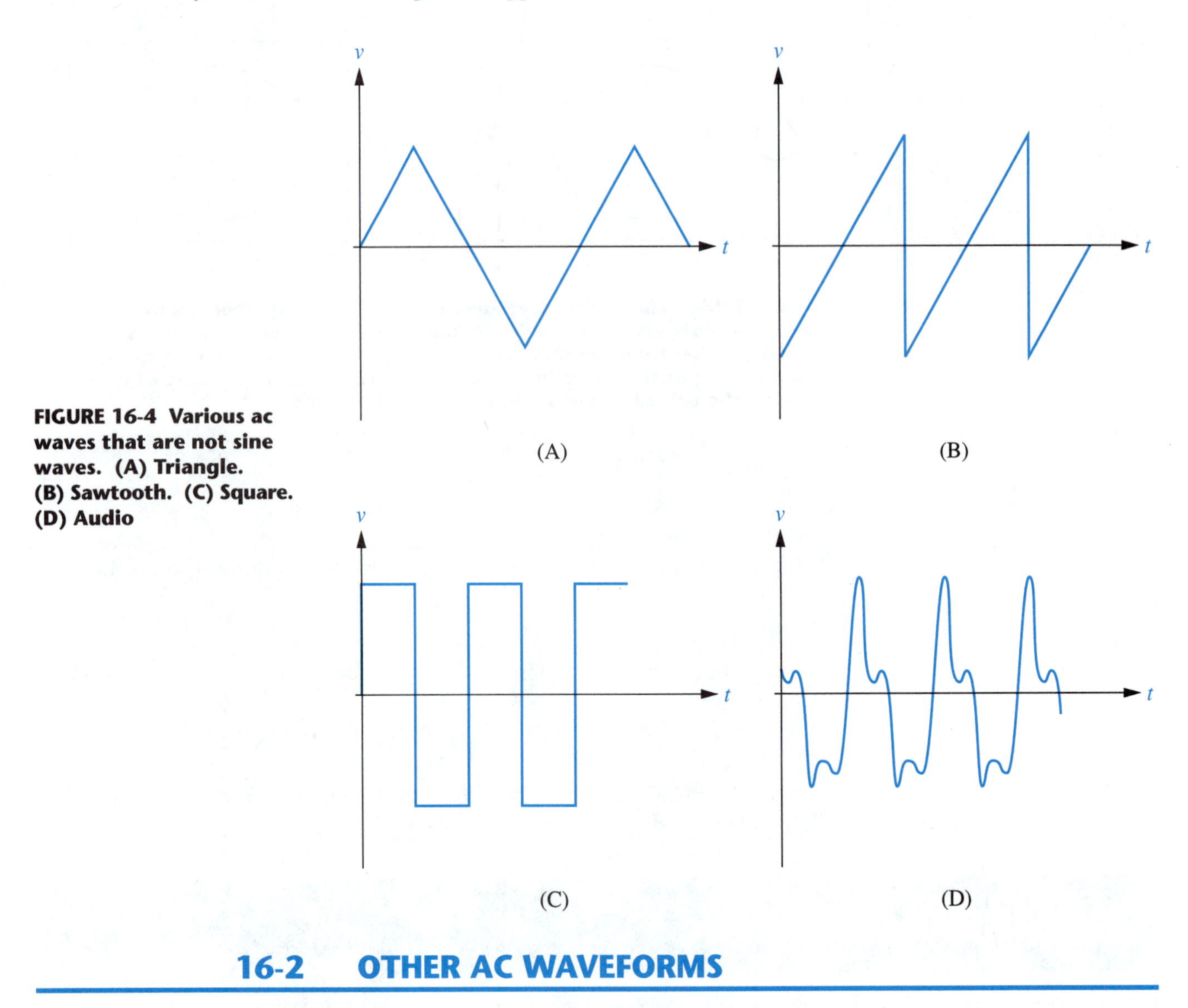

FIGURE 16-4 Various ac waves that are not sine waves. (A) Triangle. (B) Sawtooth. (C) Square. (D) Audio

16-2 OTHER AC WAVEFORMS

The most common type of ac waveform is the sine wave discussed in the previous section. However, there are many other ac waveforms that you will encounter. Several of them are shown in Figure 16-4.

Figure 16-4(A) is called a triangle waveform. It is seen in analog computing circuitry. Figure 16-4(B) is called a sawtooth wave. This waveform is applied to the deflection coils of a television picture tube to sweep the electron beam across the screen from left to right. A square wave is shown in Figure 16-4(C). It is often used to test the high-speed response of electronic devices. Figure 16-4(D) shows an audio waveform produced by a microphone. This particular audio waveform results from the sound "aah."

When we speak of ac, it is assumed that we mean sine wave ac, unless we specifically state otherwise. Thus, you wouldn't describe any of the waves in Figure 16-4 as simply ac. You would say triangle ac, sawtooth ac, and so on.

16-3 MAGNITUDE OF AC

Ac sine waves differ from each other in two ways:

1. Vertically, in their magnitude.
2. Horizontally, in their time rate of oscillation.

We will look at ac magnitude in this section.

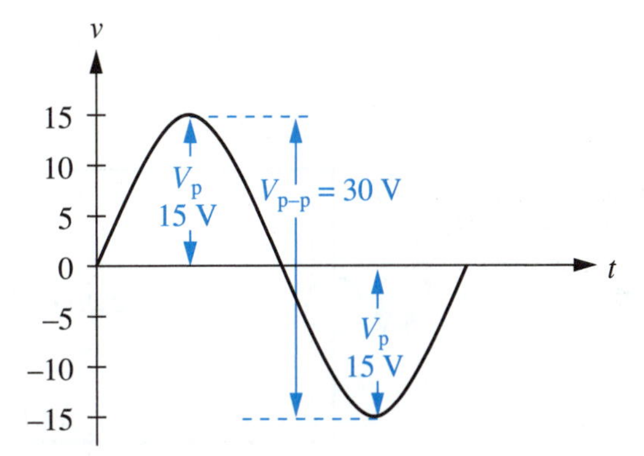

FIGURE 16-5 Illustrating peak voltage, V_p, and peak-to-peak voltage, $V_{p\text{-}p}$.

Peak and Peak-to-Peak Voltage

TECHNICAL FACT

The maximum instantaneous voltage that a sine wave reaches is called the **peak voltage**, symbolized V_p.

For the voltage sine wave in Figure 16-5, the magnitude can be expressed as

$$V_p = 15 \text{ V}$$

since the wave reaches a maximum instantaneous voltage of 15 V.

The maximum instantaneous voltage in the negative polarity is the same as for the positive polarity. Figure 16-5 shows this clearly.

TECHNICAL FACT

The overall difference between the positive peak and the negative peak is called the **peak-to-peak voltage**, symbolized $V_{p\text{-}p}$. It is twice as large as peak voltage. As a formula,

$$V_{p\text{-}p} = 2\,V_p \qquad \textbf{EQ. 16-1}$$

For the ac voltage of Figure 16-5,

$$V_{p\text{-}p} = 2\,(15 \text{ V}) = 30 \text{ V}$$

EXAMPLE 16-2

In the United States, the ac voltage from an electrical wall outlet receptacle has a peak value of about 165 V. What is its peak-to-peak value?

SOLUTION

From Equation 16-1,

$$V_{p\text{-}p} = 2\,V_p$$
$$= 2\,(165 \text{ V}) = \mathbf{330\ V}$$

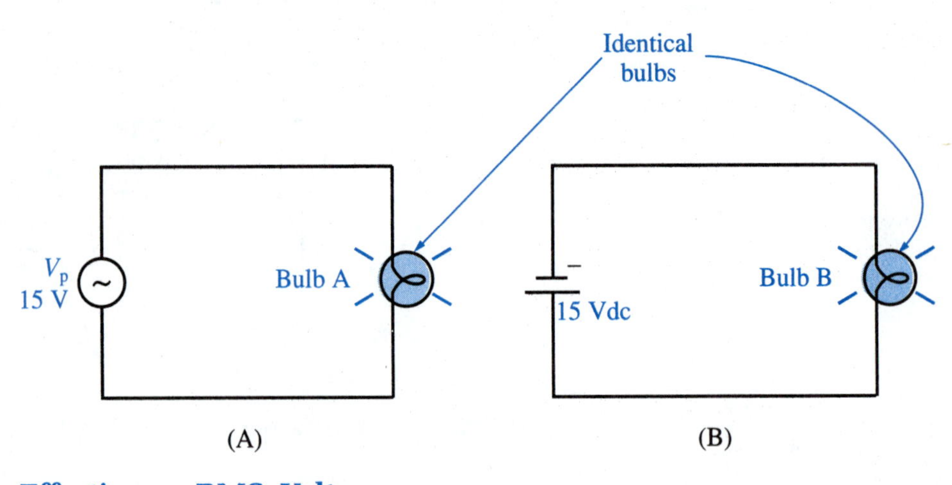

FIGURE 16-6 The ac voltage source in part A delivers less power to the bulb than the dc source in part B.

Effective, or RMS, Voltage

The magnitude of an ac sine wave can be specified in a third way. This third way is the **effective** value, also called the **rms** value. To understand the meaning of effective value, look at the two circuits in Figure 16-6.

In Figure 16-6(A), the ac voltage source has a peak value of 15 V. This ac voltage source will not make its bulb glow as brightly as the 15-V dc source in part B. The reason is that the ac voltage is less than 15 V for almost all of the time. It reaches 15 V only at the two peak moments. By contrast, the dc source is at 15 V all the time. So 15 V peak ac is not as effective as 15 V dc.

An important question is this: "How effective is 15 V peak ac? What dc value is it equivalent to?"

TECHNICAL FACT

The effective value of an ac sine wave is equal to the peak value multiplied by 0.7071. As a formula,

$$V_{eff} = 0.7071\ V_p$$

EQ. 16-2

For a 15 V peak, the effective value is

$$V_{eff} = 0.7071\ (15\ \text{V}) = 10.6\ \text{V}$$

In other words, an ac source with V_p = 15 V is equivalent to a 10.6-V dc source in its ability to deliver power to a load device. This idea is illustrated in Figure 16-7.

Equation 16-2 can be rearranged as

$$V_p = \frac{V_{eff}}{0.7071} = 1.414\ V_{eff}$$

EQ. 16-3

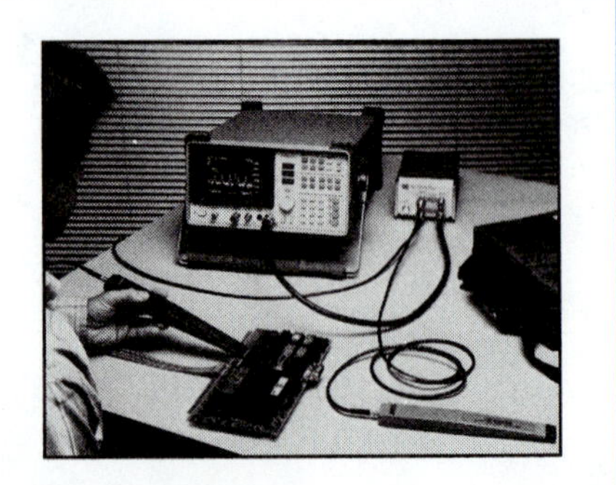

Complex ac waveforms are measured with specialized wave-analyzing instruments.
Courtesy of Hewlett-Packard Company

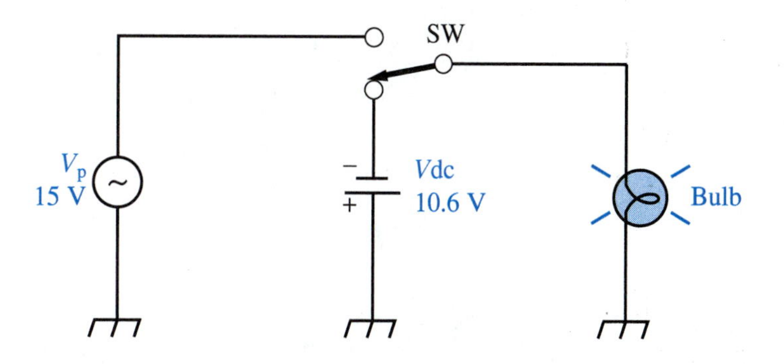

FIGURE 16-7 The bulb will glow with the same brilliance whether the switch is up or down. This is because the ac source with V_p = 15 V has the same power-delivering ability as the 10.6-V dc source.

When we speak of the value of a sine-wave voltage or current, it is assumed that we are speaking about its effective value, unless we explicitly say otherwise. For instance, we can verbally describe the ac source of Figures 16-6 and 16-7 simply as "10.6 volts," because everyone will understand that to mean 10.6 V effective value. We can also describe it as "15 volts peak." But it would be incorrect to just say "15 volts."

There is another important point regarding effective, or rms, values.

TECHNICAL FACT

Ac voltmeters and ammeters automatically read effective values.

There are rare exceptions to this rule. If so, the meter will be clearly marked as peak-indicating or peak-to-peak-indicating.

EXAMPLE 16-3

The circuit of Figure 16-8 has a 12-V ac source driving a load resistance of 48 Ω.

a) What exact numerical value will be indicated by the pointer of the VOM voltmeter?
b) What is the peak-to-peak value of the sine wave voltage?
c) What numerical value will be indicated by the digital readout of the DMM ammeter?
d) What amount of power will be dissipated by the load resistor?

SOLUTION

a) The value of the source voltage is given as simply 12 V, so we must assume this to mean 12 V rms. The VOM is an rms-indicating meter, so its pointer will show a numeric value of **12 V**.

Continued on page 288.

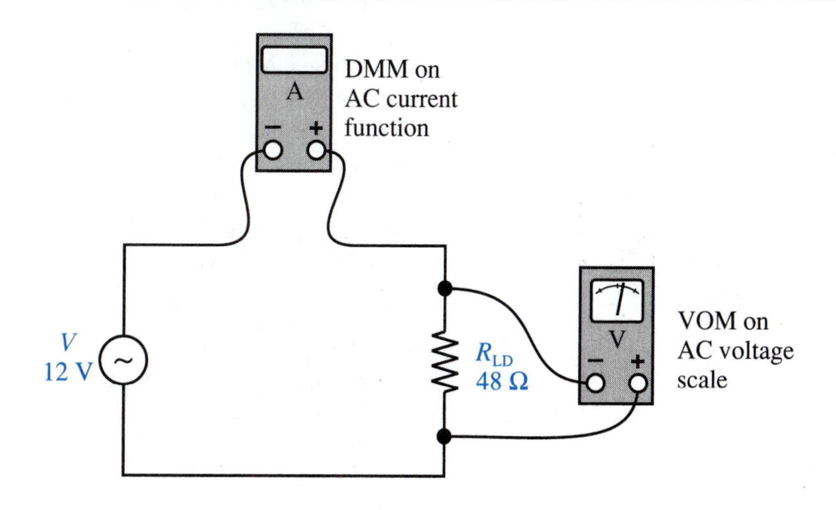

FIGURE 16-8 Single-resistor ac circuit with DMM ammeter and VOM voltmeter.

b) The peak-to-peak value can be related to the rms value by combining Equations 16-1 and 16-3.

$$V_{p\text{-}p} = 2\,V_p = 2(1.414\,V_{eff})$$

$$\boxed{V_{p\text{-}p} = 2.828\,V_{eff}}$$

EQ. 16-4

Applying Equation 16-4 to this situation gives

$$V_{p\text{-}p} = 2.828\,(12\text{ V}) = \mathbf{33.9\text{ V}}$$

c) Ohm's law applies to rms ac voltages and currents in the usual way.

$$I_{rms} = \frac{V_{rms}}{R} = \frac{12\text{ V}}{48\ \Omega} = \mathbf{0.25\text{ A}}$$

The DMM will display this numerical value, since it is a standard rms-indicating meter.

d) The usual power formulas apply to ac circuits, as long as the voltages and currents are expressed in rms units. The formulas cannot be applied with peak or peak-to-peak units.

$$P = V_{rms}\,I_{rms}$$
$$= 12\text{ V}\,(0.25\text{ A}) = \mathbf{3\text{ W}}$$

Alternatively,

$$P = \frac{(V_{rms})^2}{R_{LD}} = \frac{(12\text{ V})^2}{48\ \Omega} = \frac{144}{48} = \mathbf{3\text{ W}}$$

TECHNICAL FACT

Power in an ac circuit is not instantaneously constant like power in a dc circuit. Therefore, the calculated power value represents the **average power.**

✔ SELF-CHECK FOR SECTIONS 16-1, 16-2, AND 16-3

1. Explain the basic difference between dc and ac. [1]
2. T-F All ac waves are sine waves. [1]
3. Explain your answer to Question 2. [1]
4. For an ac waveform, the vertical axis is marked in units of________or________. [1]
5. For an ac waveform, the horizontal axis represents__________. [1]
6. In the United States, the standard residential appliance voltage is 120 V. Find the value of [2,3]
   a) V_p b) $V_{p\text{-}p}$ c) V_{rms}
7. Most flashlight bulbs are driven by 3 V dc. To get the same amount of light, what peak-to-peak value of ac voltage would be required? [3]
8. When plugged into a standard wall receptacle, a certain soldering gun carries an ac current of 4 A peak, like the waveform in Figure 16-1(B). Find its average power. [4]
9. For the soldering gun in Question 8, calculate the resistance of its heating element. [4]
10. T-F In ac terminology, the terms *effective* and *rms* mean the same thing. [3]

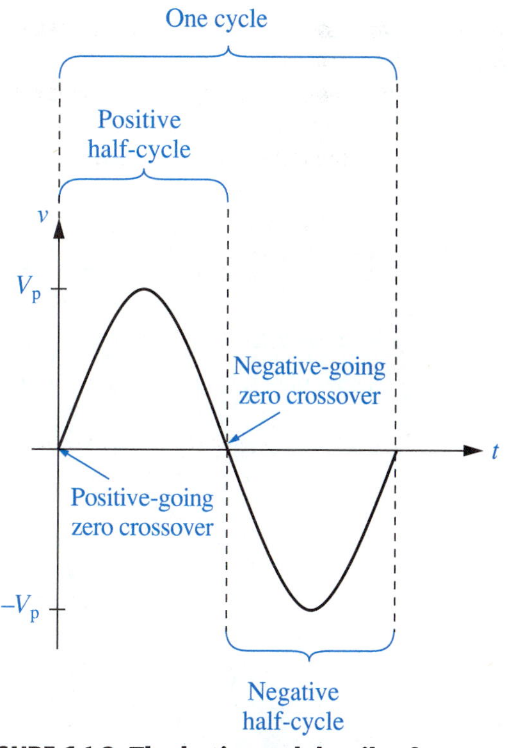

FIGURE 16-9 The horizontal details of an ac cycle

16-4 AC TIME SCALE

The horizontal time axis of an ac sine wave can be divided into sections, as shown in Figure 16-9. The voltage begins at zero, rises to its positive peak, then declines back to zero. This is called the positive half cycle. As the positive half cycle ends, the voltage passes through zero going negative. This time instant is called the negative-going zero crossover. During the negative half cycle, the voltage declines to its negative peak, then rises back to zero. At that instant, it makes a positive-going zero crossover, and starts to repeat its action.

This entire sequence of actions is called one full **cycle.** The **period** of an ac waveform is the amount of time required to complete one cycle. Period is symbolized capital T (not lower-case t, which stands for instantaneous time). The sine wave in Figure 16-10(A) has a period of 0.4 second ($T = 0.4$ s). The wave in Figure 16-10(B) has $T = 8$ ms.

The **frequency** of an ac wave is the number of cycles that it makes during an elapsed time of 1 second. Frequency is symbolized f.

A frequency meter that actually counts each oscillation is called a frequency counter. This counter can measure frequencies over the range from 10 Hz to 200 MHz with eight-digit precision (1 cycle in 100 million).
Courtesy of Simpson Electric Co.

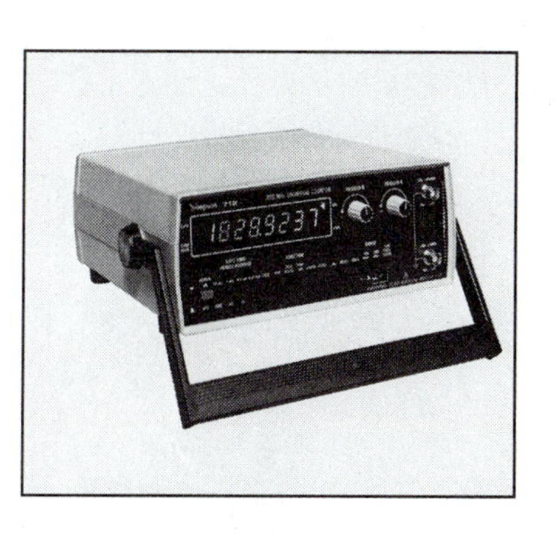

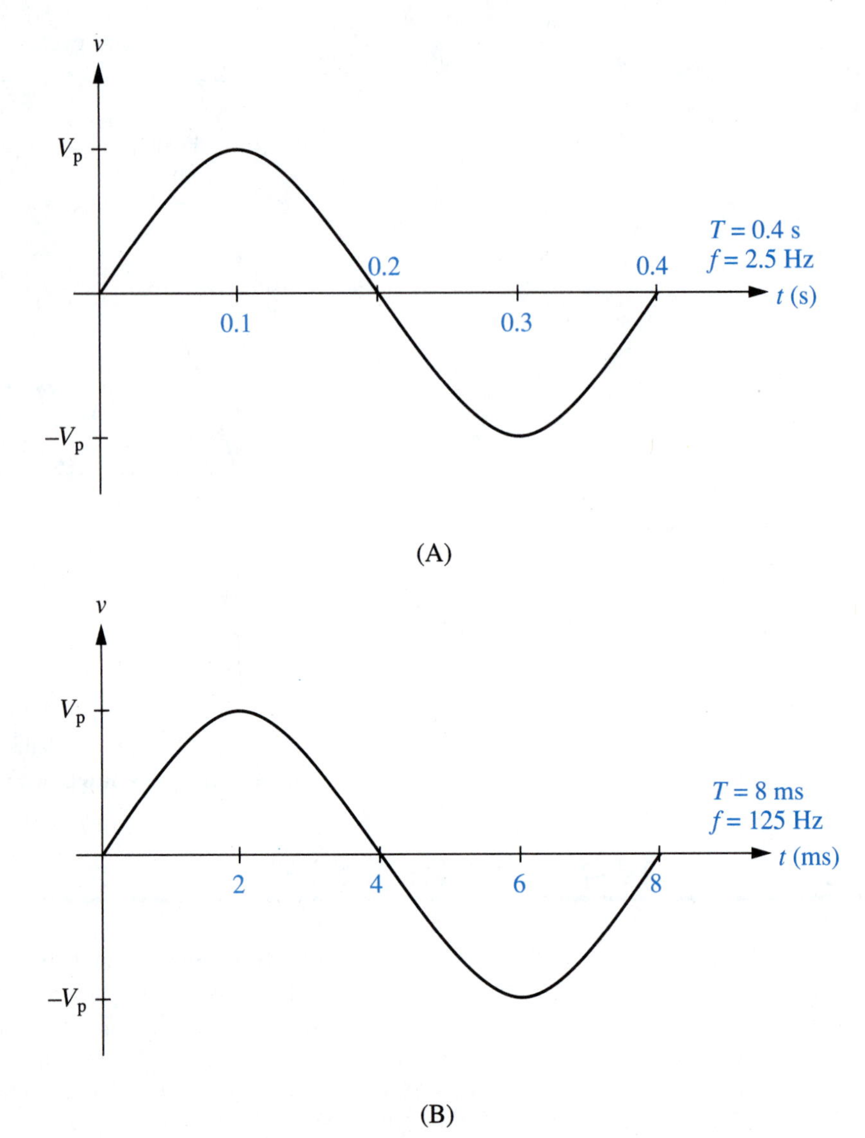

FIGURE 16-10 Specifying periods and frequencies for ac waves

TECHNICAL FACT

Frequency and period are reciprocals of each other. That is,

$$f = \frac{1}{T} \quad \text{EQ. 16-5}$$

and

$$T = \frac{1}{f} \quad \text{EQ. 16-6}$$

The units of frequency are **hertz,** symbolized Hz. One hertz is equivalent to one cycle per second. For example, in Figure 16-10(A), with $T = 0.4$ s, frequency is given by

$$f = \frac{1}{T} = \frac{1}{0.4 \text{ s}} = 2.5 \text{ Hz}$$

In Figure 16-10(B),

$$f = \frac{1}{8 \times 10^{-3} \text{ s}} = 125 \text{ Hz}$$

EXAMPLE 16-4

The ac power distribution network in the United States has a frequency of 60 Hz. Find the period of the waveform at an ac wall receptacle.

SOLUTION

Any ac wall receptacle will deliver a waveform with f = 60 Hz. From Equation 16-6,

$$T = \frac{1}{f}$$

$$= \frac{1}{60} = 0.016\ 67 \text{ s or } \mathbf{16.67\ ms}$$

16-5 SINE WAVES GRAPHED VERSUS ANGLE IN DEGREES

Sine wave voltage can be produced by a machine called an **ac alternator**. As the shaft spins in a basic alternator, it produces one cycle of sine-wave voltage for each mechanical revolution. Since a complete revolution has 360 mechanical degrees, we often think of a full cycle as having 360 **electrical degrees**. This idea is represented in Figure 16-11.

As Figure 16-11 shows, the sine wave starts at 0°, reaches its positive peak at 90°, and makes a negative-going zero crossover at 180°. The negative peak occurs at 270° and the cycle is completed at 360°.

✔ SELF-CHECK FOR SECTIONS 16-4 AND 16-5

11. For an ac waveform, what is the meaning of the term cycle? [5]
12. Define the following ac terms: [5]
    a) Positive half-cycle.
    b) Negative half-cycle.
    c) Positive-going zero crossover.
    d) Negative-going zero crossover.
13. The time of a complete cycle is called the__________of the wave. It is symbolized by the letter_______. [5]

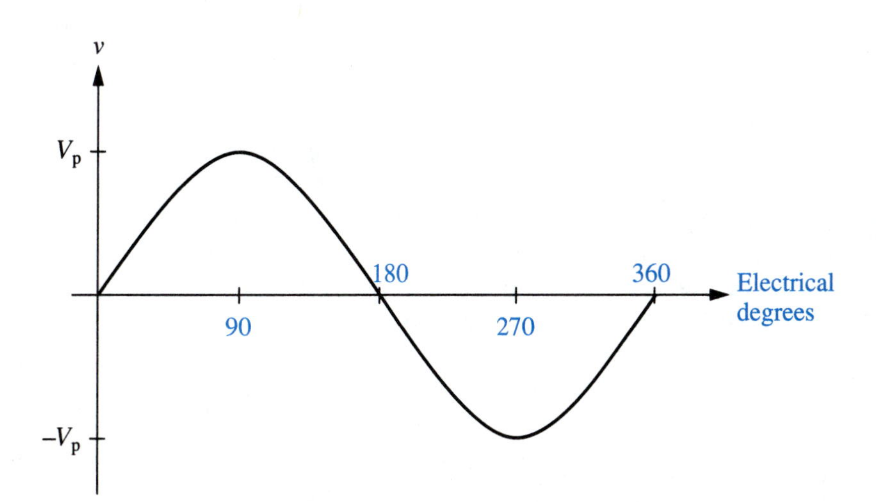

FIGURE 16-11 Graphing ac voltage with degrees on the horizontal axis

14. The basic unit of frequency is the_________. It is symbolized_____. [6]
15. One cycle per second is equivalent to 1_________. [6]
16. What is the period of a 400-Hz waveform? Repeat for a 20-kHz waveform. [6]
17. If the period of an ac wave is 5 ms, what is its frequency? Repeat for $T = 2.5\ \mu s$. [6]
18. A sine wave reaches its positive peak at_________degrees. It reaches its negative peak at_________ degrees. [1, 5]
19. T-F For a basic ac alternator, four voltage cycles are produced by each revolution of the shaft. [1, 7]

16-6 AC ALTERNATORS

7

The machine that generates a sine-wave voltage is called an ac alternator, as stated in Section 16-5. A cross-sectional diagram of an ac alternator is shown in Figure 16-12. The magnetic flux is stationary in space. It is produced by dc current through electromagnets that are wrapped around the north and south poles. These electromagnets and the dc supply that drives them are not shown in Figure 16-12.

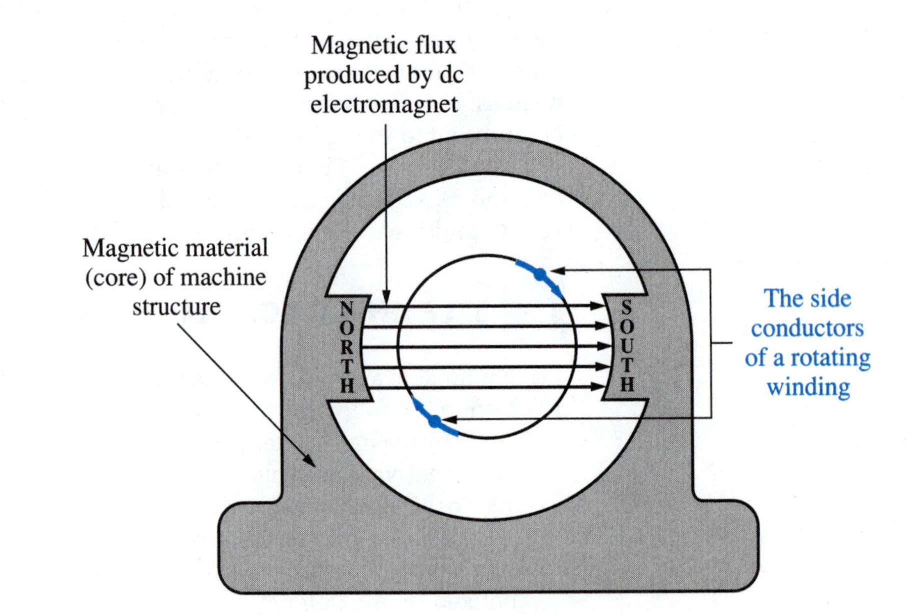

FIGURE 16-12 Simplified cross-sectional view of an ac alternator. This is the stationary-field type of construction. There is also a rotating-field type.

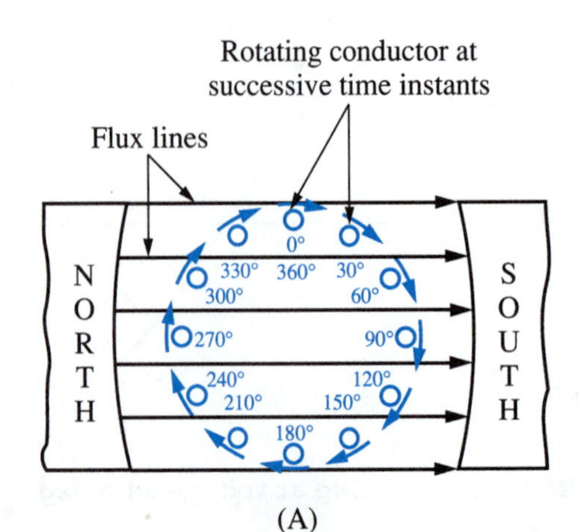

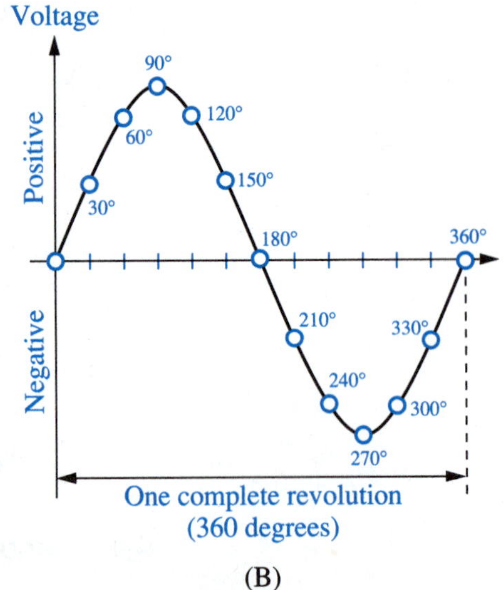

FIGURE 16-13 (A) One side of the armature winding, moving from one position to the next through the magnetic field. (B) The instantaneous voltage at each position.

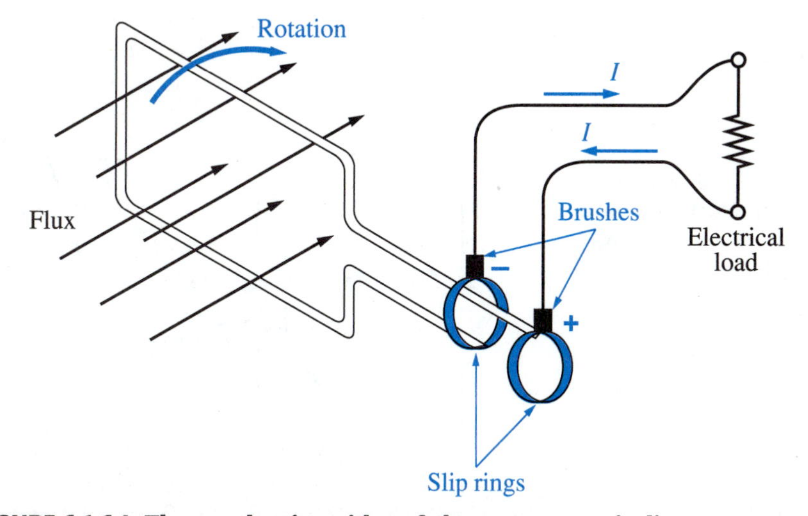

FIGURE 16-14 The conducting sides of the armature winding are connected to copper slip rings, which are mounted on the rotor structure. Carbon brushes slide on the surface of the slip rings as the rotor spins. Wires embedded in the carbon brushes carry current to and from the load. At the time instant shown, the left slip ring and brush are negative (–) and the right slip ring and brush are positive (+), as marked. When the rotor turns one-half revolution, the polarity will be reversed.

The rotating winding is supported on a cylinder called the **rotor**. The rotating winding itself is called the **armature winding**. As the spinning rotor causes the armature winding to move through the stationary flux lines, voltage is created in the armature winding. Half the time the voltage has one polarity, and the other half of the time it has the opposite polarity. For instance, at the time instant shown in Figure 16-12, the side conductor at the upper right is passing through the flux lines near the south pole. We say that the conductor is cutting south flux. At this instant, the induced voltage in the armature winding is negative on the lower-left conductor terminal, and positive on the upper-right conductor terminal. Later in time, when the rotor has spun by 180 mechanical degrees, the positions of the side conductors will be reversed, and the polarity of induced voltage will also reverse.

Figure 16-13(A) shows one conducting side of the armature winding at successive time instants as the rotor makes a complete revolution. The instantaneous voltage of this conductor terminal is plotted in Figure 16-13(B). The arrangement for carrying current from the armature winding to the load is illustrated in Figure 16-14.

Some ac alternators have more than two magnetic poles. Four, six, or more poles are possible. For such multipole machines, each revolution of the rotor produces more than one cycle of the ac sine wave. For instance, a four-pole machine produces two complete ac cycles per revolution; a six-pole machine produces three cycles per revolution, and so on.

16-7 PHASE SHIFTS

In an ac circuit with a purely resistive load, the current and voltage sine waves are perfectly synchronized. This situation is shown in Figure 16-15.

In an ac circuit with a partly capacitive load, the current is ahead of the voltage by some number of degrees between 0 and 90.

The resistive-capacitive situation is illustrated in Figure 16-16.

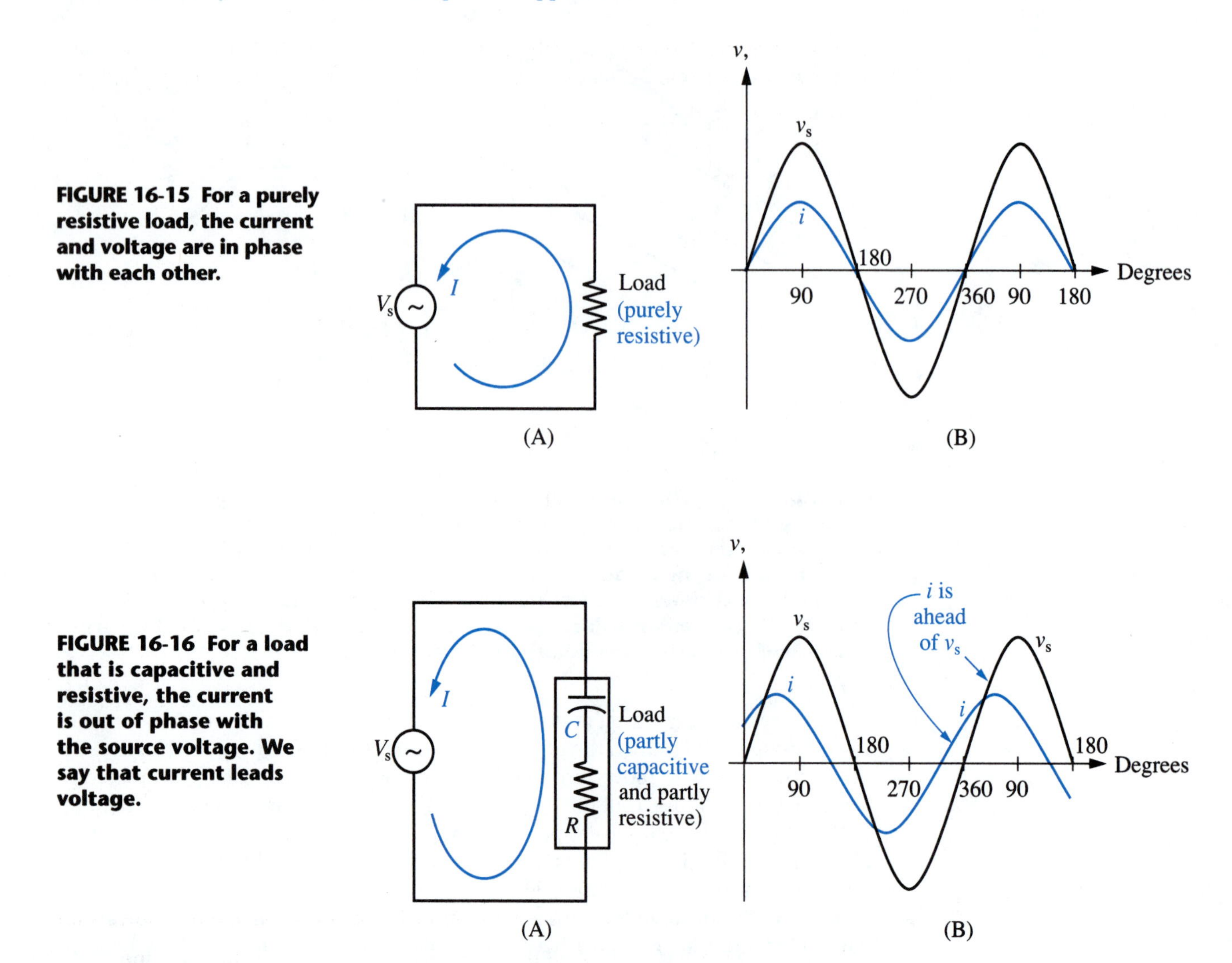

FIGURE 16-15 For a purely resistive load, the current and voltage are in phase with each other.

FIGURE 16-16 For a load that is capacitive and resistive, the current is out of phase with the source voltage. We say that current leads voltage.

TECHNICAL FACT

In an ac circuit with a partly inductive load, the current is behind the voltage by some number of degrees between 0 and 90.

The resistive-inductive situation is illustrated in Figure 16-17.

Whenever the current and voltage are out of phase, the average power is *not* given by Equation 7-5. In other words, we cannot say

$$P = V_{rms} I_{rms}$$

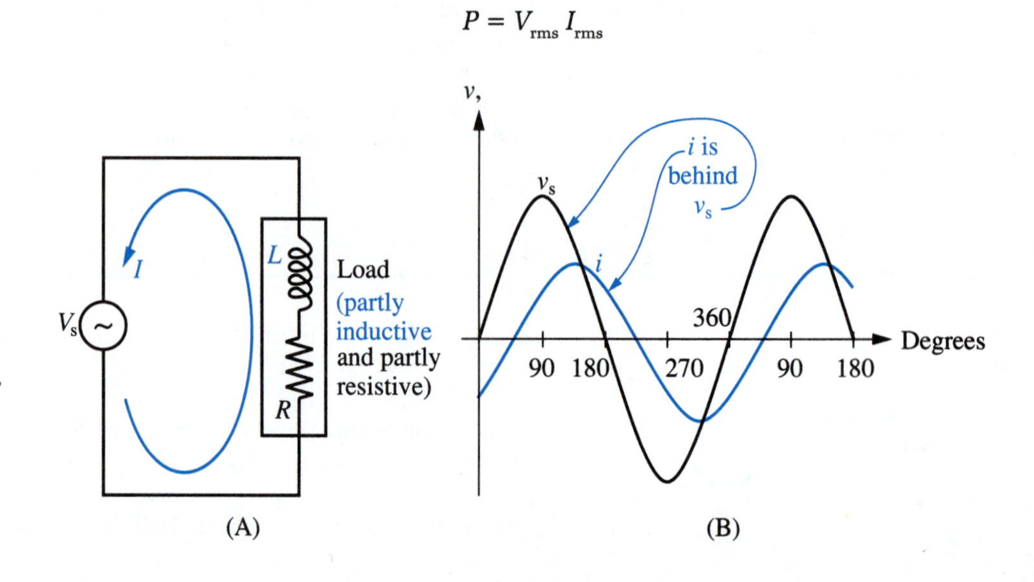

FIGURE 16-17 For a load that is inductive and resistive, the current is out of phase with the source voltage in the other direction. We say that current lags voltage.

as we were able to do for the purely resistive load in Figure 16-8. Instead, the average power must be calculated by taking into account the **power factor**, symbolized *PF*. As an equation,

$$P = V_{rms} I_{rms} (PF)$$

EQ. 16-7

Power factor depends on how far out of phase the current is from the voltage. The further out of phase they are, the smaller the power factor. The closer to being in phase they are, the larger the power factor. When they are exactly in phase (no phase shift at all), $PF = 1.0$. We will study power factor thoroughly in Chapter 19.

✔ SELF-CHECK FOR SECTIONS 16-6 AND 16-7

20. The machine that generates an ac sine-wave voltage is called an__________. [7]
21. The rotating assembly is called the__________. [7]
22. The winding that has the ac voltage generated in it is called the__________winding. [7]
23. In a two-pole alternator, the rotor must spin_____degrees to generate one complete cycle of the sine wave. [5, 7]
24. The end-terminals of the armature winding are attached to two _____ _____ which are contacted by sliding brushes. [7]
25. For a purely resistive load, ac current is_____ __________with the source voltage. [8]
26. For a partly capacitive load, ac current_______the source voltage by some amount. [8]
27. For a partly inductive load, ac current_______the source voltage by some amount. [8]

SAFETY ADVICE

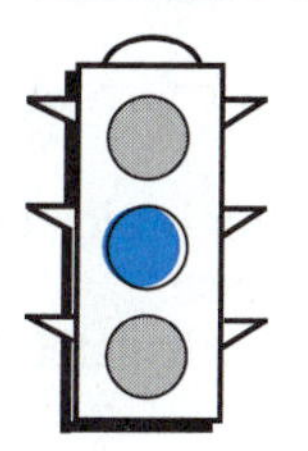

At some time, you will probably need to use an alligator-clip line cord to get access to the 120-V ac line. This would occur when you need ac power for a 120-V-rated transformer in a laboratory experiment. Here are some pieces of advice regarding the use of ac line cords with alligator clips.

1. **Always make sure that the red and black plastic boots are in place, covering the alligator clips almost to their tips.**
2. **Connect the alligator clips to their circuit points *first*, before inserting the line cord plug into an ac receptacle.**
3. **Separate and anchor the clips in a fixed position, before plugging the plug into the ac receptacle. Make certain that the exposed tips of the alligator clips can't touch each other or anything else. One good method is to place both clips between the pages of a closed book (separated by a few hundred pages of paper).**

FORMULAS

$V_{p\text{-}p} = 2\,V_p$ **EQ. 16-1**

$V_{eff} = 0.7071\,V_p$ **EQ. 16-2**

$$V_p = 1.414\ V_{eff}$$ **EQ. 16-3**

$$V_{p\text{-}p} = 2.828\ V_{eff}$$ **EQ. 16-4**

$$f = \frac{1}{T}$$ **EQ. 16-5**

$$T = \frac{1}{f}$$ **EQ. 16-6**

$$P = V_{rms} I_{rms} (PF)$$ **EQ. 16-7**

SUMMARY OF IDEAS

- Alternating current (ac) changes magnitude and direction. Ac voltage changes magnitude and polarity.
- The sine wave is the most common type of ac waveform.
- The greatest instantaneous value of an ac voltage wave is called the peak voltage, symbolized V_p. Likewise for current, I_p.
- The peak-to-peak voltage, $V_{p\text{-}p}$, is twice the peak voltage.
- The effective dc equivalent of an ac sine wave is given by $V_{eff} = 0.7071\ V_p$.
- The subscript rms is used to mean effective value.
- Direct-reading ac meters generally read in rms units.
- Ac power calculations must be made with rms values of voltage and current.
- Ac sine waves can be graphed versus time or versus angle on the horizontal axis. A 360° angle corresponds to the time of a complete cycle.
- The amount of time required for a complete cycle of oscillation is called the period, symbolized T.
- The number of oscillation cycles made in 1 second is called the frequency, symbolized f.
- Frequency is measured in hertz, symbolized Hz.
- Period and frequency are reciprocals of each other: $f = 1/T$; $T = 1/f$.
- An ac alternator is a machine that produces a sine-wave voltage by rotating its armature winding through a magnetic field.
- When an ac source drives a purely resistive load, the current and voltage are time-synchronized, or in phase.
- When an ac source drives a partially capacitive or inductive load, the current and voltage waveforms are unsynchronized, or out of phase.
- When I and V are out of phase, average power must be calculated using the power factor of the load.

CHAPTER QUESTIONS AND PROBLEMS

1. T-F The positive half-cycle of a sine wave is a mirror image of the negative half-cycle. [1]
2. T-F In an ac sine wave, the negative peak is equal to the positive peak. [1, 2]
3. What is the peak value of a sine wave with $V_{rms} = 24$ V? [3]
4. What is the peak-to-peak value of the sine wave in Problem 3? [2]
5. A certain sine wave has $V_{p\text{-}p} = 90$ V. Find its effective value. [2, 3]
6. If the ac voltage in Problem 5 were measured with a VOM, what value would the VOM indicate? [3]
7. A 50-Ω resistor is driven by an ac source with an rms voltage of 25 V. Calculate the rms current.
8. Find the peak value of current for the circuit of Problem 7. [3]
9. In the circuit of Problem 7, how much power does the resistor consume? [4]

10. Suppose that we can vary the voltage applied to the 50-Ω resistor. What value of voltage would deliver 8 W of power to the resistor? [4]
11. What would be the current for the 8-W power delivery in Problem 10? [4]
12. Define one cycle of a sine wave. [5]
13. Draw one complete cycle of a sine wave and identify the following on it: [5]
    a) Positive half-cycle
    b) Negative half-cycle
    c) Positive-going zero crossover
    d) Negative-going zero crossover
14. A certain sine wave goes from zero to its positive peak and back to zero again in 0.02 second. Find its period and frequency. [5, 6]
15. The unit hertz is a replacement for the phrase______ per______. [6]
16. What is the frequency of the ac power line in the United States? [6]
17. In the television broadcasting industry, the Channel 2 picture signal is broadcast at a frequency of 55.25 MHz. What is its period? [6]
18. In a rectified dc power supply, there are usually voltage fluctuations occurring at a frequency of 120 Hz. What is the period of these fluctuations? [6]
19. A certain AM radio station broadcasts a signal that has a period $T = 0.8$ μs. What is its frequency? [6]

Questions 20 and 21 refer to the alternator cross-section shown in Figure 16-12.

20. At the instant when the voltage is zero, where are the positions of the side conductors? [2, 7]
21. At the instant when the voltage equals V_p, where are the positions of the side conductors? [2, 7]
22. When two ac waveforms are time synchronized with each other, we say that they are_____ __________. [8]
23. When two ac waveforms are not time synchronized with each other, we say that they are_______ _____ __________. [8]
24. What does it mean to say that current leads voltage? [8]
25. What does it mean to say that current lags voltage? [8]
26. When current is out of phase with voltage, ac average power is calculated as voltage multiplied by current multiplied by__________ __________. [4, 8]

CHAPTER 17

TRANSFORMERS

The electronic instrument being used here is a gas analyzer, capable of identifying the chemical components of a gas and their degree of concentration. In this photo, a U.S. Coast Guard emergency team is using their unit to check out a barrel of unknown contents that has washed up on the shoreline. *Courtesy of NASA*

OUTLINE

17-1 Structure of Transformers
17-2 Transformer Voltage Law
17-3 Transformer Current Law
17-4 Transformer Power Law
17-5 Resistance-Matching Transformers
17-6 Other Transformer Topics
17-7 Troubleshooting Transformers
17-8 Nonideal Transformers

NEW TERMS TO WATCH FOR

step-up
step-down
primary winding
secondary winding
voltage law
turns ratio
current law
ideal transformer
power law
matching transformer
isolation
floating
tapped winding
nonideal (real) transformer
eddy-current
lamination

A transformer changes the value of ac voltage. A **step-up** transformer takes a small input voltage and converts it into a larger output voltage. A **step-down** transformer takes a large input voltage and converts it into a smaller output voltage.

In this chapter, we will study the structure and operating principles of transformers and the mathematical laws that govern their use.

After studying this chapter, you should be able to:

1. Describe the function of transformers, using the following transformer terms: primary winding, secondary winding, turns ratio, step-up, and step-down.
2. Explain how the magnetization of the transformer core creates voltage in the secondary winding.
3. Apply the transformer voltage law, the transformer current law, and the transformer power law.
4. Use a transformer to get the maximum possible power to a load resistance, R_{LD}, if the source's internal resistance, R_{int}, is known.
5. Explain how a transformer provides electrical isolation.
6. Explain the use of tapped secondary windings and multiple secondary windings.
7. Troubleshoot a transformer circuit to identify which winding failed and how it failed.
8. State the ways in which real transformers differ from the ideal model.

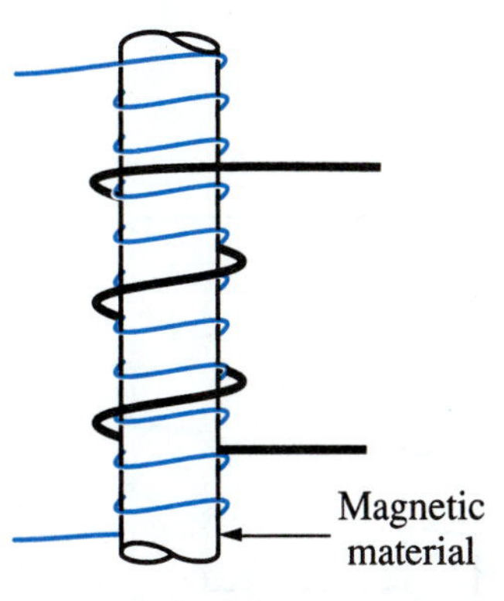

FIGURE 17-1 A transformer has two windings sharing a common magnetic core.

17-1 STRUCTURE OF TRANSFORMERS

A transformer can be thought of as two inductors wrapped around the same core material, as shown in Figure 17-1. In that figure, an inductor, or electromagnet, with many turns of thin wire has its leads going off to the left. Another inductor, with fewer turns of thicker wire, has its leads going off to the right.

In transformer terminology, the inductors are called windings. One of the two windings will be connected to an ac voltage source. The other winding will have its leads connected to a load resistance. Whichever winding is connected to the ac source is called the **primary winding.** The winding that is connected to the load is called the **secondary winding.** For example, in Figure 17-2(A), the many-turn winding to the left is connected to an ac source, while the fewer-turn winding to the right is connected to a load resistor. Therefore, the many-turn winding becomes the primary winding and the fewer-turn winding becomes the secondary. This is the step-down mode of operation, where the load voltage is less than the primary source voltage.

Figure 17-2(B) shows the opposite arrangement. The fewer-turn winding to the right receives the source voltage, while the many-turn winding to the left drives the load resistance. Here the load voltage is greater than the source voltage, so the transformer is in the step-up mode of operation.

When dealing with transformers, the terms source voltage, primary voltage, and input voltage all mean the same thing. And the terms load voltage, secondary voltage, and output voltage all mean the same thing.

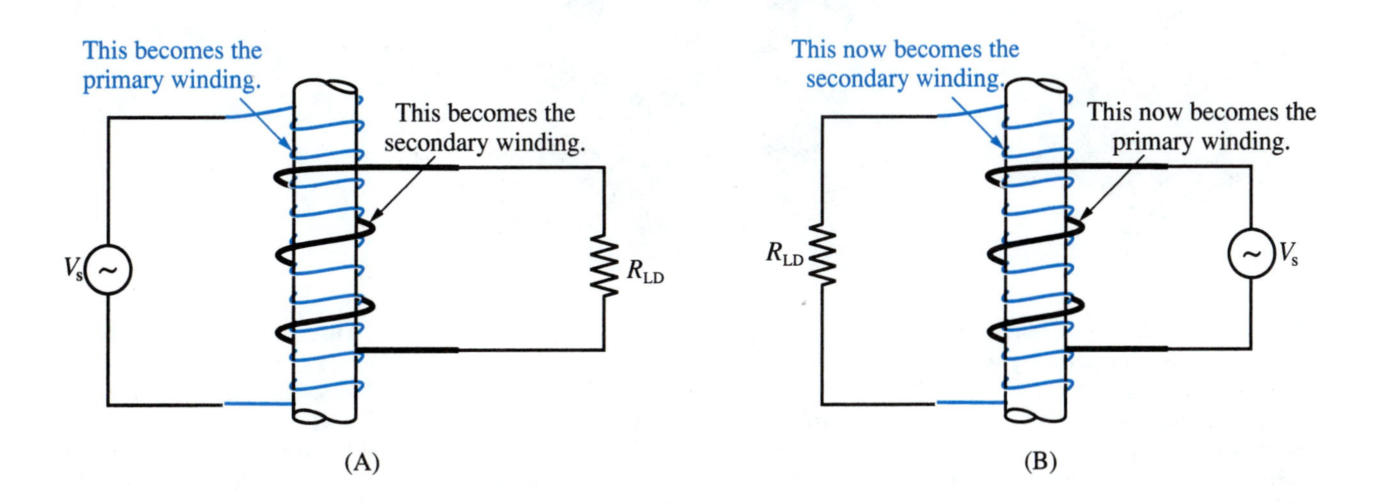

FIGURE 17-2 Telling which winding serves as the primary and which one serves as the secondary

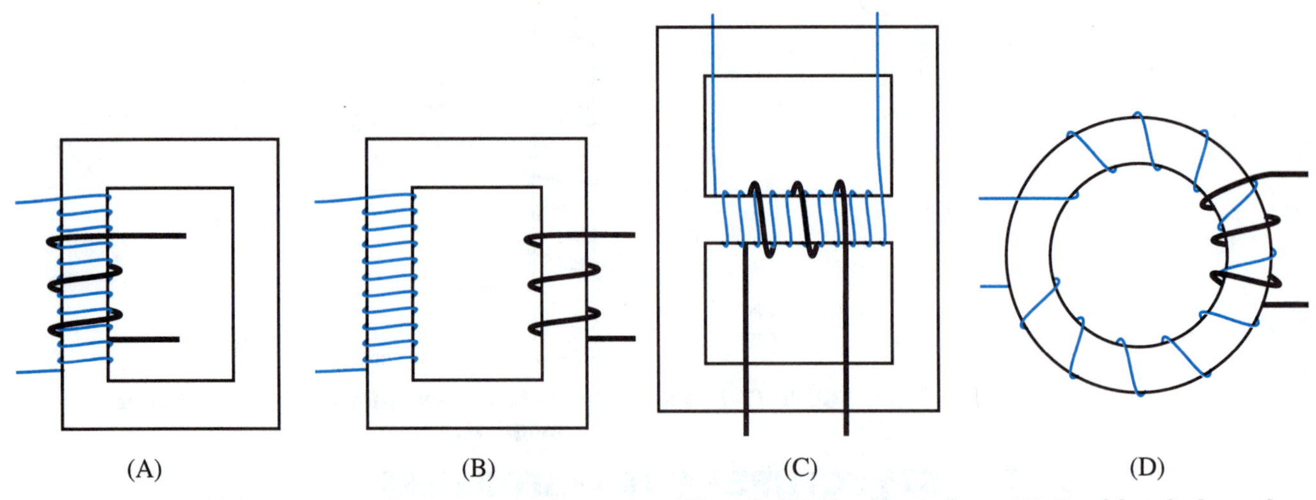

FIGURE 17-3 Common transformer core structures. (A) and (B) Single window. (C) Double window, also called E-I structure. (D) Toroid.

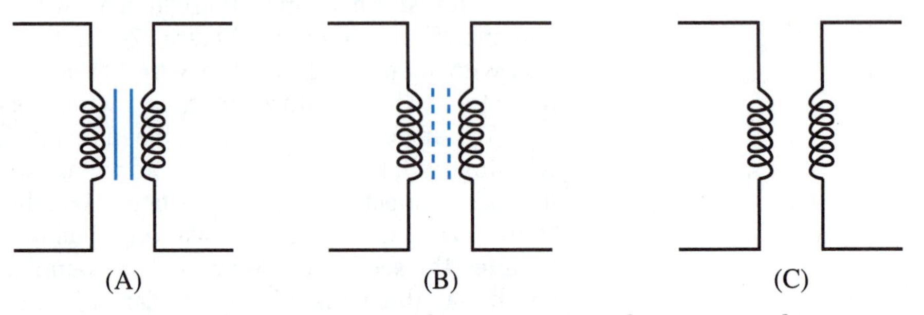

FIGURE 17-4 Transformer schematic symbols. (A) Solid parallel lines for iron core. (B) Dashed lines for ferrite core. (C) No lines for air core

The cylindrical core structure of Figures 17-1 and 17-2 is not the most common. More common are the closed-loop core structures shown in Figure 17-3.

Most transformers have a core material with high magnetic permeability. This magnifies the magnetic flux that makes the transformer work, as will be explained in Section 17-2.

If the core material contains iron metal, the transformer is symbolized schematically as shown in Figure 17-4(A). If the core is made of a substance that has iron combined with oxygen, we call it a ferrite-core transformer. Its schematic symbol is given in Figure 17-4(B). Some high-frequency transformers do not use a magnetic core at all. They have an air core, symbolized in Figure 17-4(C). Several transformers are pictured in Figure 17-5.

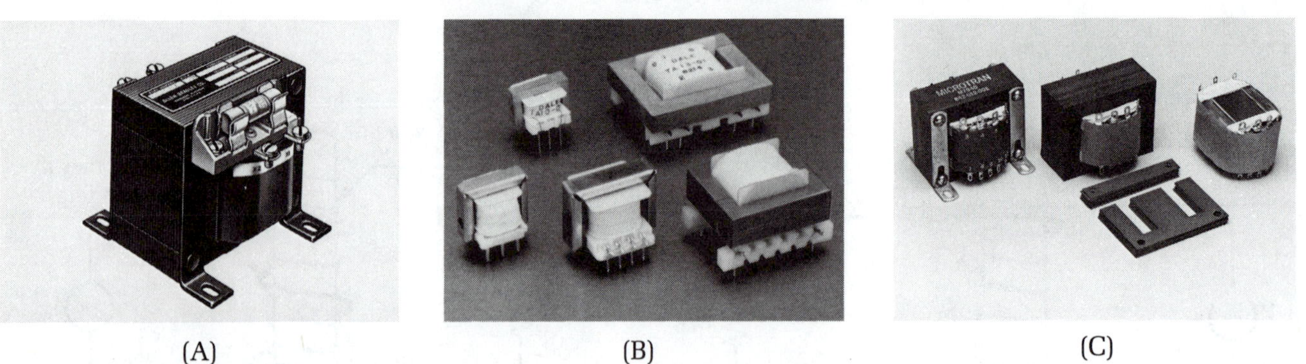

FIGURE 17-5 Various transformers. (A) Iron-core transformer with E-I structure. This unit has a built-in fuseholder for its primary winding. *Courtesy of Allen-Bradley, a Rockwell International Company.* **(B) Various double-window transformers.** *Courtesy of Dale Electronics, Inc.* **(C) Structural breakdown of a double-window (E-I frame) transformer. The preformed winding structure (containing both primary and secondary windings) is on the far right. These windings are tapped at several points. That is why they have several solder terminals. In the center, most of the E-shaped and I-shaped core slices have been inserted into the winding. The entire core structure, with mounting brackets, is held together by four screws. The finished product is on the far left.** *Courtesy of Microtran Company, Inc.*

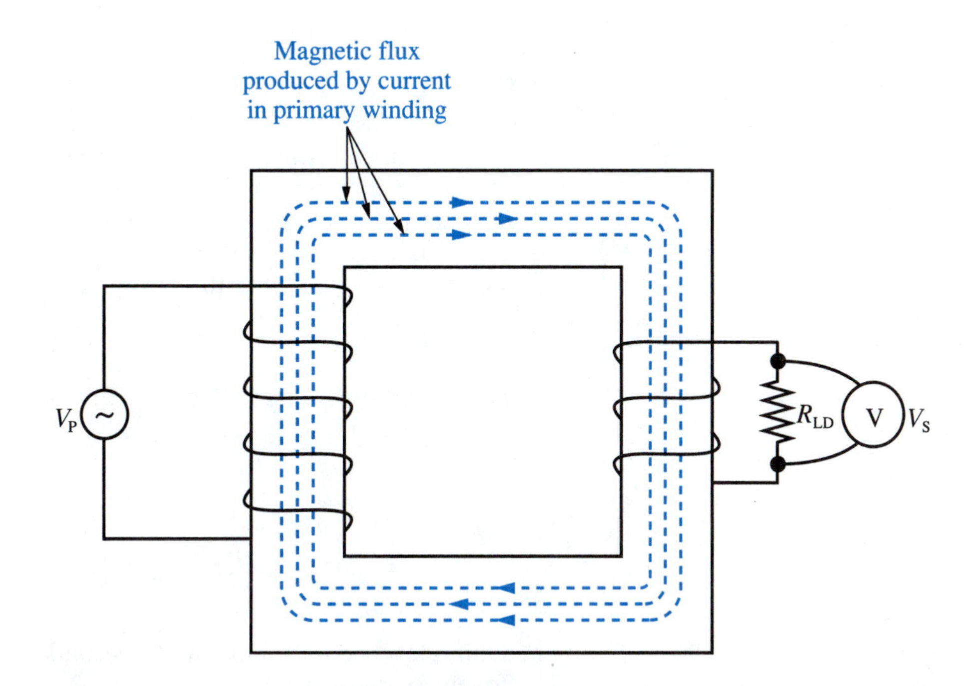

FIGURE 17-6 Flux in a transformer core. During one half-cycle of V_P, the flux is clockwise, as shown here. During the alternate half-cycle, the flux reverses to counterclockwise. The subscripts for primary and secondary are capital *P* and capital *S*, not lower-case *p* and *s*. Lower-case *p* is used for "peak"; lower-case *s* stands for "source."

17-2 TRANSFORMER VOLTAGE LAW

When the primary voltage forces current through the primary winding, magnetic flux is produced in the transformer core by electromagnet action. The flux is guided by the core so that it passes through the secondary winding. This is pictured in Figure 17-6.

As the ac current in the primary winding oscillates, it causes the flux to change in magnitude and direction, like any sine wave. The continual sine-wave oscillation of the flux causes the secondary winding to create the secondary voltage, V_S, by normal inductor action. This secondary voltage is also a sine wave, in phase with the primary voltage, V_P.

The primary winding has a certain number of turns, N_P. The number of turns on the secondary winding is symbolized N_S. Because the same amount of magnetic flux passes through both windings, this statement is true.

TECHNICAL FACT

In a transformer, the ratio of voltages is equal to the ratio of turns. As a formula,

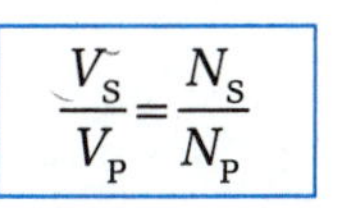

$$\frac{V_S}{V_P} = \frac{N_S}{N_P}$$

EQ. 17-1

Equation 17-1 is known as the transformer voltage law.

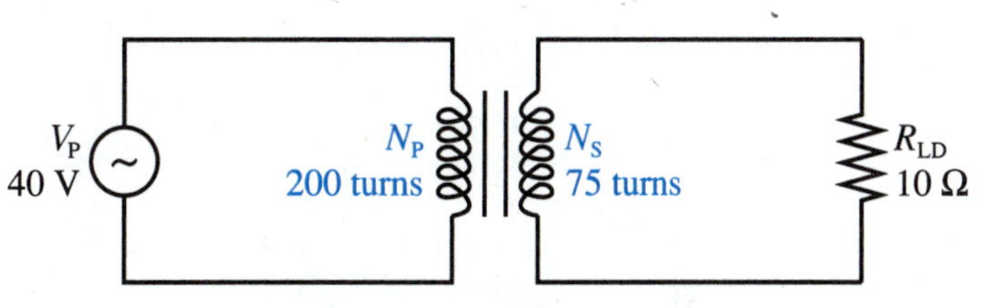

FIGURE 17-7 Transformer circuit schematic diagram

EXAMPLE 17-1

The iron-core transformer in Figure 17-7 is driven by a 40-V primary supply. The winding turns are given in that schematic as $N_P = 200$, $N_S = 75$. The secondary winding is connected to a 10-Ω load resistance.
a) Find the value of the secondary voltage, V_S.
b) Calculate the current in the secondary loop.

SOLUTION

a) From the transformer voltage law, Equation 17-1, we can say

$$\frac{V_S}{V_P} = \frac{N_S}{N_P}$$

$$V_S = V_P\left(\frac{N_S}{N_P}\right) = 40\text{ V}\left(\frac{75\text{ turns}}{200\text{ turns}}\right) = \mathbf{15\text{ V}}$$

b) Ohm's law can be applied in the secondary loop just as it can be applied to any resistive circuit.

$$I_S = \frac{V_S}{R_{LD}} = \frac{15\text{ V}}{10\ \Omega} = \mathbf{1.5\text{ A}}$$

This is an rms current, since the voltages are assumed to be given in rms units.

TECHNICAL FACT

The **turns ratio** of a transformer is defined as the ratio of secondary turns to primary turns. Turns ratio is symbolized *n*. As a formula,

$$n = \frac{N_S}{N_P}$$

EQ. 17-2

Combining Equations 17-1 and 17-2, we can write the transformer voltage law as

$$\frac{V_S}{V_P} = n$$

EQ. 17-3

EXAMPLE 17-2

a) What is the turns ratio of the transformer in Figure 17-7?
b) Suppose that the primary voltage is increased to 60 V. Find the new secondary voltage.
c) Calculate the power transferred to the load resistor.

SOLUTION

a) From Equation 17-2,

$$n = \frac{N_S}{N_P} = \frac{75\text{ turns}}{200\text{ turns}} = \mathbf{0.375}$$

Continued on page 303.

b) Writing the transformer voltage law as

$$\frac{V_S}{V_P} = n$$

we get $$V_S = V_P\,(n) = 60\text{ V }(0.375) = \mathbf{22.5\ V}$$

c) $$P = \frac{V^2}{R_{LD}} = \frac{(22.5\text{ V})^2}{10\ \Omega} = \mathbf{50.6\ W}$$

EXAMPLE 17-3

In the previous two examples, the transformer was being used in the step-down mode. Suppose the transformer is reversed so that the 60-V primary voltage is connected to the winding with only 75 turns. This is shown in Figure 17-8.

a) What is the transformer's turns ratio now?
b) Calculate the new secondary voltage.

SOLUTION

a) From Equation 17-2,

$$n = \frac{N_S}{N_P} = \frac{200\text{ turns}}{75\text{ turns}} = \mathbf{2.667}$$

b) $$\frac{V_S}{V_P} = n$$

$$V_S = 60\text{ V}(2.667) = \mathbf{160\ V}$$

Whenever a transformer is wired so that the turns ratio, n, is greater than 1, the transformer will step up the voltage. Whenever n is less than 1, voltage is stepped down.

✔ SELF-CHECK FOR SECTIONS 17-1 AND 17-2

1. The transformer winding that is driven by a separate voltage source is called the __________ winding. [1]
2. The transformer winding that connects to the load is called the__________ winding. [1]
3. For a disconnected transformer, can you state definitely which is the primary winding and which is the secondary? Explain. [1]
4. In a step-up transformer, the secondary voltage is__________than the primary voltage. [1]

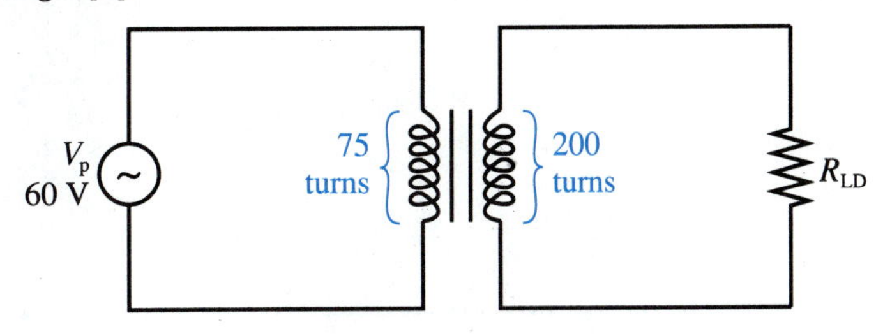

FIGURE 17-8 With the transformer windings reversed, the turns ratio becomes greater than 1. The transformer is now stepping up.

5. Explain how voltage is induced in the secondary winding of a transformer. [2]
6. What is the phase relationship between V_S and V_P? [1, 2]
7. A certain transformer has $V_P = 50$ V, $N_P = 300$ turns, and $N_S = 450$ turns. Find n. Find V_S. [3]
8. A certain transformer has $V_P = 80$ V, $V_S = 20$ V, and $N_P = 200$ turns. Find N_S. Find n. [3]

17-3 TRANSFORMER CURRENT LAW

Transformers perform a current-voltage tradeoff. The winding with the higher voltage has less current. The winding with the lower voltage has more current. This leads to the following statement.

TECHNICAL FACT

In a transformer, the ratio of currents is inversely proportional to the ratio of turns. As a formula,

$$\frac{I_P}{I_S} = \frac{N_S}{N_P} = n \qquad \text{EQ. 17-4}$$

This is called the transformer **current law.** It can also be written as

$$\frac{I_P}{I_S} = \frac{V_S}{V_P} \qquad \text{EQ. 17-5}$$

MEDICAL THERMAL IMAGING

Shown here is a medical thermal image system. It has a tripod-mounted infrared scanner that can detect body temperature differences as small as 0.1 degree Celsius. The resulting thermal images are then displayed on a video screen.

By examining such displays, medical analysts can identify specific physical problems. Reduced blood circulation, improper nerve function, and degree of internal swelling and pain all show up clearly on the image. Thus, an accurate diagnosis can sometimes be made without performing invasive tests or surgery.

Here, a patient is undergoing a nerve function test. A thermal image of two hands is shown in its color-enhanced version on page 3 of the Applications color section.

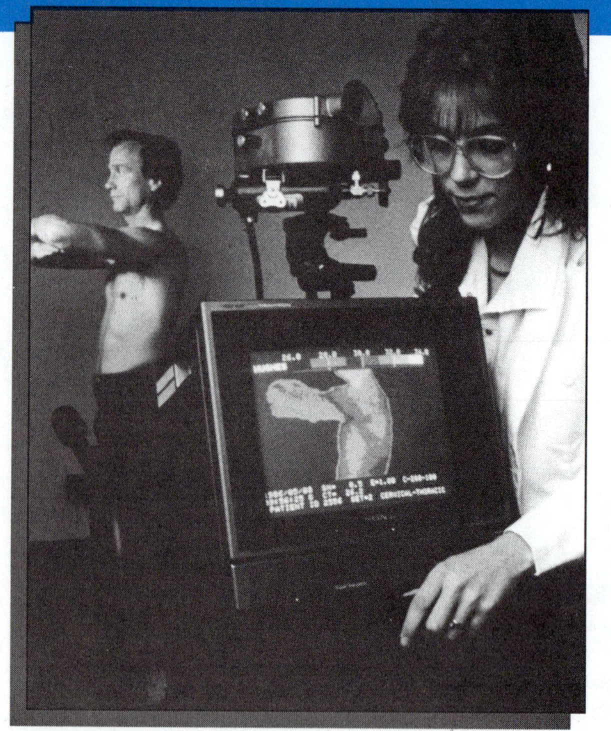

Courtesy of NASA

FIGURE 17-9 In a voltage step-up operation, current is sacrificed.

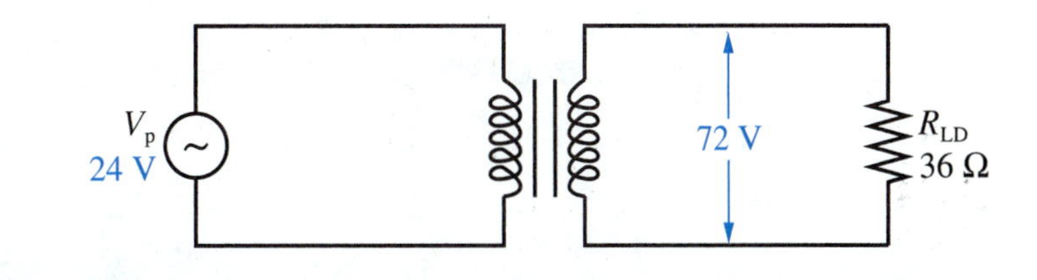

EXAMPLE 17-4

The iron-core transformer of Figure 17-9 delivers a secondary voltage of 72 V to a 36-Ω load. It is driven by a primary voltage of 24 V.

a) Find the turns ratio, n. Is this step-up or step-down operation?
b) Solve for the secondary current, I_S.
c) Solve for the primary current, I_P.

SOLUTION

a) From the transformer voltage law,

$$n = \frac{V_S}{V_P} = \frac{72\ \text{V}}{24\ \text{V}} = \mathbf{3.0}$$

This is clearly **step-up** operation.

b) Applying Ohm's law gives

$$I_S = \frac{V_S}{R_{LD}} = \frac{72\ \text{V}}{36\ \Omega} = \mathbf{2.0\ A}$$

c) From the transformer current law, Equation 17-4,

$$\frac{I_P}{I_S} = n$$

$$I_P = I_S\,(n) = 2.0\ \text{A}\,(3.0) = \mathbf{6.0\ A}$$

This example points out the current-voltage tradeoff idea of transformer operation. The increase in voltage by a factor of 3 was accompanied by a decrease in current by a factor of 3. Simply stated, to get out more voltage than we put in, we had to put in more current than we took out. The tradeoff works the other way for a step-down situation, as the next example shows.

EXAMPLE 17-5

The transformer of Figure 17-10 is stepping voltage down.

a) Find the turns ratio, n.
b) Solve for secondary current, I_S.
c) Solve for primary current, I_P. Comment on the relationship of I_P to I_S.

Continued on page 306.

FIGURE 17-10 In a voltage step-down operation, current is gained.

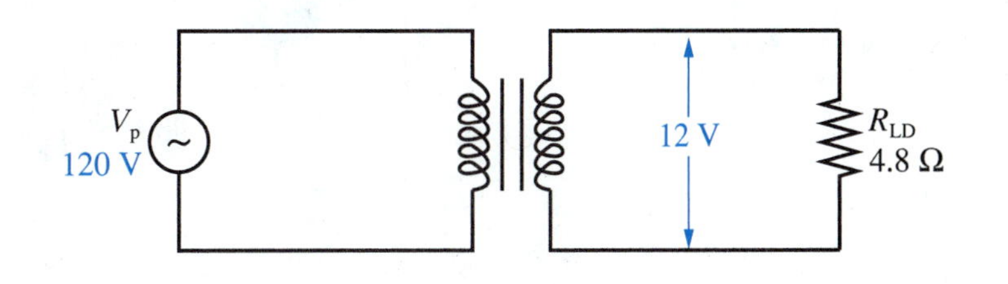

SOLUTION

a) From Equation 17-3,

$$n = \frac{V_S}{V_P} = \frac{12\text{ V}}{120\text{ V}} = \mathbf{0.1}$$

b) By Ohm's law,

$$I_S = \frac{V_S}{R_{LD}} = \frac{12\text{ V}}{4.8\ \Omega} = \mathbf{2.5\ A}$$

c) From Equation 17-4,

$$I_P = nI_S = (0.1)\ 2.5\text{ A} = \mathbf{0.25\ A}$$

Primary current is smaller than secondary current because the primary voltage is larger than secondary voltage in this step-down function.

17-4 TRANSFORMER POWER LAW

In an **ideal transformer**, no electric power is wasted. Therefore, we can say:

TECHNICAL FACT

The power delivered into the primary side by the voltage source is equal to the power delivered by the secondary side to the load resistance. In equation form,

$$P_{in(primary)} = P_{out(secondary)}$$

EQ. 17-6

$$V_P I_P = V_S I_S$$

EQ. 17-7

This is the transformer **power law.**

EXAMPLE 17-6

The ideal transformer in Figure 17-11 has a wattmeter connected in its primary side, measuring 60 W. The winding turns are given and so is the load resistance.

a) How much power is delivered to the load?
b) Calculate secondary voltage, V_S.
c) Find V_P, I_P, and I_S.

Continued on page 307.

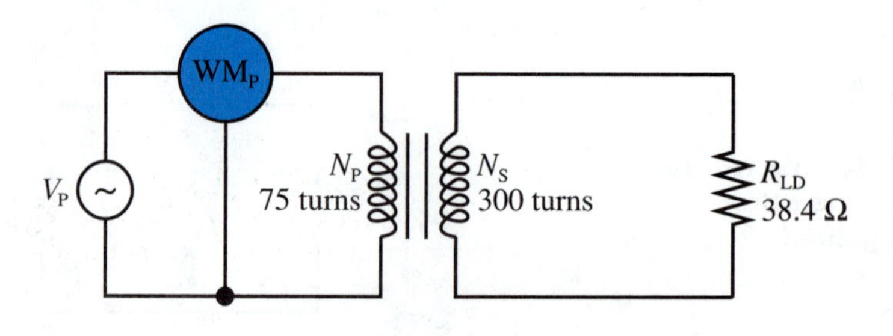

FIGURE 17-11 Installing a wattmeter to measure a transformer's primary power

SOLUTION

a) From the transformer power law,

$$P_{\text{out (to load)}} = P_{\text{in}} = \mathbf{60\ W}$$

b) The formula that relates power, resistance, and voltage is

$$P_{\text{out}} = \frac{(V_S)^2}{R_{LD}}$$

$$V_S = \sqrt{(P_{\text{out}})(R_{LD})} = \sqrt{(60\ \text{W})(38.4\ \Omega)} = \mathbf{48\ V}$$

c) Rearranging the voltage law, Equation 17-1, we get

$$\frac{V_P}{V_S} = \frac{N_P}{N_S}$$

$$V_P = V_S\left(\frac{N_P}{N_S}\right) = 48\ \text{V}\left(\frac{75\ \text{turns}}{300\ \text{turns}}\right) = \mathbf{12\ V}$$

By the power formula, Equation 7-5,

$$P_{\text{in}} = V_P I_P = 60\ \text{W}$$

$$I_P = \frac{60\ \text{W}}{12\ \text{V}} = \mathbf{5\ A}$$

Secondary current, I_S, can be found from the transformer power law, Equation 17-7.

$$V_S I_S = V_P I_P = 60\ \text{W}$$

$$I_S = \frac{60\ \text{W}}{V_S} = \frac{60\ \text{W}}{48\ \text{V}} = \mathbf{1.25\ A}$$

Alternatively, the transformer current law tells us

$$I_S = I_P\left(\frac{N_P}{N_S}\right) = 5\ \text{A}\left(\frac{75\ \text{turns}}{300\ \text{turns}}\right) = 1.25\ \text{A}$$

which checks.

In Example 17-6, the transformer receives a power of 60 watts, composed of 12 V and 5 A in the primary circuit. The transformer does not change the *amount* of power, but it does change the *composition* of the power. The power in the secondary circuit is composed of a larger voltage combined with a smaller current (48 V and 1.25 A, compared to 12 V and 5 A). This is the fundamental result of the step-up operation.

In Example 17-5, the step-down transformer changed the power composition in the opposite way. The secondary power is composed of a smaller voltage (12 V compared to 120 V primary), combined with a larger current (2.5 A compared to 0.25 A primary). This is the fundamental result of the step-down operation.

Summarizing these two situations leads to an alternative statement of the transformer power law.

TECHNICAL FACT

A transformer does not change the amount of power, but it does change the voltage and current composition of power.

All three transformer laws assume that the transformer is ideal, having 100% efficiency, with zero energy wasted as heat. In reality, all transformers waste at least a small amount of energy. The reasons for this will be discussed in Section 17-8.

✔ SELF-CHECK FOR SECTIONS 17-3 AND 17-4

9. Write the transformer current law. Be careful with the subscripts *P* and *S* referring to primary and secondary. [3]
10. The high-voltage side of a transformer carries the__________current. The low-voltage side of a transformer carries the__________current. (larger or smaller) [1, 3]
11. In a step-down transformer, the decrease in voltage is compensated for by an __________in current. [1, 3]
12. In a step-up transformer, the increase in voltage is paid for by a__________in current. [1, 3]
13. The transformer of Figure 17-9 has $n = 3.0$. Suppose it is driven by a 48-V source and is connected to a 100-Ω load resistance. [3]
    a) Find V_S.
    b) Find I_S.
    c) Find I_P.
    d) Find P_{out}.
    e) Find P_{in}.
14. The transformer in Figure 17-11 has $n = 4.0$. Suppose V_P is changed to 30 V and R_{LD} is raised to 75 Ω. What power value would the wattmeter measure? [3]
15. Figure 17-12 represents a long-distance power-transmission system from the utility company's generation site to a local substation. Explain the reason for stepping the voltage up so high before sending the power down the long-distance transmission wires. [1, 3]
16. In Figure 17-12, calculate the secondary current in the substation transformer. [3]

SAFETY ADVICE

Sometimes an electrical system is designed to operate from a high-value ac voltage, which is then stepped down by a transformer to a lower voltage. A common example of this is an industrial system with a 480 V/120 V step-down transformer. This is shown in Figure 17-13.

Systems like this are checked out in advance, before actual installation in the industrial setting. During the check-out process, there is probably no 480 V ac available to drive the transformer primary. Therefore, we must temporarily connect a 120 V ac source directly to the secondary winding, leaving the primary disconnected, as shown in Figure 17-13. Only in this way can we get ac power to the remainder of the circuit.

Be extremely careful when using this procedure. There is a tendency to think that this circuit is not very dangerous because there is not an external 480-V source driving the transformer. That thought is false. The transformer now works backward, producing 480 V across the left-side terminals in Figure 17-13. This 480 V is just as lethal as the 480-V bus line in the real industrial world.

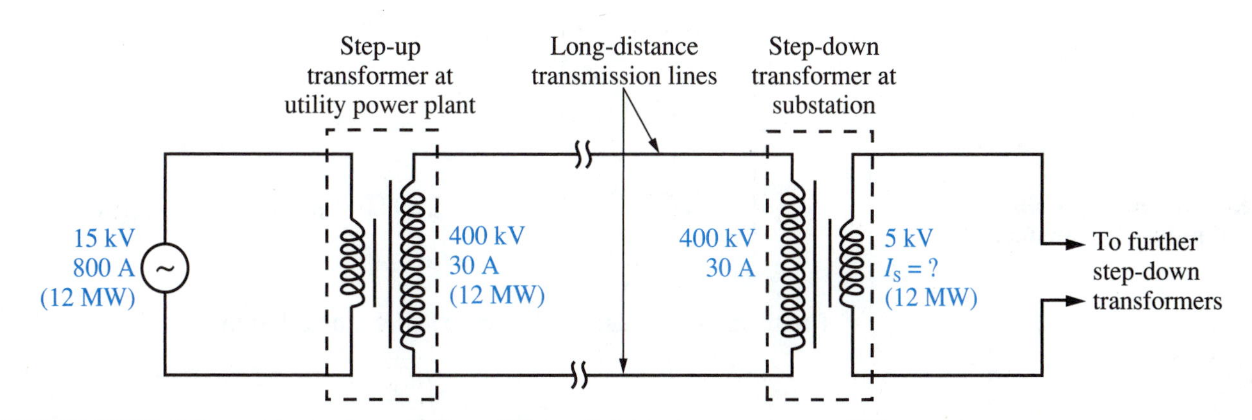

FIGURE 17-12 Schematic representation of long-distance power-transmission system

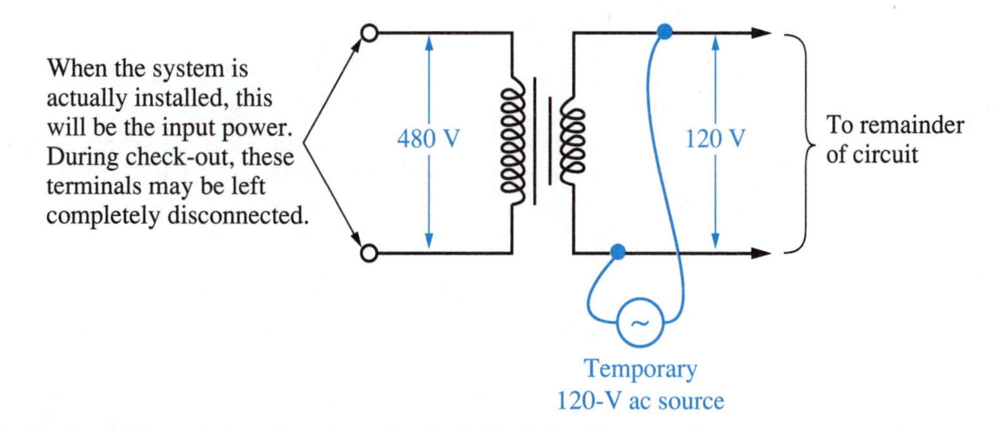

FIGURE 17-13 If a 480-V source is not available during checkout, a 120-V source may be temporarily wired directly to the secondary terminals.

17-5 RESISTANCE-MATCHING TRANSFORMERS

Most voltage sources possess internal resistance. For a fixed amount of internal resistance, R_{int}, we can make this statement.

TECHNICAL FACT

The maximum power is transferred to a load when the load resistance is equal to the source's internal resistance. That is,

$$R_{LD} = R_{int} \quad \text{(for maximum load power)} \qquad \textbf{EQ. 17-8}$$

In ac circuits, any amount of load resistance can be made to appear as if it matches the source's internal resistance. This is accomplished by installing a transformer between the source and the load. Of course, the turns ratio of the transformer must be just the proper value to apply to R_{LD} and R_{int}. This idea is illustrated in Figure 17-14.

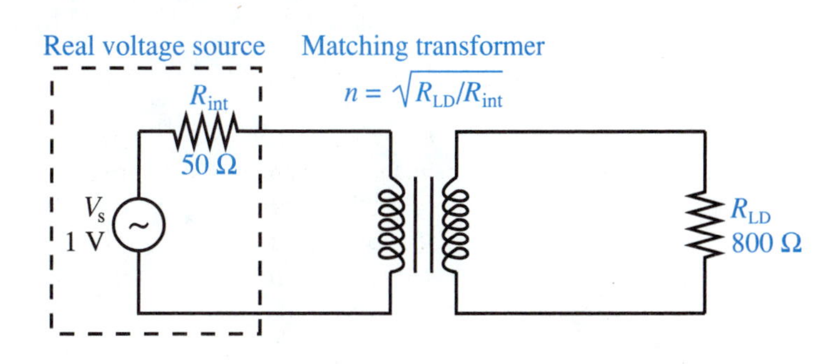

FIGURE 17-14 A matching transformer makes the load resistance appear to be equal to the source's internal resistance. This produces the maximum possible power to the load.

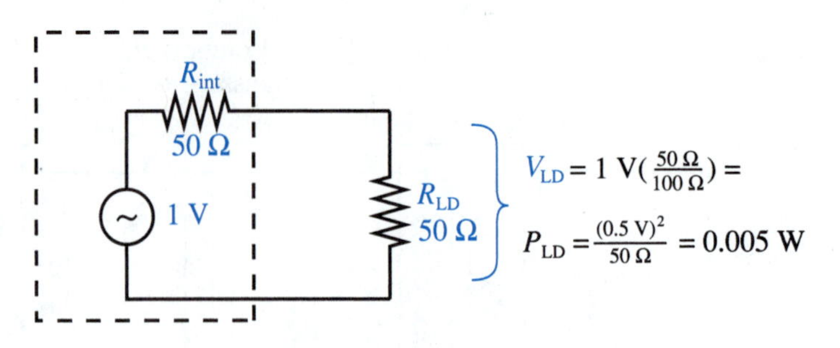

FIGURE 17-15 This is what would happen if the load resistance really did match the internal source resistance in Figure 17-13. The maximum possible load power is transferred.

The turns ratio must be chosen to satisfy the relation

$$n = \sqrt{\frac{R_{LD}}{R_{int}}}$$

EQ. 17-9

In Figure 17-14, this yields

$$n = \sqrt{\frac{800\ \Omega}{50\ \Omega}} = \sqrt{16} = 4.0$$

With $n = 4.0$, the no-load source value of 1 V is split evenly between the internal resistance and the transformer primary winding. That is, 0.5 V is dropped across R_{int} and 0.5 V is delivered to the transformer as V_P. That 0.5 V is stepped up to 2.0 V by

$$V_S = n\, V_P = (4.0)\,(0.5\text{ V}) = 2.0\text{ V}$$

With a 2.0-V secondary voltage, load power is given as

$$P_{LD} = \frac{(V_S)^2}{R_{LD}} = \frac{(2.0\text{ V})^2}{800\ \Omega} = 0.005\text{ W or 5 mW}$$

Note that this is the same amount of power that would be delivered to a 50-Ω load wired directly to the source, as pictured in Figure 17-15.

EXAMPLE 17-7

A circuit like the one in Figure 17-14 has source voltage $V_s = 5$ V, source resistance $R_{int} = 600\ \Omega$, and load resistance $R_{LD} = 14\ \Omega$.

a) What should be the turns ratio of the matching transformer in order to maximize the load power?

b) What is the amount of load power under this condition?

SOLUTION

a) From Equation 17-9,

$$n = \sqrt{\frac{R_{LD}}{R_{int}}} = \sqrt{\frac{14\ \Omega}{600\ \Omega}} = \sqrt{0.02333} = \mathbf{0.1528}$$

b) With $n = 0.1528$, the voltage source "sees" the load resistance as if it had a value of 600 Ω, the same as R_{int}. Figure 17-16 shows this.

The total power delivered by the source is given by

$$P_{total} = \frac{(V_S)^2}{R_{total}} = \frac{(5\text{ V})^2}{1200\ \Omega} = 20.83\text{ mW}$$

Half of this power is lost internally and the other half is delivered to the load.

$$P_{LD} = ½\, P_{total} = ½\,(20.83\text{ mW}) = \mathbf{10.4\text{ mW}}$$

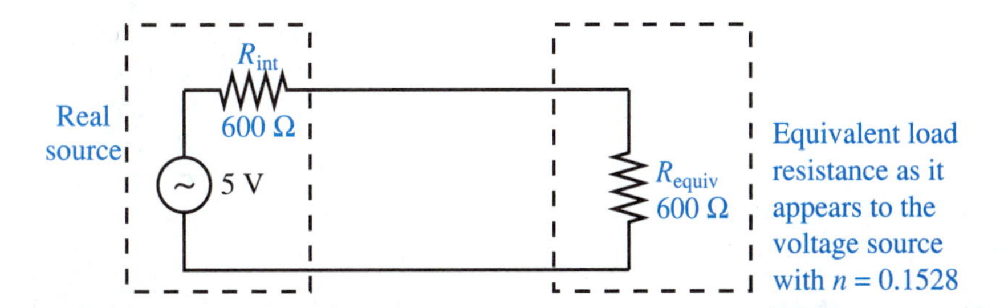

FIGURE 17-16 With a resistance-matching transformer installed, the equivalent resistance of the load becomes 600 Ω. The circuit's equivalent total resistance is then 1200 Ω. So the total power is split evenly between the source and the load.

17-6 OTHER TRANSFORMER TOPICS

Isolation

In addition to their applications already described, transformers have yet another circuit advantage.

Transformers provide electrical **isolation** between the primary and secondary circuits. Electrical isolation is the elimination of any direct electrical connection between the circuits.

Isolation is desirable for several reasons. Here are two of them:

1. The electrically isolated secondary circuit can remain completely unreferenced to earth ground. Such a circuit is said to be **floating**. It is impossible to be shocked by touching one point of a floating circuit. This safety feature is illustrated in Figure 17-17.
2. An electrically isolated circuit can be kept relatively noise-free. Noise signals are unwanted signals that interfere with proper signals. They are often injected by stray capacitance or by stray magnetic fields. Because of a transformer's isolating ability, noise signals that are injected into the primary circuit can be largely eliminated from the load-driving secondary circuit. This is illustrated in Figure 17-18.

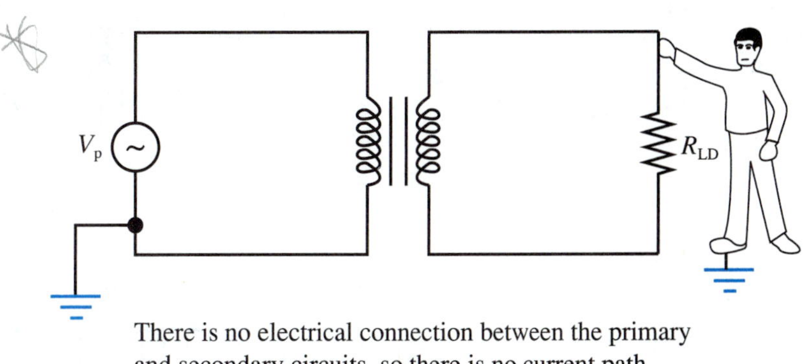

FIGURE 17-17 Transformer isolation makes the secondary circuit safe to touch.

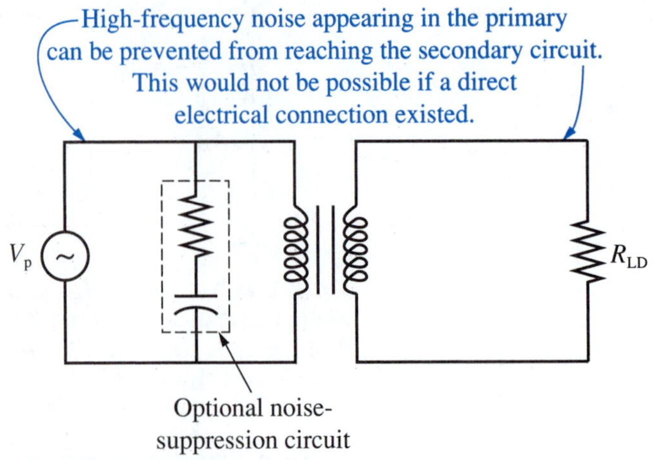

High-frequency noise appearing in the primary can be prevented from reaching the secondary circuit. This would not be possible if a direct electrical connection existed.

V_P

R_{LD}

Optional noise-suppression circuit

FIGURE 17-18 A transformer's electrical isolation feature helps suppress noise at the load.

Tapped Secondary Windings

Sometimes a secondary winding has a lead wire connected to one of its intermediate loops, as shown in Figure 17-19(A). If the lead connects to the exact middle of a winding, as shown in Figure 17-19(B), it is called a center-tapped winding. Figure 17-19(C) shows a winding with multiple tap points.

Center-tapped secondary windings are useful for making full-wave rectified dc power supplies. Such a dc power supply is shown schematically in Figure 17-20. The operation of this circuit is explained in books dealing with electronic semiconductor devices.

Transformers with center-tapped secondary windings are also used in the United States to provide two voltage levels in homes and commercial businesses. In Figure 17-21, the total secondary voltage of 240 V is used to drive heavy-demand devices like water heaters, kitchen ovens, and clothes dryers. The center-tapped voltage of 120 V is used for all other low-power loads such as lamps, kitchen countertop appliances, and home entertainment devices.

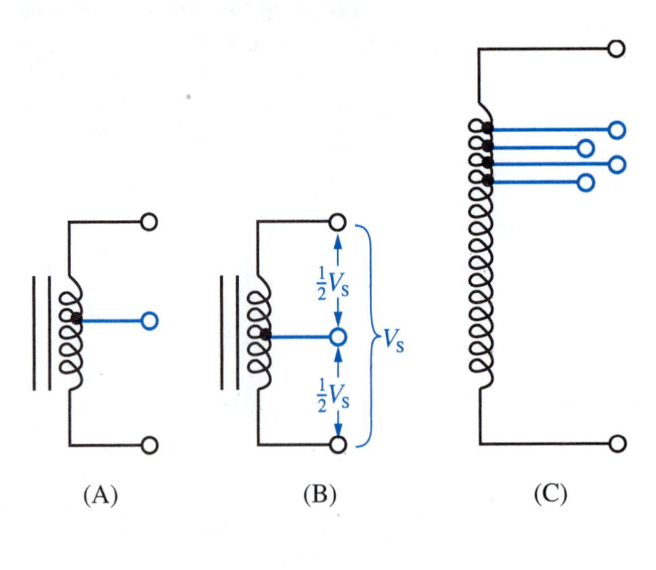

FIGURE 17-19 (A) Tapped winding. (B) Center-tapped winding, giving half the secondary voltage on each side. (C) Winding with several tap points

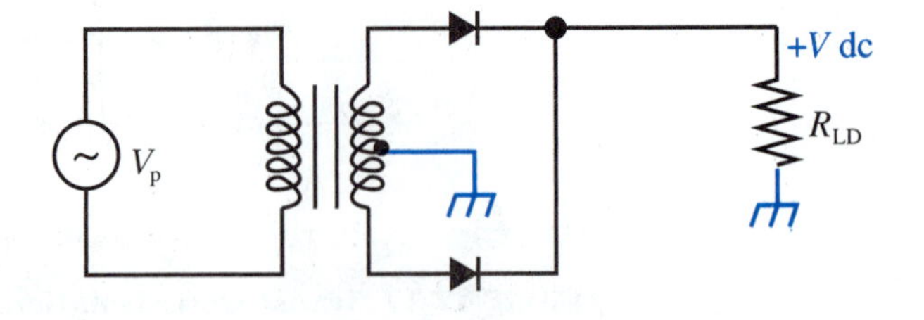

FIGURE 17-20 Full-wave rectified dc power supply using a center-tapped transformer

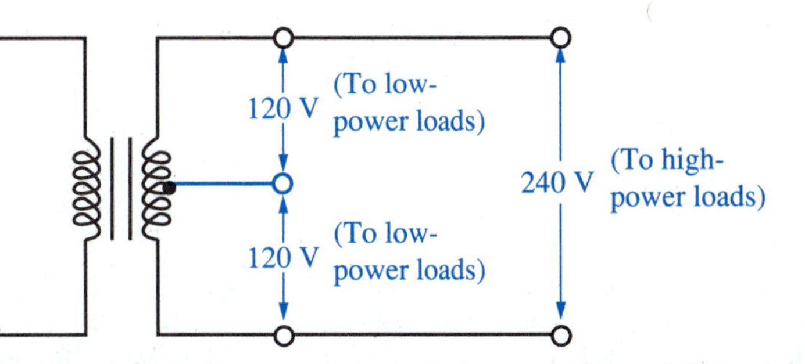

FIGURE 17-21 Residential service transformer. Primary voltage differs from one installation to another but is usually about 4–5 kV.

Multiple Secondarys

Transformers often have more than one secondary winding. Figure 17-22(A) shows the construction of a single-window transformer with two secondary windings. The schematic diagram is given in Figure 17-22(B).

The transformer voltage law applies separately to each secondary winding. That is,

$$\frac{V_{S1}}{V_P} = \frac{N_{S1}}{N_P} = n_1$$

and

$$\frac{V_{S2}}{V_P} = \frac{N_{S2}}{N_P} = n_2$$

Special-purpose transformers can have numerous secondary windings, sometimes ten or more.

✓ SELF-CHECK FOR SECTIONS 17-5 AND 17-6

17. In the practice of resistance matching, the aim is to make the load resistance appear equal to the______ __________. [4]
18. An ac source has an open-circuit value of 20 V with R_{int} = 25 Ω. It is required to drive a load resistance of 1 kΩ. What transformer turns ratio is necessary to maximize the load's power? [4]
19. In Question 18, what is that maximum value of power that will be delivered to the load? [4]
20. Name the two most common uses for center-tapped transformer windings. [6]
21. Explain the advantages of secondary circuit isolation. [5]

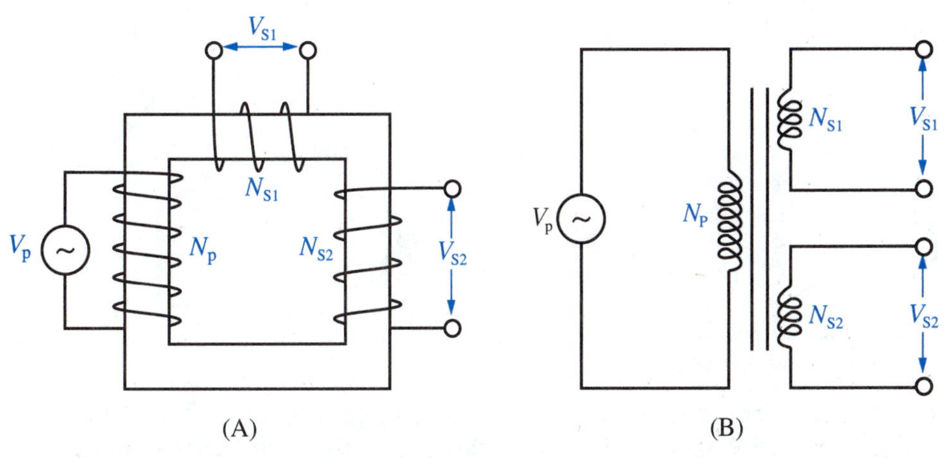

FIGURE 17-22 Transformer with two secondary windings. (A) Structure. (B) Schematic

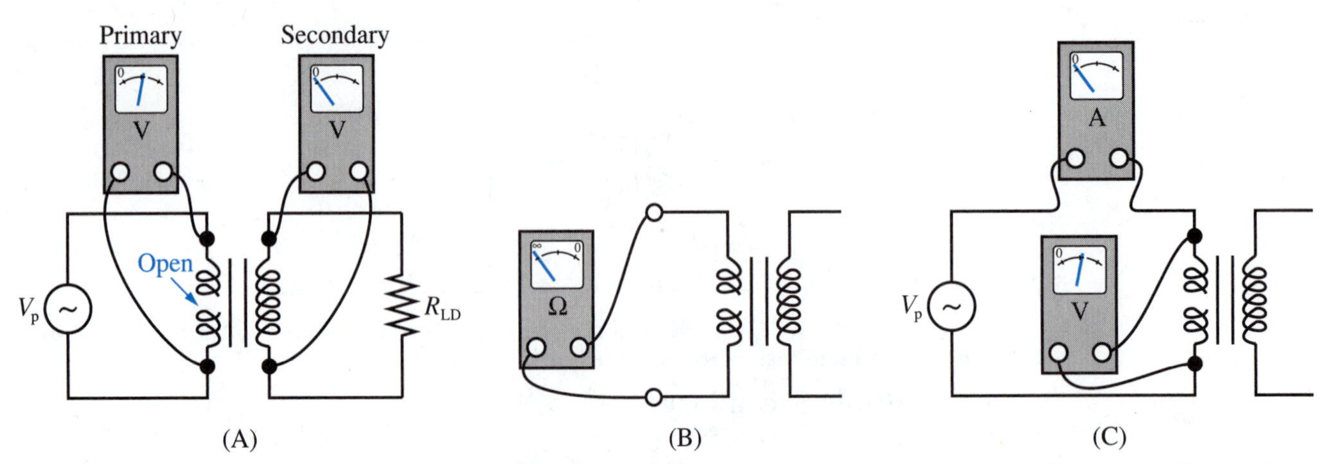

FIGURE 17-23 (A) With the primary winding failed open, these are the voltmeter readings. (B) Ohmmeter test on an isolated primary winding. (C) Ammeter test

17-7 TROUBLESHOOTING TRANSFORMERS

A transformer is essentially two inductors, so the ohmmeter tests for inductors are valid for transformers as well. Look back to Section 14-7 to review inductor troubleshoooting.

Either winding may fail open. If the primary winding fails open, the situation is shown in Figure 17-23(A). A voltmeter across the primary winding will read the normal primary voltage, but no primary current can flow through the open. Therefore, no magnetic flux is created and no secondary voltage is induced. The voltmeter in the secondary circuit reads zero.

To check a suspected open primary winding, perform the ohmmeter test shown in Figure 17-23(B). Or use the ammeter test in Figure 17-23(C) to confirm that the primary winding is open under high-voltage conditions.

If the secondary winding fails open, the situation is shown in Figure 17-24(A). The primary voltmeter reads normal voltage V_P across the primary winding, but the secondary voltmeter reads zero. Of course, this is the same voltmeter indication as in Figure 17-23(A) for an open primary winding. So if you wish to know which winding failed open, you must use an ohmmeter test or an ammeter test. These tests on the secondary winding are shown in Figure 17-24(B) and (C).

As for getting the circuit back in working order, it doesn't matter which winding failed. This is because a bad transformer is simply replaced with a new unit,

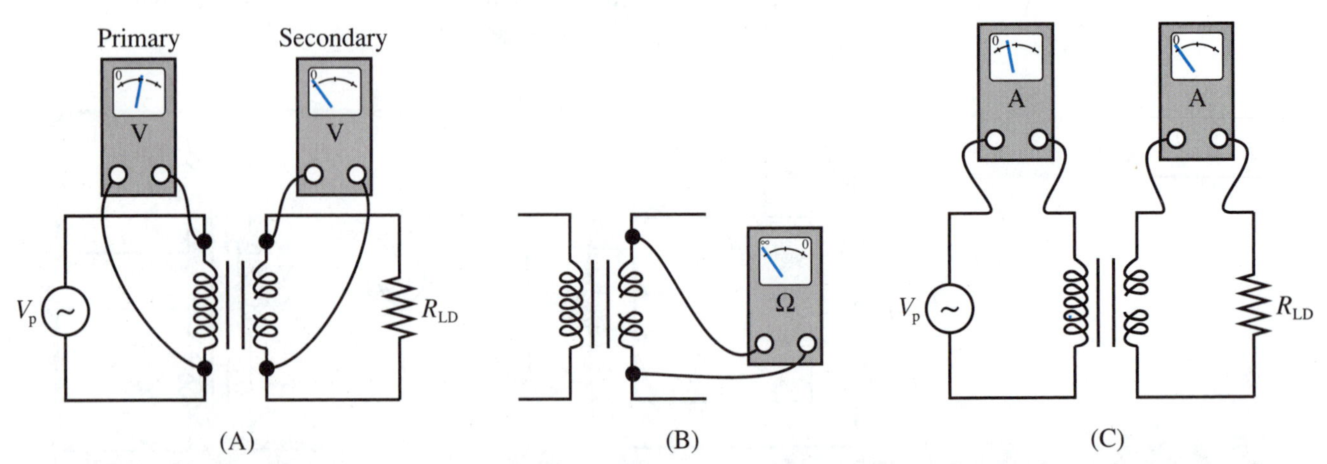

FIGURE 17-24 (A) With the secondary winding failed open, the voltmeter indications are the same as for the primary winding open [Figure 17-23(A)]. (B) Deenergize the primary and isolate the secondary winding from the load before testing for infinite ohms in the secondary. (C) Ammeter indications with the secondary winding open

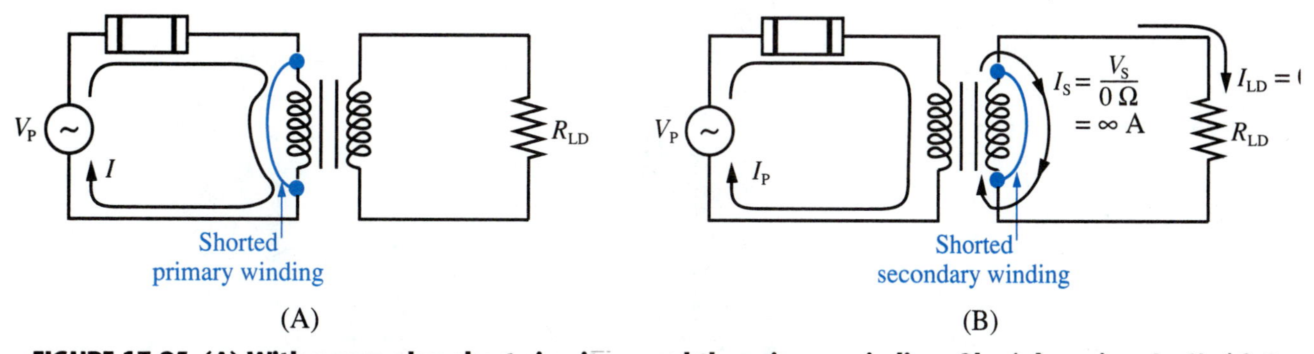

FIGURE 17-25 (A) With a zero-ohm short circuit around the primary winding, Ohm's law gives $I = V_P / 0\ \Omega = \infty$ A, so the fuse blows. (B) With a dead short-circuited secondary winding, Ohm's law predicts $I_S = V_S / 0\ \Omega = \infty$ A. The transformer current law then gives $I_P = n\ I_S = n\ (\infty\ \text{A}) = \infty$ A, so the fuse blows.

rather than being repaired. It is not cost-effective to repair a transformer unless it is a very large expensive model.

Either winding may fail shorted, either partially or completely. If either winding is dead-shorted, an overcurrent will flow in the primary circuit. This should cause the circuit protection device to blow. The situations are shown in Figure 17-25.

A short-circuited winding can be detected by an ohmmeter test. As usual, it may be that the short occurs only at the higher operating voltage, not at the ohmmeter's low test voltage.

A partially shorted primary winding is shown in Figure 17-26(A). N_P is normally 100 turns, and N_S is 500 turns. The normal turns ratio is $n = 5$. But if half the primary turns are shorted out, the number of primary turns actually working becomes only 50. Then the effective turns ratio becomes 500 turns / 50 turns = 10. This doubles V_S. By Ohm's law, secondary current, I_S, also doubles.

From the transformer current law, the primary current then becomes $I_P = 10\ I_S$. In this equation, the number 10 is twice as large as it ought to be and I_S is twice as large as it ought to be. Therefore, I_P becomes four times as large as it ought to be, which probably will blow the fuse.

A partial secondary short is shown in Figure 17-26(B). From the load's point of view, N_S is effectively lowered, so load voltage is lower than normal. The bottom (shorted) part of the winding acts like a second secondary winding. This bottom part of the secondary winding induces voltage as usual, because magnetic flux is passing through its core. But this part of the winding sees virtually zero resistance through the short, so a large current circulates via this path. By the transformer current law, this large secondary current creates a larger-than-normal primary current.

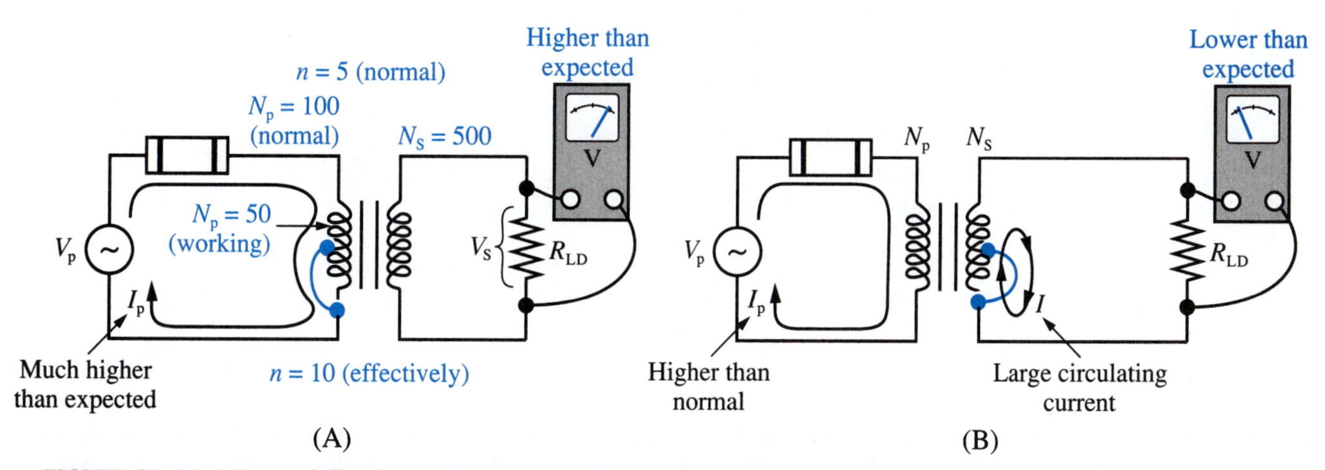

FIGURE 17-26 (A) Partially shorted primary. The result is a primary current that is much greater than normal. (B) Partially shorted secondary winding. The result is that I_P is somewhat larger than normal.

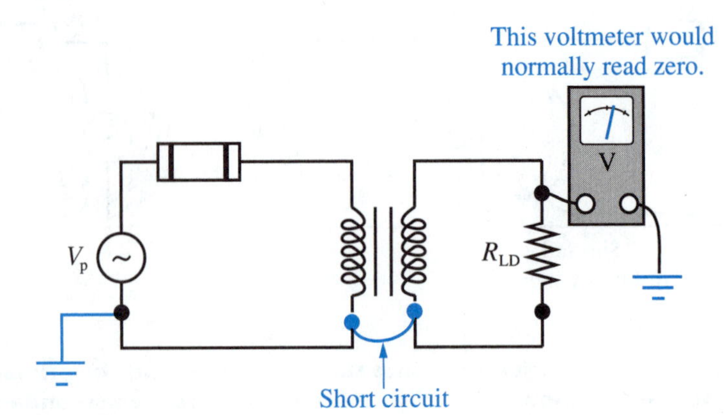

FIGURE 17-27 If a short occurs between the primary winding and the secondary winding, the secondary circuit is no longer isolated from the rest of the world.

A partially shorted transformer winding can be detected by a lower-than-normal resistance measurement on an ohmmeter test, the same as for an inductor. This assumes that you know the normal wire resistance of the winding.

The secondary and primary windings are often physically wrapped one over the other, as suggested in Figures 17-3(A), (C), and (D). Therefore, if insulation breaks down, it is possible for the two windings to make electrical contact with each other. This is shown schematically in Figure 17-27.

The short between primary and secondary windings does not necessarily cause any voltage or current value in the circuit to be abnormal. However, the secondary circuit no longer floats relative to the earth. It now becomes referenced to earth ground, as seen on the voltmeter in Figure 17-27. Therefore, the secondary circuit has lost its advantage of being free from shock hazard.

A primary-to-secondary short can be detected by an ohmmeter, with the transformer deenergized. Normally, the resistance between the windings should be ∞ Ω. As usual, a problem may not show up under low-voltage testing conditions.

17-8 NONIDEAL TRANSFORMERS

Any **nonideal (real) transformer** is bound to lose some of its input power as waste heat. Because of this waste, the output power from the secondary is slightly less than the input power to the primary. In other words, the transformer power law, Equation 17-6, is not obeyed exactly; the transformer's efficiency is less than 100%.

The wasted power occurs for three reasons:

1. Winding resistance loss (I^2R loss).
2. Core magnetization loss—also called hysteresis loss.
3. Core eddy-current loss.

Winding Resistance Loss

8

Winding resistance loss is due simply to the primary and secondary windings having nonzero resistance. As current flows in the primary winding, there is a power loss given by the formula $P_{loss} = (I_P)^2 R_P$. This is also true for the secondary winding.

Transformer manufacturers try to minimize winding resistance losses by using the heaviest gage wire possible for wrapping the windings.

Core Magnetization Loss

Every time the ac primary current reverses direction, the magnetic flux in the transformer core also reverses direction. However, most magnetic core materials have a tendency to retain their present magnetic flux direction. This tendency is called magnetic retentivity, as explained in Section 13-5.

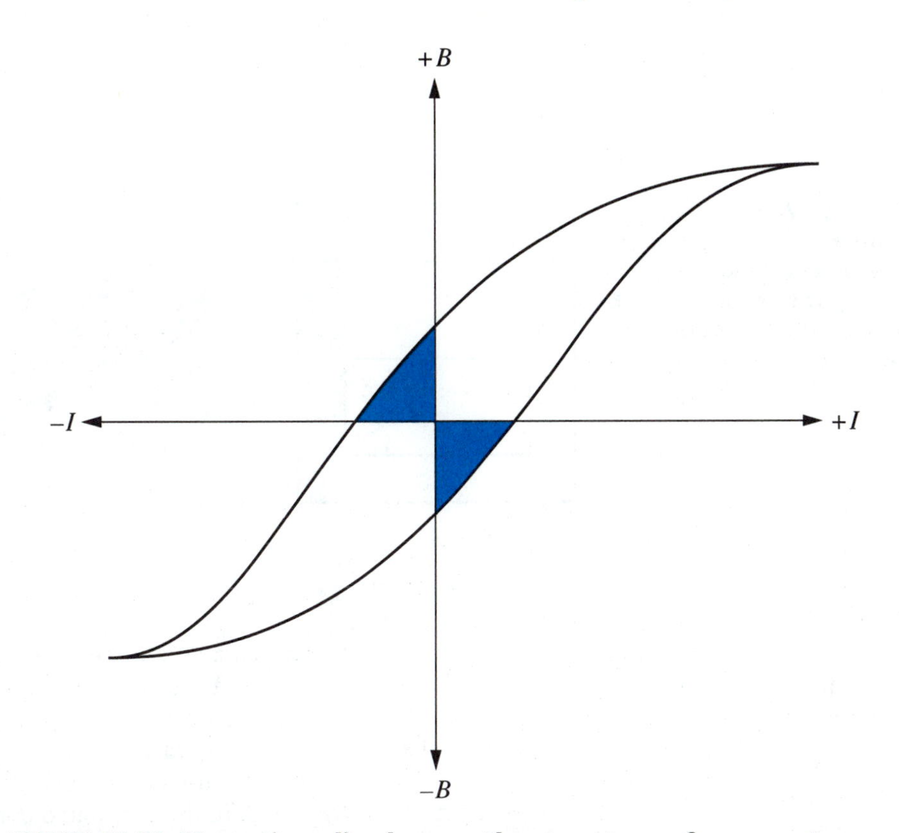

FIGURE 17-28 Magnetic cycling loop, as the current goes from zero to maximum positive, then back to zero; then repeating the process in the negative direction. Residual or retained magnetism is converted to heat energy in the shaded regions of the loop. This curve is often called a magnetic hysteresis curve, pronounced hiss-ter-é-sis.

In order to overcome the core's magnetic retentivity and force the flux to reverse, the primary winding must expend a small amount of energy. This energy is converted to heat inside the core material. Of course, this slight expenditure of energy happens twice per ac cycle, just as the primary current changes direction. Core magnetization loss is represented graphically by the magnetic cycling curve of Figure 17-28.

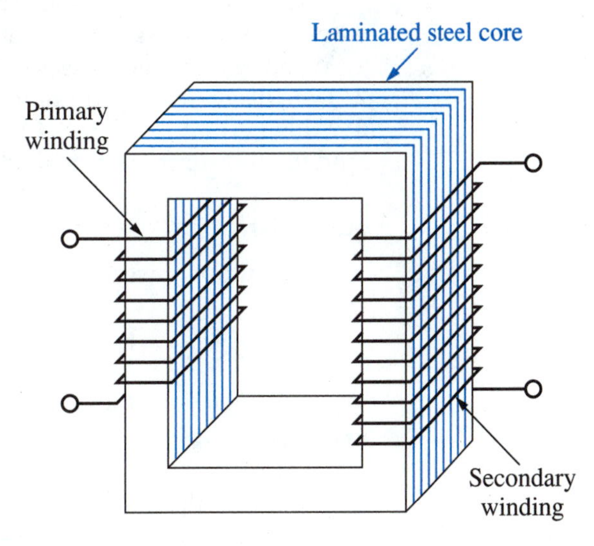

FIGURE 17-29 Laminating the core to reduce eddy-current power loss

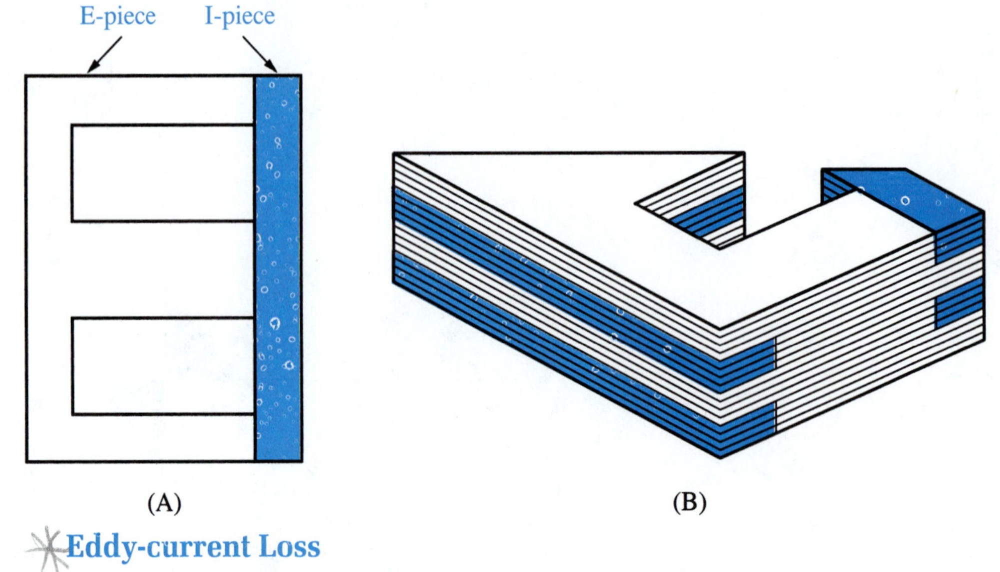

FIGURE 17-30 E-I core construction. (A) Thin E-pieces and I-pieces are brought in contact. (B) E-I lamination groups are alternated.

Eddy-current Loss

The changing magnetic flux in the core of the transformer creates voltage that tends to circulate current in the core itself, since the core is made of conducting material. This circulating current is called **eddy-current**. As it flows through the core's resistance, it produces waste heat.

Transformer manufacturers minimize eddy-currents by **laminating** the core, as shown in Figure 17-29. In the laminated core construction, layers of magnetic steel are separated by thin layers of insulation. The layers of insulation prevent the passage of current between neighboring steel layers, thus reducing the overall amount of eddy-current.

Laminated cores are often made with E-I pieces. This construction method is shown in Figure 17-30. Figure 17-30(A) shows an E-piece brought into contact with an I-piece to form one layer of a double-window core. When the core is actually assembled, there are several laminated layers having the I-piece on one side, followed by the same number of layers having the I-piece on the other side. This is illustrated in Figure 17-30(B).

EXAMPLE 17-8

A certain transformer has $V_P = 120$ V. It carries a primary current, I_P, of 0.8 A. Its power losses are as follows:

1. Winding resistance losses (primary and secondary combined) = 5.2 W.
2. Core magnetization (hysteresis) losses = 2.1 W.
3. Eddy-current losses = 0.4 W.

Find the transformer's actual efficiency.

SOLUTION

The input power to the primary side is calculated as

$$P_{in} = V_P I_P = 120 \text{ V } (0.8 \text{ A}) = 96 \text{ W}$$

The total power loss can be found by simply adding the three individual losses.

$$P_{loss\ (total)} = 5.2 \text{ W} + 2.1 \text{ W} + 0.4 \text{ W} = 7.7 \text{ W}$$

Continued on page 319.

The net output power delivered to the load by the secondary is given by

$$P_{out} = P_{in} - P_{loss} = 96\ \text{W} - 7.7\ \text{W} = 88.3\ \text{W}$$

Efficiency is calculated as

$$\eta = \frac{P_{out}}{P_{in}} = \frac{88.3\ \text{W}}{96\ \text{W}} = 0.9198 \text{ or } \mathbf{92\%}$$

✔ SELF-CHECK FOR SECTIONS 17-7 AND 17-8

22. If a transformer measures V_P normal, and a nonzero value of I_P, but $V_S = 0$, what is the probable trouble? [7]
23. Name the three causes of power loss in a real transformer. [8]
24. A certain transformer delivers 200 W to its load and operates with efficiency $\eta = 89\%$. How much power does it consume from its source? [8]
25. T-F The power loss that occurs in the windings of a transformer is sometimes called I^2R loss. [8]
26. How often must retained core magnetization be overcome in a transformer that is driven by a 60-Hz source? [8]
27. What technique is used to minimize eddy-current losses? [8]
28. T-F A narrower magnetic cycling loop results in less core magnetization loss. [8]

FORMULAS

$$\frac{V_S}{V_P} = \frac{N_S}{N_P}$$ **EQ. 17-1**

$$n = \frac{N_S}{N_P}$$ **EQ. 17-2**

$$\frac{I_P}{I_S} = \frac{N_S}{N_P} = n$$ **EQ. 17-4**

$$P_{in} = P_{out} \text{ (ideally)}$$ **EQ. 17-6**

$$V_P I_P = V_S I_S \text{ (ideally)}$$ **EQ. 17-7**

$$R_{LD} = R_{int} \text{ (for maximum power)}$$ **EQ. 17-8**

$$n = \sqrt{\frac{R_{LD}}{R_{int}}} \text{ (for matching)}$$ **EQ. 17-9**

$$\eta = \frac{P_{out}}{P_{in}} \text{ (for a nonideal transformer)}$$

SUMMARY OF IDEAS

- The primary winding of a transformer is driven by an ac voltage source. The secondary winding delivers ac power to a load.
- A transformer operates on the basis of changing magnetic flux in its core.
- In the step-down mode, secondary output voltage is less than primary input voltage. In the step-up mode, output voltage is greater than input voltage.
- The ratio of secondary voltage to primary voltage is equal to the ratio of secondary turns to primary turns, called the turns ratio. This is the transformer voltage law.

- The ratio of primary current to secondary current equals the turns ratio, n. This is the transformer current law.
- Ideally, the primary input power equals the secondary output power. This is the ideal transformer power law.
- A transformer can match a load resistance, R_{LD}, to a source internal resistance, R_{int}. This happens when $n = \sqrt{R_{LD}/R_{int}}$.
- Transformers electrically isolate the secondary circuit from the primary circuit.
- Real transformers are not 100% efficient. They waste some of the energy that is delivered to them.

CHAPTER QUESTIONS AND PROBLEMS

1. The__________winding of a transformer receives power from an external source. The_____winding delivers power to a load. [1]
2. When voltage is stepped down, current is__________. [1]
3. Draw the schematic symbols for a ferromagnetic-core transformer, a ferrite-core transformer, and an air-core transformer. [2]
4. An ideal transformer has $N_P = 250$ turns and $N_S = 100$ turns. The primary winding is driven by source voltage $V_P = 60$ V. How much secondary voltage, V_S, is induced? Is this step-up or step-down operation? [1, 3]
5. In Problem 4, if the source voltage were increased to 90 V, what would be the new value of V_S? [3]
6. T-F When the secondary load resistance is decreased, the primary current increases. [1, 3]
7. An ideal transformer is driven by $V_P = 75$ V. The secondary voltage, V_S, equals 225 V and drives a load resistance of 45 Ω. [3, 1]
   a) Find the turns ratio, n.
   b) Solve for I_S.
   c) Solve for I_P.
   d) In this transformer, the voltage has been stepped up, so the output current is__________than the input current.
8. An ideal transformer is driven by $V_P = 230$ V. V_S equals 57.5 V and drives a load resistance of 10 Ω. [3, 1]
   a) Find the turns ratio, n.
   b) Solve for I_S.
   c) Solve for I_P.
   d) In this transformer, the voltage has been stepped down, so the output current is__________than the input current.
9. A certain ideal step-down transformer has $n = 0.7$. It is driven by 120 V and is fused in the primary side with a 1-A fuse. [3]
   a) What is the maximum allowable secondary current?
   b) What is the minimum allowable load resistance?
   c) What is the maximum power that can be transferred?
10. In Figure 17-31, an ideal step-up transformer is driving a parallel group of 800-Ω resistors. How many 800-Ω resistors can be connected into the secondary circuit without blowing the fuse? [3]

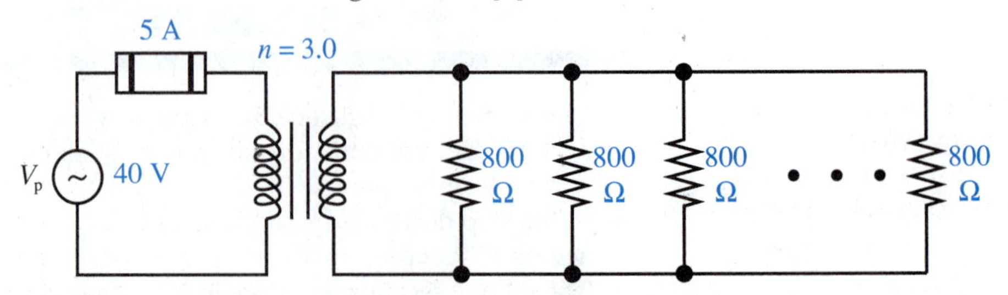

FIGURE 17-31 If too many parallel resistors are connected to the secondary side, the primary fuse will blow.

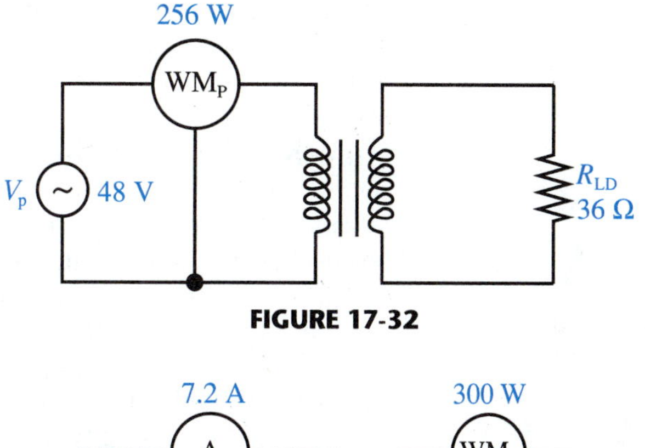

FIGURE 17-32

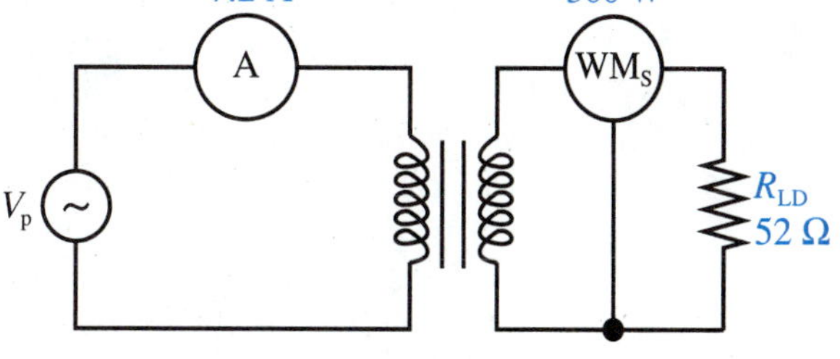

FIGURE 17-33

11. In the ideal transformer circuit of Figure 17-32, [3]
    a) What is the power delivered to the load resistor?
    b) Find V_S.
    c) Find the transformer turns ratio, n.
    d) Calculate I_S and I_P.
12. In the ideal transformer circuit of Figure 17-33, [3]
    a) How much power is delivered by the source to the primary side of the transformer?
    b) Find V_P, V_S, I_S, and n.
13. Maximum power transfer occurs when the load resistance is matched to the source's_______ ___________. [4]
14. It is desired to match a load resistance of 12 Ω to a source with $R_{int} = 50\ \Omega$. What turns ratio is needed? Is the transformer stepping up or stepping down? [4, 1]
15. It is desired to match a load resistance of 10 kΩ to a source with $R_{int} = 50\ \Omega$. What turns ratio is needed? Is the transformer stepping up or stepping down? [4, 1]
16. Describe some of the advantages that result from the isolation capability of transformers. [5]
17. A certain 120-V (primary voltage) transformer is described in a catalog as having a 12-V center-tapped winding. Explain what this means. [6]
18. In the circuit of Figure 17-34, [6, 3]
    a) Find V_{S1} and V_{S2}.
    b) Find I_{S1} and I_{S2}.
    c) Find I_P.

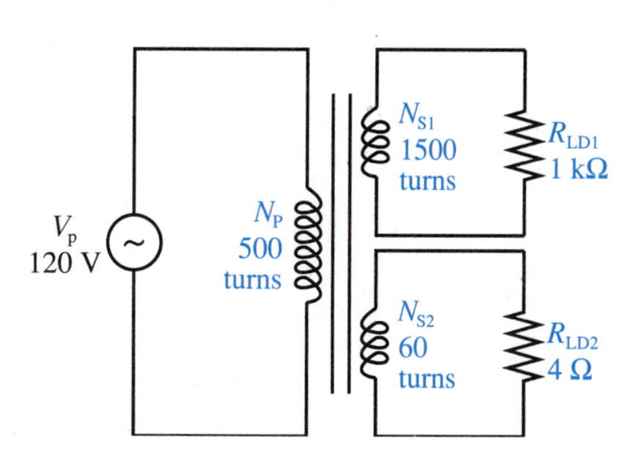

FIGURE 17-34 Dual-secondary transformer. One low-voltage and one high-voltage secondary.

19. List the reasons for power loss in a real transformer. [8]
20. A certain real transformer delivers 50 V to a 20-Ω load. If its primary power consumption is 135 W, calculate its percent efficiency. [8]

Troubleshooting Questions

21. A certain transformer has an ammeter in its primary circuit. Trouble occurs that makes the secondary output voltage become zero. The primary ammeter also reads zero. A fuse check reveals that the fuse is good and the primary voltage source is OK. What is the probable cause of this trouble? [7]
22. Trouble occurs that causes a transformer to blow its primary fuse. A measurement of the load resistance shows that the load is OK. The voltage source also checks OK. Name two possible causes of this trouble. [7]
23. For the trouble in Question 22, how would you find out which possibility actually occurred? [7]
24. A certain transformer has an ammeter in its primary circuit. Trouble occurs that makes the secondary output voltage become zero. The primary ammeter reads a nonzero value of current. Both the fuse and the voltage source are all right. What is the probable cause of this trouble? [7]
25. A technician is working in a circuit that is driven by a transformer secondary winding. Happening to touch an exposed metal surface in the circuit, the technician receives a mild shock. What is the probable trouble here? How would the technician check this suspicion? What would she do to fix the problem? [7]

CHAPTER 18

REACTANCE

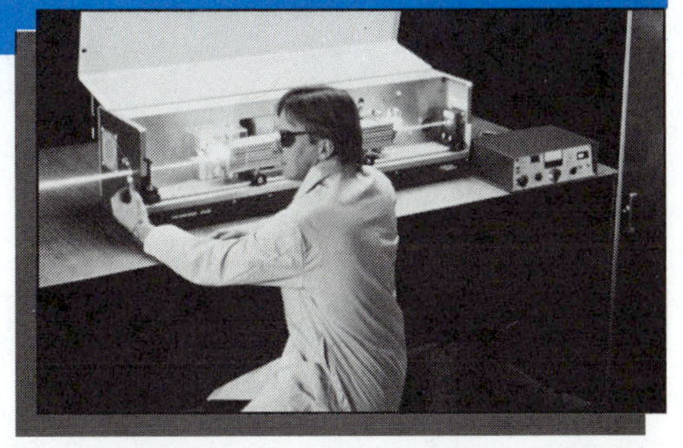

OUTLINE

18-1 Meaning of Reactance
18-2 Capacitive Reactance
18-3 Ohm's Law for Capacitors
18-4 Power in a Capacitor
18-5 Inductive Reactance
18-6 Ohm's Law for Inductors
18-7 Power in an Ideal Inductor
18-8 Real Inductors

A natural beam of light contains electromagnetic waves of varying frequencies, all of them randomly out of phase with each other. A beam of light emitted by a laser contains electromagnetic waves that are very closely matched in frequency, and perfectly in phase with each other. Laser beams like the one shown here have several medical applications. In the lower photo a laser beam is shown breaking apart a kidney stone without invasive surgery.
Courtesy of Candela Laser Corporation

NEW TERMS TO WATCH FOR

capacitive reactance
inductive reactance
leading
lagging
power pulsation
high-frequency signal
low-frequency signal
real inductor
skin effect
Q-factor

The behavior of capacitors and inductors in dc switching circuits has already been explained in Chapter 15. The time-constant principles, however, do not apply in an ac circuit. An altogether different set of ideas are needed for describing capacitors and inductors in ac sine wave circuits.

After studying this chapter, you should be able to:

1. Describe the phase relationship between current and voltage for a capacitor in an ac circuit.
2. Define capacitive reactance and calculate it from knowledge of frequency and capacitance.
3. Apply Ohm's law to a capacitive ac circuit.
4. Explain why ideal capacitors and inductors consume zero average power during a complete sine-wave cycle.
5. For an inductor in an ac circuit, describe the phase relationship between current and voltage.
6. Define inductive reactance and calculate it from knowledge of frequency and inductance.
7. Apply Ohm's law to an inductive ac circuit.
8. Explain the effect of variable frequency on capacitor behavior and inductor behavior.
9. Describe how a real inductor differs from an ideal inductor and define a real inductor's Q-factor.

18-1 MEANING OF REACTANCE

Reactance is opposition to sine-wave ac current. Reactance is measured in ohms, just like resistance. Capacitors and inductors both possess reactance. The symbol for **capacitive reactance** is X_C. The symbol for **inductive reactance** is X_L.

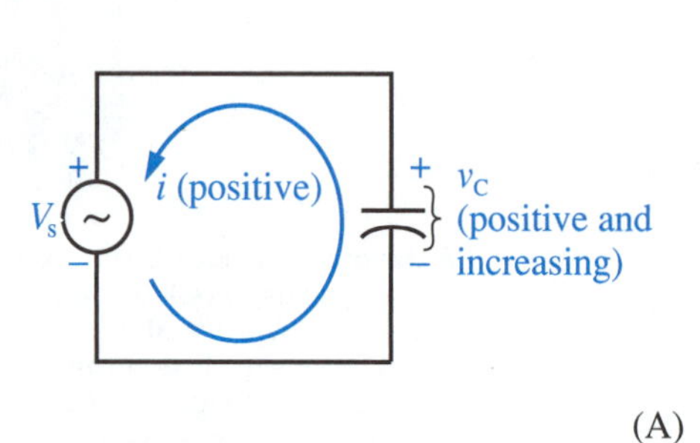

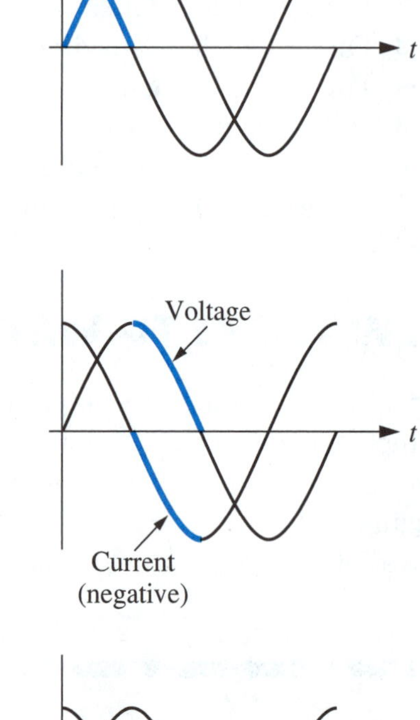

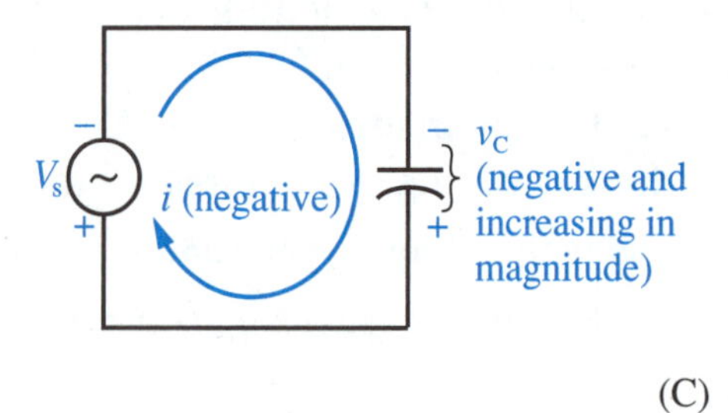

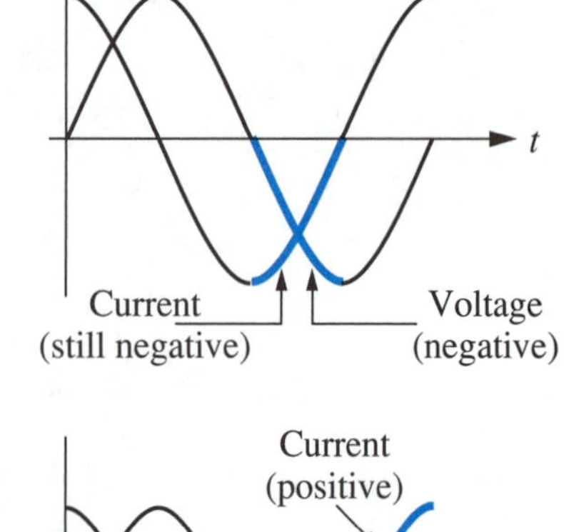

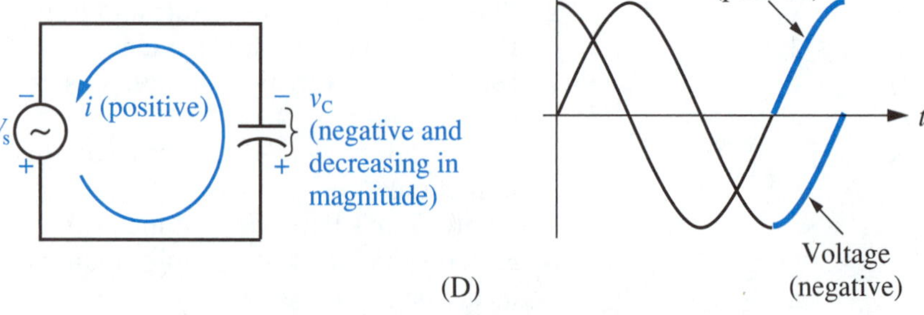

FIGURE 18-1 Explaining the relationship between capacitor voltage and current during an ac cycle.

18-2 CAPACITIVE REACTANCE

In an ac circuit, the voltage source rises toward its positive peak during the first quarter cycle, as shown in Figure 18-1(A). At the very beginning of the quarter cycle the voltage is rising quickly, so charge must be deposited on the capacitor plates quite rapidly, since

$$v_C = \frac{1}{C} q$$

The rate at which charge is deposited on the plates is instantaneous current, i, since

$$i = \frac{q}{t}$$

Therefore, the current must be large when voltage is small, near the beginning of the voltage cycle. This is shown in Figure 18-1(A). Toward the end of the first quarter cycle the voltage is larger, but it is not climbing as rapidly. Therefore, charge is not deposited on the plates as rapidly, which means that the current becomes smaller, as Figure 18-1(A) shows.

During the second quarter cycle, shown in Figure 18-1(B), the voltage is decreasing. For this to happen, charge must begin to be removed from the capacitor plates. Of course, removing charge is equivalent to reversing its flow, which is a reversal of current. Thus, the current becomes negative during the second quarter cycle, as Figure 18-1(B) indicates.

Figure 18-1(C) shows the third quarter cycle. The voltage source has become negative, so naturally the capacitor voltage, v_C, also becomes negative, as demanded by Kirchhoff's voltage law. In order for v_C to become more and more negative, charge must accumulate negative on the top terminal; therefore, the current must continue flowing in the negative direction. However, the magnitude of the current becomes smaller throughout this quarter cycle because the voltage is changing less rapidly. Figure 18-1(C) points out this relationship.

During the fourth quarter cycle the voltage remains negative, but decreases in magnitude. This means that the accumulated charge is being removed from the capacitor. Removal of charge requires a reversal of current, so we see the current returning to the positive direction in Figure 18-1(D).

Summarizing the entire sine-wave cycle, we see that:

TECHNICAL FACT

For a capacitor in an ac circuit, current and voltage are 90° out of phase, with current **leading.**

Now consider what happens if the source voltage magnitude remains the same, but its frequency is increased. With voltage the same, the charge accumulation must be the same, since

$$q = C\,v_C$$

But if frequency is higher, then the same amount of charge must be moved onto the capacitor plates in less time. This means that current, *i*, must increase, since

$$i = \frac{q}{t}$$

(same amount of charge)

(less time available)

(increased current)

The increase in current with increasing frequency is illustrated in Figure 18-2.

Of course, the size of the capacitor itself also has an effect on the amount of current. A larger capacitor requires more charge accumulation for a given voltage ($q = Cv$). A greater charge accumulation means a larger value of current. This is assuming that time does not vary (constant frequency).

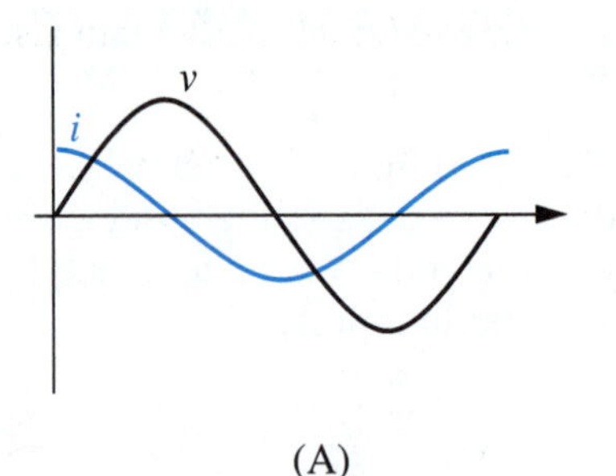

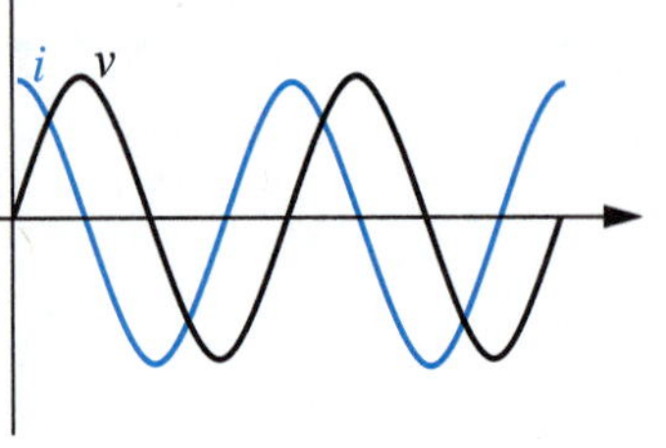

(A) (B)

FIGURE 18-2 (A) With the source at a lower frequency, a certain amount of current flows in the capacitor. (B) With the source at a higher frequency (but same voltage), a greater amount of current flows.

Since current is proportional to frequency and is also proportional to capacitance, it follows that the capacitor's opposition to current (its reactance) must be inversely proportional to frequency and inversely proportional to capacitance. Stated as a formula,

$$X_C = \frac{1}{2\pi f C}$$

EQ. 18-1

Substituting an approximate value for π gives

$$X_C = \frac{1}{6.2832\, f C}$$

EQ. 18-1

EXAMPLE 18-1

Calculate the reactance in ohms for a 0.2-μF capacitor that is driven at a frequency of 500 Hz.

SOLUTION

Use Equation 18-1. On the calculator, first multiply 2 times π , times 500 Hz, times 0.2×10^{-6} F. Then reciprocate the resulting product, using the 1/x key. The keystroke sequence is

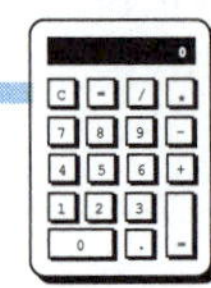

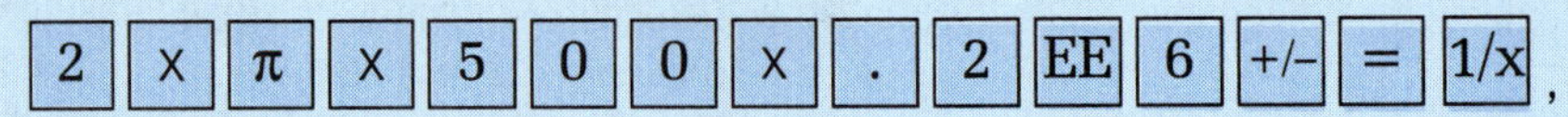

which yields

$$X_C = \frac{1}{2\pi\,(500)\,(0.2 \times 10^{-6})} = \mathbf{1592\ \Omega}$$

The number π is available on your calculator, either as its own key or as the second function of another key. The 2 times π part of the product can be replaced by the number 6.2832, as mentioned previously. For calculations that don't require great precision, the number 6.28 is all right.

EXAMPLE 18-2

What frequency value would cause the 0.2-μF capacitor to have a reactance of 700 Ω?

SOLUTION

Rearranging Equation 18-1 to solve for frequency, f, we get

$$f = \frac{1}{6.2832\, C X_C}$$

$$= \frac{1}{6.2832\,(0.2 \times 10^{-6})\,(700)} = \mathbf{1137\ Hz}$$

Of course, Equation18-1 can also be rearranged to solve for C if f and X_C are known.

✓ SELF-CHECK FOR SECTIONS 18-1 AND 18-2

1. In a purely capacitive circuit, what is the phase relationship between current and source voltage? [1]
2. A capacitor's opposition to ac current is called capacitive __________. [2]
3. In Figure 18-1, during which quarter cycle is the electron charge deposited on the bottom plate? [1]
4. During which quarter cycle is the electron charge removed from the bottom plate? [1]
5. During which quarter cycle is the electron charge deposited on the top plate? [1]
6. During which quarter cycle is the electron charge removed from the top plate? [1]
7. What is the symbol for capacitive reactance? [2]
8. Capacitive reactance is measured in units of __________. [2]
9. Calculate the capacitive reactance of a 0.01-μF capacitor at $f = 2000$ Hz. [2]
10. For the 0.01-μF capacitor of Problem 9, what frequency would produce an X_C value of 15 kΩ? [2]
11. It is desired to choose a capacitance that has a reactance of 796 Ω at a frequency of 250 Hz. What value of capacitance, C, is required? [2]

18-3 OHM'S LAW FOR CAPACITORS

In a capacitive ac circuit, the idea of reactance corresponds to the idea of resistance in a dc resistive circuit. Therefore, Ohm's law is just as valid in a capacitive ac circuit as it is in a dc circuit. We say:

$$I = \frac{V}{X_C}$$

EQ. 18–2

In Equation 18-2, voltage, V, and current, I, must be in compatible units (both rms or both peak-to-peak).

EXAMPLE 18-3

What value of current flows in the circuit of Figure 18-3?

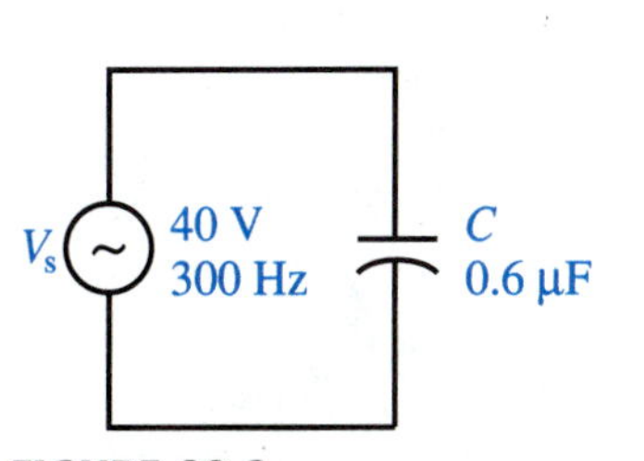

FIGURE 18-3 Demonstrating Ohm's law for a capacitive ac circuit

SOLUTION

The capacitive reactance is found from Equation 18-1 as

$$X_C = \frac{1}{6.28 fC} = \frac{1}{6.28\,(300)\,(0.6 \times 10^{-6})} = 885\ \Omega$$

Applying Ohm's law for a capacitor, Equation 18-2 gives

$$I = \frac{V_s}{X_C} = \frac{40\text{ V}}{885\ \Omega} = 0.0452\text{ A or } \mathbf{45.2\text{ mA}}$$

Since the source voltage is assumed to be expressed in rms units, the current also is in rms units.

EXAMPLE 18-4

Suppose the frequency was variable in Figure 18-3. What value of frequency would produce an rms current of 100 mA?

SOLUTION

Rearranging the Ohm's law equation, Equation 18-2, to solve for reactance, we get

$$X_C = \frac{V_s}{I} = \frac{40\ \text{V}}{100\ \text{mA}} = 400\ \Omega$$

Then, from Equation 18-1,

$$f = \frac{1}{6.28\ CX_C} = \frac{1}{6.28\ (0.6 \times 10^{-6}\ \text{F})\ (400\ \Omega)} = \mathbf{663\ Hz}$$

By combining Equations 18-1 and 18-2, we can point out the relation between current, I, and frequency, f, for a given value of capacitance.

$$I = \frac{V}{X_C} = \frac{V}{\frac{1}{2\pi fC}} = V\,(6.28\,f\,C)$$

In words, current, I, is directly proportional to frequency, f, for a given value of C. This is often stated as:

TECHNICAL FACT

Capacitors pass high-frequency signals better than they pass low-frequency signals.

EXAMPLE 18-5

Calculate the current in both of the circuits of Figure 18-4. Why are these results reasonable?

SOLUTION

In circuit A, the reactance is given by Equation 18-1 as

$$X_{CA} = \frac{1}{6.2832\,f_A\,C} = \frac{1}{6.2832\ (100\ \text{Hz})\ (2 \times 10^{-6}\ \text{F})} = 795.8\ \Omega$$

The current in circuit A is found from Ohm's law, as

$$I_A = \frac{V_s}{X_{CA}} = \frac{15\ \text{V}}{795.8\ \Omega} = \mathbf{18.85\ mA}$$

Repeating for circuit B, we get

$$X_{CB} = \frac{1}{6.2832\,f_B\,C} = \frac{1}{6.2832\ (300\ \text{Hz})\ (2 \times 10^{-6}\ \text{F})} = 265.3\ \Omega$$

$$I_B = \frac{V_s}{X_{CB}} = \frac{15\ \text{V}}{265.3\ \Omega} = \mathbf{56.55\ mA}$$

The frequency in B is three times greater than in A. Therefore, it is reasonable that the current in B is also three times as large.

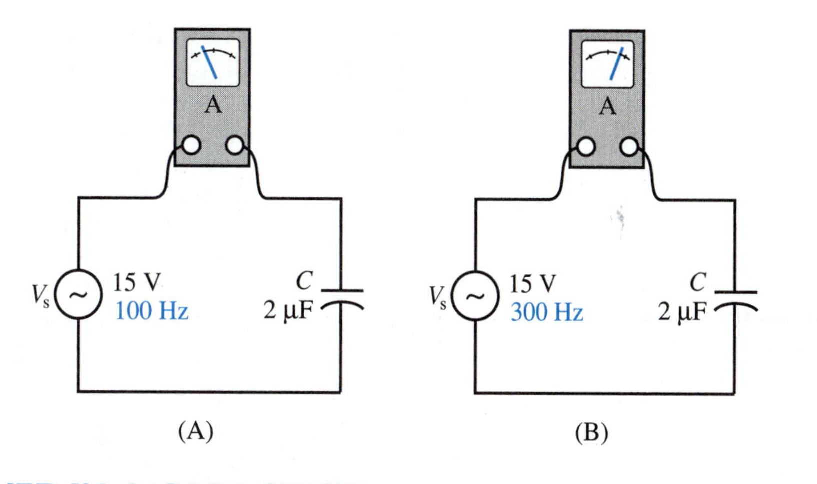

FIGURE 18-4 The effect of frequency on a capacitor. (A) Low frequency—low current. (B) Higher frequency—higher current

18-4 POWER IN A CAPACITOR

An ideal capacitor consumes zero average power in an ac circuit. Instead, it simply stores energy for one quarter cycle, then returns that energy to the source during the next quarter cycle. Refer to Figures 18-1 and 18-5 to understand this.

In the first quarter cycle, shown in part A of Figure 18-1, voltage and current are both positive. The electron current is entering the capacitor at its negative (–) terminal and exiting from the capacitor at its positive (+) terminal. This is the same way that a resistor would behave, so the capacitor is temporarily acting like a power consumer. This is indicated in Figure 18-5 as a positive **pulsation of power** during the first quarter cycle.

However, during the second quarter cycle in part B of Figure 18-1, voltage is positive but current is negative. The electron current is leaving the capacitor from its negative terminal, and entering the capacitor at its positive terminal. This is the same way that an ac voltage source would behave. Thus, the capacitor is temporarily acting like a power deliverer. This reversal of the power condition is indicated in Figure 18-5 as a negative power pulsation during the second quarter cycle.

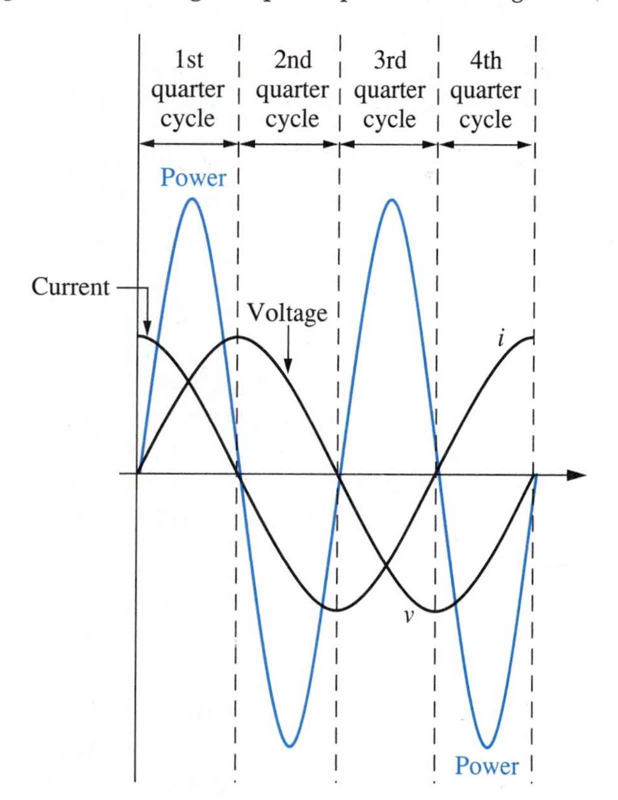

FIGURE 18-5 In a capacitive ac circuit, power pulsates once each quarter cycle. Each positive pulsation (power temporarily being consumed) is followed by a negative pulsation (power being returned).

The capacitor returns to an energy consumer during the third quarter cycle, part C of Figure 18-1. It then reverts to an energy deliverer during the fourth quarter cycle in part D of Figure 18-1. Therefore, Figure 18-5 shows another positive power pulsation in the third quarter cycle, followed by another negative power pulsation in the fourth quarter cycle of the sine wave.

So, averaged over the entire ac cycle, the same amount of energy that is temporarily stored during the first and third quarter cycles is simply returned to the source during the second and fourth quarter cycles. We can say:

TECHNICAL FACT

For an ideal capacitor in an ac circuit,

$$P = 0$$

EQ. 18-3

✓ SELF-CHECK FOR SECTIONS 18-3 AND 18-4

12. For a capacitor in an ac circuit, rms current is equal to rms voltage divided by the__________ __________. [3]
13. For a given capacitor and a fixed value of ac voltage, if the frequency is doubled, the current is__________. [2, 3]
14. Calculate the current in the circuit of Figure 18-6. [2, 3]
15. In Figure 18-6, suppose the frequency is reduced to 250 Hz, with everything else remaining the same. What is the new value of current? [2, 3]
16. In Figure 18-6, suppose frequency is variable. What value of frequency will produce current $I = 2$ mA? [2, 3]
17. Find the total current in the circuit of Figure 18-7. [2, 3]
18. T-F In an ac circuit, a capacitor spends half of its time storing energy from the source and the other half of its time returning energy to the source. [4]
19. In a capacitive ac circuit, how many power pulsations are produced by each complete cycle? [4]

18-5 INDUCTIVE REACTANCE

For an inductor, the time rate of change of current is proportional to voltage, as expressed in Equation 14-1.

$$v_{\mathrm{L}} = L\left(\frac{\Delta i}{\Delta t}\right)$$

We can use this fact to explain the relationship between current and voltage in an ac inductive circuit. Refer to Figure 18-8.

5

In part A of Figure 18-8, we begin by looking at the second quarter cycle, when voltage has reached its positive peak and is returning to zero. The time rate of change of current is greatest when its waveform is the steepest. This occurs at the moment when the current sine wave is crossing through zero, about to become

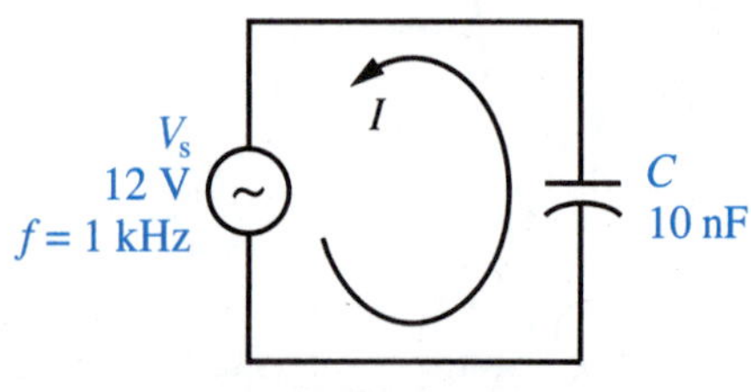

FIGURE 18-6

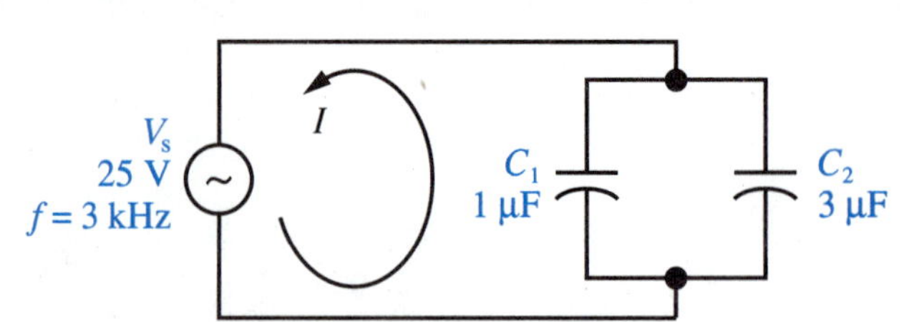

FIGURE 18-7 Ohm's law for a two-capacitor parallel circuit

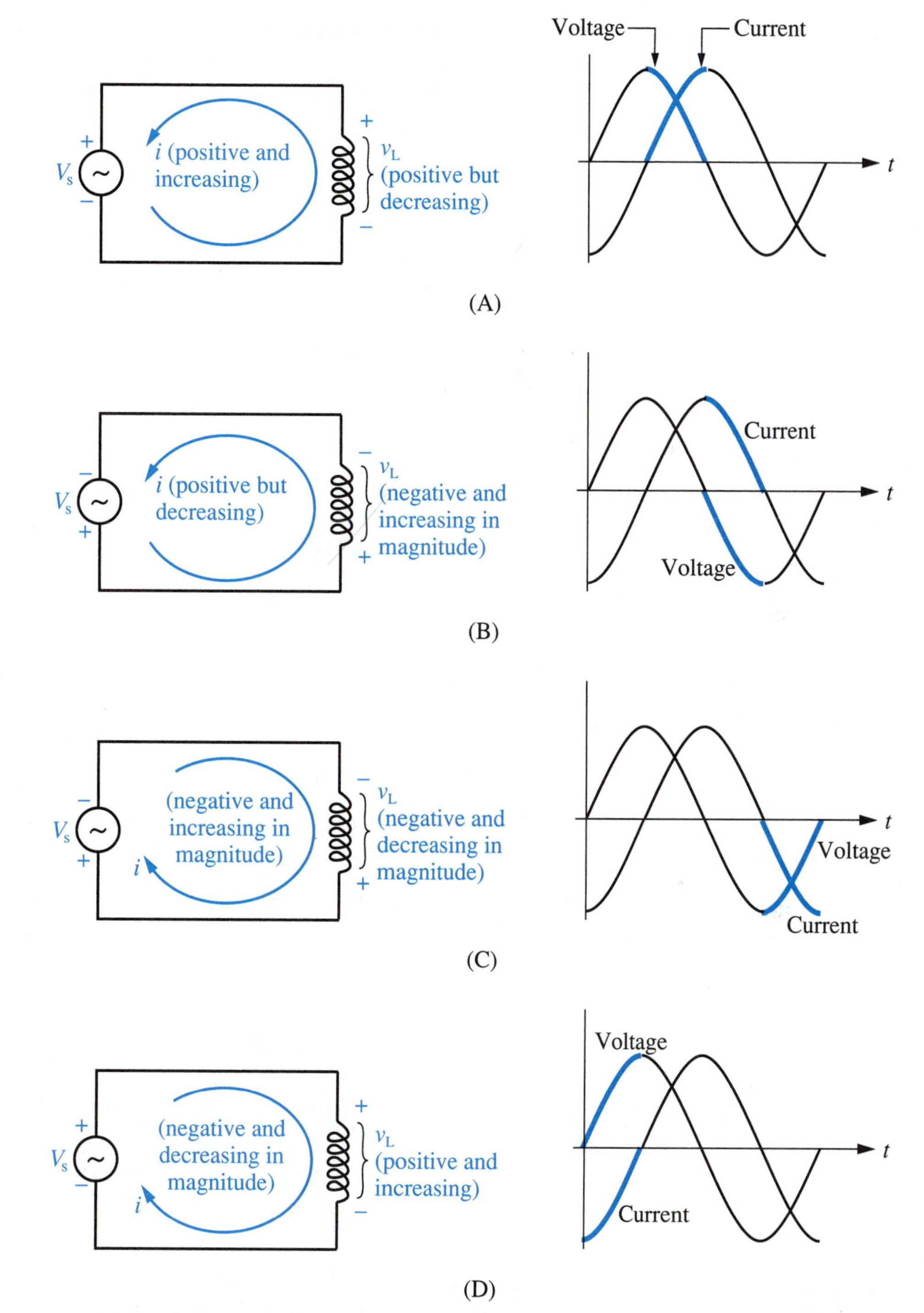

FIGURE 18-8 Explaining the relationship between inductor voltage and current during an ac cycle

positive. Thus, according to Equation 14-1, the sine wave of current must just be crossing through zero at the moment when the voltage sine wave is at its peak value. This is shown in Figure 18-8(A).

Figure 18-8(B) shows the next quarter cycle. Voltage becomes negative, so the time rate of change of current must also become negative—that is, heading downward rather than upward. This is shown in Figure 18-8(B).

During the next quarter cycle, shown in Figure 18-8(C), voltage remains negative so current must continue heading downward. But since the magnitude of voltage is becoming less, the current waveform must become less steep. This is shown in Figure 18-8(C).

In Figure 18-8(D), which is actually the first quarter cycle of voltage, the current returns to an upward (positive) slope, since the voltage returns to its positive region.

Summarizing the entire sine-wave cycle, we see that

TECHNICAL FACT

In an inductive ac circuit, current and voltage are 90° out of phase, with current **lagging** voltage.

Now consider what happens if the source voltage magnitude remains the same, but its frequency is increased. The higher frequency tends to make the zero crossover points on the current waveform steeper than before. But according to Equation 14-1, the actual steepness of the current waveform ($\Delta i/\Delta t$) cannot change if the voltage remains the same. Therefore, the current magnitude must decrease in order to compensate for the more rapid oscillation, and in that way hold the value of $\Delta i/\Delta t$ at a constant value.

In other words, if Δt becomes smaller because f increases, Δi must also become smaller in order to balance Equation 14-1.

$$v_L = L\frac{\Delta i}{\Delta t}$$

(same amount of voltage) → v_L

Δi: (less current to compensate for less time)

Δt: (less time because of frequency increase)

The decrease in current with increasing frequency is illustrated in Figure 18-9.

The size of the inductor itself also has an effect on the amount of current. A larger inductor can create a given amount of voltage with a smaller rate of change of current.

$$v_L = L\left(\frac{\Delta i}{\Delta t}\right)$$

L: If this is larger

$\left(\frac{\Delta i}{\Delta t}\right)$: this term can be smaller and still result in a constant v_L.

Therefore, a larger inductor results in a proportionately smaller current, assuming a constant frequency and constant voltage.

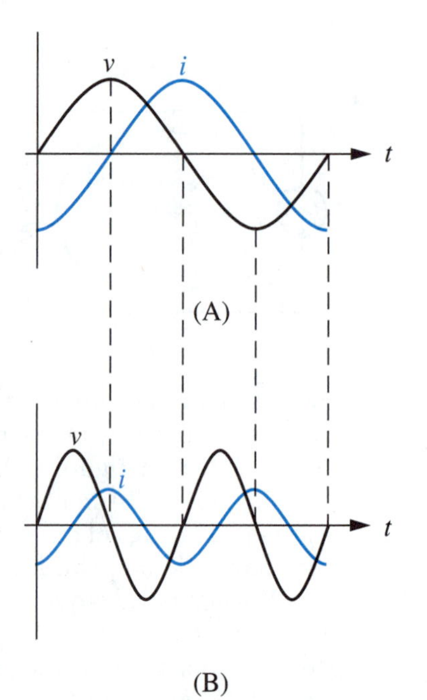

FIGURE 18-9 (A) At a lower source frequency, a certain amount of inductive current flows. (B) At a higher source frequency (same voltage), less current flows.

FILM SPECIAL EFFECTS

Special effects in films are often created by combining film frames. Here a single frame in the movie *GHOST* has been created by combining, or compositing, two separately shot frames. See page 7 in the Applications color section.

- The still frame, (A), shows simply two walls and a door.
- In frame (B), the door has been removed from its hinges and a blue screen has been erected to cover the door opening and the glass wall to the corner. In this frame the actor is extending his arm through the open door in normal fashion. The blue screen provides clear photographic delineation against his body.
- A black-and-clear copy, called a matte, is then made of this action frame. The electronically controlled matte-making process causes the film space occupied by the actor's body to convert to black. The blue screen backing and the rest of the action frame to the left of the blue screen convert to clear. Special-effects technicians then alter the matte frame by changing the actor's arm below the elbow from black to clear. A reverse of this altered matte pattern is then produced. It is called a counter-matte and is shown in (C). Note that the arm below the elbow is missing from the counter-matte image.
- The counter-matte filmstrip is then physically overlaid on the film that shows the actor in motion. The physical alignment between the counter-matte film and the action film must be virtually perfect. For example, the tip of the actor's nose on the counter-matte copy must lie precisely on the tip of his nose on the action film. This exact alignment is called registration. It cannot be checked automatically, but requires careful visual inspection by a technician. These two overlaid (called bi-packed) film frames are then rephotographed. The resulting film shows the actor and everything to the right of the actor, but not his arm below the elbow. This new film is called the dupe. The dupe filmstrip is rewound to be used again later. Then the altered matte, which is not shown here, is physically overlaid on still frame (A). This bi-pack combination is rephotographed onto the dupe film. Using optoelectronic sensors, the dupe's exposure from this still bi-pack is matched to the earlier exposure from the action bi-pack. This final rephotography produces the movie frame shown in (D).

Repeating this process for many frames, we see the ghost thrusting his arm straight through the solid door.

(A)

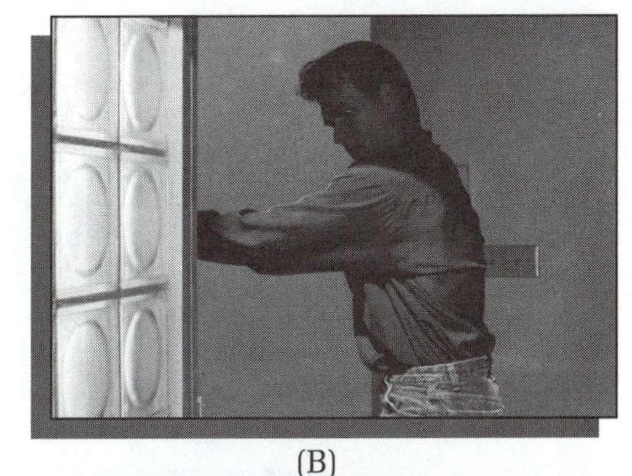

(B)

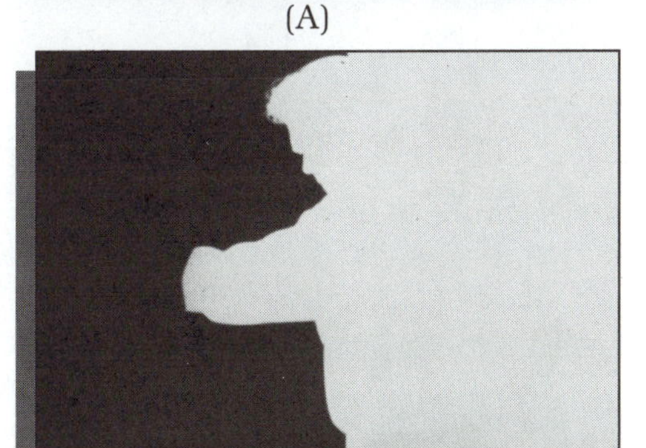

(C)

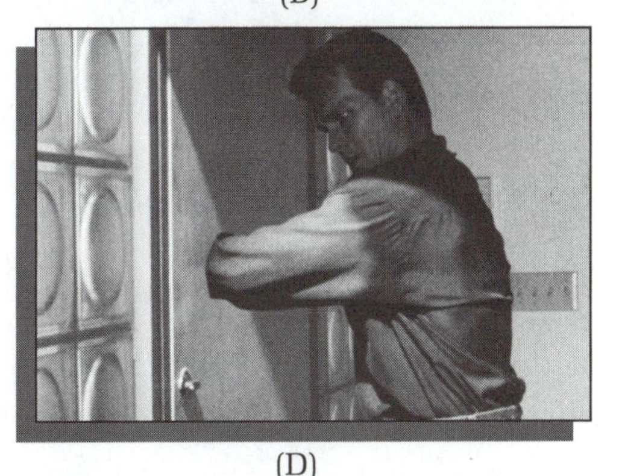

(D)

Creating the illusion of ghostly powers. (A) Background. (B) Action photographed against blue screen. (C) Altered counter-matte. (D) Final composite frame. *Courtesy of Paramount Pictures, copyright 1990,*

Since current is inversely proportional to frequency and inversely proportional to inductance, *L*, it follows that the opposition to current (reactance) is directly proportional to frequency and to inductance. Stated as a formula,

$$X_L = 2\pi fL$$

EQ. 18-4

EXAMPLE 18-6

a) What is the reactance of a 1.5-henry inductor at 100 Hz?
b) Repeat for a frequency of 1 kHz. Why is this result reasonable?

SOLUTION

a) From Equation 18-4,

$$X_L = 2\pi fL = 6.2832\ (100\ \text{Hz})\ (1.5\ \text{H}) = \mathbf{942.5\ \Omega}$$

b) $X_L = 6.2832\ (1 \times 10^3\ \text{Hz})\ (1.5\ \text{H}) = \mathbf{9425\ \Omega}$

The reactance is 10 times larger since the frequency is 10 times as great.

18-6 OHM'S LAW FOR INDUCTORS

TECHNICAL FACT

Ohm's law applies to inductors in ac circuits just as it applies to capacitors. We write for an inductor

$$I = \frac{V}{X_L}$$

EQ. 18-5

EXAMPLE 18-7

Figure 18-10 shows a constant inductance, 0.5 H, driven by different frequencies. Find the current in each case and explain why these results are reasonable.

SOLUTION

For circuit A,

$$X_{LA} = 2\pi f_A L = 6.2832\ (100\ \text{Hz})\ (0.5\ \text{H}) = 314.2\ \Omega$$

From Equation 18-5, Ohm's law for an inductor,

$$I_A = \frac{V_s}{X_{LA}} = \frac{15\ \text{V}}{314.2\ \Omega} = \mathbf{47.75\ mA}$$

Repeating for circuit B,

$$X_{LB} = 6.2832\ (300\ \text{Hz})\ (0.5\ \text{H}) = 942.5\ \Omega$$

$$I_B = \frac{15\ \text{V}}{942.5\ \Omega} = \mathbf{15.92\ mA}$$

This is reasonable because circuit B, with three times the frequency, allows only one-third as much current.

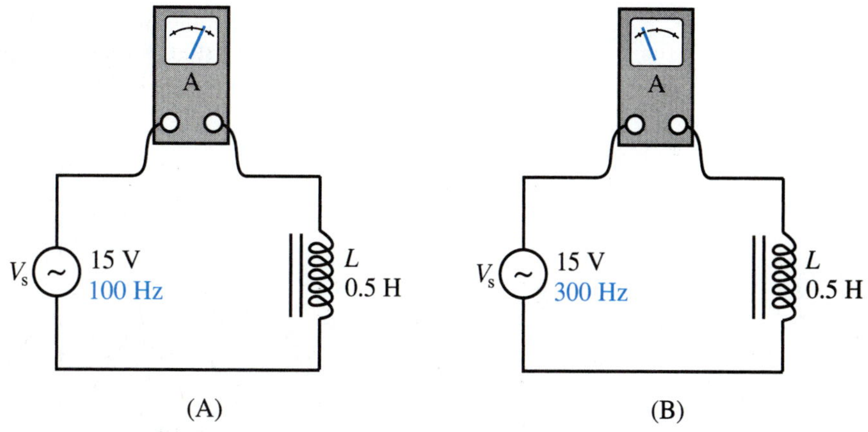

FIGURE 18-10 The effect of frequency on an inductor. (A) Low frequency, larger current. (B) Higher frequency, less current

As a general rule, we say

Inductors pass low-frequency signals better than they pass high-frequency signals.

EXAMPLE 18-8

In Figure 18-11, suppose the frequency to be variable. What frequency will produce a total source current of 200 mA? The inductors are magnetically shielded from one another so they do not interact.

SOLUTION

If two parallel inductors do not interact magnetically, they can be combined by the reciprocal formula.

$$\frac{1}{L_T} = \frac{1}{L_1} + \frac{1}{L_2} = \frac{1}{200\text{ mH}} + \frac{1}{300\text{ mH}}$$

$$L_T = 120\text{ mH}$$

Rearranging Equation 18-5 gives

$$X_{LT} = \frac{V_s}{I_T} = \frac{30\text{ V}}{200\text{ mA}} = 60\ \Omega$$

Then rearranging the inductive reactance formula, Equation 18-4, gives

$$f = \frac{X_{LT}}{2\pi L_T} = \frac{60\ \Omega}{6.2832\ (120 \times 10^{-3}\text{ H})} = \mathbf{79.6\ Hz}$$

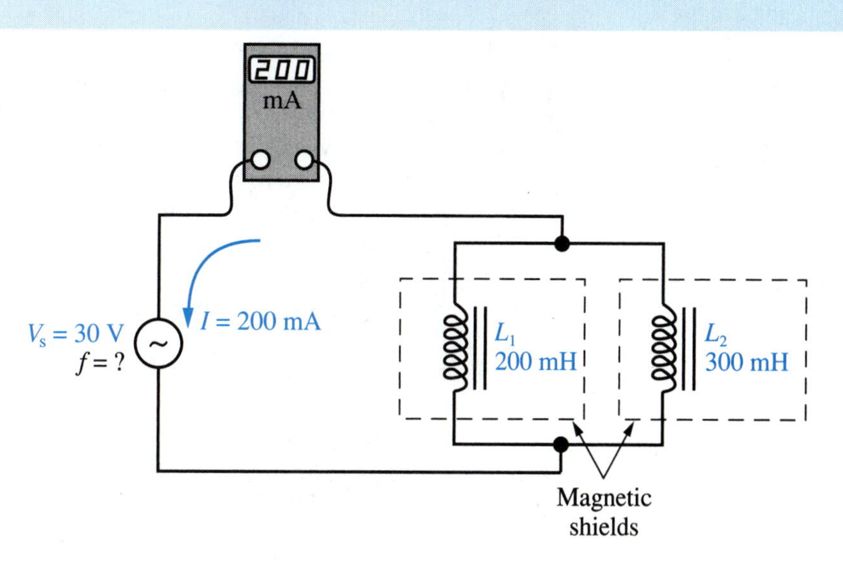

FIGURE 18-11 Shielded inductors connected in parallel. Applying Ohm's law.

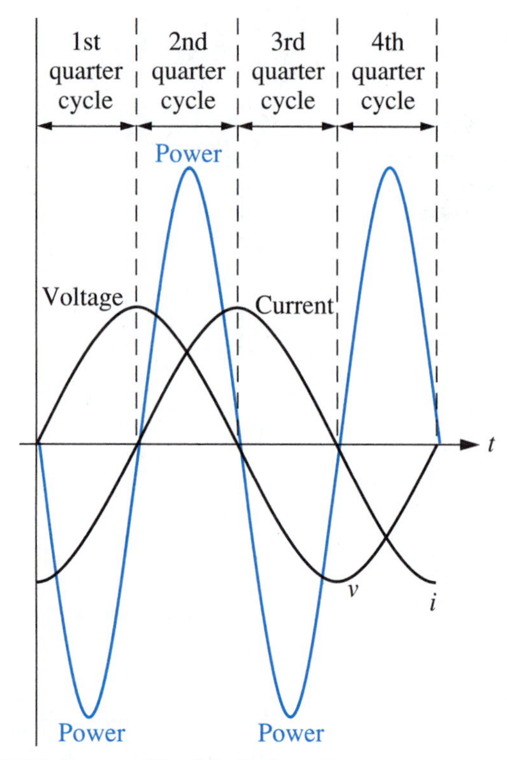

FIGURE 18-12 For an ideal inductor the average power is zero.

18-7 POWER IN AN IDEAL INDUCTOR

An ideal inductor is like an ideal capacitor in that it consumes zero average power. Energy is temporarily stored in the magnetic field of the inductor during one quarter cycle, but is then returned to the source during the next quarter cycle.

In Figure 18-12, the inductor stores energy when current and voltage are both positive together during the second quarter cycle, and also when they are both negative together during the fourth quarter cycle. But power becomes negative when current and voltage are opposite each other in the first and third quarter cycles. Negative power pulsations represent a return of stored energy to the source.

Averaging over the complete cycle, for an ideal inductor, $P = O$.

18-8 REAL INDUCTORS

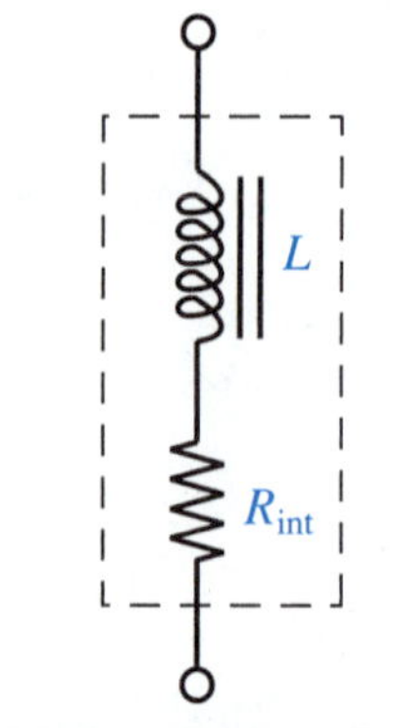

FIGURE 18-13 A real inductor is usually visualized as an ideal inductor in series with an internal resistance.

Real-life inductors differ from the ideal model in that they always possess some internal resistance. We think of a **real inductor** as shown in Figure 18-13.

The internal resistance of a real inductor is due in part to the ohmic resistance of the wire in its winding. However, there are other ac effects that cause the actual internal resistance to be greater than the ohmic resistance that would be measured by a dc ohmmeter. This is pointed out in Figure 18-14.

There are three ac effects that account for this difference.

1. Magnetic hysteresis losses in the inductor's core. This is like a transformer.
2. Eddy-current losses in the core; also like a transformer.
3. Skin effect in the winding.

Skin effect is the tendency for ac current to concentrate near the surface, or "skin," of a piece of wire. Ac current does not flow evenly across the wire's entire cross section, the way dc current does. This is illustrated in Figure 18-15. Because ac current does not use the center of the wire, the cross-sectional area is effectively reduced. This effectively increases the wire's resistance.

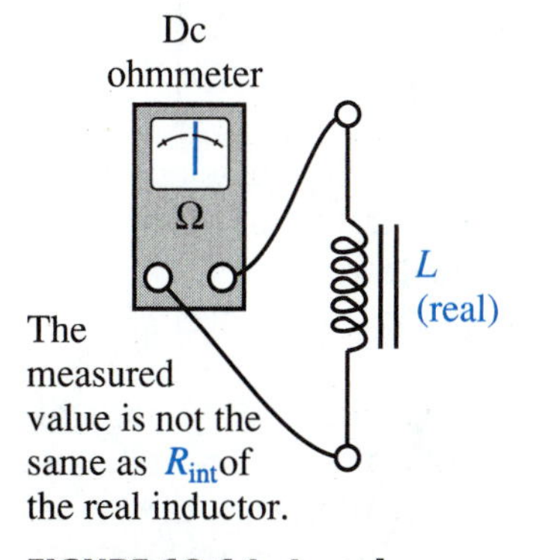

FIGURE 18-14 A real inductor's R_{int} is always greater than the wire resistance measured by a dc ohmmeter.

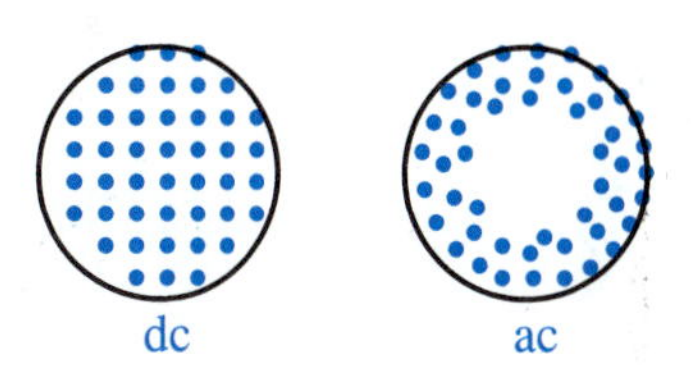

FIGURE 18-15 With ac current, the moving charge tends to concentrate near the surface of the wire.

9

A real inductor's internal resistance must be measured with an ac instrument that is designed to take these three effects into account. Often, a real inductor's internal resistance is specified by stating its quality factor, or **Q-factor.**

TECHNICAL FACT

An inductor's Q-factor is defined as the ratio of its reactance to its internal resistance. As an equation,

$$Q = \frac{X_L}{R_{int}}$$

EQ. 18-6

Since Q is a ratio of ohms to ohms, it is a pure number. It has no units.

EXAMPLE 18-9

An impedance bridge is an ac instrument that is often used to measure real inductors. The operator selects the operating frequency of the impedance bridge, and the instrument reads out inductance, L, and Q-factor.

Suppose a particular measurement is made at $f = 3$ kHz, with the bridge indicating $L = 450$ mH, and $Q = 4.6$. What is the inductor's effective internal resistance?

SOLUTION

Knowing that $f = 3$ kHz and $L = 450$ mH, first find the reactance, X_L, from Equation 18-4.

$$X_L = 6.28fL = 6.28\,(3 \times 10^3 \text{ Hz})\,(450 \times 10^{-3} \text{ H}) = 8482\ \Omega$$

Then, by rearranging Equation 18-6, we get

$$R_{int} = \frac{X_L}{Q} = \frac{8482\ \Omega}{4.6} = \mathbf{1844\ \Omega}$$

For any specific inductor, it is clear that Q varies with the operating frequency, since X_L is frequency dependent.

SELF-CHECK FOR SECTIONS 18-5 THROUGH 18-8

20. T-F For an ideal inductor in an ac circuit, current leads voltage by 90°. [5]
21. T-F For a fixed inductance, as frequency increases, inductive reactance increases proportionately. [6]
22. T-F For a fixed frequency, reactance, X_L, is directly proportional to inductance, L. [6]
23. Inductive reactance is measured in units of__________. [6]
24. Find the reactance of a 100-mH inductor at 2 kHz. [6]
25. Find the reactance of a 100-mH inductor at 4 kHz. [6]
26. If a 100-mH inductor is driven by a 2-kHz, 10-V source, what is the current? [6, 7]
27. If a 100-mH inductor is driven by a 4-kHz, 10-V source, what is the current? [6, 7]
28. Inductors pass__________frequency signals better than they pass__________ frequency signals. [8]
29. The inductor of Problem 24 has $Q = 3.2$. Find its effective internal resistance. [9]

FORMULAS

$$X_C = \frac{1}{2 \pi f C}$$ **EQ. 18-1**

$$X_L = 2 \pi f L$$ **EQ. 18-4**

$$I = \frac{V}{X_C}$$ for a capacitor **EQ. 18-2**

$$I = \frac{V}{X_L}$$ for an inductor **EQ. 18-5**

For an ideal capacitor or inductor,
$P = 0$ **EQ. 18-3**

$$Q = \frac{X_L}{R_{int}}$$ **EQ. 18-6**

SUMMARY OF IDEAS

- Reactance is the resistance-like opposition to ac current that is shown by capacitors and inductors.
- Capacitive reactance, X_C, is inversely proportional to frequency and inversely proportional to capacitance ($X_C = 1 / 2 \pi f C$).
- Ohm's law for a capacitor enables us to calculate a capacitor's ac current ($I = V / X_C$).
- In an ideal capacitor, current and voltage are out of phase by 90°, with current leading voltage.
- For an ideal capacitor, average power is zero.
- Capacitors are frequency-sensitive; they pass high-frequency signals better than they pass low-frequency signals.
- Inductive reactance, X_L, is proportional to frequency and proportional to inductance ($X_L = 2 \pi f L$).
- Ohm's law for an inductor enables us to calculate an inductor's ac current ($I = V / X_L$).
- In an ideal inductor, current and voltage are out of phase by 90°, with current lagging voltage. Therefore, average power is zero.
- Inductors are frequency-sensitive; they pass low-frequency signals better than they pass high-frequency signals.
- Real inductors have internal resistance that depends partly on ac effects.
- An inductor's Q-factor is the ratio of its reactance to its internal resistance ($Q = X_L / R_{int}$).

CHAPTER QUESTIONS AND PROBLEMS

1. When an ac capacitor voltage is at its peak, the capacitor's current is __________. [1]
2. When an ac capacitor voltage is zero, the capacitor's current is__________. [1]
3. For an ideal capacitor, current__________voltage by_______degrees. [1]
4. Which will produce more current through a 0.3-μF capacitor, a 12-V ac source at 60 Hz or a 12-V ac source at 400 Hz? Explain why. [3, 8]
5. Calculate the reactance of the 0.3-μF capacitor at 60 Hz. Then calculate the capacitor's current with the 12-V, 60-Hz source. [2, 3]
6. Repeat Problem 5 for the 12-V, 400-Hz source. [2, 3]
7. If f is doubled, X_C will be__________. [2, 8]
8. If C is doubled, X_C will be__________. [2]
9. a) At what frequency does a 0.005-μF capacitor have a reactance of 100 Ω? [2]
   b) Repeat for $X_C = 40$ kΩ.
10. a) What amount of capacitance has a reactance of 600 Ω at 1 kHz? [2]
    b) Repeat for $X_C = 14$ kΩ.
11. We want to drive a current of 8 mA through a 0.05-μF capacitor at $f = 400$ Hz. What voltage is necessary? [2, 3]
12. Capacitors pass__________frequency signals better than they pass__________ frequency signals. [8]
13. In an ideal inductor, ac current__________voltage by_______degrees. [5]
14. Inductors pass__________frequency signals better than they pass__________ frequency signals. [8]
15. Which will produce more current through a 0.7-H inductor, a 12-V ac source at 60 Hz, or a 12-V ac source at 400 Hz? Explain why. [8]
16. Calculate the reactance of the 0.7-H inductor at 60 Hz. Then calculate the inductor's current with the 12-V, 60-Hz source. [6, 7]
17. Repeat Problem 16 for the 12-V, 400-Hz source. [6, 7]
18. If f is doubled, X_L will be__________. [6]
19. If L is doubled, X_L will be__________. [6]
20. a) At what frequency does a 300-mH inductor have a reactance of 200 Ω? [6]
    b) Repeat for $X_L = 4$ kΩ.
21. a) What amount of inductance has a reactance of 80 Ω at 1 kHz? [6]
    b) Repeat for $X_L = 2.5$ kΩ.
22. We want to drive a current of 10 mA through a 0.8-H inductor at 400 Hz. What voltage is necessary? [6, 7]
23. In Problem 22, suppose that frequency is halved to 200 Hz. The value of current will__________by a factor of 2. (increase or decrease) [8]
24. A particular inductor is measured on an impedance bridge at 2 kHz , giving $L = 28$ mH, $Q = 4.1$. [9]
    a) What is its reactance?
    b) What is the value of its internal resistance, R_{int} ?
25. T-F An inductor with a high Q-factor has low winding resistance. [9]

CHAPTER 19

IMPEDANCE

OUTLINE

Silicon crystals produced in zero-gravity conditions can be made much larger and with fewer defects than crystals produced on earth. This is a great advantage in the building of advanced microelectronic devices.
Here, a payload specialist aboard a NASA Space Shuttle is preparing a crystal-growing apparatus. Virtually perfect crystals that were grown in space are shown in the magnified photo. *Courtesy of NASA*

NEW TERMS TO WATCH FOR

phasor (arrow)
resolve (a phasor)
in-phase component
out-of-phase component
right triangle
hypotenuse
cosine
arc cosine
impedance
power factor
apparent power
voltampere
true power
power triangle
quadrature power
var
coupling capacitor

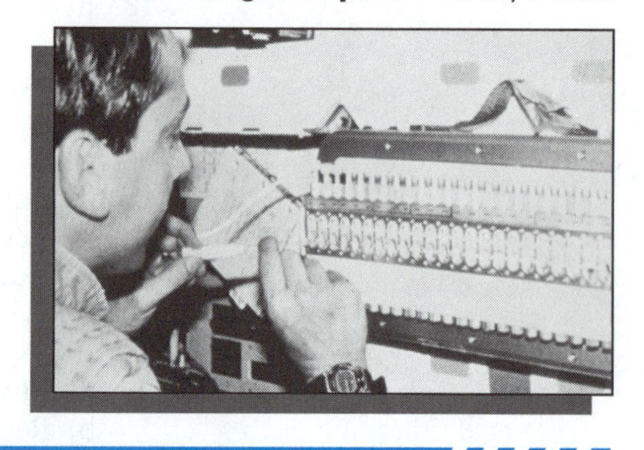

Many ac circuits have both resistance and reactance. Because the voltages and currents are out of phase with one another, we must use a new analysis tool called the phasor diagram.

After studying this chapter, you should be able to:

1. Represent the magnitude and phase of two or more sine waves on a phasor diagram.
2. Graphically resolve ac phasor variables into horizontal and vertical components.
3. Combine out-of-phase ac phasor variables by the Pythagorean theorem to find magnitude and phase angle.
4. Combine out-of-phase ac phasor variables graphically to find magnitude.
5. Analyze parallel ac circuits, both resistor-capacitor and resistor-inductor.
6. Define impedance. Calculate any one of the three variables, impedance, source voltage, and total current, given the other two.
7. Troubleshoot parallel ac circuits to find the location of a short or an open.
8. Explain the importance of power factor in an ac circuit.
9. Define apparent power and quadrature power and explain how they are different from true power.
10. Relate the impedance of a series ac circuit to resistance and reactance.
11. Analyze series ac circuits, both *RC* and *RL*.
12. Troubleshoot series ac circuits to find the location of a short or an open.

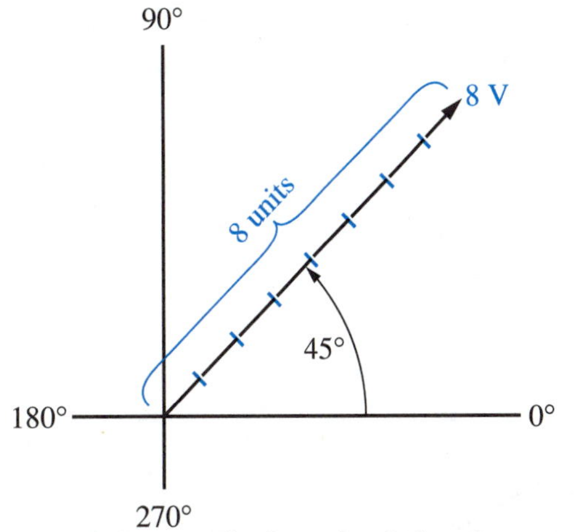

FIGURE 19-1 A voltage phasor. The length of the phasor arrow represents the magnitude of voltage (8 V in this case). The phasor's position is 45 degrees above the x axis. This means that the voltage is out of phase (leading) by 45 electrical degrees from some other voltage or current, which is regarded as the reference variable. If the reference variable's phasor were shown, it would lie in the 0° position.

19-1 PHASORS

Drawing actual sine waveforms to show magnitude and phase relationship is difficult. We have a simpler method of showing the same information. It is the phasor method. A **phasor** is a straight-line arrow drawn on a set of x-y axes. The length of the arrow represents the magnitude of the electrical variable, and its position represents the phase angle of the electrical variable.

For example, in Figure 19-1, the length is 8 units, representing 8 volts. The position is 45° above the x axis. This means that this voltage is 45° out of phase with whatever electrical variable is directly on the x axis. The electrical variable that is

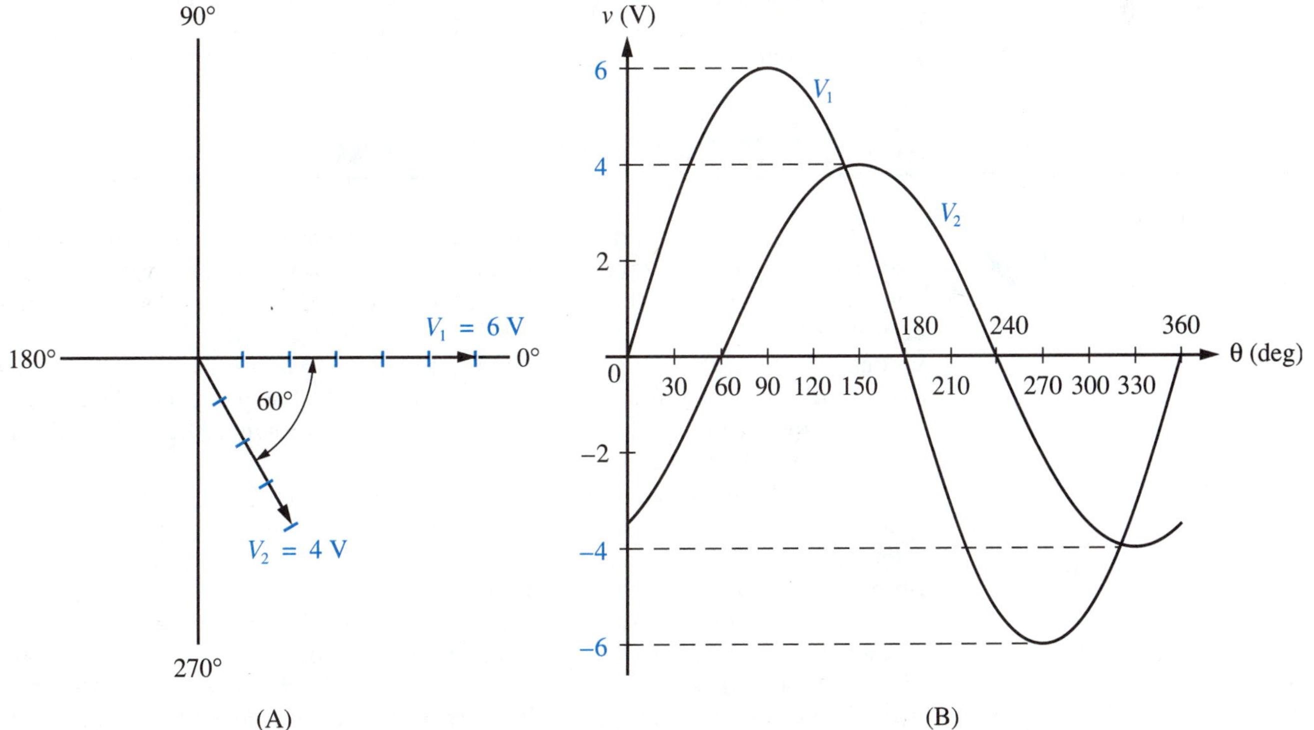

FIGURE 19-2 (A) A phasor diagram containing two phasors, V_1 (6 V) and V_2 (4 V). The reference variable is V_1. Because the V_2 phasor is drawn 60° below the x axis, it means that V_2 lags V_1 by 60 electrical degrees. (B) Sine waveforms of V_1 and V_2, showing that V_2 crosses through zero 60° after V_1 crosses through zero (V_2 lags V_1 by 60°.)

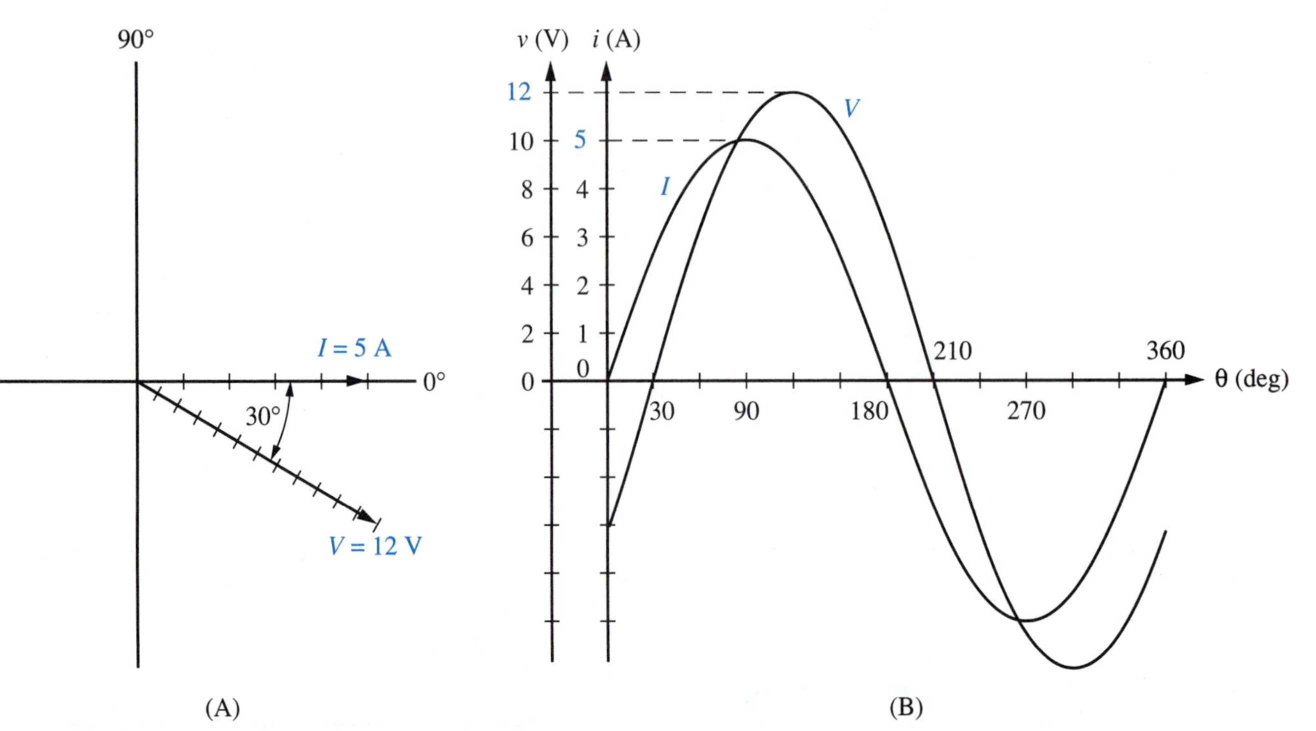

FIGURE 19-3 (A) Current and voltage on the same phasor diagram, with current as the reference. Voltage lags current by 30°. The scale factor for current is ¼ inch = 1 ampere. For voltage, ⅛ inch = 1 volt. (B) Sine waveforms of current and voltage shown in (A).

directly on the x axis (the 0° position) is called the reference variable.

Figure 19-2(A) shows a phasor diagram that represents two voltages, V_1 and V_2. The diagram indicates that V_1 is considered the reference voltage, and it has a magnitude of 6 V. Voltage V_2 has a magnitude of 4 V and it lags V_1 by 60°. The actual sine waveforms of V_1 and V_2 are shown in Figure 19-2(B).

A current phasor and a voltage phasor can appear on the same diagram. Since they are two different quantities, they can have different magnitude scales, as shown in Figure 19-3(A). Their waveforms are shown in Figure 19-3(B).

19-2 RESOLVING PHASORS GRAPHICALLY

2

A phasor that is not directly on an axis can be broken up, or **resolved,** into two parts. One part lies on the x axis and the other part lies on the y axis. This is demonstrated in Figure 19-4. Figure 19-4(A) shows a current of 5 A leading the reference variable by 36.9°. The x-axis part is found in Figure 19-4(B) by dropping a vertical line from the arrowhead down to the x axis. Measuring along the x axis shows a distance of 4 units (1 unit = ¼ inch), or 4 A. The y-axis part is found in Figure 19-4(C) by projecting a horizontal line from the arrowhead over to the y axis. Measurement shows 3 units (¾ inch), or 3 A.

The parts of the phasor are called its components. The horizontal x-axis component is often referred to as the **in-phase component,** meaning in phase with the reference variable. The vertical y-axis component is referred to as the **out-of-phase component,** meaning 90° out of phase with the reference variable.

In Figures 19-4 and 19-5, with current as the phasor and source voltage as the reference variable, the in-phase component is the resistive current and the out-of-phase component is the reactive current. For the example of Figures 19-4 and 19-5, the reactive current is capacitive reactive, since it leads voltage by 90°.

Figure 19-5 shows the overall situation as it would be in an actual circuit. Combining a 4-A resistive current with a 3-A reactive current produces a total current of 5 A, which is out of phase with the source voltage by 36.9°.

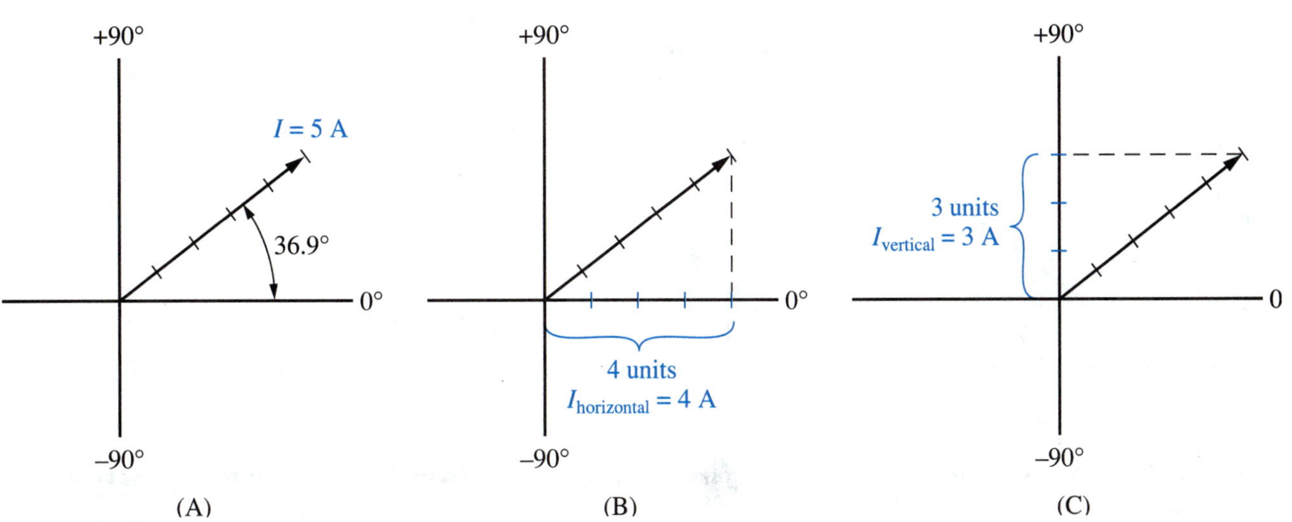

FIGURE 19-4 Resolving a phasor with a scale factor of ¼ inch = 1 A. (A) Initial phasor. (B) Finding the horizontal (x axis) part. (C) Finding the vertical (y axis) part.

✓ SELF-CHECK FOR SECTIONS 19-1 AND 19-2

1. The length of a phasor arrow shows the__________of the electrical variable. [1]
2. The position of a phasor arrow shows the______ __________of the electrical variable. [1]
3. T-F In a phasor diagram, all the voltage phasors must have the same scale. [1]
4. T-F In a phasor diagram, a voltage phasor and a current phasor must have the same scale. [1]
5. Draw a phasor for 20 V leading the reference variable by 60°. [1]
6. Draw a phasor for 3 A lagging the reference variable by 70°. [1]
7. A current of 4 A lags a 25-V source voltage by 30°. Draw the phasor diagram, assuming that source voltage is the reference variable. [1]
8. A circuit is driven by a source voltage of 50 V. It carries a total current of 2 A, lagging the voltage by 30°. Using careful scaling, find: [2]
   a) The resistive (horizontal) component of current.
   b) The reactive (vertical) component of current.

19-3 COMBINING PHASORS

In Figure 19-5, the reactive current phasor can be moved to the right, where it joins the tips of the other two phasors. This has been done in Figure 19-6(A). It is clear that this forms a triangle containing a 90° angle, called a **right triangle**.

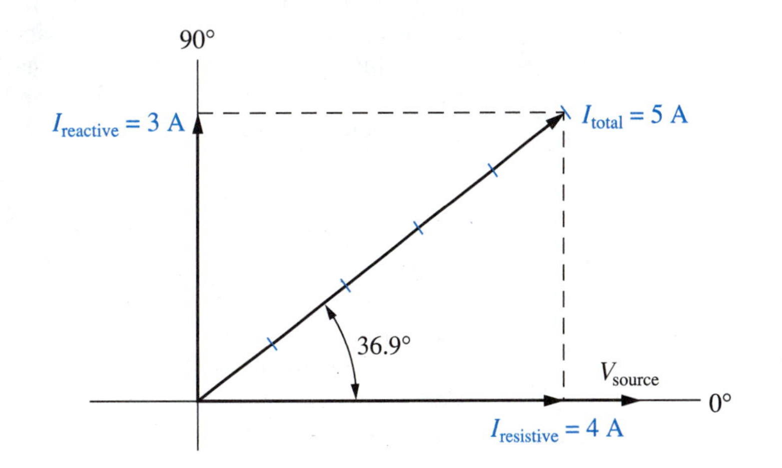

FIGURE 19-5 Current phasor and its components. The scale factor is ½ inch = 1 A.

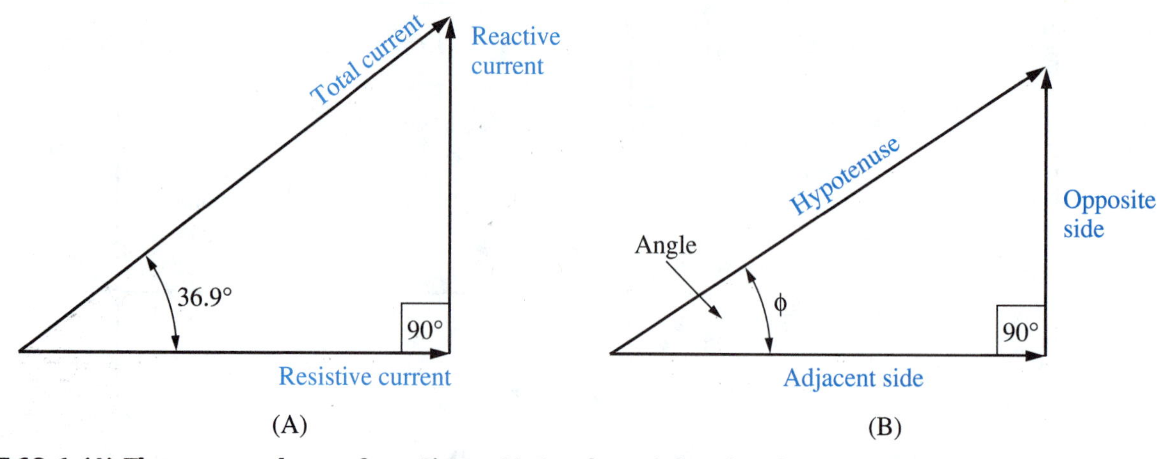

FIGURE 19-6 (A) The current phasors from Figure 19-5 make a right triangle. (B) Terms for a right triangle

In Figure 19-6(B), the right triangle has been redrawn with the standard mathematical labels. The symbol ø (Greek phi) symbolizes the angle that we use to analyze the triangle. The **hypotenuse** is the side of the triangle that is across from the 90° angle. It is the longest side of the right triangle. The adjacent side is the other side that forms the angle ø. It is alongside ø, or adjacent to it. The side across from ø is called the opposite side.

The three sides are related to each other by the Pythagorean theorem. Using the symbols *hyp* for hypotenuse, *adj* for adjacent side, and *opp* for opposite side, the Pythagorean theorem is written as

$$hyp^2 = adj^2 + opp^2$$

EQ. 19-1

The ratio of the adjacent side to the hypotenuse is called the **cosine** of ø, or cos ø. As an equation,

$$\cos ø = \frac{\text{adjacent side}}{\text{hypotenuse}}$$

EQ. 19-2

Suppose that the sides of the triangle represent electric currents, as they do in Figures 19-5 and 19-6. This is the common way to represent an ac circuit whenever the total current is out of phase with the source voltage. Figure 19-6 makes it clear that the hypotenuse represents the total current and the triangle's angle ø represents the electrical phase angle between total current and source voltage V_s. The triangle's opposite side is the reactive component of current, which leads (or lags) the voltage by 90°. The adjacent side is the resistive component of current, which is in phase with the voltage. We will symbolize the total current as I_T, the resistive current as I_R, and the reactive current as I_X.

Then Equations 19-2 and 19-1 can be written in realistic electrical terms as

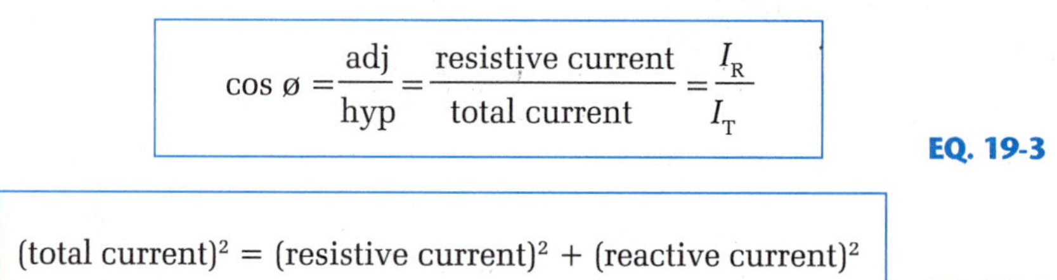

$$\cos ø = \frac{\text{adj}}{\text{hyp}} = \frac{\text{resistive current}}{\text{total current}} = \frac{I_R}{I_T}$$

EQ. 19-3

and

$$(\text{total current})^2 = (\text{resistive current})^2 + (\text{reactive current})^2$$

EQ. 19-4

or

$$I_T^2 = I_R^2 + I_X^2$$

EQ. 19-4

Equation 19-4 can be rearranged as

$$I_T = \sqrt{I_R^2 + I_X^2}$$

EQ. 19-5

Now if we know resistive current, I_R, and reactive current, I_X, we can use Equation 19-5 to solve for the total current, I_T .

EXAMPLE 19-1

A certain ac circuit carries resistive current I_R = 9 A and a leading (capacitive) reactive current I_X = 12 A. What is the value of total current?

SOLUTION

Applying Equation 19-5, we have

$$I_T = \sqrt{I_R^{\,2} + I_X^{\,2}}$$
$$= \sqrt{9^2 + 12^2} = \sqrt{81 + 144} = \sqrt{225}$$
$$= \mathbf{15\ A}$$

Notice that the simple arithmetic sum of $I_R + I_X$ does *not* equal I_T. This is the way that ac circuits always work. Whenever ac variables are not exactly in phase with each other, they cannot be simply summed arithmetically.

Combining two phasors into one resultant is the reverse of resolving one phasor into two components. Therefore, we can combine two phasors graphically by reversing the technique in Section 19-2.

EXAMPLE 19-2

Using careful scaling, graphically combine I_R = 9 A with I_X = 12 A to find the total current. This should duplicate the Pythagorean theorem result from Example 19-1.

SOLUTION

Using the scale ⅛ inch = 1 A, Figure 19-7(A) shows I_R = 9 A on the x axis and I_X = 12 A on the y axis. Extend a horizontal line to the right from the tip of the I_X phasor. Then extend a vertical line up from the tip of the I_R phasor. These extensions have been done in Figure 19-7(B).

The point where the lines cross is the tip of the I_T phasor. Therefore, we can draw the I_T phasor as shown in Figure 19-7(C). Marking it in ⅛-inch units, we find its length to be 15 units, or **15 A**.

If we know the hypotenuse and the adjacent side of a right triangle, we can find the angle ø. Electrically, this is the same as knowing the total current, I_T, and the resistive current, I_R . For any right triangle in general, we use Equation 19-2. For an ac electrical circuit in particular, we use Equation 19-3, which is equivalent to Equation 19-2.

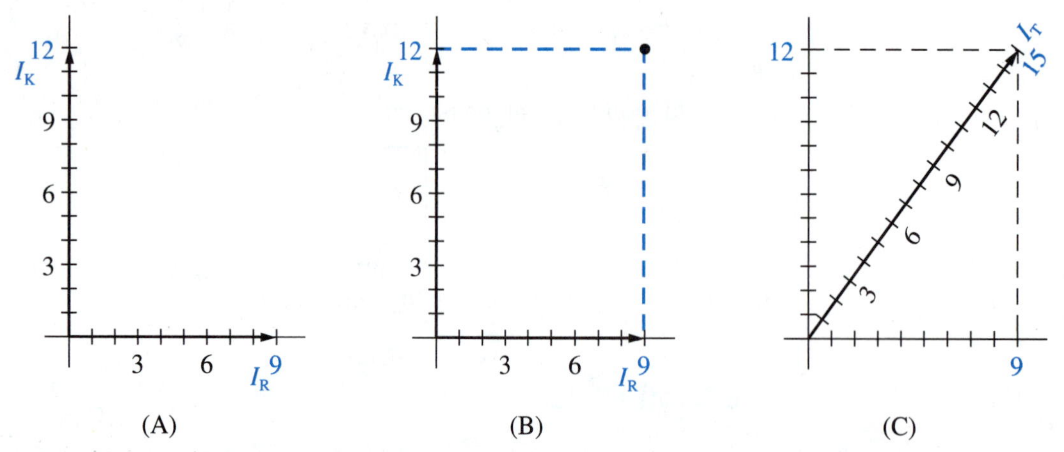

FIGURE 19-7 Graphical combination of two phasors. (A) Showing the horizontal and vertical components, to scale. (B) Extending from the tips. This is the reverse of what was done in Figure 19-4. (C) The resultant, or phasor sum.

$$\cos ø = \frac{\text{adjacent side}}{\text{hypotenuse}}$$

EQ. 19-2

$$\cos ø = \frac{I_R}{I_T}$$

EQ. 19-3

The angle ø is the phase angle of the circuit. It tells how far out of phase the total current is from the source voltage, as indicated in Figures 19-5 and 19-6. To find the phase angle ø, we follow what Equation 19-3 tells us. We first divide I_R by I_T. This always gives a number less than 1. For instance, in Examples 19-1 and 19-2 the number is

$$\frac{I_R}{I_T} = \frac{9 \text{ A}}{15 \text{ A}} = 0.6$$

This number is the cosine of the angle ø. That is what Equation 19-3 means. Therefore, in this example, ø is whatever angle has a cosine of 0.6. Using a scientific calculator, it is very easy to find out this angle in degrees.

To explain how to do this, we first must have a way of symbolizing the statement "ø is the angle whose cosine is 0.6." One way to symbolize that statement is

$$ø = \arccos (0.6)$$

EQ. 19-6

This would be stated in speech as " phi equals **arc cosine** of zero point six." An alternative symbol, which is actually more common on hand-held calculators, is

$$ø = \cos^{-1} (0.6)$$

EQ. 19-7

which is spoken exactly the same way. The symbol $\cos^{-1}$ in Equation 19-7 means the same thing as the symbol arccos in Equation 19-6.

On a scientific calculator, the $\cos^{-1}$ function is the 2nd function of the cos key. So, to solve for ø using Equation 19-7, here is the keystroke sequence:

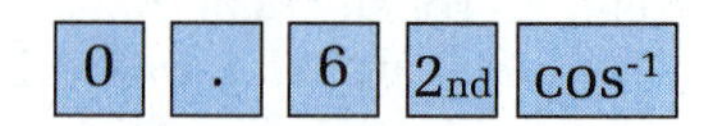

which produces a display of 53.13 degrees. Therefore, the angle of the triangle in

Figure 19-7 is 53.13°. Electrically, it means that I_T is out of phase with V_s by 53.13°.

As a general equation for any values of I_R and I_T, we have

3

$$ø = \cos^{-1}\left(\frac{I_R}{I_T}\right)$$

EQ. 19-8

EXAMPLE 19-3

Suppose that you didn't know the phase angle in Figures 19-4 and 19-5. Knowing that I_R = 4 A and I_T = 5 A, find phase angle ø.

SOLUTION

To apply Equation 19-8, first divide I_R by I_T.

$$\frac{I_R}{I_T} = \frac{4\text{ A}}{5\text{ A}} = 0.8$$

Then push the 2nd key, followed by the cos key. The result is **36.87°**, which is ø.

See Appendix—Using a Scientific Hand-held Calculator for further advice on using the cos and $\cos^{-1}$ functions.

✓ SELF-CHECK FOR SECTION 19-3

9. T-F A right triangle is one that contains a 90° angle.
10. In a right triangle, cos ø is equal to the__________side divided by the hypotenuse.
11. For an ac circuit with total current out of phase with the source voltage, the resistive current is represented by the__________side of the triangle. The symbol for resistive current is_______. [2]
12. For an ac circuit with total current out of phase with the source voltage, the reactive current is represented by the__________side of the triangle. The symbol for reactive current is_______. [2]
13. A 120-V ac circuit has I_R = 4.2 A and I_X = 1.8 A, leading. Draw the circuit's phasor diagram. Then: [3]
    a) Find the magnitude of total current, I_T.
    b) Describe the phase angle of I_T relative to V_s. Is I_T leading or lagging?

19-4 SOLVING PARALLEL AC CIRCUITS

Parallel RC Circuits

In a parallel resistor-capacitor circuit like Figure 19-8(A), the entire source voltage appears across both components. Because the source voltage is common to both circuit components, it is used as the reference variable. It is placed at the 0° position on the circuit's phasor diagram. This is indicated in Figure 19-8(C).

The individual branch currents can be calculated with Ohm's law, as indicated in Figure 19-8(B). For the resistor,

$$I_R = \frac{V_s}{R} = \frac{30\text{ V}}{12\text{ k}\Omega} = 2.5\text{ mA}$$

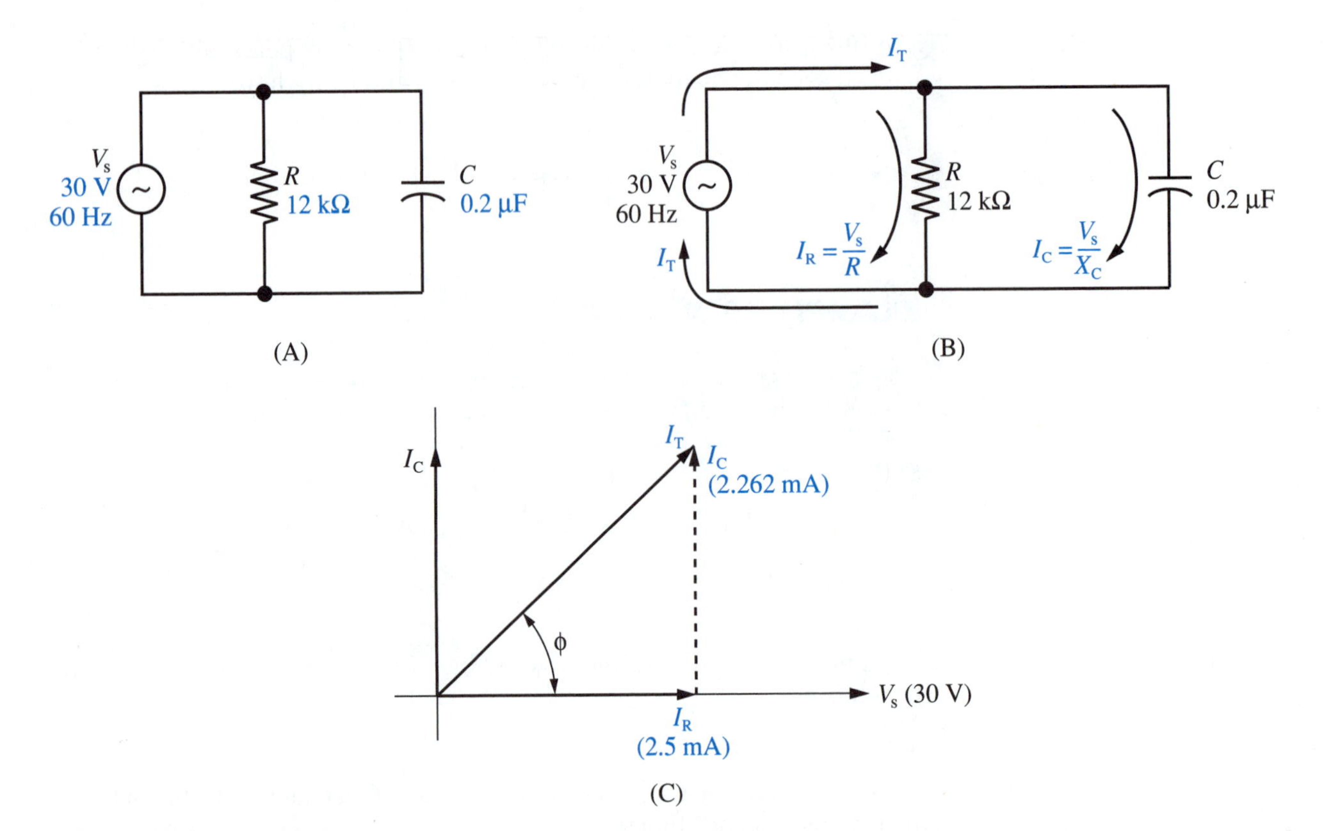

FIGURE 19-8 (A) Parallel RC circuit. (B) Individual branch currents are calculated by Ohm's law. (C) Total current and phase angle are found from the phasor diagram. I_C (shown dashed) is moved to the tip of I_R to form a right triangle. I_T = 3.37 mA; ø = 42.1°.

This resistive current is in phase with the voltage, so it is shown on the horizontal axis in Figure 19-8(C), on top of the voltage phasor.

The capacitive reactance is given by

$$X_C = \frac{1}{6.28\ fC} = \frac{1}{6.28\ (60)\ 0.2 \times 10^{-6}} = 13.26\ \text{k}\Omega$$

Applying Ohm's law for a capacitor, we get

$$I_C = \frac{V_s}{X_C} = \frac{30\ \text{V}}{13.26\ \text{k}\Omega} = 2.262\ \text{mA}$$

This current leads the voltage by 90°, so it is shown in the upward vertical direction in the phasor diagram of Figure 19-8(C).

In Section 19-3, we extended lines from the tips of I_R and I_X. A quicker way to visualize this is to just move the I_X phasor so that its tail is at the tip of I_R. Then draw I_T to form a right triangle, as shown in Figure 19-8(C).

Using Equation 19-5 to combine I_R and I_C, we have

$$I_T = \sqrt{I_R^{\,2} + I_C^{\,2}}$$
$$= \sqrt{2.5^2 + 2.262^2} = \sqrt{6.25 + 5.117}$$
$$= 3.371\ \text{mA}$$

A reliable and efficient keystroke sequence for the Pythagorean theorem is to use the calculator's memory to sum the squares.

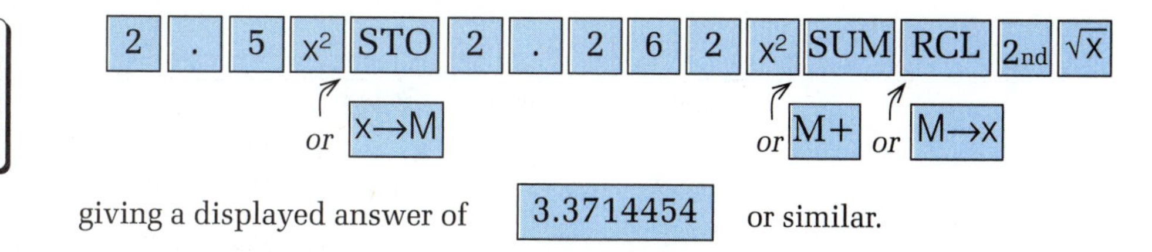

giving a displayed answer of 3.3714454 or similar.

The phase angle ø can be found from Equation 19-8.

$$\begin{aligned} ø &= \cos^{-1}\left(\frac{I_R}{I_T}\right) \\ &= \cos^{-1}\frac{2.5\text{ mA}}{3.371\text{ mA}} = \cos^{-1}(0.7416) \\ &= 42.14° \end{aligned}$$

Rounding to three significant figures, we conclude that the total current in Figure 19-8 has a magnitude of 3.37 mA, and it leads the source voltage by 42.1 degrees.

When an ac circuit has a combination of resistance and reactance, we define its **impedance** as follows:

TECHNICAL FACT

A resistive/reactive circuit's overall opposition to current is called its impedance, symbolized *Z*. It is measured in basic units of ohms.

When dealing with a resistive/reactive circuit, Ohm's law is written as

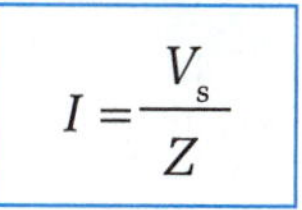

$$I = \frac{V_s}{Z}$$

EQ. 19-9

Of course, Equation 19-9 can be rearranged to solve for any one of the variables.

EXAMPLE 19-4

Find the impedance of the circuit in Figure 19-8. Use the information that we have found out about that circuit.

SOLUTION

Rearranging Equation 19-9 gives

$$\begin{aligned} Z &= \frac{V_s}{I_T} \\ &= \frac{30\text{ V}}{3.371\text{ mA}} \\ &= 8.899\text{ k}\Omega \text{ or } \mathbf{8.90\text{ k}\Omega} \end{aligned}$$

The parallel circuit's overall ohmic value, its impedance, is less than the ohmic value of either individual branch ($R = 12$ kΩ and $X_C = 13.3$ kΩ). This is like the situation in a parallel dc circuit, where total resistance is less than any individual branch resistance.

The impedance of a parallel RC circuit can be calculated directly from the resistance and capacitive reactance values.

$$Z = \frac{R X_C}{\sqrt{R^2 + X_C^2}}$$

EQ. 19-10

Applying this formula to the Figure 19-8 circuit with kilohm units, we get

$$Z = \frac{(12)(13.26)}{\sqrt{12^2 + 13.26^2}} = \frac{159.12}{\sqrt{319.83}}$$

$$= \frac{159.12}{17.884} = 8.897 \text{ or } 8.90 \text{ k}\Omega$$

which agrees with the result from Example 19-4.

Parallel RL Circuits

A parallel RL circuit is solved the same way as a parallel RC circuit. The only difference is that the reactive component of current lags the source voltage by 90°, rather than leads it.

EXAMPLE 19-5

Figure 19-9(A) shows an ideal parallel *RL* circuit.

a) Find the individual branch currents I_R and I_L using Ohm's law.
b) Draw the phasor diagram and use it to find the magnitude of I_T and its phase angle relative to V_s.
c) Find the circuit impedance.

SOLUTION

a) Applying Ohm's law to the resistor branch gives

$$I_R = \frac{V_s}{R} = \frac{50 \text{ V}}{600\ \Omega} = \mathbf{83.33\ mA}$$

The inductive reactance is calculated from

$$X_L = 6.2832 \text{ fL} = 6.2832\ (400)(0.35) = 879.6\ \Omega$$

Ohm's law for the inductor branch gives

$$I_L = \frac{V_s}{X_L} = \frac{50 \text{ V}}{879.6\ \Omega} = \mathbf{56.84\ mA}$$

These results are recorded in Figure 19-9(B).

b) I_R is in phase with reference phasor V_s. I_L lags V_s by 90°, as shown in the diagram of Figure 19-9(C). Applying the Pythagorean theorem, Equation 19-5, gives us

$$I_T = \sqrt{I_R^2 + I_L^2}$$

$$= \sqrt{83.33^2 + 56.84^2}$$

$$= 100.87 \text{ mA}$$

Phase angle ø is found by using Equation 19-8:

$$ø = \cos^{-1}\left(\frac{I_R}{I_T}\right)$$

Continued on page 351.

$$= \cos^{-1}\left(\frac{83.33 \text{ mA}}{100.87 \text{ mA}}\right) = \cos^{-1}(0.8261)$$

$$= 34.30°$$

Rounding to three significant figures, we would describe the total current as **101 mA,** lagging the source by **34.3°**.

c) The circuit impedance, Z, can be found by either of the two methods as before. From Ohm's law, we have

$$Z = \frac{V_s}{I_T} = \frac{50 \text{ V}}{100.87 \text{ mA}} = \mathbf{495.7\ \Omega}$$

Substituting X_L for X_C in Equation 19-10 gives us

$$Z = \frac{RX_L}{\sqrt{R^2 + X_L^2}}$$

EQ. 19-11

$$= \frac{600 \times 879.6}{\sqrt{600^2 + 879.6^2}} = \mathbf{495.7\ \Omega}$$

In Example 19-5, the inductive reactive current was considered to be exactly 90° out of phase with the source. In a more realistic situation, the inductor would be nonideal, containing some internal resistance. This would cause the inductor branch current to be out of phase by some angle less than 90°. Then the phasor diagram would not form a right triangle, and the solution would be more difficult.

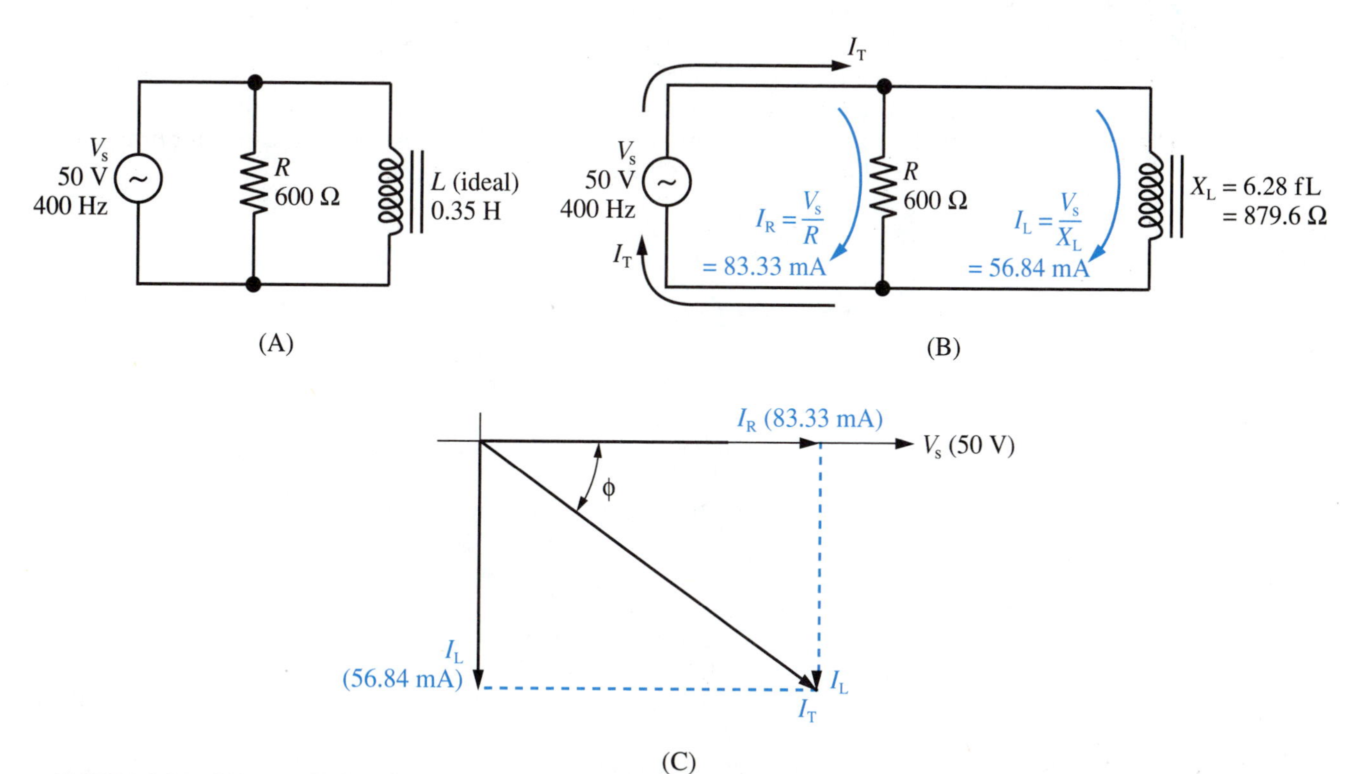

FIGURE 19-9 (A) Parallel RL circuit containing an ideal inductor. (B) Individual branch currents are calculated by Ohm's law. (C) I_T and ø are found from the phasor diagram: I_T = 101 mA ; ø = 34.3°.

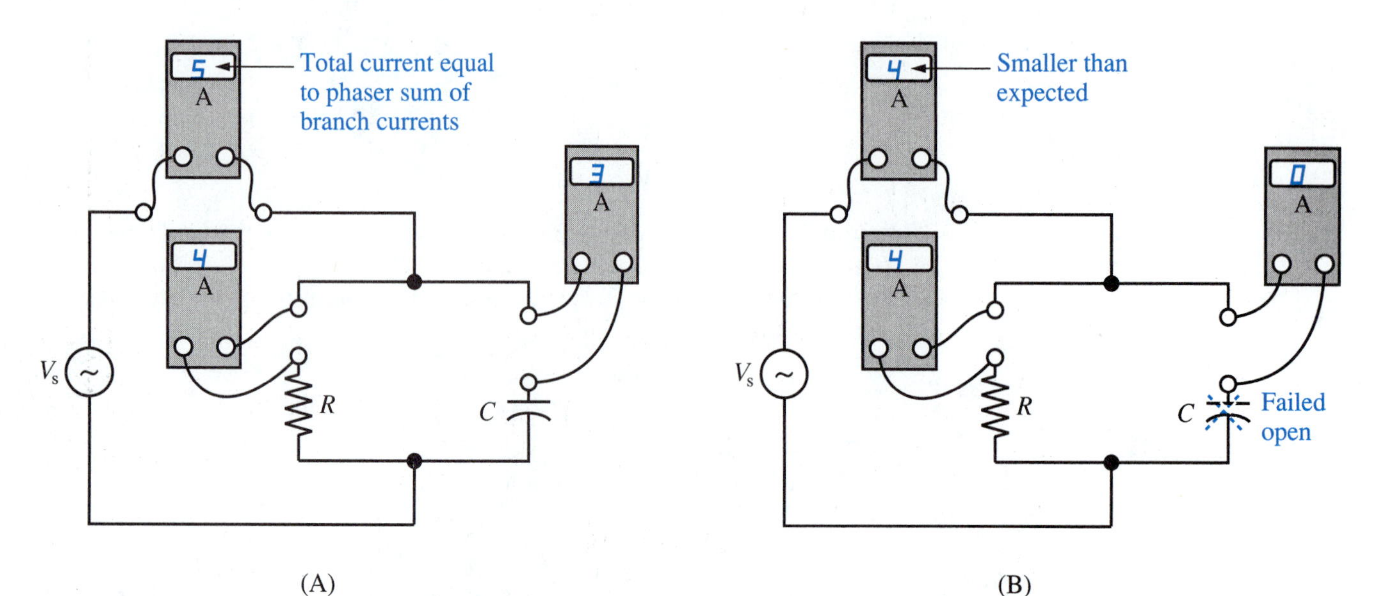

FIGURE 19-10 Parallel RC circuit. (A) Working properly. (B) With capacitive branch failed open, I_T is equal to just I_R.

Troubleshooting Parallel Ac Circuits

Shorts. A dead short across any branch of a parallel ac circuit produces an overcurrent that will blow the fuse. A partial short may not blow the fuse, but it will cause the total supply current to be greater than normal. You can find out which branch component has shorted by the same basic methods used for dc circuits. These methods were illustrated in Figures 10-14 and 10-15.

Opens. If an ac circuit branch fails open, the total line current becomes smaller than normal. This is illustrated in Figure 19-10.

In a circuit with many branches, use an ac voltmeter to show whether the open is in the load component itself or in the wiring joining the load component to the supply leads. A final check of a reactive component, capacitor or inductor, can be done with an LCR meter.

✔ SELF-CHECK FOR SECTION 19-4

14. In a parallel *RC* or *RL* circuit, the reference phasor is the__________ __________ phasor. [5]
15. The total current, I_T,__________the source voltage, V_s, in a parallel *RC* circuit. (leads or lags) [5]
16. In a parallel resistive/reactive (*RX*) circuit, write the equation that gives the phase angle ø if I_R and I_T are known. [5]
17. A parallel *RC* circuit has $V_s = 12$ V, $R = 800\ \Omega$, and $X_C = 600\ \Omega$. [5, 6]
    a) Find I_R and I_C.
    b) Draw the phasor diagram. From it find I_T and ø.
    c) Solve for the total circuit impedance, *Z*.
18. An ideal parallel *RL* circuit has $V_s = 120$ V, $R = 3$ kΩ, and $X_L = 2.5$ kΩ. [5, 6]
    a) Find I_R and I_L.
    b) Draw the phasor diagram and use it to find I_T and ø.
    c) Solve for the circuit's *Z*.

EXPLODING FIRECRACKER

These photos show a firecracker just before, during, and just after the moment of explosion. They were obtained by an ingenious method of electronically storing and shifting photographic images. A video camera continually photographs the firecracker (or any object that is going to have something drastic happen to it). The camera may take photos at the rate of 10 frames per second, say. The time interval between photos would then be 0.1 second. Each photographic frame is electronically divided into a large number of tiny areas. The brightness or darkness of each tiny area is then represented by a coded group of electronic data bits.

One electronic memory bank stores an entire collection of data bits that represents the appearance of each photographic frame. There are 16 such memory banks, so they can store the most recent 16 frames. In our example, that would mean the most recent 1.6 seconds, looking backward in time.

We think of these 16 memory banks as lined up, from left to right. Memory bank Number 16 is on the far left, and memory bank Number 1 is on the far right. Memory bank 16 stores the frame that was shot 16 time intervals ago (1.6 seconds ago). Memory bank 1 stores the frame that was shot most recently (within the last 0.1 second).

Each time a new frame is shot, it is quickly coded and stored in memory bank 1. At that same instant, the electronic data that was in memory bank 1 is shifted into memory bank 2. Also at that same instant, the electronic data that was in memory bank 2 is shifted into memory bank 3. And so on, to memory bank 15 being shifted into memory bank 16. The data that was in memory bank 16 gets thrown away.

This process continues until some drastic event happens. A drastic event always produces some detectable change, such as a change in light intensity. An electronic sensor is placed so that it is able to detect that change. When the drastic change occurs, the electronic sensor sends a signal to the video system, telling it to continue for eight more photographs, then stop.

Therefore, when the photography stops, the memory banks contain eight frame images that occurred before the event, and eight more that occurred during and after the event. By straddling the event in this way, the system is guaranteed to capture the details of the event itself.

An electronic system that uses this shifting idea is called a shift register. Shift registers are studied in a course on digital electronics.

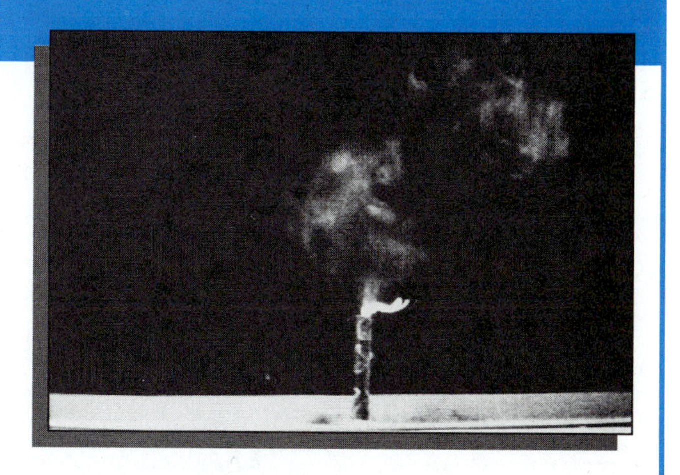

Courtesy of Colorado Video, Inc.

19-5 POWER FACTOR

We know from Chapter 18 that reactive current, whether inductive or capacitive, transfers zero power. When current and voltage are 90° out of phase, energy is temporarily stored in the reactor during one quarter cycle, then immediately returned to the source during the following quarter cycle.

For resistive current, on the other hand, the average power, P, over a complete ac cycle is simply

$$P = V_s I_R$$

EQ. 19-12

In Equation 19-12, V_s and I_R must both be expressed in rms units.

EXAMPLE 19-6

Calculate the average power in the parallel *RL* circuit of Figure 19-9.

SOLUTION

The inductive reactive current, I_L, contributes nothing to the circuit's power. Only I_R matters. From Equation 19-12,

$$P = V_s I_R$$
$$= 50 \text{ V } (83.33 \times 10^{-3} \text{ A}) = \mathbf{4.17 \text{ W}}$$

We often don't know the resistive component of the total current. Instead, we know the overall magnitude of I_T, and its phase angle relative to the voltage source. In that case, we write the average power formula as

$$P = V_s I_T (\cos ø)$$

EQ. 19-13

where I_R in Equation 19-12 is replaced with I_T (cos ø) . This replacement is gotten by rearranging cos ø = (I_R / I_T), Equation 19-3.

TECHNICAL FACT

The factor cos ø in Equation 19-13 is called the **power factor**, symbolized *PF*. We often write the ac power formula as

$$P = V_s I_T (PF)$$

EQ. 19-14

where

$$PF = \cos ø$$

EQ. 19-15

EXAMPLE 19-7

A certain *RL* circuit has V_s = 120 V, I_T = 2.3 A, lagging the source by 25°.

a) Calculate this circuit's power factor.
b) What is the circuit's average power?

Continued on page 355.

SOLUTION

a) Power factor, *PF*, is given by Equation 19-15 as

$$PF = \cos ø$$
$$= \cos 25°$$

Here, use the plain COS function of your calculator, not the 2nd function, $\cos^{-1}$. The keystroke sequence is 2 5 COS, giving a display of **0.8192**.

Power factors are often converted to percentages by multiplying the decimal times 100%. In this case, *PF* = **81.9%.**

b) From Equation 19-14,

$$P = V_s I_T (PF)$$
$$= 120 \text{ V } (2.3 \text{ A}) (0.8192) = \mathbf{226 \text{ W}}$$

TECHNICAL FACT

The simple product of $V_s I_T$ in Equation 19-14 is called the **apparent power,** symbolized *S*. Its measurement units are **voltamperes,** symbolized VA.

$$S = V_s I_T \text{ (voltamperes)}$$

EQ. 19-16

When dealing with the idea of apparent power, we sometimes make the distinction clear by describing power, *P*, as **true power.** Therefore, for any ac circuit, we can write

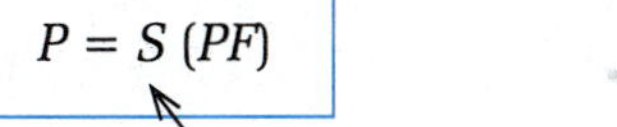

$$P = S (PF)$$

EQ. 19-17

power factor, equal to cos ø

true power in watts

apparent power in voltamperes

EXAMPLE 19-8

A certain ac motor appears to the source as a parallel *RL* circuit, with *R* = 11.5 Ω and *L* = 0.05 H. This equivalence is shown in Figure 19-11.

For these branch values, the motor's total current I_T is 11.71 A, lagging the voltage by 31.39°.

a) What is the apparent power?

b) What is the true power consumed by the motor?

SOLUTION

a) From Equation 19-16,

$$S = V_s I_T$$
$$= 115 \text{ V } (11.71 \text{ A}) = \mathbf{1347 \text{ VA}} \text{ or } \mathbf{1.35 \text{ kVA}}$$

b) True power is given by Equation 19-17,

$$P = S (PF)$$
$$= 1347 \text{ VA } (\cos 31.39°)$$
$$= 1347 (0.8536) = \mathbf{1150 \text{ W}} \text{ or } \mathbf{1.15 \text{ kW}}$$

TECHNICAL FACT

Ac sources, alternators and transformers, are rated in terms of apparent power, rather than true power.

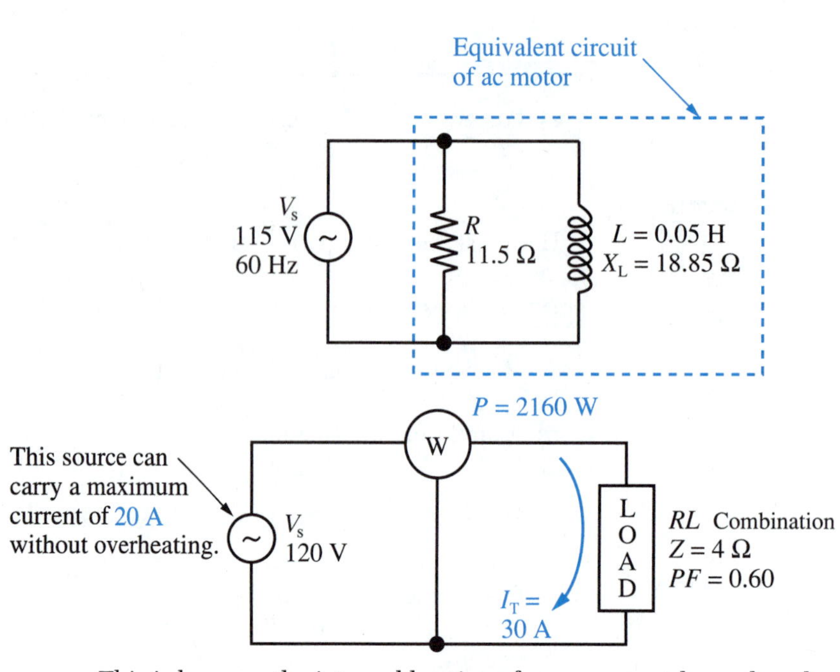

FIGURE 19-11 Schematic of an ac motor operating from the standard ac line. All motors have inductance associated with their windings. They can be viewed as equivalent to a parallel RL circuit.

FIGURE 19-12 For rating an ac source, apparent power is the important quantity, not true power.

This is because the internal heating of an ac source depends only on the magnitude of its current, not on the phase of that current.

Look at Figure 19-12 to understand apparent power rating. The ac source (perhaps a 120-V transformer secondary winding) is designed to carry a maximum current of 20 A safely. If the current exceeds 20 A, the winding's I^2R losses will cause the transformer to overheat. Now if we mistakenly rated the transformer as

$$P_{max} = 120\text{ V }(20\text{ A}) = 2400\text{ W (true power)}$$

the situation in Figure 19-12 would seem to be safe. The wattmeter reads less than 2400 W, since

$$I_T = \frac{V_s}{Z} = \frac{120\text{ V}}{4\ \Omega} = 30\text{ A}$$

$$P = V_s I_T (PF) = 120\text{ V }(30\text{ A})(0.60) = 2160\text{ W (true power)}$$

However, this situation is clearly not safe because 30 A far exceeds the 20-A maximum limit.

The transformer should properly be rated at

$$S_{max} = 120\text{ V }(20\text{ A}) = 2400\text{ VA (apparent power)}$$

Then, ignoring the load's power factor, the actual S value is given by

$$S = 120\text{ V }(30\text{ A}) = 3600\text{ VA (apparent power)}$$

which is clearly dangerous, since it exceeds the 2400 VA rating.

In summary, it is not correct to rate an ac source in terms of true power (watts). If the circuit load has a low PF, it is possible for the ac source to be delivering a rather small amount of true power even while its windings are overheating due to excessive current. Instead, constant-voltage ac sources are correctly rated in terms of apparent power (voltamperes).

The **power triangle** of Figure 19-13 shows the relationship between apparent power, true power, and a third variable called **quadrature power,** symbolized Q. As Figure 19-13 indicates, the relationship among these three is given by the Pythagorean theorem as

$$S^2 = P^2 + Q^2$$

EQ. 19-18

The basic unit of quadrature power is the **var** (voltampere reactive).

The power triangle is very similar to the phasor diagram of the ac circuit, as Figure 19-13(B) points out.

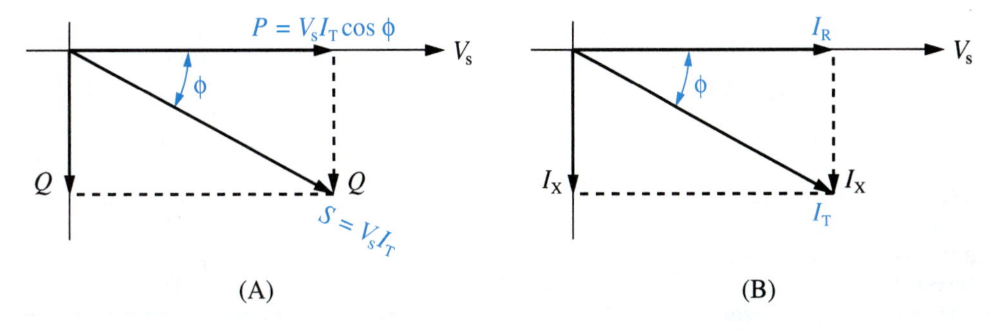

FIGURE 19-13 (A) The power triangle. (B) Similarity of current phasor diagram to power triangle

EXAMPLE 19-9

For the ac motor circuit of Figure 19-11, find the quadrature power, Q, in vars.

SOLUTION

From Equation 19-18,

$$
\begin{aligned}
S^2 &= P^2 + Q^2 \\
Q^2 &= S^2 - P^2 \\
Q &= \sqrt{S^2 - P^2} \\
&= \sqrt{(1347\ \text{VA})^2 - (1150\ \text{W})^2} = \sqrt{491\,909} \\
&= \mathbf{701.4\ var}
\end{aligned}
$$

The preceding rearrangement of the Pythagorean theorem is always allowed, whenever you must solve for one of the sides of a right triangle, rather than the hypotenuse. The calculator keystroke sequence is just like the sequence shown earlier, except that after squaring the second number, you must change its sign to negative, by pushing the +/– key before summing to the memory. The +/– key is called the CS key on some calculators.

A load's quadrature power is important to heavy industrial users of electric energy. Such users often seek to eliminate the quadrature power of inductive loads by connecting capacitors in parallel.

✔ SELF-CHECK FOR SECTION 19-5

19. The cosine of the current-voltage phase angle ø is called______ __________. [8]
20. A circuit has rms voltage of 9 V and rms current of 2.5 A, lagging the voltage by 27°. [8]
    a) What is the circuit's power factor?
    b) What is the average power consumption?
21. In Question 20, would the answers be any different if current were leading voltage by 27°? [8]
22. A certain electric motor takes 5 A from the 120-V ac line. A wattmeter in the circuit reads 520 W. What is the motor's power factor? [8]
23. The basic unit of apparent power is the__________. [9]
24. T-F The maximum rating of a transformer is commonly expressed in units of watts. [9]

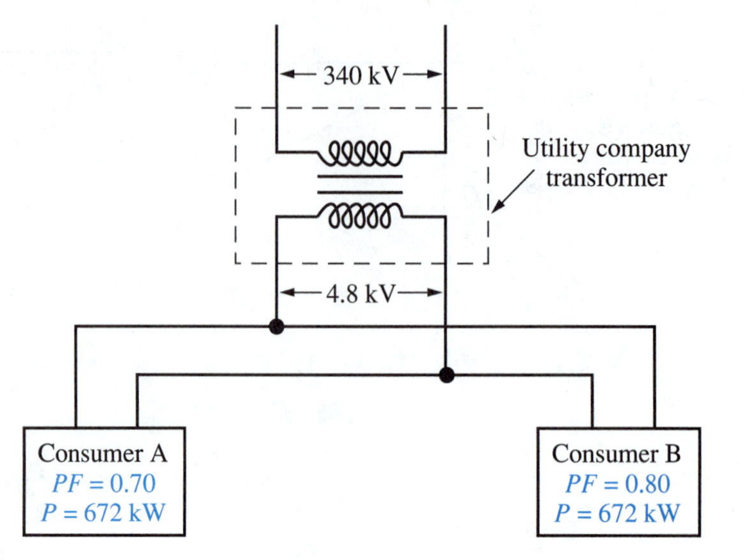

FIGURE 19-14 The consumer with the better power factor needs less current to receive the same amount of power.

25. T-F In the power triangle, the relationship between apparent power, S, true power, P, and quadrature power, Q, is $S^2 = P^2 + Q^2$. [9]
26. In Figure 19-14, an electric utility company transformer is supplying energy to consumers A and B. Both consume 672 kW of true power, but consumer B has a more resistive load than consumer A (PF = 0.80 for B versus 0.70 for A). [8, 9]
    a) Find the current in the lines that go to consumer A.
    b) Find the current in the lines that go to consumer B.
    c) Which consumer causes greater I^2R power loss along the lines? (Assume they have the same wire gage and are the same distance from the transformer).
27. Electric utility companies charge their customers for the true energy consumed. In Problem 26, which consumer is likely to get a price discount? (Dollar cost per kilowatthour of energy.) Explain why. [8, 9]

19-6 SOLVING SERIES AC CIRCUITS

To solve parallel ac circuits, we combined currents. The justification for this was Kirchhoff's current law. In series ac circuits, we combine voltages. Kirchhoff's voltage law is the justification.

For example, look at the series *RL* circuit in Figure 19-15, where current, I, is known in advance. The 0.1-A current is common to both components in a series arrangement. Therefore, applying Ohm's law twice, we get

$$V_R = IR = (0.1\text{ A})\ 200\ \Omega = 20\text{ V}$$
$$V_C = IX_C = (0.1\text{ A})\ 318.3\ \Omega = 31.83\text{ V}$$

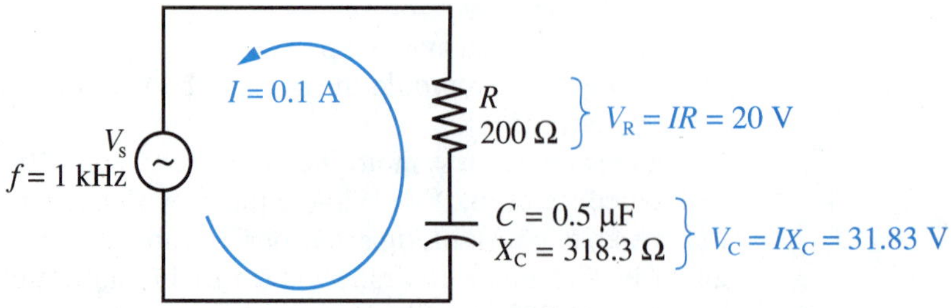

V_s
$f = 1$ kHz
$I = 0.1$ A
R 200 Ω — $V_R = IR = 20$ V
$C = 0.5$ μF, $X_C = 318.3$ Ω — $V_C = IX_C = 31.83$ V

FIGURE 19-15 Series RC circuit with known current

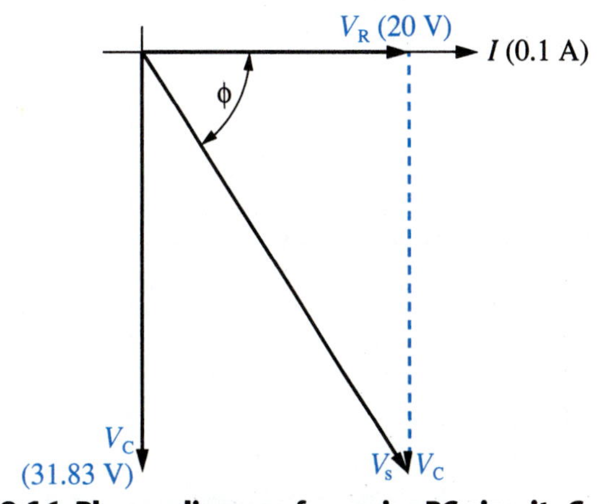

FIGURE 19-16 Phasor diagram for series RC circuit. Current is the reference phasor.

Of course, we can't add V_R and V_C by simple arithmetic, because they aren't in phase with each other. That is, we cannot say

$$V_s = 20\text{ V} + 31.83\text{ V} = 51.83\text{ V (incorrect)}$$

Instead, we must show V_R and V_C on a phasor diagram and combine them using the Pythagorean theorem.

The current is used as the reference (0°) phasor in Figure 19-16 since it is common to everything in the circuit. V_R is in phase with the current so it is also drawn in the 0° position. Capacitor voltage V_C lags the current by 90°, so it is shown pointing down in Figure 19-16.

Kirchhoff's voltage law tells us that the phasor combination of V_R and V_C must produce the total source voltage V_s. Applying the Pythagorean theorem, we have

$$V_s^2 = V_R^2 + V_C^2$$
$$V_s = \sqrt{V_R^2 + V_C^2}$$
$$= \sqrt{20^2 + (31.83)^2} = \sqrt{1413.15}$$
$$= 37.59\text{ V}$$

The phase angle ø is found from the ratio of the adjacent side to the hypotenuse, as we did before. The only difference now is that the adjacent side and the hypotenuse represent voltages, rather than currents.

$$ø = \cos^{-1}\left(\frac{\text{adj}}{\text{hyp}}\right) = \cos^{-1}\left(\frac{V_R}{V_s}\right)$$
$$= \cos^{-1}\left(\frac{20\text{ V}}{37.59\text{ V}}\right) = \cos^{-1}(0.5321)$$
$$= 57.86°$$

The source voltage is 37.6 V, lagging the current by 57.9°. Alternatively, current leads V_s by 57.9°.

Combining Ohmic Values

In a series circuit containing both resistance and reactance, the ohmic values can be combined by phasor techniques to find the total impedance. An ohmic value has the same location on a phasor diagram that its voltage would have. Therefore, resistance R lies on the x axis, in phase with current I, just like V_R. This is shown in Figure 19-17(A). In Figure 19-17(B), capacitive reactance, X_C, points down, lagging I by 90°, just like V_C. In Figure 19-17(C), inductive reactance, X_L, points up, leading I by 90°, just like V_L.

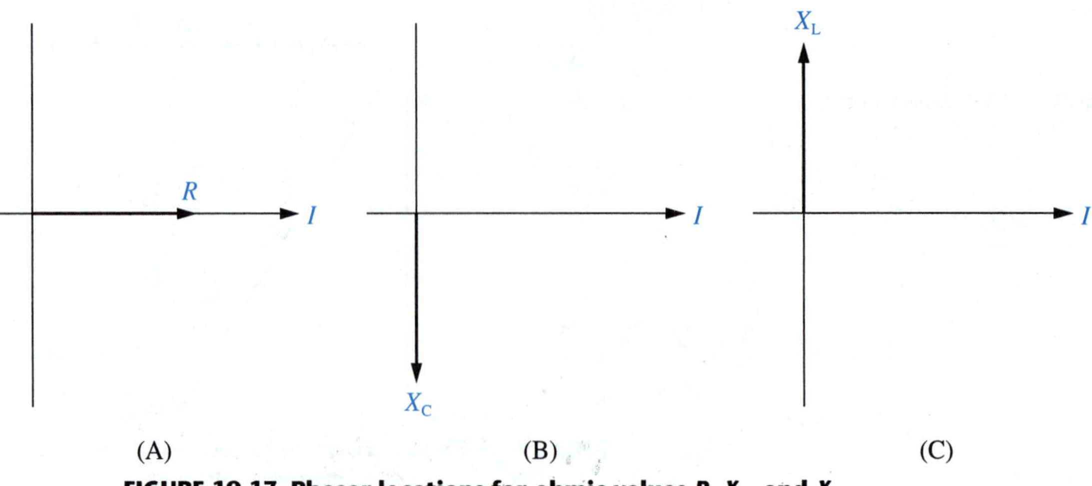

FIGURE 19-17 Phasor locations for ohmic values R, X_C, and X_L.

EXAMPLE 19-10

For the series *RC* circuit of Figure 19-15, draw the phasor diagram of ohmic values and then find the circuit's total impedance.

SOLUTION

Figure 19-18 is the phasor diagram. Using the Pythagorean theorem, we have

$$\begin{aligned} Z &= \sqrt{R^2 + X_C^{\,2}} \\ &= \sqrt{200^2 + (318.3)^2} = \sqrt{141\,315} \\ &= \mathbf{375.9\ \Omega} \end{aligned}$$

Using this information with the known current value of $I = 0.1$ A, we can apply Ohm's law to verify the source voltage calculated earlier.

$$V_s = IZ = (0.1\text{ A})\,(375.9\ \Omega) = 37.59\text{ V},$$

which agrees with the earlier result.

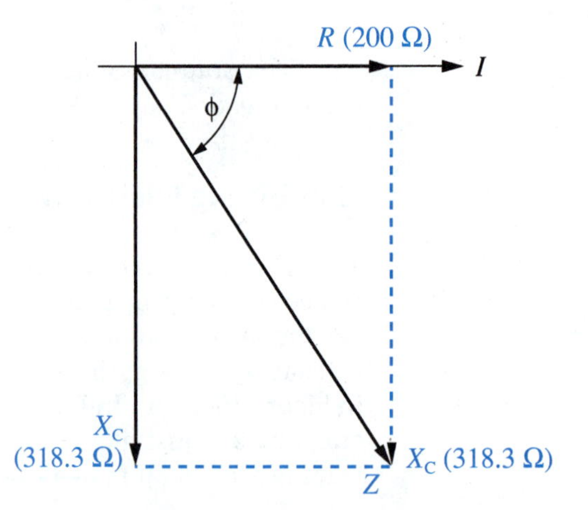

FIGURE 19-18 Phasor diagram of ohmic values. This triangle has the same proportions as the triangle in Figure 19-16.

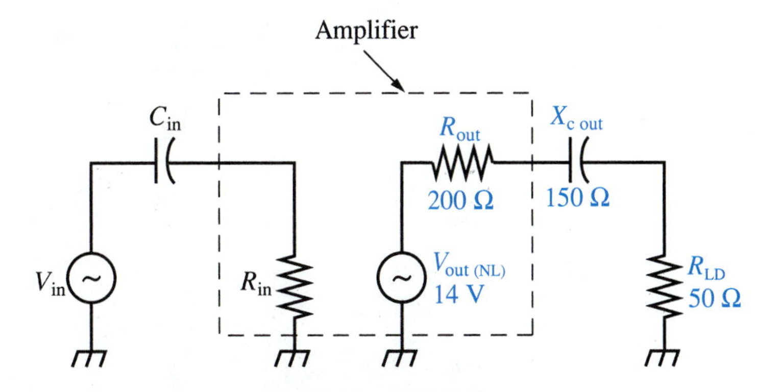

FIGURE 19-19 Equivalent circuit of an electronic amplifier with a significant amount of capacitive reactance due to the output coupling capacitor

EXAMPLE 19-11

An electronic amplifier is a circuit for boosting small voltages into larger voltages, while at the same time increasing the current capability. Amplifiers usually have their output signal connected through a **coupling capacitor** to the load, as shown in Figure 19-19. The purpose of the output coupling capacitor is to block any dc voltage within the amplifier circuitry from appearing at the load.

In many situations, the reactance of the output coupling capacitor is negligible. However, in Figure 19-19, X_{Cout} is 150 Ω, which is *not* negligible.

Amplifiers have a certain amount of output resistance, R_{out}, which is like the internal resistance of a dc source. In Figure 19-19, R_{out} = 200 Ω. The no-load (load disconnected) output voltage is 14 V for this amplifier [$V_{out\ (NL)}$ = 14 V].

a) Find the total impedance of the output circuit in Figure 19-19.
b) Calculate the current in the output circuit.
c) Find the value of V_{LD}.
d) Show the phase relationship between V_{LD} and $V_{out\ (NL)}$ on a phasor diagram.

SOLUTION

11

a) The output circuit is a series combination of two resistances and one capacitive reactance. Adding the resistances by simple arithmetic gives

$$R_T = R_{out} + R_{LD}$$
$$= 200\ \Omega + 50\ \Omega = 250\ \Omega$$

R_T and X_{Cout} are placed on a phasor diagram in Figure 19-20.

$$Z = \sqrt{R_T^2 + (X_{Cout})^2}$$
$$= \sqrt{250^2 + 150^2} = \mathbf{291.5\ \Omega}$$

Continued on page 362.

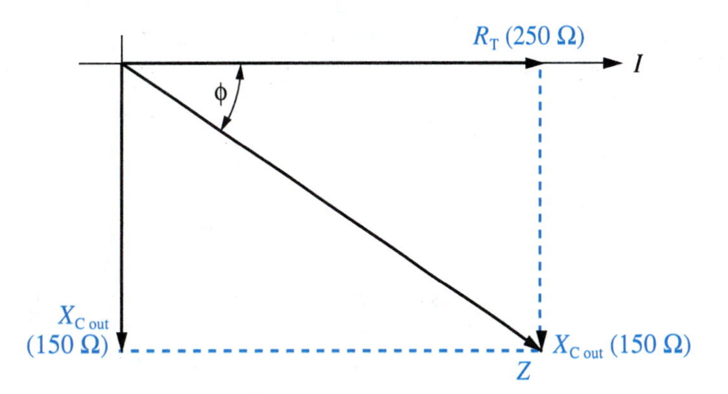

FIGURE 19-20 Phasor diagram of ohmic values. Current, *I*, is the reference phasor, even though its value is not known yet.

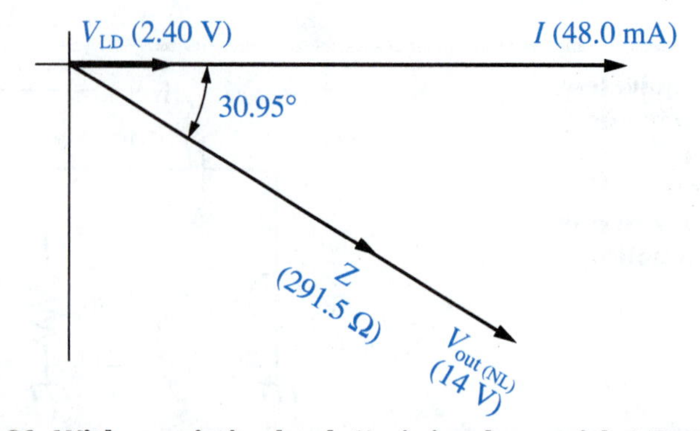

FIGURE 19-21 With a resistive load, V_{LD} is in phase with I. But because the overall circuit contains some capacitance, the source voltage [$V_{out(NL)}$] lags current, I.

b) Applying Ohm's law to the entire series *RC* output circuit,

$$I = \frac{V_{out\,(NL)}}{Z} = \frac{14\text{ V}}{291.5\ \Omega} = \mathbf{48.0\text{ mA}}$$

c) Applying Ohm's law to the load resistance alone,

$$V_{LD} = IR_{LD} = (48.0\text{ mA})\ 50\ \Omega = \mathbf{2.40\text{ V}}$$

d) The source voltage of this circuit is $V_{out\,(NL)}$. It has the same phasor position as total impedance, Z. (Total voltage and total impedance go together in the same position on a phasor diagram.) From Figure 19-20, we can see that

$$\cos\varnothing = \frac{R_T}{Z}$$

$$\varnothing = \cos^{-1}\left(\frac{R_T}{Z}\right) = \cos^{-1}\left(\frac{250\ \Omega}{291.5\ \Omega}\right)$$

$$= \cos^{-1}(0.8576)$$

$$= 30.95°$$

Therefore, $V_{out\,(NL)}$ is drawn 30.95° below the horizontal in Figure 19-21. Naturally V_{LD} is in phase with the current, I, since the load is purely resistive. It is shown in the 0° position in Figure 19-21. It is clear that V_{LD} leads $V_{out\,(NL)}$ by 30.95°.

Series RL Circuit With an Ideal Inductor

In a series resistor-inductor circuit with an ideal inductor, the solution is very similar to a series *RC* circuit. The only difference is that the ohmic phasors form a triangle above the x axis, rather than below it.

Self-Check problems 34 and 35 will give you practice in solving series *RL* circuits.

Nonideal Inductor in a Series RL Circuit

Real inductors often have a considerable amount of internal resistance, R_{int}, as we know from Section 18-8. When a real inductor is in series with a component resistor, the R_{int} value is simply added arithmetically to the component resistance, R. The resulting total resistance is symbolized R_T. Then X_L and R_T are combined on a phasor diagram to get impedance, Z.

SAFETY ADVICE

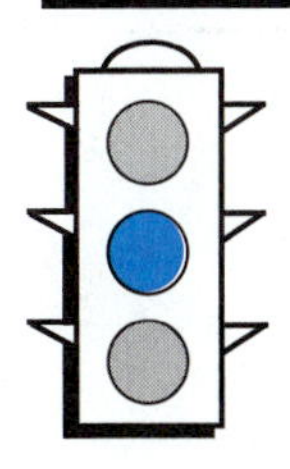

Any inductive load device that has a particular ac voltage rating must not be subjected to that same value of dc voltage. For example, a relay coil that is rated at 24 V, 60 Hz ac, must not be driven by a 24-V dc source. If 24 V dc is applied to the coil, its current will be too large. Unless the coil is protected by a fuse, this will overheat the winding, damaging its insulation and ruining the coil. Of course, any continued overcurrent always brings the risk of fire.

The reason for the overcurrent is that the coil has no reactance under dc conditions. Its impedance now consists only of its resistance. The coil designers were counting on the 60-Hz reactance to combine with the resistance. The resistance-reactance combination was supposed to provide the proper amount of impedance to limit the current to a correct value.

For the same reason, the coil must not be driven by an ac source with a frequency that is very much lower than 60 Hz.

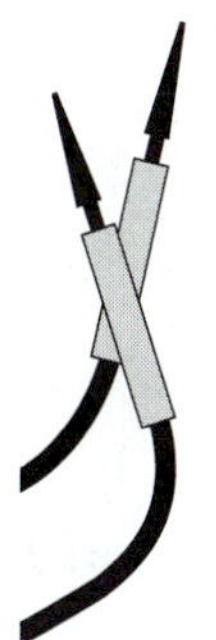

Troubleshooting Series Ac Circuits

Opens. An open in a series ac circuit has the same effect as an open in a series dc circuit—all current stops and all the load components become deenergized. You can find the exact location of the open by probing around the circuit with an ac voltmeter, in the same way that was shown for a dc circuit in Figure 9-20.

Shorts. In a series ac circuit, a shorted resistor causes the current to become larger, in the usual way.

The series ac circuits studied so far have been plain *RC* and plain *RL*. They have had one kind of reactance or the other, but not both (not both X_C and X_L). For such series circuits, plain *RC* or plain *RL*, a shorted reactor also causes current to become larger, as shown in Figure 19-22.

In that figure, the reactance is inductive due to a combination of L_1 and L_2. If one inductor fails shorted, the net inductive reactance decreases. In this case, X_{net} decreases from 100 Ω to 60 Ω. This causes the total impedance to decrease, and current rises.

12

To locate a shorted component, measure the voltage across each one. The shorted component will measure 0 V if it is a dead short. It will measure a less-than-expected value for a partial short. This was explained in Figure 9-22 for dc circuits.

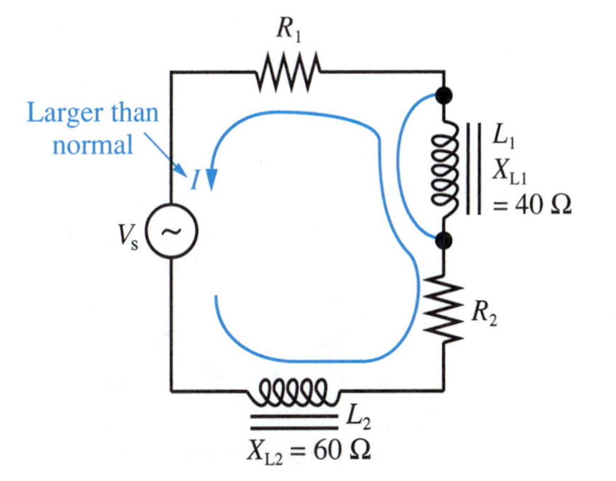

FIGURE 19-22 Effect of a shorted inductor in a series RL circuit.

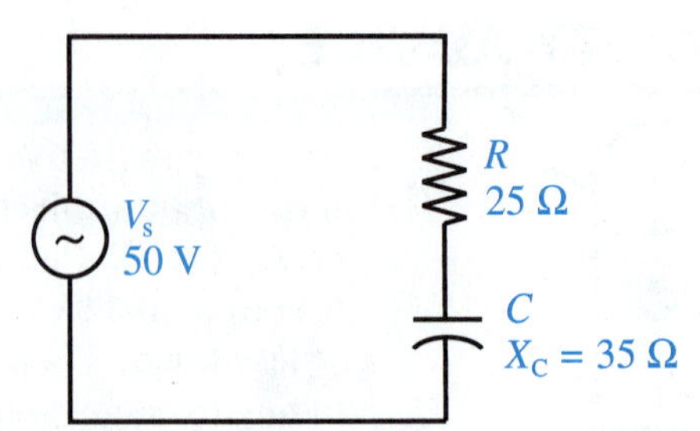

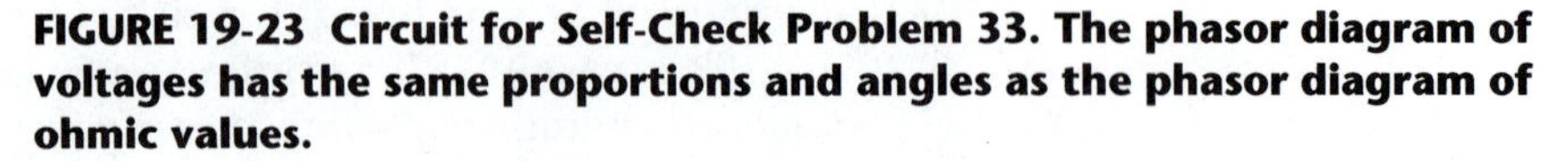
FIGURE 19-23 Circuit for Self-Check Problem 33. The phasor diagram of voltages has the same proportions and angles as the phasor diagram of ohmic values.

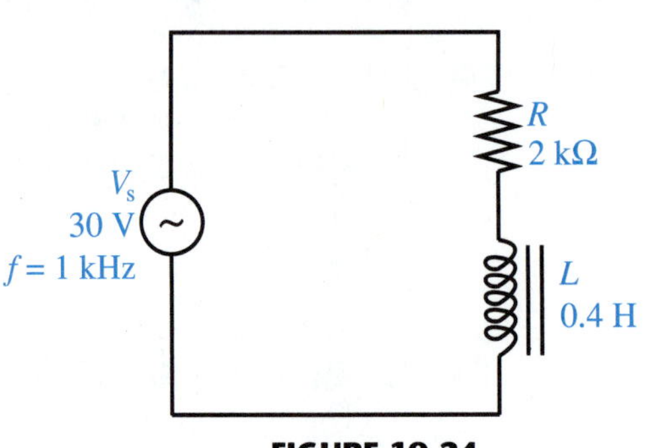

FIGURE 19-24

SELF-CHECK FOR SECTION 19-6

28. In a series ac circuit, the__________is the reference variable because it is the same for all circuit components. [11]
29. In a phasor diagram for a series ac circuit, resistance is placed on the________axis. [10, 11]
30. In a phasor diagram for a series ac circuit, where is capacitive reactance placed? Explain why. [10, 11]
31. Repeat Question 30 for inductive reactance. [10, 11]
32. T-F In a phasor diagram for a series circuit, the position of the impedance is the same as the position of the source voltage. [11]
33. For the series *RC* circuit of Figure 19-23, draw the phasor diagram of ohmic values. Then, [10, 11]
    a) Find the circuit impedance, *Z*.
    b) Find the current, *I*, and its phase relation to V_s .
    c) Calculate V_R and V_C.
    d) Draw the phasor diagram of voltages. Compare it to the phasor diagram of ohmic values.
34. For the ideal series *RL* circuit of Figure 19-24, [11]
    a) Draw a phasor diagram showing current, *I*, as the reference phasor.
    b) Show the proper positions of the ohmic values—resistance, *R*, and inductive reactance, X_L .
    c) What is the circuit's total impedance, *Z*?
    d) Find the value of current, *I*, and its phase angle relative to V_s.
35. For the series RL circuit in Problem 34, [11]
    a) Find V_R , the voltage across the resistor.
    b) Find the voltage V_L across the ideal inductor.
    c) Prove that Kirchhoff's voltage law is satisfied. That is, show that the phasor addition, or combination, of V_R and V_L gives a total voltage of 30 V.

FORMULAS

$\text{hypotenuse}^2 = \text{adjacent}^2 + \text{opposite}^2$ **EQ. 19-1**

$\cos ø = \frac{\text{adjacent}}{\text{hypotenuse}}$ **EQ. 19-2**

For parallel ac circuits:

$\cos ø = \frac{\text{resistive current}}{\text{total current}} = \frac{I_R}{I_T}$ **EQ. 19-3**

$I_T^2 = I_R^2 + I_X^2$ **EQ. 19-4**

$I_T = \sqrt{I_R^2 + I_X^2}$ **EQ. 19-5**

$Z = \frac{RX}{\sqrt{R^2 + X^2}}$ **EQS. 19-10 AND 19-11**

For all ac circuits:

$I = \frac{V_s}{Z}$ **EQ. 19-9**

$PF = \cos ø$ **EQ. 19-15**

$P = V_s I_R$ **EQ. 19-12**

$S = V_s I_T$ (voltamperes) **EQ. 19-16**

$P = V_s I_T \cos ø$ **EQ. 19-13**

$P = S\,(PF)$ (watts) **EQ. 19-17**

$P = V_s I_T\,(PF)$ **EQ. 19-14**

$S^2 = P^2 + Q^2$ (Q in vars) **EQ. 19-18**

For series ac circuits:

$V_s = \sqrt{V_R^2 + V_X^2}$

$Z = \sqrt{R^2 + X^2}$

SUMMARY OF IDEAS

- A phasor diagram is a simple method of representing the magnitude and phase of ac current or voltage.
- An ac current or voltage phasor can be resolved into its horizontal and vertical components.
- Two ac phasors that are displaced by 90° can be combined using the Pythagorean theorem.
- In parallel ac circuits, the source voltage is the reference phasor and the branch current phasors are combined to give the total current.
- For a circuit that is partly resistive and partly reactive, the overall opposition to current is called impedance, symbolized Z.
- For such a circuit, Ohm's law is written $V_s = I Z$. Any one of these three variables can be found if the other two are known.
- A circuit's power factor is the cosine of the phase angle between total current and source voltage.
- Power factor can be thought of as the ratio of the in-phase component of current to the total current (ratio of I_R to I_T) .
- Ac sources are rated in terms of their maximum apparent power, not their true power.
- In series ac circuits, the current is the reference phasor, and the component voltages are combined to give the total (source) voltage.
- In a series ac circuit, the ohmic values of resistance and reactance can be placed on a phasor diagram and combined to give the circuit's impedance (total opposition ability).

CHAPTER QUESTIONS AND PROBLEMS

1. Every phasor diagram for an ac circuit has a reference variable. Where is the reference variable's phasor placed ? [1]
2. Draw a phasor diagram with source voltage as the reference variable and total current lagging by 45°. [1]
3. Repeat Question 2 for I_T leading by 45°. [1]
4. When a phasor is resolved into two components, the component on the__________axis is called the in-phase component; the component on the__________axis is called the out-of-phase component. [2]
5. In a certain ac circuit, suppose $I_R = 4$ A and $I_X = 6$ A. Use the graphical method shown in Figure 19-7 to find I_T. Use a scale factor of ¼ inch = 1 A. [4]
6. For a right triangle, the hypotenuse, opposite side, and adjacent side are related by the__________theorem. [3]
7. Write the formula for the theorem in Question 6. [3]
8. In a phasor diagram of electric currents, a right triangle is formed by I_R, I_X, and I_T. In that triangle, the cosine of ø is equal to the__________current divided by the total current.
9. An ac circuit has an in-phase current $I_R = 2.0$ A, and a leading out-of-phase current $I_X = 2.7$ A. The reference variable is V_s. [3]
   a) What is the magnitude of total current I_T?
   b) Find the phase angle ø.
   c) Describe the phase relation of I_T to V_s.
10. For the parallel *RC* circuit of Figure 19-25, [5]
    a) Find individual branch currents I_R and I_C.
    b) Using a phasor diagram, find total current, I_T, and its phase angle relative to V_s.
11. In a parallel *RC* circuit, the total current always__________the voltage. (leads or lags) [5]
12. In a parallel *RC* circuit, if I_C is greater than I_R, the total current, I_T, will be out of phase with V_s by__________than 45°. (Answer more than or less than.) [5]
13. In Figure 19-8(A), suppose the frequency is increased to 120 Hz. Everything else stays the same. Find the following: [5, 6]
    a) I_R
    b) I_C
    c) I_T
    d) ø
    e) *Z* (from Ohm's law)
    f) *Z* (from Equation 19-10)

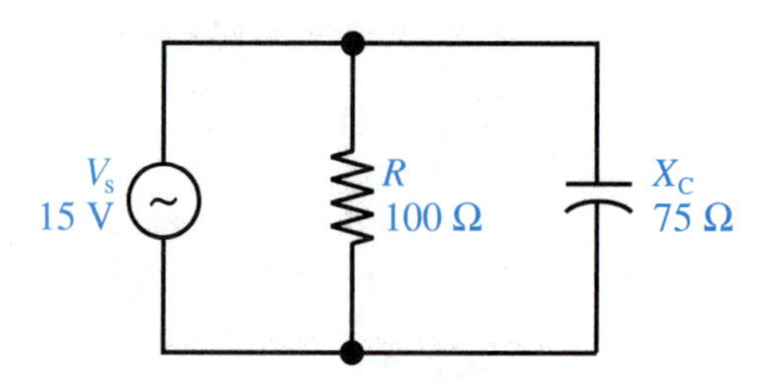

FIGURE 19-25

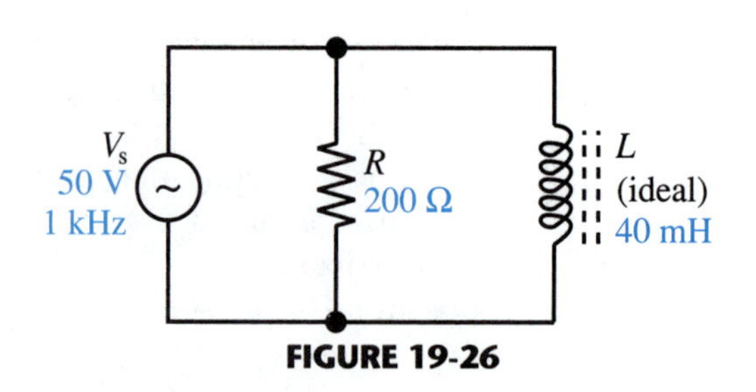

FIGURE 19-26

14. In a parallel *RC* circuit, if the frequency is increased, the total current will __________. [5]
15. For the parallel *RL* circuit of Figure 19-26, [5]
    a) Find individual branch currents I_R and I_L.
    b) Using a phasor diagram, find total current, I_T, and its phase angle relative to V_s.
16. In a parallel *RL* circuit, I_T always__________V_s. [5]
17. In a parallel *RL* circuit like Figure 19-26, suppose the frequency is increased. For each of the following variables, tell whether it will increase, decrease, or stay the same. [5]
    a) I_R will__________.
    b) X_L will__________.
    c) I_L will__________.
    d) I_T will__________.
    e) ø will__________.
18. From Problem 10, Figure 19-25, V_s, I_R, and I_C are known. Find the true power, *P*. [9]
19. Find the power factor, *PF*, for the circuit in Figure 19-25. [8, 9]
20. Apparent power is symbolized __________. It is measured in basic units of__________. [9]
21. Find the apparent power, *S*, for the circuit in Figure 19-25. [9]
22. For the parallel *RL* circuit of Figure 19-27, [5, 9]
    a) Find I_R, I_L, I_T and ø.
    b) What is the circuit's apparent power, *S*?
    c) What is the circuit's true power, *P*?
    d) Find the circuit's power factor.
23. T-F An electric load with a high power factor is more economical than one with a low power factor, all other things being equal. [8]
24. In a series ac circuit, the reference variable is the__________. [1, 11]
25. For the series *RC* circuit of Figure 19-28, [10, 11]
    a) Calculate X_C.
    b) Combine *R* and X_C on a phasor diagram to find *Z*.
    c) Find *I* and its phase relationship to V_s.
26. T-F In a series *RC* circuit, *Z* is always larger than X_C. [10]
27. For the circuit in Problem 25 (Figure 19-28), you can calculate V_R as 44.1 V and V_C as 66.8 V. If you add $V_R + V_C$ by simple arithmetic, you get a result of 110.9 V, which is larger than the source voltage of 80 V. Why doesn't this work? [11]
28. Show the correct way to combine V_R and V_C for the circuit of Figure 19-28. [11]

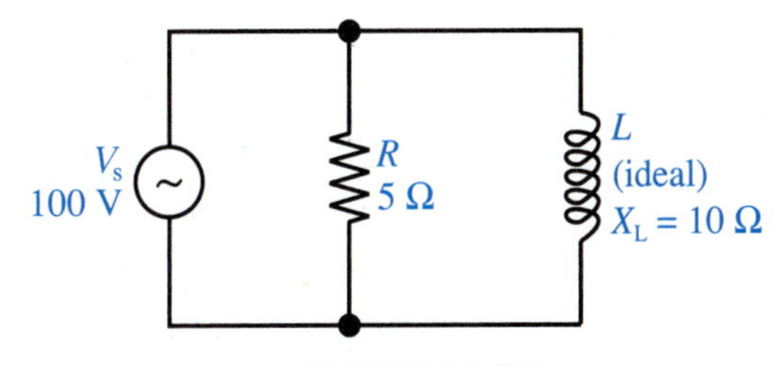

FIGURE 19-27

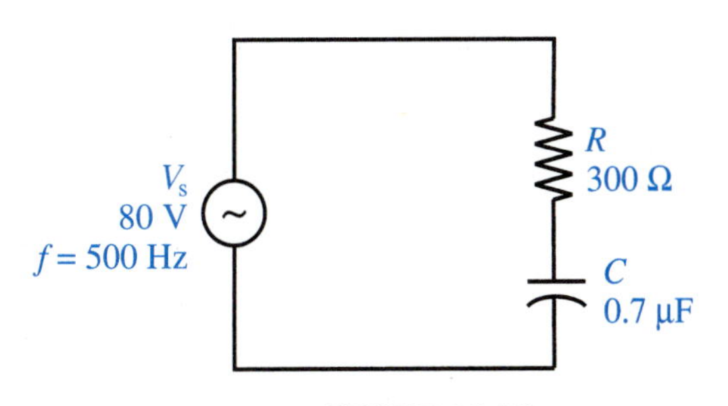

FIGURE 19-28

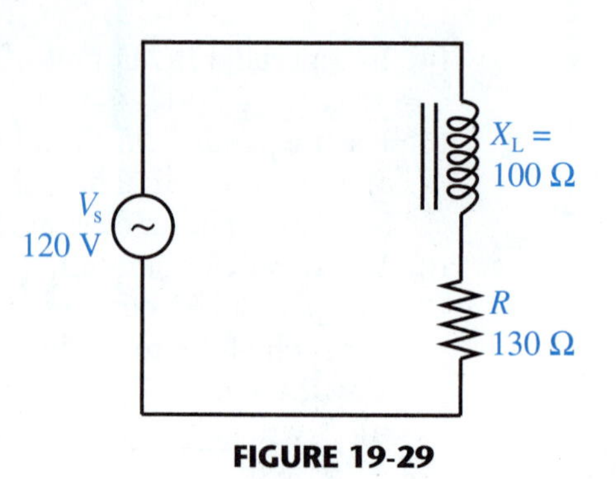

FIGURE 19-29

29. For the ideal series *RL* circuit in Figure 19-29, [10, 11]
    a) Find *Z* using a phasor diagram.
    b) Find *I*.
    c) Calculate V_L and V_R.
30. For the *RL* circuit of Problem 29, find the following: [8, 9]
    a) Apparent power.
    b) True power.
    c) Power factor.
    d) Quadrature power.

CHAPTER 20

RESONANCE

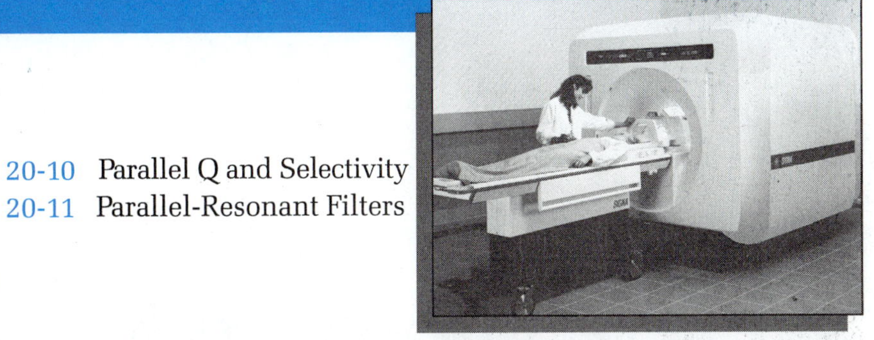

With a magnetic resonance imaging (MRI) machine, doctors can study a patient's brain and spinal condition without surgery or injection of dyes. The image shown here is of the lower spine looking from the person's side. It clearly shows a ruptured disc at the bend of the spine.
Courtesy of GE Medical Systems

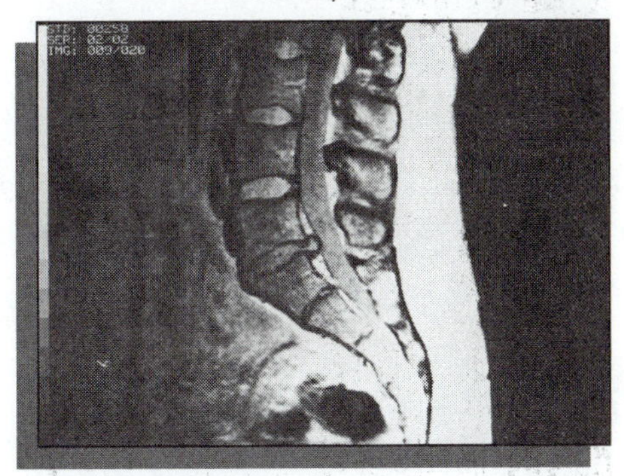

OUTLINE

NEW TERMS TO WATCH FOR

resonant frequency
frequency-response curve
bandwidth
low cutoff frequency
high cutoff frequency
selectivity
circuit Q (quality)
filter
band-pass
band-stop
dropping resistor
television sound trap
video (picture) amplifier
audio (sound) amplifier
low-pass
high-pass
parallel resonance

In an ac circuit containing both an inductor and a capacitor, there is one particular frequency that produces a vigorous response from the circuit. This is called the resonant frequency.

After studying this chapter, you should be able to:

1. Use Ohm's law to analyze the behavior of a series-resonant circuit.
2. Find the resonant frequency for a series *LCR* circuit.
3. From the frequency-response curve for a series *LCR* circuit, identify the high cutoff frequency, the low cutoff frequency, and the bandwidth.
4. Describe the relationship between circuit *Q* and circuit selectivity, or bandwidth, for a series *LCR* circuit.
5. Calculate any one of the three variables, f_r, Q_T, or *Bw*, given the other two.
6. Show how to obtain band-pass filtering and band-stop filtering using series *LCR*.
7. Show how to obtain low-pass and high-pass filtering using *RC* and *RL* series combinations.
8. Find the resonant frequency for an ideal parallel *LCR* circuit, and analyze its behavior using Ohm's law.
9. Interpret the frequency-response for a parallel *LCR* circuit, including high and low cutoff frequencies, bandwidth, selectivity, and circuit *Q*. Calculate any one of the three variables, f_r, $Q_{T(parallel)}$, or *Bw*, given the other two.
10. Show how to obtain band-pass and band-stop filtering, using parallel *LCR*.

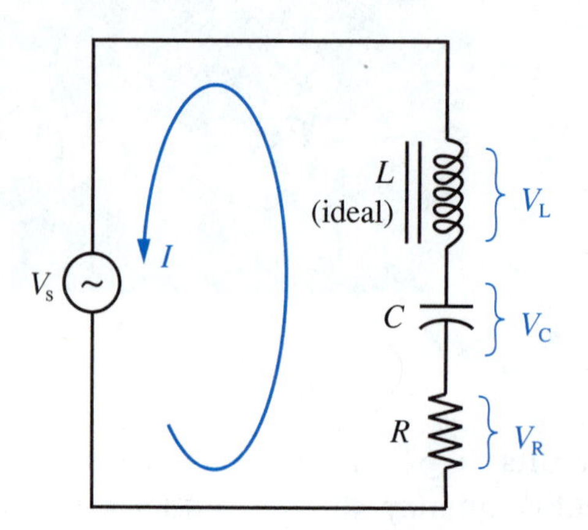

FIGURE 20-1 Series LCR circuit

20-1 LCR SERIES CIRCUITS

A series circuit containing an inductor, a capacitor, and a resistor is shown in Figure 20-1. The voltage across the resistor, V_R, is in phase with the current. This is shown in the waveform graph of Figure 20-2(A) and in the phasor diagram of Figure 20-2(B).

The inductor voltage, V_L, leads the current by 90°. The waveform graph in Figure 20-2(A) shows this, and the same fact is presented in the phasor diagram of Figure 20-2(B). Capacitor voltage, V_C, lags the current by 90°, as Figures 20-2(A) and 20-2(B) make clear.

By inspecting the waveforms of Figure 20-2(A), it can be seen that the instantaneous value of inductor voltage, v_L, is always the opposite of the instan-

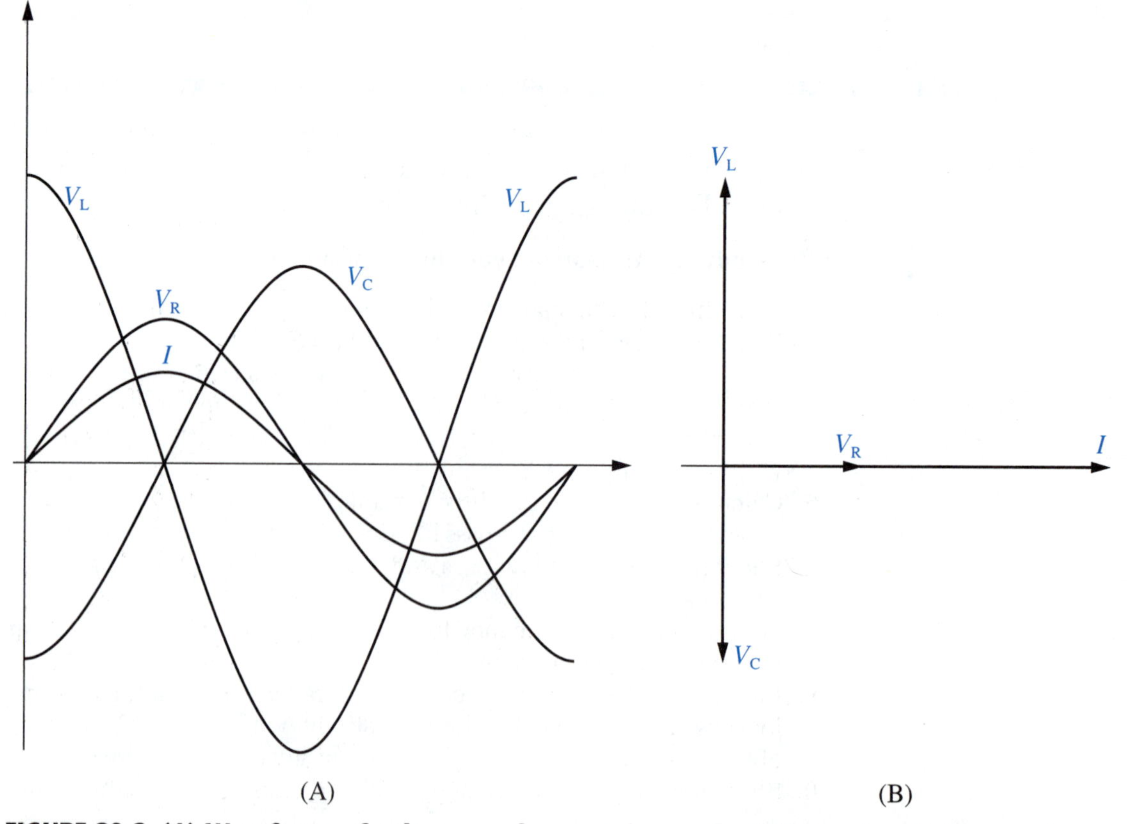

FIGURE 20-2 (A) Waveforms of voltages and current in a series LCR circuit. (B) Phasor diagram

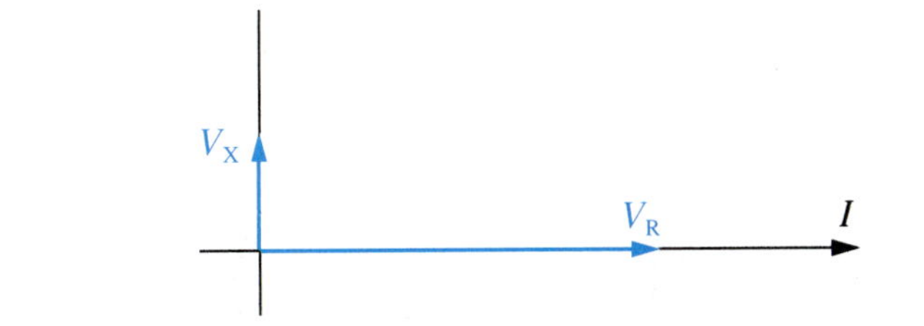

FIGURE 20-3 V_L and V_C subtract from one another to give the net reactive voltage, V_X.

taneous capacitor voltage, v_C. When v_L is instantaneously positive (meaning positive on the top terminal in the schematic of Figure 20-1), capacitor voltage, v_C, is instantaneously negative (negative on the top terminal of C in Figure 20-1). When v_L goes instantaneously negative, v_C reverses to instantaneously positive. Verify this by checking the waveforms carefully.

Thus, as far as Kirchhoff's voltage law is concerned, V_L and V_C don't add, they subtract. With V_L larger than V_C, as shown in Figure 20-2, the inductor-capacitor combination has a net voltage of

$$V_X = V_L - V_C$$

EQ. 20-1

In Equation 20-1, the smaller voltage has been subtracted from the larger to give the net reactive voltage in the circuit.

On a phasor diagram, this means that the shorter phasor is subtracted from the longer phasor (they point in opposite directions). As a general rule, we can say:

TECHNICAL FACT

In a series LCR circuit, the *difference* between V_L and V_C becomes the net reactive voltage phasor, V_X. This V_X phasor points in the direction of the longer phasor (larger voltage).

Applying this rule to Figure 20-2 gives the phasor diagram shown in Figure 20-3.

EXAMPLE 20-1

For the LCR circuit of Figure 20-1, suppose V_L = 12 V, V_C = 8 V, and V_R = 6 V.

a) Find the combination of these three voltages, which equals V_s.
b) What is the phase relationship between I and V_s?
c) If current I = 0.15 A, find the circuit's impedance.

SOLUTION

a) The voltages across these three components are shown in the phasor diagram of Figure 20-4(A). With V_L larger than V_C,

$$V_X = V_L - V_C = 12\text{ V} - 8\text{ V} = 4\text{ V}$$

V_X points upward, as shown in Figure 20-4(B). Using the Pythagorean theorem,

$$V_T = V_s = \sqrt{V_R^2 + V_X^2}$$
$$= \sqrt{6^2 + 4^2} = \mathbf{7.21\ V}$$

b)
$$\varnothing = \cos^{-1}\left(\frac{\text{adj}}{\text{hyp}}\right) = \cos^{-1}\left(\frac{V_R}{V_T}\right) = \cos^{-1}\left(\frac{6\text{ V}}{7.2\text{ V}}\right) = \cos^{-1}(0.8321)$$
$$= \mathbf{33.7°}\ (I \text{ lagging } V_s)$$

c)
$$Z = \frac{V_s}{I} = \frac{7.21\text{ V}}{0.15\text{ A}} = \mathbf{48.1\ \Omega}$$

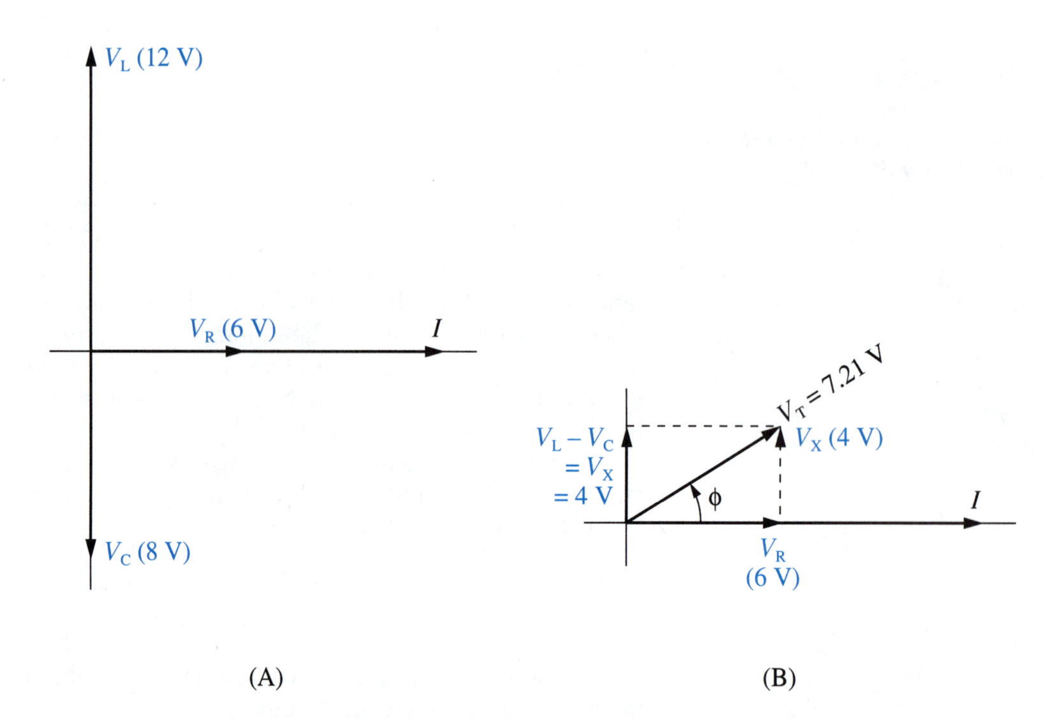

FIGURE 20-4 (A) Phasor diagram for Example 20-1. (B) V_X points in the direction of V_L, since V_L is larger than V_C. The overall circuit is therefore inductive/resistive, with I lagging V_s.

20-2 SERIES-RESONANT FREQUENCY

If the source frequency is variable in Figure 20-1, there will be some frequency at which $V_C = V_L$. Then these two voltages exactly cancel each other, causing the circuit to appear totally resistive, with no reactive nature.

TECHNICAL FACT

Every series LCR circuit has a certain **resonant frequency,** f_r. At $f = f_r$,

1. Z is at its minimum possible value, since X_L and X_C cancel each other. Therefore,

$$Z_{min} = R$$

EQ. 20-2

2. I is at its maximum possible value, since Z is minimum.
3. $ø = 0°$, because the circuit becomes totally resistive.

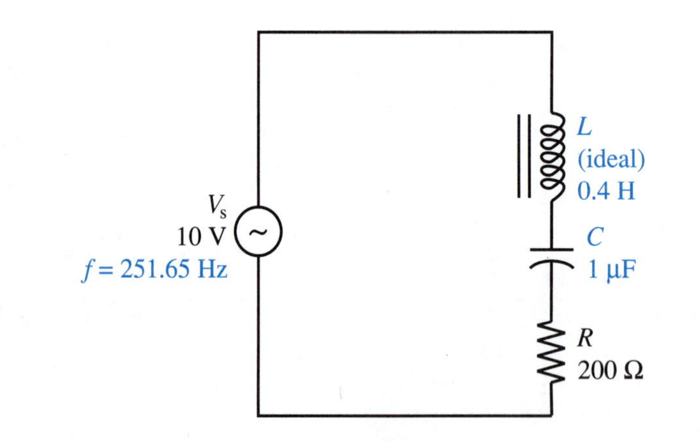

FIGURE 20-5 Series LCR combination being driven at its resonant frequency

EXAMPLE 20-2

The series *LCR* circuit of Figure 20-5 is being driven at its resonant frequency, which is 251.65 Hz.

a) Prove that this is the resonant frequency by calculating X_L and X_C.
b) Draw a phasor diagram of ohmic values to show that $Z = R$.
c) Find the current, I.
d) Apply Ohm's law to the inductor and to the capacitor to find V_L and V_C.

SOLUTION

a) $$X_L = 2\,\pi f L = 6.2832\ (251.65\text{ Hz})\ (0.4\text{ H}) = \mathbf{632.46\ \Omega}$$

$$X_C = \frac{1}{2\,\pi f C} = \frac{1}{6.2832\ (251.65\text{ Hz})\ (1 \times 10^{-6}\text{ F})} = \frac{1}{1.5811 \times 10^{-3}} = \mathbf{632.46\ \Omega}$$

With X_L and X_C being exactly equal, V_L must exactly equal V_C. Therefore, the two voltages cancel out, making this the resonant frequency.

b) The initial phasor diagram of Figure 20-6(A) quickly reduces to Figure 20-6(B).

$$Z = \sqrt{R^2 + X_{net}^{\ 2}} = \sqrt{R^2 + 0}$$
$$= R$$

c) $$I = \frac{V_s}{Z} = \frac{V_s}{R} = \frac{10\text{ V}}{200\ \Omega} = \mathbf{0.05\ A}$$

d) For the inductor,

$$V_L = I X_L = (0.05\text{ A})(632.46\ \Omega) = \mathbf{31.6\ V}$$

For the capacitor,

$$V_C = I X_C = (0.05\text{ A})(632.46\ \Omega) = \mathbf{31.6\ V}$$

Note that the individual inductor and capacitor voltages are much larger than the source voltage itself. This is an unusual feature of series-resonant circuits.

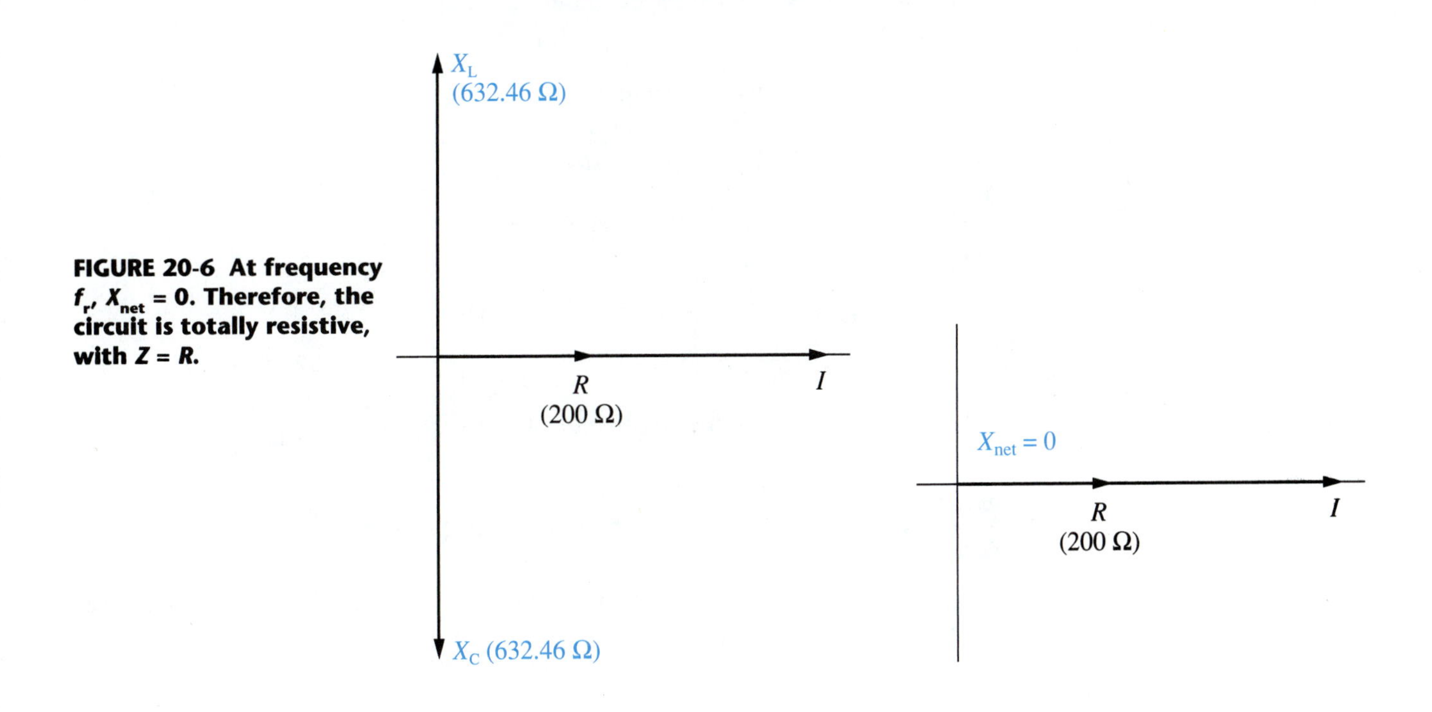

FIGURE 20-6 At frequency f_r, $X_{net} = 0$. Therefore, the circuit is totally resistive, with $Z = R$.

In order for resonance to occur, it is necessary that $X_L = X_C$. That is,

$$2\pi f L = \frac{1}{2\pi f C}$$

EQ. 20-3

There is only one frequency that will satisfy Equation 20-3, the **resonant frequency** f_r. Rearranging Equation 20-3 to solve for frequency yields

$$f_r = \frac{1}{2\pi\sqrt{LC}}$$

EQ. 20-4

With L expressed in henrys and C expressed in farads, Equation 20-4 gives the resonant frequency in hertz.

EXAMPLE 20-3

The series LCR circuit of Figure 20-7 has $L = 0.7$ H, $C = 2$ µF, and $R = 120\ \Omega$.

a) Solve for the resonant frequency, f_r.
b) Prove that $X_L = X_C$ at this frequency.

SOLUTION

a) Applying Equation 20-4, we get

$$f_r = \frac{1}{2\pi\sqrt{LC}} = \frac{1}{6.2832\sqrt{(0.7) \times (2 \times 10^{-6})}}$$

The best way to solve Equation 20-4 with a hand-held calculator is by the following sequence of operations:

1. Multiply L times C.
2. Take the square root of that result, using the $\sqrt{x}$ key.
3. Multiply the square root times 2 times π, or 6.2832.
4. Use the 1 / x key to reciprocate the previous result.

Following this sequence for the values above gives

$$f_r = \mathbf{134.5\ Hz}$$

The exact keystroke sequence would be

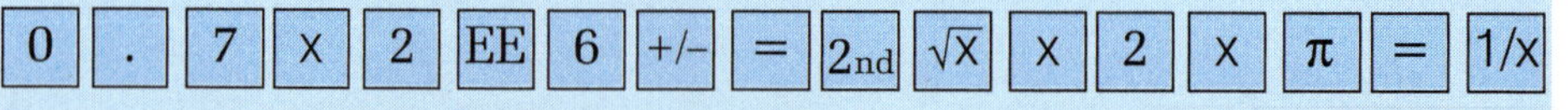

producing a display of 134.51, or similar.

b) At the resonant frequency of 134.5 Hz,

$$X_L = 6.2832\, f_r\, L = 6.2832\,(134.5)(0.7) = 591.6\ \Omega$$

and $$X_C = \frac{1}{6.2832\, f_r\, C} = \frac{1}{6.2832\,(134.5)(2 \times 10^{-6})} = 591.6\ \Omega$$

which demonstrates that the two reactances are exactly equal at f_r.

SELF-CHECK FOR SECTIONS 20-1 AND 20-2

1. In a series LCR circuit, the net reactance is equal to the smaller reactance subtracted from the__________ __________. [2]

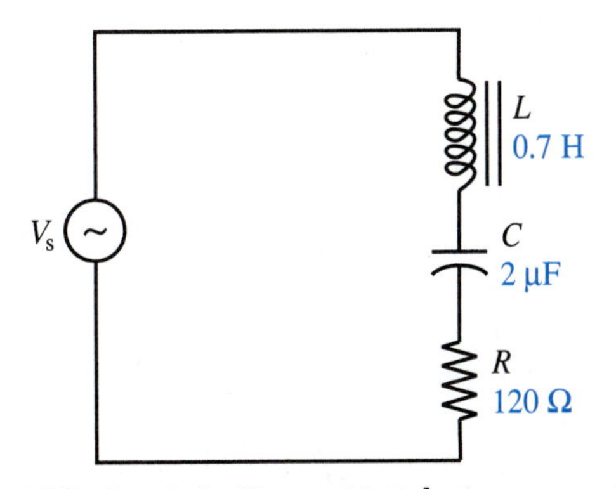

FIGURE 20-7 Series LCR circuit whose resonant frequency is to be found in Example 20-3

2. Take the series *LCR* circuit in Figure 20-7, but suppose it is driven at a nonresonant frequency of 200 Hz. Find X_C, X_L, and net reactance, X_{net}.
3. For the circuit in Problem 2, what is the impedance, *Z*?
4. For a series *LCR* circuit operating at the resonant frequency, f_r, the__________ __________exactly cancels the__________ __________. [1]
5. At the resonant frequency, the impedance of a series *LCR* circuit is minimum and is equal to__________. [2]
6. A series *LCR* circuit has $L = 0.2$ H, $C = 0.05$ μF, and $R = 600\ \Omega$. Calculate the resonant frequency. [1]
7. If the circuit in Problem 6 is driven by a voltage of 12 V at the resonant frequency, what is the value of current? [2]
8. For the conditions in Problem 7, what is the phase angle between current and source voltage, V_s? What is the circuit's power factor? [2]
9. For the conditions in Problem 7, calculate V_L. Repeat for V_C. Compare these voltages to the 12-V source voltage V_s, and explain what has occurred. [2]

20-3 FREQUENCY RESPONSE

A phasor diagram can describe the action of a series *LCR* circuit at only one frequency, namely the frequency at which X_L and X_C are calculated. To completely describe the action of a series *LCR* circuit at many frequencies, we plot a **frequency-response curve** of current, *I*, versus frequency, *f*. *I* goes on the y axis and *f* goes on the x axis. This idea is shown in Figure 20-8.

Sometimes a frequency-response curve is obtained by experimentally testing the circuit, taking actual measurements of current at many different frequencies. Other times a frequency-response curve is obtained strictly by calculation. The calculations are a lot of work, since every data point on the curve must be found

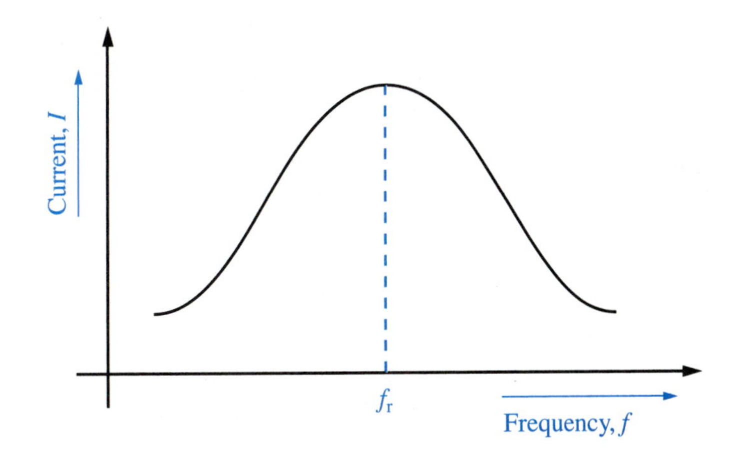

FIGURE 20-8 Graph of current versus frequency for series LCR circuit. Maximum current occurs when $f = f_r$. At all other frequencies, above or below f_r, there is less current.

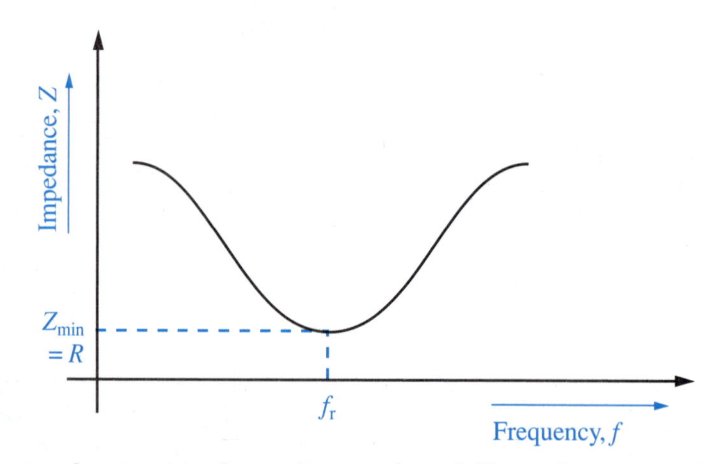

FIGURE 20-9 Alternative way of showing frequency response of series LCR. At $f = f_r$, the impedance is minimal.

by redrawing the circuit's phasor diagram for a different frequency. This involves recalculating X_L, X_C, X_{net}, Z, and I, over and over again. Such lengthy repetitive calculations are best performed by a computer.

A frequency-response curve may also be presented as a plot of impedance, Z, versus frequency, f. This is shown in Figure 20-9.

SAFETY ADVICE

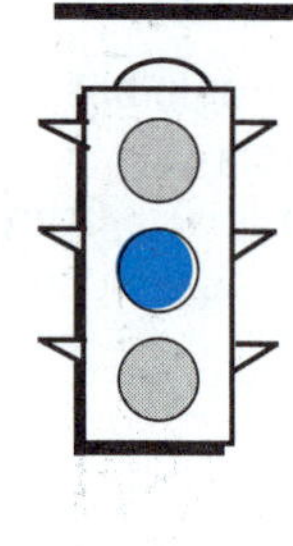

In a series-resonant circuit, the individual component voltages V_C and V_L can be much greater than the source voltage, V_s. This happens because the reactances of the capacitor and inductor are much greater than the circuit's resistance, which is the only current-limiter in the circuit. See Figure 20-5 and the solution to Example 20-2, part d.

The V_C and V_L voltages can therefore deliver a shock or damage the capacitor or inductor. This is true even though the circuit would seem harmless, since it has a low-value source voltage.

Bandwidth

TECHNICAL FACT

The **bandwidth** (symbolized *Bw*) of a series *LCR* circuit is the range of frequencies that allows the current to be at least 70.7% of the maximum value.

The bandwidth idea is illustrated in Figure 20-10.

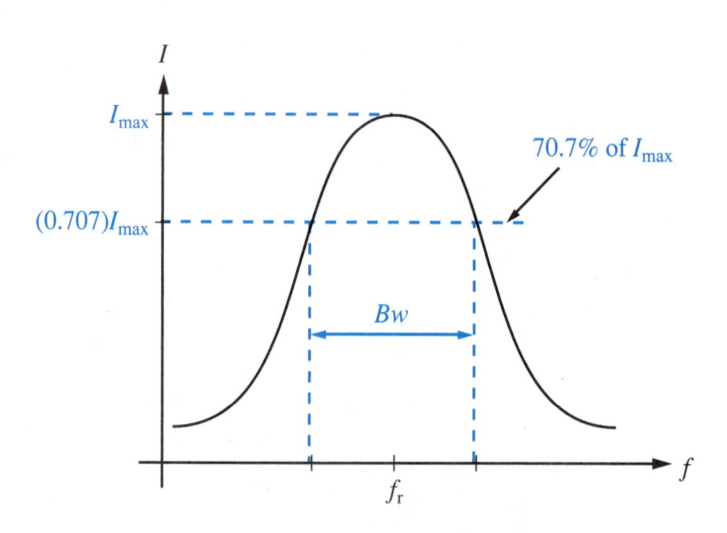

FIGURE 20-10 Bandwidth is the range of frequencies around f_r that permit the current to be at least 70.7% of the maximum value.

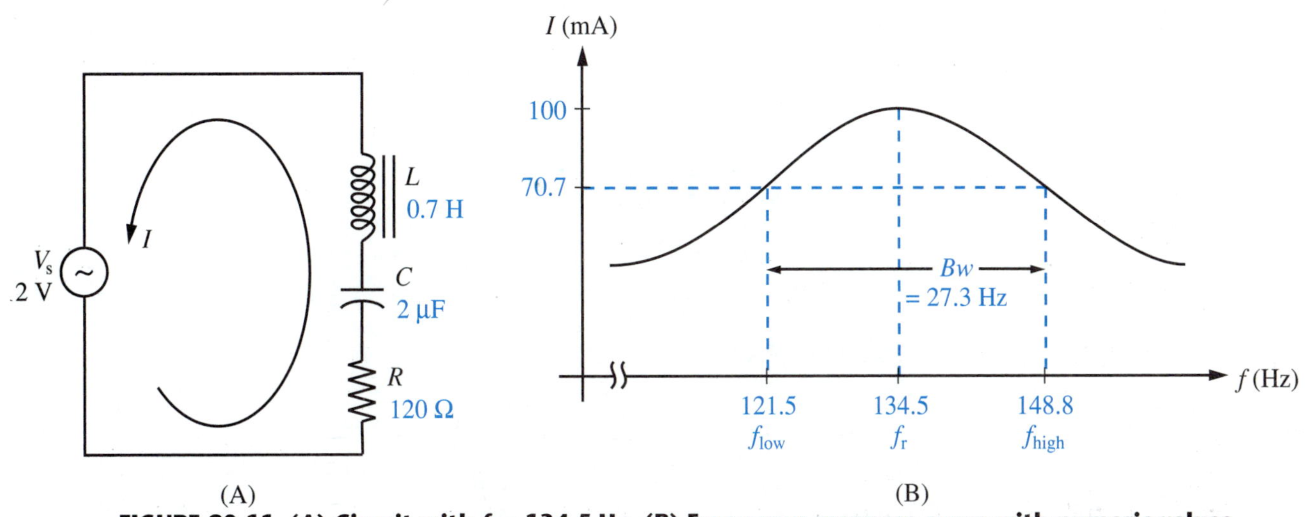

FIGURE 20-11 (A) Circuit with f_r = 134.5 Hz. (B) Frequency-response curve with numeric values

The circuit from Example 20-3 has been redrawn in Figure 20-11(A), with V_s = 12 V. At the 134.5-Hz resonant frequency,

$$I_{max} = \frac{V_s}{Z_{min}} = \frac{V_s}{R} = \frac{12\ \text{V}}{120\ \Omega} = 0.1\ \text{A or } 100\ \text{mA}$$

It can be proved that the current would decline to 70.7 mA at f = 121.56 Hz, below resonance. It can also be proved that I = 70.7 mA at f = 148.84 Hz, above resonance. (To prove these facts for yourself, see Chapter Questions and Problems 41 and 42 at the end of this chapter.) These actual values are specified in the circuit's frequency-response curve, shown in Figure 20-11(B).

TECHNICAL FACT

The frequency below resonance that produces $I = 0.707\ I_{max}$ is called the **low cutoff frequency**, symbolized f_{low}. The frequency above resonance that produces $I = 0.707\ I_{max}$ is called the **high cutoff frequency**, symbolized f_{high}.

Bandwidth is the difference between the high cutoff frequency and the low cutoff frequency. As an equation,

$$Bw = f_{high} - f_{low}$$

EQ. 20-5

In Figure 20-11,

$$Bw = 148.84\ \text{Hz} - 121.56\ \text{Hz} = 27.3\ \text{Hz}$$

Predicting Bandwidth. If we can actually measure the circuit current experimentally, it is not too hard to find its bandwidth. We proceed as follows:

1. Adjust the source frequency until the current reaches its maximum value; measure that value as I_{max}.
2. Multiply I_{max} by 0.7071. This is the cutoff value of current.
3. Adjust the source frequency above resonance until the measured value of current equals the cutoff value. Record that frequency as f_{high}.
4. Repeat part 3 below resonance, to find f_{low}.
5. Subtract $f_{high} - f_{low}$ to find Bw.

The preceding method is fine for a real circuit, but what if we are just working with paper and pencil? We need a formula for predicting Bw from knowledge of the circuit's components. By a lengthy mathematical process, this formula has been derived:

$$Bw = \frac{R}{2\pi L} = \frac{R}{6.2832\,L}$$

EQ. 20-6

EXAMPLE 20-4

Predict the bandwidth of the circuit in Figure 20-11(A) from knowledge of the components only.

SOLUTION

Applying Equation 20-6, we get

$$Bw = \frac{R}{2\pi L} = \frac{120}{6.2832\,(0.7)} = 27.28 \text{ or } \mathbf{27.3\ Hz}$$

which agrees with the earlier result.

20-4 SELECTIVITY AND QUALITY (*Q*)

Figure 20-12 shows two frequency-response curves that are quite different. The one in part A is steep-sided, while the one in part B has gently sloped sides. We say that the steep-sided curve represents a very **selective** circuit. This is because frequencies very near f_r produce a rather large current, but frequencies slightly different from f_r have a drastic reduction in current. The circuit *selects* the range of frequencies near f_r.

On the other hand, the gently sloped curve of part B means that there is not such an abrupt change in current. Therefore, the circuit is not very frequency-selective.

We quantify selectivity by stating the bandwidth. A small or narrow bandwidth means a quite selective frequency-response from the circuit. A large or wide bandwidth means a not-so-selective frequency-response.

There are some applications where a narrow bandwidth is required. For instance, in an AM radio receiver the signal from the antenna is a mixture of all the frequencies from the radio broadcasting stations in the vicinity. These frequencies range from 540 kHz at the low end to 1600 kHz at the upper end of the AM frequency band. The individual stations may have frequencies that differ by only

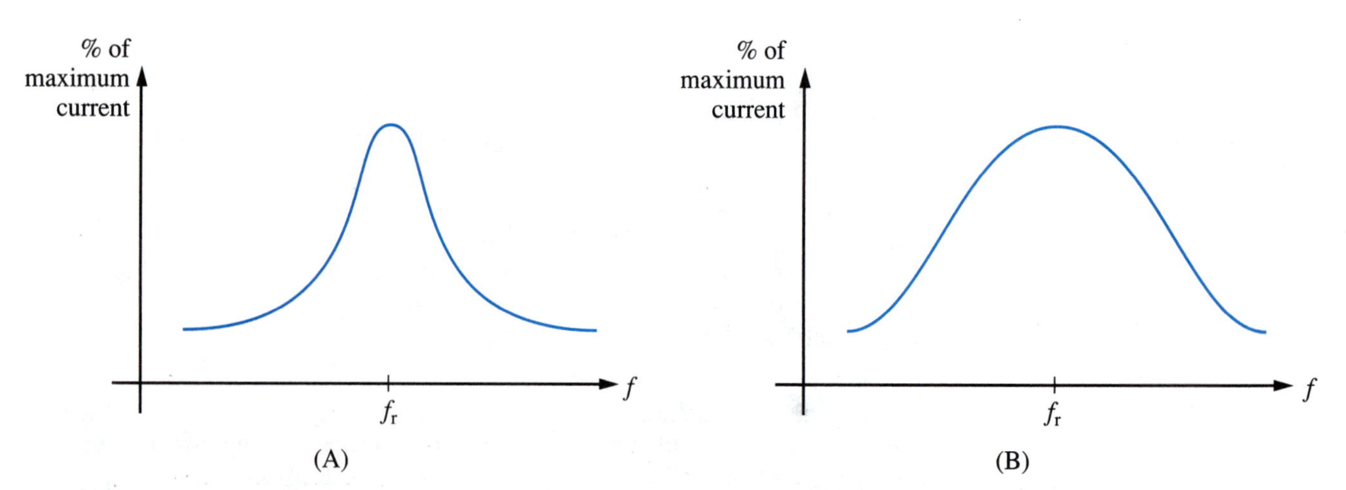

FIGURE 20-12 (A) Selective frequency response. (B) Not-so-selective frequency response

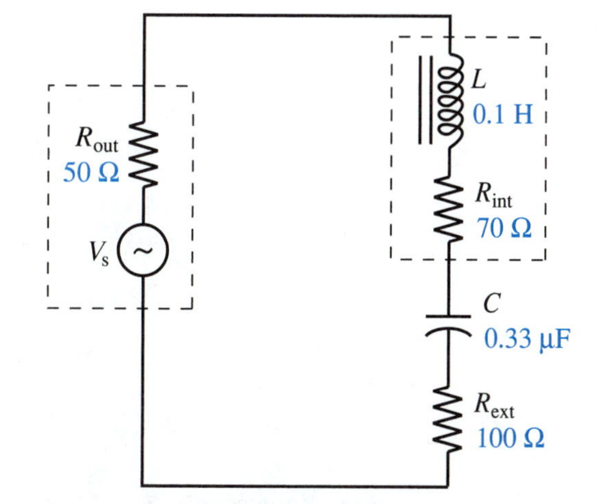

FIGURE 20-13 LCR circuit with real components. The circuit's Q_T value indicates its selectivity. Be careful to distinguish between Q of the inductor and Q_T of the total circuit. Q_T for the total circuit is always less than inductor Q.

10 kHz. Therefore, the series-resonant tuning circuits in the radio must have a bandwidth of less than 10 kHz, while the signals themselves are several hundreds of kHz, to prevent two stations being heard at the same time.

When you tune a radio or TV, you are changing a capacitance or inductance value in an *LCR* circuit, thereby changing its resonant frequency. This will be discussed further in Section 20-5.

Relationship of Bandwidth to Circuit Q. For a series-resonant *LCR* circuit, the reactances of the inductor and capacitor are equal; both can be symbolized as X_r. The subscript *r* signifies the resonant condition.

TECHNICAL FACT

For a series-resonant circuit, **circuit Q** or total Q is equal to the resonant reactance, X_r, divided by the circuit's total resistance. As a formula,

$$Q_T = \frac{X_r}{R_T}$$

EQ. 20-7

In Equation 20-7, the total resistance, R_T, includes the component resistance, the internal resistance of a nonideal inductor, and the internal resistance of the voltage source.

EXAMPLE 20-5

Figure 20-13 shows an LCR circuit containing a real inductor and a real voltage source. The real inductor has an internal resistance, R_{int}, of 70 Ω. The real voltage source has an internal resistance or output resistance, R_{out}, of 50 Ω. The 100-Ω component resistance is symbolized R_{ext} to emphasize that it is external to the inductor and the voltage source.

a) Find the resonant frequency, f_r.
b) What is the value of resonant reactance, X_r?
c) Calculate the total circuit quality, Q_T.

SOLUTION

a) Resistances have no effect on the resonant frequency; it depends on *L* and *C* only. From Equation 20-4,

Continued on page 380.

$$f_r = \frac{1}{2\pi\sqrt{LC}} = \frac{1}{6.2832\sqrt{(0.1)(0.33 \times 10^{-6})}}$$
$$= \mathbf{876.1\ Hz}$$

b) Use either the capacitive reactance or inductive reactance formula. Choosing inductive reactance,

$$X_r = 2\pi f_r L = 6.2832\,(876.1)\,(0.1) = \mathbf{550.5\ \Omega}$$

c) Add all resistances to obtain R_T.

$$R_T = R_{ext} + R_{int} + R_{out}$$
$$= 100 + 70 + 50 = 220\ \Omega$$

From Equation 20-7,

$$Q_T = \frac{X_r}{R_T} = \frac{550.5\ \Omega}{220\ \Omega} = \mathbf{2.50}$$

Q_T is a unitless number, being a ratio of ohms to ohms.

4

Q_T is useful for predicting selectivity. Higher-Q circuits tend to be more selective (narrower bandwidths). Low-Q circuits tend to be less selective (wider bandwidths). In fact, bandwidth can be calculated directly from knowledge of f_r and Q_T.

$$Bw = \frac{f_r}{Q_T}$$

EQ. 20-8

The effect of circuit Q on bandwidth is illustrated in Figure 20-14.

EXAMPLE 20-6

For the circuit of Figure 20-13, used in Example 20-5, do the following:

a) Calculate bandwidth from Equation 20-8.

b) Calculate bandwidth from Equation 20-6, and compare.

SOLUTION

a) $Bw = \frac{f_r}{Q_T} = \frac{876.1\ \text{Hz}}{2.50} = \mathbf{350\ Hz}$

b) Using the circuit's total resistance in Equation 20-6, we have

$$Bw = \frac{R_T}{2\pi L} = \frac{220\ \Omega}{6.2832\ (0.1\ \text{H})} = \mathbf{350\ Hz}$$

which agrees with part a.

In general, the bandwidth of a series-resonant circuit is not spread equally on both sides of the resonant frequency, f_r. Instead, f_{high} tends to be further away from f_r, with f_{low} closer to f_r. We can check this out by referring back to the cutoff frequencies given for the circuit of Figure 20-11(A).

The distance from f_{low} to f_r is

$$f_r - f_{low} = 134.5\ \text{Hz} - 121.6\ \text{Hz} = 12.9\ \text{Hz}$$

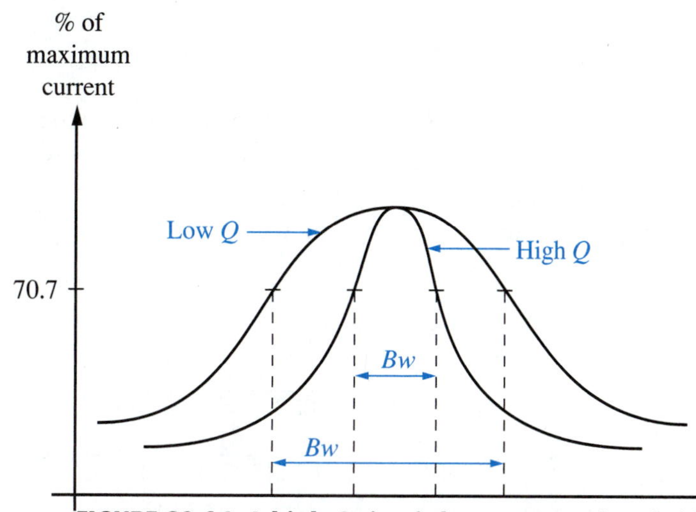

FIGURE 20-14 A high-*Q* circuit has a narrow bandwidth (greater selectivity). A low-*Q* circuit has a wider bandwidth (less selectivity).

The distance from f_r to f_{high} is

$$f_{high} - f_r = 148.8\text{ Hz} - 134.5\text{ Hz} = 14.3\text{ Hz}$$

The idea here is that you cannot exactly locate the cutoff frequencies by just dividing the bandwidth in half, and saying:

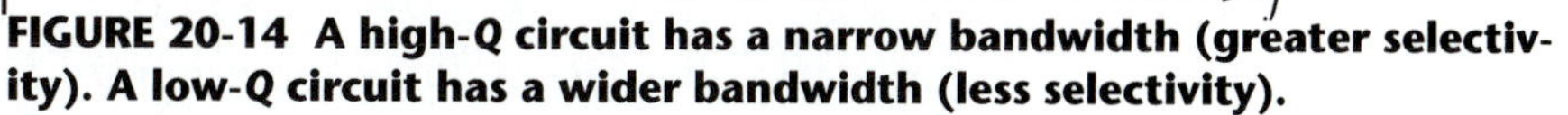

$$f_{high} = f_r + \tfrac{1}{2}\,Bw$$

EQ. 20-9

$$f_{low} = f_r - \tfrac{1}{2}\,Bw$$

EQ. 20-10

These equations are *not* exactly correct.

However, for high-*Q* circuits, Equations 20-9 and 20-10 give nearly correct answers. As a general rule, if Q_T is greater than 10, Equations 20-9 and 20-10 can be used to get very close approximations of f_{low} and f_{high}.

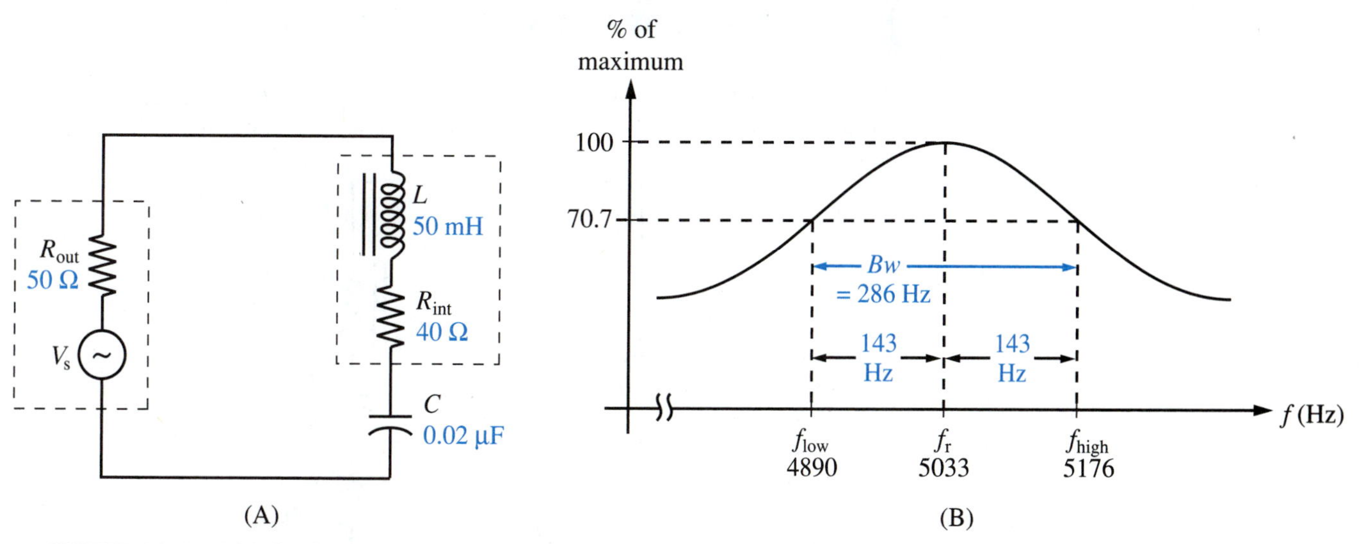

FIGURE 20-15 This high-*Q* circuit has a nearly symmetrical frequency response (f_r nearly in the center of the bandwidth).

EXAMPLE 20-7

The circuit of Figure 20-15 has no component resistance. The only resistances are the internal resistances in the inductor and the voltage source.

a) Find f_r.
b) Find X_r.
c) Find Q_T.
d) Find Bw.
e) Estimate f_{low} and f_{high}.

SOLUTION

a) $f_r = \dfrac{1}{2\pi\sqrt{LC}} = \dfrac{1}{6.2832\sqrt{(50 \times 10^{-3})(0.02 \times 10^{-6})}} = \mathbf{5033\ Hz}$

b) $X_r = \dfrac{1}{2\pi f_r C} = \dfrac{1}{6.2832\ (5033)(0.02 \times 10^{-6})} = \mathbf{1581\ \Omega}$

c) $Q_T = \dfrac{X_r}{R_T} = \dfrac{X_r}{R_{out} + R_{int}} = \dfrac{1581\ \Omega}{(50 + 40)\ \Omega} = \mathbf{17.6}$

d) $Bw = \dfrac{f_r}{Q_T} = \dfrac{5033}{17.6} = \mathbf{286\ Hz}$

e) Since $Q_T > 10$, we can use Equations 20-9 and 20-10.

$$\tfrac{1}{2}\,Bw = \tfrac{1}{2}\,(286\ \text{Hz}) = 143\ \text{Hz}$$

$$f_{low} \cong f_r - \tfrac{1}{2}\,Bw$$

$$= 5033\ \text{Hz} - 143\ \text{Hz} = \mathbf{4890\ Hz}$$

$$f_{high} \cong f_r + \tfrac{1}{2}\,Bw$$

$$= 5033\ \text{Hz} + 143\ \text{Hz} = \mathbf{5176\ Hz}$$

The approximate frequency-response curve is shown in Figure 20-15(B).

Using exact calculation techniques, it could be proved that the actual values of the cutoff frequencies are:

$$f_{low} = 4892\ \text{Hz}$$

$$f_{high} = 5178\ \text{Hz}$$

It is clear that the approximate values in Example 20-7 are very close indeed.

SELF-CHECK FOR SECTIONS 20-3 AND 20-4

10. In a frequency-response curve, the horizontal (x axis) variable is__________. [3]
11. The highest point on a frequency-response curve occurs at the__________ frequency. [3]

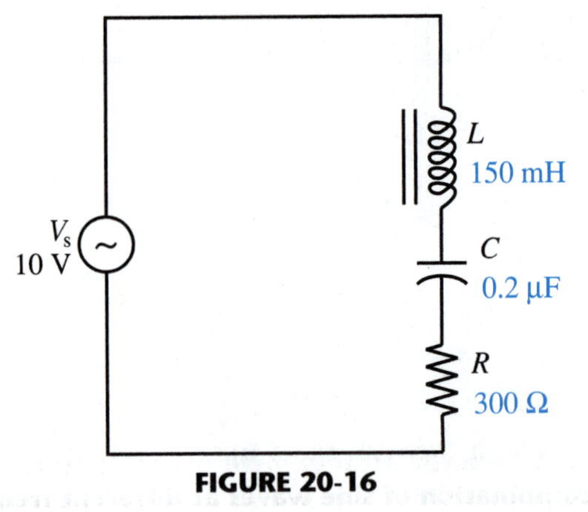

FIGURE 20-16

12. Explain the difference between a highly selective circuit and a not-very-selective circuit. How do the two circuits respond to various signal frequencies? [4]
13. Explain the meaning of the term *low cutoff frequency* (f_{low}). Repeat for f_{high}. [3]
14. The difference between f_{high} and f_{low} is called the_________. [3]
15. A selective circuit has a_________bandwidth, while a nonselective circuit has a_________bandwidth. [4]
16. What is the bandwidth of the *LCR* circuit of Figure 20-16? [5]
17. For the circuit of Figure 20-16, calculate the following: [5]
    a) f_r
    b) X_r
    c) Q_T
    d) *Bw*, using Equation 20-8. (Check against the answer from Problem 16.)
18. To decrease the bandwidth of a series-resonant circuit, Q_T must be_________. Therefore, the total resistance must be_________. (increased or decreased) [4]
19. T-F Obtaining a very narrow bandwidth requires a high-*Q* inductor. [4]

20-5 SERIES-RESONANT FILTERS

A **filter** is a circuit that passes signals at certain frequencies to a load. Signals at other frequencies are rejected—they cannot pass to the load.

Band-pass Filters

The basic structure of a series-resonant **band-pass** filter is shown in Figure 20-17(A). Think of the input voltage, V_{in}, as a combination of many signals, all at different frequencies. This is illustrated in Figure 20-17(B). Some of those signals will have frequencies close to the resonant frequency, and some will have frequencies far from the resonant frequency.

For those signals close to f_r, the *LC* combination has a small value of net reactance; in fact, $X_{net} = 0\ \Omega$ right at f_r. Therefore, for signals close to f_r, very little voltage will be dropped across the *LC* combination. Most of the voltage will make it through to R_{LD}.

But for signals far from f_r, the *LC* combination has a large value of net reactance (net inductive for frequencies far above f_r, and net capacitive for frequencies far below f_r). Therefore, most of the signal voltage will be dropped across the *LC* combination. Very little of the voltage will make it through to R_{LD}.

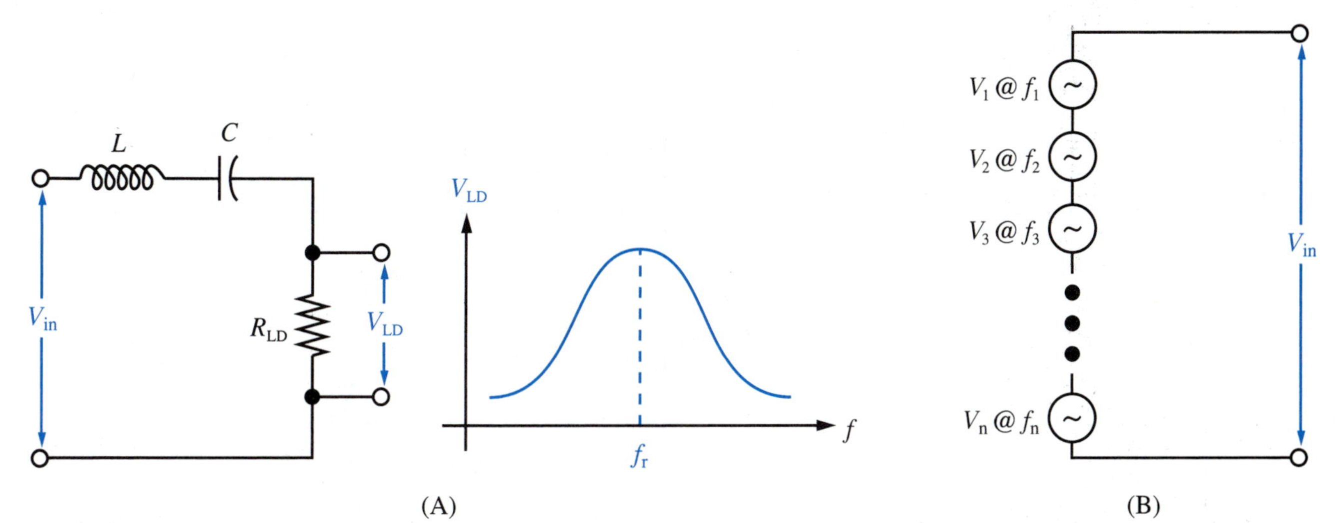

FIGURE 20-17 (A) Schematic arrangement of series-resonant band-pass filter. (B) The input voltage can be visualized as a combination of sine waves at different frequencies.

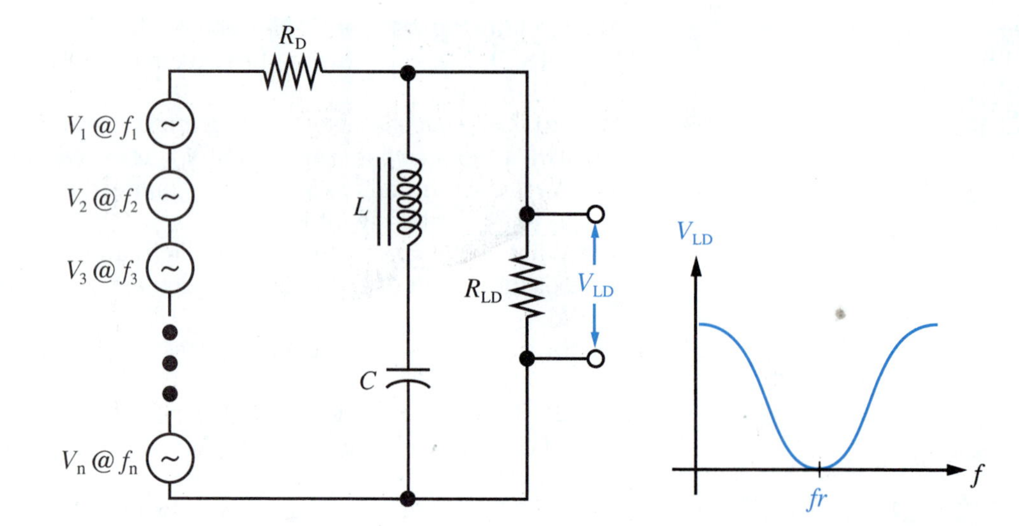

FIGURE 20-18 Schematic arrangement of series-resonant band-stop filter

The general sketch of V_{LD} versus frequency is drawn alongside the load terminals in Figure 20-17(A). As that sketch shows, the filter allows the band of frequencies near f_r to pass to the load, but frequencies outside that band are blocked, or filtered out.

Band-stop Filters

6

The basic structure of a series-resonant **band-stop** filter is shown in Figure 20-18. The *LC* combination will have a certain resonant frequency, f_r. Again the overall input, V_{in}, is viewed as a mixture of signals, some with frequencies close to f_r and some with frequencies far away from f_r.

For those signals with frequencies close to f_r, the combination has a low net reactance; in fact, $X_{net} = 0\ \Omega$ right at f_r. This low reactance "shorts out" the load. Therefore, very little voltage makes it through to the load: instead, most of the signal voltage will be dropped across **dropping resistor** R_D.

But for signals far from f_r, the *LC* combination has a large value of net reactance, so R_{LD} is not shorted out. Instead, the two resistors R_{LD} and R_D share the voltage by normal resistive voltage division. If R_{LD} is greater than R_D, then most of the signal voltage appears across R_{LD}.

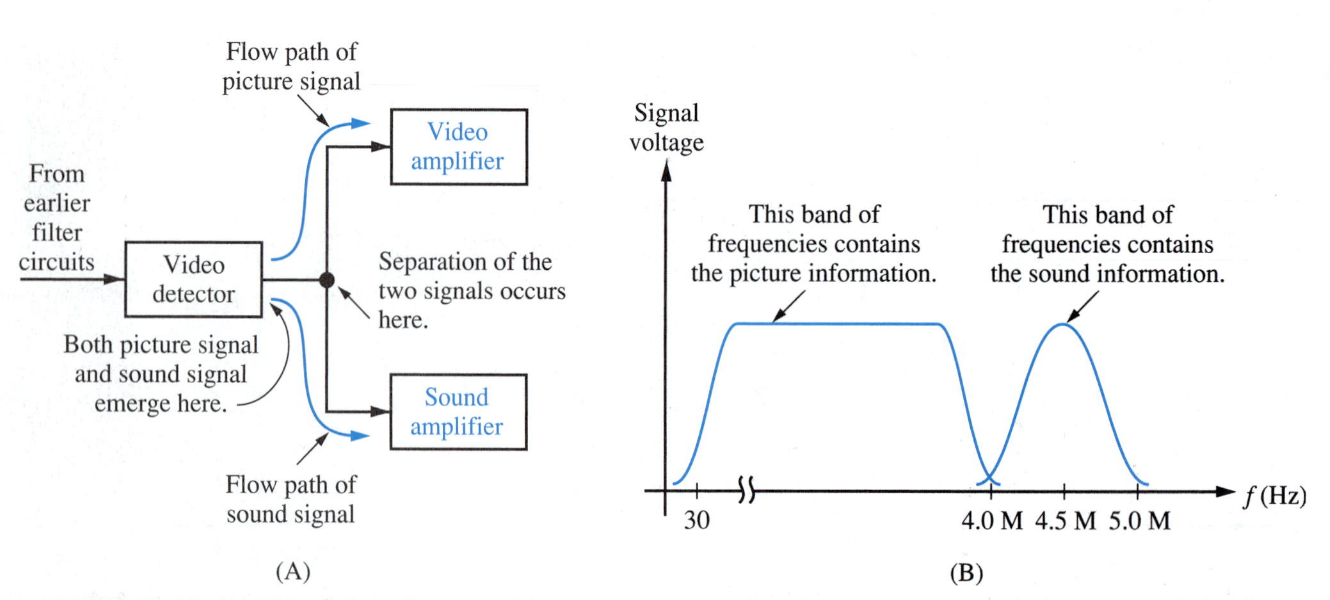

FIGURE 20-19 (A) Visualizing the separation of picture (video) and sound (audio) signals in a television. (B) Frequency bands of picture signal and sound signal.

The general sketch of V_{LD} versus frequency is drawn alongside the load terminals in Figure 20-18. As that sketch shows, the filter stops the band of frequencies near f_r. Frequencies outside that band make their way through the filter to the load.

Television Sound Trap. A good example of a series-resonant band-stop filter is the **sound trap** in a television receiver. As Figure 20-19 shows, the signal in a TV receiver contains both picture information and sound information. It can be thought of as a mixture of signals from about 30 Hz to about 5 MHz. The frequencies in the range from 30 Hz to 4.0 MHz contain the picture information. The frequencies from 4.0 MHz to 5.0 MHz contain the sound information, as indicated in Figure 20-19(B).

These two bands of frequencies must be separated inside the TV. The picture frequency band must be sent to the **video (picture) amplifier**, where it will eventually reach the picture tube. The sound frequency band must be sent to the **audio (sound) amplifier**, where it will eventually reach the loudspeaker. The separation is diagrammed in Figure 20-19(A).

STRUCTURAL DIAGNOSIS

Many older buildings are deteriorating due to environmental contamination of their structural materials. To help preserve such buildings, it is necessary to know just which contaminants have gotten into the materials. Of course, it is possible to drill into the material to get a core sample. This sample can then be analyzed in a laboratory. However, drilling for a core sample is itself harmful to the building.

Shown at top is a noninvasive method of finding out the contaminants. A radioactive neutron source is placed on one side of the material. A gamma-ray detector is placed on the opposite side. As the high-velocity neutrons pass through the building material, they occasionally collide with the atomic nuclei of both the building material and the contaminants. These collisions cause gamma-rays to be produced. Gamma-rays are electromagnetic waves that are higher in frequency than X-rays.

The specific frequency of a gamma-ray depends on the atomic number of the element that the neutron has collided with. The detector counts how many gamma-rays have been produced at each specific frequency. Those numbers are then carried within the detector. Later, they can be visually displayed on an electronic instrument called a spectrum analyzer. In the lower photo, scientists are using the spectrum analyzer display to see which contaminants are present and what their quantities are.

Courtesy of NASA

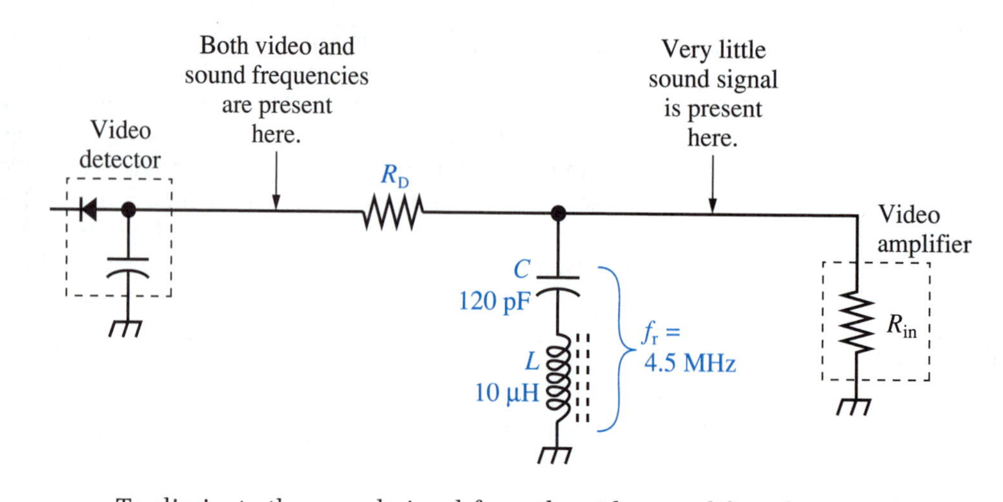

FIGURE 20-20 Series-resonant band-stop filter for stopping the sound band, around 4.5 MHz

To eliminate the sound signal from the video amplifier, the manufacturer places a series-resonant circuit, with f_r = 4.5 MHz, in parallel with the input to the video amplifier. This is shown in simplified form in Figure 20-20.

In that figure, the LC circuit presents a very low net reactance to frequencies near 4.5 MHz. This low net reactance, combined with dropping resistance, R_D, causes most of the sound signal to be dropped across R_D. Very little of the sound band reaches the input of the video amplifier.

20-6 LOW-PASS AND HIGH-PASS FILTERS

There are two additional filtering modes, besides the band-pass and band-stop operations described in Section 20-5. These two modes are low-pass filtering and high-pass filtering.

Low-pass Filters

All filter circuits receive an input signal that is a mixture of many frequencies. This was illustrated in Figures 20-17(B) and 20-18. A **low-pass** filter passes the low-frequency signal components to the load, but it prevents the high-frequency signal components from reaching the load. This frequency-discrimination is pictured in Figure 20-21.

An elementary low-pass filter circuit can be built with just a resistor and a capacitor. This is shown in Figure 20-22(A). In that circuit, capacitance, C, is chosen so that its reactance is much less than R_D (1000 Ω) for the high-frequency signals that are being rejected (far to the right on the frequency-response graphs in Figure 20-21.) With X_C much less than 1000 Ω, most of the high-frequency input voltage is dropped across R_D, by voltage division. Very little appears across the C terminals with X_C so small. Thus, V_{out} has very little high-frequency content.

However, the capacitive reactance of C is much greater than 1000 Ω in the low-frequency range (far to the left on the frequency-response graphs of Figure 20-21.) With X_C much greater than 1000 Ω, voltage division demands that most of the low-frequency input voltage appear across the C terminals. Thus, V_{out} contains most of the low-frequency content of V_{in}.

A low-pass filter can also be built with a resistor-inductor combination. This is shown in Figure 20-22(B). At low frequencies, inductive reactance, X_L, is much less than resistance, R. So, by voltage division, most of the low-frequency signal appears across R, as V_{out}.

At high frequencies, inductive reactance, X_L, is much greater than R. Therefore, by voltage division, most of the high-frequency input voltage is dropped across X_L. Very little appears across R. V_{out} has very little high-frequency content.

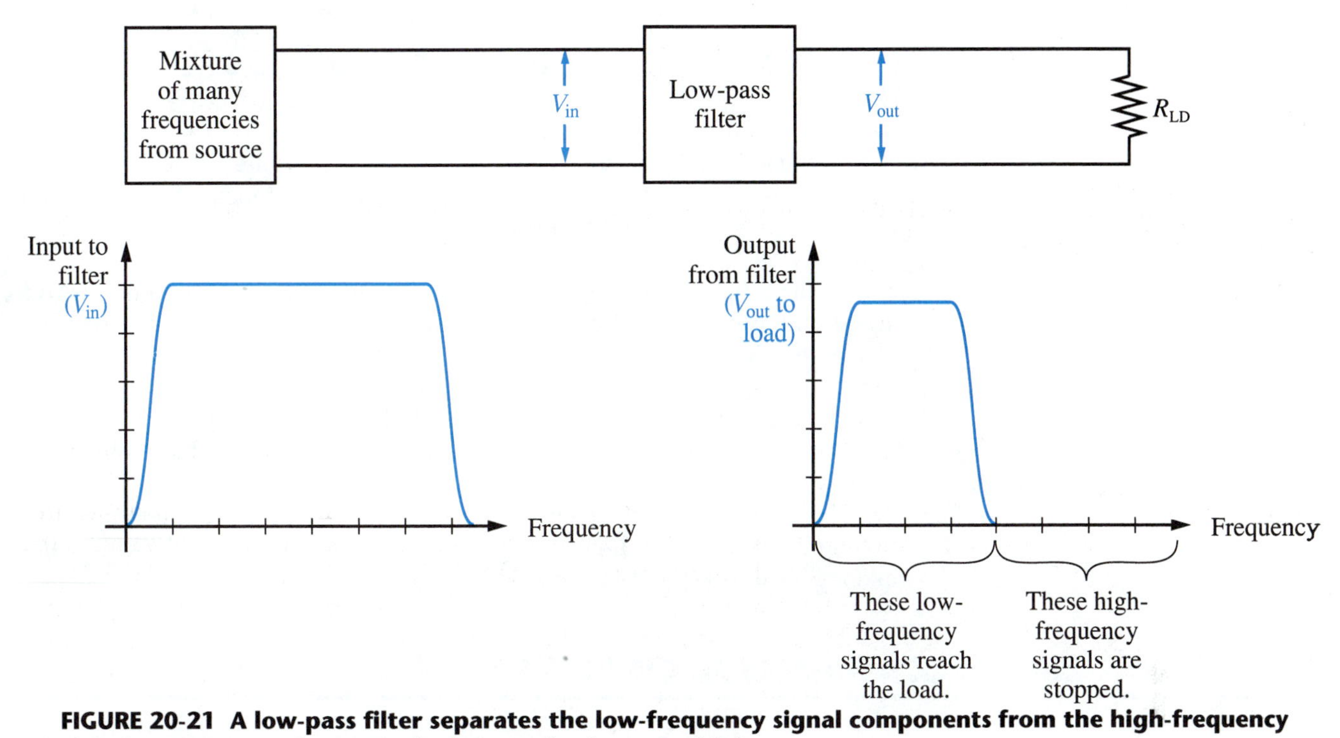

FIGURE 20-21 A low-pass filter separates the low-frequency signal components from the high-frequency signal components. It passes the lows and rejects the highs.

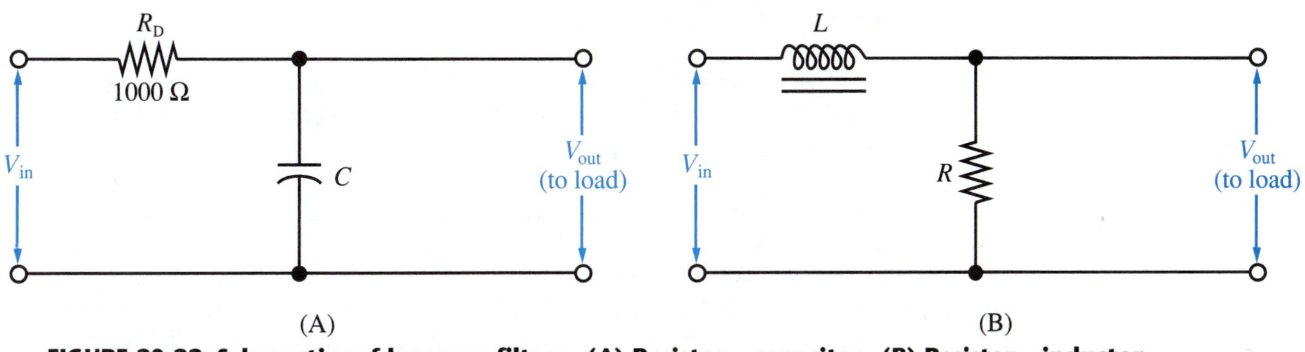

FIGURE 20-22 Schematics of low-pass filters. (A) Resistor – capacitor. (B) Resistor – inductor

High-pass Filters

A **high-pass** *RC* filter is built by just reversing the positions of the resistor and the capacitor. The *R-C* arangement is shown in Figure 20-23(A).

The high-pass *R-L* arrangement is shown in Figure 20-23(B). It has the resistor and inductor reversed, compared to the low-pass filter of Figure 20-22(B).

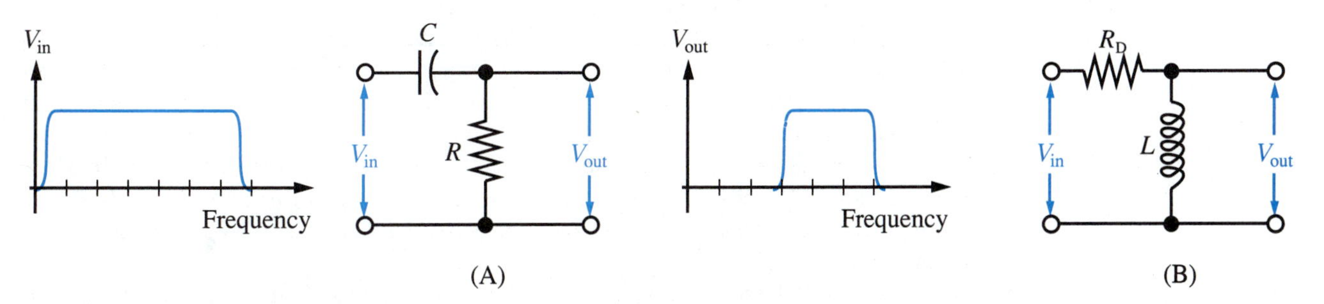

FIGURE 20-23 (A) Schematic of resistor-capacitor high-pass filter, and graph of output response. (B) Resistor-inductor high-pass filter

✔ SELF-CHECK FOR SECTIONS 20-5 AND 20-6

20. Explain what a band-pass filter does. [6]
21. Sketch the schematic arrangement of an *LC* band-pass filter. Explain how this *LC* combination passes close-to-resonant-frequency signals to the load but blocks the far-frequency signals. [6]
22. Explain what a band-stop filter does. [6]
23. Sketch the schematic arrangement of an *LC* band-stop filter. Explain how this *LC* combination passes far-from-resonant-frequency signals to the load but blocks the close-frequency signals. [6]
24. Explain what a low-pass filter does. [7]
25. Sketch the schematic arrangement of an *RC* low-pass filter. Explain how this *RC* combination passes low-frequency signals to the load but blocks high-frequency signals. [7]
26. Sketch the schematic arrangement of an *RC* high-pass filter. Explain how this *RC* combination passes high-frequency signals to the load but blocks low-frequency signals. [7]

20-7 LCR PARALLEL CIRCUITS

A parallel circuit containing an ideal inductor, capacitor, and resistor is shown in Figure 20-24. Source voltage, V_s, is regarded as the reference variable in this parallel circuit for the usual reason—it is common to each component.

Resistor current, I_R, is in phase with V_s, of course. Inductor current, I_L, lags V_s by 90°, and capacitor current, I_C, leads V_s by 90°. These phase relationships are indicated in the waveform graphs of Figure 20-25(A) and the phasor diagram of Figure 20-25(B).

The waveforms of Figure 20-25(A) make it clear that the instantaneous value of inductor current, i_L, is always opposite from the instantaneous capacitor current, i_C. That is, whenever i_L is flowing from top to bottom, i_C is flowing from bottom to top, and vice versa.

Therefore, as far as Kirchhoff's current law is concerned, I_L and I_C subtract from each other, rather than add. With I_L larger than I_C, as shown in Figure 20-25, the *LC* parallel pair has a net reactive current of

$$I_X = I_L - I_C$$

EQ. 20-11

This is shown in the phasor diagram of Figure 20-26.

This net reactive current flows in the leads between the resistor and the *LC* combination in Figure 20-24.

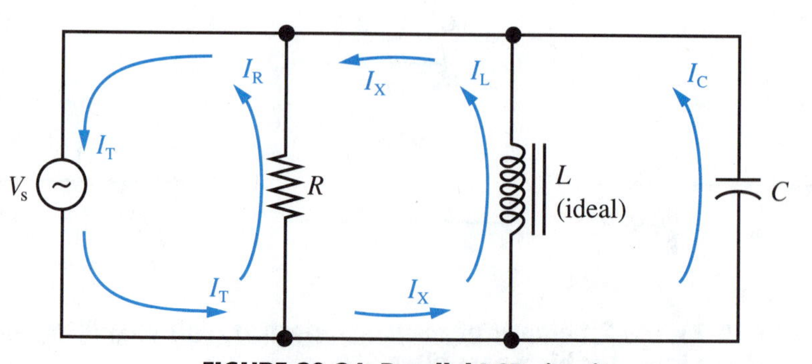

FIGURE 20-24 Parallel LCR circuit

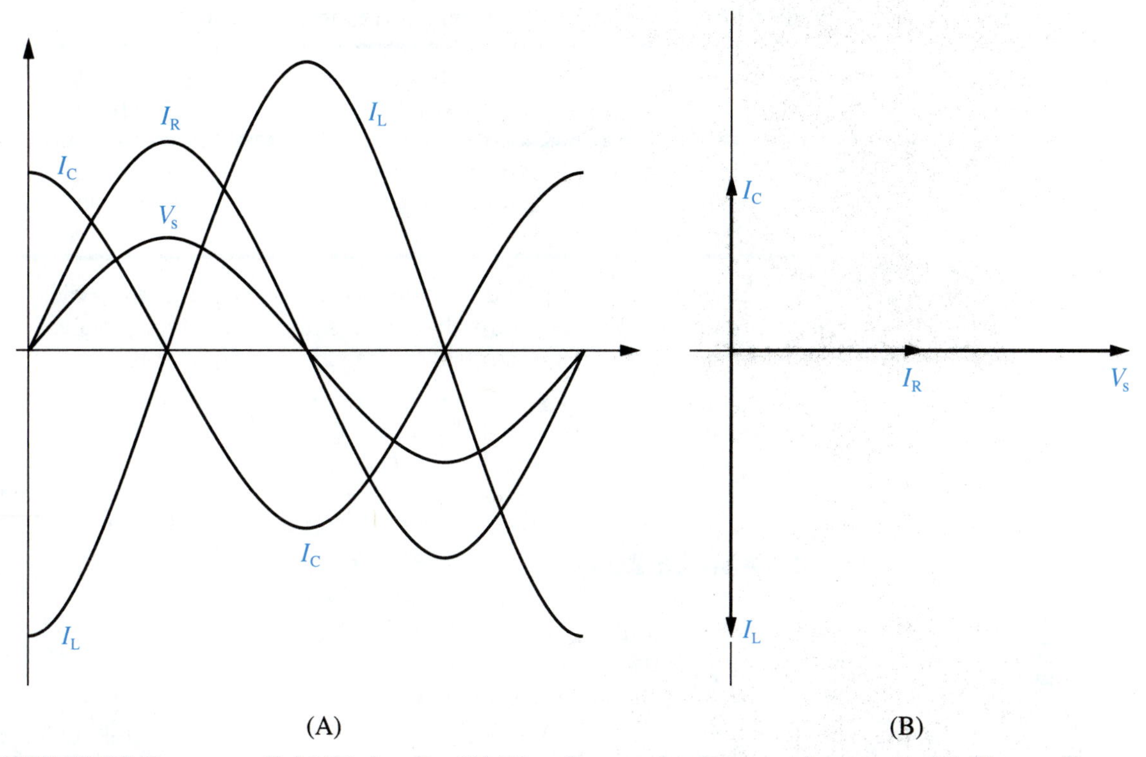

FIGURE 20-25 For a parallel LCR circuit. (A) Waveforms of voltage and current. (B) Phasor diagram

FIGURE 20-26 I_C is subtracted from I_L to give the net reactive current, I_X.

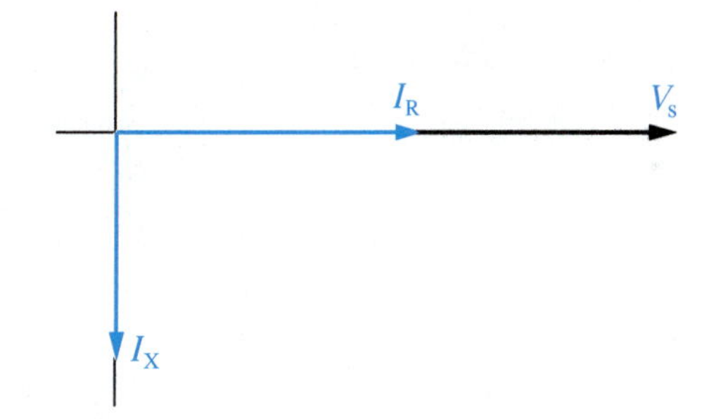

20-8 PARALLEL-RESONANT FREQUENCY

If the source frequency is variable in Figure 20-24, there will be some frequency that causes X_L to equal X_C. Under that condition, I_L and I_C exactly cancel each other. This makes the circuit appear totally resistive, with no reactive nature.

TECHNICAL FACT

Every parallel LCR circuit has a certain **parallel-resonant** frequency, f_r. At $f = f_r$:

1. Z is at its maximum possible value, with

$$Z_{max} = R$$

EQ. 20-12

2. I_T is at its minimum possible value, since Z is maximum.
3. $ø = 0°$, because the circuit becomes totally resistive.

Comparisons between parallel resonance and series resonance are pointed out in Table 20-1.

TABLE 20-1 Comparing Series Resonance and Parallel Resonance

	SERIES RESONANCE	PARALLEL RESONANCE
Z	minimum	maximum
I	maximum	minimum
ø	0°	0°

The condition for parallel resonance is the same as for series resonance, namely that $X_L = X_C$. Therefore, the formula for resonant frequency, f_r, is also the same. For a parallel *LCR* circuit containing ideal components

$$f_r = \frac{1}{6.28\sqrt{LC}}$$

EQ. 20-13

EXAMPLE 20-8

The parallel *LCR* circuit of Figure 20-27 has $L = 170$ mH, $C = 0.05$ µF, and $R = 2000\ \Omega$.

a) Solve for the resonant frequency, f_r.
b) Calculate the resonant reactance, X_r, of both the inductor and the capacitor.
c) Calculate the individual branch currents I_L, I_C, and I_R. Place them on a phasor diagram and combine them to find I_T and ø.
d) What is the circuit's total impedance, Z?

SOLUTION

a) Equation 20-13 gives us

$$f_r = \frac{1}{6.28\sqrt{LC}} = \frac{1}{6.28\sqrt{(170 \times 10^{-3})(0.05 \times 10^{-6})}}$$
$$= \mathbf{1726\ Hz}$$

b) Using the capacitive reactance formula,

$$X_r = \frac{1}{6.28 f_r C} = \frac{1}{6.28\,(1726)\,(0.05 \times 10^{-6})}$$
$$= \mathbf{1844\ \Omega}$$

c) Applying Ohm's law to each branch individually,

$$I_R = \frac{V_s}{R} = \frac{20\ \text{V}}{2\ \text{k}\Omega} = \mathbf{10.0\ mA}$$

$$I_L = \frac{V_s}{X_L} = \frac{V_s}{X_r} = \frac{20\ \text{V}}{1844\ \Omega} = \mathbf{10.85\ mA}$$

$$I_C = \frac{V_s}{X_C} = \frac{V_s}{X_r} = \mathbf{10.85\ mA}$$

Figure 20-28(A) is the phasor diagram for this situation. It quickly reduces to Figure 20-28(B). With $I_X = 0$, there is zero current flowing in the wires that join the resistor to the *LC* combination in Figure 20-27.

d) From Ohm's law,

$$Z = \frac{V_s}{I_T} = \frac{20\ \text{V}}{10\ \text{mA}} = \mathbf{2\ k\Omega} = R$$

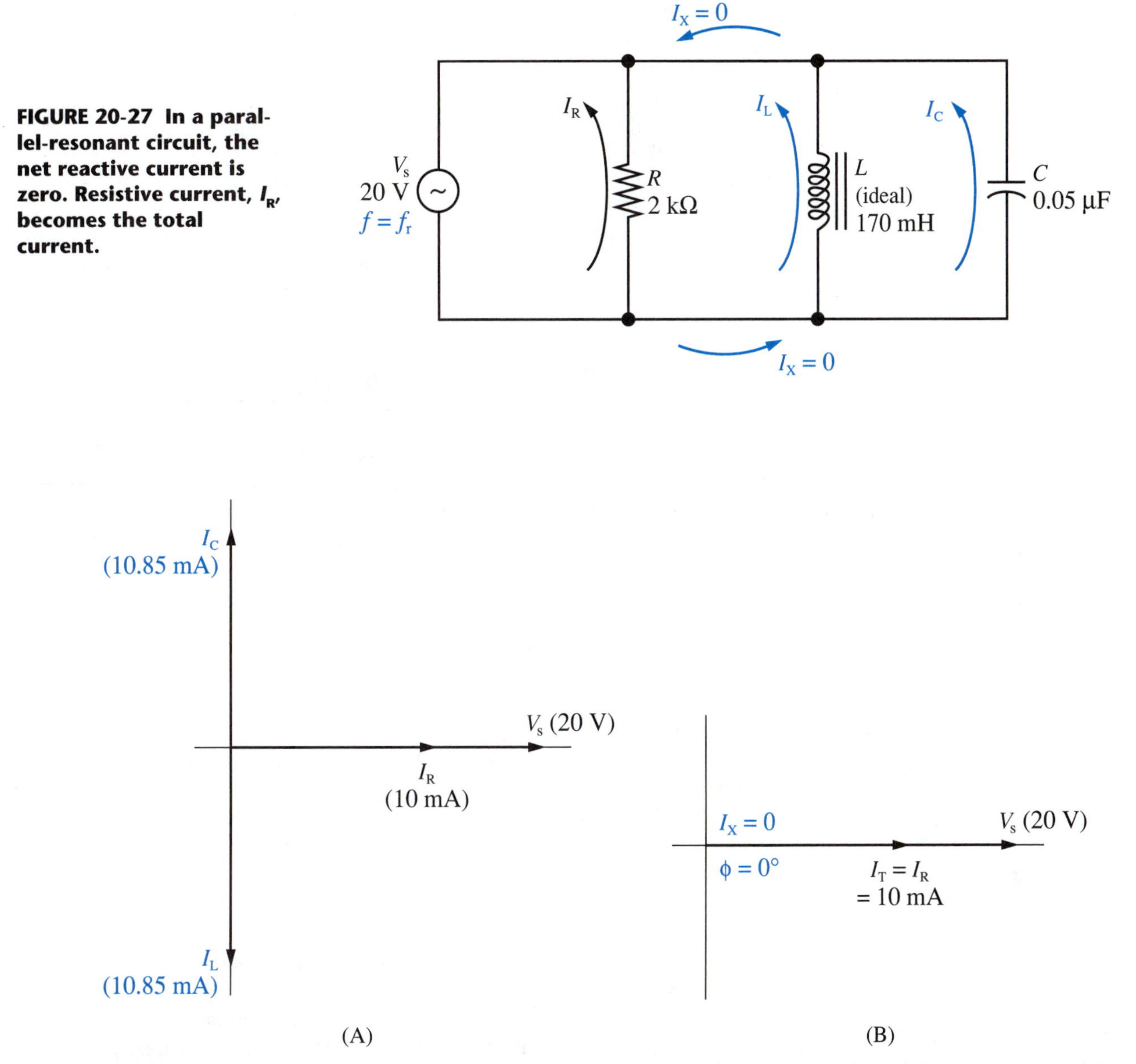

FIGURE 20-27 In a parallel-resonant circuit, the net reactive current is zero. Resistive current, I_R, becomes the total current.

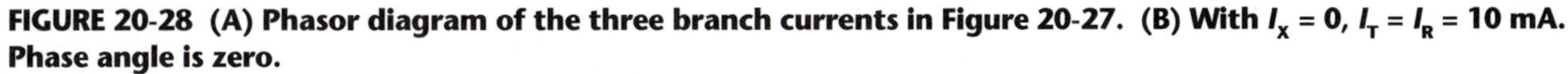

FIGURE 20-28 (A) Phasor diagram of the three branch currents in Figure 20-27. (B) With $I_X = 0$, $I_T = I_R = 10$ mA. Phase angle is zero.

✔ SELF-CHECK FOR SECTIONS 20-7 AND 20-8

27. T-F In an ideal parallel *LCR* circuit operating at its resonant frequency, the inductive current and capacitive current are exactly equal and cancel each other. [8]
28. T-F In an ideal parallel *LCR* circuit driven at its resonant frequency, the total current equals the current through the resistive branch. [8]
29. In an ideal parallel *LCR* circuit operating at resonance, the net reactive current, I_X, equals__________. [8]
30. In a parallel *LCR* circuit, the__________rises to a maximum at the resonant frequency. [8]
31. Referring to the answer to Question 30, explain the important difference between parallel resonance and series resonance. [2, 8]

32. In an ideal parallel-resonant circuit with $R = 2.5\ k\Omega$, $L = 300\ mH$, and $C = 0.75\ \mu F$, the impedance at resonance equals__________. [8]
33. In a series-resonant circuit, current becomes maximum at $f = f_r$. In a parallel-resonant circuit, total current becomes__________ at $f = f_r$. [2, 8]
34. For the component values in Question 32, driven by a voltage source with $V_s = 12\ V$, find: [8, 9]
   a) f_r
   b) X_r
   c) I_L
   d) I_C
   e) I_T
35. In Problem 34, compare the individual branch current, I_L, to the total current, I_T. How can it be that the total current is less than the current through a single branch? [8]

20-9 PARALLEL FREQUENCY-RESPONSE

A general Z-versus-f curve is drawn in Figure 20-29(A). As that curve shows, Z reaches its maximum value at the resonant frequency, f_r. There are two frequencies where $Z = (0.7071)\ Z_{max}$. As before, these are called the low cutoff frequency, f_{low}, and the high cutoff frequency, f_{high}.

Also as before, the bandwidth of a parallel LCR circuit is given by

$$Bw = f_{high} - f_{low}$$

EQ. 20-14

Parallel frequency-response can also be expressed by a curve of total current, I_T, versus frequency, as shown in Figure 20-29(B). As that graph makes clear, I_T dips to its minimum value at the resonant point, given by

$$I_{T(min)} = \frac{V_s}{Z_{max}} = \frac{V_s}{R}$$

EQ. 20-15

At the two cutoff frequencies, I_T grows to 1.414 times as large as $I_{T(min)}$, since $1/0.7071 = 1.414$.

Figure 20-29 further demonstrates the oppositeness of parallel-resonant frequency response (maximum Z, minimum I) and series-resonant frequency response (minimum Z, maximum I).

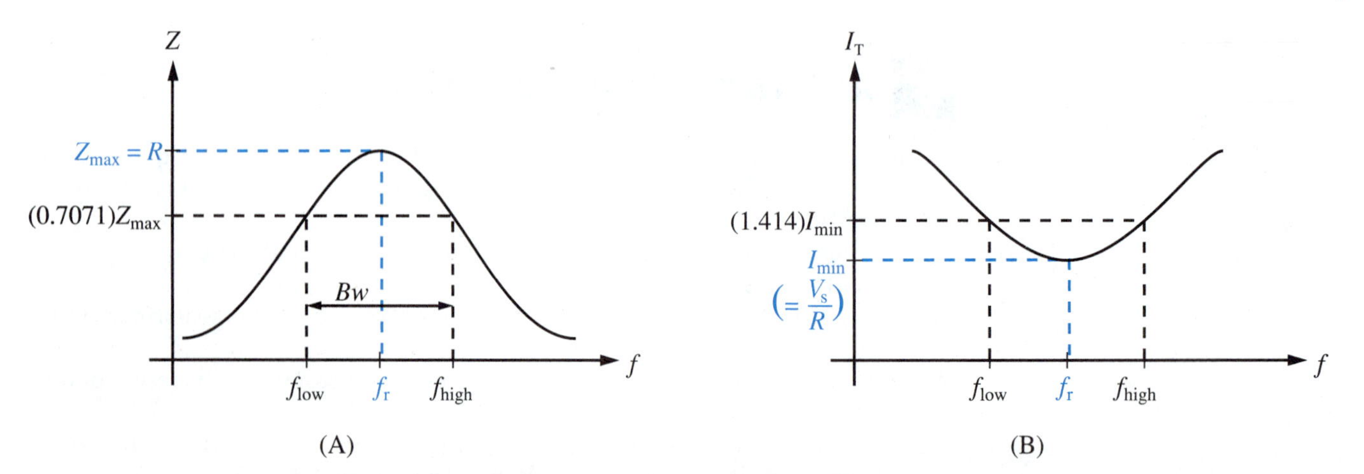

FIGURE 20-29 Frequency-response curves for parallel LCR circuit. (A) *Z* versus *f*. (B) *I* versus *f*.

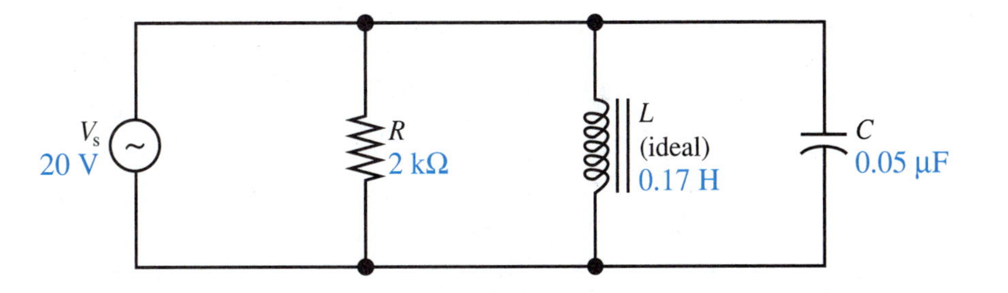

FIGURE 20-30 Circuit for finding out the frequency response of the LCR circuit in Figure 20-27. Bandwidth is 1592 Hz, not symmetrical.

Predicting Bandwidth. We can predict parallel-resonant bandwidth just from knowledge of the electrical component values, as we did for series resonance in Section 20-3. It can be shown by mathematical derivation that the bandwidth of a parallel *LCR* circuit is given by

$$Bw = \frac{1}{6.28\ RC}$$

EQ. 20-16

EXAMPLE 20-9

The circuit of Figure 20-27 has been redrawn in Figure 20-30 for convenience. Predict its bandwidth.

SOLUTION

From Equation 20-16,

$$Bw = \frac{1}{6.2832\ RC} = \frac{1}{6.2832\ (2 \times 10^3)\ (0.05 \times 10^{-6})}$$

$$= \mathbf{1591.5\ Hz}$$

For this circuit, the cutoff frequencies are:

$$f_{\text{low}} = 1105.1\text{ Hz}$$
$$f_{\text{high}} = 2697.6\text{ Hz}$$

(You can prove these cutoff values by calculating the individual reactances and all three of the individual branch currents. Then combine them on phasor diagrams to demonstrate that $I / I_{\text{min}} = 1.414$.)

Again, the frequency-response curve is nonsymmetrical around the resonant frequency of 1726 Hz. The distance from f_{low} to f_r is 1726 Hz – 1105 Hz = 621 Hz. There is a greater distance from f_r to f_{high}: 2697 Hz – 1726 Hz = 971 Hz.

20-10 PARALLEL Q AND SELECTIVITY

For parallel resonance, we define the circuit's *Q* differently than for series resonance. For an ideal parallel-resonant circuit, shown in Figure 20-31(A), total circuit *Q* is defined as

$$Q_{\text{T(parallel)}} = \frac{R}{X_r}$$

EQ. 20-17

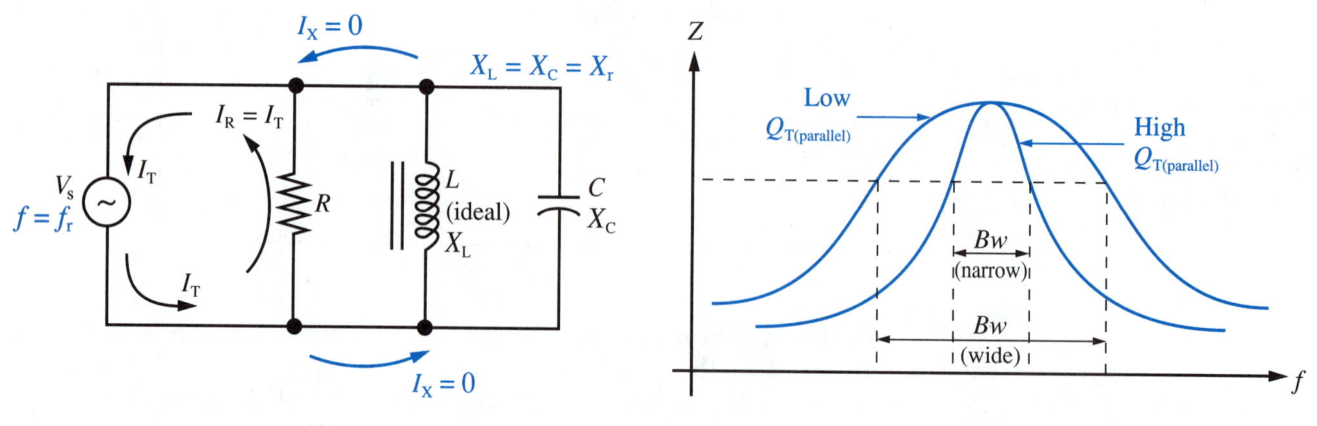

FIGURE 20-31 (A) General parallel LCR circuit at resonance. (B) Higher Q produces greater selectivity (narrower *Bw*).

The subscript T indicates that this is for the total circuit, not just a single component. As before, X_r symbolizes the reactance of either the inductor or the capacitor at f_r.

Once $Q_{T(parallel)}$ is known, it again enables us to predict bandwidth:

$$Bw = \frac{f_r}{Q_{T(parallel)}}$$

EQ. 20-18

Thus, a higher-Q parallel-resonant circuit has greater selectivity than a lower-Q circuit, the same as for series resonance. This is illustrated in Figure 20-31(B).

EXAMPLE 20-10

For the circuit of Figures 20-27 and 20-30,

a) Find $Q_{T(parallel)}$.
b) Calculate the bandwidth and compare to the earlier result.

SOLUTION

a) From Example 20-8, resonant reactance, X_r, is known as

$$X_r = 1844\ \Omega$$

From Equation 20-17,

$$Q_{T(parallel)} = \frac{R}{X_r} = \frac{2000\ \Omega}{1844\ \Omega} = \mathbf{1.0846}$$

9

As before, there are no units on Q since it is a ratio of ohms to ohms.

b) From previous calculations, $f_r = 1726$ Hz. Using Equation 20-18,

$$Bw = \frac{f_r}{Q_{T(parallel)}} = \frac{1726\ \text{Hz}}{1.0846} = \mathbf{1591.5\ Hz}$$

which agrees with the result in Example 20-9.

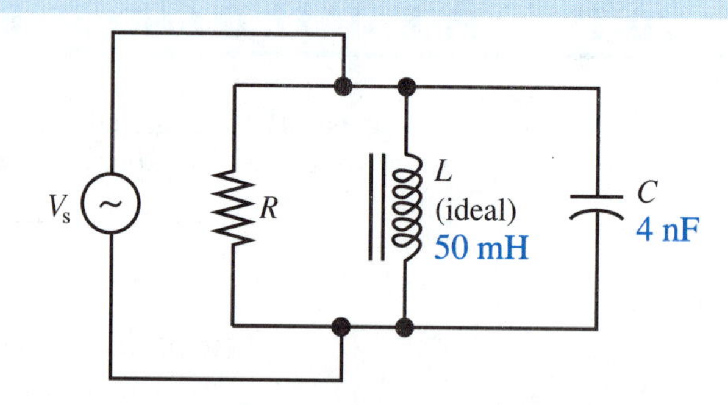

FIGURE 20-32 Selecting *R* to produce a desired bandwidth

EXAMPLE 20-11

The parallel circuit of Figure 20-32 has L and C already specified. We are to choose R.

a) What is the resonant frequency?
b) If we want the bandwidth to be 800 Hz, what should the circuit Q be?
c) Choose the right value of R to give that value of circuit Q.
d) Plot an approximate frequency-response curve of Z versus f.

SOLUTION

a) $$f_r = \frac{1}{6.28\sqrt{LC}} = \frac{1}{6.2832\sqrt{(50 \times 10^{-3})(4 \times 10^{-9})}} = \mathbf{11\ 254\ Hz}$$

b) Rearranging Equation 20-18,

$$Q_{T(\text{parallel})} = \frac{f_r}{Bw} = \frac{11\ 254\ \text{Hz}}{800\ \text{Hz}} = \mathbf{14.07}$$

c) From Equation 20-17,

$$R = Q_{T(\text{parallel})}(X_r) = (14.07)\,X_r$$

Using inductance to find X_r, we get

$$X_r = X_L = 6.2832\ (11\ 254\ \text{Hz})\ (50 \times 10^{-3}\ \text{H}) = 3536\ \Omega,$$

so $$R = (14.07)\ 3536\ \Omega = \mathbf{49.75\ k\Omega}$$

d) This is a situation where we can approximate f_{low} and f_{high} by assuming that the frequency-response curve is nearly symmetrical around f_r. We did the same thing for a high-Q series LCR circuit in Example 20-7. Since $Q_{T(\text{parallel})}$ is greater than 10 in this example, the approximation will work quite well.

$$f_{low} \cong f_r - \tfrac{1}{2}\,Bw = 11\ 254 - \tfrac{1}{2}\,(800) = 10\ 854\ \text{Hz}$$

$$f_{high} \cong f_r + \tfrac{1}{2}\,Bw = 11\ 254 + \tfrac{1}{2}\,(800) = 11\ 654\ \text{Hz}$$

The approximate frequency-response curve is sketched in **Figure 20-33.**

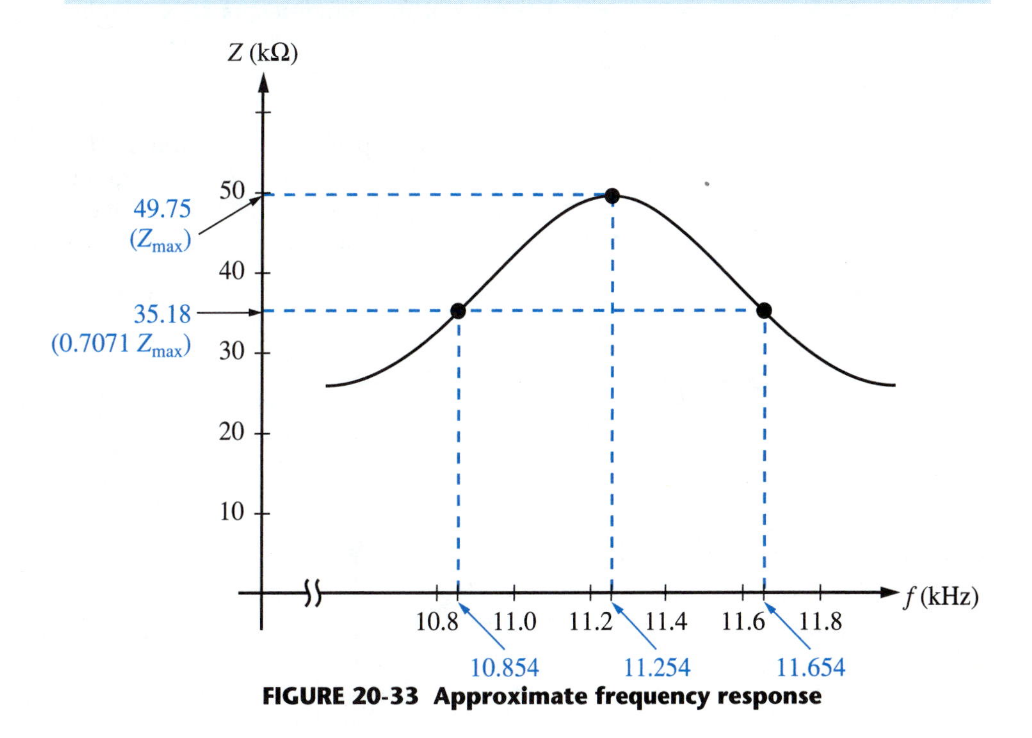

FIGURE 20-33 Approximate frequency response

SELF-CHECK FOR SECTIONS 20-9 AND 20-10

36. Which two circuit components determine the bandwidth of a parallel LCR circuit? [9]
37. An ideal parallel LCR circuit has R = 1.5 kΩ, L = 700 mH, and C = 50 nF. Calculate the circuit's bandwidth. [8, 9]
38. In a parallel LCR circuit, larger R results in a__________bandwidth. [9]
39. T-F For a parallel-resonant circuit, a lower value of resistance results in a higher value of total circuit Q. [9]
40. T-F For a parallel-resonant circuit, higher Q produces greater selectivity. [9]
41. A certain parallel LCR circuit has a resonant frequency of 10 kHz and a resonant reactance X_r = 2.5 kΩ. [9]
    a) What value of $Q_{T(parallel)}$ will give a bandwidth of 4 kHz?
    b) What value of resistance, R, is required for this?

20-11 PARALLEL-RESONANT FILTERS

Parallel-resonant LC filters can produce the same results as series-resonant filters.

Band-pass

For band-pass operation, the parallel-resonant LC combination is placed in parallel with the load, as shown in Figure 20-34. Dropping resistor R_D is located before the LC combination. At frequencies near f_r, the LC combination presents a very high impedance. Therefore, within that band of frequencies R_{LD} and R_D divide the input voltage proportionately. If R_{LD} is much greater than R_D, then most of the input appears across the load.

However, at frequencies far from resonance, the LC combination presents a low impedance, much less than R_D. Therefore, at frequencies outside the pass band, most of the input voltage is dropped across R_D. Very little gets through to the load. We can think of the LC combination as "shorting out" the load.

The general frequency-response is sketched alongside R_{LD} in Figure 20-34.

Band-stop

For band-stop operation, the parallel LC combination is placed in series with the load, as shown in Figure 20-35. At frequencies near f_r, the LC combination has a very high impedance. Therefore, most of the signal is dropped across LC, with very little getting through to R_{LD}, for frequencies in the stop band.

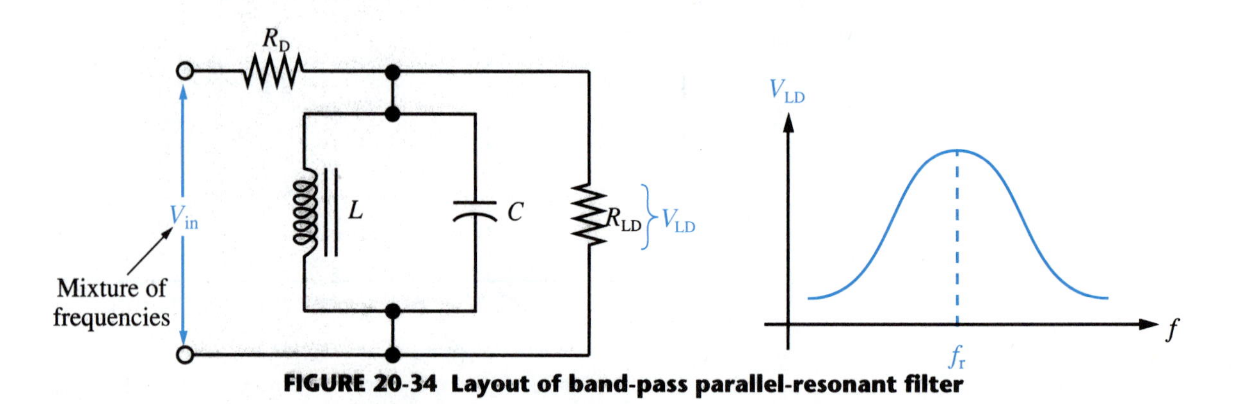

FIGURE 20-34 Layout of band-pass parallel-resonant filter

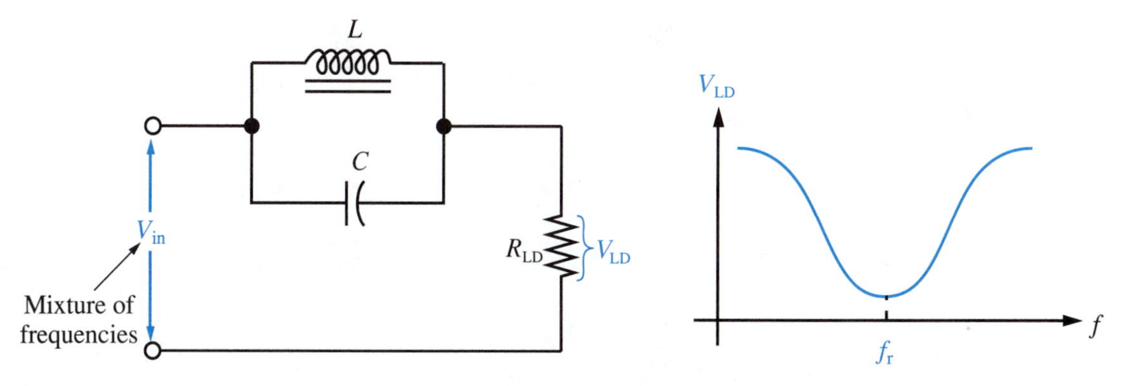

FIGURE 20-35 Layout of band-stop parallel-resonant filter

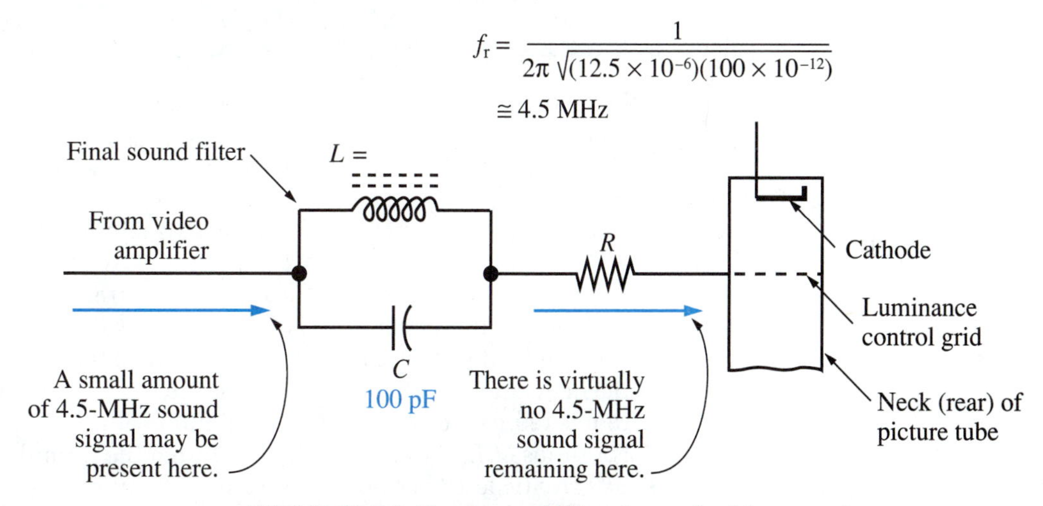

FIGURE 20-36 Final sound filter in a television receiver

At frequencies far away from f_r, the LC combination has a very low impedance, much less than R_{LD} itself. Therefore, most of the signal gets through to R_{LD}, for frequencies outside the stop band.

The band-stop response is sketched alongside R_{LD} in Figure 20-35.

Sound-signal Band-stop Filter Between a TV's Video Amplifier and Picture Tube. As we saw in Section 20-5, most of a television's sound signal is blocked from the input of its video amplifier by the action of a series LC band-stop filter. This is shown schematically in Figure 20-19.

However, this band-stop filter never works perfectly. A small amount of the 4.5-MHz-band sound signal is bound to sneak by the input filter and enter the video amplifier. From there, it will be amplified and will eventually reach the luminance control grid of the picture tube, causing sound interference in the TV picture.

To eliminate this problem, a parallel-resonant band-stop filter is sometimes placed in the signal line between the video amplifier and the picture tube. This is shown in Figure 20-36.

FORMULAS

For series resonance:

$$Z_{minimum} = R \quad \textbf{EQ. 20-2}$$

$$f_r = \frac{1}{6.28\sqrt{LC}} \quad \textbf{EQ. 20-4}$$

$$Bw = f_{high} - f_{low} \quad \textbf{EQ. 20-5}$$

$$Bw = \frac{R}{6.28\,L} \quad \textbf{EQ. 20-6}$$

$$Q_T = \frac{X_r}{R_T} \quad \textbf{EQ. 20-7}$$

$$Bw = \frac{f_r}{Q_T} \quad \textbf{EQ. 20-8}$$

For high-Q circuits only:

$f_{high} \cong f_r + ½\ Bw$ **EQ. 20-9**

$f_{low} \cong f_r - ½\ Bw$ **EQ. 20-10**

For parallel resonance:

$Z_{maximum} = R$ **EQ. 20-12**

$f_r = \dfrac{1}{6.28\sqrt{LC}}$ **EQ. 20-13**

$Bw = \dfrac{1}{6.28\,R\,C}$ **EQ. 20-16**

$Q_{T(parallel)} = \dfrac{R}{X_r}$ **EQ. 20-17**

$Bw = \dfrac{f_r}{Q_{T(parallel)}}$ **EQ. 20-18**

SUMMARY OF IDEAS

- In a series LCR circuit, inductor voltage and capacitor voltage tend to cancel each other.
- V_L and V_C completely cancel each other at the series-resonant frequency, symbolized f_r.
- At frequency = f_r, circuit impedance, Z, is minimum, and is equal to resistance, R.
- A frequency-response curve is a graph of current (or percent of maximum current) versus frequency.
- The bandwidth (Bw) is the difference between the low cutoff frequency, f_{low}, and the high cutoff frequency, f_{high}. These are the two frequencies at which the current is reduced to 70.7% of its maximum value.
- For series LCR, Bw can be calculated from the formula $Bw = R / 6.28\ L$.
- Selectivity describes how effectively a circuit rejects nonresonant frequencies.
- High-Q circuits are more selective (have narrower bandwidths). Low-Q circuits are less selective (have wider bandwidths).
- Bandwidth is equal to resonant frequency divided by total circuit Q ($Bw = f_r / Q_T$).
- A band-pass filter allows frequencies close to f_r to reach the load ; but it rejects frequencies far above f_r or far below f_r.
- A band-stop filter stops frequencies close to f_r from reaching the load, but it allows frequencies far from f_r to reach the load.
- A series LC band-pass filter goes in series with the load. A series LC band-stop filter goes in parallel with the load.
- In an ideal parallel LCR circuit, inductor current and capacitor current cancel each other completely at the parallel-resonant frequency, f_r.
- At $f = f_r$, circuit impedance, Z, is maximum and is equal to resistance, R. Total current, I_T, is at its minimum value and has a phase angle $ø = 0°$.
- For parallel LCR, a frequency-response curve is usually a graph of percentage of maximum impedance versus frequency.
- For parallel resonance, Bw, f_{low}, and f_{high} are defined the same as for series resonance.
- For parallel resonance, total circuit Q is given by $Q_{T(parallel)} = R / X_r$.
- Bandwidth can be calculated from $Bw = f_r / Q_{T(parallel)}$.
- A parallel-resonant LC band-pass filter is connected in parallel with the load. A parallel LC band-stop filter is connected in series with the load.

CHAPTER QUESTIONS AND PROBLEMS

1. For a series LCR circuit operating at resonance, V_L and V_C are out of phase with each other by_______degrees. Their magnitudes are__________. [2]
2. For a series LCR circuit operating at resonance, what is the circuit's power factor? [2]

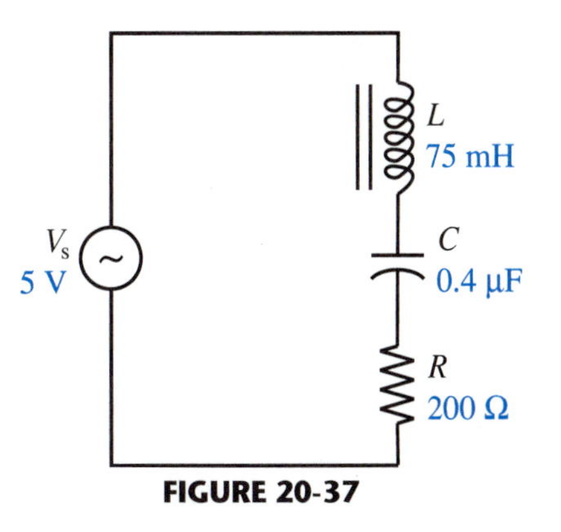

FIGURE 20-37

3. If the inductance, L, of a series LCR circuit is increased, the resonant frequency, f_r, is__________. [1]
4. If the capacitance, C, of a series LCR circuit is increased, the resonant frequency, f_r, is__________. [1]
5. If the resistance, R, of a series LCR circuit is increased, the resonant frequency, f_r, is__________. [1]
6. In a series LCR circuit operating at a frequency higher than f_r, the reactor voltages do not cancel each other because V_L is__________than V_C.
7. In a frequency-response curve, the horizontal (x-axis) variable is__________. [3]
8. The highest point on a frequency-response curve occurs at the__________ frequency. [3]
9. A frequency-response curve with very steep sides indicates that the circuit has __________selectivity. If a curve has gently-sloping sides, the circuit has __________selectivity. (Answer greater or lesser.) [3, 4]
10. Explain the meaning of the terms low cutoff frequency and high cutoff frequency. [3]
11. Bandwidth is defined as the difference between__________and__________. [3]
12. The bandwidth of a series LCR circuit is determined by the component values of the __________and the__________. [2]
13. In Figure 20-37: [1,5]
    a) Calculate the resonant frequency, f_r.
    b) What is the bandwidth?
14. If you wanted to make the circuit of Figure 20-37 more selective without changing its resonant frequency, which component would you change? Would you make it larger or smaller? [4, 5]
15. As Problem 14 suggests, let us reduce the bandwidth of Figure 20-37 to 150 Hz. [4, 5]
    a) What new value of Q_T is needed?
    b) What should the new R value be?
16. In Figure 20-37, are f_{low} and f_{high} approximately the same distance from f_r? Explain why. [4]
17. What does the symbol X_r stand for? [4]
18. In a real (nonideal) series-resonant circuit, what other resistances, besides the component resistor, make up the total resistance, R_T? [4]
19. T-F To get a very narrow bandwidth you must use a high-Q inductor. [4]
20. For a series-resonant circuit, the minimum possible bandwidth is gotten by removing the component resistance, R, altogether. Then what factors determine the minimum Bw? [4]
21. What type of filter prevents a band of frequencies close to the resonant frequency from passing through to the load? [6]
22. When a series LC combination is connected in series with the load, it creates a__________filter. [6]
23. When a series LC combination is connected in parallel with the load, it creates a__________filter. [6]

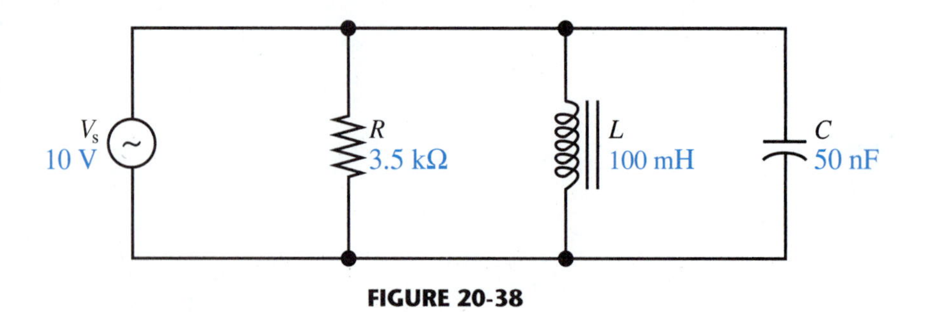

FIGURE 20-38

24. For the circuit of Figure 20-37, with the source at the resonant frequency, what is the power factor? [2]
25. In Problem 24, with the source at f_r, find the value of maximum current. [2]
26. In Problem 24, calculate the circuit's power by using $P = V_s I (PF)$. [2]
27. In Problem 24, calculate the circuit's power using $P = I^2R$. Compare to the answer from Problem 26. [2]
28. T-F In an ideal parallel-resonant circuit, the total current equals the current through the resistive branch. [9]
29. In an ideal parallel-resonant circuit, the net reactive current, I_X, equals __________. [9]
30. In a parallel *LCR* circuit, the__________reaches its maximum value at the resonant frequency, f_r; therefore, the__________reaches its minimum value. [9]
31. According to your answers from Question 30, what is the major difference between series-resonant and parallel-resonant circuits? [2, 9]
32. In a parallel *LCR* circuit operating at a frequency higher than f_r, the reactor currents do not cancel each other out because I_L is__________than I_C.
33. For the circuit of Figure 20-38: [8, 9]
    a) Calculate f_r.
    b) Find bandwidth from $Bw = 1/6.28\ RC$.
    c) Calculate X_r.
    d) Find $Q_{T(parallel)}$.
    e) Find Bw from $Bw = f_r / Q_{T(parallel)}$.
34. In Problem 33, if you wanted to change the *Bw* of the circuit without changing f_r, which component would you change? Would you make it larger or smaller? [8, 9]
35. As suggested in Question 34, let us reduce the bandwidth to 500 Hz. [9]
    a) What new value of $Q_{T(parallel)}$ is needed?
    b) What should be the new *R* value?
36. T-F For a parallel *LCR* circuit operating at one of the cutoff frequencies, $Z = (0.7071)\ Z_{max}$, and $I_T = (1.414)\ I_{min}$. [9]
37. For the circuit of Problem 33 operating at the resonant frequency, find: [9]
    a) Power factor, *PF*.
    b) Total current, I_T.
    c) Inductor current, I_L.
    d) Capacitor current, I_C.
38. For the circuit of Problems 33 and 37, find: [9]
    a) Total circuit power, from $P = V_s I_T (PF)$.
    b) Resistor power from $P = (I_R)^2 R$. Comment on your answers.
39. When a parallel *LC* combination is connected in series with the load, it creates a __________filter. [10]
40. When a parallel *LC* combination is connected in parallel with the load, it creates a __________filter. [10]
41. The circuit in Figure 20-11(A) resonates at 134.5 Hz, with $I_{max} = 100$ mA. For $f = 121.56$ Hz, calculate X_L and X_C and draw the circuit's phasor diagram of ohmic values. Work out the phasor triangle's solution to find *Z* and *I*. Does this prove that $f_{low} = 121.56$ Hz?
42. Repeat Problem 41 for $f = 148.84$ Hz. Does this prove that $f_{high} = 148.84$ Hz?

CHAPTER 21

SOLID-STATE ELECTRONICS

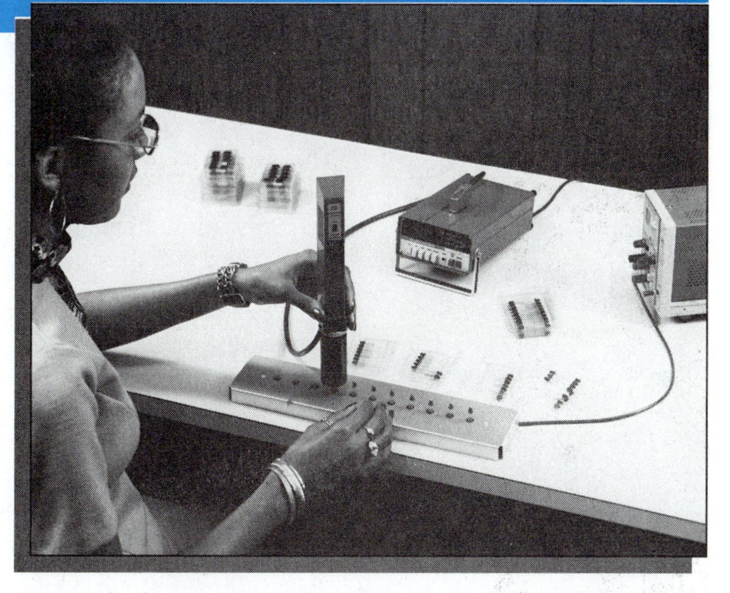

Test probe for measuring the efficiency of a light-emitting diode.
Courtesy of Tektronix, Inc.

OUTLINE

NEW TERMS TO WATCH FOR

diode
forward direction
bias
reverse direction
rectify
passive device
active device
bipolar junction transistor (BJT)
current gain
power gain
voltage gain
collector
base
emitter
common-emitter
current source
p-n junction
dc bias currents
coupling capacitor
field-effect transistor (FET)
drain
gate
source
input resistance
distortion
integrated circuit
discrete
dual-in-line package
chip

Electricity and magnetism have to do with physical events that occur more-or-less naturally. For example, resistance is a physical quality that exists in nature without human creation. It is certainly true that we manipulate the earth's material to alter natural resistance. We do this to get specific, predictable results. The same thing can be said about capacitance, and inductance, and natural magnets, and almost everything else that we have studied so far.

However, in the final three decades of the 1800s, human experimentation began to cause events that were not natural. The first clear-cut example of this was the vacuum tube, a sealed glass enclosure containing almost no air. It was originally developed by Thomas Edison as a means of producing artificial light. Within a few years, though, Edison realized that the vacuum tube could perform as a one-way path of electric current. This one-way action was not natural. Nothing even similar had ever happened before. Since then, of course, hundreds of new devices have been invented that produce never-before-seen results. Electronics, as distinguished from electricity, is the field that deals with such non-natural devices and their applications.

Until 1948, all electronic devices were vacuum tubes. In that year, the first electronic device made out of solid material was invented. Solids have great advantages over vacuum tubes. That is why today, solid electronic devices have replaced vacuum tubes in all but a few applications.

In this chapter, we will give an overview of three important solid-state electronic devices. Then we will touch on the topic of integrated circuits. You will study the ideas of solid-state electronics in greater detail in a later course.

After studying this chapter, you should be able to:

1. Draw the schematic symbol for a diode and label the terminals.
2. Given a circuit schematic containing a diode, tell whether the diode is blocking or conducting.
3. Apply Ohm's law and Kirchhoff's voltage law to diode/resistor circuits.
4. Describe the rectifying process. Sketch the output waveform of a sine-wave-rectified circuit.
5. Sketch a half-wave-rectified output waveform that would be obtained from a three-phase automobile alternator. Identify the battery-charging intervals.
6. Explain the meaning of power amplification.
7. Draw the schematic symbols for *npn* and *pnp* bipolar transistors, and label the terminals.
8. Using the current source idea, explain the current-amplifying process for a bipolar transistor in the common-emitter mode.
9. Given a complete circuit schematic diagram of a basic BJT amplifier, calculate its current gain, A_I, its voltage gain, A_V, and its power gain, A_P.
10. Show how to connect the dc power supplies to correctly bias a *pnp* transistor and an *npn* transistor.
11. Sketch the combined ac and dc input and output waveforms for a bipolar transistor amplifier.
12. Explain the operation of input and output coupling capacitors for a transistor amplifier.
13. Draw the schematic symbols for *p*-channel and *n*-channel junction field-effect transistors, and label the terminals.
14. Show how to connect the dc power supplies to correctly bias a *p*-channel JFET and an *n*-channel JFET.
15. Using the current source idea, explain the amplifying operation of a JFET in the common-source mode.
16. Explain why a dc supply terminal acts like a short circuit to ground for an ac signal.
17. State the main practical advantage of a JFET over a BJT.
18. Calculate the equivalent ac drain resistance for a JFET amplifier when the load is not placed directly in the drain lead, but is coupled to the drain through a capacitor. Do the same for equivalent ac collector resistance in a BJT amplifier.
19. Given a complete circuit schematic diagram of a basic JFET amplifier, calculate its current gain, A_I, its voltage gain, A_V, and its power gain, A_P, and sketch the combined ac and dc input and output waveforms.
20. State the basic imperfection of a JFET amplifier and explain why it occurs.

21-1 DIODES

A **diode** is a one-way conductor. Its schematic symbol is shown in Figure 21-1, with the names of its terminals. If a source tries to force current through the

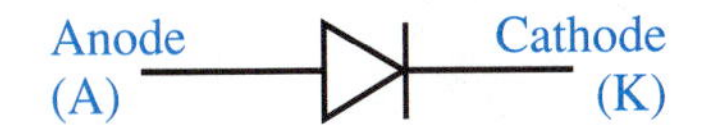

FIGURE 21-1 Diode schematic symbol. The terminal names are anode (symbolized A) and cathode (symbolized K).

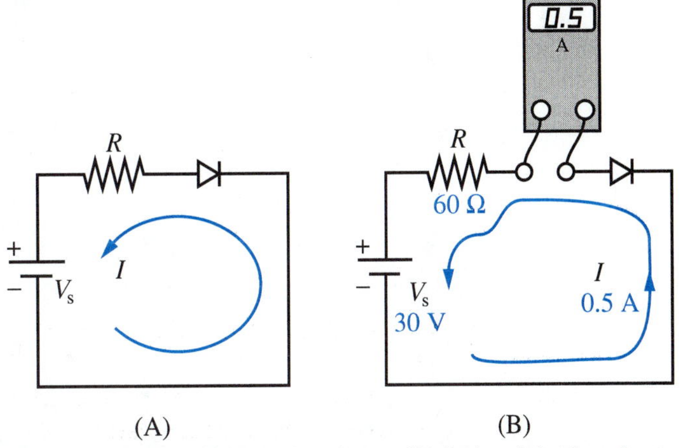

FIGURE 21-2 (A) Forward direction for a diode is with the electron current entering on the pointed side (the cathode) and exiting on the flat side (the anode). We say that the diode is *conducting* current. Another term that means the same thing is that the diode has a forward bias. (B) Since the diode has almost 0 Ω of resistance in the forward direction, the approximate current, *I*, is given by just Ohm's law. $I = V_s / R = 30\ V / 60\ \Omega = 0.5\ A$.

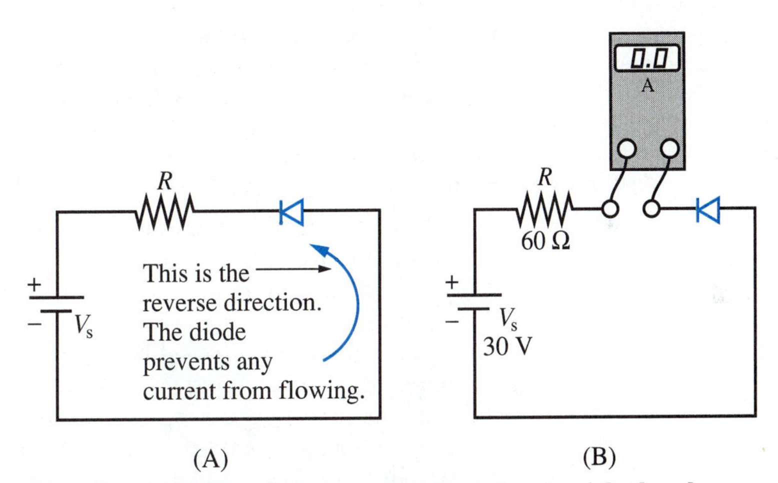

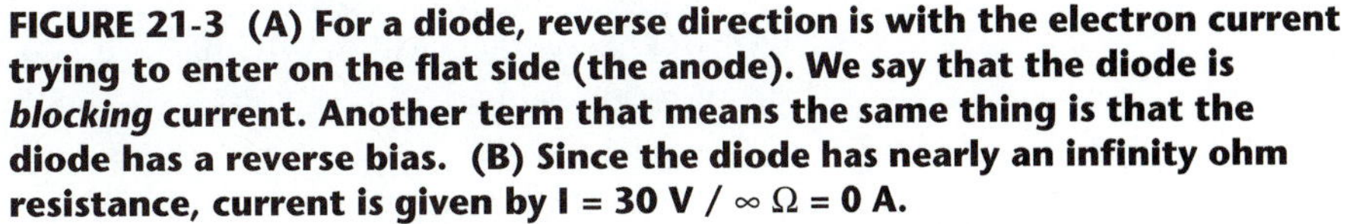
FIGURE 21-3 (A) For a diode, reverse direction is with the electron current trying to enter on the flat side (the anode). We say that the diode is *blocking* current. Another term that means the same thing is that the diode has a reverse bias. (B) Since the diode has nearly an infinity ohm resistance, current is given by $I = 30\ V / \infty\ \Omega = 0\ A$.

diode in the **forward direction**, the diode presents very little resistance, ideally 0 Ω. Current flows freely. The forward-current direction is the direction *against the arrow*, as shown in Figure 21-2. In that figure, resistor *R* is present in the circuit to limit the current flow. This is necessary because the diode itself offers almost zero resistance. It acts like a nearly perfect conductor.

But if a source tries to force current through the diode in the **reverse direction**, the diode presents a great resistance. Virtually no current is able to flow. The reverse current direction is the direction *with the arrow*. This is shown in Figure 21-3. In that figure, virtually zero current flows because the diode presents almost infinite resistance. It acts like a nearly perfect insulator.

Figure 21-4 shows the physical appearance of several diodes. In general, physically larger diodes tend to have greater current capability than smaller diodes.

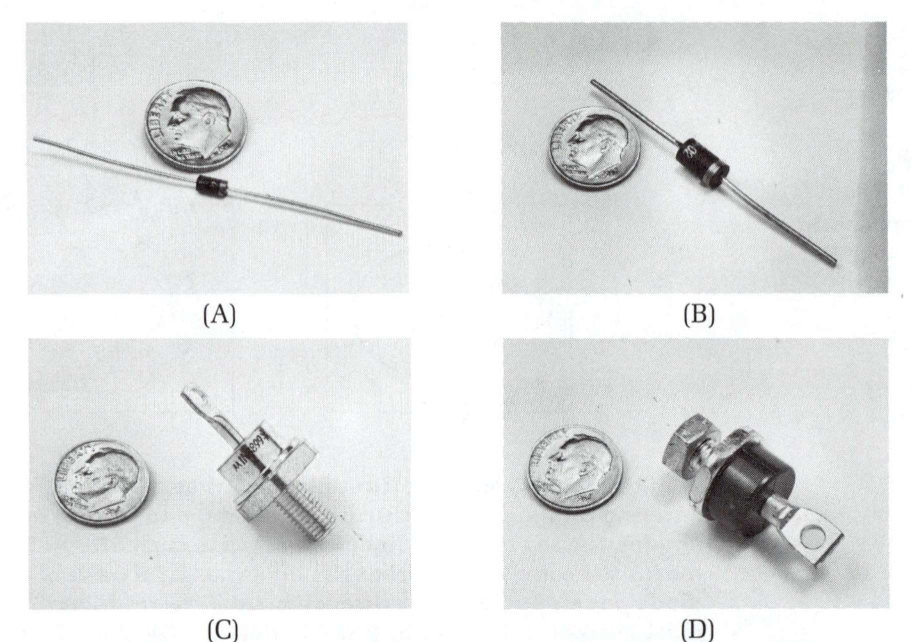

FIGURE 21-4 Various diodes. (A) A type No. 1N4004 diode, a popular diode for rectifying ac to dc. Diode type numbers often start with the characters 1N; this is because diodes have one internal junction where different kinds of silicon crystals are joined together. The No. 1N4004 diode can carry a maximum average current of 1 A without overheating, and can block a maximum reverse voltage of 400 V. (B) When a cylindrical diode body has a single stripe, that stripe marks the cathode end. Here the cathode lead is on the right. The anode lead is on the left. This particular diode has a 3-A maximum current rating. (C) Stud diodes are bolted to a heat sink. For some types, the bolt side is the cathode and the solder terminal lead is the anode. For other types, anode and cathode are the other way around. When this particular diode is properly attached to a heat sink in ambient air at 25°C (77°F), it can handle a maximum average current of 20 A. (D) This stud diode has a maximum average current rating of 35 A, but it can carry momentary surges of 500 A.

EXAMPLE 21-1

2

Figure 21-5 shows a diode in a switchable circuit. A digital voltmeter is connected to indicate the diode's voltage. Its current is indicated by a digital ammeter. Both meters are dual-polarity.

a) With the switch in the up position, as shown, does the diode conduct, or does it block?
b) What voltage value will be indicated by the voltmeter? (Assume the diode is ideal.)
c) What current value will be indicated by the ammeter?

SOLUTION

a) Voltage source V_{s1} is forcing electron current to flow against the arrow, through the diode from bottom to top. The diode is **conducting**.
b) An ideal forward-biased diode acts like a 0-Ω short circuit. Therefore, its voltage is **0 V**.

c) With the ideal diode having zero voltage drop, Kirchhoff's voltage law requires that the entire source voltage appears across the resistor. That is,

$$V_R = V_{s1} = 15\text{ V}$$

From Ohm's law,

$$I_R = \frac{V_R}{R} = \frac{15\text{ V}}{100\ \Omega} = \mathbf{0.15\ A}$$

The circuit is a series circuit, so this 0.15-A current is the only current. It will be the value read by the ammeter.

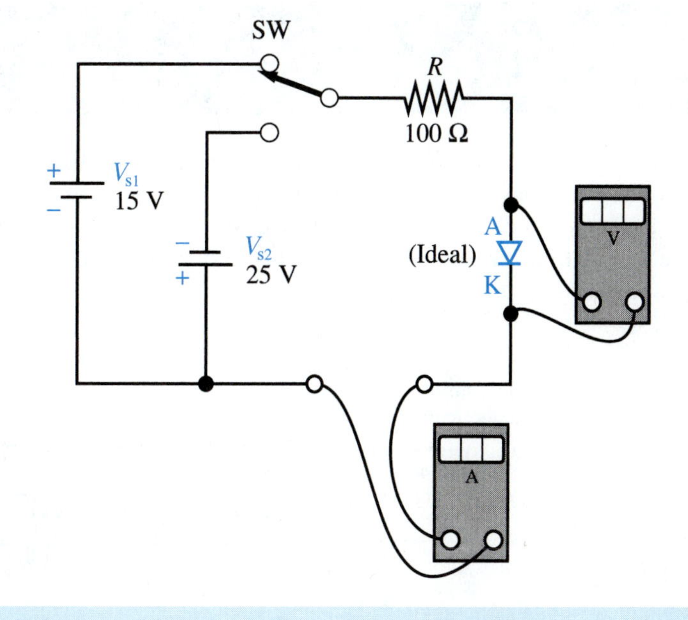

FIGURE 21-5 Switched diode circuit. The ammeter and voltmeter are DMMs with dual polarity—they can correctly measure either polarity.

EXAMPLE 21-2

Repeat Example 21-1, but this time with the switch moved to the down position in Figure 21-5.

SOLUTION

a) Voltage source V_{s2} is trying to force electron current to flow in the direction of the arrow. The diode will not allow the current to pass; it **blocks.**

b) and c) Answer current first. With the ideal diode blocking, $I = \mathbf{0\ A}$. This is what the ammeter will read.

Answer voltage second. With the diode blocking, it acts like an open circuit. Therefore, the entire source voltage appears across the diode terminals. The voltmeter reads **25 V**, positive on the bottom (K) and negative on the top (A).

Examples 21-1 and 21-2 make the following facts clear.

TECHNICAL FACT

A diode conducts when its arrow points toward the more negative terminal in a circuit.

Said the other way,

TECHNICAL FACT

A diode blocks when its arrow points toward the more positive terminal in a circuit.

21-2 HALF-WAVE RECTIFIERS

One of the important uses for diodes is to convert ac into pulsating dc. This process is called **rectifying**. It is illustrated in Figure 21-6. In Figure 21-6, the diode is connected in series with a load resistor. This series combination is driven by a sine wave ac voltage source. Often a transformer secondary winding is the ac voltage source for a rectifier.

RADAR

Radar is an electrical and electronic system that sends out bursts of electromagnetic waves from a transmitting antenna. The radar system's electronic circuitry measures the time for the wave bursts to travel to an object, reflect off the object's surface, and return to a receiving antenna. A longer turnaround time indicates that the object is farther away. A shorter turnaround time indicates that the object is closer. The basic system layout is illustrated in the figure below.

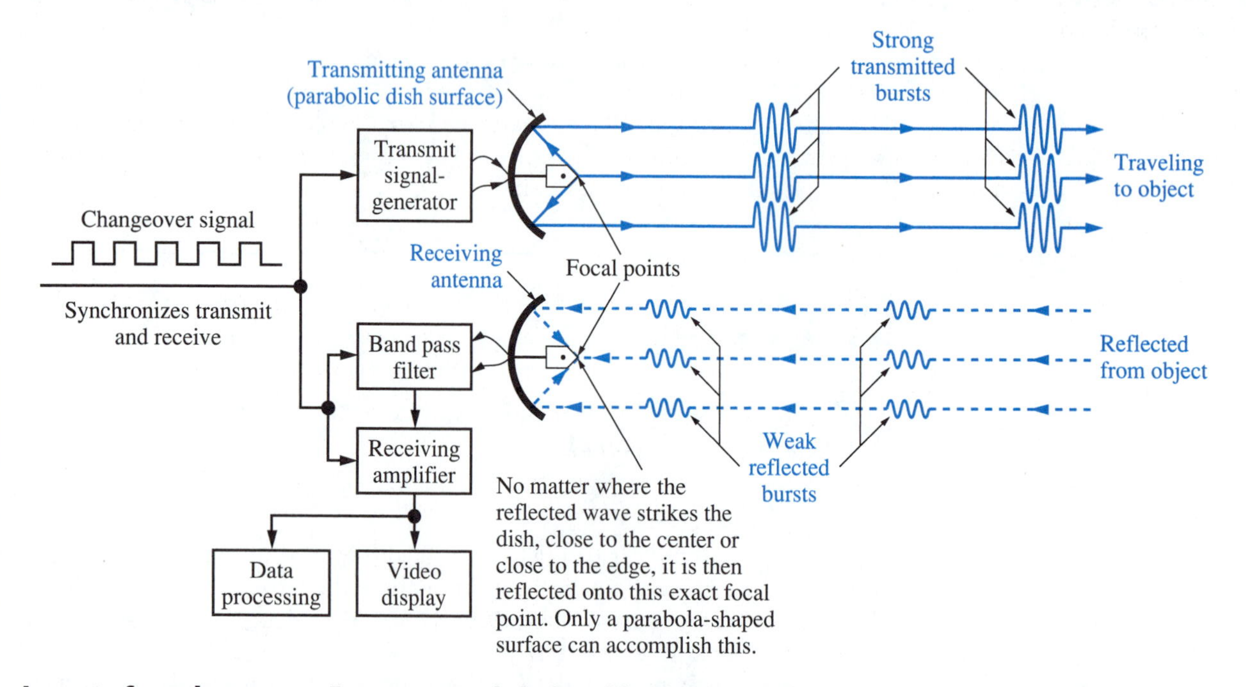

Layout of a radar system. For conceptual clarity, this diagram shows two antennas. In reality, a single antenna is used for transmitting and receiving, since the two operations do not occur simultaneously. Here, the transmitted waves are shown perfectly parallel to each other. Actually, they spread slightly apart, at an angle of about 1–3 degrees.

The wave bursts (often called pulses) travel at nearly the speed of light in a vacuum—3×10^8 meters / second, or about 186 000 miles / second. So even at the considerable distances encountered in military applications, the turnaround time is quite short. For example, at a distance of 100 km (62 miles), the turnaround time is on the order of 0.3 millisecond (100×10^3 m / 3×10^8 m/s = 0.3×10^{-3} s). At a typical civilian traffic-control distance, say 1/2 mile, turnaround time is on the order of 2 μs.

By comparing distances at successive bursts, the radar's electronic circuitry is able to measure the speed of the object, and to tell its direction of motion. These are the only important variables in traffic-control radar.

Military radar is much more sophisticated. It takes into account the portion of the transmitted energy that is reflected back and the details of the reflection pattern. Also factored in are the antenna's angle of elevation and its angle of azimuth (angle that it is turned relative to a line on the earth's surface). Military radar is therefore able to detect and measure finer variables, such as the object's size, shape, exact compass bearing, and, if it is moving through the air, its altitude and rate of descent or ascent.

The photograph on page 8 of the Applications color section shows a pair of high-performance military radar units, which can be used on land or mounted on the deck of a warship. This particular model transmits wave bursts in the frequency range of 2.2 to 2.4 gigahertz. The 30-foot-diameter (9.14-meter) dish is made of aluminum. The interior dish surface deviates from the shape of a perfect parabola by no more than 1.5 mm (0.06 inch). The dish is supported in back by 16 welded aluminum grids or "trusses," which are clearly visible in the closer unit. The dish and its supporting members weigh about 32 000 newtons (7000 pounds).

Continued on page 407.

Its range of elevation is from –7 degrees to +178 degrees (tipped 7 degrees below horizontal to the left, to tipped 2 degrees above horizontal to the right). The range of azimuth travel is from –368 degrees to +368 degrees (starting at midpoint, turn counterclockwise one full revolution plus 8 degrees, then back to midpoint, then clockwise one full revolution plus 8 degrees).

Each motion axis is driven by two 22.4-kW (30-hp) dc motors, operating through transmissions with gear ratios of 248:1 (elevation) and 236:1 (azimuth). This gearing boosts the elevation torque to 53 500 N-m maximum (about 72 500 lb-ft). The elevation driven gear, or elevation wheel, is the large circular gear whose bottom is even with the head of the technician standing on the service platform. Maximum torque to the azimuth axis is 50 800 N-m (about 70 000 lb-ft). Azimuth rotation occurs at the break in the pedestal just below the service platform. The entire service platform rotates with the antenna.

Due to the ability of the dc motors' drive circuits to control and regulate these great torques, the system's mechanical tracking capabilities are astounding. Its maximum speed of tracking is 30 deg/s for both azimuth and elevation. From a standstill, it can accelerate to maximum speed in 0.4 s. This enables it to start from a standstill, turn a full azimuth revolution, and stop in 12.7 s. In elevation, it can start from a standstill, move from left horizontal to nearly right horizontal, and stop in 6.8 s.

Half the time the secondary voltage, v_S, has the positive polarity, which is shown in Figure 21-7(A). During such a positive half-cycle, the diode arrow points to the negative source terminal. Therefore, the diode conducts. Ideally, the v_{LD} waveform, shown on the right, is exactly the same as the v_S source waveform, shown on the left.

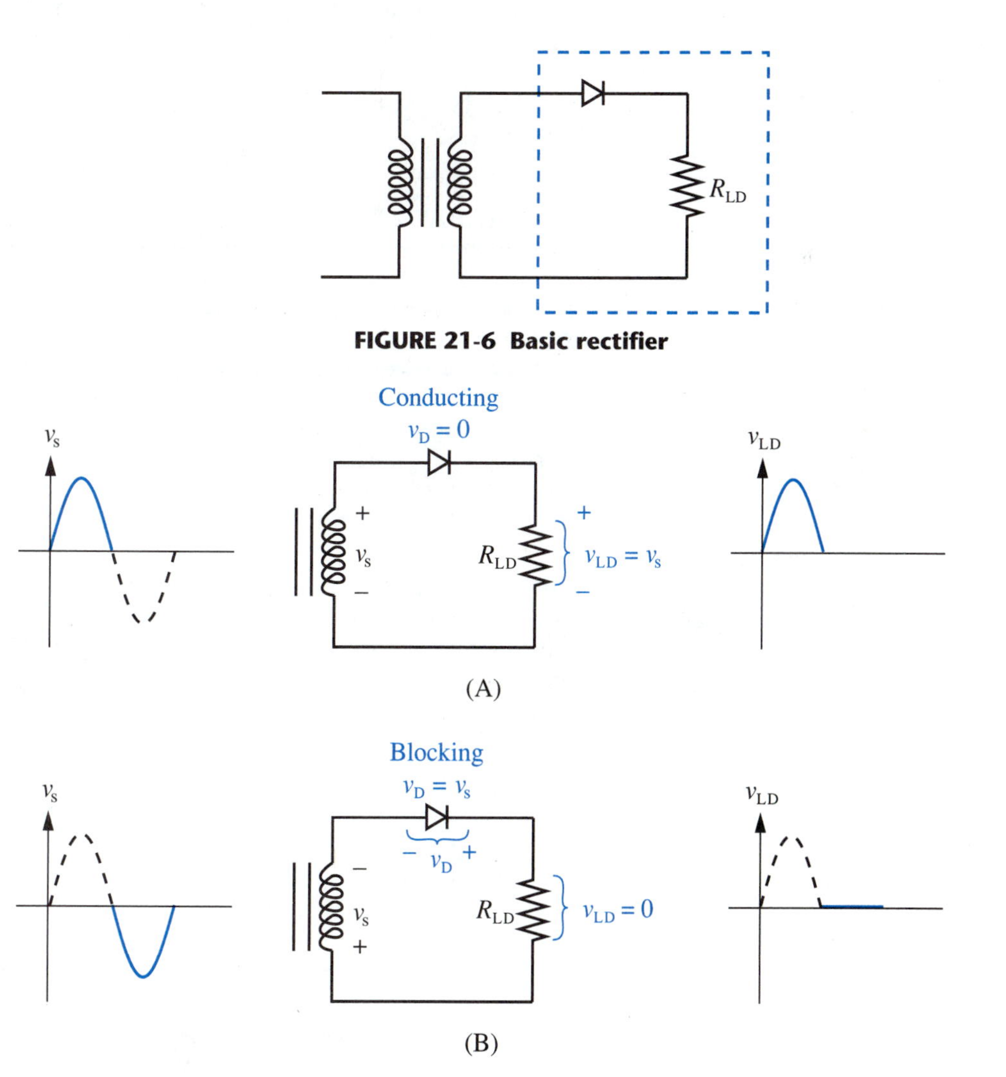

FIGURE 21-6 Basic rectifier

FIGURE 21-7 Operation of a half-wave rectifier. (A) During the positive half-cycle of ac. (B) During the negative half-cycle of ac

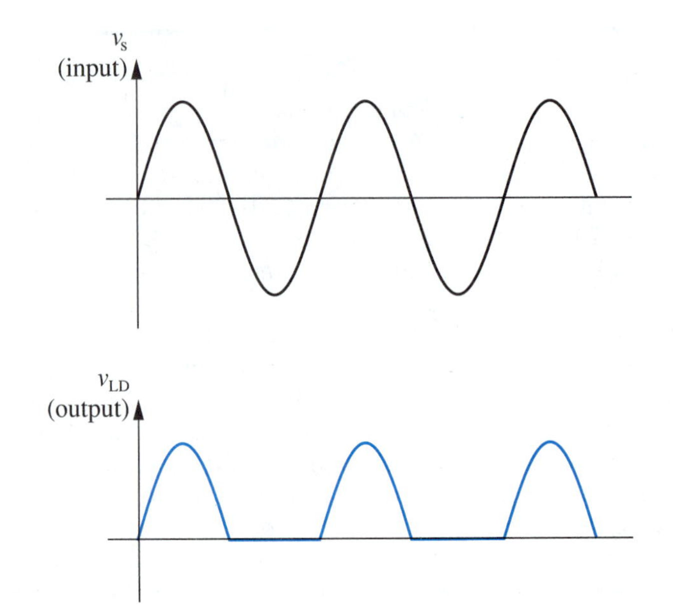

FIGURE 21-8 Waveforms for a half-wave rectifier. During positive half-cycles, v_{LD} equals the transformer v_S. During negative half-cycles, $v_{LD} = 0$ V.

The other half of the time, secondary voltage, v_S, has the negative polarity, shown in Figure 21-7(B). During such a negative half-cycle, the diode arrow points toward the positive source terminal. Therefore, the diode blocks. The entire v_S source voltage waveform, shown on the left, appears across the diode. Zero voltage appears across the load, as shown in the waveform on the right.

If you find it helpful, think of the ideal diode as a fast-acting switch. Every time the source goes positive, the diode "switch" quickly closes. Every time the source reverses to negative, the diode switch quickly opens.

The instant-by-instant operation of a half-wave rectifier is summarized in Table 21-1.

TABLE 21-1 Summarizing Events in a Half-wave Rectifier

SOURCE POLARITY	DIODE CONDITION	DIODE IS LIKE THIS SWITCH STATE	VOLTAGE v_D ACROSS IDEAL DIODE	LOAD VOLTAGE v_{LD}
Positive	Conducting (forward-biased)	Closed	0 V	v_S
Negative	Blocking (reverse-biased)	Open	v_S	0 V

4

Sketching these events over several cycles of the ac source, we get the waveforms shown in Figure 21-8. They make it clear that the ac input waveform is converted to a single-polarity, therefore dc, output waveform. This dc is not smooth dc, like we would get from a battery or high-quality power supply. It is *pulsating* dc.

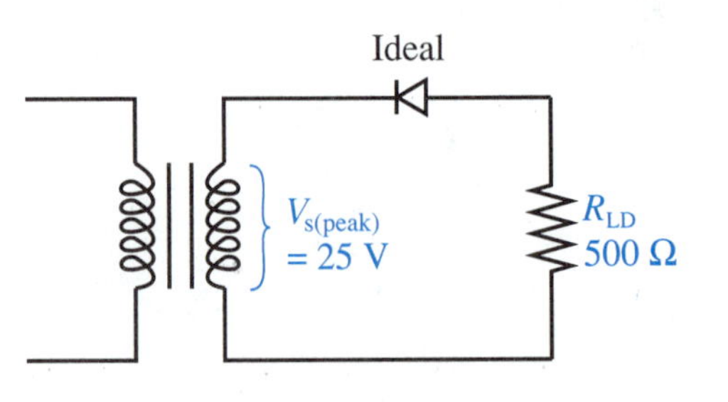

FIGURE 21-9 Negative half-wave rectifier

EXAMPLE 21-3

In Figure 21-9, the peak value of the secondary voltage is 25 V.

a) Sketch the v_{LD} waveform.
b) What is the peak value of current?

SOLUTION

a) The diode is reversed from Figures 21-6 and 21-7. Now it points toward the top terminal of the secondary winding. Therefore, it will conduct when the winding is negative on top. In other words, the diode will conduct during the negative half-cycle. This will apply a pulsating dc voltage to the load that is negative on top, positive on bottom. We regard that polarity as negative, so the waveform is as shown in **Figure 21-10.**
b) Apply Ohm's law at the peak instant.

$$i_p = \frac{V_p}{R_{LD}}$$

$$= \frac{-25\text{ V}}{500\ \Omega} = \mathbf{-50\ mA}$$

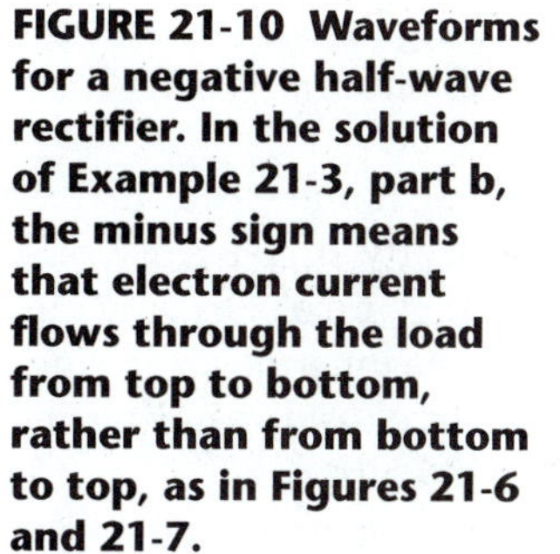

FIGURE 21-10 Waveforms for a negative half-wave rectifier. In the solution of Example 21-3, part b, the minus sign means that electron current flows through the load from top to bottom, rather than from bottom to top, as in Figures 21-6 and 21-7.

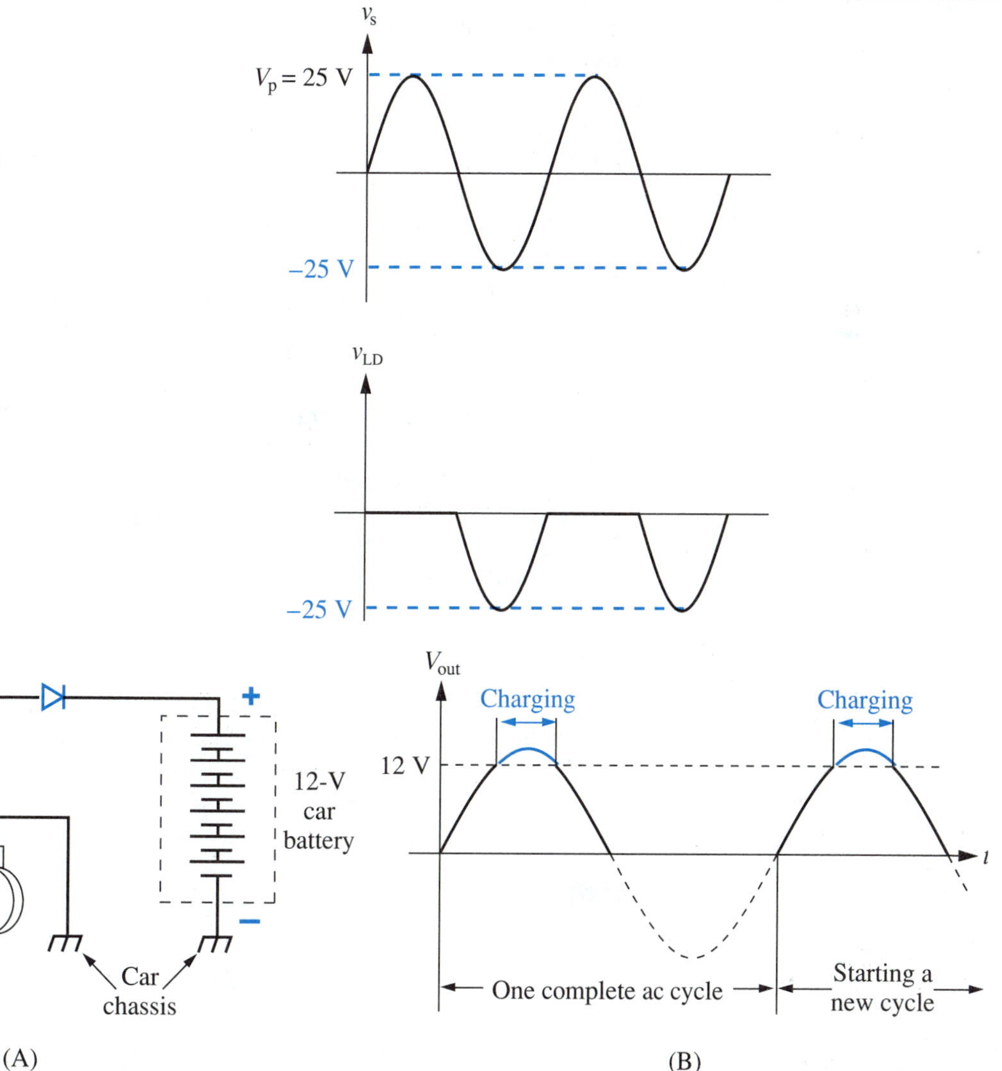

FIGURE 21-11 (A) Half-wave rectified single-diode charging circuit. (B) Output waveform. Battery charging occurs only during the indicated interval within the ac cycle.

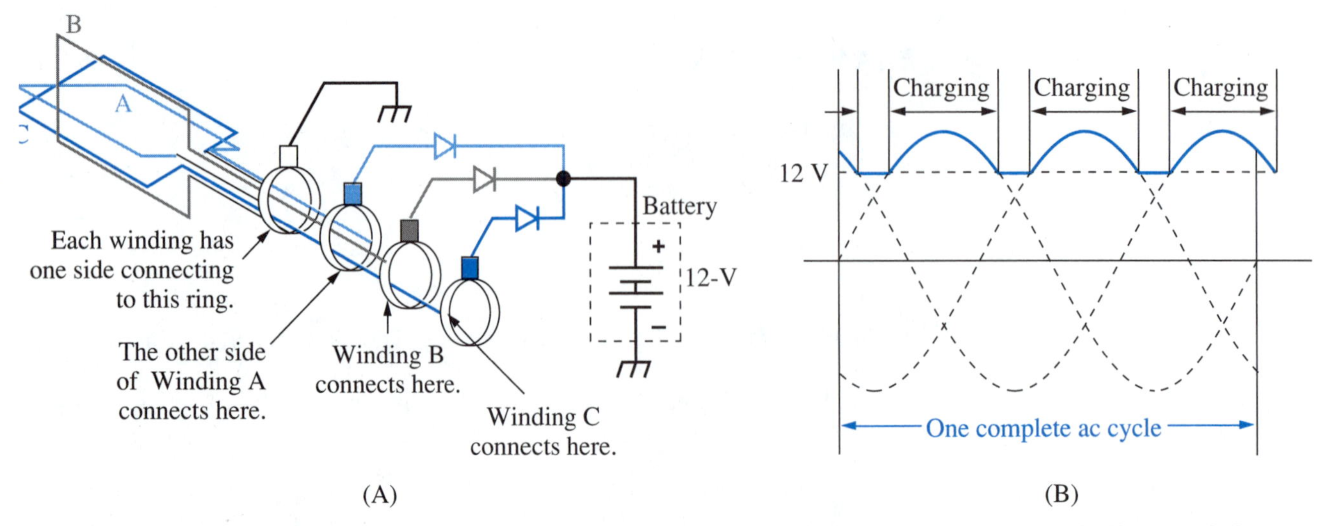

FIGURE 21-12 (A) Simplified view of an automotive charging circuit. There are three windings and three rectifying diodes. (B) Output voltage waveform. This circuit is called a three-phase half-wave rectifier.

Diode Application—The Rectifiers in an Automobile Alternator. We saw in Section 16-6 how an alternator induces a sine wave ac voltage. An alternator's ac output must be rectified to dc before it can be used to recharge a car's battery. This is shown schematically in Figure 21-11(A). The half-wave-rectified output waveform is sketched in Figure 21-11(B). Battery charging actually occurs only during those intervals of the ac cycle when the output voltage is greater than 12 V, the present battery voltage.

Figure 21-12 shows an improved alternator. It has not just one winding, but three windings, labeled A, B, and C. They are mechanically offset from one another by 120 degrees. Therefore, they induce sine waves that are electrically out of phase by 120 degrees. All three windings are connected to the innermost slip ring, the one farthest to the left in Figure 21-12(A). The other sides of each winding are connected to the three remaining slip rings, one winding per slip ring.

From the three outer slip rings, each brush leads out to a separate diode, connected to the diode's anode side. All three diodes have their cathodes tied together, as Figure 21-12(A) shows. The cathode tie-point is then connected to the positive terminal of the car battery.

The advantage of this three-winding / three-diode design is shown in Figure 21-12(B). Now there are three intervals during a single ac cycle when the rectified output voltage is greater than the 12-V battery voltage. Therefore, the battery is actually being recharged for a greater overall portion of the cycle time.

5

The arrangements in Figures 21-11 and 21-12 are half-wave rectifiers because they use only the positive half-cycle, not the negative half-cycle. A full-wave rectifier would use both half-cycles. You will study full-wave rectifiers in a course devoted to solid-state electronics.

✓ SELF-CHECK FOR SECTIONS 21-1 AND 21-2

1. Draw the schematic diagram of a 10-V dc voltage source, a 5-Ω resistor, and a diode, all in series. Label the terminals of the diode (A and K). [1]
2. Based on how you oriented the diode in problem 1, will it conduct, or will it block? [2]
3. Turn the diode around. Now what does it do, conduct or block? [2]

4. One of your two circuits makes the diode conduct. For that circuit: [3]
   a) What is the current value? Show direction.
   b) What amount of voltage V_R is dropped across the resistor? Show polarity.
   c) What amount of voltage V_D is dropped across the diode? Show polarity.
5. One of your two circuits makes the diode block. For that circuit: [3]
   a) What is the current value? Show direction.
   b) What amount of voltage V_R is dropped across the resistor? Show polarity.
   c) What amount of voltage V_D is dropped across the diode? Show polarity.

21-3 CURRENT-OPERATED TRANSISTORS

A diode is a **passive device.** It receives an input waveform and allows a portion of that input to appear at the output. A diode cannot create its own output waveform. An **active device** creates its own output waveform, which is greater than the input waveform. The first solid-state active device was the current-operated transistor, usually called the **bipolar junction transistor,** or **BJT**.

Before we get involved in terminal names, voltage polarities, and other circuit details, let us try to get an overall view of what a bipolar transistor accomplishes. In simplest terms, a bipolar transistor receives a small input current waveform and creates a larger output current waveform. This action is illustrated in Figure 21-13. In that figure, a signal source supplies a small ac input signal to the transistor's input terminal. The ac input voltage, V_{in}, exists between the input terminal and the chassis ground terminal. The ac input current, I_{in}, flows back and forth between the input terminal and the ground terminal. Both V_{in} and I_{in} are small, as shown by the input waveforms on the left of Figure 21-13.

The transistor is connected to a dc power source, which is not shown in the block-type diagram of Figure 21-13. The transistor draws on the power reserves of the dc source to produce an output current, I_{out}, that is much larger than the input current, I_{in}. For a particular transistor, output current, I_{out}, is larger than I_{in} by a certain factor. That factor is called the **current gain** of the transistor. It is symbolized by the Greek letter β, pronounced báy-tah. A modern bipolar transistor usually has a β factor that is some value between 100 and 300.

Output voltage, V_{out}, may be larger than V_{in}, or it may not be. That depends on the resistance of the load and other circuit variables that we are not concerned with right now. The output waveforms on the right of Figure 21-13 show the larger output current and an unchanged output voltage ($V_{out} = V_{in}$). The output waveform

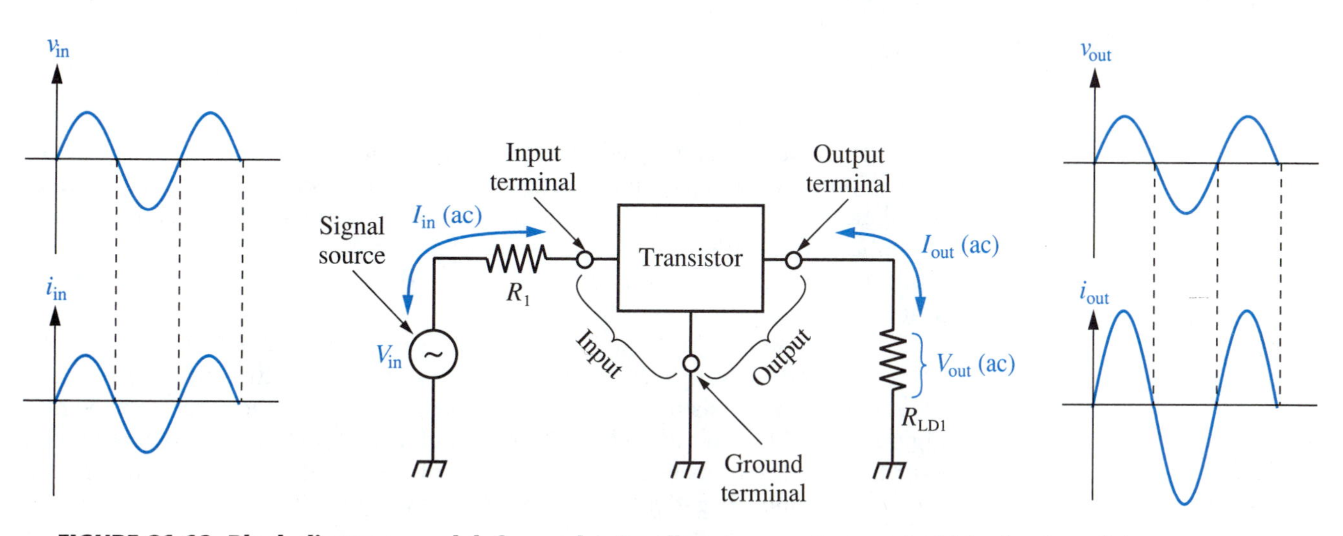

FIGURE 21-13 Block diagram model, for understanding a current-operated bipolar transistor

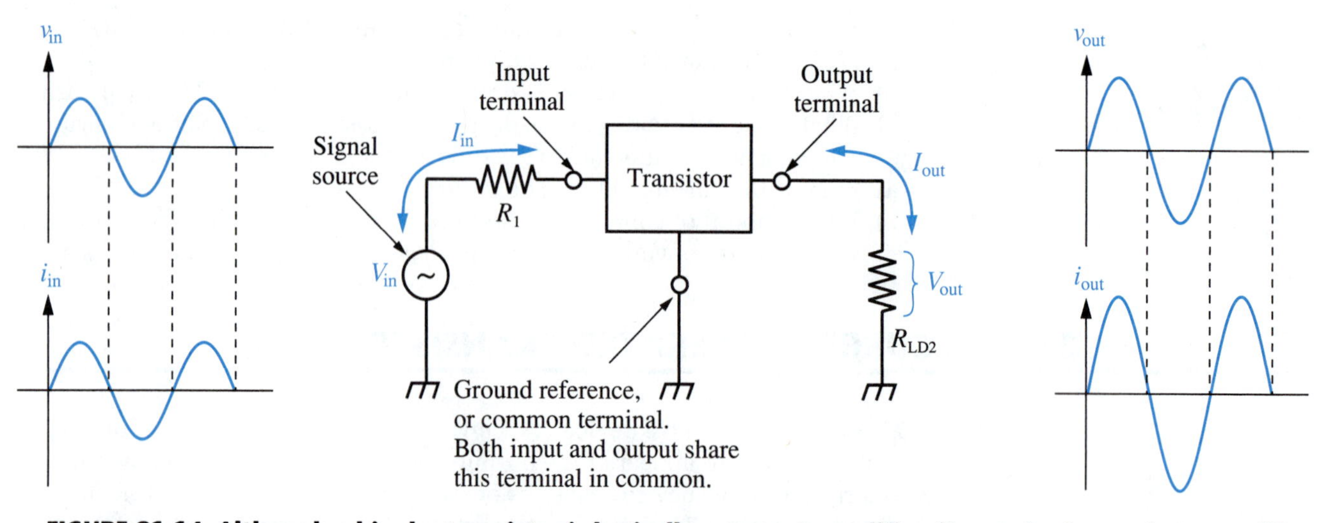

FIGURE 21-14 Although a bipolar transistor is basically a current amplifier, it can also be made to amplify voltage. Compare these output waveforms to the output waveform in Figure 21-13.

shapes are duplicates of the input waveform shapes. In this figure, the transistor receives a sine wave output, so it produces a sine wave output. If the input were a triangle wave, the transistor would produce a triangle wave output. The same holds true for any waveshape.

In Figure 21-13, V_{out} is the same magnitude as V_{in}. You should think of a bipolar junction transistor as a current amplifier, not necessarily a voltage amplifier. But even if the voltage just holds steady, notice carefully that:

TECHNICAL FACT

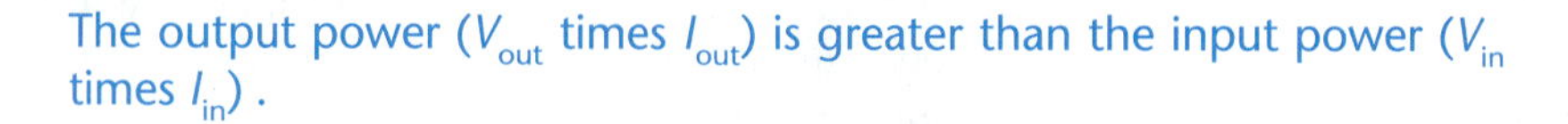
The output power (V_{out} times I_{out}) is greater than the input power (V_{in} times I_{in}) .

This **power gain** is the essence of a true amplifier. This is something that a transformer can never do.

6

The larger amount of output power that is delivered to the load actually comes from the dc power source. The transistor controls how power is drawn from the dc source and delivered to the load.

Figure 21-14 shows a situation where V_{out} is larger than V_{in}. This demonstrates the following:

TECHNICAL FACT

The power gain of a bipolar transistor amplifier circuit can be due to a combination of current gain and voltage gain.

Power gain is also called power amplification. It is symbolized A_P. Current gain, or amplification, is symbolized A_I. (For the overall amplifier circuit, A_I can be less than the transistor's β. This is explained later.) **Voltage gain,** or amplification, is symbolized A_V. In each case, the *A* stands for amplification, and the subscript tells what is being amplified, or made larger. For all three variables, the words *gain* and *amplification* are used interchangeably.

Several bipolar junction transistors are shown in Figure 21-15.

Transistor Symbols and Terminals. There are two different polarity types of bipolar transistors. One polarity type is called *npn*; the other is called *pnp*. Figure 21-16(A) shows the schematic symbol for the *npn* type. The letters *npn* are used because the transistor has three layers of semiconducting material, with the top layer having a *n*egative charge, the middle layer having a *p*ositive charge, and the bottom layer having a *n*egative charge. This layered structure is illustrated in Figure 21-16(B).

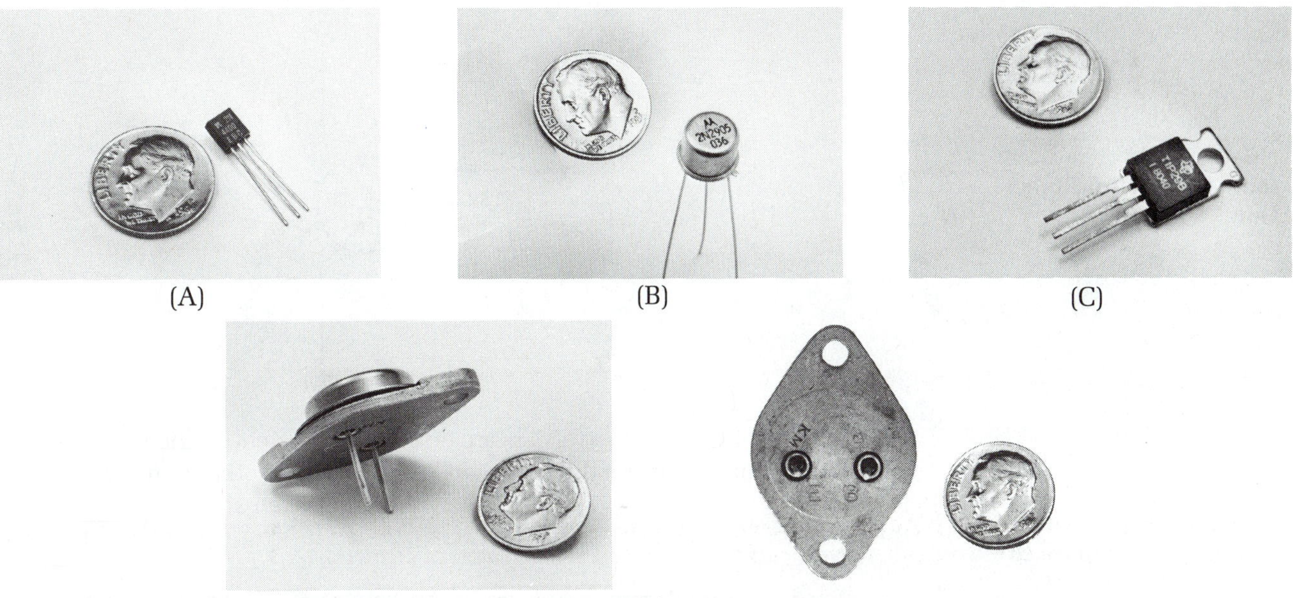

FIGURE 21-15 Physical appearance of bipolar transistors. (A) Small plastic package. Transistors in this package typically have these maximum ratings: power dissipation = 0.3 W; collector current = 0.5 A. The lead configuration is usually E, B, C, from left to right. (B) Larger metal-can package. Typical maximum ratings are: P_{max} = 5 W (heat sunk); I_C = 3 A. The tab (shown here on the right) marks the emitter, E. Usually, B is the middle terminal and collector, C, is opposite the tab. (C) Plastic-and-metal package. The metal surface is usually connected to the middle lead, which is the collector, C. Usually E is on the left and B is on the right. Typical maximum ratings are: P_{max} = 10 W (heat sunk); I_C = 5 A. (D) High-power all-metal case. Typical maximum ratings when heat-sunk are: P_{max} = 100 W; $I_{C(max)}$ = 15 A. The metal case is usually the collector, C. In the bottom view, the left terminal is the emitter, E, and the right terminal is the base, B.

Figure 21-17(A) shows the *pnp* schematic symbol. The only difference is the direction of the arrow. A *pnp* transistor has three layers that contain the opposite impurity from an *npn* transistor.

As indicated in Figures 21-16 and 21-17, each layer is connected to an external terminal. The top terminal is called the **collector,** labeled C. The middle is called the **base,** labeled B. The bottom is called the **emitter,** labeled E. The solid layers themselves can be referred to as the collector region, the base region, and the emitter region.

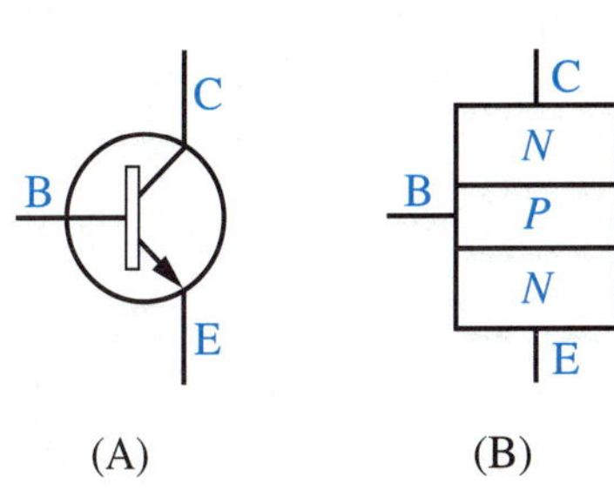

FIGURE 21-16 (A) Schematic symbol for an *npn*-type bipolar transistor. (B) In the actual construction of a bipolar transistor, there are three layers of silicon. Silicon is a semiconductor material, that has rather low conductivity (high resistivity) at room temperatures. Each silicon layer contains some impurity that is purposely placed there during the transistor's manufacture. If the impurity is arsenic, the silicon layer contains extra electrons in the outer-shell arrangement. This is a surplus of negative charge-carriers, so the layer is called an *n*-layer. If the impurity is boron, the layer has a shortage of electrons in the outer-shell arrangement. This shortage of negative charge-carriers is like having positive charge-carriers, or holes. Therefore the layer is called a *p*-layer.

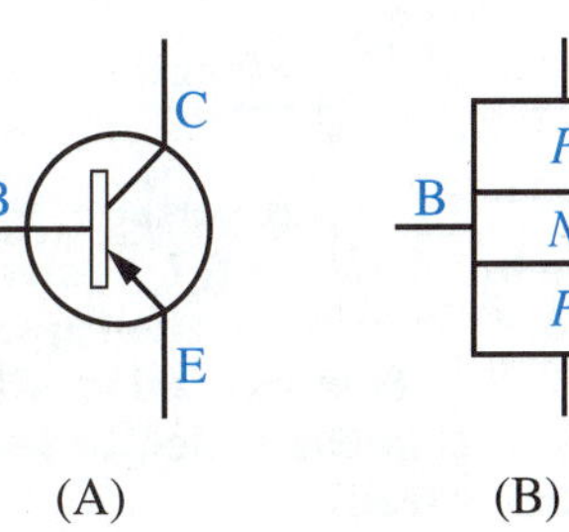

FIGURE 21-17 *Pnp* bipolar transistor. (A) Schematic symbol. (B) Three-layer structure

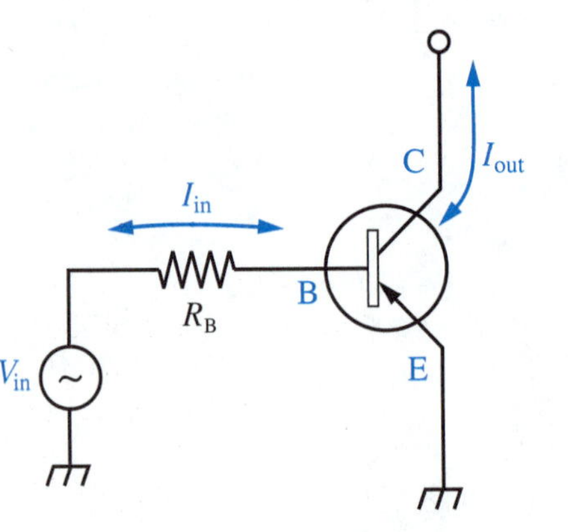

FIGURE 21-18 The common-emitter configuration.

Transistor Operation. There are three quite different methods for operating a bipolar transistor. The methods are different from each other in regard to which lead is connected to circuit ground. We will study the most popular one of these three methods. The other two methods are covered in a course dedicated to solid-state electronics.

The most popular method is called **common-emitter.** In it, the emitter, *E*, is connected to circuit ground. The input current is injected into the base, B. The output current appears at the collector, C. Figure 21-18 shows this arrangement. This method is called common-emitter because the emitter, being connected to ground, is the terminal that both the input and the output have in common. This commonality is shown more clearly in Figure 21-19.

The ac input current flows back and forth between base and emitter, through the interior solid of the transistor. In Figure 21-19, I_{in} passes through the base region and through the emitter region. The input voltage is applied between the outside of the base resistor and the emitter, as indicated in that figure.

The ac output current flows back and forth between the collector and emitter. I_{out} passes through the collector region, through the base region, and through the emitter region. The output voltage appears across the load resistor, which is placed in series with the collector lead in Figure 21-19. As that figure indicates, V_{out} appearing across R_{LD} is the same as V_{out} appearing between collector and emitter.

With the output variables involving the emitter and the input variables also involving the emitter, it is clear that the emitter is common to both input and output.

A bipolar transistor always guarantees that the current in the collector is β times larger than the current in the base lead. The transistor's collector becomes a **current source.**

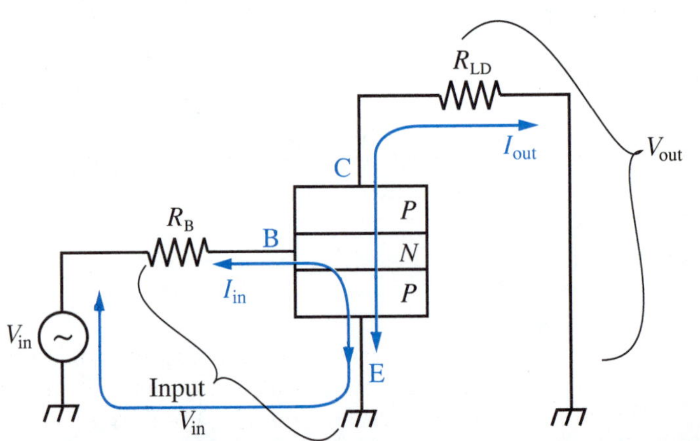

FIGURE 21-19 Input voltage, V_{in}, is measured relative to the emitter terminal, and the input current, I_{in}, emerges from the inside of the transistor on the emitter lead. I_{in} returns to the signal source through the ground connection. Output voltage, V_{out}, is measured relative to the emitter terminal. The output current, I_{out}, which is in the collector lead, must complete its path by passing through the emitter lead.

TECHNICAL FACT

An ideal current source maintains a given amount of current regardless of the load resistance that is connected to it. If the load resistance is low, a current source develops only a small terminal voltage. If the load resistance is high, a current source develops a larger terminal voltage, whatever value is required to satisfy Ohm's law.

Compare this mentally to an ideal voltage source, which maintains a given output voltage regardless of the load resistance connected to it.

So, in Figure 21-19, if R_{LD} is a small value, V_{out} will be a small value. This means that the voltage amplification will not be great, even though the current amplification is great. On the other hand, if R_{LD} is a large value of resistance, V_{out} will be a large voltage. This means that the voltage amplification, as well as the current amplification, will be great. This is pictured in Figure 21-14.

You should think of the order of events like this:

1. A certain amount of base (input) current is forced to flow between base and emitter. This current is ac and is given approximately by

$$I_b \cong \frac{V_{in}}{R_B}$$

EQ. 21-1

Equation 21-1 is approximately correct because the transistor itself has very low internal resistance between its base lead and emitter lead. In most cases, this internal resistance is negligible compared to the external resistance, R_B.

2. The transistor is operated by the base current. Whatever the base current value from Equation 21-1, the collector current, I_c, will be β times as large. That is,

8

$$I_c = \beta I_b$$

EQ. 21-2

TECHNICAL FACT

I_c is independent of load resistance, independent of any voltage that appears in the collector circuit, and independent of everything in the entire circuit *except* I_b. This is what is meant by the phrase *current-operated transistor.*

3. The ac output voltage, V_{out}, is developed because the transistor forces ac current, I_c, to flow through R_{LD}. V_{out} becomes whatever value is required by the Ohm's law formula

$$V_{out} = I_c R_{LD}$$

EQ. 21-3

(As a practical matter, there is a certain maximum limit on V_{out}, but we are not concerning ourselves with that now.)

EXAMPLE 21-4

In a circuit configuration like the one in Figures 21-18 and 21-19, suppose the transistor's beta factor has been measured on an instrument called a transistor curve-tracer, giving $\beta = 120$. For the input circuit, $V_{in} = 0.5$ V, and $R_B = 10$ kΩ. The load resistance in the collector is $R_{LD} = 300$ Ω.

a) Find the base input current, I_b (also symbolized I_{in}).
b) Find the collector output current, I_c (also symbolized I_{out}).
c) What amount of output voltage, V_{out}, is produced across R_{LD}?

Continued on page 416.

SOLUTION

a) Ignoring the internal resistance between the base and emitter of the transistor, we can use Equation 21-1, Ohm's law, to find I_b.

$$I_b = \frac{V_{in}}{R_B}$$

$$= \frac{0.5\text{ V}}{10\text{ k}\Omega} = \mathbf{50 \times 10^{-6}\ A\ \text{or}\ 50\ \mu A}$$

Here the ac base input current is quite small. This is typical for bipolar transistors in the common-emitter configuration.

b) The transistor becomes an ac current source with output current given by Equation 21-2,

$$I_{out} = I_c = \beta\, I_b$$

$$= 120\ (50 \times 10^{-6}\text{ A}) = \mathbf{6 \times 10^{-3}\ A\ \text{or}\ 6\ mA}$$

This value of current must flow in the collector lead regardless of what amount of resistance is placed in the collector lead (within limits, as mentioned before).

c) Applying Ohm's law, Equation 21-3, to the load resistor, we get

$$V_{out} = I_c\, R_{LD}$$

$$= (6 \times 10^{-3}\text{ A})\ (300\ \Omega)$$

$$= \mathbf{1.8\ V}$$

In an amplifier circuit, current amplification, A_I, is defined as output current divided by input current.

$$A_I = \frac{I_{out}}{I_{in}}$$

EQ. 21-4

Voltage amplification is output voltage divided by input voltage.

$$A_V = \frac{V_{out}}{V_{in}}$$

EQ. 21-5

Power amplification is output power divided by input power.

$$A_P = \frac{P_{out}}{P_{in}}$$

EQ. 21-6

Substituting $P = VI$ into Equation 21-6, we get

$$A_P = \frac{P_{out}}{P_{in}} = \frac{(V_{out})\ (I_{out})}{(V_{in})\ (I_{in})} = \left(\frac{V_{out}}{V_{in}}\right) \times \left(\frac{I_{out}}{I_{in}}\right)$$

$$A_P = A_V\, A_I$$

EQ. 21-7

Equation 21-7 tells us that we can find the power gain of an amplifier by multiplying the amplifier's voltage gain times its current gain.

EXAMPLE 21-5

For the circuit in Example 21-4,

a) Find the current amplification, A_I.
b) Find the voltage amplification, A_V.
c) Find the power amplification, A_P.

SOLUTION

a) Using Equation 21-4,

$$A_I = \frac{I_{out}}{I_{in}} = \frac{I_c}{I_b}$$

$$= \frac{6 \times 10^{-3}\ \text{A}}{50 \times 10^{-6}\ \text{A}}$$

$$= \mathbf{120}$$

In this case, A_I is the same as the transistor's beta factor.

b) From Equation 21-5,

$$A_V = \frac{V_{out}}{V_{in}}$$

$$= \frac{1.8\ \text{V}}{0.5\ \text{V}}$$

$$= \mathbf{3.6}$$

c) Using Equation 21-7,

$$A_P = A_V A_I$$

$$= 3.6\ (120)$$

$$= \mathbf{432}$$

Or, we could find both the input power and the output power, and use Equation 21-6.

$$P_{in} = V_{in} I_{in} = V_{in} I_b$$

$$= (0.5\ \text{V})\ (50 \times 10^{-6}\ \text{A})$$

$$= 25 \times 10^{-6}\ \text{W} = 25\ \mu\text{W}$$

$$P_{out} = V_{out} I_{out} = V_{out} I_c$$

$$= (1.8\ \text{V})\ (6 \times 10^{-3}\ \text{A})$$

$$= 10.8 \times 10^{-3}\ \text{W} = 10.8\ \text{mW}$$

These calculations assume that the values of voltages and currents are rms values. Then, from Equation 21-6

$$A_P = \frac{P_{out}}{P_{in}}$$

$$= \frac{10.8 \times 10^{-3}\ \text{W}}{25 \times 10^{-6}\ \text{W}}$$

$$= \mathbf{432}$$

There are no measurement units on current amplification, voltage amplification, or power amplification. Amplifications are just ratios of identical units, so the results are unitless numbers.

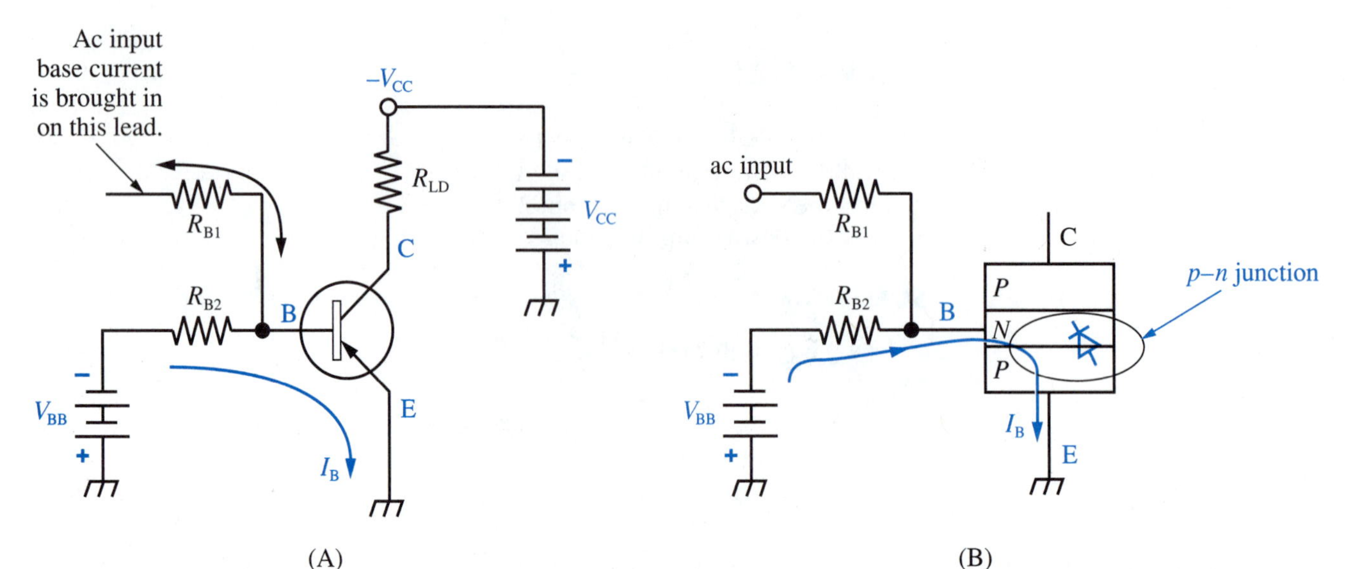

FIGURE 21-20 (A) Dc supply connections for a *pnp* transistor. (B) The base-emitter junction is just like the diode studied in Section 21-1. Current can pass in only one direction, from *n* (base) to *p* (emitter).

Be sure you appreciate the newness of what is happening here. Up till now, not a single electrical device that we have studied has been able to deliver more output power than input power. For example, consider a transformer. There, our ideal model could only break even. Any real transformer actually loses power, giving out less than was put in. To gain power, we need an active device like a bipolar transistor. Working with the energy stored in a dc power supply, the transistor controls the delivery of power to the load, making power gain possible.

Connecting the Dc Power Supplies. So far, we have regarded the transistor as an ac signal-amplifier. This is perfectly correct. As brought out above though, a transistor cannot boost the ac power level all by itself. Instead, its collector circuit must be connected to a dc power source so that it can extract power from that source to produce the output signal.

In addition to the collector dc requirement, it is also necessary for the base circuit to be connected to a dc supply. The proper polarities of base dc supply voltage and collector dc supply voltage for a *pnp* transistor are shown in Figure 21-20. The base dc supply voltage is symbolized V_{BB}. It must have a negative polarity relative to the emitter for a *pnp* transistor. This is because the base-emitter pair of leads is equivalent to a diode. There must be some flow of dc current through the transistor's internal "diode" before the ac base input current can be applied. This is

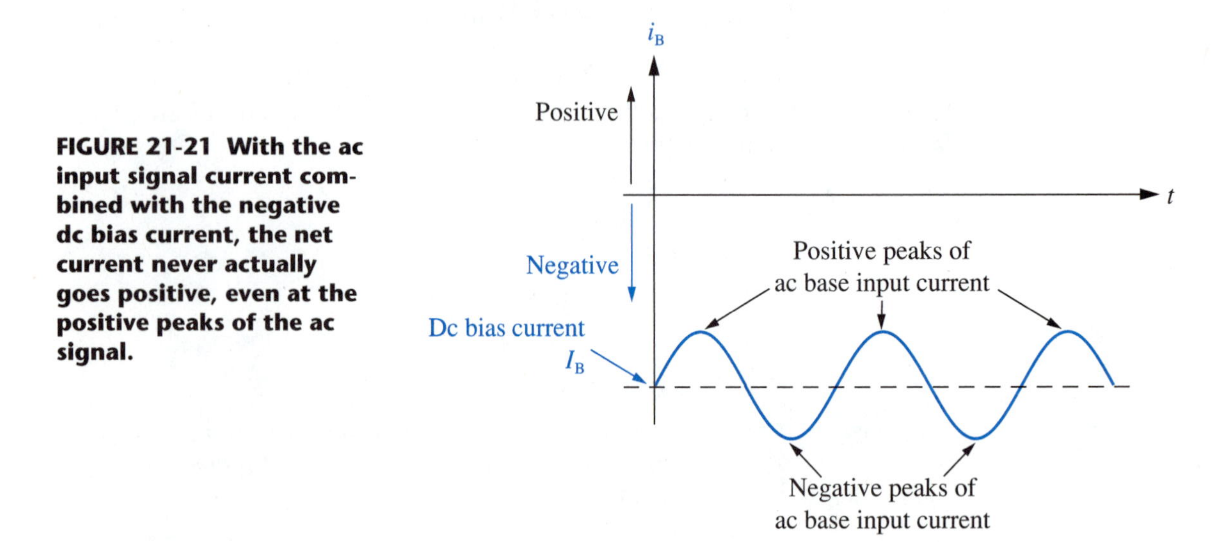

FIGURE 21-21 With the ac input signal current combined with the negative dc bias current, the net current never actually goes positive, even at the positive peaks of the ac signal.

shown in Figure 21-20(A), with the base electron current flowing against the arrow, the same as for a diode.

Inside the transistor, a junction exists where the bottom of the base *n*-type layer meets the top of the emitter *p*-type layer. We call this a ***p-n* junction**. The solid-state diode from Sections 21-1 and 21-2 contains just this, a single *p-n* junction. The schematic arrow points from the *p*-layer to the n-layer, as shown in Figure 21-20(B). Electron current can pass through the junction from *n* to *p*, against the arrow. But it cannot pass from *p* to *n*, in the same direction as the arrow.

This explains why the dc base voltage must be used to forward-bias the base-emitter *p-n* junction in Figure 21-20 before the ac signal current can be injected. If the ac current were applied by itself, it would reverse-bias the *p-n* junction half of the time. This would block every positive half-cycle of current, destroying the sine wave shape of the input signal. But the arrangement in Figure 21-20 allows the ac input signal to combine with the dc bias current. This is shown in the waveform of Figure 21-21. That waveform makes it clear that the net current remains negative at all times. It never reverse-biases the base-emitter *p-n* junction.

Going back to Figure 21-20, we see that the collector dc supply voltage is symbolized V_{CC}.

TECHNICAL FACT

For a *pnp* transistor, V_{CC} must be more negative than V_{BB}.

For example, if V_{BB} is –6 V, then a V_{CC} value of –10 V or –15 V would be all right. But a V_{CC} value of –5 V would not be all right, because –5 V is not as negative as –6 V.

There is another *p-n* junction inside the transistor where the *n*-type base layer meets the *p*-type collector layer. This junction is called the collector-base junction. When we say that V_{CC} must be more negative than V_{BB}, that is the same as saying that the collector-base junction must be reverse-biased. This is made clearer in Figure 21-22.

The "diode" arrow always points toward the *n*-type layer. In Figure 21-22, the diode points down. With –6 V applied to the pointed side of the diode arrow, and –10 V dc applied to the flat side, the more negative voltage is on the flat side. Therefore, the diode does not conduct. None of the dc base current that enters the transistor on the base lead flows out by the collector lead. All of it flows out by the emitter lead. This is a requirement for proper operation of the transistor.

One of the internal junctions has forward-bias polarity and the other junction has reverse-bias polarity. This is the reason for the name bipolar junction transistor.

Now here is the point where the ingeniousness of the transistor idea comes into play. The forward bias of the base-emitter junction in Figure 21-22 causes a certain amount of base current, I_B, to cross the junction. This current is due entirely to excess electrons from the *n*-type base layer being forced to move into the *p*-type emitter layer. Once they enter the emitter layer, they remain as free electrons on the silicon outer-shell crystalline structure. Being always repelled by the negative

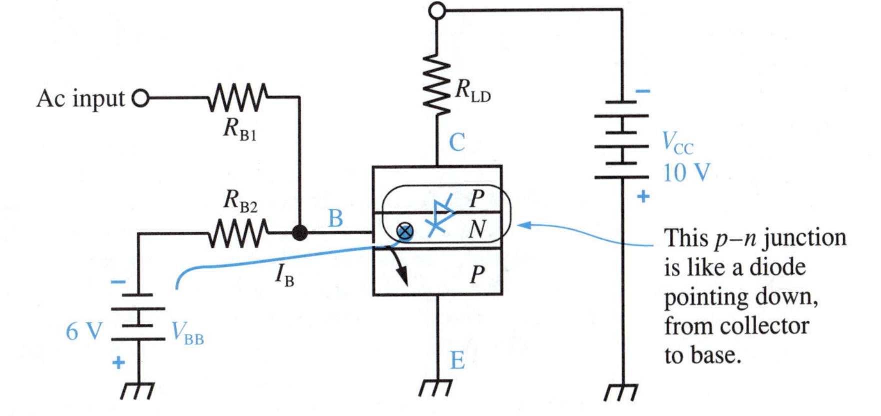

FIGURE 21-22 The collector- base junction must be reverse-biased by the dc supplies. For a *pnp* transistor, this means that V_{CC} must be more negative than V_{BB}.

voltage on the base, they jump from one atom to another, and eventually make their way through the entire thickness of the emitter region. They exit by the emitter lead and return through the ground connection to the positive terminal of the V_{BB} supply.

At the same time that this is happening, positive-charge-carrying holes from the *p*-type emitter layer are being attracted by the negative voltage on the base. Therefore, they cross the base-emitter junction in the opposite direction, moving *up* in Figure 21-22.

TECHNICAL FACT

The number of holes moving up across the base-emitter junction is much greater (β times greater) than the number of electrons moving down across the base-emitter junction in a *pnp* transistor. This is because the transistor manufacturer placed many more boron atoms (holes) in the emitter region than arsenic atoms (free electrons) in the base region.

Once the positive charge-carrying holes arrive in the base layer in Figure 21-22, they have two choices:

1. Move horizontally to the left, toward the negative potential on the base terminal. The transistor manufacturer has made the horizontal dimension of the base region rather wide, referring to Figure 21-22. Therefore, the center of the base-emitter junction (the "average" crossing point) is rather far away from the base terminal.
2. Move vertically up toward the even more negative potential on the collector terminal. The transistor manufacturer has made the vertical height of the base region quite short, referring to Figure 21-22. The holes don't have very far to go to reach the collector-base junction.

As you would expect, virtually all the positively charged holes take choice number 2. Once they have been sucked into the collector region, they don't turn back. They just keep on moving through the collector layer until they emerge from

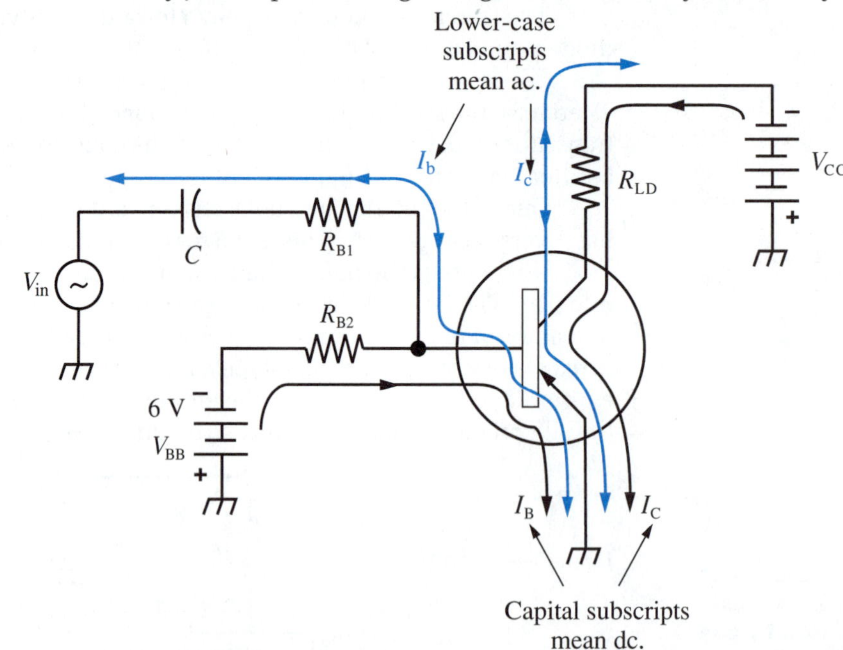

FIGURE 21-23 A complete common-emitter transistor amplifier. The overall base current consists of a dc part, I_B, and an ac part, I_b. The ac part is the input signal. The overall collector current consists of a dc part, I_C, and an ac part, I_c. The ac part is the output signal.
The input coupling capacitor acts as a short circuit for the high-frequency ac input signal. But it completely blocks any of the dc base bias current from going backward through R_{B1} and passing through the input source to reach ground. All the dc bias current from V_{BB} must flow into the transistor's base, as it is supposed to.

the transistor on the collector lead. As we discussed in Section 4-5, hole current moving out on the collector lead is no different from electron current moving in on the collector lead. Thus, a much larger dc electron current, symbolized I_C, passes through the transistor from the collector to the emitter, then passes through the transistor from the base to the emitter (I_B). This is the breakthrough idea that ignited solid-state electronics.

Dc and Ac Operating Together—The Complete Amplifier Circuit. With both junctions properly biased and both **dc bias currents** flowing, the ac input current is passed through the input coupling capacitor, then through R_{B1} to the base terminal, as shown in Figure 21-23. When the transistor receives the ac base input current, it produces an amplified ac collector current by the internal action just described.

Ac base current is symbolized I_b. The lower-case *b* subscript distinguishes this ac base current from the dc bias base current, which is symbolized I_B, with a capital *B* subscript. Likewise, ac collector current is symbolized I_c, with a lower-case *c*

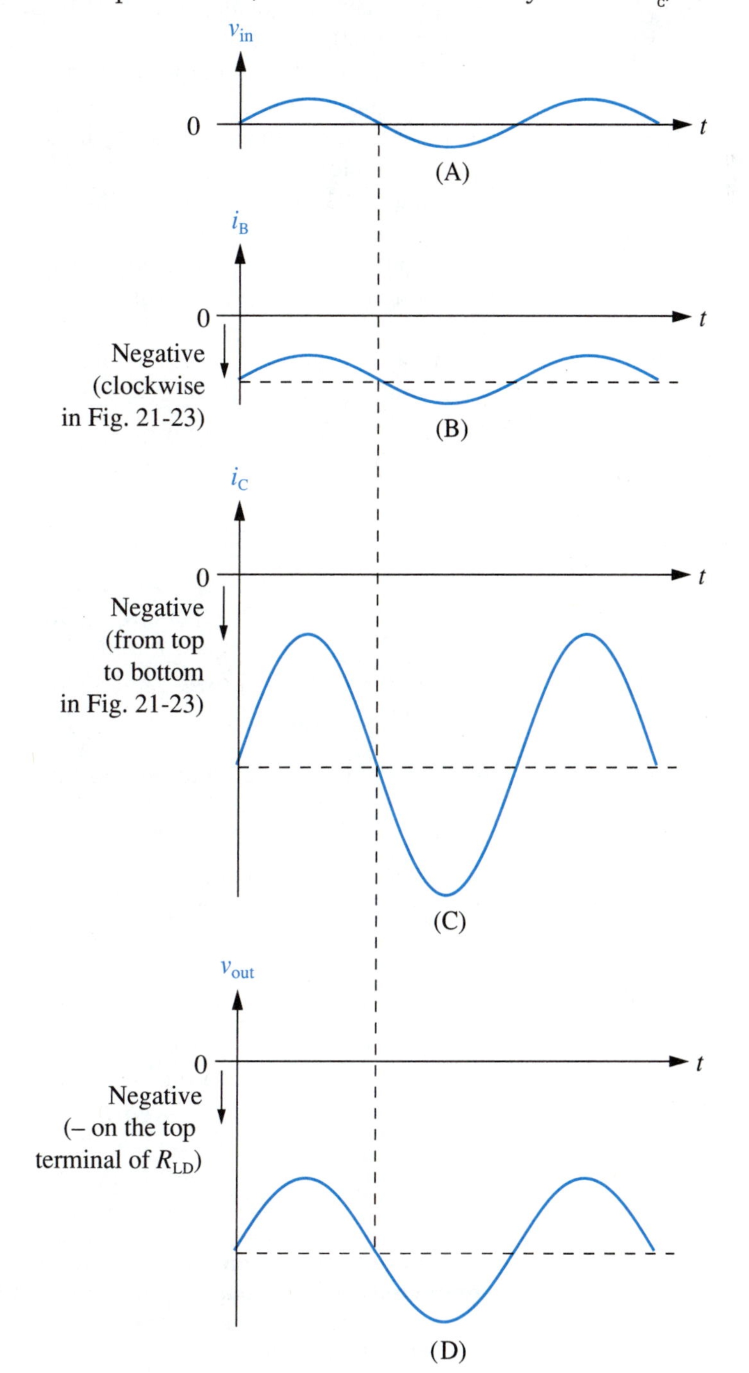

FIGURE 21-24 Waveforms of voltage and current for the complete amplifier of Figure 21-23. (A) Input voltage. (B) Total base current (ac combined with dc bias). (C) Total collector current. (D) Total output voltage

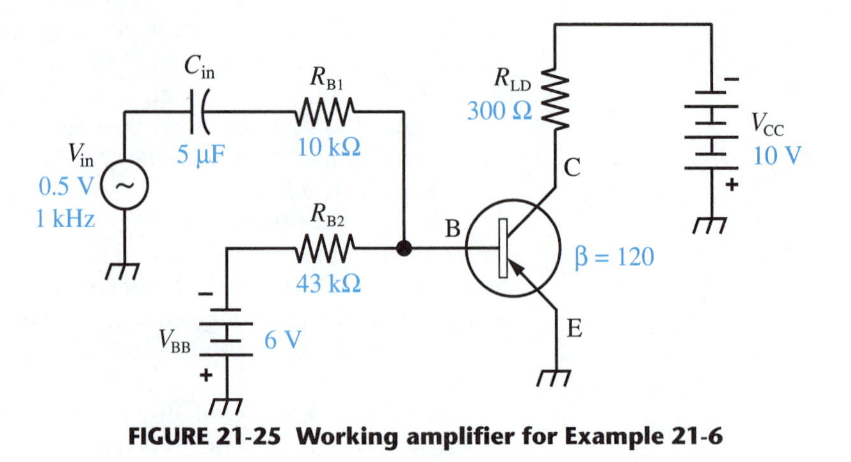

FIGURE 21-25 Working amplifier for Example 21-6

subscript. This distinguishes it from dc bias collector current, which is symbolized I_C, with a capital *C* subscript.

The waveforms of total base current, i_B, and total collector current, i_C, are shown in Figure 21-24. Also shown are the v_{in} and v_{out} waveforms. The output voltage waveform has been drawn assuming that R_{LD} is large enough to cause v_{out} to be somewhat greater than v_{in}.

In Figure 21-24, the output voltage waveform is not pure ac. Instead, it is ac superimposed on (combined with) a negative dc bias voltage. This happens because the load resistor is placed directly in the collector lead. In that position, R_{LD} is bound to experience both the dc bias collector current, I_C, and the ac output signal collector current, I_c.

An output voltage that is a combination of ac and dc is acceptable in some applications. In many applications, it is not acceptable. In the next section, we will show how to remove the dc part from the output voltage.

EXAMPLE 21-6

In Figure 21-25, suppose that we have the same transistor and the same values of R_{B1} (10 kΩ) and R_{LD} (300 Ω) that we had in Examples 21-4 and 21-5. With V_{in} = 0.5 V, the ac circuit's operation will be just the same as before in Examples 21-4 and 21-5. That is, I_b = 50 μA, I_c = 6 mA, and V_{out} = 1.8 V.

The dc supplies are V_{BB} = −6 V and V_{CC} = −10 V, with the base bias resistor R_{B2} = 43 kΩ.

a) Find the amount of dc base bias current, I_B.
b) Find the amount of dc collector bias current, I_C.
c) Find the amount of dc voltage across R_{LD} (the dc voltage that would exist even if there were no ac signal present).
d) Draw the waveform of v_{RLD}, the total voltage across R_{LD}.

SOLUTION

a) The dc base bias circuit has V_{BB} driving R_{B2}, followed by the base-emitter junction. For approximation, we can think of the B-E junction as an ideal conducting diode, having zero voltage drop. Then, by Ohm's law,

$$I_B \cong \frac{V_{BB}}{R_{B2}} \qquad \textbf{EQ. 21-8}$$

$$= \frac{6\text{ V}}{43\text{ k}\Omega}$$

= **0.1395 mA** or **0.140 mA**

Continued on page 423.

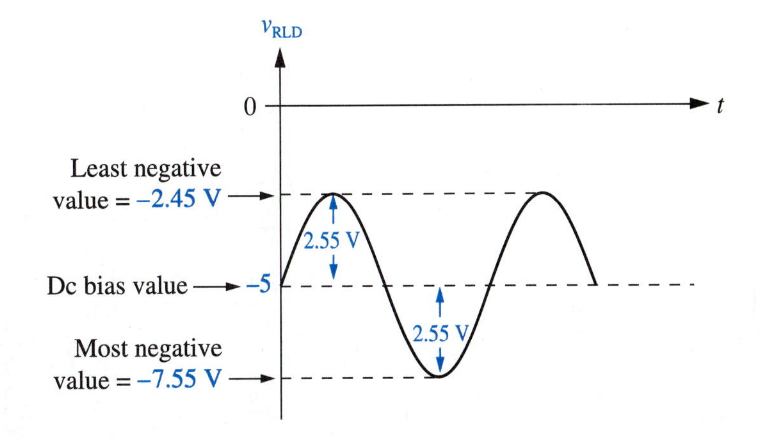

FIGURE 21-26 Waveform of v_{RLD}. An ac sine wave with peak value 2.55 V is superimposed on –5V dc.

b) The dc version of Equation 21-2 is

$$I_C = \beta I_B$$

EQ. 21-9

Here, we are using the approximation that the transistor's dc current gain is the same factor (β) as its ac current gain.

From Equation 21-9,

$$I_C = 120\ (0.1395\ \text{mA}) = 16.74\ \text{mA or } \mathbf{16.7\ mA}$$

c) Using Ohm's law again,

$$V_{RLD} = I_C R_{LD}$$
$$= (16.74\ \text{mA})\ (300\ \Omega)$$
$$= \mathbf{5.02\ V}$$

This is a negative voltage, since its polarity is negative on the top of R_{LD} and positive on the bottom.

d) From Example 21-4, $V_{out(rms)}$ = 1.8 V.

$$V_{out(peak)} = (1.414)\ V_{out(rms)}$$
$$= (1.414)\ (1.8) = 2.55\ \text{V}$$

Therefore, the total voltage across R_{LD} oscillates between:

1. Its most negative value, which is –5 V dc – 2.55 $V_{(peak)}$ = –7.55 V, and
2. Its least negative value, which is –5 V dc + 2.55 $V_{(peak)}$ = –2.45 V.

This waveform is drawn in **Figure 21-26.**

Dc Bias for an npn Transistor. An *npn* transistor must be connected to its dc supplies as shown in Figure 21-27(A). The two key requirements are:

1. The base must be positive with respect to the emitter. This forward-biases the base-emitter *p-n* junction as shown in Figure 21-27(B), establishing the dc base bias current, I_B.
2. The collector must be even more positive than the base. This reverse-biases the collector-base *p-n* junction, the same as was done for the *pnp* transistor.

Once the biasing supplies are connected, an *npn* transistor exhibits the same basic internal action as a *pnp* transistor, but with the charge-carriers reversed. Here, repeated, is the basic description.

1. A certain number of positive charge-carrying holes from the *p*-type base region move across the base-emitter junction in the downward direction in Figure 21-27(B). Hole-current moving down is equivalent to electron current moving up. This is the origin of dc base current, I_B, in that figure.
2. A much greater number of negative charge-carrying free electrons moves across the base-emitter junction in the up direction. This number is much greater

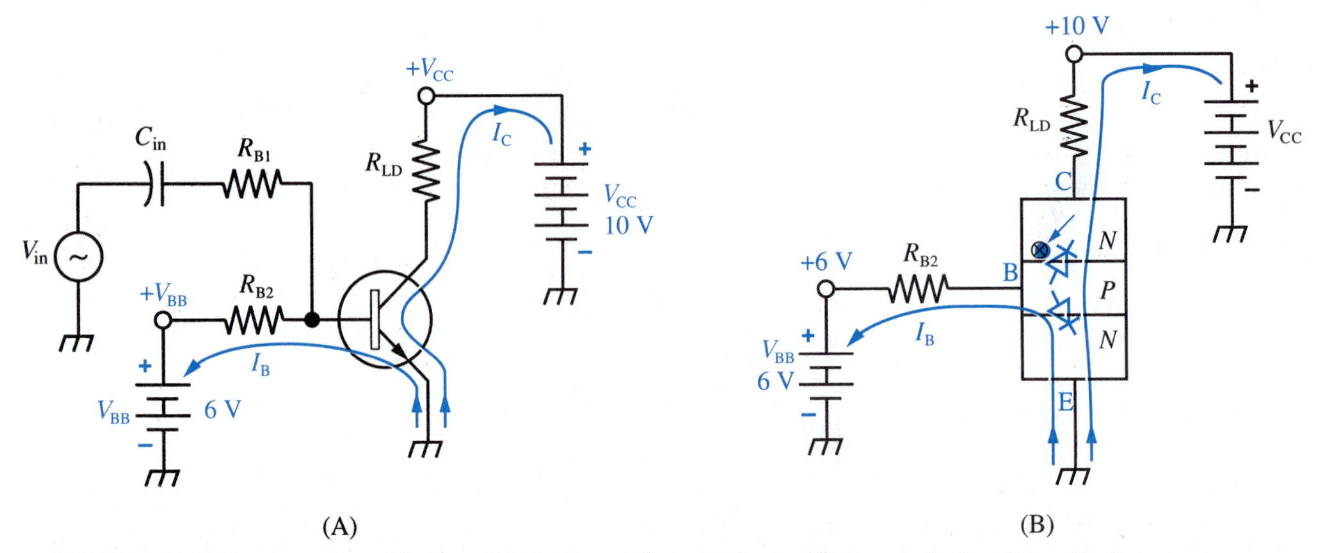

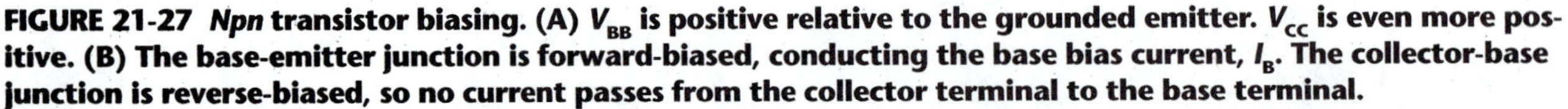
FIGURE 21-27 ***Npn*** **transistor biasing. (A)** V_{BB} **is positive relative to the grounded emitter.** V_{CC} **is even more positive. (B) The base-emitter junction is forward-biased, conducting the base bias current,** I_B**. The collector-base junction is reverse-biased, so no current passes from the collector terminal to the base terminal.**

because the transistor manufacturer has placed many more arsenic atoms (free electrons) in the emitter region than boron atoms (holes) in the base region. The ratio of densities of these impurity atoms is β.

3. The great number of free electrons have a difficult horizontal path to the base terminal, but an easier vertical path to the collector region. The collector is attracting them more strongly than the base anyway, because it is more positive than the base.
4. Virtually all the free electrons that have been emitted from the emitter region are collected by the collector region. They move through the collector region and exit from the collector lead, producing I_C.

The voltage and current waveforms for the *npn* transistor amplifier are shown in Figure 21-28. The dc part of the v_{out} waveform can be removed by making a change in the circuit of Figure 21-27(A). Instead of connecting the load resistance directly in the collector lead, we can do this:

1. Remove R_{LD} and place a new resistor, R_C, in the collector lead. The R_C resistance must be present so that collector current, I_C, is able to develop a voltage. If R_C were not present, the output current would flow, but the transistor current source would develop zero output voltage.
2. Tap the collector terminal with an output **coupling capacitor,** C_{out}.
3. Connect the load resistance from the outside terminal of C_{out} to ground, as shown in Figure 21-29.

12 In Figure 21-29, capacitance, C_{out}, charges to the value of dc voltage that exists on the collector. This is the positive dc voltage level shown in Figure 21-28(D). The capacitor charges positive on the inside and negative on the outside. Therefore, if C_{out} is an electrolytic capacitor, it must be connected with that polarization. The same remark applies to C_{in}. If it is electrolytic, its terminal polarization must be positive to the inside of the schematic diagram, negative to the outside.

The output voltage waveform across R_{LD} now becomes pure ac. The dc that exists at the collector terminal, V_C, is blocked by the output coupling capacitor. The new v_{out} waveform is shown in Figure 21-30(D).

For the Figure 21-29 amplifier, here is the reason why the v_{out} waveform is inverted, or phase-shifted by 180° compared to the v_{in} waveform. At a time instant when v_{in} is positive, total base current, i_B, and total collector current, i_C, in Figure 21-29 are more positive than average. This is shown by the waveforms in Figure 21-28(B) and (C) and also by Figure 21-30 (B) and (C). When i_C is more positive than average, it causes the voltage across R_C to become more negative on its

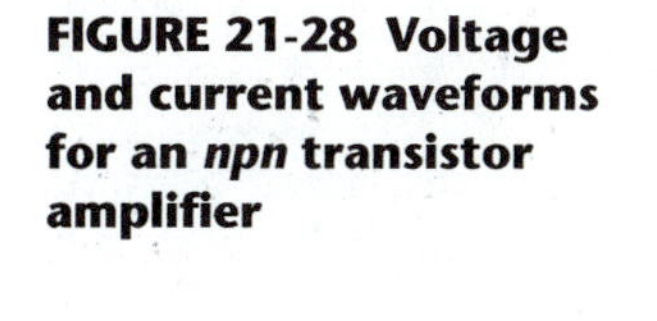

FIGURE 21-28 Voltage and current waveforms for an *npn* transistor amplifier

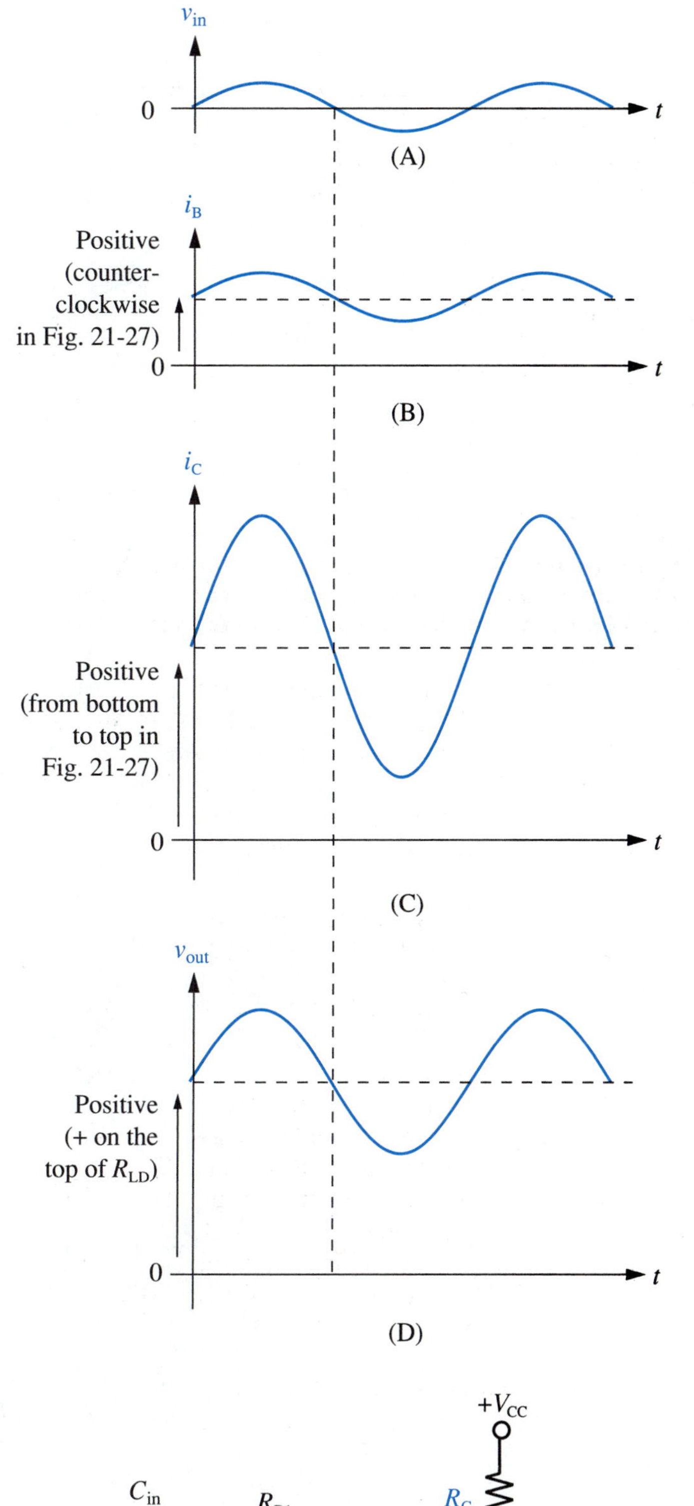

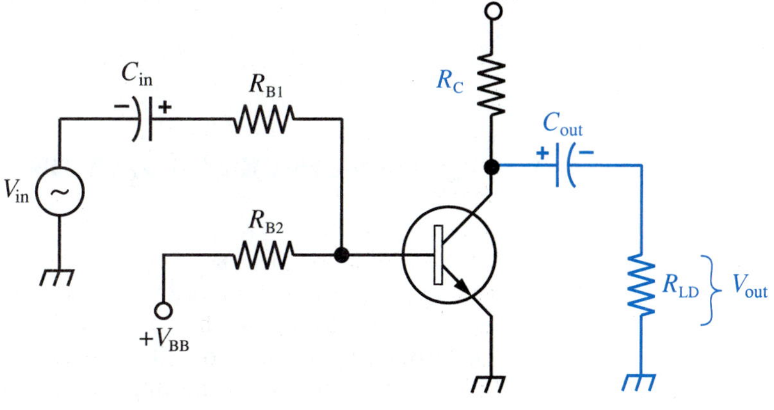

FIGURE 21-29 Removing the dc component of load voltage by changing the location of R_{LD}

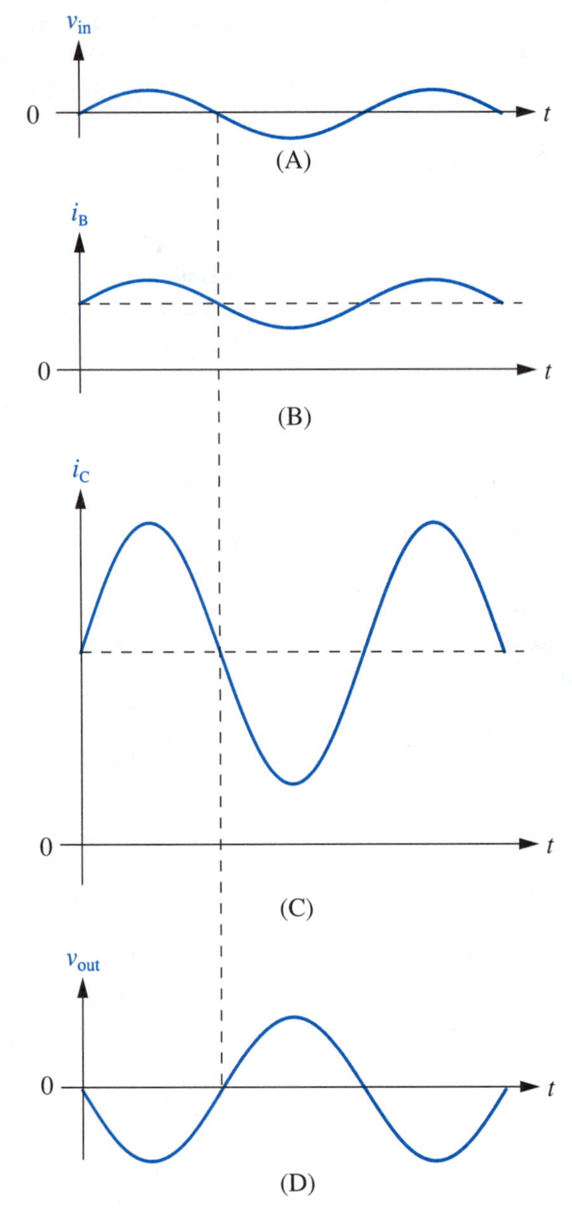

FIGURE 21-30 Waveforms for common-emitter *npn* amplifier with capacitor-coupled output. The v_{out} ac waveform is inverted with respect to the input.

bottom terminal. This is because a more positive than average value for i_C means a greater electron current flowing through R_C from bottom to top. As R_C becomes more negative on its bottom terminal, that ac variation is passed right through C_{out}, pulling the top terminal of R_{LD} negative. Therefore, the positive half-cycle of v_{in} produces a negative half-cycle of v_{out}. And vice-versa.

We say that a common-emitter amplifier produces an inverted output signal. In most applications, this inversion is acceptable. In those applications where it is not acceptable, we can pass the first output signal through a second common-emitter amplifier to reinvert it.

✔ SELF-CHECK FOR SECTION 21-3

6. What is the difference between an active device and a passive device? [6]
7. In a bipolar junction transistor, the output current is controlled by the __________. [8]
8. The words *gain* and__________mean the same thing. [6, 11]
9. Draw the schematic symbol for an *npn* transistor; label the three terminals. [7]
10. Repeat Problem 9 for a *pnp* transistor. [7]

11. The most popular operating method for a bipolar junction transistor is with the __________ connected to circuit ground. This is called common-__________. [8]
12. T-F In a bipolar junction transistor, the density of impurity atoms in the emitter region is much greater than the density of impurity atoms in the base region. [8]
13. In Figure 21-25, C_{in} has a capacitance of 5 μF. We have regarded this as a 0-Ω direct connection, for 1 kHz ac. Calculate X_{Cin} and compare it to R_{B1}. Is the 0-Ω assumption proper? [12]
14. In Figure 21-25, suppose that a different transistor is substituted, with $\beta = 175$. Repeat Examples 21-4, 21-5, and 21-6. [11]
15. For an *npn* transistor, the dc bias voltage on the base must be__________with respect to the emitter. (positive or negative) [9]
16. For an *npn* transistor, the dc bias on the collector must be__________with respect to the base. (positive or negative) [9]
17. In a common-emitter amplifier, how do we remove the dc component of voltage from the load waveform? [12]
18. When the method of Question 17 is used, the output voltage waveform is __________ phase with the input voltage waveform. [11, 12]

21-4 VOLTAGE-OPERATED TRANSISTORS

For the bipolar transistors in the previous section, the base-to-emitter *current* controlled what happened in the output circuit. The amount of i_B controlled the amount of i_C, and the interaction of i_C with the resistance in the collector circuit determined v_{OUT}, by Ohm's law.

In this section, we will study transistors in which the *voltage* between the input terminal and ground controls what happens in the output circuit. Such transistors are called **field-effect transistors,** or **FETs.** This can be pronounced as the letters themselves, eff-ee-tee, or it can be pronounced fet, as in get.

There are two fundamentally different kinds of FETs. They are called:

1. The junction-FET, or JFET.
2. The metal-oxide semiconductor FET, or MOSFET. Another name for MOSFET is insulated-gate FET, or IGFET.

We will look only at JFETs. You will study MOSFETs in a later course. A JFET has three terminals, like a BJT. The terminals are called **drain** (D), **gate** (G), and **source** (S). There is a close equivalence between the three JFET terminals and the three BJT terminals. This is shown in Table 21-2.

TABLE 21-2 Equivalences between FET Terminals and BJT Terminals

FIELD-EFFECT TRANSISTOR FET	BIPOLAR JUNCTION TRANSISTOR BJT
drain (D)	collector (C)
gate (G)	base (B)
source (S)	emitter (E)

The schematic symbols for the two polarities of JFETs are given in Figure 21-31.

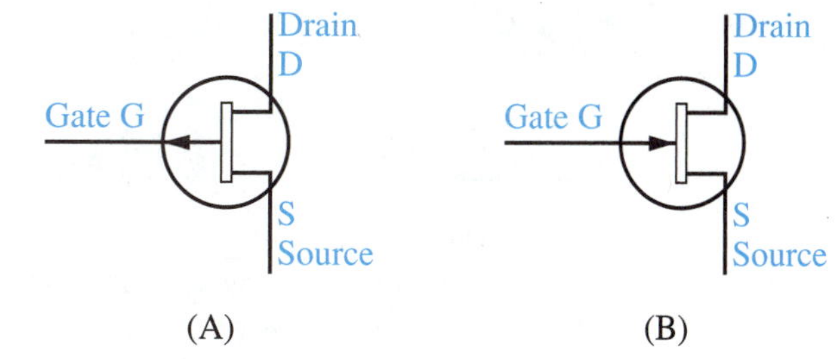

FIGURE 21-31 JFET schematic symbols and terminal names. (A) *p*-channel (B) *n*-channel

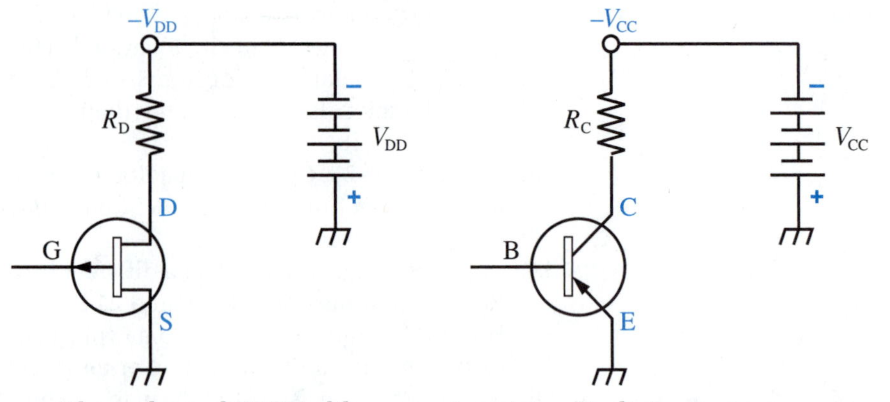

FIGURE 21-32 A *p*-channel FET is dc-biased on its main current terminals like a *pnp* BJT.

The *p*-channel JFET is like a *pnp* transistor. Its drain-source power-supply polarity is the same as a *pnp*'s. Namely, it is negative on the top (drain) terminal, and positive ground on the bottom (source) terminal. This similarity is pointed out in Figure 21-32. In Figure 21-32, both transistor symbols are drawn in the usual way, with the input terminal on the left side. One way to remember that a *p*-channel FET is like a *pnp* BJT is that their arrows both point from right to left. Of course, this only works when the input terminals are in their usual positions, on the left.

Inside a JFET, there is one *p-n* junction. For now, it is best to think of the *p-n* junction as existing between the gate and the source. We will not draw the actual internal structure of a JFET at this time. The arrow in the JFET schematic symbol indicates which side of the junction is a *p*-type region, and which side is *n*-type. Thus, in Figure 21-31(A), the p-channel JFET has a *p*-type source and an *n*-type gate, since the "diode" arrow points from source to gate.

In Figure 21-31(B), the *n*-channel JFET has a *p*-type gate and an *n*-type source, since the arrow points from gate to source. It is helpful to remember that the channel name of the JFET gives the type of the source region. That is, in a *p*-channel JFET, the source is a *p*-type region. In an *n*-channel JFET, the source is an *n*-type region.

An *n*-channel JFET is like an *npn* transistor, in terms of dc power supply polarity. This is shown in Figure 21-33. Again, both arrows point in the same direction, left to right, when the schematic symbols are drawn in the usual way.

Gate Biasing. A junction FET is quite different from a bipolar transistor when it comes to the dc bias voltage required on its gate input terminal.

TECHNICAL FACT

For a JFET, the gate-source *p-n* junction must be reverse-biased (blocking). It must never be forward-biased (conducting) like a bipolar transistor.

Thus, the *p*-channel FET in Figure 21-32 must have a positive dc voltage applied to its gate in order to make the gate-source junction reverse-biased. This is shown in Figure 21-34(A).

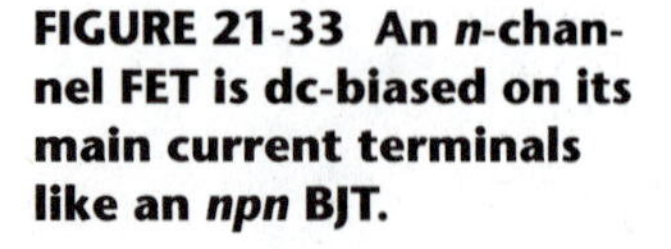

FIGURE 21-33 An *n*-channel FET is dc-biased on its main current terminals like an *npn* BJT.

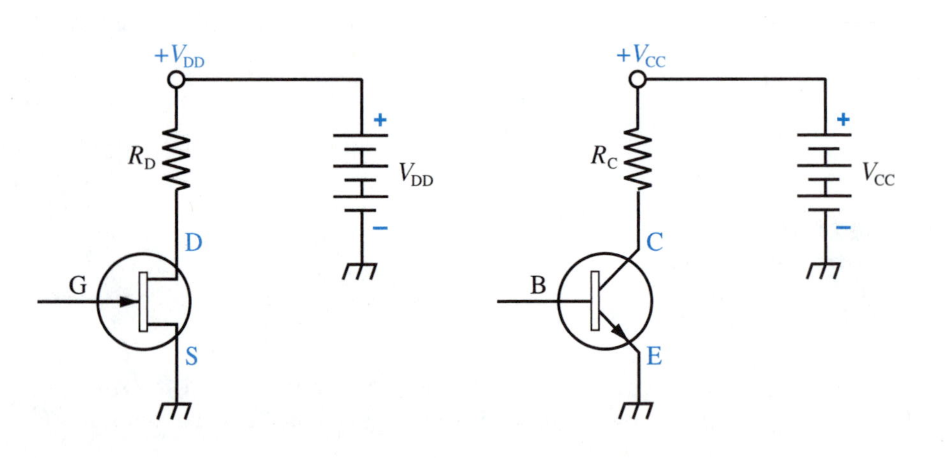

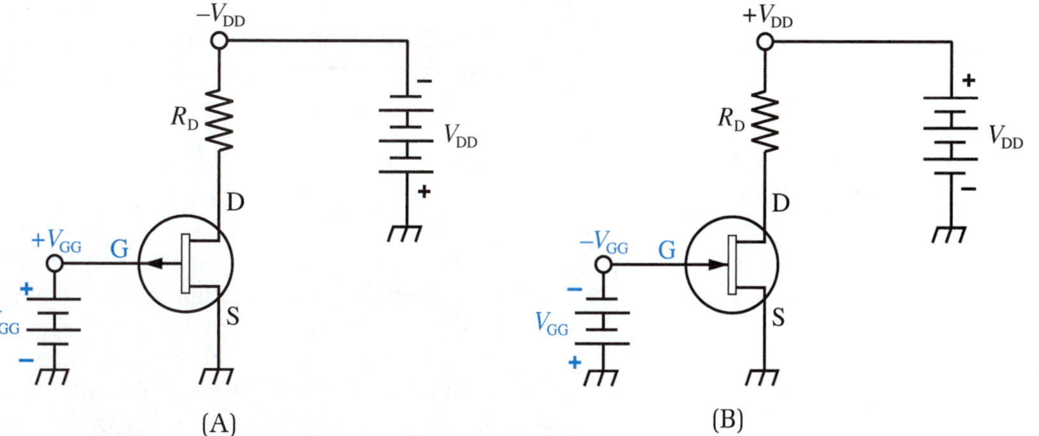

FIGURE 21-34 Showing the gate dc bias along with the drain dc bias for a JFET. (A) *p*-channel JFET. (B) *n*-channel JFET

14

The *n*-channel FET of Figure 21-33 must have a negative dc voltage applied to its gate. This makes its gate-source junction reverse-biased, as indicated in Figure 21-34(B).

Input-Output Relationship. An FET is a voltage-operated transistor. This means that the voltage between the gate and the source (V_{GS}) controls the current in the output circuit.

TECHNICAL FACT

An FET becomes a current source, like a BJT. The FET's drain current, I_D, is completely determined by V_{GS}. I_D is independent of drain resistance, R_D, independent of any voltage that appears in the drain circuit (within limits), and independent of everything else in the entire circuit, except V_{GS}. This is what is meant by a voltage-operated transistor.

15

Note the fundamental difference between an FET and a BJT. For a BJT, it was the current I_B from the base to the common terminal that controlled the current source in the output circuit. Now, for an FET, it is the voltage V_{GS} between the gate and the common terminal that controls the current source in the output circuit.

For an FET, the equation that relates the output current, I_D, to the controlling voltage, V_{GS}, is not a simple one. This makes FETs a bit harder to grasp than bipolar transistors, where the $I_C = \beta I_B$ equation is very simple.

Refer to Figure 21-35 to understand the relationship between V_{GS} and drain current, I_D. The FET in Figure 21-35(A) is an *n*-channel unit, so V_{DD} is positive and

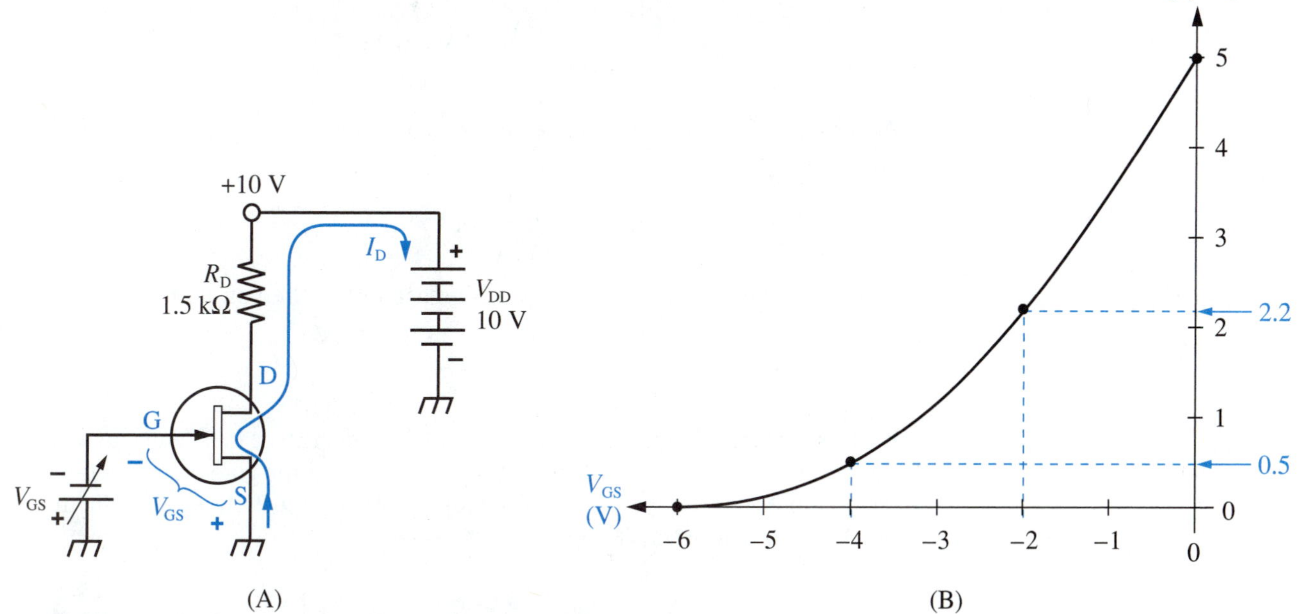

FIGURE 21-35 Showing the effect of input voltage, V_{GS}, on output current, I_D. (A) Test circuit. (B) Graph of I_D versus V_{GS}. The more negative V_{GS} becomes, the smaller I_D becomes.

V_{GS} is negative. Both voltages are measured with respect to the grounded source. The schematic diagram shows a variable V_{GS} voltage source.

TECHNICAL FACT

In a JFET, a small magnitude of V_{GS} (a value close to zero) produces a large amount of drain current, I_D. A large magnitude of V_{GS} (a value further from 0 V) produces a smaller amount of I_D.

A typical example of this relationship between I_D and V_{GS} is graphed in Figure 21-35(B). Four specific points are identified on that graph. Let us explain them.

1. $V_{GS} = -2$ V.
   At $V_{GS} = -2$ V, the drain current is 2.2 mA. For the *n*-channel FET of Figure 21-35(A), the drain current flows by this path: From ground into the source terminal, S, of the FET. It then passes through the body of the FET and emerges at the drain terminal, *D*. From there, it passes through resistor R_D to the V_{DD} source. I_D develops a certain voltage across R_D that is negative on the bottom, positive on the top. In this example, the voltage V_{RD} is given by Ohm's law as

$$V_{RD} = I_D R_D$$
$$= 2.2 \text{ mA } (1.5 \text{ k}\Omega)$$
$$= 3.3 \text{ V}$$

   By Kirchhoff's voltage law, the voltage on the drain relative to the source, V_{DS}, can be found.

$$V_{DS} = V_{DD} - V_{RD}$$
$$= 10 \text{ V} - 3.3 \text{ V}$$
$$= 6.7 \text{ V}$$

2. $V_{GS} = -4$ V.
   If V_{GS} changes to −4 V, the drain current decreases to $I_D = 0.5$ mA. This is shown clearly on the graph of Figure 21-35(B). The smaller drain current causes a smaller voltage to be developed across R_D.

$$V_{RD} = I_D R_D$$
$$= 0.5 \text{ mA } (1.5 \text{ k}\Omega)$$
$$= 0.75 \text{ V}$$

   V_{DS} is then given by

$$V_{DS} = V_{DD} - V_{RD}$$
$$= 10 \text{ V} - 0.75 \text{ V}$$
$$= 9.25 \text{ V}$$

   As the controlling voltage changed by 2 volts (from −2 V to −4 V), the output variables changed by these amounts:

$$\Delta I_D = \text{change in } I_D = 0.5 \text{ mA} - 2.2 \text{ mA}$$
$$= -1.7 \text{ mA}$$

   where the minus sign just means a decrease rather than an increase.

$$\Delta V_{DS} = 9.25 \text{ V} - 6.7 \text{ V}$$
$$= 2.55 \text{ V}$$

3. $V_{GS} = -6$ V.
   For this particular FET, if V_{GS} changes to −6 V, I_D decreases to 0. The value of V_{GS} that forces the drain current to stop ($I_D = 0$) is called the *gate-source cutoff voltage*, symbolized $V_{GS(off)}$. It is one of the important specifications that distinguishes one FET from another.

   Now,

$$V_{RD} = (0 \text{ mA})(1.5 \text{ k}\Omega) = 0 \text{ V}$$

$$V_{DS} = 10 \text{ V} - 0 \text{ V} = 10 \text{ V}$$

Changing from V_{GS} = –4 V to V_{GS} = –6 V is the same amount of change that we went through a moment ago, as V_{GS} went from –2 V to –4 V. In both cases, V_{GS} became more negative by 2 V. But the output variables have not changed by the same amount. Table 21-3 demonstrates this.

TABLE 21-3 Changing the Input Controlling Variable of an FET

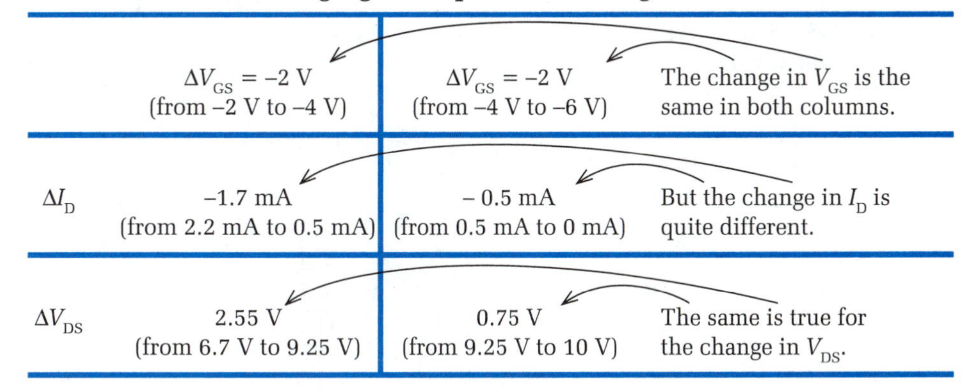

	ΔV_{GS} = –2 V (from –2 V to –4 V)	ΔV_{GS} = –2 V (from –4 V to –6 V)	The change in V_{GS} is the same in both columns.
ΔI_D	–1.7 mA (from 2.2 mA to 0.5 mA)	– 0.5 mA (from 0.5 mA to 0 mA)	But the change in I_D is quite different.
ΔV_{DS}	2.55 V (from 6.7 V to 9.25 V)	0.75 V (from 9.25 V to 10 V)	The same is true for the change in V_{DS}.

Note: For an FET, a given amount of change (2 V) in the input controlling variable, V_{GS}, does not produce a constant amount of charge in the output variable, I_D. Output voltage V_{DS} then follows I_D.

When we see an input-output relationship like the one in Table 21-3, we say that the device is *nonlinear*. For a nonlinear device, graphing the output versus the input does not produce a straight line. Of course, we knew this coming in, as soon as we saw Figure 21-35(B).

A circuit containing a nonlinear device like an FET is harder to understand and analyze than one containing a linear device like a bipolar transistor. In many applications, nonlinearity is a practical disadvantage. But there are some applications where it can actually be an advantage.

4. V_{GS} = 0 V.

A JFET requires that the gate-source junction must never be forward-biased. V_{GS} is allowed to become 0 V though, because that gives no bias at all. For the FET in Figure 21-35, V_{GS} = 0 V causes an I_D value of 5 mA. This is the FET's maximum possible value of drain current. It is called *shorted-gate drain current,* symbolized I_{DSS}. It is another one of the specifications that distinguishes one FET from another.

If you want, you can prove that the 2-volt change from V_{GS} = 0 V to V_{GS} = –2 V produces much more drastic changes in the output variables than either of the 2-volt changes in Table 21-3.

For a *p*-channel JFET, the polarities of the power supplies in Figure 21-35(A) will be reversed. Naturally, I_D will flow in the opposite direction, through the FET from top to bottom.

Applying the Ac Input Signal to the Gate. To combine an ac input signal with the gate's dc bias voltage, we must do two things.

1. Place a gate resistor, R_G, between the V_{GG} source and the G terminal.
2. Couple the ac input voltage through an input capacitor to the G terminal.

These steps have been taken in Figure 21-36.

Input coupling capacitor C_{in} accomplishes the same things that it did for a bipolar transistor amplifier. It blocks any dc current from V_{GG} that might otherwise flow through the V_{in} voltage source, while it passes the ac input voltage directly to the gate input terminal.

Resistor R_G is necessary so that V_{in} is not shorted to ground through V_{GG}. This issue of how a dc voltage source appears to an ac signal is a very important issue. It will come up again and again in your further study of electronic amplifiers.

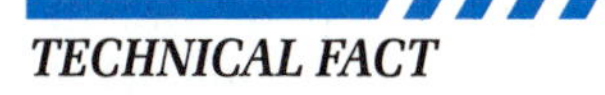

TECHNICAL FACT

An ac signal views the ungrounded terminal of a dc voltage source (power supply) as if it were shorted to ground.

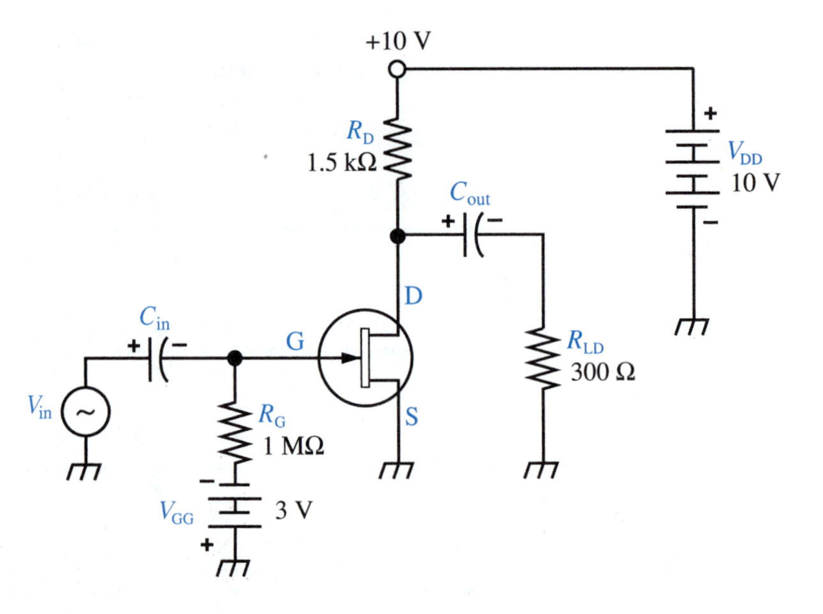

FIGURE 21-36 Complete FET common-source amplifier

There are several different ways of explaining this fact. One explanation is this: Since the voltage at the ungrounded source terminal is guaranteed to be a certain unchanging dc value, it is impossible for any ac voltage to exist on that terminal. The only circuit arrangement that absolutely prevents any voltage from existing at a point is a short circuit from that point to ground.

In Figure 21-36, a large value has been chosen for R_G. Here is the reason for that.

TECHNICAL FACT

The chief advantage of an FET amplifier is that its **input resistance** is very high. This means that the amplifier's input terminal presents a great amount of resistance to the signal source, thereby taking very little current from the signal source.

17

Many real-life signal sources are incapable of delivering very much current. Such signal sources need an amplifier with high input resistance.

In Figure 21-36, the input signal, V_{in}, views C_{in} as virtually a 0-Ω connection to the G terminal. This is because the size of the capacitor is selected to make reactance, X_{Cin}, very low. When the signal reaches the right side of C_{in}, it has two possible paths to ground:

1. Through the gate-source junction of the JFET.
2. Through R_G, then straight through V_{GG} to ground.

Remember, V_{GG} acts like a short circuit to ground, for an ac signal.

These two paths are in parallel with each other, from the viewpoint of the ac signal. Therefore, the amplifier's input resistance, R_{in}, is given by the reciprocal formula

$$\frac{1}{R_{in}} = \frac{1}{R_{\text{p-n junction}}} + \frac{1}{R_G}$$

EQ. 21-10

The first path is a reverse-biased *p-n* junction, which ideally has infinite resistance. In reality, its resistance is usually several thousand million ohms (several gigohms). This is such a huge value that it might as well be ∞.

Because R_G is in parallel with virtually ∞ Ω, the amplifier's overall R_{in} is virtually equal to R_G. As stated earlier, one of the main reasons for having an FET amplifier is to get a high value of R_{in}. So in Figure 21-36, we choose a high value of R_G, 1 MΩ.

Producing the Ac Output Signal in the Drain Circuit. We now realize that a dc supply behaves like a short circuit to ground for high-frequency ac.

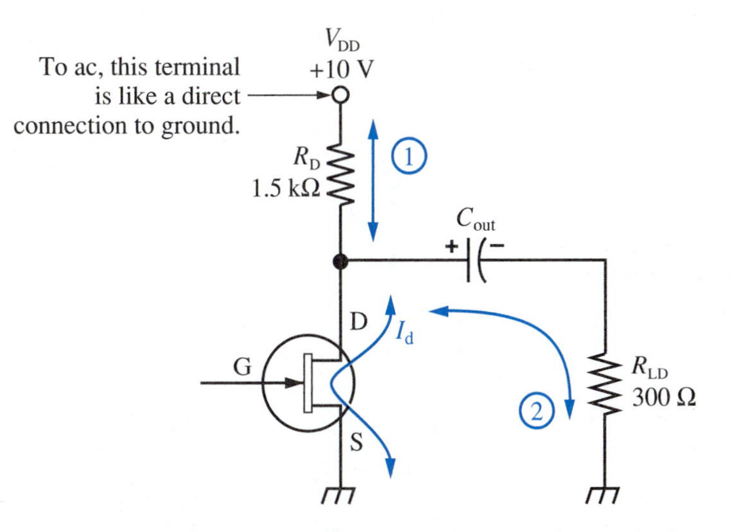

FIGURE 21-37 Developing V_{out} by passing I_d through the equivalent ac resistance seen from the drain, symbolized r_D.

Knowing that, let us take a close look at what happens in the drain circuit of Figure 21-36. Drain current, i_D, is a combination of ac and dc, just like i_C in a bipolar transistor amplifier. Only the ac part of i_D is shown in the partial schematic of Figure 21-37.

As ac current I_d emerges from the drain terminal, it sees two parallel paths to ground. They are labeled in Figure 21-37 as:

1. Through R_D, then right through the V_{DD} supply to ground. In fact, you can think of the V_{DD} terminal itself as an "ac ground."
2. Through C_{out} (X_{Cout} is very low, virtually 0 Ω), then through R_{LD} to ground.

TECHNICAL FACT

For an FET, the drain ac current source "sees" an equivalent ac resistance of $R_D \| R_{LD}$. This equivalent ac resistance is symbolized r_D, with a lower-case r. Therefore, $r_D = R_D \| R_{LD}$, or

$$\frac{1}{r_D} = \frac{1}{R_D} + \frac{1}{R_{LD}} \qquad \text{EQ. 21-11}$$

Equation 21-11 can also be written

$$r_D = \frac{R_D(R_{LD})}{R_D + R_{LD}} \qquad \text{EQ. 21-12}$$

Since the ac I_d current is not all passing through R_D, it does not produce an output voltage given by $V_{out} = I_d R_D$. Instead, the ac output voltage is given by

$$V_{out} = I_d\, r_D \qquad \text{EQ. 21-13}$$

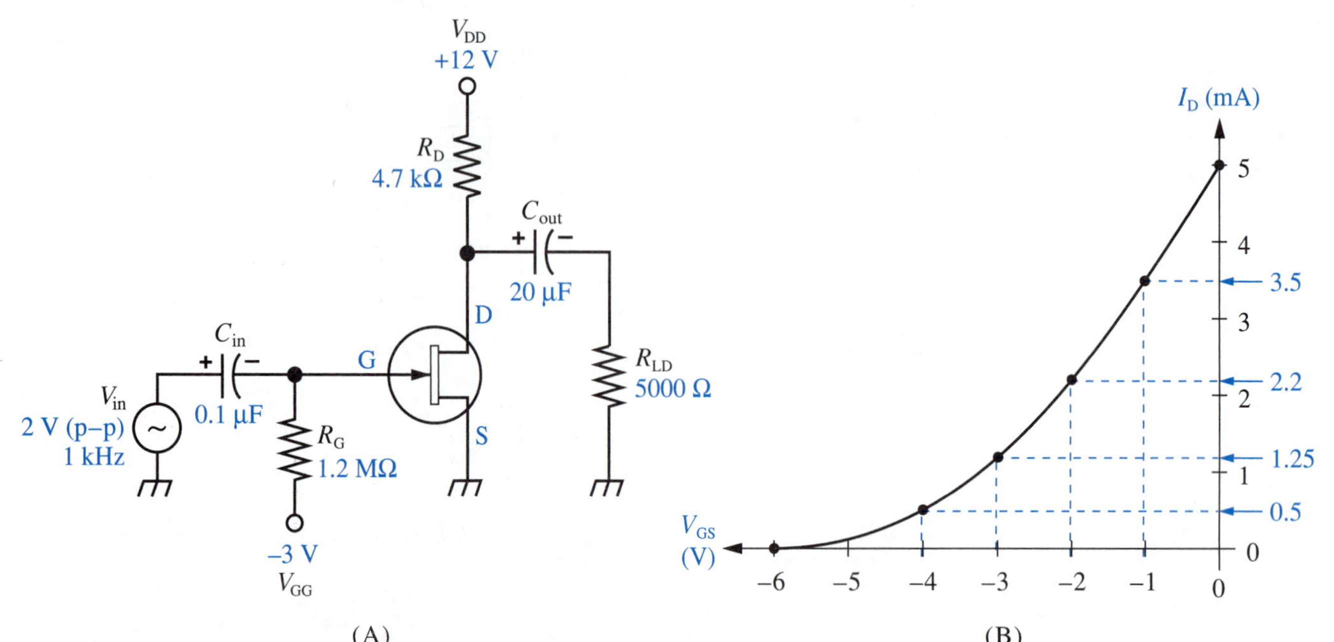

FIGURE 21-38 (A) Complete schematic diagram of a JFET common-source amplifier. A JFET, like a BJT, has two other operating modes, common-gate and common-drain. (B) Graph of I_D versus V_{GS} for this FET.

EXAMPLE 21-7

For the FET amplifier shown in Figure 21-38(A):

a) Find the amplifier's input resistance, R_{in}.
b) Describe the v_{GS} voltage oscillations. What actual values does v_{GS} oscillate back and forth between?
c) Describe the total i_D oscillations. What actual values does i_D oscillate back and forth between?
d) Describe I_d, the ac part of total i_D .
e) What value of ac output voltage, V_{out}, is produced?

SOLUTION

a) With V_{GG} = –3 V, V_{GS} reverse-biases the gate-source *p-n* junction. Therefore, its resistance is virtually infinite. From Equation 21-10,

$$\frac{1}{R_{in}} = \frac{1}{R_{\text{p-n junction}}} + \frac{1}{R_G} = \frac{1}{\infty} + \frac{1}{R_G}$$

$$R_{in} = R_G = \mathbf{1.2\ M\Omega}$$

b) V_{in} is 2 volts peak-to-peak. This is superimposed on (combined with) the V_{GG} value of –3 V. This –3 V is the V_{GS} bias value. It produces a dc bias drain current of 1.2 mA, as indicated in Figure 21-38(B). When v_{in} swings instantaneously positive 1 V, v_{GS} swings to –2 V, since –3 V + 1 V = –2 V. When v_{in} swings instantaneously negative 1 V, v_{GS} swings to –3 V – 1 V = –4 V. Thus, v_{GS} oscillates back and forth **between –2 V and –4 V.**

c) From the FET characteristic curve of Figure 21-38(B), we see that i_D = 2.2 mA for v_{GS} = –2 V, and i_D = 0.5 mA for v_{GS} = –4 V. Therefore, i_D oscillates back and forth **between 2.2 mA and 0.5 mA.**

d) I_d has a peak-to-peak value given by

$$2.2\text{ mA} - 0.5\text{ mA} = \mathbf{1.7\ mA}$$

Continued on page 435.

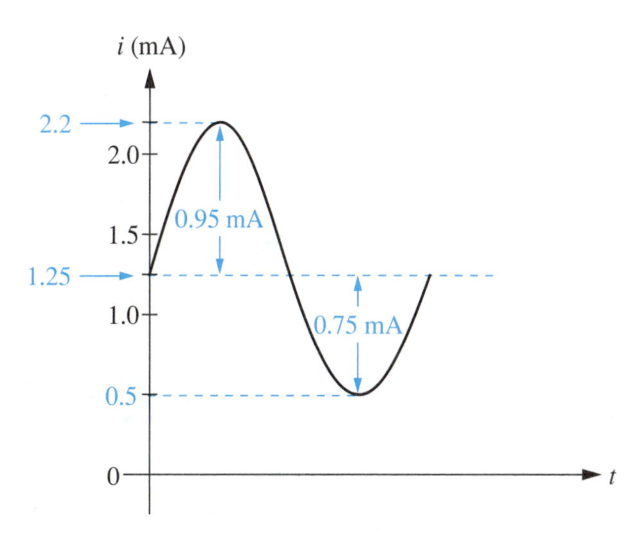

FIGURE 21-39 Waveform of i_D in Example 21-7. The dc-bias value of I_D (the value that would occur if no input signal were applied) is 1.25 mA. This can be found from the graph of Figure 21-38(B), using $V_{GS} = V_{GG} = -3$ V.

However, I_d **is not a sine wave.** Instead, it will have a positive peak that is greater than its negative peak, as shown in Figure 21-39. This is caused by the nonlinear input-output behavior of the FET. As mentioned before, this may not be acceptable, depending on the nature of the application.

e) V_{out} will have the same shape as I_d in Figure 21-39. Its peak-to-peak value is given by Ohm's law, Equation 21-13.

$$V_{out} = I_d \, r_D = (1.7 \text{ mA}) \, r_D$$

The equivalent ac drain resistance can be found by Equation 21-11 or Equation 21-12. Using Equation 21-12,

$$r_D = \frac{R_D \, R_{LD}}{R_D + R_{LD}} = \frac{(4700)\,(5000)}{4700 + 5000} = 2423 \ \Omega$$

Then

$$V_{out} = (1.7 \text{ mA})\,(2423 \ \Omega)$$
$$= \mathbf{4.12 \ V \ (peak\text{-}to\text{-}peak)}$$

20

The output of a bipolar junction transistor would not have the signal **distortion** that is shown in the JFET waveform of Figure 21-39. This is because the BJT is a linear device. In truth though, linearity is always a matter of degree. If a bipolar transistor were forced to go through very large output swings, it might exhibit some small amount of distortion. On the other hand, if the signal were very small in the FET amplifier, the output distortion would be very slight, perhaps negligible. We often distinguish between these two operating conditions by the terms *large-signal operation* (some distortion is likely) and *small-signal operation* (any distortion will be minimal).

SELF-CHECK FOR SECTION 21-4

19. What is the fundamental operating difference between a field-effect transistor and a bipolar junction transistor? [14]
20. Draw the schematic symbol for a *p*-channel JFET and label the terminals. Repeat for an *n*-channel JFET. [13]
21. For a JFET, the__________terminal is like the collector terminal of a BJT. [13]
22. For a JFET, the__________terminal is like the base terminal of a BJT. [13]
23. For a JFET, the__________terminal is like the emitter terminal of a BJT. [13]
24. For a JFET, what variable controls the output current, I_D? [14]

25. Define gate-source cutoff voltage, $V_{GS(off)}$. [14]
26. When V_{GS} becomes smaller in magnitude (closer to 0 V), I_D becomes __________(larger or smaller). [14]
27. For a__________device, graphing the output versus the input produces a straight line. [8]
28. For a__________device, graphing the output versus the input produces a curve, not a straight line. [14, 20]
29. Define shorted-gate drain current, I_{DSS}. [14]
30. In an amplifier circuit, a dc voltage source appears to an ac signal like a __________ to ground. [16]

21-5 INTEGRATED CIRCUITS

Many bipolar junction transistors and field-effect transistors are manufactured and sold as single units, one in a package. All the transistors in Figure 21-15 are like that. Using microphotography techniques, manufacturers can also construct multiple-transistor circuits in a single package. Entire circuits are connected internally in the package, complete with all resistors, capacitors, and diodes. Such circuits are called **integrated circuits** (ICs).

Elaborate multiple-transistor circuits are rather expensive to build using individually packaged transistors (called **discrete** transistors), along with individual supporting components. Also, the greater the number of electrical connections

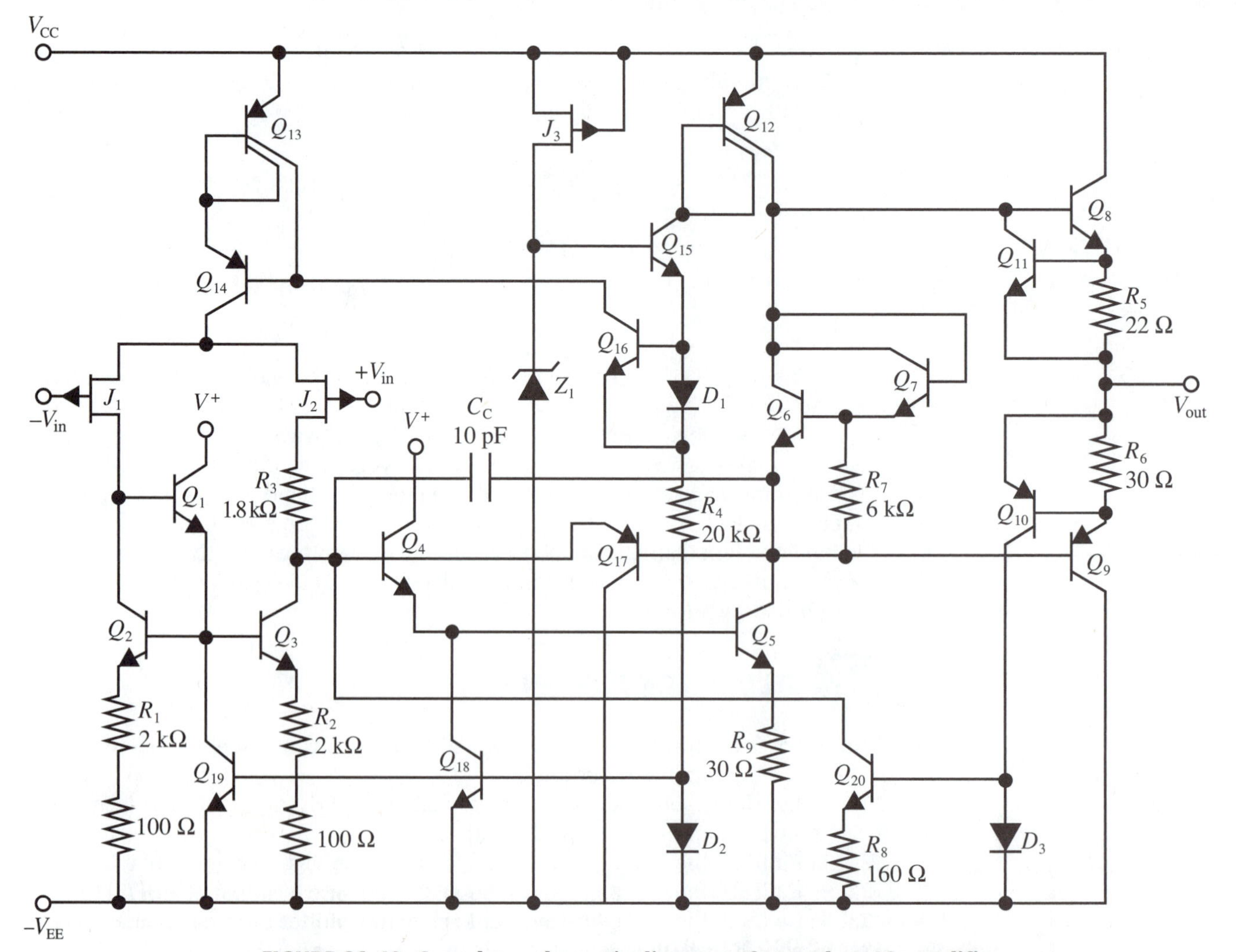

FIGURE 21-40 Complete schematic diagram of a modern IC amplifier

involved, the greater the chances for trouble. But such circuits can be built very inexpensively and reliably as integrated circuits.

For example, the circuit shown in Figure 21-40 contains 20 bipolar transistors, 3 *p*-channel JFETs, 4 diodes, 11 resistors, and 1 capacitor. It is an amplifier with pretty impressive specifications: voltage gain, $A_V = 100\ 000$; input resistance, $R_{in} = 1000\ G\Omega$; bandwidth 0 Hz (dc) to 4 MHz; in other words, there is no f_{low}. The frequency-response curve of A_V versus f is flat all the way down to 0 Hz.

This amplifier circuit can be bought at a price of two for forty-nine cents, as an IC. Figure 21-41(A) shows the eight-terminal IC package that contains two of these amplifiers. The user simply connects the power supplies and ground to the proper terminals, or pins, and connects the input signal between its proper pins. The amplified output signal immediately appears at the output pin. These IC amplifiers are also available in the round package shown in Figure 21-41(B).

All of the amplifier's components are integrated into a single silicon crystal, called a **chip**. Inside the eight-pin packages of Figures 21-41(A) or (B), very little of the space is taken up by the chip itself. Most of the space is taken up by the eight connecting wires that join the silicon chip to the eight pins. In this example, the chip itself is only about 2 mm x 2 mm (about 0.08 in x 0.08 in) in dimension. A chip of this complexity, containing fewer than 50 transistors in total, is classified as small-scale integration (SSI).

By enlarging the chip size, it is possible to get many more than 50 transistors onto the chip and into the package. Packages containing from 50 to 400 transistors are called medium-scale integrated circuits (MSI). Many digital circuits are MSI. They usually appear in 14-pin or 16-pin DIPs, like Figure 21-41(C).

(Digital circuits handle only two voltage levels, one level for 0 and one level for 1. This is unlike the amplifiers that we have studied in this chapter, which have continuously variable voltages.) Integrated circuits that handle sine waves and other continuously variable voltages are often called linear ICs, to distinguish them from digital ICs.

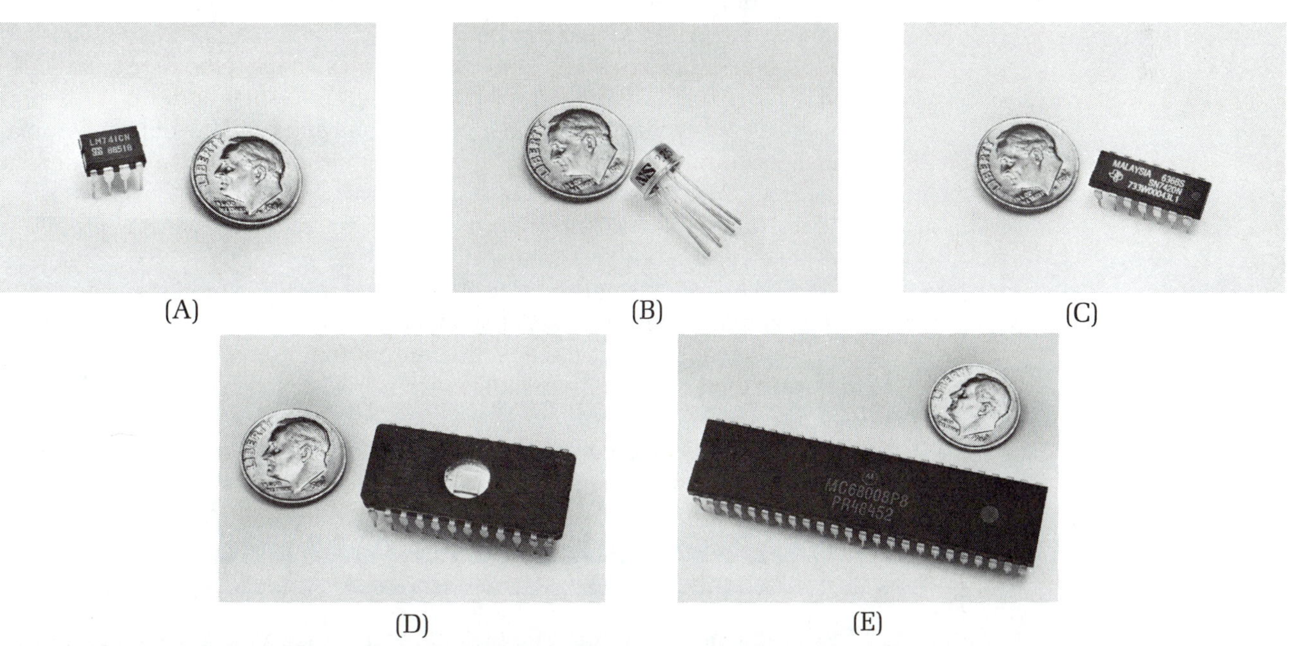

FIGURE 21-41 Integrated circuits. (A) Eight-pin package. Four pins are in line along one side, and four more along the other side. A package with this pin layout is called a dual-in-line package, or DIP. (B) Round eight-pin package, often called the "top-hat" package. (C) Fourteen-pin DIP. (D) Twenty-four-pin DIP containing an Erasable Programmable Read-Only Memory, or EPROM. This particular unit stores 16 384 bits of information (0 or 1), organized into groups of 8 bits each. The contents of this memory can be erased by shining ultraviolet light through the built-in window. (E) Forty-eight-pin DIP. This particular IC is a microprocessor, containing over 68 000 transistors.

As technology got better, it became possible to squeeze more and more transistors onto a chip. Now we refer to chips containing 400 to 10 000 transistors as large-scale integration (LSI). The digital memory device shown in Figure 21-41(D) is an example of LSI. Chips with over 10 000 transistors are referred to as very large-scale integration (VLSI). An example is shown in Figure 21-41(E).

FORMULAS

$I_c = \beta I_b$ **EQ. 21-2**

$A_I = \dfrac{I_{out}}{I_{in}}$ **EQ. 21-4**

$A_V = \dfrac{V_{out}}{V_{in}}$ **EQ. 21-5**

$A_P = \dfrac{P_{out}}{P_{in}}$ **EQ. 21-6**

$A_P = A_V A_I$ **EQ. 21-7**

$\dfrac{1}{r_D} = \dfrac{1}{R_D} + \dfrac{1}{R_{LD}}$ **EQ. 21-11**

$V_{out} = I_d r_D$ **EQ. 21-13**

SUMMARY OF IDEAS

- A diode conducts current in one direction only. If a diode is reverse-biased, it blocks current.
- A diode is able to rectify ac into pulsating dc.
- A bipolar junction transistor, or BJT, is an active device. When properly connected to dc power supplies, it can amplify power.
- The output current of a BJT is controlled by its input current.
- In the common-emitter operating mode, input current, I_b, flows between base and emitter ; output current, I_c, flows between collector and emitter. I_c is proportional to I_b, making the BJT a linear device.
- Coupling capacitors can be used to carry the ac signals at the input and at the output of a BJT or FET amplifier, while blocking the dc bias currents. This eliminates any dc bias voltage from the input terminal and from the output terminal.
- A transistor amplifier (BJT or FET) provides either voltage gain, or current gain, or both. In all cases, it provides power gain.
- A junction field-effect transistor, or JFET, is an active device whose output current, I_d, is controlled by its input voltage, V_{gs}.
- For an FET, I_d is not proportional to V_{gs}, making the FET a nonlinear device.
- A dc power supply terminal acts like an ac ground.
- The equivalent ac resistance seen at the output terminal of a transistor is equal to the load resistance, R_{LD}, in parallel with the resistance in the output lead (R_C or R_D).
- Integrated circuits, ICs, contain many transistors and supporting components connected together on a single chip of silicon.

CHAPTER QUESTIONS AND PROBLEMS

1. Draw a schematic diagram of a 10-V dc voltage source connected to a series combination of a 200-Ω resistor and an ideal diode, with the diode forward-biased. [1,2]
2. For the diagram in Problem 1: [3]
   a) Find the circuit current.
   b) Find the magnitude and polarity of V_R, the voltage across the resistor.
   c) Find the magnitude and polarity of V_D, the voltage across the diode.

3. Repeat Problems 1 and 2, but with the diode reverse-biased. [1, 2, 3]
4. Draw a schematic diagram of a transformer secondary winding with $V_{S(rms)} = 24$ V, driving an ideal diode and load resistor. Then draw the load voltage waveform, showing its peak value and its synchronization with the sine wave V_S waveform. [2, 4]
5. Repeat Problem 4, but with the diode reversed. [2, 4]

Problems 6 through 32 refer to bipolar junction transistors.

6. Draw the schematic symbol of a *pnp* transistor and give the terminal names and letter symbols. Repeat for an *npn* transistor. [7]
7. In the common-emitter mode, which transistor terminal receives the input signal? Which one delivers the output signal? [7, 8]
8. In any electronic amplifier, it will always be true that output__________ is greater than input__________. [6]
9. Another word for amplification is__________. [6]
10. When a bipolar transistor is properly biased, the base-emitter *p-n* junction will be__________-biased, but the collector-base *p-n* junction will be __________-biased. (Answer forward or reverse.) [9]
11. The collector output circuit of a transistor acts like a__________source. (voltage or current). [8]
12. For a BJT, the ratio of collector current, I_C, to base current, I_B, is symbolized ______. [8]
13. Explain the specific operating details of a current source. In your explanation, make it clear how a current source is different from a voltage source. [8]
14. Draw a complete schematic diagram of a transistor amplifier that fits the following description: [7, 9]
    - Transistor type — *npn*;
    - Configuration — common emitter;
    - $V_{CC} = 20$ V;
    - $V_{BB} = 6$ V, coupled through $R_{B2} = 91$ kΩ;
    - Input circuit — 1-kHz sine wave voltage source with $V_{in(p-p)} = 4$ V, coupled through $C_{in} = 0.27$ μF and $R_{B1} = 56$ kΩ;
    - Output circuit — $R_C = 820$ Ω; $R_{LD} = 1.5$ kΩ coupled through $C_{out} = 10$ μF (electrolytic).

Problems 15 through 32 refer to the amplifier that you have drawn for Problem 14.

15. Assuming that the base-emitter *p-n* junction is ideal, find the dc base bias current, I_B. [10]
16. Find the dc collector bias current, I_C. [10]
17. Find the dc voltage drop across R_C. [10]
18. What is the dc bias voltage at the collector terminal relative to ground? [10]
19. Prove that C_{in} has negligible reactance, compared to R_{B1}. [12]
20. Calculate the peak-to-peak ac base input current. Use Ohm's law, $I_b = V_{in} / R_{B1}$. [10]
21. Find the ac collector output current (p-p). Use $I_c = \beta I_b$. [10]
22. Calculate the equivalent ac collector resistance, r_C. [17]
23. Find V_c, the ac output voltage (p-p). Use Ohm's law, $V_c = I_c r_C$. [10]
24. Prove that C_{out} has virtually negligible reactance, compared to R_{LD}. [12]
25. T-F Output load voltage, symbolized V_{out} or V_{LD}, is the same as V_c found in Problem 23. [10, 11]
26. T-F At an instant when V_{in} is positive, V_{LD} is negative. [10]
27. What is the voltage gain, A_V, for this amplifier? [11]
28. Find the peak-to-peak ac load current, I_{LD}. Use Ohm's law. [10]
29. What is the current gain, A_I for this amplifier? [11]
30. Calculate the power gain for this amplifier, using Equation 21-7. [11]

31. Calculate the load power. Be sure to use rms values of current and/or voltage. You should get about 3.9 mW. [6, 10]
32. What is the true origin of this 3.9 mW output power? [6, 10]

Problems 33 through 54 refer to junction field-effect transistors.

33. Draw the schematic symbol of a *p*-channel JFET, and give the terminal names and letter symbols. Repeat for an *n*-channel JFET. [13]
34. In the common-source mode, which transistor terminal receives the input signal? Which one delivers the output signal? [13, 14]
35. When a JFET is properly biased, the *p-n* junction is__________– biased. (forward or reverse) [15]
36. The drain output circuit of a JFET acts like a__________source. (voltage or current) [14]
37. T-F For an FET, the controlling variable is gate-to-source voltage, V_{GS}. [14]
38. Referring to Question 37, describe the operating difference between an FET and a bipolar junction transistor. [8, 14]
39. T-F For a BJT, the input-output relationship between I_b and I_c is linear. [8]
40. T-F For a JFET, the input-output relationship between V_{gs} and I_d is linear. [14]

For Problems 41 through 54, use the circuit diagram of Figure 21-36, and the JFET characteristic curve of Figure 21-38(B), but with the following new component values:

- V_{DD} = 24 V;
- V_{GG} = –2 V;
- $V_{in(p-p)}$ = 2 V;
- R_G = 2.2 MΩ;
- R_D = 5.1 kΩ;
- R_{LD} = 250 Ω;
- C_{in} and C_{out} both have negligible reactances.

41. What is the amplifier's input resistance seen by the input voltage source? [18]
42. Find the dc drain bias current, I_D . [19]
43. Find the dc voltage drop across R_D. [19]
44. What is the dc bias voltage at the drain terminal relative to ground? [19]
45. Find the maximum and minimum instantaneous values of drain current, i_D . [19]
46. Draw a waveform graph of i_D versus time. Show the oscillations about the I_D value that you found in Problem 42. [19]
47. T-F The result from Problem 46 shows a true sine wave ac superimposed on dc. [20]
48. Calculate the equivalent ac drain resistance, r_D. [17]
49. Find V_d, the ac output voltage (p-p). Use Ohm's law, $V_d = I_d\, r_D$. [19]
50. T-F At an instant when V_{in} is positive, V_{LD} is negative. [19]
51. Using the peak-to-peak value from Problem 49, what is the voltage gain, A_V, for this amplifier? [19]
52. Find the current demand from the signal source by applying Ohm's law with the R_{in} value from Problem 41. [18, 19]
53. Get the peak-to-peak drain current value from Problem 45. Using that value and the peak-to-peak current from Problem 52, find the current gain, A_I, for this amplifier. [19]
54. Calculate the power gain for this amplifier, using Equation 21-7. [19]

GLOSSARY

Ac current: Current that changes direction periodically.

Ac voltage: Voltage that changes polarity periodically.

Alternator: A machine that uses magnetism to generate a sine-wave ac voltage when its shaft is rotated.

Ammeter: The electrical instrument that measures current.

Ampere (A): The basic measurement unit for current. One ampere is one coulomb per second.

Amplifier: A circuit that provides power gain.

Analog: Circuit or measuring instrument that allows continuously variable values of voltage and current. As distinguished from digital, which allows changes in voltage to occur only in definite jumps.

Anode (A): The side of a diode that must be positive to make the diode conduct. Electron current exits from the anode.

Apparent power (*S*): The product of rms voltage times rms current, without regard to their phase relation. Measured in units of voltamperes.

Arc: A short-lived burst of current through the air between two switch surfaces, occurring at the moment the switch opens.

Arc cosine: The angle that has a particular given value as its cosine.

Armature winding: The winding in an alternator that induces the ac output voltage. In general, for any electromechanical machine, it is the winding that produces the machine's output product (voltage for a generator, or torque for a motor). Sometimes the armature is on the rotor and sometimes it is on the stationary part of the machine.

Atomic particle: Tiny part of an atom. The three particles are electrons, protons, and neutrons.

Audio: Referring to frequencies within the range of human hearing, from about 100 Hz to about 15 kHz.

AWG number: American wire gage. A number indicating the gage, or thickness, of a conductor. Lower numbers indicate heavier (thicker) conductors.

Balanced: The condition of a Wheatstone bridge that causes the output voltage to be zero. Means the same thing as nulled.

Band-pass filter: A filter circuit that passes a signal to the load if the signal frequency is between f_{low} and f_{high}.

Band-stop filter: A filter circuit that prevents a signal from reaching the load if the signal frequency is between f_{low} and f_{high}.

Bandwidth (*Bw*): The range of frequencies that produce at least 70.7% of the maximum response from the circuit. Equal to the difference between f_{high} and f_{low}.

Base (B): The lead of a BJT that receives the input signal in the standard common-emitter configuration.

Base bias current (I_B): The dc current that passes into the base lead of a BJT, before any signal current occurs.

Base-emitter junction: The *pn* junction between the base region and the emitter region of a BJT.

Battery: A voltage source that works by chemical reactions taking place at its electrodes.

Battery charger: An electric source that can recharge a battery made of secondary cells.

Beta (β): The ratio of collector current, I_C, to base current, I_B, for a BJT. Beta is the current gain of a BJT in common-emitter configuration.

Bipolar junction transistor (BJT): The original current-operated transistor, consisting of three alternately doped semiconductor regions.

Branch: A piece of a larger circuit containing either a single resistor or a group of series resistors.

Capacitance (*C*): The electrical quality that specifies how much charge must be stored to produce a given value of voltage.

Capacitive reactance (X_C): A capacitor's opposition to current in an ac circuit.

Capacitor: Electrical device having capacitance. It consists of two metal plates separated by a dielectric layer.

Cathode (K): The side of a diode that must be negative to make the diode conduct. Electron current enters on the cathode.

Cell: An individual chemical unit of a battery, which usually contains several cells in series.

Charge (*Q*): The electric property of electrons and protons that produces attraction and repulsion forces.

Charging, a capacitor: The process of moving charge from one plate to the other of a capacitor, thereby establishing voltage across the capacitor.

Chassis: Usually, the metal enclosure that holds a circuit and often serves as the electrical ground point. For an automobile, the chassis is the structural metal that furnishes the return path for current to the battery or alternator.

Chip: The piece of silicon that is the base material for the integration of transistors and their support devices, forming an integrated circuit.

Circuit breaker: A device that protects against overcurrents by opening a reclosable contact if an overcurrent is sensed.

Closed: The condition of a switch when the switch is providing a current path between two terminals.

Coaxial cable: Cable with a single solid conductor in the center, surrounded by a dielectric material with carefully controlled capacitive characteristics, surrounded by metal braid, which serves as the common conductor.

Collector (C): The terminal of a BJT from which the output signal is taken, in standard common-emitter configuration.

Collector-base junction: The *pn* junction between the collector region and the base region of a BJT.

Common: The electrical point in a circuit with respect to which all voltages are measured.

Common-emitter configuration, or amplifier: A BJT amplifier in which the emitter is at ground potential, and is common to both the input voltage and the output voltage.

Common-source configuration, or amplifier: An FET amplifier in which the source terminal is at ground potential, and is common to both the input voltage and the output voltage.

Common terminal: The terminal of a double-throw switch contact that is shared by the N.O. contact and the N.C. contact.

Common wire: In a 120-V ac circuit, the wire that is at approximately earth potential and that is carrying the load current. It has white insulation.

Conductance (*G*): The measure of a device's ability to pass current. It is measured in basic units of siemens (S). Conductance is the reverse (reciprocal) of resistance.

Conductor: A material that allows electric current to pass. Opposite of insulator.

Contact: The two surfaces that touch each other to make a switch closed, and that separate from each other to make a switch open.

Core: The material that the windings of an inductor or transformer are wrapped around.

Core losses: Power losses in the magnetic core of an inductor or transformer that are converted into waste heat inside the core. There are two kinds of core losses, hysteresis losses and eddy-current losses.

Cosine: The ratio of the adjacent side of a triangle to the hypotenuse. In an ac parallel circuit, the cosine of the phase angle between total current and voltage is the ratio of the resistive current, I_R, to the total current, I_T.

Coulomb (C): The basic measurement unit of charge. One coulomb is the charge contained in 6.24×10^{18} electrons.

Coupling capacitor: A capacitor that is used to feed the input signal into an amplifier or take the output signal from an amplifier, while blocking dc bias current.

Current (*I*): The movement of electric charge through a conductor.

Current coil, of a wattmeter: The coil in a wattmeter that senses the current through the device whose power is being measured. It is an electromagnet, containing a small number of turns of low-resistance wire.

Current gain (A_I): The factor by which the output current is greater than the input current for an amplifier.

Current source: A source that maintains a constant output current in the face of varying load resistance. The output voltage changes to satisfy Ohm's law.

Cutoff frequency: One of the two frequencies that causes a circuit's response to decline to 70.7% of its maximum response. The two cutoff frequencies are symbolized f_{low} and f_{high}.

Cycle: A complete oscillation of an ac wave.

Dc: Current or voltage that has a constant direction or polarity. It does not periodically change from positive to negative and back to positive.

Dc power supply: An electronic device that provides smooth dc by rectifying ac.

Dielectric: The material that separates the plates of a capacitor.

Digital: Circuitry in which only two voltage levels occur. As distinguished from analog, which has continuous values of voltage.

Diode: An electronic device that conducts current in one direction, but blocks current in the other direction.

Discharging, a capacitor: The process of removing charge from the plates of a capacitor, thereby eliminating voltage across the capacitor.

Discrete: A way of making electronic devices in which one package contains just a single device. Different from integrated, in which one package contains an entire circuit consisting of many individual electronic devices.

DMM: Digital multimeter, capable of measuring current, voltage, or resistance.

Doped silicon: Silicon that has had impurities scattered throughout it.

Double-pole: A switch containing two electrically separate contacts with a common operating mechanism. Idea extends to three-pole, four-pole, and so on.

Double-throw: A switch that changes the states of two contacts when it is actuated. One contact goes from open to closed, and the other goes from closed to open.

Drain (D): One of the terminals of a field-effect transistor. It is the terminal that produces the output when the FET is used in the standard common-source configuration.

Dropping resistor: A resistor placed in series with a load for the purpose of reducing the voltage across the load.

Eddy current: Whirlpool-like current that is induced in the cross-sectional plane of a magnetic core by time-varying flux in the core.

Effective value: The dc-equivalent value of an ac variable in terms of power transfer. Also called rms value.

Efficiency (η): The ratio of output power to input power for any electrical or electromechanical device. Usually expressed as a percentage.

Electrode: One of the electrical conducting terminals of a chemical cell, or battery.

Electrolyte: The chemical compound that a cell's electrodes react with to produce an electromotive force.

Electrolytic capacitor: A capacitor in which the dielectric layer is a very thin layer of aluminum or tantalum oxide that does not form until the capacitor is energized.

Electromagnet: A magnet that consists of a coiled winding carrying electric current.

Electron: The subatomic particle, near the outside of the atom, that carries negative charge.

Electrostatic: Referring to static charge, which has been concentrated in fixed positions.

Emitter: One of the three terminals of a BJT. It is the terminal that is connected to chassis ground in the standard common-emitter configuration.

Energize: To cause voltage to be applied to an electrical device, so that it starts working.

Energy (*W*): The ability to do something useful, some work. Electrical energy is the ability to move charge against an opposing voltage.

Engineering units: Multiples or submultiples of the basic measurement unit that change by a factor of one thousand from one engineering unit to the next. The most common ones in electrical work are kilo, mega, milli, micro, nano, and pico.

Equivalent resistance (or circuit): A single resistance that makes the circuit behave in the same way as the actual combination of two or more resistors.

Farad (F): Basic measurement unit for capacitance.

Faraday's law: The law that relates voltage to the time rate of change of magnetic flux. $V = N\,(\Delta\Phi / \Delta t)$.

Fast-acting fuse: A fuse that blows very quickly, even for a moderate overcurrent.

Ferrite: A material containing iron oxide that has fairly high magnetic permeability, and that tends to have lower core losses than ferromagnetic material.

Ferromagnetic: A material containing iron or an iron alloy that has high magnetic permeability.

Field-effect transistor (FET): A transistor in which the input voltage controls the output's constant-current source.

Filter: A circuit that discriminates among ac signals of different frequencies, passing some frequencies and blocking others.

Floating: Having no electrical connection to an earth-grounded circuit. A floating circuit cannot shock you if you touch just one point. A nonfloating circuit *can* shock you that way.

Flux (Φ): The number of magnetic field lines present, without regard to their area of containment. Measured in basic units of webers (Wb).

Flux density (*B*): The number of magnetic flux lines per square meter of containment area. Measured in basic units of teslas (T).

Forward bias: An applied voltage that makes a diode conduct.

Frequency (*f*): The number of cycles, or oscillations, that occur in one second. Its basic measurement unit is the hertz (Hz).

Frequency-response curve: A graph of a circuit's response (on the *y* axis) versus the frequency at which the circuit is operated (on the *x* axis).

Full load (FL): The loading condition that causes a voltage source to deliver its maximum possible or allowable current to the load.

Fuse: A device that protects against overcurrents by overheating and melting its link.

Gain: The factor by which some electrical variable (power, voltage, or current) is increased by an amplifier. Also called amplification, symbolized *A*.

Gate (G): One of the three terminals of an FET. It is the terminal that receives the input signal in the standard common-source configuration.

Gate-source voltage (V_{GS}): The input voltage that operates an FET amplifier in the standard common-source configuration.

Ground: Generally, the electrical point in a circuit that all voltages are measured relative to. In some circuits, this point may actually be the earth, the ground we walk on.

Ground-fault interrupter: Safety device that opens a circuit contact if it senses a difference in the currents in the hot wire and the common wire.

Grounding wire: In a 120-V (or higher) ac circuit, the noninsulated wire that connects all the circuit enclosures to earth ground potential. It does not carry any current if the circuit is functioning properly.

Half-wave rectifier: A rectifier in which one half-cycle of the ac input is passed to the load, and the other half-cycle is simply blocked. Distinguished from full-wave rectifier.

Henry (H): The basic measurement unit of inductance.

Hertz (Hz): The basic measurement unit of frequency. One Hz is one cycle per second.

High-cutoff frequency (f_{high}): The higher of the two cutoff frequencies for a frequency-selective circuit. Or, the cutoff frequency that is higher than the resonant or midband frequency.

High-pass filter: A filter that is designed to pass high-frequency signals to the load, while blocking low-frequency signals.

Hole: A place where an electron ought to be.

Hole current: Current that is visualized as flowing from the positive terminal of the source, through the circuit, and back to the negative terminal of the source.

Horizontal, or *x*, component: The portion of a phasor variable that is in phase with the reference variable in an ac circuit.

Hot: Having a voltage that is rather large, referenced to the circuit ground. The hot wire in a 120-V ac circuit is about 120 V away from ground potential, and is therefore dangerous. Its insulation color is usually red or black.

Hysteresis: The behavior of a magnetic core that causes its flux density to be different for a given amount of current, depending on whether the current is increasing or decreasing.

IC (integrated circuit): An entire circuit consisting of many active electronic devices, transistors, completely interconnected and contained in a single package.

Ideal: A perfect device that obeys the theoretical laws exactly. It usually implies that there is no wasted heat energy produced. Used to describe inductors, capacitors, transformers, voltage sources, diodes, and many other devices.

Ignition coil: A transformer whose secondary voltage is large enough to arc across spark plug electrodes, thereby igniting a fuel mixture. The secondary voltage is induced when the current in the primary winding is suddenly established or interrupted.

Impedance (Z): The combined current-opposing effect of resistance and reactance in an ac circuit.

Impedance bridge: An electrical measuring instrument that uses the balanced bridge idea to obtain measurements of inductance or capacitance, and their associated Q or D factors. The Q or D factor is an indirect measure of the associated resistance.

Incandescent lamp: A lamp that produces light by raising a metal filament to a very high temperature inside a vacuum or inert gas-filled enclosure.

Inductance (L): The electrical quality that relates voltage to the time rate of change of current. Another way to look at inductance is as the quality that relates a coil's magnetic flux to the amount of current in the coil.

Inductive kickback: The short-lived high voltage that is induced by an inductor when its current is suddenly stopped.

Inductive reactance (X_L): An inductor's opposition to current in an ac circuit.

Instantaneous value: The value of a varying voltage or current at a specific time instant.

Insulator: A material that does not allow current to flow. Opposite of conductor.

Integrated circuit: *See* IC.

Internal resistance (R_{int}): The resistance that is associated with the internal construction of a voltage source. Also called output resistance.

Isolation: The feature of transformers that means there is no electrical connection between the primary circuit and the secondary circuit.

Joule (J): The basic measurement unit of energy.

Kilowatthour (kWh): The popular commercial unit for measuring electrical energy. One kWh is equal to 3 600 000 J.

Kirchhoff's current law: The law that tells us that the total amount of current entering an electrical point must equal the total amount of current leaving that electrical point.

Kirchhoff's voltage law: The law that tells us that the sum of the voltage drops in a circuit (around a circuit loop) must equal the applied source voltage.

Lagging: Reaching a point on an ac waveform later in time than the other variable. Contrasted with leading.

Laminated core: A magnetic core that is made of thin pieces of ferromagnetic material alternating with thin pieces of insulating material. Used to reduce the eddy-current effect.

LCR meter: An electronic instrument that can measure inductance, capacitance, or resistance. Most LCR meters can also measure combinations of R and L, or R and C.

Leading: Reaching a point on an ac waveform earlier in time than the other variable. Contrasted with lagging.

Leakage: (1) The imperfection of a capacitor that enables some of the charge to return to the original plate through the dielectric. (2) The imperfection in wire insulation that allows current to pass from the hot wire directly to earth ground, bypassing the load.

Left-hand rule: The rule that tells the direction of circulation of magnetic flux, given the direction of electron current in a straight wire.

Lenz's law: The law that indicates the polarity of an inductor's induced voltage.

Linear: A circuit or circuit component that has a proportional relationship between its input and its output (or its cause and its effect).

Load: The part of an electric circuit that actually converts electrical energy into a useful form.

Low-cutoff frequency (f_{low}): The lower of the two cutoff frequencies for a frequency-selective circuit.

Low-pass filter: A filter that is designed to pass low-frequency signals to the load, while blocking high-frequency signals.

Magnetic field: The "force" field in the space near a magnet.

Magnetic flux: *See* flux.

Magnetic flux density: *See* flux density.

Magnetic polarity: The identification of one point on a magnet as north and another point as south.

Magnetic saturation: The tendency of any magnetic core to become perfectly aligned with the externally applied magnetic field if the magnetizing force becomes large enough. Once the saturation point is reached, any further increase in H no longer produces a magnified increase in B.

Magnetizing force (H): The number of ampere-turns per unit of core length. Therefore, the tendency of an electromagnet to establish magnetic flux, without regard to the core effect. This variable is sometimes called magnetic field intensity or magnetic field strength. These names are misleading, and should not be used.

Magnetomotive force (MMF): The number of ampere-turns in an electromagnet, without regard for the length of the core.

Magnitude: The amount of a variable, without regard to its sign (positive or negative).

Matching: The practice of using a transformer to make a load's resistance appear to be the same value as the source's internal (output) resistance. Also called resistance matching, and, not really correctly, impedance matching.

Maximum power theorem: The idea that for a fixed internal supply resistance, R_{int}, the load resistance must equal R_{int} to get the maximum possible load power.

Multimeter: An electric measuring instrument that can serve several functions, measuring voltage, current, or resistance.

Mutual inductance: Interaction that sometimes occurs between two individual inductors.

Negative half-cycle: The half of an ac voltage oscillation in which the voltage is negative on top and positive on bottom (assumed). Or, for current, the half of the oscillation in which the electron current passes through the load from top to bottom (assumed).

Node: An electrical point in a circuit.

Noise: An undesired electrical voltage that is mixed in with the desired signal voltage.

No-load (NL): The operating condition in which there is no current being taken from a source, because there is nothing connected to its output terminals. The load impedance is infinite.

Nominal value: The value of a component that the manufacturer was attempting to produce. Distinguished from actual value.

Nonideal: An imperfect device that does not obey the theoretical laws exactly. It usually implies that the device produces some waste heat energy. Also called real.

Normally closed (N.C.): A switch that is closed when not actuated, opening when actuated.

Normally open (N.O.): A switch that is open when not actuated, closing when actuated.

North: The end of a magnet from which flux lines emerge.

***Npn* transistor:** A BJT that has an *n*-type collector and emitter, and a *p*-type base.

***N*-type semiconductor:** Silicon that is doped with an impurity that provides free electrons.

Nulled: The condition of a Wheatstone bridge that causes the output voltage to be zero.

Ohm (Ω): The basic measurement unit of resistance. Also the basic measurement unit of reactance and impedance.

Ohmmeter: An electric meter for measuring resistance in ohms.

Ohm's law: The law that expresses the relationship among current, voltage, and resistance. In an ac circuit, resistance is replaced by reactance or impedance.

Open: The condition of a switch when the switch is not providing a current path between its terminals.

Open circuit: A break, perhaps accidental, that prevents a complete current flow path.

Overcurrent: A current that is too large for a circuit's conductors to carry safely. It results from a malfunction in the circuit, or a human error.

Parallel: Connected together between the same two electrical points. Connected so that voltages must be equal.

Peak-to-peak value, of voltage or current ($V_{p\text{-}p}$ or $I_{p\text{-}p}$): The difference between the positive peak and the negative peak values. For a true ac wave, $V_{p\text{-}p}$ is twice V_p.

Peak value, of voltage or current (V_p or I_p): The maximum instantaneous value that an ac waveform reaches.

Period (T): The amount of time required for an ac wave to go through a complete cycle.

Permanent magnet: A metal magnet that does not require

electric current through a coil.

Permeability, relative (μ_r): The factor by which a magnetic material magnifies the magnetic flux density, compared to air.

Phase: The time relation between two (or more) ac variables.

Phase angle (ø): The time-based relation between two ac variables, assuming that the time for a full cycle (T) is equivalent to a 360-degree angle.

Phasor: An arrow that conveys information about the magnitude and phase of an ac variable.

***Pnp* transistor:** A BJT that has a *p*-type collector and emitter, and an *n*-type base.

Polarity: The identification of one point as negative (–) and the other point as positive (+) for a voltage.

Positive half-cycle: The half of an ac voltage oscillation in which the voltage is positive on top and negative on bottom (assumed). Or, for current, the half of the oscillation in which the electron current passes through the load from bottom to top (assumed).

Potentiometer: A three-terminal variable resistor, which can also serve as a variable voltage divider.

Power (P): The time rate of energy use.

Power factor (PF): The portion of the total current in an ac circuit that actually contributes to power transfer from source to load. Power factor is equal to the cosine of the phase angle between current and voltage in the circuit.

Power gain (A_P): The factor by which the output power is greater than the input power for an amplifier.

Primary cell: A chemical cell that cannot be safely recharged.

Primary winding: The winding of a transformer that receives input power from the source.

Printed circuit (PC) board: A thin board of insulating material that has copper tracks adhered to it. Electrical components are mounted on the PC board, and are electrically interconnected by the copper tracks.

Proton: The subatomic particle that carries positive charge. It is in the nucleus, and is not movable by electrical means.

Pythagorean theorem: The formula that gives the relationship of the adjacent side, the opposite side, and the hypotenuse in a right triangle. For ac circuits, it gives the relationship of the horizontal (in-phase) component, the vertical (out-of-phase) component, and the total value.

***P*-type semiconductor:** Silicon that is doped with an impurity that provides holes.

***Q* (quality factor):** The figure of merit regarding frequency selectivity. In a series ac circuit, Q is the ratio of reactance to resistance . In a parallel ac circuit, Q is the ratio of resistance to reactance (resistance divided by reactance).

Reactance (X): The ability of a capacitor or inductor to oppose the passage of ac current. Capacitive reactance is symbolized X_C. Inductive reactance is symbolized X_L.

Reciprocal: The relationship of one idea being the number one (1) divided by the other idea. For example, resistance and conductance are reciprocals because $R = 1/G$ and $G = 1/R$.

Rectifier: A circuit that converts ac to dc.

Relay: An electromagnetically operated switch.

Residual flux: The flux that is retained by the core when the magnetizing force has been removed.

Resistance (R): The ability of a device to oppose the passage of dc current. Its basic measurement unit is the ohm.

Resistivity: The atomic characteristic of a material that represents its ability to oppose dc current.

Resistor: An electrical component that is manufactured for the deliberate purpose of having resistance.

Resonance: The ability of inductor/capacitor combinations to give vigorous circuit response at a particular frequency, called the resonant frequency.

Resonant frequency (f_r): The particular frequency at which resonance occurs for a given LC combination.

Retentivity: The ability or tendency of some magnetic core materials to retain a small amount of flux even when the magnetizing force has been removed.

Reverse bias: An applied voltage that causes a *pn* junction, or diode, to block current.

Rheostat: A two-terminal variable resistor.

Rms value: The dc-equivalent value of an ac variable in terms of power transfer. Also called effective value.

Rotor: The cylindrical rotating part of an alternator, or of any electromechanical machine.

Scale: A set of numbers on a measuring instrument that gives meaning to a particular pointer position.

Schematic diagram: A diagram of an electric circuit that does not show the electric devices with their realistic appearance.

Secondary cell: A chemical cell that can be safely recharged by having the chemical reactions reversed at the surfaces of its electrodes.

Secondary winding: The winding of a transformer that delivers output power to the load.

Selectivity (frequency): The ability of a particular circuit to favor frequencies near resonance but to discriminate against frequencies far from resonance.

Semiconductor: A material that is neither a good conductor, nor a good insulator. Usually refers to silicon that has been doped with an impurity.

Series: Connected together in line with one another. Connected so that currents must be equal.

Shielding, electrical: Protecting from undesired noise voltage pickup arising from stray capacitance between nearby conductors and the point that is shielded.

Shielding, magnetic: Protecting from undesired noise voltage pickup arising from stray time-varying magnetic fields in the vicinity.

Short circuit: A path by which current can flow directly to ground. Usually means an improper path due to a circuit malfunction.

Siemens (S): The basic measurement unit of conductance.

Sine wave: An ac wave that follows the mathematical sine function.

Single-throw: A switch that changes the state of only one contact per pole when it is actuated. The contact either goes from open to closed or goes from closed to open.

Skin effect: The tendency for ac current to concentrate near the surface of a conductor, rather than being evenly distributed over the conductor's cross section, like dc current.

Slip rings: Insulated copper rings that encircle the shaft of an electromechanical machine, carrying current between the winding on the rotor and the outside world.

Slow-blow fuse: A fuse that takes a longer-than-usual time to melt, for a moderate overcurrent.

Solar cell: A solid-state electronic device that can produce electric power from sunlight.

Solenoid: A cylindrical electromagnet. Often having a movable core that is pulled into the hollow cylinder by magnetic force.

Solid-state: Referring to electronic devices that are made of solid silicon, rather than vacuum tubes.

Source (S): One of the three terminals of an FET. It is the terminal that is connected to chassis ground in the standard common-source configuration.

Source (of voltage): The voltage-providing device that drives a circuit.

Source voltage value (V_s): The amount of voltage provided by a voltage source.

South: The end of a magnet at which flux lines reenter the core.

Speaker, or loudspeaker: A mechanically vibrating device that receives electrical energy from an ac source and converts it partly into sound energy.

Step-down mode: The mode of operating a transformer that causes the secondary voltage to be less than the primary voltage.

Step-up mode: The mode of operating a transformer that causes the secondary voltage to be greater than the primary voltage.

Stray capacitance: Capacitance that is present as an unavoidable outcome of normal circuit construction. It is often undesirable. Distinguished from component capacitance.

Switch (SW): A device that enables us to complete a circuit's current path, or break open the path.

Tachometer: Instrument for measuring rotational speed of a shaft.

Tapped: Having an internal point(s) accessible to the outside world, not just the two end points.

Temperature stability: The ability of an electric device to maintain its original character at very high or very low temperatures.

Tesla (T): The basic measurement unit of flux density.

Time constant (τ): A certain amount of real time, during which a transient goes through 63% of its change.

Time constant curve: The particular shape of the waveform of voltage or current during an electrical transient. Usually plotted as a percentage of maximum on the *y* axis, versus number of elapsed time constants on the *x* axis. Also called universal time constant curve.

Tolerance: The percentage deviation from the nominal value that is allowed in the manufacturing process.

Toroid: Doughnut-shaped magnetic core, used because it has no abrupt corners to disrupt the flux pattern.

Transformer: A two-inductor device that can raise or lower the value of an ac voltage. It also raises or lowers the value of current, in the opposite direction from voltage.

Transformer current law: The formula that tells us that, ideally, the ratio of primary current to secondary current is equal to the turns ratio of the transformer.

Transformer voltage law: The formula that tells us that, ideally, the ratio of secondary voltage to primary voltage is equal to the turns ratio of the transformer.

Transient: A temporary variation in voltage and/or current produced by a capacitor or inductor following a switching action.

Transistor: A solid-state electronic device that provides power amplification.

Troubleshooting: The process of finding the reason for a circuit malfunction.

True power: Actual power in an ac circuit, as distinguished from apparent power.

Turns ratio (*n*): The ratio of the number of secondary winding turns to the number of primary winding turns in a transformer.

Vertical, or *y*, component: The portion of a phasor variable that is 90° out of phase with the reference variable in an ac circuit.

Volt (V): The basic measurement unit for voltage.

Voltage (*V*): The force-like electrical variable that causes current in a conductor.

Voltage coil, of a wattmeter: The coil in a wattmeter that senses the voltage across the device whose power is being measured. It is an electromagnet, containing a large number of turns of high-resistance wire.

Voltage divider: A group of series resistors (or a pot) that causes a certain portion of an overall source voltage to be available.

Voltage gain (A_V): The factor by which the output voltage is greater than the input voltage for an amplifier.

Voltage source: The voltage-providing device that drives a circuit.

Voltampere (VA): The basic measurement unit for apparent power.

Voltmeter: The electrical instrument that measures voltage.

VOM (Volt-ohm-milliammeter): Analog instrument capable of measuring all three basic variables.

Watt (W): The basic measurement unit for power.

Wattmeter: The electrical instrument that measures power.

Waveform: A graph of instantaneous voltage or current versus time.

Weber (Wb): The basic measurement unit for magnetic flux.

Wheatstone bridge: A circuit containing four resistances, at least one of which is variable. The variable resistance is adjusted until the voltage division is the same on the right and left sides.

Winding: A spiral coil of wire wrapped around a core. Inductors, transformers, and motors contain windings.

APPENDIX

USING A SCIENTIFIC HAND-HELD CALCULATOR

Hand-held scientific calculators have certainly taken the drudgery out of mathematical calculations. Computations that used to require several minutes of concentrated, tiring effort can now be accomplished with ease, in a few seconds, and with much better confidence in the result. This frees us to concentrate our mental efforts in more profitable ways, namely on understanding the ideas of electricity. You owe it to yourself to become skilled in the use of such a calculator.

For an introductory course in electricity, the functions that you need are the following:

- The four arithmetic functions, +, –, x, and ÷.
- The scientific notation function. Its key is usually marked EXP or EE.
- The square function, marked x^2, and square root function, marked $\sqrt{x}$.
- The reciprocal, 1/x, function.

Besides these basic functions, memory capability is very handy. The three memory functions are 1) "Place into memory," marked STO or X→M, or similar; 2) "Add to memory," usually marked SUM or M+; 3) "Recall from memory," marked RCL or M→X, or similar.

The grouping function (the two parentheses keys) is useful, especially when applying the product-over-the-sum formula. If your calculator has a π key, that saves you the trouble of entering the actual value of π (3.1416) or 2π (6.2832) in every formula involving π.

And finally, to avoid trigonometry look-up tables, the trig functions SIN, COS, and TAN, and their inverses SIN^{-1}, COS^{-1}, and TAN^{-1} are very helpful. In this book, we have avoided the SIN and TAN functions and their inverses. We used only COS and COS^{-1}.

Addition

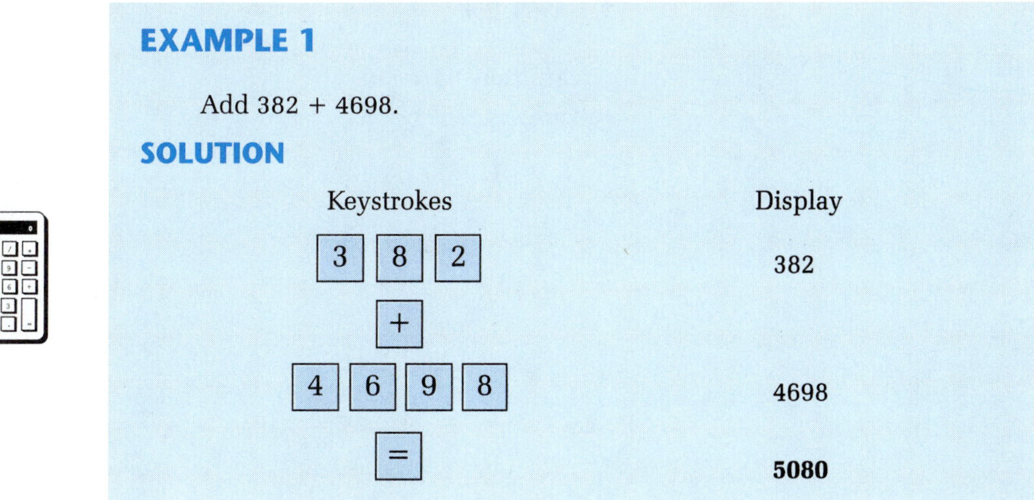

EXAMPLE 1

Add 382 + 4698.

SOLUTION

Keystrokes	Display
[3] [8] [2]	382
[+]	
[4] [6] [9] [8]	4698
[=]	**5080**

EXAMPLE 2

Add $3.41 \times 10^2 + 1.93 \times 10^3$.

SOLUTION

Keystrokes	Display
3 . 4 1 EE 2	3.41 02 or similar
+	
1 . 9 3 EE 3	1.93 03 or similar
=	**2.271 03** or similar

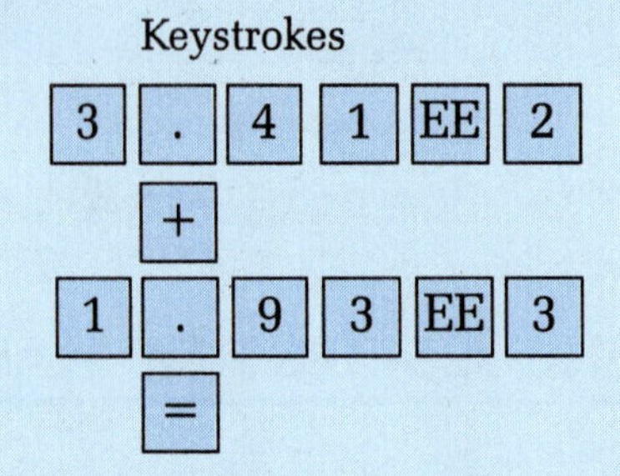

which is equivalent to the number **2271.**

Subtraction

EXAMPLE 3

Subtract 702 – 128.

SOLUTION

Keystrokes	Display
7 0 2	702
–	
1 2 8	128
=	**574**

Multiplication

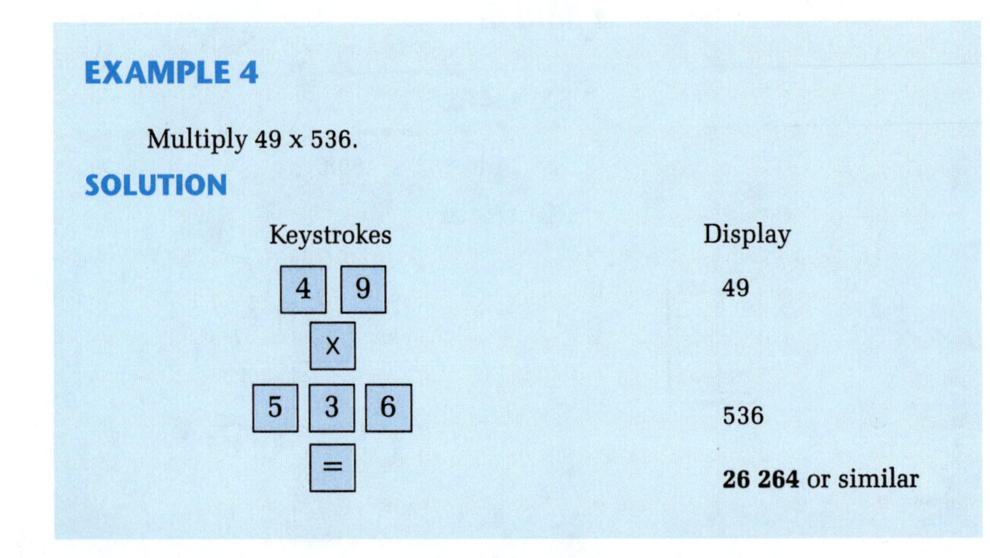

EXAMPLE 4

Multiply 49 x 536.

SOLUTION

Keystrokes	Display
4 9	49
x	
5 3 6	536
=	**26 264** or similar

Division

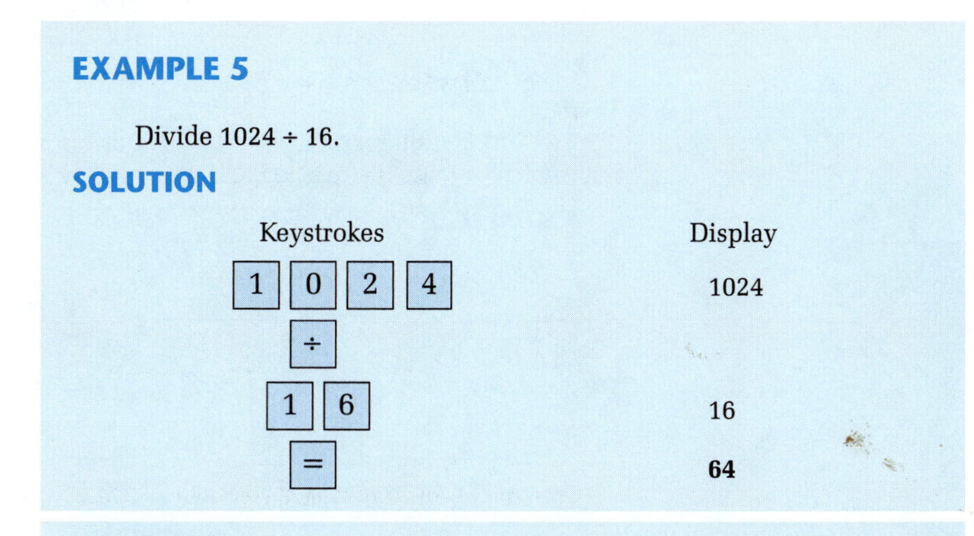

EXAMPLE 5

Divide 1024 ÷ 16.

SOLUTION

Keystrokes	Display
1 0 2 4	1024
÷	
1 6	16
=	**64**

EXAMPLE 6

Divide 4.56 x 10^{-2} ÷ 1.2 x 10^{-3}.

SOLUTION

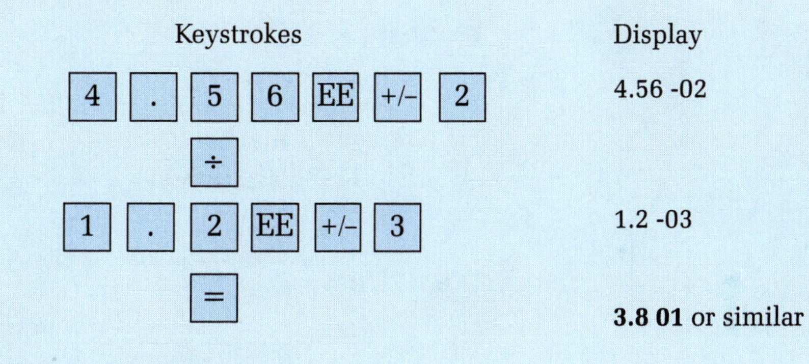

Keystrokes	Display
4 . 5 6 EE +/– 2	4.56 -02
÷	
1 . 2 EE +/– 3	1.2 -03
=	**3.8 01** or similar

which is equivalent to **38.**

Square

EXAMPLE 7

If current, *I*, equals 0.15 A and resistance, *R*, equals 400 Ω, find power, *P*. Use the formula $P = I^2R$.

SOLUTION

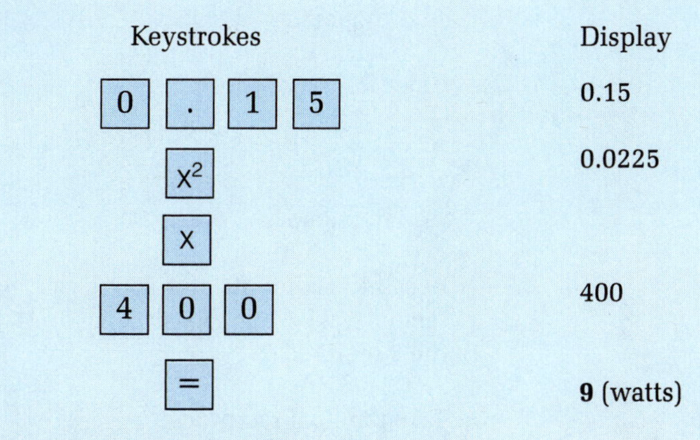

Keystrokes	Display
0 . 1 5	0.15
x^2	0.0225
x	
4 0 0	400
=	**9** (watts)

Square root

EXAMPLE 8

A certain resistor with $R = 28\ \Omega$ is burning 7 W of power. What is the voltage across the resistor? Use the formula $V = \sqrt{PR}$.

SOLUTION

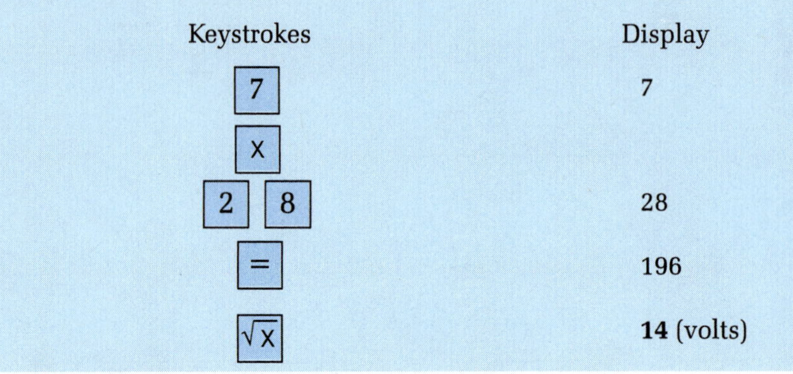

Keystrokes	Display
7	7
X	
2 8	28
=	196
√X	**14** (volts)

1/x and Memory Functions

EXAMPLE 9

These three resistors are in parallel: $R_1 = 24\ \Omega$, $R_2 = 40\ \Omega$, and $R_3 = 15\ \Omega$. Find the equivalent total resistance, R_T. Use the formula

$$\frac{1}{R_T} = \frac{1}{R_1} + \frac{1}{R_2} + \frac{1}{R_3}.$$

SOLUTION

	Keystrokes	Display
Enter R_1	2 4	24
Reciprocal of R_1	1/x	0.0416667
Store in memory	STO	0.0416667
Enter R_2	4 0	40
Reciprocal of R_2	1/x	0.025
Add to memory	SUM	0.025
Enter R_3	1 5	15
Reciprocal of R_3	1/x	0.0666667
Add to memory	SUM	0.0666667
Recall the contents of the memory, which is the sum of the 3 reciprocals	RCL	0.1333333
Take the final reciprocal	1/x	**7.5** (ohms)

If your calculator does not have memory, you can attempt to accumulate reciprocals in the display. Not all calculators can do this. If yours cannot, you will have to write down each reciprocal on paper as it is found. Then add them by reentering their values from the paper. Then take the final reciprocal.

Grouping, or Parentheses, Function

EXAMPLE 10

There are two resistors in parallel: $R_1 = 680\ \Omega$, and $R_2 = 470\ \Omega$. Find the equivalent total resistance, R_T, using the product-over-the-sum formula $R_T = \frac{R_1 \times R_2}{R_1 + R_2}$.

SOLUTION

	Keystrokes	Display
Enter the first resistance	[6] [8] [0]	680
Product	[x]	
Enter the second resistance	[4] [7] [0]	470
Divided by	[÷]	319600
Open parenthesis to start the sum	[(]	
Enter the first resistance value in the group (R_1)	[6] [8] [0]	680
Sum	[+]	
Enter the second resistance value in the group (R_2)	[4] [7] [0]	470
Close parenthesis to end the sum	[)]	1150
The final result	[=]	**277.9** (ohms)

COS Function

When using the cos and $\cos^{-1}$ functions, your calculator must be in the DEG mode, since we are dealing with angles in degrees (as opposed to radians or gradians). On most scientific calculators, this is set by the DRG key. Check out your calculator's instruction booklet. Also, when using the cos and $\cos^{-1}$ functions, it may take your calculator a short while (less than two seconds) to display the answer.

EXAMPLE 11

A certain ac circuit has a current-voltage phase angle of $ø = 35°$. Use the formula $PF = \cos ø$ to find the circuit's power factor.

SOLUTION

	Keystrokes	Display
With the calculator in DEG mode, enter the angle	[3] [5]	35
Use the cosine function	[COS]	0.81915

This could be rounded to $PF =$ **0.82** or **82%.**

COS^{-1} Function

EXAMPLE 12

A certain ac circuit has $PF = 0.91$. Find the phase angle between current and voltage.

SOLUTION

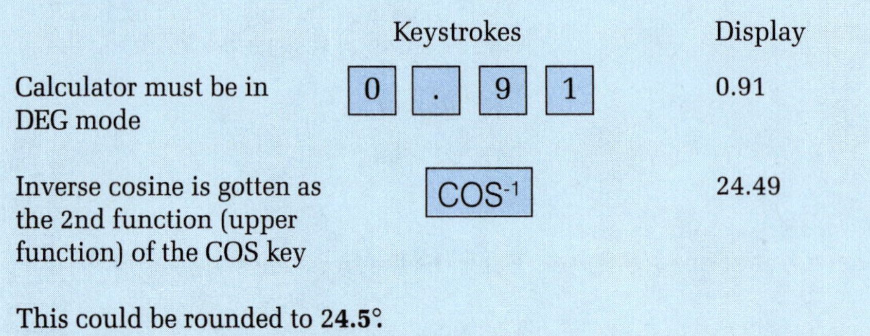

	Keystrokes	Display
Calculator must be in DEG mode	0 . 9 1	0.91
Inverse cosine is gotten as the 2nd function (upper function) of the COS key	COS^{-1}	24.49

This could be rounded to **24.5°**.

ANSWERS

CHAPTER 1 Charge and Current

Self-Check for Sections 1-1, 1-2, and 1-3

1. Electrons, protons, neutrons
2. protons, neutrons
3. Electrons
4. Electrons negative, protons positive, neutrons have no charge
5. attract, repel
6. positive, negative
7. Outer shell
8. Outer shell
9. positive

Self-Check for Sections 1-4 and 1-5

10. Voltage. No voltage was applied.
11. Outer-shell electrons. They are held more loosely.
12. direct, alternating
13. Direct current
14. Alternating current
15. I
16. Ampere. A
17. Q
18. Coulomb. C
19. one ampere
20. 31 200 000 000 000 000 000
21. 0.25 A
22. 0.5 C, 2.5 A

Chapter Questions and Problems (Odd numbers)

1. Electrons, protons, neutrons
3. the same as
5. negative
7. attract
9. Two negatively charged objects
11. current
13. two, one
15. F
17. ampere, A
19. 3 A
21. 0.04 A
23. Q
25. quickly

CHAPTER 2 Voltage and Voltage Sources

Self-Check for Sections 2-1, 2-2, and 2-3

1. higher
2. electromotive force
3. volt
4. T
5. Current
6. Current
7. Yes, voltage may be present without a path for current flow.
8. No, current cannot flow without voltage to force it.
9. Voltage does not travel "through" wires.

Self-Check for Section 2-4

10. Positive electrode, negative electrode, and electrolyte
11. primary, secondary
12. Some chemical reactions can be undone, some cannot be undone safely.
13. A wire moving through a magnetic field
14. rectified power supplies
15. Nonpolluting, renewable, practical in remote areas

Chapter Questions and Problems (Odd numbers)

1. V. V
3. T
5. T
7. It causes loosely held electrons to become free.
11. 12.6 V
13. Alkaline batteries
15. Lead-acid batteries
17. It causes chemical reactions to become undone.
19. negative, positive
21. Solar
23. Electronic power supplies
25. Generator

CHAPTER 3 Resistance

Self-Check for Sections 3-1 and 3-2

2. ohm
3. T
4. 1 A

Self-Check for Sections 3-3, 3-4, and 3-5

5. Carbon-composition, wirewound, cermet, and deposited-film
6. ± 5%, ± 10%
7. a) 1600 Ω, ± 10%. b) 680 Ω, ± 5%. c) 3.9 Ω, ± 10%. d) 47 Ω, ± 5%. e) 8450 Ω, ± 2%
8. a) 10.8–13.2. b) 351–429. c) 7790–8610
11. Violet, green, brown, gold

Self-Check for Sections 3-6 and 3-7

12. A variable resistor can be adjusted.
13. A potentiometer has three leads.
14. decreases
15. increase, 50
16. The resistance from the top to the tap would increase an unknown amount.
17. 1.5
18. 0.28
19. The resistance of the tungsten will be 3.2 times higher.
20. 0.023 85 Ω

Chapter 3 Questions and Problems (Odd numbers)

1. less
3. The ohm. Ω
5. A thin film is sprayed onto an insulating base or deposited by vaporization methods.
7. potentiometer
9. F
13. ± 10%
15. 0 Ω tap-to-top, 1000 Ω tap-to-bottom
17. 500 Ω tap-to-top, 500 Ω tap-to-bottom
19. 8.4 Ω
21. high
23. medium
25. Heat produced by the filament's resistance

CHAPTER 4 Schematic Diagrams

Self-Check for Sections 4-1 and 4-2

1. Schematics make drawings as simple as possible.
3. Switch location does not affect operation.
4. Up

Self-Check for Sections 4-3 and 4-4

6. R_1 is negative on top, R_2 is positive on top, and current is counterclockwise.
7. All lamps are positive on top.
8. Current flows from bottom to top in each lamp.
9. Current through lamp 2 stops. Lamp 2 will be positive on both sides.
10. Current direction and voltage polarity will remain the same in lamps 1 and 3. Switch 2 will not have any effect there.

Chapter Questions and Problems (Odd numbers)

3. open
5. knife, toggle
7. The voltage source causes current flow.
9. Insulators prevent contact with conductors.
11. Switches turn circuits on and off.
13. Speaker
15. Welder
17. 20 V
19. 15 V
21. 7 V
23. a) and b) No. There is no complete current path.
25. conventional

CHAPTER 5 Measuring Electric Circuits

Self-Check for Sections 5-1 and 5-2

1. Disconnect power, open the circuit above or below R_1, connect the ammeter to close the circuit, reapply power.
2. Disconnect power, open the circuit above or below R_2, connect the ammeter to close the circuit, reapply power.
3. Disconnect power, open the circuit near the source, connect the ammeter to close the circuit, reapply power.
4. Since $I_{R_1} = I_T$, any ammeter connection will do.
5. Anywhere in the circuit
6. $I_{R_1} = I_{R_2}$
7. $I_{R_1} = I_{R_2} = I_{R_3}$
8. Connect the voltmeter leads across R_1 with the black lead on top.
9. Connect the voltmeter leads across R_2 with the black lead on top.
10. Connect the negative lead above R_1. Connect the positive lead below R_2.
11. Connect the negative lead above R_2, the positive lead below R_3.
12. No.

Self-Check for Sections 5-3 and 5-4

13. T
14. F
15. Remove the lamp, connect the meter across R_2.
16. Open the circuit between R_2 and the lamp.
17. analog
18. digital multimeter

Chapter Questions and Problems (Odd numbers)

1. T
3. T
5. F
7. digital
9. Across an isolated, deenergized device
11. Other parallel paths in the circuit may be measured unintentionally.
13. T
15. VOM, DMM
17. Disconnect power, open the circuit near R_1, connect the ammeter, reapply power.
19. Disconnect power, open the circuit near R_3, connect the ammeter, reapply power.
21. Connect the voltmeter across R_1.
23. Connect the voltmeter across R_4.
25. Deenergize the circuit, connect the ohmmeter across R_4.

CHAPTER 6 Ohm's Law

Self-Check for Sections 6-1, 6-2, and 6-3

1. 3 A
2. 1.5 A. Reducing the voltage by half reduces the current by half.
3. 16 Ω
4. 32 Ω. Doubling the resistance halves the current.
5. 60 V
6. a) 0.662 A. b) 0.375 A. c) lower. d) 0 A. There is no complete current path.

Self-Check for Section 6-4

7. a) 4.7×10^3 b) 6.2×10^6 c) 3.6×10^{-3} d) 1.5×10^{-5}
8. a) 572 b) 6100 c) 0.083 d) 0.000 049
9. a) 3.645×10^6 b) 6.072×10^3 c) 7.98×10^{-5} d) 2.928×10^{-1}
10. a) 2.27×10^2 b) 6.31×10^{-3} c) 9.22×10^{-6} d) 4.76×10^6
11. a) 9.5×10^2 b) 2.1×10^4 c) 4.66×10^3 d) 6.38×10^4

Self-Check for Sections 6-5 and 6-6

12. 4 kΩ
13. 159.1 V
14. 50 nA
15. a) 400 mA. b) 200 μA. c) 4.26 mA
16. R_1 = 5 kΩ, R_2 = 500 Ω, R_3 = 50 Ω

Chapter Questions and Problems (Odd numbers)

1. lower
3. 411 mA
5. 48 Ω
9. 600 mV
11. a) 3.9×10^2 b) 1.6×10^3 c) 1.25×10^7 d) 5.3×10^5
13. a) 8.5×10^{-3} b) 9.2×10^{-4} c) 4.31×10^{-1} d) 6.4×10^{-6}
15. a) 2.158×10^6 b) 9.69×10^2 c) 1.107×10^{-1} d) 3.026×10^{-4}
17. a) 3.83×10^3 b) 4.22×10^4 c) -2.72×10^6 d) 4.6×10^{-2}
19. 6.67 μA
21. 1.5 mV
25. 50 V

CHAPTER 7 Power and Energy

Self-Check for Sections 7-1, 7-2, and 7-3

1. energy
2. newton. N
3. 225 J
4. 157.5 J
5. 67.5 J less than it started with
6. 270 J
7. Power is the time rate of energy.
8. watt. W
9. 0.9 W
10. 1 W
11. 12 kWh, 43.2 MJ
12. 78 cents

Self-Check for Section 7-4

13. 384 W
14. 327 Ω
15. 135 W
16. 27.4 V
17. 38.7 V
18. Increasing voltage will increase current which will also increase power.
19. 3.6 A
20. 35 W. It converts it to sound and heat.
21. 115 Ω
22. 7 mA

Self-Check for Sections 7-5 and 7-6

23. waste heat
24. efficiency
25. Torque, rotational speed, shaft power
26. It is a high-efficiency device.
27. Light
28. Low efficiency
29. newton-meter, N-m
30. radian per second, rad/s
31. watts
32. pound-foot
33. rpm
34. horsepower, HP
35. current
36. voltage

Chapter Questions and Problems (Odd numbers)

1. F
3. joule, J
5. 1 volt, 1 coulomb
7. W
9. kilowatt hour, kWh
11. 1 joule, time, 1 second
13. $I = \frac{P}{V}$, $V = \frac{P}{I}$
15. 109 A
17. $P = \frac{V^2}{R}$, $R = \frac{V^2}{P}$, $V = \sqrt{PR}$
19. 25 W
21. $P = I^2R$, $R = \frac{P}{I^2}$, $I = \sqrt{\frac{P}{R}}$
23. 247 mΩ
25. 30.24 W
27. No.
29. 3.9%

CHAPTER 8 Safety Devices, Wires, Switches, and Relays

Self-Check for Sections 8-1, 8-2, and 8-3

1. overcurrent
2. Nearly zero ohms
3. Part of the load is shorted.
4. T
5. T
6. F
7. infinite
8. thermal, magnetic
9. A circuit breaker does not have to be replaced.
10. F

Self-Check for Sections 8-4 and 8-5

11. The voltage source could be shorted to the enclosure, and a person could touch the enclosure while also in contact with the earth ground.
12. An overcurrent occurs, blowing the fuse.
13. An overcurrent may not occur.
14. A partial short to the enclosure causes a difference in the supply-wire currents.

Self-Check for Sections 8-6, 8-7, and 8-8

15. thinner
16. millimeter, mil
17. 20 A
18. 15 A
19. T
20. A double-throw switch has 2 contacts sharing a common terminal.

Self-Check for Sections 8-9 and 8-10

23. A relay is operated by current through its coil.
24. T
25. coil, contact
26. N.O. is open when coil is deenergized and closes when coil is energized. N.C. is opposite.

Chapter Questions and Problems (Odd numbers)

1. A circuit breaker
3. Moderate
5. 1–10 ms
7. To ensure that they can extinguish an arc.
9. A short to the frame blows the fuse.
11. T
21. Connect the coil to the switch and low-voltage supply. Connect the contacts to the high voltage.
23. deenergized, open, closed
25. Contact RA-1 closed to provide another path for current to the coil.

CHAPTER 9 Series Circuits

Self-Check for Sections 9-1 and 9-2

1. T
2. F
3. No, their position has nothing to do with it.
7. a) $R_T = 3750\ \Omega$ b) $R_1 = 575\ \Omega$ c) $R_2 = 1.2\ k\Omega$ d) $R_4 = 1275\ \Omega$ e) $V_4 = 15.3$ V

Self-Check for Section 9-3

8. T
9. T
10. $V_{R_1} = 25$ V, $V_{R_2} = 37.5$ V, $V_{R_3} = 17.5$ V, $V_S = 80$ V
11. 8 V. No.
12. 27 V
13. 35 V

Self-Check for Sections 9-4 and 9-5

14. Real voltage sources have internal resistances.
15. a) 14.4 V b) 13.5 V
16. 3.6 V
17. 767 Ω
18. 12.7 V

Chapter Questions and Problems (Odd numbers)

1. $V = IR$. With current the same, different resistance requires different voltage.
3. a) 8 V b) 11 V
5. a) 700 Ω b) 200 Ω
7. a) $V_1 = 12$ V b) $V_2 = 17.6$ V c) $V_3 = 8.4$ V d) $R_3 = 5.25\ \Omega$
9. a) $R_T = 32.5\ \Omega$ b) $V_1 = 12.3$ V, $V_2 = 24.6$ V c) 36.9 V

11. a) $V_{OUT(FL)} = 20$ V b) $I_{FL} = 0.2$ A
    c) $V_{OUT(NL)} = 24$ V
13. a) 14 V b) 26.6 V
15. a) $V_1 = 13.5$ V b) $V_2 = 10.4$ V, $V_3 = 16.1$ V
    c) Yes
17. 26.5 V
19. R_4 is open.
21. $R_{int} = 150\ \Omega$
23. Between 50% and 70%, the wiper is losing its mechanical contact with the resistance element.
25. V_S is actually greater than 120 V. The dial is incorrect.

CHAPTER 10 Parallel Circuits

Self-Check for Sections 10-1 and 10-2

1. equal, unequal
2. lower, lowest
3. a) $R_T = 2.99\ k\Omega$ b) $I_T = 20$ mA
4. a) $R_T = 1.83\ k\Omega$ b) $I_T = 32.8$ mA
5. A lower resistance will result in a higher current.
6. $R_T = 60\ \Omega$
7. 150 Ω
8. 100 Ω
9. $R_T = \frac{R}{N}$. The total resistance for a parallel circuit of N equal resistors will equal the individual resistance divided by the number of resistors, N.
10. $G_T = 0.17$ S

Self-Check for Section 10-3

11. current
12. $I_T = 6.75$ A
13. $I_T = 1.05$ A
14. $R_4 = 60\ \Omega$
15. 4 A leaving

Chapter Questions and Problems (Odd numbers)

1. a) $R_T = 1.26\ k\Omega$. b) $I_T = 14.3$ mA.
   c) Above or below the source.
3. 666.7 Ω
5. 30 Ω
7. $G_T = \frac{1}{R_T} = 795\ \mu S$
9. $I_2 = 21$ mA
19. a) 6 A, 72 W b) 16 A, 192 W. c) 36 A, 432 W
21. A single load burn-out will not affect the other loads in a parallel circuit.
23. a) $I_2 = 12.5$ mA, $I_3 = 27.5$ mA, $I_T = 65$ mA.
    b) $R_1 = 2\ k\Omega$, $R_3 = 1.82\ k\Omega$, $R_T = 769\ \Omega$.
25. $I_1 = 67$ mA, $I_2 = 133$ mA, $I_3 = 0.2$ A, $I_4 = 0.1$ A
27. $R_1 = 90\ \Omega$, $R_2 = 60\ \Omega$
29. Yes, total resistance will increase, lowering total current.
31. $I_1 = 1$ A, $I_2 = 0.8$ A, $I_3 = 2$ A, $I_4 = 0.5$ A. R_2 is open.
33. The short is in R_5, R_6, R_7, or R_8.
35. The short occurs only under high-voltage operating conditions.

CHAPTER 11 Series-Parallel Circuits

Self-Check for Sections 11-1 and 11-2

1. $V_1 = V_2 = V_3 = 10$ V, $V_4 = 40$ V, $I_1 = 0.2$ A, $I_2 = 0.13$ A, $I_3 = 0.17$ A, $I_4 = 0.5$ A
2. $I_T = 0.25$ A, $I_1 = 200$ mA, $I_2 = 67$ mA, $I_3 = 133$ mA, $I_4 = 50$ mA, $V_1 = 8$ V, $V_{2-3} = 10$ V, $V_4 = 18$ V
3. $P_1 = 1.6$ W, $P_2 = 0.67$ W, $P_3 = 1.3$ W, $P_4 = 0.9$ W
4. $I_T = 400$ mA, $I_1 = I_2 = 200$ mA, $I_3 = 133$ mA, $I_4 = 67$ mA, $V_1 = 9$ V, $V_2 = 5$ V, $V_{3-4} = 4$ V
5. $I_T = 26.4$ mA, $I_1 = 13.9$ mA, $I_2 = 8.3$ mA, $I_3 = 5.5$ mA, $I_4 = 12.5$ mA, $V_1 = 16.7$ V, $V_{2-3} = 8.3$ V, $V_4 = 25$ V

Self-Check for Section 11-3

6. 333 Ω
7. $V_2 = 7.05$ V, $V_4 = 7.05$ V
8. No. Since R_1 and R_2 are fixed, V_2 is constant.
9. Yes. As R_4 changes, V_4 changes until $V_4 = V_2$.
10. No. 333 Ω will balance this bridge at any voltage.

Chapter Questions and Problems (Odd numbers)

1. Simplify the circuit.
3. $I_T = I_1 = I_3 = 4$ mA, $I_2 = 1.5$ mA, $I_4 = I_5 = 2.5$ mA, $V_1 = 12$ V, $V_2 = 7.5$ V, $V_3 = 8$ V, $V_4 = 2.5$ V, $V_5 = 5$ V
5. R_T and I_T only
17. $I_1 = 10$ A, $I_2 = 15$ A, $I_{neutral} = 5$ A, left to right.
19. No. No.
21. $V_1 = 6.9$ V, $V_2 = 23.1$ V, $V_3 = 10.8$ V, $V_4 = 19.2$ V
23. 6 kΩ
25. Yes. V_4 changes from 19.2 V to 23.1 V. V_3 changes from 10.8 V to 6.9 V.
27. R_2 is open.
29. R_2 is open.
31. No, V_{R_1} should equal 16.7 V.
33. No, V_{R_2} should equal 7.5 V.
35. No, V_{R_1} should equal 20 V and V_{R_6} should equal 12.8 V.

CHAPTER 12 Capacitance

Self-Check for Sections 12-1 and 12-2

1. The capacitor must have the charge moved from one plate to the other.
2. coulomb, volt
3. 200 V
4. 400 nF
5. microfarads, nanofarads, and picofarads
6. voltage source
7. 2 nF, 2000 pF
8. 0.004 μF

Self-Check for Sections 12-3 and 12-4

9. Plate area, distance between plates, and dielectric constant
10. increased
11. decreased
12. Neoprene because it has a higher dielectric constant.
13. 1.15 nF
14. maximum usable voltage
15. It has a large plate area in a small package.
16. – 55℃ to 150℃
17. High leakage, poor tolerance, not very temperature stable, polarity is restricted.
18. 15 000 pF, ± 5% tolerance, – 55℃ to 85℃ temperature range, ± 3.3% change over temperature range

Self-Check for Sections 12-5 and 12-6

19. Intermeshing plate and compression types
20. series
21. $C_T = 2.7$ nF
22. parallel
23. $C_T = 248$ nF

Self-Check for Sections 12-7, 12-8, and 12-9

24. one
25. open
26. shorted
27. F
28. T
29. outside
30. Near the peak of the pulsation
31. When the voltage of the pulsation becomes less than the voltage of the capacitor
32. Some of the charge on the capacitor must drain off to provide current to the load.
33. T
34. The capacitor takes time to charge.

Chapter Questions and Problems (Odd numbers)

1. 2.5 V
3. 500 μF
5. increased
7. T
9. 8 pF
11. outside, inside
13. The dielectric (leakage) resistance is not infinite.
15. parallel
17. The combination in (A) which has $C_T = 14$ μF. The combination in (B) has $C_T = 12$ μF.
19. It cannot detect high-voltage problems. It cannot test small capacitors.
21. They can become shorted or open, or their value can change.
23. After surging toward zero ohms, the pointer will return to a medium value, not a high value.
25. Yes, it is out of tolerance.
27. The 6 μF capacitor may be open.

CHAPTER 13 Magnetism

Self-Check for Sections 13-1 and 13-2

1. flux
2. Flux comes out of the north pole and re-enters at the south pole.
3. The strength of the field
4. F
5. south
6. north

Self-Check for Sections 13-3, 13-4, and 13-5

7. Iron atoms are tiny magnets themselves.
8. T
9. North is on top, south in on the bottom.
10. South is now on top.
11. 0.16 tesla
12. Gradually
13. Point R will be higher.

Self-Check for Sections 13-6 and 13-7

14. Permeable materials are used to divert magnetic fields from their natural path.
15. Yes, the coil will attract the core regardless of current direction.
16. increasing
17. A higher current flow will produce a stronger magnetic field.

Self-Check for Section 13-8

18. 5.56 mT
19. 0.125 mWb (one fourth of the total flux)

20. 60 000 A t/m
21. 0.076 T
22. 3.78 T
23. 1800 A t

Chapter Questions and Problems (Odd numbers)

1. T
3. An arrowhead
5. Weber, tesla
7. 0.556 T
9. T
11. The factor by which the material's permeability is greater than air's permeability.
13. 8250 A • t/m
15. The *B* axis (vertical axis) intercept is higher, at 0.4 T instead of 0.25 T.
17. The material in Problem 14
19. $H = 468$ A • t/m, $B = 0.590$ mT. (Both are doubled.)
21. 200 mA
23. F
25. 4000 turns. Increasing *N* increases *H* and *B*. Increased B means greater magnetic attraction force.

CHAPTER 14 Inductance

Self-Check for Sections 14-1 and 14-2

1. Inductance
2. This is because the inductor attempts to force current to flow CCW, in the opposite direction to the change being produced by the dc voltage source.
3. As the inductor voltage slowly decreases toward zero, the difference between V_s and v_L slowly increases. As this net driving voltage increases, *i* increases according to Ohm's law.
4. T
5. a) Negative on top, positive on bottom
   b) Negative on bottom, positive on top
6. henry; H
7. 0.625 H

Self-Check for Sections 14-3 and 14-4

8. Core length, number of winding turns, magnetic permeability of the core material
9. 303 mH
10. 360 mH
11. Ferromagnetic, ferrite, air (or nonpermeable)
13. Ferromagnetic
14. T
15. toroid
16. low
17. greater
18. To the right, in the center of the winding length.

Self-Check for Sections 14-5 through 14-8

19. 1.0 H
20. 0.24 H
21. a and c and d
22. No. There is no way to relate *L* to *R* without knowing the exact details of construction.
23. A noise burst tends to change current through the load momentarily. The inductor quickly induces a voltage with a polarity opposing the noise burst.
24. Series
25. 1800 rpm

Chapter Questions and Problems (Odd numbers)

1. henry; H
3. It induces a voltage that tends to maintain the current at its previous value.
5. 12 V
7. 1189 turns
9. decreased
11. Ferromagnetic
13. Air
15. It is not possible to get large values of inductance.
17. T
19. 1) Placing the inductors physically farther apart.
    2) Rotating one inductor by 90°.
    3) Shielding one or both inductors.
21. Voltage magnitude
23. 0 Ω
25. The inductor is functioning properly; it is not the cause of the trouble.

CHAPTER 15 Time Constants

Self-Check for Section 15-1

1. Capacitors and inductors
2. rising, falling
3. falling, rising
4. fast, slow

Self-Check for Sections 15-2, 15-3, and 15-4

5. 63
6. 86
7. 5
8. 30 ms
9. 30 ms
10. 150 ms
11. 20 μs
12. 95%

Chapter Questions and Problems (Odd numbers)

1. T
3. 2 τ
5. T
7. a) 100 mA b) 37 mA, 5 mA c) 5 ms

9. a) 14.8 mA, 2 mA b) 12.5 ms
11. T
13. a) 1.5 H b) 20 ms c) 400 mA d) 252 mA, 344 mA, 380 mA, 400 mA
15. same
17. Pole 1, because its load is partially inductive, not totally resistive.

CHAPTER 16 Alternating Current and Voltage

Self-Check for Sections 16-1, 16-2, 16-3

1. Dc is instantaneously constant in value and never changes direction (polarity). Ac is continually changing in value and does change direction.
2. F
3. Any waveform that oscillates between positive and negative is ac. It isn't necessary for it to have the distinctive sine shape.
4. voltage (volts), current (amps)
5. time
6. a) 170 V b) 340 V c) 120 V
7. 8.48 $V_{p\text{-}p}$
8. 339 W
9. 42.4 Ω
10. T

Self-Check for Sections 16-4 and 16-5

11. Passing through the entire range of values, both positive and negative.
12. a) Passing through all the positive values, from zero to peak and back to zero.
    b) Passing through all the negative values.
    c) The moment at which the variable is at zero, beginning a positive half-cycle.
    d) The moment beginning a negative half-cycle.
13. period, T
14. hertz, Hz
15. hertz
16. 2.5 ms, 50 μs
17. 200 Hz, 400 kHz
18. 90, 270
19. F

Self-Check for Sections 16-6 and 16-7

20. alternator
21. rotor
22. armature
23. 360
24. slip rings
25. in phase
26. leads
27. lags

Chapter Questions and Problems (Odd numbers)

1. T
3. 33.9 V
5. 31.8 V
7. 0.5 A
9. 12.5 W
11. 0.4 A
15. cycles, second
17. 18.1 ns
19. 1.25 MHz, usually labeled 1250 kHz on the dial
21. 3 and 9 o'clock
23. out of phase
25. Current reaches its positive-going zero crossover after voltage does.

CHAPTER 17 Transformers

Self-Check for Sections 17-1 and 17-2

1. primary
2. secondary
3. No, either winding can serve as the primary.
4. larger
5. The primary ac current produces flux in the core that is continually changing. This changing flux causes the secondary winding to induce voltage.
6. In-phase
7. 1.5 V, 75 V
8. 50 turns, 0.25

Self-Check for Sections 17-3 and 17-4

9. $I_P / I_S = N_S / N_P$
10. smaller, larger
11. increase
12. decrease
13. a) 144 V b) 1.44 A c) 4.32 A d) 207 W e) 207 W
14. 192 W
15. Stepping the voltage up to a high level causes the current in the transmission wires to be much lower, for a given amount of power transfer. Lower current reduces the I^2R power loss (waste) due to wire resistance.
16. 2.4 kA

Self-Check for Sections 17-5 and 17-6

17. source's internal resistance
18. 6.32
19. 4 W
20. 1) For providing two voltage levels, one for driving high-power loads and one for driving low-power loads. 2) For full-wave rectifying
21. Isolation makes it easier to filter out, or prevent the appearance of, high-frequency electrical noise. Under certain conditions, isolation can make the circuit safer, less likely to produce electrical shock.

Self-Check for Sections 17-7 and 17-8

22. The secondary winding is open.
23. 1) I^2R loss due to winding resistance. 2) I^2R loss due to eddy-current in the core material. 3) Core magnetization loss due to the core's tendency to retain its present magnetic field.
24. 225 W
25. T
26. 120 times per second
27. laminating the core with thin layers of insulation
28. T

Chapter Questions and Problems (Odd numbers)

1. primary, secondary
5. 36 V
7. a) 3 b) 5 A c) 15 A d) less
9. a) 1.43 A b) 58.8 Ω c) 120 W
11. a) 256 W b) 96 V c) 2
    d) I_S = 2.67 A, I_P = 5.33 A
13. internal resistance
15. n = 14.14, up
17. There are 12 V across the entire winding. There are 6 V between the center-tapped lead and either end lead.
19. I^2R loss in windings, I^2R eddy current loss in core, magnetization reversal loss in core
21. The primary winding has failed open.
23. Perform an ohmmeter test of both windings.
25. There is a short between the grounded primary circuit and the secondary circuit. Perform an ohmmeter test between the primary and secondary windings (or a high-voltage test). Replace the transformer.

CHAPTER 18 Reactance

Self-Check for Sections 18-1 and 18-2

1. Current leads voltage by 90°
2. reactance
3. First quarter cycle
4. Second quarter cycle
5. Third quarter cycle
6. Fourth quarter cycle
7. X_C
8. ohms
9. 7.96 kΩ
10. 1061 Hz
11. 0.8 μF

Self-Check for Sections 18-3 and 18-4

12. capacitive reactance
13. doubled
14. 0.754 mA
15. 0.189 mA (one-fourth the previous value)
16. 2653 Hz
17. 1.89 A
18. T
19. 4

Self-Check for Sections 18-5 through 18-8

20. F
21. T
22. T
23. ohms
24. 1257 Ω
25. 2513 Ω (twice the previous value)
26. 7.96 mA
27. 3.98 mA (half the previous value)
28. low, high
29. 393 Ω

Chapter Questions and Problems (Odd numbers)

1. zero
3. leads, 90
5. 8842 Ω, 1.36 mA
7. halved
9. a) 318 kHz
   b) 796 Hz
11. 63.7 V
13. lags, 90
15. A 12-V ac source at 60 Hz. Low frequency produces less inductive reactance, therefore more current.
17. 6.82 mA
19. doubled
21. a) 12.7 mH
    b) 0.398 H
23. increase
25. T

CHAPTER 19 Impedance

Self-Check for Sections 19-1 and 19-2

1. magnitude
2. phase angle
3. T
4. F
8. a) 1.7 A
   b) 1 A

Self-Check for Section 19-3

9. T
10. adjacent
11. adjacent, I_R
12. opposite, I_X
13. a) 4.57 A
    b) 23.2°, leading

Self-Check for Section 19-4

14. source voltage
15. leads
16. $ø = \cos^{-1}(I_R / I_T)$

17. a) $I_R = 15$ mA, $I_C = 20$ mA
    b) 25 mA, $ø = 53°$
    c) 480 Ω
18. a) $I_R = 40$ mA, $I_L = 48$ mA
    b) $I_T = 62.5$ mA, $ø = 50.2°$
    c) $Z = 1.92$ kΩ

Self-Check for Section 19-5

19. power factor
20. a) $PF = 0.89$
    b) 20 W
21. No
22. 0.87
23. voltampere
24. F
25. T
26. a) 200 A
    b) 175 A
    c) Consumer A, since their current is greater.
27. Consumer B because with less power lost in the transmission lines, the utility company can produce less power at the generating site, thereby reducing their expense, while still delivering the same amount of power at the consumer's load site.

Self-Check for Section 19-6

28. current
29. horizontal
30. It is placed vertically downward because capacitive voltage lags the current by 90°.
31. Inductive reactance is placed vertically upward because inductive voltage leads the current by 90°.
32. T
33. a) 43.0 Ω
    b) 1.16 A, leading V_s by 54.5°
    c) $V_R = 29.0$ V, $V_C = 40.6$ V
    d) Similar triangles
34. c) $Z = 3.21$ kΩ
    d) $I = 9.34$ mA, lagging by 51.5°
35. a) $V_R = 18.7$ V
    b) $V_L = 23.5$ V
    c) $\sqrt{(18.7)^2 + (23.5)^2} = 30.0$ V

Chapter Questions and Problems (Odd numbers)

1. On the x axis, to the right
5. $I_T = 7.2$ A
7. $adj^2 + opp^2 = hyp^2$
9. a) 3.36 A
   b) 53.5°
   c) I_T leads V_s by 53.5°.
11. leads
13. a) 2.5 mA d) 61.1°
    b) 4.52 mA e) 5.80 kΩ
    c) 5.17 mA f) 5.80 kΩ
15. a) $I_R = 250$ mA, $I_L = 199$ mA
    b) $I_T = 319$ mA, lagging V_s by 38.5°
17. a) stay the same c) decrease e) decrease
    b) increase d) decrease
19. $PF = 0.60$
21. $S = 3.75$ VA
23. T
25. a) $X_C = 455$ Ω b) $Z = 545$ Ω c) $I_T = 0.147$ A, leading V_s by 56.6°
27. Because it is not correct to add two out-of-phase voltages by simple arithmetic. Instead, they must be combined by phasor methods.
29. a) $Z = 164$ Ω
    b) $I = 0.732$ A
    c) $V_L = 95.1$ V, $V_R = 73.2$ V

CHAPTER 20 Resonance

Self-Check for Sections 20-1 and 20-2

1. larger reactance
2. $X_C = 397.9$ Ω, $X_L = 879.7$ Ω, $X_{net} = 481.8$ Ω
3. $Z = 496.5$ Ω
4. inductive reactance, capacitive reactance
5. resistance, R
6. 1591 Hz
7. 0.1 A
8. $ø = 0°$, $PF = 1.0$
9. $V_L = 40$ V, $V_C = 40$ V. These individual component voltages are much greater than the source voltage because the resonant impedance is so low, allowing the resonant current to be large. These two 40-V values combine to equal 0 V.

Self-Check for Sections 20-3 and 20-4

10. frequency
11. resonant
12. A highly selective circuit causes a great change in current as the frequency varies from the f_r value. A not-so-selective circuit causes less change in current for the same frequency variation.
13. f_{low} is the frequency below the resonant frequency, f_r, that causes the circuit current to decline to 0.7071 times the resonant current. f_{high} is the frequency above the resonant frequency that causes the same response.
14. bandwidth
15. narrow, wide
16. 318 Hz
17. a) 918.9 Hz c) 2.89
    b) 866.0 Ω d) 318 Hz. It checks.
18. increased, decreased
19. T

Self-Check for Sections 20-5 and 20-6

20. It passes to the load those signals whose frequencies are within a band near f_r. It blocks signals outside that band.

21. At frequencies close to f_r, the LC combination has low net reactance; therefore, the signal voltage divides with a large portion across the load. At frequencies far from f_r, the reverse occurs.
22. It passes to the load those signals whose frequencies are outside a band near f_r. It blocks signals within that band.
23. It blocks the close-frequency signals because the LC combination has low net reactance; therefore, the parallel combination of the LC combination with R_{LD} has a net impedance that is low. The signal voltage then divides with most of the signal across dropping resistor R_D, and very little across the load combination.
24. It passes low-frequency signals to the load, but blocks high-frequency signals.
25. C is in parallel with the load, so at low frequencies where X_C is high, the parallel combination has a large net impedance. Therefore, the signal divides with most of its voltage across the load parallel combination. But at high frequencies, X_C is low, so the parallel combination has low net impedance. Therefore, most of the signal voltage drops across the resistor, R.
26. C is in series with the load. At low frequencies, where X_C is large, most of the signal is dropped across C. But at high frequencies, where X_C is low, the signal divides with most of its voltage appearing across the load.

Self-Check for Sections 20-7 and 20-8

27. T
28. T
29. zero
30. total impedance
31. In parallel resonance, Z_T increases to maximum at f_r. In series resonance, Z_T decreases to minimum at f_r.
32. 2.5 kΩ
33. minimum
34. a) 335.5 Hz d) 19.0 mA
    b) 632.5 Ω e) 4.8 mA
    c) 19.0 mA
35. I_L is larger than I_T. This can happen because the large individual branch current I_L is cancelled out by an equal-magnitude current I_C in the neighboring branch that is 180° out of phase with I_L.

Self-Check for Sections 20-9 and 20-10

36. The resistor and the capacitor
37. 2122 Hz
38. smaller (narrower)
39. F
40. T
41. a) $Q_{T(parallel)} = 2.5$ b) $R = 6.25$ kΩ

Chapter Questions and Problems (Odd numbers)

1. 180°, equal
3. decreased
5. unchanged
7. frequency, f
9. greater, lesser
11. high cutoff frequency, f_{high}, low cutoff frequency, f_{low}
13. a) 919 Hz b) 424 Hz
15. a) $Q_T = 6.13$ b) $R = 70.6\ \Omega$
17. Reactance of either the inductor or the capacitor, at the resonant frequency.
19. T
21. Band-stop
23. band-stop
25. 25 mA
27. 0.125 W. It matches.
29. zero
31. In a series-resonant circuit, I is maximum when $f = f_r$. In a parallel-resonant circuit, I is minimum when $f = f_r$, because Z is maximum.
33. a) 2251 Hz d) 2.47
    b) 910 Hz e) 910 Hz
    c) 1414 Ω
35. a) 4.50 b) 6366 Ω
37. a) $PF = 1.0$ b) $I_T = 2.86$ mA
    c) $I_L = 7.07$ mA d) $I_C = 7.07$ mA
39. band-stop
41. $X_L = 534.65\ \Omega$, $X_C = 654.64\ \Omega$, $X_{net} = 120.0\ \Omega$, $Z = 169.71\ \Omega$, $I = 70.71$ mA. Yes, this proves that $f_{low} = 121.56$ Hz, since $I / I_{max} = 0.7071$.

CHAPTER 21 Solid-State Electronics

Self-Check for Sections 21-1 and 21-2

3. Opposite of your answer to Question 2.
4. a) 2 A b) 5 V c) 0 V, no polarity
5. a) 0 A, no direction b) 0 V, no polarity
   c) 5 V, + on K, – on A

Self-Check for Section 21-3

6. An active device extracts energy from a dc power supply and delivers it to the load in an output waveform, with the output waveform shape determined by an input signal waveform. A passive device just reacts to the input signal in accordance with Ohm's law and Kirchhoff's laws.
7. input current
8. amplification
11. emitter, emitter
12. T

13. X_{Cin} = 31.8 Ω, R_{B1} = 10 000 Ω. This *RC* combination has a total impedance, *Z*, of 10 000.05 Ω, which is so extremely close to the value of R_{B1} that X_{Cin} can be regarded as zero.
14. *Example 21-4*
    a) I_b (I_{in}) = 50 μA, the same as before
    b) I_c (I_{out}) = 8.75 mA
    c) V_{out} = 2.625 V (3.71 V_{peak})
    Example 21-5
    a) A_I = 175
    b) A_V = 5.25
    c) A_P = 919
    Example 21-6
    a) I_B = 0.14 mA, the same as before
    b) I_C = 24.5 mA
    c) V_{RLD} = 7.35 V
    d) Oscillating between −11.0 V and −3.61 V
15. positive
16. positive
17. By connecting a capacitor between the collector and the load, with the other side of the load grounded.
18. 180° out of

Self-Check for Section 21-4

19. An FET is operated by its input voltage, while the BJT is operated by its input current.
21. drain
22. gate
23. source
24. V_{GS}
25. The value of V_{GS} that causes I_D to decrease to zero.
26. larger
27. linear
28. nonlinear
29. The current that flows through the source-drain path when V_{GS} = 0 V.
30. short circuit, or direct connection

Chapter Questions and Problems (Odd numbers)

3. a) *I* = 0
   b) V_R = 0
   c) V_D = 10 V, + on *K*
5. Half-wave-rectified waveform, with V_{peak} = 34 V
7. base, collector
9. gain
11. current
13. A current source maintains its current at a constant value even though the load resistance changes. A current source will vary its output voltage, raising it for large values of load resistance and lowering it for small values of load resistance. A voltage source maintains its voltage at a constant value, allowing the current to vary inversely with the load resistance.
15. I_B = 65.9 μA
17. V_{RC} = 8.11 V
19. X_{Cin} = 589 Ω, which raises the *RC* input circuit's impedance to 56 003 Ω, a negligible difference.
21. $I_{c(p-p)}$ = 10.7 mA
23. $V_{c(p-p)}$ = 5.67 V
25. T
27. A_V = 1.42
29. A_I = 52.9
35. reverse
37. T
39. T
41. R_{in} = 2.2 MΩ
43. V_{RD} = 11.0 V
45. $i_{D(max)}$ = 3.47 mA, $i_{D(min)}$ = 1.20 mA
47. F
49. V_d = 3.81 V
51. A_V = 1.90
53. A_I = 2497

INDEX